Inverse and Ill-Posed Problems Series

Operator Theory and Ill-Posed Problems

Also available in the Inverse and Ill-Posed Problems Series:

Well-posed, Ill-posed, and
Intermediate Problems with Applications
Yu.P. Petrov and V.S. Sizikov
Direct Methods of Solving
Multidimensional Inverse Hyperbolic Problems
S.I. Kabanikhin, A.D. Satybaev and M.A. Shishlenin
Characterisation of Bio-Particles from
Light Scattering
V.P. Maltsev and K.A. Semyanov
Carleman Estimates for Coefficient Inverse
Problems and Numerical Applications
M.V. Klibanov and A.A Timonov
Counterexamples in Optimal Control Theory
S.Ya. Serovaiskii
Inverse Problems of Mathematical Physics
*M.M. Lavrentiev, A.V. Avdeev, M.M. Lavrentiev, Jr.,
and V.I. Priimenko*
Ill-Posed Boundary-Value Problems
S.E. Temirbolat
Linear Sobolev Type Equations and Degenerate
Semigroups of Operators
G.A. Sviridyuk and V.E. Fedorov
Ill-Posed and Non-Classical Problems of Mathematical
Physics and Analysis
Editors: M.M. Lavrent'ev and S.I. Kabanikhin
Forward and Inverse Problems for Hyperbolic,
Elliptic and Mixed Type Equations
A.G. Megrabov
Nonclassical Linear Volterra Equations of the First Kind
A.S. Apartsyn
Poorly Visible Media in X-ray Tomography
D.S. Anikonov, V.G. Nazarov, and I.V. Prokhorov
Dynamical Inverse Problems of Distributed Systems
V.I. Maksimov
Theory of Linear Ill-Posed Problems and its Applica-
tions
V.K. Ivanov, V.V. Vasin and V.P. Tanana
Ill-Posed Internal Boundary Value Problems for the
Biharmonic Equation
M.A. Atakhodzhaev
Investigation Methods for Inverse Problems
V.G. Romanov
Operator Theory. Nonclassical Problems
S.G. Pyatkov
Inverse Problems for Partial Differential Equations
Yu.Ya. Belov
Method of Spectral Mappings in the Inverse Problem
Theory
V. Yurko
Theory of Linear Optimization
I.I. Eremin
Integral Geometry and Inverse Problems for Kinetic
Equations
A.Kh. Amirov
Computer Modelling in Tomography and
Ill-Posed Problems
M.M. Lavrent'ev, S.M. Zerkal and O.E. Trofimov
An Introduction to Identification Problems via F
unctional Analysis
A. Lorenzi
Coefficient Inverse Problems for Parabolic Type
Equations and Their Application
P.G. Danilaev

Inverse Problems for Kinetic and
Other Evolution Equations
Yu.E. Anikonov
Inverse Problems of Wave Processes
A.S. Blagoveshchenskii
Uniqueness Problems for Degenerating
Equations and Nonclassical Problems
S.P. Shishatskii, A. Asanov and E.R. Atamanov
Uniqueness Questions in Reconstruction of
Multidimensional Tomography-Type
Projection Data
V.P. Golubyatnikov
Monte Carlo Method for Solving
Inverse Problems of Radiation Transfer
V.S. Antyufeev
Introduction to the Theory of Inverse Problems
A.L. Bukhgeim
Identification Problems of Wave
Phenomena - Theory and Numerics
S.I. Kabanikhin and A. Lorenzi
Inverse Problems of Electromagnetic Geophysical Fields
P.S. Martyshko
Composite Type Equations and Inverse Problems
A.I. Kozhanov
Inverse Problems of Vibrational Spectroscopy
*A.G. Yagola, I.V. Kochikov, G.M. Kuramshina and
Yu.A. Pentin*
Elements of the Theory of Inverse Problems
A.M. Denisov
Volterra Equations and Inverse Problems
A.L. Bughgeim
Small Parameter Method in Multidimensional Inverse
Problems
A.S. Barashkov
Regularization, Uniqueness and Existence of Volterra
Equations of the First Kind
A. Asanov
Methods for Solution of Nonlinear Operator Equations
V.P. Tanana
Inverse and Ill-Posed Sources Problems
Yu.E. Anikonov, B.A. Bubnov and G.N. Erokhin
Methods for Solving Operator Equations
V.P. Tanana
Nonclassical and Inverse Problems for Pseudoparabolic
Equations
A. Asanov and E.R. Atamanov
Formulas in Inverse and Ill-Posed Problems
Yu.E. Anikonov
Inverse Logarithmic Potential Problem
V.G. Cherednichenko
Multidimensional Inverse and Ill-Posed Problems
for Differential Equations
Yu.E. Anikonov
Ill-Posed Problems with A Priori Information
V.V. Vasin and A.L. Ageev
Integral Geometry of Tensor Fields
V.A. Sharafutdinov
Inverse Problems for Maxwell's Equations
V.G. Romanov and S.I. Kabanikhin

INVERSE AND ILL-POSED PROBLEMS SERIES

Operator Theory and Ill-Posed Problems

M.M. Lavrent'ev and L.Ya. Savel'ev

LEIDEN • BOSTON
2006

A C.I.P. record for this book is available from the Library of Congress

ISBN-13: 978-90-6764-448-8
ISBN-10: 90-6764-448-X

Printed and bound in The Netherlands.

The book is based on the course of lectures on calculus and functional analysis and several special courses given by the authors at Novosibirsk State University. It also includes results of research carried out at the Institute of Mathematics of the Siberian Branch of the Russian Academy of Sciences. A brief introduction to the language of set theory and elements of abstract, linear, and multilinear algebra is provided. The language of topology is introduced and fundamental concepts of analysis for vector spaces and manifolds are described in detail. The most often used spaces of smooth and generalized functions, their transformations, and the classes of linear and nonlinear operators are considered. Special attention is given to spectral theory and the fixed point theorems. A brief presentation of degree theory is provided. The part devoted to ill-posed problems includes a description of partial differential equations, integral and operator equations, and problems of integral geometry.

The book can serve as a textbook or reference on functional analysis. It contains many examples. It can also be of interest to specialists in the above fields.

Contents

BASIC CONCEPTS   1

Chapter 1. Set theory   2

1.1. Sets 2

    1.1.1. Elements and subsets 2

    1.1.2. The algebra of sets 3

    1.1.3. Cartesian product 5

1.2. Correspondences 7

    1.2.1. Images and inverse images 7

    1.2.2. Functions 8

    1.2.3. Collections of sets 11

1.3. Relations 14

    1.3.1. Reflexivity, transitivity, and symmetry 14

    1.3.2. Equivalence 15

    1.3.3. Order 17

1.4. Induction 23

    1.4.1. Well-ordered sets 23

    1.4.2. Discrete sets 24

    1.4.3. Zorn's lemma 26

1.5. Natural numbers 27

    1.5.1. Decimal natural numbers 27

    1.5.2. The isomorphism theorem 28

    1.5.3. Countable sets 29

Chapter 2. Algebra   30

2.1. Abstract algebra 30

     2.1.1. Semigroups 30

     2.1.2. Groups 37

     2.1.3. Rings and fields 55

     2.1.4. Lattices 64

     2.1.5. Numbers 69

2.2. Linear algebra 75

     2.2.1. Vector spaces 76

     2.2.2. Linear operators 89

     2.2.3. Linear functionals 102

     2.2.4. Scalar products 114

     2.2.5. Normed spaces 121

     2.2.6. Euclidean spaces 130

2.3. Multilinear algebra 141

     2.3.1. Tensor product 141

     2.3.2. Exterior product 146

Chapter 3. Calculus   150

3.1. Limit 150

     3.1.1. Topological spaces 150

     3.1.2. Directed sets 170

     3.1.3. Convergence 177

3.2. Differential 197

     3.2.1. The definition of the differential 197

     3.2.2. Differentiation rules 206

     3.2.3. Lagrange's theorem 210

     3.2.4. Termwise differentiation 215

     3.2.5. Total differentials 217

     3.2.6. Solution of functional equations 226

     3.2.7. Taylor's formula 241

     3.2.8. Local minima 250

     3.2.9. Smooth curves 258

     3.2.10. A simplest variational problem 261

3.3. Integral 263
    3.3.1. Measures 264
    3.3.2. Classical definition of the integral 272
    3.3.3. Limit theorems 289
    3.3.4. Measurable functions 295
    3.3.5. The Fubini and Tonelli theorems 298
    3.3.6. Indefinite integrals 316
3.4. Analysis on manifolds 321
    3.4.1. Manifolds 321
    3.4.2. The rank theorem 326
    3.4.3. Sard's theorem 327
    3.4.4. Differential forms 328
    3.4.5. The Poincare theorem 331
    3.4.6. Change of variables 334
    3.4.7. Integral over a manifold 335
    3.4.8. The Stokes formula 342
    3.4.9. Map degree 348
    3.4.10. Applications 351

OPERATORS   353

Chapter 4. Linear operators   **354**
4.1. Hilbert spaces 354
    4.1.1. Orthogonal projection 354
    4.1.2. Continuous linear functionals 356
    4.1.3. The spaces $\mathcal{L}^2 = \mathcal{L}^2(U, \mu)$ 359
4.2. Fourier series 363
    4.2.1. Fourier coefficients 363
    4.2.2. Isomorphism of Hilbert spaces 368
4.3. Function spaces 369
    4.3.1. Metric spaces 369
    4.3.2. Smooth functions 371
    4.3.3. Lebesgue spaces 376
    4.3.4. Distributions 380
    4.3.5. Sobolev spaces 388

4.4. Fourier transform 391
    4.4.1. Transforms of rapidly decreasing functions 391
    4.4.2. Transforms of slowly increasing distributions 393
    4.4.3. The Fourier–Plancherel transform 395
    4.4.4. The Fourier–Stieltjes transform 395
    4.4.5. The Radon transform 396
4.5. Bounded linear operators 397
    4.5.1. Extensions of functionals 397
    4.5.2. Uniform boundedness of operators 399
    4.5.3. Inversion of operators 401
    4.5.4. Closedness of the graph of an operator 403
    4.5.5. Weak compactness 405
4.6. Compact linear operators 408
    4.6.1. Examples of compact operators 408
    4.6.2. Properties of compact operators 410
    4.6.3. Adjoint operators 411
    4.6.4. Fredholm operators 414
    4.6.5. Fredholm theorems 417
4.7. Self-adjoint operators 419
    4.7.1. Banach adjoint operators 419
    4.7.2. Hilbert adjoint operators 420
    4.7.3. Hermitian and normal operators 422
    4.7.4. Unitary operators 423
    4.7.5. Positive operators 424
4.8. Spectra of operators 426
    4.8.1. Classification of spectra 426
    4.8.2. The spectrum of a closed operator 431
    4.8.3. The spectrum of a bounded operator 433
    4.8.4. The spectrum of a compact operator 435
    4.8.5. The spectrum of a self-adjoint operator 435
4.9. Spectral theorem 444
    4.9.1. Projection measures 445
    4.9.2. Integrals of bounded functions 451
    4.9.3. Integrals of unbounded functions 459
    4.9.4. Spectral theorem 462
    4.9.5. Operator functions 466

4.10. Operator exponential 468

    4.10.1. Problem formulation 468

    4.10.2. Semigroups of operators 470

    4.10.3. The Laplace transform 475

    4.10.4. Stone's theorem 476

    4.10.5. Evolution equations 478

Chapter 5. Nonlinear operators **483**

5.1. Fixed points 483

    5.1.1. The Brouwer theorem 483

    5.1.2. The Tikhonov theorem and the Schauder theorem . . . 487

5.2. Saddle points 490

    5.2.1. Kakutani's theorem 490

    5.2.2. von Neumann theorem 493

5.3. Monotonic operators 497

    5.3.1. Definition and properties 497

    5.3.2. Equations with monotonic operators 499

5.4. Nonlinear contractions 501

    5.4.1. Contracting semigroups of operators 501

    5.4.2. Approximation 503

5.5. Degree theory 504

    5.5.1. Finite-dimensional spaces 504

    5.5.2. The Leray–Schauder degree 508

ILL-POSED PROBLEMS 511

Chapter 6. Classic problems **512**

6.1. Mathematical description of the laws of physics 512

6.2. Equations of the first order 518

6.3. Classification of differential equations of the second order . . . 519

6.4. Elliptic equations 521

6.5. Hyperbolic and parabolic equations 527

6.6. The notion of well-posedness 529

Chapter 7. Ill-posed problems   531

7.1. Ill-posed Cauchy problems 531

7.2. Analytic continuation and interior problems 534

7.3. Weakly and strongly ill-posed problems. Problems of differentiation. 536

7.4. Reducing ill-posed problems to integral equations 537

Chapter 8. Physical problems leading to ill-posed problems   541

8.1. Interpretation of measurement data from physical devices . . . 541

8.2. Interpretation of gravimetric data 543

8.3. Problems for the diffusion equation 546

8.4. Determining physical fields from the measurements data   . . . 547

8.5. Tomography 548

Chapter 9. Operator and integral equations   552

9.1. Definitions of well-posedness 552

9.2. Regularization 555

9.3. Linear operator equations 559

9.4. Integral equations with weak singularities 564

9.5. Scalar Volterra equations 565

9.6. Volterra operator equations 568

Chapter 10. Evolution equations   571

10.1. Cauchy problem and semigroups of operators 571

10.2. Equations in a Hilbert space 573

10.3. Equations with variable operator 577

10.4. Equations of the second order 578

10.5. Well-posed and ill-posed Cauchy problems 580

10.6. Equations with integro-differential operators 581

Chapter 11. Problems of integral geometry   584

11.1. Statement of problems of integral geometry 584

11.2. The Radon problem 584

11.3. Reconstructing a function from spherical means 588

11.4. Planar problem of the general form 594

11.5. Spatial problems of the general form 602

11.6. Problems of the Volterra type for manifolds invariant with
respect to the translation group 614

11.7. Planar problems of integral geometry with a perturbation . . 618

Chapter 12. Inverse problems   626

12.1. Statement of inverse problems 626

12.2. Inverse dynamic problem. A linearization method 628

12.3. A general method for studying inverse problems for hyperbolic equations 637

12.4. The connection between inverse problems for hyperbolic, elliptic, and parabolic equations 644

12.5. Problems of determining a Riemannian metric 651

Chapter 13. Several areas of the theory of ill-posed problems, inverse problems, and applications   659

Bibliography   662

Index   673

Preface

The book consists of three major parts. The first two parts, written by L. Ya. Savel'ev, deal with general mathematical concepts and certain areas of operator theory. The third part, written by M. M. Lavrent'ev, is devoted to ill-posed problems. It can be read independently of the first two parts and presents a good example of applying the methods of calculus and functional analysis. The book is based on the lectures given by the authors at Novosibirsk State University.

The part "Basic Concepts" briefly introduces the language of set theory and concepts of abstract, linear and multilinear algebra. We also introduce the language of topology and consider fundamental concepts of calculus: the limit, the differential, and the integral. A special section is devoted to analysis on manifolds.

The part "Operators" describes the most important function spaces and operator classes for both linear and nonlinear operators. Different kinds of generalized functions and their transformations are considered. Elements of the theory of linear operators are presented. Spectral theory is given a special focus. We prove main theorems on stationary points of nonlinear transformations and briefly introduce the theory of mapping degree.

The part "Ill-Posed Problems" is devoted to problems of mathematical physics, integral and operator equations, evolution equations and problems of integral geometry. It also deals with problems of analytic continuation.

Detailed coverage of the subjects and numerous examples and exercises make it possible to use the book as a textbook on some areas of calculus and functional analysis. It can also be used as a reference textbook because of the extensive scope and detailed references with comments.

Several sections contain new research results, namely, a description of a general model of linear continuous extension of a vector measure to the integral. A uniqueness theorem is proved for a new type of equations with integro-differential operators. Examples in which the Radon problem is ill-posed are given. The problem is proved to be weakly ill-posed. The problem with incomplete initial data is analyzed. A uniqueness theorem for a special problem of integral geometry is proved. The above results were discussed at the seminars in the Sobolev Institute of Mathematics (Novosibirsk).

The authors wish to thank their colleagues for valuable criticism and advice.

BASIC CONCEPTS

This part of the book describes the areas of set theory, abstract and linear algebra, and geometry of spaces that are widely used in calculus and functional analysis.

The chapter "Set Theory" presents the abstract mathematical language used in modern mathematical analysis. The necessary notions are introduced and useful facts from set theory are provided. The abstract mathematical language makes it possible to formulate propositions in a simple form and make arguments concise and clear. The abstract character of the language is compensated by providing specific examples. The logical connectives *if ... then, only if, if and only if* (or *equivalent*) are often denoted by the arrows $\Rightarrow$, $\Leftarrow$, $\Leftrightarrow$, respectively. The symbols $\forall$ and $\exists$ often stand for the expressions *for all* and *for some*. The general principle of induction and the equivalent propositions used in the proofs of many important theorems are formulated.

The chapter "Algebra" includes elements of abstract, linear, and multilinear algebra. It focuses on the areas of these theories that are often used in calculus and functional analysis. Considerable attention is given to algebraic structures, namely semigroups, groups, rings, fields, vector spaces, and algebras. Normed and Euclidean spaces and their geometry are described. A separate section is devoted to elements of multilinear algebra and tensor calculus.

The chapter "Calculus" is devoted to limit theory, differential and integral calculus. Directional convergence in topological spaces, differentiation in vector spaces, and integration with respect to a measure are described. A separate section is devoted to analysis on manifolds.

Chapter 1.

Set theory

Throughout the book we will use naive set theory, which assumes that sets with all the necessary properties exist. The chapter describes the formal language of set theory and operations on sets.

1.1.   SETS

There is no formal definition of a set. It is assumed that properties used to construct sets are not contradictory and the sets with these properties are well defined.

1.1.1.   Elements and subsets

Each *nonempty set* is defined by its *elements*. It is also assumed that there exists an *empty set*, which is a set with no elements. Sets are usually denoted by capital letters, while their elements are denoted by small letters.

1. We indicate that *an element a belongs to a set E* by writing $a \in E$. The negation of this statement is written as $a \notin E$. A set E defined by enumerating all of its elements a, b, c, ... is said to *consist of elements a, b, c, ...*, which is denoted by $E = \{a, b, c, \dots\}$. The order in which the elements are put together does not matter.

Two sets are considered to be equal if they consist of the same elements. If sets A and B are equal, we write $A = B$. The negation of this statement is written as $A \neq B$.

Sets consisting of exactly one element are called *singletons* and are often identified with their elements.

2. If every element of a set A belongs to a set B, then A is called a *subset* of B. Alternatively, we say that B *includes* A ($B \supseteq A$), or A *is included in* B ($A \subseteq B$). Otherwise, if B contains no elements of A, we write $B \not\supseteq A$ or $A \not\subseteq B$. If $A \subseteq B$ and $A \neq B$, then the inclusion of A in B is called *proper* and designated as $A \subset B$ or $B \supset A$.

Every set E includes the empty set, which is denoted by $\varnothing$.

From the definitions, it follows that $A = B$ if and only if $A \subseteq B$ and $B \subseteq A$.

For a set E, it is convenient to designate its subsets by $\{x \in E \mid \ldots\}$, where the vertical bar precedes the description of a certain property defining a given subset. Some notations use colon instead of the vertical bar. If it is obvious from the context that x is an element of a set E, then we write x instead of $x \in E$.

Nonempty subsets of a set E that are not equal to E are called *proper subsets of E.*

For any set E, there exists a *class* $\mathcal{P} = \mathcal{P}(E)$ of all subsets of E. Here the term *class* means sets whose elements are sets. It is introduced for the sake of convenience in order to avoid phrases such as "a set of sets" or "a set of subsets".

Example. Consider the set of digits $F = \{0, 1, 2, 3, 4, 5, 6, 7, 8, 9\}$. The set of even digits $A = \{0, 2, 4, 6, 8\}$ and the set of odd digits $B = \{1, 3, 5, 7, 9\}$ are subsets of F. The set $A = \{2, 4, 6, 8\}$ is a proper subset of A. Neither A nor B includes the set $D = \{0, 1\}$.

1.1.2.   The algebra of sets

The operations of union, intersection, and set difference are defined on the class $\mathcal{P} = \mathcal{P}(E)$ of all subsets of a set E.

1. Let X, Y, and Z be sets of class $\mathcal{P}$. The set $X \cup Y = \{x \mid x \in X \text{ or } y \in Y\} \in \mathcal{P}$, consisting of elements that belong to X or Y is called the *union of X and Y.*

The set $X \cap Y = \{x \mid x \in X \text{ and } y \in Y\} \in \mathcal{P}$, consisting of elements that belong to both X and Y is called the *intersection of X and Y.*

The set $Y \setminus X = \{y \mid y \in Y \text{ and } y \notin X\} \in \mathcal{P}$, consisting of elements that belong to Y and do not belong to X is called the *complement of X with respect to Y.*

The complement of X with respect to E is called simply the *complement X* and denoted by X' or X^c. Obviously, $Y \setminus X = Y \setminus (X \cap Y) = Y \cap X'$.

2. If $X \cap Y = \varnothing$, then the sets X and Y said to be *non-intersecting* or *disjoint*. It follows from the definitions that $X \cap Y = \varnothing$ is equivalent to $Y \subseteq X'$.

A class is called *disjoint* or *pairwise disjoint* if any two distinct sets in it are disjoint. From the definition, it follows that the empty class and the class containing exactly one set are disjoint.

3. The operations of union, intersection and set difference satisfy the following easily verifiable conditions:

$$X \cup X' = E, \quad X \cap X' = \varnothing \qquad (1)$$

— *partition;*

$$X \cup X = X, \quad X \cap X = X \qquad (2)$$

— *idempotency;*

$$X \cup Y = Y \cup X, \quad X \cap Y = Y \cap X \qquad (3)$$

— *commutativity;*

$$(X \cup Y) \cup Z = X \cup (Y \cup Z), \quad (X \cap Y) \cap Z = X \cap (Y \cap Z) \qquad (4)$$

— *associativity;*

$$\begin{aligned}(X \cup Y) \cap Z = (X \cap Z) \cup (Y \cap Z), \\ (X \cap Y) \cup Z = (X \cup Z) \cap (Y \cup Z)\end{aligned} \qquad (5)$$

— *distributivity;*

$$X \cup (X \cap Y) = X, \quad X \cap (X \cup Y) = X \qquad (6)$$

— *absorption;*

$$(X \cup Y)' = X' \cap Y', \quad (X \cap Y)' = X' \cup Y' \qquad (7)$$

— *duality;*

$$(X')' = X \qquad (8)$$

— *involution.*

The class $\mathcal{P}$ with operations $\cup$, $\cap$, $'$ is a *boolean algebra of sets.*

4. The operations of union and intersection can be generalized. Consider an arbitrary class $C \subseteq \mathcal{P}$. The set

$$\cup \mathcal{C} = \{x \mid x \in C \text{ for some } C \in \mathcal{C}\} \in \mathcal{P}$$

of all elements that belong to at least one set of class $\mathcal{C}$ is called the *union of sets of the class* $\mathcal{C}$.

For a set A, every class $\mathcal{C}$ such that $\cup \mathcal{C}$ includes A is called a *cover of* A.

A disjoint class $\mathcal{D}$ such that $\cup \mathcal{D} = A$ is called a *partition of* A (the usual assumption being that all sets in $\mathcal{D}$ are nonempty). The set

$$\cap \mathcal{C} = \{x \mid x \in C \text{ for every } C \in \mathcal{C}\} \in \mathcal{P},$$

is called the *intersection of sets of the class* $\mathcal{C}$.

For example, for a class $\mathcal{C}$ consisting of sets $A, B, C, \dots$, their union and intersection are written as

$$A \cup B \cup C \cup \dots, \quad A \cap B \cap C \cap \dots.$$

In particular, if $\mathcal{C} = \{A, B\}$, then

$$\cup \{A, B\} = A \cup B, \quad \cap \{A, B\} = A \cap B,$$

which follows from the definition of union and intersection, with $X = A$, $Y = B$.

5. Example. For the sets A, B, C, D, F in the example of Section 1.1.1, the following statements are true:

$$A \cup B = F, \quad A \cap B = \varnothing, \quad A' = B, \quad B' = A,$$
$$A \setminus C = \{0\}, \quad A \setminus D = C, \quad A \cap C = C, \quad A \cap D = \{0\},$$
$$B \cup C \cup D = F, \quad B \cap C \cap D = \varnothing.$$

Remark. We do not justify the well-posedness of the above definitions since we use naive set theory. In axiomatic theories, it is ensured by the axioms (see Kuratowski and Mostowski, 1970, Chapter. II).

1.1.3. Cartesian product

The notion of the Cartesian product is associated with geometry rather than algebra. It formalizes the transition from the line to the plane and the multidimensional space. The definition of the Cartesian product is based on the definition of an ordered pair.

1. Consider arbitrary elements $a \in A$ and $b \in B$, singletons $\{a\}$ and $\{b\}$, and the set $\{a, b\}$.

The class $(a, b) = \{\{a\}, \{a, b\}\}$ with elements $\{a\}$ and $\{a, b\}$ is called the *ordered pair whose first element is a and second is b.* .

We emphasize that this definition uses only the notion of a set. The element $a \in A$, $a \neq b$, is labeled as the *first* because $\{a\} \in \{\{a\}, \{a, b\}\}$, but $\{b\} \notin \{\{a\}, \{a, b\}\}$. For brevity, we will call an ordered pair simply a *pair* and write ab instead of (a, b).

Let $x \in A$, $y \in B$. Then the ordered pairs (x, y) and (a, b) are equal if and only if their first elements x and a are equal and the same is true for their second elements y and b.

Exercise. Prove the foregoing statement.

If $a \neq b$, then $(a, b) \neq (b, a)$. By definition, $(a, a) = \{\{a\}\}$ for every $a \in A \cap B$.

2. The set $A \times B = \{(x, y) \mid x \in A,\ y \in B\}$ consisting of all ordered pairs whose first elements belong to A and whose second elements belong to B is called the *Cartesian product of the sets A and B*. If $A \neq B$, where $A \neq \varnothing$ and $B \neq \varnothing$, then $A \times B \neq B \times A$.

Example. Let $A = B = F = \{0, 1, 2, 3, 4, 5, 6, 7, 8, 9\}$. Then the Cartesian product $A \times B = F \times F$ consists of the ordered pairs $00, 01, \ldots, 09$, $10, 11, \ldots, 19, \ldots, 90, 91, \ldots, 99$.

3. Nonempty classes whose elements are nonempty will be called *nondegenerate*. Consider a nondegenerate disjoint class $\mathcal{D}$. *A sample of $\mathcal{D}$ is a set Y consisting of elements of sets in $\mathcal{D}$ such that for every set $X \in \mathcal{D}$ there is exactly one element $y \in Y$ that belongs to X.*

We assume that the Axiom of choice holds.

Axiom of Choice. *There exists a sample of any nondegenerate disjoint class of sets.*

The set $\prod \mathcal{D}$ of all samples of $\mathcal{D}$ is called the *Cartesian product of the sets of the class $\mathcal{D}$*. The axiom of choice states that $\prod \mathcal{D} \neq \varnothing$ for every nondegenerate disjoint class $\mathcal{D}$. For a class $\mathcal{D}$ whose elements are sets $A, B, C, \ldots$, the Cartesian product $\prod \mathcal{D}$ is also written as $A \times B \times C \times \cdots$.

Example. Let $F = \{0, 1, 2, 3, 4, 5, 6, 7, 8, 9\}$ and let $\mathcal{D}$ be the class consisting of the sets $A = \{1\} \times F$, $B = \{2\} \times F$, $C = \{3\} \times F$. Then

$$\prod \mathcal{D} = \{\{(1, x), (2, y), (3, z)\} \mid x, y, z \in F\}.$$

Elements of $F \times F \times F$ can be conveniently written in the form of a sequence $xyz = 000, 001, \ldots, 999$.

1.2.   CORRESPONDENCES

Correspondences are formally defined as sets of ordered pairs. The meaning of this notion is intuitively clear.

1.2.1.   Images and inverse images

For any sets A and B, every subset S of the Cartesian product $A \times B$ is called a *correspondence* between A and B. If $(x, y) \in S$, this is often written in the form xSy or $S : x \to y$. Similarly, $S \subseteq A \times B$ is often written as $S : A \to B$.

1. For $x \in A$, the set $S(x) = \{y \mid (x, y) \in S\} \subseteq B$ is called the *image of the element* x under the correspondence S. This image may be empty. For $X \subseteq A$, the union $S(X) = \cup\{S(x) \mid x \in X\}$ of images $S(x)$ of all elements $x \in X$ is called the *image of the set* X under the correspondence S. In particular, the set $S(A) \subseteq B$ is called the *image* or the *range of the correspondence* S, also denoted by Ran S.

2. The correspondence $S^{-1} = \{(y, x) \mid (x, y) \in S\} \subseteq B \times A$ is called the *inverse* of S. The image $S^{-1}(y) \subseteq A$ of an element $y \in B$ under S^{-1} is called the *inverse image of* y *under* S. This inverse image may be empty. The image $S^{-1}(Y) \subseteq A$ of a set $Y \subseteq B$ under S^{-1} is called the *inverse image of* Y *under* S. It is obvious that $S^{-1}(Y) = \cup\{S^{-1}(y) \mid y \in Y\}$. The set $S^{-1}(B) \subseteq A$ is called the *domain of definition of* S. The correspondence S is said to be *defined on* $S^{-1}(B)$, this set being denoted by Dom S and called simply the *domain of* S.

3. It follows from the definitions that

$$\text{Dom } S = \{x \in A \mid y \in S(x) \text{ for some } y \in B\},$$
$$\text{Ran } S = \{y \in B \mid y \in S(x) \text{ for some } x \in A\}.$$

The condition in the first equality means that $S(x) \neq \varnothing$, and the condition in the second one means that $S^{-1}(y) \neq \varnothing$.) If Dom $S = A$, then the correspondence S is said to be *defined everywhere*. If $\operatorname{Ran} S = B$, then the correspondence S is called a *covering correspondence*.

Let $S \subseteq T \subseteq A \times B$. Then S is called a *restriction* of the correspondence T, and T is called an *extension* of the correspondence S. The inclusion $S \subseteq T$ is equivalent to the inclusions $S \subseteq \operatorname{Dom} T$ and $S(x) \subseteq T(x)$ for every $x \in \operatorname{Dom} S$.

4. Consider sets A, B, C and correspondences $S \subseteq A \times B$, $T \subseteq B \times C$. The correspondence $TS \subseteq A \times C$ defined by the formula $(TS)(x) = T(S(x))$ ($x \in A$) is called the *composition of correspondences* S and T. The domain $\operatorname{Dom}(TS)$ is empty if $\operatorname{Ran} S$ and $\operatorname{Dom} T$ do not intersect.

It follows from the definitions that $(TS)^{-1} = S^{-1}T^{-1}$.

The composition of the correspondences S and T can be represented in a diagram. Diagrams are especially convenient to represent compositions of more than two correspondences. The term *product* is often used as a synonym to *composition*.

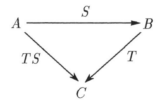

Example. Let $F = \{0, 1, 2, 3, 4, 5, 6, 7, 8, 9\}$ and $F \times F = \{00, 01, \ldots, 09;$ $10, 11, \ldots, 19; \ldots 90, 91, \ldots, 99\}$. The correspondence $S = \{01, 02, \ldots, 09,$ $12, \ldots, 19, \ldots, 89\} \subseteq F \times F$ is a *strict order* on F. In this case, $y \in S(x)$ means that $y > x$. The correspondence $S^{-1} = \{10, 21, 20, \ldots, 98, \ldots, 91, 90\}$ $\subseteq F \times F$ is the *inverse of the strict order* on F. Then $y \in S^{-1}(x)$ means that $y < x$.

The correspondence $T = \{00, 11, 24, 39\} \subseteq F \times F$ associates every number with its square: $z \in T(y)$ means $z = y^2$. We have $T(S(0)) = T(\{1, 2, 3,$ $\ldots, 9\}) = \{1, 4, 9\}$, $T(S(1)) = T(\{2, 3, \ldots, 9\}) = \{4, 9\}$, $T(S(2)) =$ $T(\{3, \ldots, 9\}) = \{9\}$, $T(S(3)) = \cdots = T(\{9\}) = \varnothing$. Therefore, $TS = \{01, 04, 09, 14, 19, 29\} \subseteq F \times F$.

1.2.2.   Functions

Correspondences for which the image of every element is either empty or consists of exactly one element are called functions.

1. A correspondence $f \subseteq A \times B$ such that every $x \in \operatorname{Dom} f \subseteq A$ has a unique image $y \in \operatorname{Ran} f \subseteq B$ is said to be a *many-to-one correspondence*. A many-to-one correspondence $f \subseteq A \times B$ is also called a *function from A to B* or a *function defined on A with values in B*. It is conventional to write $y = f(x)$ instead of $\{y\} = f(x)$. The element y is called the *value of f* corresponding to x.

Exercise. Prove that the composition of functions is a function.

Remark. By definition, a function is a set of ordered pairs. From the definition, it is intuitively clear that a function is equivalent to its graph. Sometimes a graph $f \subseteq A \times B$ is defined independently and a correspondence from A to B is defined as a triple (A, B, f). This provides a way to treat functions as sets, which often appears to be convenient.

2. Functions defined everywhere are called *maps*. A function $f \subseteq A \times B$ with $\operatorname{Dom} f = A$ is said to map A *to B*. In addition, if $\operatorname{Ran} f = B$, then f is said to map A *onto B*. If f maps A to B, it is conventional to write f: $A \to B$ or $A \xrightarrow{f} B$ instead of $f \subseteq A \times B$. The set of all maps of A to B is denoted by B^A.

3. The correspondence $f^{-1} \subseteq B \times A$ inverse to $f \subseteq A \times B$ is not necessarily a function. If f is a many-to-one correspondence and so is its inverse, then f is called a *one-to-one correspondence*. A one-to-one map from A to B is called an *injection* or *embedding*. A one-to-one map from A onto B is called a *surjection*. It is also called an *isomorphism from A onto B*. If there exists an isomorphism between two sets, these sets are said to be *isomorphic*. An isomorphism from A onto B is also called a *substitution of A into B*, and an isomorphism from A onto itself is called an *automorphism* or a *permutation* of A. All isomorphisms are one-to-one surjections.

Examples. 1) Let $F = \{0, 1, 2, 3, 4, 5, 6, 7, 8, 9\}$. Then $F \times F = \{00, 01, \ldots, 09; 10, 11, \ldots, 19; \ldots 90, 91, \ldots, 99\}$. The correspondence $f = \{00, 12, 24, 36, 48\}$ is a function with domain $\operatorname{Dom} f = \{0, 1, 2, 3, 4\}$ and range $\operatorname{Ran} f = \{0, 2, 4, 6, 8\}$. The inverse correspondence $f^{-1} = \{00, 21, 42, 63, 84\}$ is also a function with domain $\operatorname{Dom} f^{-1} = \operatorname{Ran} f$ and range $\operatorname{Ran} f^{-1} = \operatorname{Dom} f$. It is clear that f is a one-to-one correspondence. It maps the set $A = \{0, 1, 2, 3, 4\}$ onto $B = \{0, 2, 4, 6, 8\}$. These sets are isomorphic.

2) The map $X \to y$ ($X \in \mathcal{D}$, $y \in X$) of a disjoint class $\mathcal{D}$ onto its sample Y is an isomorphism (see Section 1.1.3).

Remark. We emphasize the importance of specifying the correspondences together with the sets involved. The set of pairs f in the preceding example is included in both Cartesian products $A \times B$ and $F \times F$. At the same time, the correspondence f in $A \times B$ is an isomorphism, whereas it is not so in $F \times F$.

4. For sets A, B, C, consider embeddings $f : A \to B$, $g : B \to C$, and their composition $h = gf : A \to C$.

Proposition. *The composition h of embeddings f and g is an embedding.*

Proof. We now prove that h is an embedding. Let $h(x) = h(y)$ for some $x, y \in A$. This means that $g(f(x)) = g(f(y))$. We have $f(x) = f(y)$ because g is a one-to-one correspondence. Since f is a one-to-one correspondence, it follows that $x = y$. The composition h is a one-to-one correspondence. $\square$

Corollary. *The composition h of isomorphisms f and g is an isomorphism.*

Proof. Based on the above proposition, it suffices to show that the composition $h = gf$ of the surjections g and f is a surjection. Take any $z \in C$. Since g is a surjection, there exists an element $y \in B$ such that $g(y) = z$. Since f is a surjection, there exists an element $x \in A$ such that $f(x) = y$. Consequently, $h(x) = g(f(x)) = z$ and thus h is a surjection. $\square$

5. It is often convenient to represent sets by functions called indicators.

Consider a set U, the class $\mathcal{P} = \mathcal{P}(U)$ of all subsets of U, and the set $\mathbb{B} = \{0, 1\}$. For every $X \in \mathcal{P}$, the function $\operatorname{ind} X : U \to \mathbb{B}$ is defined, with values $\operatorname{ind} X(u) = 1$ for $u \in X$ and $\operatorname{ind} X(u) = 0$ for $u \notin X$.

Functions defined on U with values in $\mathbb{B}$ are called *indicators on* U. The function $\operatorname{ind} X$ is called the *indicator of the subset* X *of* U. The set of all indicators on U is denoted by $\mathcal{I} = \mathcal{I}(U)$.

The map ind from $\mathcal{P}$ to $\mathcal{I}$, which associates every set $X \in \mathcal{P}$ with its indicator $\operatorname{ind} X \in \mathcal{I}$, is an isomorphism from $\mathcal{P}$ onto $\mathcal{I}$. The isomorphism $\operatorname{ind} : \mathcal{P} \to \mathcal{I}$ associates sets in $\mathcal{P}$ with their indicators in $\mathcal{I}$.

Exercise. Prove that $\operatorname{ind} : \mathcal{P} \to \mathcal{I}$ is an isomorphism.

Remark. We have demonstrated that functions can be thought of as sets of a specific kind, while sets can be thought of as functions of a specific kind.

6. Functions can be thought of as operators of a general kind, or *abstract operators*. All operators considered in the following sections have specific domains and ranges, as well as specific properties of the correspondences between the domain and the range.

All general definitions and statements for correspondences and functions are valid in the operator theory. The properties of images, inverse images, and compositions are of special importance in this theory. Complex compositions can be conveniently represented using diagrams with arrows.

1.2.3. Collections of sets

The term *collection* is a synonym for the term *function*. It is used in a specific notation for values of functions that involves indices, which often appears to be convenient.

1. Consider a set I of elements called *indices*. Any map $f : I \to B$ that maps I into B will be called a *collection of elements of B indexed by I*. We will use the designation f_i $(i \in I)$ for $f(i)$, and (f_i) for f. Thus, $f = (f_i)$. Whenever necessary, the index set is specified: (f_i) $(i \in I)$.

For every collection of sets $X_i \in \mathcal{P} = \mathcal{P}(E)$, we define the *union* $\cup X_i$ and the *intersection* $\cap X_i$:

$$\cup X_i = \{x \mid x \in X_i \text{ for some } i \in I\},$$
$$\cap X_i = \{x \mid x \in X_i \text{ for every } i \in I\}.$$

These definitions agree with those for the union and intersection of a class of sets given in Section 1.1.2. The index set is specified whenever necessary: $\bigcup_{i \in I}, \bigcap_{i \in I}$.

From the definitions, it follows that

$$\bigcup X_i = \varnothing, \quad \bigcap X_i = E \quad (i \in \varnothing)$$

for subsets of the collection E indexed by the empty set I. A collection whose index set is empty is called an *empty collection*.

For a set A, a collection of nonempty sets X_i such that A is included in $X = \bigcup X_i$ is called a *cover of A*. If $X = A$ and the sets X_i are pairwise disjoint, then the cover (X_i) is called a *set partition of A*.

Every pairwise disjoint class $\mathcal{D}$ of nonempty sets $X \subseteq E$ can be thought of as a collection — the identity map in $\mathcal{D}$, which maps $\mathcal{D}$ onto itself. In this case, the sets $X \in \mathcal{D}$ serve as indices of themselves. If $\bigcup \mathcal{D} = E$, then $\mathcal{D}$ is also called a *set partition of E*. For the sake of simplicity, the requirement for elements of set partitions to be nonempty is sometimes omitted.

The Cartesian product of the disjoint class $\mathcal{D}$ consisting of sets $\{i\} \times X_i = \{(i, x_i) \mid x_i \in X_i\}$ $(i \in I)$ is called the *Cartesian product of the collection* X_i $(i \in I)$ and denoted by $\prod X_i$ or $\prod_{i \in I} X_i$. Its elements are collections $x = (x_i)$ of $x_i \in X_i$ $(i \in I)$:

$$\prod X_i = \{x = (x_i) \mid x_i \in X_i \text{ for every } i \in I\};$$

or, equivalently, functions $f : I \to X$ such that $f(i) \in X_i$ for every index $i \in I$:

$$\prod X_i = \{f : I \to X \mid f(i) \in X_i \text{ for every } i \in I\}.$$

From the definition, it follows that the Cartesian product of an empty collection is empty.

2. The main properties of operations defined for collections of sets are easy to verify. They can be represented as identities. If A, B, and C are index sets, X_α $(\alpha \in A)$ and Y_β $(\beta \in B)$ are collections of sets, $p : A \to A$ is a permutation, $A(\gamma)$ $(\gamma \in C)$ is a collection of subsets of A that is a cover of A, then the following properties hold:

$$\bigcup_{\alpha \in A} X_{p(\alpha)} = \bigcup_{\alpha \in A} X_\alpha, \quad \bigcap_{\alpha \in A} X_{p(\alpha)} = \bigcap_{\alpha \in A} X_\alpha \tag{1}$$

— *commutativity*;

$$\bigcup_{\alpha \in A} X_\alpha = \bigcup_{\gamma \in C}\left(\bigcup_{\alpha \in A(\gamma)} X_\alpha\right), \quad \bigcap_{\alpha \in A} X_\alpha = \bigcap_{\gamma \in C}\left(\bigcap_{\alpha \in A(\gamma)} X_\alpha\right) \tag{2}$$

— *associativity*;

$$\left(\bigcup_{\alpha \in A} X_\alpha\right) \cap \left(\bigcup_{\beta \in B} Y_\beta\right) = \bigcup_{(\alpha,\beta) \in A \times B} (X_\alpha \cap Y_\beta),$$

$$\left(\bigcap_{\alpha \in A} X_\alpha\right) \cup \left(\bigcap_{\beta \in B} Y_\beta\right) = \bigcap_{(\alpha,\beta) \in A \times B} (X_\alpha \cup Y_\beta) \tag{3}$$

— *distributivity*;

$$\left(\bigcup_{\alpha \in A} X_\alpha\right)' = \bigcap_{\alpha \in A} X_\alpha', \quad \left(\bigcap_{\alpha \in A} X_\alpha\right)' = \bigcup_{\alpha \in A} X_\alpha' \tag{4}$$

— duality;

$$\left(\bigcup_{\alpha \in A} X_\alpha\right) \times \left(\bigcup_{\beta \in B} Y_\beta\right) = \bigcup_{(\alpha,\beta) \in A \times B} (X_\alpha \times Y_\beta),$$

$$\left(\bigcap_{\alpha \in A} X_\alpha\right) \times \left(\bigcap_{\beta \in B} Y_\beta\right) = \bigcap_{(\alpha,\beta) \in A \times B} (X_\alpha \times Y_\beta) \tag{5}$$

— cartesian distributivity.

If $A = B$, then the foregoing identity can be simplified as follows:

$$\left(\bigcap_{\alpha \in A} X_\alpha\right) \times \left(\bigcap_{\alpha \in A} Y_\alpha\right) = \bigcap_{\alpha \in A} (X_\alpha \times Y_\alpha). \tag{6}$$

Exercise. Prove the above identities.

4. Consider sets E, F, and $Y \subseteq F$, collections $X_i \subseteq E$ and $Y_j \subseteq F$, and a map $f : E \to F$. The following equalities hold for images and inverse images:

$$f(\cup X_i) = \cup f(X_i); \tag{7}$$
$$f^{-1}(\cup Y_j) = \cup f^{-1}(Y_j), \quad f^{-1}(\cap Y_j) = \cap f^{-1}(Y_j); \tag{8}$$
$$f^{-1}(Y') = f^{-1}(Y)'. \tag{9}$$

For the image of intersection, only the following inclusion is true:

$$f(\cap X_i) \subseteq \cap f(X_i). \tag{10}$$

Exercise. Prove equalities (7)–(10).

Example. Let $E = F = \{0,1,2,3,4,5,6,7,8,9\}$, $X_1 = \{1,3,5,7,9\}$, $X_2 = \{0,2,4,6,8\}$ and let $f : E \to F$ be defined by the equalities $f(0) = f(1) = 1$, $f(2) = f(3) = 3$, $f(4) = f(5) = 5$, $f(6) = f(7) = 7$, $f(8) = f(9) = 9$. Then $f(X_1) = f(X_2) = X_1$, $X_1 \cap X_2 = \varnothing$ and $f(X_1 \cap X_2) = f(\varnothing) = \varnothing \subset X_1 = f(X_1) = f(X_2) = X_1$.

Let $Y_1 = \{1, 2, 3, 4, 5\}$, $Y_2 = \{2, 3, 4, 5, 6\}$, $Y_3 = \{3, 4, 5, 6, 7\}$. We have $f^{-1}(Y_1) = \{0, 1, 2, 3, 4, 5\}$, $f^{-1}(Y_2) = \{2, 3, 4, 5\}$, $f^{-1}(Y_3) = \{2, 3, 4, 5, 6, 7\}$ and

$$Y_1 \cap Y_2 \cap Y_3 = \{3, 4, 5\}, \quad f^{-1}(Y_1) \cap f^{-1}(Y_2) \cap f^{-1}(Y_3) = \{2, 3, 4, 5\}.$$

Therefore, $f^{-1}(Y_1 \cap Y_2 \cap Y_3) = \{2, 3, 4, 5\} = f^{-1}(Y_1) \cap f^{-1}(Y_2) \cap f^{-1}(Y_3)$.

This example illustrates identities (8) and inclusion (10).

6. The union $\cup$ and intersection $\cap$ of a collection of sets can be viewed as abstract operators. Their range is $\mathcal{P}$ and their domain of definition is the set $F(I, \mathcal{P})$ of maps that map an index set I to the class $\mathcal{P} = \mathcal{P}(E)$ of all subsets of a given set E. For every map $f : I \to \mathcal{P}$, its images under the operators of union and intersection are $\cup f = \bigcup_i f_i \in \mathcal{P}$ and $\cap f = \bigcap_i f_i \in \mathcal{P}$, respectively.

Equalities (1)–(10) represent the properties of the operators $\cup$ and $\cap$.

1.3.   RELATIONS

Correspondences that are subsets of $E \times E$ are called *relations* on E. Important examples of relations are equivalence and order. Note that any map from A to B is a relation on $E = A \cup B$.

1.3.1.   Reflexivity, transitivity, and symmetry

The properties of reflexivity, transitivity, and symmetry are very common and can be considered to be the most important.

1. A relation $R \subseteq E \times E$ such that $x \in R(x)$ for every $x \in E$ is said to be *reflexive*. In a reflexive relation, every element of E is related to itself $(x \to x)$ and, maybe, other elements.

A relation $R \subseteq E \times E$ such that $y \in R(x)$ and $z \in R(y)$ implies $z \in R(x)$ for every $x, y, z \in E$ is said to be *transitive*. In other words, in a transitive relation, if x is related to y, then x is related to all elements that y is related to: $x \to y$, $y \to z \Rightarrow x \to z$.

A relation $R \subseteq E \times E$ such that $y \in R(x)$ implies $x \in R(y)$ for all $x, y \in E$ is said to be *symmetric*: $x \to y \Rightarrow y \to x$. Every symmetric relation coincides with its inverse $R = R^{-1}$.

A relation $R \subseteq E \times E$ such that $y \in R(x)$ and $x \in R(y)$ implies $x = y$ for all $x, y \in E$ is said to be *antisymmetric*: $x \to y$, $y \to x \Rightarrow x = y$. Under an antisymmetric relation, only equal elements are related to each other.

2. Examples. The relations S and T in the example of Subsection 1.2.1 are transitive and antisymmetric, and so is their composition TS.

Consider the relation $R = \{00, 11, 22, 33, 44, 55, 66, 77, 88, 99\}$, which is the equality relation for digits. It is reflexive, transitive, and symmetric. The order relation $R \cup S$ for digits is reflexive and transitive.

The relation f in Example 1 of Subsection 1.2.2 is not transitive. Indeed, if $y = 2x$ and $z = 2y$, then $z \neq 2x$ for $x \neq 0$.

1.3.2. Equivalence

A reflexive, transitive, and symmetric relation is called an *equivalence relation*. If R is an equivalence relation on E, then the notation $x \sim y$ is often used instead of $y \in R(x)$ and $x \to y$.

1. The following theorem provides a criterion for equivalence relations.

Theorem. *Every equivalence relation on a given set defines a partition of this set. Conversely, every partition of a given set introduces an equivalence relation on it.*

Proof. Let R be an equivalence relation on E and $\mathcal{D}$ is the class of all images $R(x)$ of elements $x \in E$. It is required to prove that $\mathcal{D}$ is a partition of E. Since R is reflexive, $x \in R(x)$ and therefore $\mathcal{D}$ is a cover of E. Let $a, b \in E$, $x \in R(a)$, $y \in R(b)$, and suppose that there exists an element $c \in R(a) \cap R(b)$, i.e., $x \sim a$, $y \sim b$ and $c \sim a$, $c \sim b$. Since R is symmetric and transitive, we have $c \sim a$, $c \sim b \Rightarrow a \sim c$, $c \sim b \Rightarrow a \sim b$ and $x \sim a$, $a \sim b \Rightarrow x \sim b \Leftrightarrow x \in R(b) \Rightarrow R(a) \subseteq R(b)$. Similarly, $y \sim b$, $a \sim b \Rightarrow y \sim b$, $b \sim a \Rightarrow y \sim a \Leftrightarrow y \in R(a) \Rightarrow R(b) \subseteq R(a)$. It follows that $R(a) = R(b)$. Consequently, if $R(a) \neq R(b)$, then $R(a) \cap R(b) = \varnothing$, and therefore $\mathcal{D}$ is a partition of E.

We now prove the converse. Let $\mathcal{D}$ be a partition of E such that all subsets in the partition are nonempty. Consider the relation R on E introduced by $\mathcal{D}$ such that every element $x \in E$ is related to the set $X \in \mathcal{D}$ that covers x. By definition, R is the union of cartesian squares $X \times X$ of all sets $X \in \mathcal{D}$. Since $x \in X = R(x)$, it follows that R is reflexive. If $y \in X = R(x)$, then $X = R(y)$, $x \in R(y)$ and therefore R is symmetric. Since $R(x) = R(y)$ for any $y \in R(x)$, we have $z \in R(x)$ for any $z \in R(y)$, which means that R is transitive. Thus, R is an equivalence relation. $\square$

The partition of E introduced by the equivalence relation R is called a *factor class* obtained by factoring E with respect to R and is denoted by

E/R. The elements of E/R are called *factor sets*. Any two elements that belong to the same factor set are equivalent.

2. Examples. 1) The simplest example of an equivalence relation is equality on a set. It defines a partition of the set consisting of one-element subsets:

$$R = \{(x, x) \mid x \in E\}, \quad \mathcal{D} = \{\{x\} \mid x \in E\}.$$

2) The relation that associates any two digits whenever they are both even or both odd is an equivalence relation introduced by the partition

$$\mathcal{D} = \{\{0, 2, 4, 6, 8\}, \{1, 3, 5, 7, 9\}\}$$

of the set $F = \{0, 1, 2, 3, 4, 5, 6, 7, 8, 9\}$.

3) Consider a set of statements associated with subsets of a set U so that the implication $A \Rightarrow B$ means $A \subseteq B$ for the corresponding sets. Then the logical equivalence $A \Leftrightarrow B$ of the statements A and B means $A = B$ for the corresponding sets. The equality of sets is an equivalence relation on the class $\mathcal{P} = \mathcal{P}(U)$ of all subsets of U.

3. A less trivial example of an equivalence relation uses the definition of cardinality, which is a formal generalization of the intuitive notion of the number of elements in a set.

Let $\mathcal{P} = \mathcal{P}(U)$ be the class of all subsets of a set U. By $\leftrightarrow$ we denote the relation on $\mathcal{P}$ induced by an isomorphism of two sets: for $X, Y \in \mathcal{P}$, $X \leftrightarrow Y$ means that X and Y are isomorphic (i.e., there exists a one-to-one map from X onto Y). The relation $\leftrightarrow$ is an equivalence relation. Indeed, it is reflexive (the identity map from X onto itself is an isomorphism), transitive (the composition of isomorphisms from X onto Y and from Y onto Z is an isomorphism), and symmetric (the inverse of any isomorphism from X onto Y is an isomorphism from Y onto X).

Thus, $\mathcal{P}$ can be partitioned into a collection of pairwise disjoint nonempty classes, each consisting of sets isomorphic to each other. Isomorphic sets are said to be *equipollent*, and classes of sets isomorphic to each other are called *cardinals*. For example, the class $0 = \{\varnothing\}$, whose only member is the empty set, and the class $1 = \{\{x\} \mid x \in U\}$ of all singletons $\{x\} \in \mathcal{P}$ are cardinals.

For a set X, the class of all sets isomorphic to X is called the *power* of X or the *cardinal number* of X, and is denoted by $|X|$. This class depends on the given universal set U whose subsets are elements of $|X|$. To emphasize this dependence, the notation $|X|_U$ can be used instead of $|X|$. By Dedekind's definition, any set X that is not isomorphic to any of its proper subsets is said to be *finite* . If X is finite, we write $|X| < \infty$. Otherwise, X is said to be *infinite* ($|X| = \infty$). Obviously, every subset of a finite set is finite. Operations defined for cardinal numbers are described in detail in Kuratowski and Mostowski (1970, Chapter 5), and Bourbaki (1965, Chapter 3).

Exercise. Prove that if A is a finite set and B is an infinite set, then $|A \cup B| = |B|$.

The following theorem provides a criterion for two sets to be equipollent.

The Schröder-Bernstein Theorem. *Sets A and B are equipollent if and only if there are injections of the set A into the set B and of B into A.*

The proof can be found in Kuratowski and Mostowski (1970, (Chapter 1, Corollary 4.4).

Remark. In naive set theory, which is employed here, it is assumed that the introduction of the universal set helps to avoid some paradoxes associated with the set of all sets (Kuratowski and Mostowski, 1970, Chapter 2, Section 3). The existence of the class of all subsets of a given set is postulated by the Axiom of power set (Kuratowski and Mostowski, 1970, Chapter 2, Section 2). For this reason, in the present book, by the term a *class of sets* we usually mean a *class of subsets of a certain universal set*. These notions are assumed to be self-evident no formal definition is provided for them.

1.3.3. Order

Any reflexive and transitive relation is called an *order*.

Most orders considered in practice possess some additional properties. For example, any symmetric order is an equivalence relation. The union of any transitive relation and the equality relation is an order. Order is not assumed to be defined on the entire set and therefore is also called *partial order*.

1. Let R be an order on a set E. We will write $y \succeq x$ instead of $(x, y) \in R$ and say that the element y *succeeds to* the element x.

An important type of order is an antisymmetric order. Moreover, the terms *order* and *partial order* usually imply antisymmetry. A relation that is only reflexive and transitive is called a *preorder*. For antisymmetric orders, we write $y \geq x$ instead of $y \succeq x$ and say that y is *greater than* x. By the definition of an antisymmetric relation, $x \geq y$ and $y \geq x$ implies $y = x$.

For an order R on E, its inverse R^{-1} is an order on E. We will write $y \preceq x$ instead of $(x, y) \in R^{-1}$ and say that the element y *precedes* the element x. If R is antisymmetric, then R^{-1} is also antisymmetric. For antisymmetric orders, we write $y \leq x$ instead of $y \preceq x$ and say that y is *less than* x.

The negation of a relation statement will be denoted by a slash mark across the relation sign.

An order can be viewed as "direct" and denoted by $\succeq$ in one case, whereas in other contexts it may be convenient to view the same order as inverse and denote it by $\preceq$, since these terms and designations are relative.

If $y \succeq x$ and $y \not\preceq x$, we write $y \succ x$ and say that y *strictly succeeds* to x. If $y \geq x$ and $y \not\leq x$, we write $y > x$ and say that y *is strictly greater than* x. Note that the condition $y \not\leq x$ is equivalent to $y \neq x$ by antisymmetry. The designations $y \prec x$, $y < x$, and the terms *strictly precede* and *strictly less* are defined similarly.

Any elements x and y such that $y \succeq x$ or $x \succeq y$ are said to be *comparable*. An order such that any two elements of the ordered set in question are comparable is called a *linear order* or a *linear order*. For a total antisymmetric order, the trichotomy law holds: for any y and x, either $y > x$, or $y = x$, or $y < x$.

Example. For any set, the set of cardinals of its subsets is linearly ordered (see Kuratowski and Mostowski, 1970, Chapter 1, Corollary 5.5).

Any two elements x and y of a partially ordered set such that $y \succeq x$ and $x \succeq y$ are called *order-equivalent* or simply *equivalent*. It is easy to verify that this relation is reflexive, transitive, and symmetric. This equivalence will be denoted by $\equiv$.

Exercise. Prove that $\equiv$ is an equivalence relation.

Any set E taken together with an order $\succeq$ on it, i.e., the pair $(E, \succeq)$ is an *ordered set*. It is often denoted simply by E whenever the meaning is clear from the context.

Examples. The usual order on the set of digits can be written as $0 < 1 < 2 < 3 < 4 < 5 < 6 < 7 < 8 < 9$. It is a total antisymmetric order.

The set of all ordered pairs of digits can be ordered with respect to their first elements. For any digits a, b, x, y, let $ab \leq xy$ mean that $a \leq x$. This order is not antisymmetric. For example, $12 \leq 13$ and $13 \leq 12$ although $12 \neq 13$.

The inclusion relation $\supseteq$ is an order on the class $\mathcal{P} = \mathcal{P}(U)$ of all subsets of a set U. This order is antisymmetric, but it is not total if there exist at least two distinct elements in U.

2. Let X be a subset of a set E and $\succeq$ be an order on E.

If $b \in E$ succeeds to every $x \in X$, then b is said to *succeed to the set* X, which is denoted by $b \succeq X$. If an element $a \in E$ precedes every $x \in X$, then a is said to *precede the set* X, which is denoted by $a \preceq X$.

Any ordered set E such that for any $x, y \in E$ there is a $z \in E$ that succeeds to x and y is called a *directed set*. Directed sets are sometimes called *nets* (because of the graphical representation of elements of directed sets in the form of net points).

We emphasize that the order that defines a directed set is not necessarily antisymmetric.

3. Assume that an order on E is antisymmetric. For $b \in E$ and $X \subset E$ such that *b succeeds to X*, we say that b is an *upper bound of X*, or X is *majorized by b*, or b is a *majorant of X*. If an element a precedes X, then a is called an *upper bound of X*, or a *minorant of X*. An upper bound of X that is contained within X is called the *greatest element of X* and denoted by $\max X$. A lower bound of X that is contained within X is called the *least element of X* and denoted by $\min X$. The least element of a set is sometimes called its *initial* element, and the greatest element is called its *last* or *terminal* element.

The least upper bound of X is called the *supremum of X* and is denoted by $\sup X$. The greatest lower bound of X is called the *infimum of X* and is denoted by $\inf X$. The supremum and infimum of X do not necessarily belong to X. The notions of infimum and supremum are extensions of the notions of the least element and the greatest element.

An element that is not smaller than any other element in E is called a *maximal element*. An element that is not greater than any other element in E is called a *minimal element*. These notions are extensions of the notions of the smallest and the greatest elements of an ordered set.

The greatest and the least elements of an ordered set are necessarily unique (if they exist). This is not the case for maximal and minimal elements. For any two comparable elements x and y, however, the greatest element and the least element of the set $\{x, y\}$ do coincide with its maximal and minimal elements, respectively. The least of them is denoted by $\min\{x, y\}$ or $x \wedge y$, and the greatest by $\max\{x, y\}$ or $x \vee y$. As these notions are sometimes easy to confuse, special care must be taken when using them for partial orders.

Examples. Consider the class $\mathcal{P} = \mathcal{P}(U)$ of all subsets of a set U and the order relation $\supseteq$ for $\mathcal{P}$. Clearly, $U = \max \mathcal{P}$ and $\varnothing = \min \mathcal{P}$.

It follows from the definitions that $\sup\{X, Y\} = X \cup Y$ and $\inf\{X, Y\} = X \cap Y$ for any $X, Y \in \mathcal{P}$. Furthermore, for any class $C \subseteq \mathcal{P}$, $\sup C = \cup C$ and $\inf C = \cap C$.

The set $E = \{00, 01, \ldots, 09, 10, 20, \ldots, 90\}$ with order relation $00 < 01 < \cdots < 09$; $10 < 20 < \cdots < 90$ has two maximal and two minimal elements.

4. Let E and F be ordered sets. We will take the liberty of using the same designation $\succeq$ for both orders. A map $f : E \to F$ such that $f(y) \succeq f(x)$ whenever $y \succeq x$ $(x, y \in E)$ is called *monotonic*. A monotonic map is also said to be *order-preserving*. If $f(y) \preceq f(x)$ whenever $y \succeq x$ $(x, y \in E)$, then f is called *antimonotonic*. Antimonotonic maps reverse the order.

Monotonic and antimonotonic maps defined on ordered sets with antisymmetric orders are called *increasing* and *decreasing*, respectively. Let E and F be ordered sets with antisymmetric orders denoted by $\geq$ in both cases. A map $f : E \to F$ such that $f(y) > f(x)$ whenever $y > x$ $(y, x \in E)$ is said to be *strictly increasing*. A map $f : E \to F$ such that $f(y) < f(x)$ whenever $y > x$ $(y, x \in E)$ is said to be *strictly decreasing*.

If E and F are ordered sets, a strictly increasing one-to-one map f from E to F such that f^{-1} is strictly increasing is called an *embedding from E into F*. If $f(E) = F$, then f is called an *isomorphism from E onto F* and the ordered sets E and F are called *isomorphic*. These definitions of embedding and isomorphism are extensions of similar definitions for sets in general, the added requirement being the preservation of order.

If there exists an isomorphism from an ordered set E onto an ordered set F, then the corresponding orders are called *equivalent*.

Examples. For $E = \{\alpha, \beta, \gamma, \delta, \varepsilon, \zeta, \eta, \theta, \iota, \varkappa\}$ and $F = \{0, 1, 2, 3, 4, 5, 6, 7, 8, 9\}$, we define the orders $\alpha < \beta < \gamma < \delta < \varepsilon < \zeta < \eta < \theta < \iota < \varkappa$ and $0 < 1 < 2 < 3 < 4 < 5 < 6 < 7 < 8 < 9$, respectively. Then the map $f : E \to F$ such that $f(\alpha) = 0$, $f(\beta) = 1$, $f(\gamma) = 2$, $f(\delta) = 3$, $f(\varepsilon) = 4$, $f(\zeta) = 5$, $f(\eta) = 6$, $f(\theta) = 7$, $f(\iota) = 8$, and $f(\varkappa) = 9$ is an order-preserving isomorphism from E onto F. These ordered sets are isomorphic.

The alphabetical order on the set of letters E and the conventional order of numbers in F are equivalent.

5. Let A and B be ordered sets. It is possible to introduce an order on the cartesian product $A \times B$ based on the orders on A and B. The two main types of order that can be introduced on $A \times B$ are coordinatewise order and lexicographic order.

The *coordinatewise order* is defined as follows: $(x, y) \succeq (a, b)$ in $A \times B$ if $x \succeq a$ in A and $y \succeq b$ in B. The *lexicographic order* is defined as follows: $(x, y) \succeq (a, b)$ in $A \times B$ if $(1) x \succ a$ in A or (2) $x \equiv a$ in A and $y \succeq b$ in B (if the first coordinates are order-equivalent, then the order of two pairs is determined by the order of the second coordinates). It is easy to verify that the orders defined this way on $A \times B$ are reflexive and transitive.

Exercise. Prove that the coordinatewise and lexicographic orders are reflexive and transitive.

The ordered set $A \times B$ with coordinatewise order is called the *Cartesian product of the ordered sets A and B*. The ordered set $A \times B$ with lexicographic order is called the *lexicographic product of the ordered sets A and B*. It is obvious that the relation $(x, y) \succeq (a, b)$ holds in the lexicographic sense if it holds coordinatewise. The converse is not true.

Example. Let $A = \{1, \ldots, 9\}$, $B = \{0, 1, \ldots, 9\}$ and $0 < 1 < 2 < 3 < 4 < 5 < 6 < 7 < 8 < 9$. Then $A \times B = \{10, 11, \ldots, 19; \ldots; 90, 91, 99\}$ and the lexicographic order is defined by the inequalities $10 < 11 < \cdots < 19 \ldots 90 < 91 < 99$. For the coordinatewise order, the inequalities $10 < 11 < \cdots < 19$ and $< 90 < 91 < 99$ hold, whereas the inequality $19 < 90$ does not hold. The pairs 19 and 90 are not comparable coordinatewise.

We now define an order relation for the cardinal numbers $|X|$ and $|Y|$ of the sets X and Y: by $|X| \leq |Y|$ we will mean that X can be embedded into Y. The relation $\leq$ is obviously reflexive. By the proposition on the composition of embeddings and the Schröder-Bernstein Theorem, this relation is also transitive and antisymmetric.

By definition, the power of the cartesian product $A \times B$ is equal to the product of the powers of A and B:

$$|A \times B| = |A| \cdot |B|.$$

If A or B is infinite, then $|A \times B| = |A| \vee |B|$. In particular, if A is infinite and B is countable, then $|A \times B| = |A|$ (see Bourbaki, 1965, Chapter 3, Section 6.3, Corollary 4).

Consequently, the class $\mathcal{D}(A)$ of all countable subsets of any infinite set A ($|A| = \infty$) is equipotent to A. The same is true for the class $\mathcal{K}(A) \subseteq \mathcal{D}(A)$ of all finite subsets of A:

$$|A| = |\mathcal{K}(A)| = |\mathcal{D}(A)| \quad (|A| = \infty).$$

Let $Y^X = \mathcal{F}(X, Y)$ be the set of all maps from X into Y. If $|X| \geq 1$ and $|Y| \geq 2$, then $|Y^X| > |X|$. For example, for $Y = \{0, 1\}$, the set of all indicators on X is equipotent to the class $2^X = \mathcal{P}(X)$ of all subsets of X, and $|2^X| = |\mathcal{P}(X)| > |X|$ by the Cantor theorem (Kuratowski and Mostowski, 1970, Chapter 5, Section 3, Theorem 2; Bourbaki, 1965, Chapter 3, Section 3, Theorem 2).

It is easy to verify that if the orders on A and B are antisymmetric, then so are the coordinatewise and the lexicographic orders on $A \times B$. In this case, the condition $x \equiv a$ in the definition of the lexicographic order can be replaced with the condition $x = a$. If the orders on A and B are total, then the same is true for the lexicographic order on $A \times B$, but not necessarily for the coordinatewise order, which was illustrated in the previous example.

It is clear that the cartesian product and the lexicographic product of directed sets is a directed set.

6.   Let $\mathcal{D}$ be a class of pairwise disjoint ordered sets. It is possible to introduce an order on $\cup \mathcal{D}$ based on the orders on sets in $\mathcal{D}$ and an antisymmetric order on $\mathcal{D}$. For $x \in X$ and $y \in Y$, where $X, Y \in \mathcal{D}$, we put $y \succeq x$ if (1) $Y > X$ in $\mathcal{D}$ or (2) $Y = X$ and $y \succeq x$ in X. The union $\cup \mathcal{D}$ ordered this way is called the *order sum* of the sets of the class $\mathcal{D}$. It is denoted by $\sum \mathcal{D}$.

Example. Let $\mathcal{D}$ be the class containing the sets $X = F = \{0, 1, \ldots, 9\}$, $Y = (F \setminus \{0\}) \times F = \{10, 11, \ldots, 99\}$, and $Z = Y \times X = \{100, 101, \ldots, 999\}$. Let the orders on X, Y, and Z be defined by the inequalities $0 < 1 < \cdots < 9$, $10 < 11 < \cdots < 99$, and $100 < 101 < \cdots < 999$, respectively. Let the order on $\mathcal{D}$ be defined by the inequalities $X < Y < Z$. The union $\cup \mathcal{D} = X \cup Y \cup Z$ becomes the order sum $\sum \mathcal{D} = X + Y + Z$ if we introduce two more inequalities: $9 < 10$ and $99 < 100$.

It is easy to verify that if the orders on the component sets are antisymmetric, then so is the order on the union. Furthermore, if $\mathcal{D}$ and each set in it are linearly ordered, then so is the union.

Exercise. Prove the foregoing statements.

Remark. The order on $\cup\mathcal{D}$ used in the definition of the order sum $\sum\mathcal{D}$ is related to the lexicographic order. Indeed, for any two elements, we first compare the sets that contain these elements. Then, if these sets are equal, the elements themselves are compared. That is, we use the lexicographic order for pairs of the form $(X, x) \in \mathcal{D}\times\cup\mathcal{D}$ such that $X \in \mathcal{D}$ and $x \in X$. The pairs (X, y) such that $X \in \mathcal{D}$ and $y \notin X$ are not taken into consideration.

1.4.   INDUCTION

In this section, we prove several theorems for specific ordered sets. These theorems are often used in mathematical argument. All the orders considered in this section are assumed to be antisymmetric.

1.4.1.   Well-ordered sets

An ordered set E with order $\geq$ is said to be *well ordered* if every nonempty subset of E has a least element. The order $\geq$ is called a *well-order relation*. Obviously, every well-order is a linear order.

If $\geq$ is a well-order, its inverse $\leq$ is not necessarily a well-order. In particular, if there is no greatest element in a well-ordered set E with well-order $\geq$, then $\leq$ is not a well-order.

The empty with empty order relation is a well-ordered set.

1. Every two elements $a, b \in E$ with linear order $\geq$ such that $a \leq b$ define the following *intervals* in E with *endpoints* a and b:

$$]a, b[= \{x \mid a < x < b\}, \quad [a, b[= \{x \mid a \leq x < b\},$$
$$]a, b] = \{x \mid a < x \leq b\}, \quad [a, b] = \{x \mid a \leq x \leq b\}.$$

Intervals of the form $[a, b]$ are also called *closed intervals*, and intervals of the form $]a, b[$ are called *open intervals* or *segments*. An interval of either of these types with endpoints a and b will be denoted by $|a, b|$. The elements a and b are also called the *left endpoint* and the *right endpoint* of the interval $|a, b|$.

Note that if $a = b$, then the interval $[a, a] = \{a\}$ consists of a single point, while $]a, a[= [a, a[=]a, a] = \varnothing$. Intervals of the form $]a, b[$ may be empty even if $a \neq b$.

For any $a \in E$, the following intervals in E with endpoint a are defined:

$$]a, \rightarrow[= \{x \mid a < x\}, \quad [a, \rightarrow[= \{x \mid a \leq x\}.$$

An interval of either of these types with endpoint a will be denoted by $|a, \rightarrow|$. Note that if E has a greatest element, $b = \max E$, then $|a, \rightarrow|$ is the same as $[a, b]$ or $]a, b]$. Similarly, we introduce intervals of the form

$$]\leftarrow, b[= \{x \mid x < b\}, \quad]\leftarrow, b] = \{x \mid x \leq b\}$$

with endpoint b.

2. Let E be a nonempty well-ordered set with $a = \min E$ and let X be a subset of E.

Theorem. *For any $x \in E$, let $[a, x[\subseteq X$ imply $x \in X$. Then $X = E$.*

Proof. Consider the complement set $X' = E \setminus X$. If $X \neq E$, then $X' \neq \varnothing$ and there is an element b such that $b = \min X' \in X'$ because E is well ordered. At the same time, we have $[a, b[\subseteq X$ and therefore $b \in X$ by the assumption of the theorem. A contradiction.   □

This theorem represents the *principle of induction for well-ordered sets*.

Remark. Since $[a, a[= \varnothing \subseteq X$, in the assumption of the principle of induction it is required that $a \in X$.

1.4.2.   Discrete sets

In the most general definition, a well-ordered set E is called *discrete* if every $x \in E$ such that $x \neq a = \min E$ and $x \neq b = \max E$ has an immediate predecessor $x_- = \max]\leftarrow, x[$ and an immediate successor $x_+ = \min]x, \rightarrow[$. In a discrete set, if $a \neq b$, then there exist elements a_+ and b_-. The empty set and singletons are discrete sets. If E has no least and greatest elements, then the definition applies to all $x \in E$.

We now give the definition of a discrete well-ordered set for the sake of convenience. In what follows, whenever we need the most general definition in order to include linearly ordered discrete sets, this will be specified explicitly.

1. If E is a well-ordered set, then every element $x \in E$ that is not its greatest element has an *immediate successor* $x_+ = \min]x, \rightarrow[$.

A well-ordered set E is called *discrete* if every $x \in E$ such that $x \neq a = \min E$ has an immediate predecessor $x_- = \max\,]\leftarrow, x[$. By definition, in a discrete set E, every $x \in E$ such that $x \neq a = \min E$ and $x \neq b = \max E$ has an immediate predecessor x_- and an immediate successor x_+. Moreover, if $a \neq b$, then there exist a_+ and b_-. If a set is discrete with respect to a certain order, this order is called *discrete*.

Exercise. Prove the Denjoy theorem, which states that all linear orders on a given set are discrete and equivalent to each other if and only if this set is finite (see Savel'ev, 1969a, Introduction, subsection 2.11.1)

If the last element (endpoint) exists in a discrete set, this set is called a *finite discrete set*. The empty set with empty order is discrete. Nonempty discrete sets with no last element are called *infinite discrete sets*.

For simplicity, x_+ and x_- are called the elements *preceding* x and *succeeding to* x whenever no other elements are considered and there is no confusion.

Lemma. *For any finite discrete sets A and B, the lexicographic product $A \times B$ is finite.*

Proof. Let $Z \subseteq A \times B$, $Z \neq \varnothing$, and $X = \{x \mid xy \in Z$ for some $y \in B\} \subseteq A$. There exist elements $a = \min X$ and $c = \max X$. Consider the sets $Y(a) = \{y \mid ay \in Z\}$, $Y(c) = \{y \mid cy \in Z\}$ and the elements $b = \min Y(a)$, $d = \max Y(c)$. We have $ab = \min Z$ and $cd = \max Z$ in the lexicographic order. Therefore, $A \times B$ with the lexicographic order is a finite discrete set. □

Example. The set $E = \{0, 1, 2, 3, 4, 5, 6, 7, 8, 9\}$ with the conventional order $0 < 1 < 2 < 3 < 4 < 5 < 6 < 7 < 8 < 9$ is discrete. From the definitions, it follows that $0 = \min E$, $9 = \max E$, $0_+ = 1$, and $9_- = 8$. For the rest of the numbers in E, the respective immediate predecessors and successors are $1_- = 0$, $1_+ = 2$, $2_- = 1$, $2_+ = 3, \ldots$.

2. Let E be a nonempty discrete set with order $\geq$, $a = \min E$, and $X \subset E$.

Theorem. *Let (1) $a = \min E \in X$ and (2) for every $x \neq b = \max E$, $x \in X$ implies $x_+ \in X$. Then $X = E$.*

Proof. The assertion of this theorem follows from the theorem that was proved in Section 1.4.1, 2. Indeed, the condition $a \in X$ for $x = a$ in the assumption of that theorem is satisfied by condition (1). For every $x > a$, $[a, x[= [a, x_-] \subseteq X$ implies $x_- \in X$ and, by condition (2), $x = (x_-)_+ \in X$. $\square$

This theorem represents the *principle of induction for discrete sets*.

The requirement $x \neq b = \max E$ in the assumption of the theorem can be omitted if E is infinite.

3. On any infinite discrete set E, we can define the *shift operator* T that maps every element $x \in E$ to its immediate successor x_+. The image of T is $T(E) = E \setminus \{a\}$, where $a = \min E$.

The operator S that maps every element $y \in E$, $y \neq a$, to its immediate predecessor y_- is defined on $E \setminus \{a\}$. The image of S is $S(E \setminus \{a\}) = E$.

It follows from the definitions that $S(T(x)) = (x_+)_- = x$ $(x \in E)$, $T(S(y)) = (y_-)_+ = y$ $(y \in E, y \neq a)$, i.e., the operator S is the inverse of T. For this reason, S is called the *backward shift operator*.

1.4.3. Zorn's lemma

Zorn's lemma is often used in proofs by induction. It states that, under certain conditions, every element in an ordered set is majorized by some maximal element.

1. Consider a set E with order $\geq$. This order induces an order on every $D \subset E$. If the induced order is a linear order on D, then D is called a *chain* in E. A chain that is not contained within any other chain is called *maximal*. It is a maximal element in the set of all chains in E ordered by inclusion.

The following statement is equivalent to the axiom of choice.

The Hausdorff maximality theorem. *In any ordered set, every chain is contained in a maximal chain.*

The theorem does not provide any specific method of extending the chain to a maximal chain.

2. An ordered set is called *linearly majorized* if every chain in it has an upper bound. The Hausdorff maximality theorem implies the following statement.

Zorn's lemma. *Every ordered set in which every chain has an upper bound contains a maximal element.*

Proof. Let E be an ordered set with order $\geq$. Let $a \in E$. Then $L = \{a\}$ is a chain in E. By the Hausdorff maximality theorem, there exists a maximal chain M in E such that $M \supseteq L$. Assume that every chain in E has an upper bound, and let $b \in E$ be an upper bound for the chain M. Then $a \leq b$.

It follows that the element b is maximal. Indeed, suppose $b < c$ for some $c \in E$. Then $x \leq b < c$ for all $x \in M$ and $N = M \cup \{c\}$ is a chain in E, which is a contradiction because M is a maximal chain. $\square$

Remark. Axiom of choice, the Hausdorff maximality theorem, and Zorn's lemma are equivalent. A brief proof of this statement is given in Kelley (1975).

1.5. NATURAL NUMBERS

Natural numbers provide a way to count and specify the order of elements in well-ordered sets. The role of natural numbers in mathematics is described in the famous phrase of Leopold Kronecker: "God made the integers, and all the rest is the work of man". The natural numbers are formally defined by a system of axioms. For example, see Kuratowski and Mostowski (1970), Chapter 3, where a brief description of the natural numbers in terms of naive set theory is provided. For details, see also Savel'ev (1969), Introduction to Chapter 2.

1.5.1. Decimal natural numbers

The most common representation of natural numbers is the one that uses the set of digits $F = \{0, 1, \ldots, , 9\}$.

We will also consider the set $F_0 = \{1, 2, \ldots, 9\}$ of nonzero digits. The set $\mathbb{N}$ of decimal natural numbers will be defined by induction as a discrete infinite set.

First, we specify an infinite discrete class of sets using the sets F and F_0. The initial element of this class is the set $a = F_0$. For an element x that is already defined, its immediate successor x_+ is defined as the cartesian product of x and F: $x_+ = x \times F$ $(x \in \mathbb{N})$. In particular, we have

$$a_+ = F_0 \times F = \{10, 11, \ldots, 19; 20, 21, \ldots, 29; \ldots 90, 91, \ldots, 99\},$$
$$(a_+)_+ = F_0 \times F \times F, \quad ((a_+)_+)_+ = F_0 \times F \times F \times F.$$

Thus, the initial element a and the rule of obtaining x_+ from x defines the well-ordered class

$$\mathcal{N} = \{F_0, F_0 \times F, F_0 \times F \times F, \dots\}.$$

Clearly, this class consists of pairwise disjoint sets. Every set in $\mathcal{N}$ can be ordered with the lexicographic order. The order sum

$$\mathbb{N} = \sum \mathcal{N}$$

represents *ordered set of decimal natural numbers.*

1.5.2.   The isomorphism theorem

Aside from the decimal representation, there are many other representations of natural numbers. This is justified by the following theorem.

Theorem. *For any two infinite discrete sets A and B, there exists a unique monotonic isomorphism from A onto B.*

Proof. 1) We now prove the existence part. Assume $a = \min A$, $b = \min B$ and consider the correspondence $\varphi : A \to B$ such that $\varphi(a) = b$ and $\varphi(x_+) = \varphi(x)_+$ $(x \in A)$. Let X be the set of all $x \in A$ such that φ is defined and is strictly monotone on the interval $[a, x]$. Put $\varphi(x) = Y$. It follows from the definition of φ that $a \in X$, $b \in Y$. Furthermore, for any $x \in A$ and $y \in B$, if $x \in X$ and $y \in Y$, then $(x_+) \in X$ and $(y_+) \in Y$. By induction, $X = A$, $Y = B$, and φ is the required isomorphism.

2) We now prove the uniqueness of φ. Let ψ be a monotonic isomorphism from A onto B. We have $\psi(a) = b = \varphi(a)$ and the equality $\psi(x) = \varphi(x)$ implies $\psi(x_+) = \psi(x)_+ = \varphi(x)_+ = \varphi(x_+)$ for every $x \in A$. By induction, it follows that $\psi = \varphi$.   □

Corollary. *Every infinite discrete set is isomorphic to the ordered set of decimal natural numbers.*

Therefore, for the ordered set of natural numbers we can take an arbitrary infinite discrete set, for example, the set of binary natural numbers defined in a similar way on the basis of the sets $B_0 = \{1\}$ and $B = \{0, 1\}$.

When mentioning the natural numbers, the particular way of their representation is usually not specified, and $\mathbb{N}$ is called the *ordered set of natural numbers.* The elements of $\mathbb{N}$ are also called *indices* when used for ordering a set.

1.5.3. Countable sets

Depending on their cardinality, sets are classified into finite and infinite, countable and uncountable.

Every finite set is isomorphic to an interval of the set of natural numbers. If a finite set K is isomorphic to an interval $[1, n] \subseteq \mathbb{N}$, then K is said to be *consisting of n elements*, which is written as $|K| = n$.

A set is said to be *countable* if it is isomorphic to the set of natural numbers. The power of any countable set D, which is equal to the power of $\mathbb{N}$, is denoted by ω: $|D| = |\mathbb{N}| = \omega$. The symbol ∞ is also used instead of ω.

A set is said to be *uncountable* if it is not isomorphic to any subset of $\mathbb{N}$. An example of an uncountable set is the set $F(\mathbb{N})$ of all maps from $\mathbb{N}$ to itself (see Kuratowski and Mostowski, 1970, Section 5.5).

Remark. A common property of finite and countable sets is their being isomorphic to subsets of the set $\mathbb{N}$ of integer numbers. The elements of finite or countable set can be enumerated. It is therefore convenient to consider finite sets to be countable as well and classify countable sets into finite and infinite.

Exercises. Prove the following statements: 1) The image of every countable set under any map is countable. 2) The image of every finite set under any map is finite. 3) The set $\mathbb{N} \times \mathbb{N}$ is isomorphic to $\mathbb{N}$. 4) The union of a countable collection of countable sets is countable.

Chapter 2.

Algebra

This chapter describes the language of abstract algebra, which provides an efficient and precise tool to formulate many important results of operator theory.

2.1. ABSTRACT ALGEBRA

This section examines some of the most common algebraic structures.

2.1.1. Semigroups

Semigroups are the simplest of the most common algebraic structures, since they have only one operation, which is assumed to be associative. The algebraic theory of semigroups is examined in detail in Clifford and Preston (1967).

1. A *semigroup* is a pair (H, h), where H is a set and h is a map from $H \times H$ to H (called *operation*). The operation h is usually called *addition* or *multiplication* and is denoted by $+$ or $\cdot$, respectively. The results of addition or multiplication are called the *sum* or the *product*, respectively. The standard designations are $h(x, y) = x + y$ and $h(x, y) = x \cdot y$. It is customary to write simply xy instead of $x \cdot y$. A semigroup with addition is called *additive*, while a semigroup with multiplication is called *multiplicative*.

We will write H instead of (H, h) for brevity. By definition, the operation in a semigroup is *associative*:

$$h(h(x, y), z) = h(x, h(y, z)) \quad (x, y, z \in H).$$

For additive and multiplicative semigroups, this property is written as

$$(x+y)+z = x+(y+z), \quad (xy)z = x(yz) \quad (x,y,z \in H).$$

The associativity means that the way the terms are grouped does not affect the result of summation and multiplication. Pairs (H, h) that satisfy all conditions in the definition of a semigroup except the associativity of the operation h are called *groupoids*.

Examples. 1) The additive semigroup $(\mathbb{N}, +)$ and the multiplicative semigroup $(\mathbb{N}, \cdot)$, where $\mathbb{N}$ is the set of natural numbers. The analogous semigroups for the sets of integer, rational, real, and complex numbers ($\mathbb{Z}$, $\mathbb{Q}$, $\mathbb{R}$, $\mathbb{C}$).

2) The semigroup $(\mathcal{F}, \circ)$, where $\mathcal{F} = \mathcal{F}(E)$ is a set of all transformations of a set E, and $\circ$ is the composition of transformations. Multiplicative notation is standard for this semigroup, i.e., the composition $\beta \circ \alpha$ of transformations β and α is written as $\beta\alpha$.

3) The semigroups $(\mathcal{P}, \cup)$ and $(\mathcal{P}, \cap)$, where $\mathcal{P} = \mathcal{P}(E)$ is a set of all subsets of a set E and the operations are the union and the intersection, respectively.

2. A semigroup (H, h) is said to be *commutative* or *Abelian* if its operation h is symmetric, i.e.,

$$h(x, y) = h(y, x) \quad (x, y \in H).$$

For additive and multiplicative semigroups this property is written as follows:

$$x + y = y + x, \quad xy = yx \quad (x, y \in H).$$

The sum $\sum x_i$ and the product $\prod x_i$ of a finite collection of elements x_i $(1 \leq i \leq n)$ is defined by induction.

The semigroups in Examples 1 and 3 are Abelian, whereas the semigroup of transformations in Example 2 is not Abelian.

For a semigroup (H, h), an element $e \in H$ is said to be *neutral* if

$$h(e, x) = h(x, e) = x \quad (x \in H).$$

It is necessarily unique. In additive semigroups, the neutral element is called *zero*, written 0, whereas in multiplicative semigroups it is called the *identity*, written 1. We have

$$x + 0 = 0 + x = x, \quad 1 \cdot x = x \cdot 1 = x \quad (x \in H).$$

In a multiplicative semigroup $(H, \cdot)$, zero is defined by the equality

$$0 \cdot x = x \cdot 0 = 0 \quad (x \in H).$$

Assume that there exists a neutral element e in a semigroup (H, h). For $x \in H$, its *inverse* $x^{-1} \in H$ is defined by the equalities

$$h(x, x^{-1}) = h(x^{-1}, x) = e.$$

If an inverse element x^{-1} exists for $x \in H$, then it is unique. In particular, $e^{-1} = e$. In multiplicative notation, the equalities that define an inverse element are written as $xx^{-1} = x^{-1}x = 1$. In additive notation, $-x$ stands for x^{-1} and the equalities defining an inverse element are written as $x + (-x) = (-x) + x = 0$.

Examples. 1) There is no neutral element in $(\mathbb{N}, +)$.

2) The neutral element of the semigroup $(\mathcal{F}, \circ)$ is the identity transformation id defined by the equality $\mathrm{id}\,(x) = x$ $(x \in E)$. The inverse of the automorphism $f \in \mathcal{F}$ is the automorphism f^{-1} defined by the equalities $f(f^{-1}(x)) = f^{-1}(f(x)) = x$ $(x \in E)$.

3) The neutral elements in $(\mathcal{P}, \cup)$ and $(\mathcal{P}, \cap)$ are the empty set $\varnothing$ and the entire set E, respectively. Only the neutral elements in these semigroups have inverse elements.

3. Let (G, g) and (H, h) be semigroups. If $G \subseteq H$ and $g \subseteq h$ (g is a restriction of h), then (G, g) is said to be a *subsemigroup* of (H, h), which is written as $(G, g) \subseteq (H, h)$. Thus, by definition,

$$(G, g) \subseteq (H, h) \Longleftrightarrow G \subseteq H, g \subseteq h.$$

The set $(G, g) \subseteq G \times G \times G \times G = G^4$ is a subset of $(H, h) \subseteq H \times H \times H \times H = H^4$.

If $h(H \times G) \subseteq G$ or $h(G \times H) \subseteq G$, then the semigroup (G, g) is called a *left ideal* or a *right ideal* of the semigroup (H, h), respectively, in accordance with the position of H. If a left ideal is also a right ideal, then it is called a *two-sided ideal*, or simply an *ideal*. In commutative semigroups, the notions of a right ideal and a left ideal coincide.

In additive and multiplicative semigroups, the inclusion that defines a right ideal can be replaced with the conditions

$$x + y \in G, \quad xy \in G \quad (x \in G, \ y \in H).$$

Examples. 1) Let $(2\mathbb{N}, \cdot)$ and $(\mathbb{N}, \cdot)$ be multiplicative semigroups of even and all natural numbers, respectively. Clearly, $(2\mathbb{N}, \cdot)$ is an ideal of $(\mathbb{N}, \cdot)$, but the additive semigroup $(2\mathbb{N}, +)$ is not an ideal of $(\mathbb{N}, +)$.

2) Let $c \in E$ and let a transformation $\gamma \in \mathcal{F} = \mathcal{F}(E)$ be such that $\gamma(E) = c$. It is obvious that $\gamma \circ \alpha = \gamma$ for any $\alpha \in \mathcal{F}$ and $(\gamma, \circ)$ is a right ideal of $(\mathcal{F}, \circ)$. At the same time, if there is an element $b \in E$ such that $b \neq c$, then $\beta \circ \gamma = \beta \neq \gamma$ for $\beta \in \mathcal{F}$ such that $\beta(E) = b$, and $(\gamma, \circ)$ is not a left ideal of $(\mathcal{F}, \circ)$.

3) Let $C \subseteq E$ and let $\mathcal{P}(C)$ be the class of all subsets of C. It is clear that $C \cap X \subseteq C$ for any $X \subseteq E$ and $(\mathcal{P}(C), \cap)$ is an ideal of $(\mathcal{P}(E), \cap)$. At the same time, $(\mathcal{P}(C), \cup)$ is not an ideal of $(\mathcal{P}(E), \cup)$.

4. An ideal (G, g) of a semigroup (H, h) is said to be *proper* whenever $\varnothing \subset G \subset H$ (i.e., $G \neq \varnothing$ and $G \neq H$). A proper ideal is called a *maximal ideal* if it is not included in any other proper ideal of (H, h).

Proposition. *Every proper ideal of a semigroup with the neutral element is included in a maximal ideal.*

Proof. Let $(\mathcal{E}, \subseteq)$ be the ordered set of all proper ideals of a semigroup (H, h) that include a given ideal (G, g). The union $M = \cup\mathcal{L}$ of all ideals $L \in \mathcal{L}$ is an upper bound for every chain $\mathcal{L} \subseteq \mathcal{E}$. Indeed, we have $L \subseteq M$ and if $a_1 \in L_1$, $a_2 \in L_2$, then, since $L_1 \subseteq L_2$ or $L_1 \supseteq L_2$, it follows that $h(a_1, a_2) \in L_1$ or $h(a_1, a_2) \in L_2$. Hence $h(a_1, a_2) \in M$. The neutral element does not belong to any proper ideal $L \in \mathcal{L}$ and therefore it does not belong to M. We conclude that $M \in \mathcal{E}$. Application of Zorn's lemma completes the proof. $\square$

Remark. The proof of the preceding proposition is very typical. Similar argument is used in the proofs of many analogous propositions.

Example. Let $\mathbb{B} = \{0, 1\}$ and let addition and multiplication be defined by the tables

+	0	1
0	0	1
1	1	0

·	0	1
0	0	0
1	0	1

The semigroup $(\mathbb{B}, +)$ has no proper ideals. The only proper ideal of the semigroup $(\mathbb{B}, \cdot)$ is $(\{0\}, \cdot)$, which is obviously maximal.

5. Any element x of a semigroup (H, h) such that $h(x, x) = x$ is called an *idempotent*. This term is associated with multiplicative notation $x^2 = x \cdot x = x$. The neutral element e is obviously an idempotent. All elements of the semigroups $(\mathcal{P}, \cup)$ and $(\mathcal{P}, \cap)$ are idempotents:

$$X \cup X = X, \quad X \cap X = X \quad (X \in \mathcal{P}).$$

Let $I = I(H)$ be the set of idempotents of the semigroup (H, h). We introduce a *natural order* on I as follows:

$$x \le y \Leftrightarrow h(x, y) = h(y, x) = x \quad (x, y \in J).$$

Exercise. Prove that this relation is reflexive, transitive, and antisymmetric.

Natural orders for the semigroups $(\mathcal{P}, \cup)$ and $(\mathcal{P}, \cap)$ are the relations $\supseteq$ and $\subseteq$:

$$X \cup Y = X \Leftrightarrow X \supseteq Y, \quad X \cap Y = X \Leftrightarrow X \subseteq Y.$$

6. For semigroups (G, g) and (H, h), any map $\varphi : G \to H$ that preserves the semigroup operation, i.e.,

$$\varphi(g(x, y)) = h(\varphi(x), \varphi(y)) \quad (x, y \in G),$$

is called a *homomorphism* from (G, g) into (H, h), written $\varphi : (G, g) \to (H, h)$. If φ is surjective (i.e., $\varphi(G) = H$), then (H, h) is said to be *homomorphic* to (G, g).

An *additive homomorphism* $\alpha : (G, +) \to (H, +)$ is defined by the condition

$$\alpha(x + y) = \alpha(x) + \alpha(y) \quad (x, y \in G).$$

A *multiplicative homomorphism* $\beta : (G, \cdot) \to (H, \cdot)$ is defined by the formula

$$\beta(xy) = \beta(x)\beta(y) \quad (x, y \in G).$$

Homomorphisms of the form $\varphi : (G, +) \to (H, \cdot)$ and $\psi : (G, \cdot) \to (H, +)$ have the following properties:

$$\varphi(x + y) = \varphi(x)\varphi(y), \quad \psi(xy) = \psi(x) + \psi(y) \quad (x, y \in G).$$

All definitions for arbitrary maps apply to homomorphisms: one-to-one homomorphisms are called *embeddings*, surjective embeddings are called *isomorphisms*, and the isomorphisms of a semigroup onto itself are called *automorphisms*. For isomorphic semigroups, we write $(G, g) \simeq (H, h)$ or $G \simeq H$, meaning that there exists an isomorphism $\varphi : G \to H$ and its inverse $\varphi^{-1} : H \to G$.

Exercise. Prove that the map inverse to an isomorphism from a semi-group to a semigroup is an isomorphism.

Examples. 1) The map $\varphi(n) = 2n$ is an isomorphism from $(\mathbb{N}, +)$ onto $(2\mathbb{N}, +)$.

2) The map $\varphi(X) = X'$ is an isomorphism from $(\mathcal{P}, \cup)$ onto $(\mathcal{P}, \cap)$ and of $(\mathcal{P}, \cap)$ onto $(\mathcal{P}, \cup)$. This follows from the property of duality

$$(X \cup Y)' = X' \cap Y', \quad (X \cap Y)' = X' \cup Y' \quad (X, Y \in \mathcal{P}).$$

The isomorphism φ coincides with the isomorphism φ^{-1} because the operation of taking the complement of a set is involutive.

7. Any semigroup (H, h) is isomorphic to a subsemigroup $(\mathcal{T}(H), \cdot)$ of the semigroup $(\mathcal{F}(H), \cdot)$ of transformations of H (see Clifford and Preston, 1967).

The elements of $\mathcal{T}(H)$ are the left-shift operators defined for any $x \in H$ by the formula $\tau_x(z) = h(x, z)$ $(z \in H)$. By the associativity of h,

$$\tau_x(\tau_y(z)) = h(x, h(y, z)) = h(h(x, y), z) = \tau_{h(x,y)}(z) \quad (x, y, z \in H),$$

and therefore $(\mathcal{T}(H), \cdot)$ is a semigroup. Clearly, $\mathcal{T}(H) \subseteq \mathcal{F}(H)$. The preceding equalities for the operators τ_x, τ_y, and $\tau_{h(x,y)}$ imply $\tau_{h(x,y)} = \tau_x \cdot \tau_y$ $(x, y \in H)$. Consequently, the map $\tau : H \to \mathcal{T}(H)$ is a homomorphism from (H, h) *onto* $(\mathcal{T}(H), \cdot)$ or *into* $(\mathcal{F}(H), \cdot)$.

Exercise. Determine the conditions under which this homomorphism becomes an isomorphism.

Example. The operator representations of $(\mathbb{N}, +)$ and $(\mathbb{N}, \cdot)$ are isomorphic to these semigroups. For this reason, natural numbers can be viewed as operators.

We could consider the *right-shift* operators instead of the left-shift operators. For semigroups with commutative operation, they coincide and are therefore called simply *shift operators*.

8. For any set E, the set $\mathcal{R} = \mathcal{R}(E)$ of all relations on E with the operation $\circ$ of composition of relations is a semigroup. We will use multiplicative notation and write TS (or ST) instead of $T \circ S$ and Rx (or xR) instead of $R(x)$ for $R, S, T \in \mathcal{R}$. Whether to use the left-side or the right-side notation is a matter of convention. The variants TS and Rx are more common.

Note that $\mathcal{R} = \mathcal{P}(E \times E)$, i.e., every relation R is a subset of $E \times E$. A relation R can be viewed as a map from E to $\mathcal{P}(E)$ such that $R(x) = \{y \mid (x, y) \in \mathcal{R}\}$ for $x \in E$, so that $\mathcal{R} = \mathcal{F}(E, \mathcal{P}(E))$.

We now verify that the composition of relations is associative. Let $R, S, T \in \mathcal{R}$ and $x \in E$. From the definition of the composition of correspondences, it follows that

$$((TS)R)(x) = (TS)(R(x)) = T(S(R(x))) = T((SR)(x)),$$

and thus $(TS)R = T(SR)$. For this reason, $(\mathcal{R}, \circ) = (\mathcal{R}, \cdot)$ is a semigroup.

The semigroup $(\mathcal{F}, \circ) = (\mathcal{F}, \cdot)$ of transformations of E can be embedded into the semigroup $(\mathcal{R}, \cdot)$ and is therefore a subsemigroup of $(\mathcal{R}, \cdot)$. The identity transformation I is the identity element in $(\mathcal{R}, \cdot)$. The empty relation $\varnothing$ is the zero element in $(\mathcal{R}, \cdot)$.

The semigroup $(\mathcal{R}, \cdot)$ is noncommutative since $ST = TS$ is not necessarily true.

Example. Let $E = \mathbb{B} = \{0, 1\}$, $\mathcal{R}(E) = \mathcal{R}(\mathbb{B}) = \mathcal{P}(\mathbb{B} \times \mathbb{B}) = \mathcal{P}\{00, 01, 10, 11\}$ ($2^4 = 16$ relations in total), $S = \{00, 01\}$, $T = \{00, 10\}$. Then $ST = \{00\} \neq \{00, 01, 10, 11\} = TS$. Such compositions can be illustrated by the following diagrams:

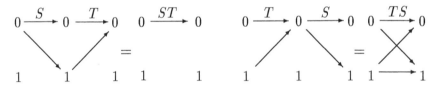

The above example also illustrates the duality of the functional notation $(ST)(x) = S(T(x))$ and the operator notation $x(ST) = (xS)T$.

Note that S and T are inverses of each other as correspondences: $(0, 0) \leftrightarrow (0, 0)$, $(0, 1) \leftrightarrow (1, 0)$. In the semigroup $(\mathcal{R}(E), \cdot)$, however, S and T are not inverses of each other. The algebraic inverse of an element may not coincide with its inverse as a correspondence.

Correspondences and, particularly, relations are sometimes called *multivalued functions* or *multivalued abstract operators*. Their values are sets.

9. The following statements hold for compositions of homomorphisms.

Proposition. *The composition of homomorphisms of semigroups is a homomorphism.*

Proof. Let F, G, and H be semigroups, $\alpha : F \to G$ and $\beta : G \to H$ be homomorphisms, and $\gamma = \beta\alpha : F \to H$ be the composition of α and β. It is required to prove that γ is a homomorphism. In additive notation, we have

$$\gamma(x + y) = \beta(\alpha(x + y))$$
$$= \beta(\alpha(x) + \alpha(y)) = \beta(\alpha(x)) + \beta(\alpha(y)) = \gamma(x) + \gamma(y)$$

for all $x, y \in F$. Consequently, γ is a homomorphism. $\square$

For sets, the composition of isomorphisms is an isomorphism (see Section 1.2.3).

Corollary. *The composition of isomorphisms of semigroups is an isomorphism.*

Exercises. Prove the following statements for relations $R, S, T \in \mathcal{R}(E)$:

1) $R \subseteq S \Rightarrow RT \subseteq ST$, $TR \subseteq TS$.

2) $RS \subseteq SR \Rightarrow RS = SR$ if R and S are symmetric.

3) If S^{-1} is the inverse of S as an automorphism of E, then S^{-1} is the algebraic inverse of S as an element of the semigroup $(\mathcal{R}(E), \cdot)$.

2.1.2. Groups

A semigroup is called a *group* if it has a neutral element and every element of this semigroup has an inverse.

Groups are the main subject of Chapters 2 and 7 in the monograph van der Waerden (1991). Different versions of the definition of a group are discussed in Clifford and Preston (1967).

1. All general definitions and propositions for semigroups can be extended to groups.

A subsemigroup of a group is its *subgroup* if it is itself a group. The neutral element of a group belongs to its every subgroup. Indeed, for any element of a subgroup, its inverse also belongs to it and the group operation applied to them yields the neutral element.

Examples. 1) The semigroups $(\mathcal{P}, \cup)$ and $(\mathcal{P}, \cap)$ are not groups: $X \cup Y = \varnothing$ only if $X = Y = \varnothing$, and $X \cap Y = U$ only if $X = Y = U$ ($\mathcal{P} = \mathcal{P}(U)$, $U \neq \varnothing$).

2) The semigroup $(\mathcal{R}, \cdot)$ of all transformations of the universal set U and their compositions is not a group. At the same time, the semigroup $(\mathcal{A}, \cdot)$ of all automorphisms of U is a group since for any $R \in \mathcal{A}$ its inverse element in the semigroup is its inverse automorphism $R^{-1} \in \mathcal{A}$.

3) The additive semigroup $(\mathbb{B}, +)$ with $\mathbb{B} = \{0, 1\}$ is a group. We have $0 + 0 = 0$, $1 + 1 = 0$. We call $(\mathbb{B}, +)$ the *binary group*. The multiplicative semigroup $(\mathbb{B}, \cdot)$ is not a group, whereas the semigroup $(\mathbb{B} \setminus \{0\}, \cdot) = (\{1\}, \cdot)$ is a group.

2. A classical example of an Abelian group is the additive group of integers $(\mathbb{Z}, +)$. It is obtained from the semigroup $(\mathbb{N}, +)$ by adding zero and the inverse elements.

Every natural number n is associated with the *negative integer number* $-n$. The set of all negative integers is denoted by $-\mathbb{N}$. The set of *integers* is the union $\mathbb{Z} = (-\mathbb{N}) \cup \{0\} \cup \mathbb{N}$. The operation of addition in $\mathbb{Z}$ is defined in the usual way. The set $\mathbb{Z}$ with addition, written $(\mathbb{Z}, +)$, is an Abelian group.

The order on $-\mathbb{N}$ is determined by the order on $\mathbb{N}$: $-m \leq -n$ if $m \leq n$. In addition, $-m < 0 < n$ for all $-m \in -\mathbb{N}$ and $n \in \mathbb{N}$. The order on $\mathbb{Z}$ is consistent with the addition operation: $x \leq y \Rightarrow x + z \leq y + z$ ($x, y, z \in \mathbb{Z}$). By adding the order to the group of integers, we obtain the *ordered group* $(\mathbb{Z}, +, \leq)$.

For every even integer $2n$ there is a negative number $-2n$. The union of all such numbers, $2\mathbb{Z} = (-2\mathbb{N}) \cup \{0\} \cup (2\mathbb{N})$, together with the induced addition operation constitute the subgroup $(2\mathbb{Z}, +)$ of *even* integers.

The set $\mathbb{Z}$ with the multiplication operation defined in the usual way constitutes the multiplicative semigroup $(\mathbb{Z}, \cdot)$. This semigroup is commutative, but it is not a group.

Exercise. Prove that the set of integers $\mathbb{Z}$ is countable.

3. In what follows, for the sake of clarity, we will use either additive or multiplicative notation instead of general notation.

It is easy to verify the uniqueness of a neutral element e in a group $(G, \cdot)$: if $ax = xa = x$ for every $x \in G$, then $a = ae = e$. For every $x \in G$, its inverse x^{-1} is also unique. Indeed, if $yx = xy = e$ for some $y \in G$, then $y = y(xx^{-1}) = (yx)x^{-1} = x^{-1}$. In any group $(G, \cdot)$, the equations $ax = b$ and $ya = b$ have unique solutions $x = a^{-1}b$ and $y = ba^{-1}$ for any $a, b \in G$. We say that these equations are *uniquely solvable*. The solvability of equations $ax = b$ and $ya = b$ is a distinguishing feature of groups as compared to semigroups.

The existence and uniqueness of inverses in groups make it possible to define the *inverse operation*, called *subtraction* in additive groups and *division* in multiplicative groups). The results of subtraction and division are called the *difference* and the *quotient*, respectively: $x + (-y) = x - y$, $x \cdot y^{-1} = x/y$.

Because of the existence of inverses, the only idempotent element in a group is the neutral element.

In additive and multiplicative groups, there are automorphisms $\alpha : x \to -x$ and $\beta : x \to x^{-1}$ ($x \in G$), respectively. In the group $(\mathbb{Z}, +)$, the former automorphism takes positive numbers into negative ones and vice versa. The automorphisms α and β are *involutions*, i.e., $\alpha^2 = \alpha \circ \alpha = \mathrm{id}$, $\beta^2 = \beta \circ \beta = \mathrm{id}$.

Integer numbers are used to define *integer multiples* and *integer powers* of elements in a group. This is done by induction:

$$0 \cdot x = 0, \quad 1 \cdot x = x, \quad (n+1) \cdot x = n \cdot x + x, \quad (-n) \cdot x = -(n \cdot x);$$
$$x^0 = 1, \quad x^1 = x, \quad x^{n+1} = x^n \cdot x, \quad x^{-n} = (x^n)^{-1}$$

for all elements x of the group and integer numbers n. In this context, elements of the group $(\mathbb{Z}, +)$ are *operators* acting on the groups $(G, +)$ and $(G, \cdot)$. The following properties of multiples and powers are also proved by induction: $(m + n)x = mx + nx$, $x^{m+n} = x^m x^n$ for all elements x of the group and integers m and n.

4. *Normal subgroups* $(F, +)$ and $(F, \cdot)$ of groups $(G, +)$ and $(G, \cdot)$ are subgroups such that

$$x + F = F + x, \quad x \cdot F = F \cdot x$$

for all $x \in G$. All subgroups of an Abelian group are normal.

Example. The element $xyx^{-1}y^{-1}$ is called the *commutator* of elements x and y of a multiplicative group $(G, \cdot)$. The subgroup $(K, \cdot)$ generated by the set of all commutators in $(G, \cdot)$ is called the *commutator subgroup* of $(G, \cdot)$. The commutator subgroup is a normal subgroup.

Remark. Since every subgroup of a group contains its neutral element, it follows that it has no proper ideals, i.e., the equalities $F + G = F$ ($FG = F$) or $G + F = F$ ($GF = F$) hold only when $F = G$.

5.   Homomorphisms of groups preserve the operation together with the neutral element and the inverses. For an *additive homomorphism* α : $(F, +) \to (G, +)$, we have

$$\alpha(x + y) = \alpha(x) + \alpha(y), \quad \alpha(0) = 0, \quad \alpha(-x) = -\alpha(x) \quad (x \in F).$$

The first property (postulated in the definition) implies the other two. Indeed,

$$\alpha(0) + \alpha(x) = \alpha(0 + x) = \alpha(x) = \alpha(x + 0) = \alpha(x) + \alpha(0),$$
$$\alpha(-x) + \alpha(x) = \alpha(-x + x) = \alpha(0) = \alpha(x - x) = \alpha(x) + \alpha(-x).$$

A *multiplicative homomorphism* β : $(F, \cdot) \to (G, \cdot)$ has analogous properties:

$$\beta(xy) = \beta(x)\beta(y), \quad \beta(1) = 1, \quad \beta(x^{-1}) = ((\beta(x))^{-1} \quad (x \in F).$$

Alongside with homomorphism, the notions of *embedding, isomorphism,* and *automorphism* defined for semigroups are extended to groups.

Examples. 1) The map α : $n \to 2n$ $(n \in \mathbb{Z})$ is an isomorphism from $(\mathbb{Z}, +)$ onto $(2\mathbb{Z}, +)$.

2) The map α : $\mathbb{Z} \to \mathbb{B}$ whose range consists of two elements, $\alpha(2n) = 0$ and $\alpha(2n + 1) = 1$ $(n \in \mathbb{Z})$, is a homomorphism from $(\mathbb{Z}, +)$ onto $(\mathbb{B}, +)$.

3) The map β : $x \to x^{-1}$ $(x \in G)$ is a one-to-one map, but $(xy)^{-1} = y^{-1}x^{-1}$. If $(G, \cdot)$ is an Abelian group, then β is an automorphism.

The inverse image $\varphi^{-1}(e) \subseteq F$ of the neutral element under the homomorphism φ : $F \to G$ is called the *kernel of the homomorphism* φ, written Ker φ.

For an additive homomorphism α and a multiplicative homomorphism β, their kernels are determined by the equalities

$$\text{Ker } \alpha = \alpha^{-1}(0) = \{x \in F \mid \alpha(x) = 0\},$$
$$\text{Ker } \beta = \beta^{-1}(1) = \{x \in F \mid \beta(x) = 1\}.$$

In Examples 1)–3), Ker $\alpha = \{0\}$, Ker $\alpha = 2\mathbb{Z}$, Ker $\beta = \{1\}$.

Lemma. *A homomorphism is an embedding if and only if its kernel contains only the neutral element.*

Proof. Let $\alpha : (F, +) \to (G, +)$ and $x, y, z \in F$. If $\alpha(x) = \alpha(y) \Rightarrow x = y$, then $\alpha(z) = 0 = \alpha(0) \Rightarrow z = 0$ and $\operatorname{Ker} \alpha = \{0\}$. Conversely, if $\alpha(z) = 0 \Rightarrow z = 0$, then $\alpha(x) = \alpha(y) \Rightarrow 0 = \alpha(x) - \alpha(y) = \alpha(x - y) \Rightarrow x - y = 0 \Rightarrow x = y$ and therefore α is an embedding. $\square$

Exercise. Prove this lemma for a homomorphism $\beta : (F, \cdot) \to (G, \cdot)$.

6. We now prove a lemma on a connection between homomorphisms and normal subgroups.

Lemma. *The kernel of a homomorphism is a normal subgroup.*

Proof. Let $\alpha : (F, +) \to (G, +)$, $N = \operatorname{Ker} \alpha$, $u \in N$, $v \in N$, $x \in G$.

1) Since $\alpha(u + v) = \alpha(u) + \alpha(v) = 0 + 0 = 0$, we have $u + v \in N$. Furthermore, $\alpha(0) = 0$ implies $0 \in N$, and $\alpha(-u) = -\alpha(u) = -0 = 0$ implies $-u \in N$. It follows that $(N, +)$ is a subgroup of F.

2) Let $x + u = y$ be an arbitrary element of $x + N$. Then $u = -x + y$ and $0 = \alpha(u) = -\alpha(x) + \alpha(y)$. Hence $\alpha(x) = \alpha(y)$. We have $0 = \alpha(y) - \alpha(x) = \alpha(y - x)$ and $y - x = v \in N$. Then $y = v + x \in N + x$ and $x + N \subseteq N + x$. The inclusion $x + N \supseteq N + x$ is established in a similar fashion. Thus, $x + N = N + x$ and N is a normal subgroup. $\square$

Exercise. Prove that every normal subgroup is a kernel of a homomorphism.

7. Every normal subgroup N of a group G defines an equivalence relation on G that is consistent with the group operation. This provides a way to define a new group G/N that is homomorphic to G. We will examine this kind of groups in detail because of its importance.

Lemma A. *The relation $x \sim y \Leftrightarrow x - y \in N$ $(x, y \in G)$ is an equivalence on $(G, +)$ consistent with the addition operation.*

Proof. 1) We have $0 = x - x \in N \Rightarrow x \sim x$, $x \sim y \Rightarrow y - x = -(x - y) \in N \Rightarrow y \sim x$ and $x \sim y, y \sim z \Rightarrow x - y, y - z \in N \Rightarrow x - z = (x - y) + (y - z) \in N \Rightarrow x \sim z$. It follows that $\sim$ is an equivalence relation.

2) Let $a, b, x, y \in G$ and $x \sim a$, $y \sim b$. We set $x - a = u \in N$, $y - b = v \in N$, and $v - a = -a + w$ for some $w \in N$ (since $N + (-a) = (-a) + N$). Then $(x + y) - (a + b) = x + y - b - a = x + v - a = x - a + w = u + w \in N$ and $x + y \sim a + b$. This means that the relation $\sim$ is consistent with the addition operation. $\square$

Lemma B. *The relation* $\sim$ *defined by the condition* $x \sim y \Leftrightarrow xy^{-1} \in N$ $(x, y \in G)$ *is an equivalence on* $(G, \cdot)$ *consistent with the multiplication operation.*

Exercise. Prove Lemma B using arguments similar to those used in the proof of Lemma A.

We introduce the following notation for the sake of convenience:

$$\bar{a} = a + N \ (a \in (G, +)), \quad \bar{a} = aN \ (a \in (G, \cdot)), \quad \bar{G} = G/N.$$

From the definitions, it follows that elements a and b of the group G are equal if and only if the corresponding elements $\bar{a}$ and $\bar{b}$ of the class $\bar{G}$ are equal, i.e., $a \sim b \Leftrightarrow \bar{a} = \bar{b}$.

Corollary A. *For the group* $(G, +)$, *the class* $\bar{G} = G/N$ *of sets* $\bar{a} = a + N$ $(a \in G)$ *is a partition of the set* G. *An addition operation in* $\bar{G}$ *is defined as follows:*

$$\bar{a} + \bar{b} = \overline{a + b} \quad (a, b \in G)$$

This operation is associative and there exist a neutral element and inverse elements with respect to it.

Proof. 1) From Lemma A and the theorem of Section 1.3.2 (1), it follows that $\bar{G}$ is a partition of G.

2) The addition operation in $\bar{G}$ is well defined because the equivalence relation $\sim$ is consistent with addition in $(G, +)$. Indeed, $\bar{x} = \bar{a}, \bar{y} = \bar{b} \Rightarrow x + y = a + b \Rightarrow \overline{x + y} = \overline{a + b} \ (a, b, x, y \in G)$.

3) The associativity of addition in $(G, +)$ ensures the associativity of addition in $\bar{G}$: $\bar{a} + (\bar{b} + \bar{c}) = \bar{a} + \overline{(b + c)} = \overline{a + (b + c)} = \overline{(a + b) + c} = \overline{(a + b)} + \bar{c} = (\bar{a} + \bar{b}) + \bar{c} \ (a, b, c \in G)$.

Since $\bar{a} + \bar{0} = \overline{a + 0} = \bar{a} = \overline{0 + a} = \bar{0} + \bar{a} \ (a \in G)$, it follows that $\bar{0}$ is the neutral element with respect to addition in $\bar{G}$. Since $\bar{a} + \overline{(-a)} = \overline{a + (-a)} = \bar{0} = \overline{(-a) + a} = \overline{(-a)} + \bar{a}$, we have $\overline{-a} = -\bar{a} \ (a \in G)$. $\square$

Corollary B. *For the group* $(G, +)$, *the class* $\bar{G} = G/N$ *of sets* $\bar{a} = aN$ $(a \in G)$ *is a partition of the set* G. *A multiplication operation in* $\bar{G}$ *is defined as follows:*

$$\bar{a}\bar{b} = \overline{ab} \quad (a, b \in G)$$

This operation is associative and there exist a neutral element and inverse elements with respect to it.

Exercise. Prove Corollary B using arguments similar to those used in the proof of Corollary A.

Corollaries A and B imply that $(G/N, +)$ and $(G/N, \cdot)$ are groups. They are called *factor groups* of the groups $(G, +)$ and $(G, \cdot)$ over the normal subgroups $(N, +)$ and $(N, \cdot)$, respectively.

Examples. 1) Let $(G, +) = (\mathbb{Z}, +)$ and $(N, +) = (2\mathbb{Z}, +)$. Then the factor group $(G/N, +) = (\mathbb{Z}/2\mathbb{Z}, +)$ is isomorphic to the binary group $(\mathbb{B}, +)$.

2) For every normal subgroup F of a group G there exists a factor group G/F. In particular, for the subgroup $F = \{e\}$ containing only the neutral element, the factor group $G/\{e\}$ is isomorphic to G. The factor group G/G is isomorphic to the trivial group $\{e\}$.

3) For every homomorphism $\varphi : F \to G$, there exists a factor group $F/\mathrm{Ker}\,\varphi$. An example of such a group is $(\mathbb{Z}/2\mathbb{Z}, +)$.

Remark. From the definitions, it follows that the factor sets that make up the factor group G/N are equipollent to the normal subgroup N. Assume that the sets G, N, and $\bar{G}$ consist of p, n, and $\bar{p}$ elements, respectively. Then $p = \bar{p} \cdot n$ and $\bar{p} = p/n$.

8. For the group G and the factor group $\bar{G}$, there is a *natural homomorphism* $\bar{\varphi} : x \to \bar{x}$ $(x \in G)$. The following equalities hold in the additive and the multiplicative cases, respectively:

$$\bar{\varphi}(x + y) = \overline{x + y} = \bar{x} + \bar{y} = \bar{\varphi}(x) + \bar{\varphi}(y) \quad (x, y \in (G, +)),$$
$$\bar{\varphi}(xy) = \overline{xy} = \bar{x}\bar{y} = \bar{\varphi}(x)\bar{\varphi}(y) \quad (x, y \in (G, \cdot)).$$

The natural homomorphism provides a convenient way to describe the properties of equivalent elements of the group. For example, the homomorphism from $(\mathbb{Z}, +)$ onto the group $(\mathbb{Z}/2\mathbb{Z}, +)$, which is isomorphic to $(\mathbb{B}, +)$, describes the property of being even.

The following proposition is often used when dealing with natural homomorphisms.

Proposition. *The composition of homomorphisms of groups is a homomorphism.*

Proof. Let F, G, and H be groups, $\alpha : F \to G$ and $\beta : G \to H$ be homomorphisms, and let $\gamma = \beta\alpha : F \to H$ be the composition of α and β. From the proposition in Section 2.1.1 (10), it follows that the map γ is a homomorphism between the corresponding semigroups. As was proved in part 5 of this subsection, the homomorphism γ preserves the neutral element and inverses.   □

Corollary. *The composition of isomorphisms of groups is an isomorphism.*

Example. Let $F = G = H = \mathbb{Z}$ and $\alpha(x) = 2x$, $\beta(y) = 2y$. Then $\gamma(x) = 2(2x) = 4x$ for $x \in \mathbb{Z}$.

9. Let G and $\bar{G}$ be groups, N be a normal subgroup of G, and G/N be the factor group of G over N. If isomorphic groups are considered equivalent, then we can say that any group is homomorphic to G if and only if it is a factor group of G.

Theorem. 1) *Let $\varphi : G \to \bar{G}$ be a homomorphism from G onto $\bar{G}$ such that $N = \ker \varphi$. Then $\bar{G}$ is isomorphic to G/N.*

2) *Let $\bar{G}$ be isomorphic to G/N. Then there exists a homomorphism from G onto $\bar{G}$ such that $N = \ker \varphi$.*

Proof. As follows from the lemma in part 6 of this subsection, the kernel $N = \ker \varphi$ is a normal subgroup and therefore it determines the factor group G/N. Consider the natural homomorphism $\bar{\varphi} : G \to G/N$, the inverse correspondence $\varphi^{-1} : \bar{G} \to G$, and the composition $\gamma = \bar{\varphi}\varphi^{-1} : \bar{G} \to G/N$.

The correspondence γ is injective. Indeed, by assumption, for every $\bar{a} \in \bar{G}$ there is an $a \in G$ such that $\varphi(a) = \bar{a}$. If $x \in G$ and $\varphi(x) = \bar{a}$, then $\varphi(x) = \varphi(a)$, $x \sim a$ and $\bar{\varphi}(x) = \bar{\varphi}(a)$. Hence, $\gamma(\bar{a}) = \{\bar{\varphi}(x) \mid \varphi(x) = \bar{a}\} = \bar{\varphi}(a)$.

We now prove that γ is a homomorphism. Let $\bar{a} = \varphi(a)$ and $\bar{b} = \varphi(b)$. Then $\bar{a} + \bar{b} = \varphi(a) + \varphi(b) = \varphi(a + b) = \overline{a + b}$ and $\gamma(\bar{a} + \bar{b}) = \gamma(\overline{a + b}) = \bar{\varphi}(a + b) = \bar{\varphi}(a) + \bar{\varphi}(b)$.

The kernel of the homomorphism γ is trivial. We have $\ker \gamma = \gamma^{-1}(N) = \varphi(\bar{\varphi}^{-1}(N)) = \varphi(0) = \bar{0}$. From the lemma in part 5 of this subsection, it follows that γ is an isomorphism from $\bar{G}$ onto G/N.

2) Let $\gamma : \bar{G} \to G/N$ be an isomorphism and $\gamma^{-1} : G/N \to \bar{G}$ be the isomorphism inverse to γ. Then, by the proposition of Subsection 2.1.1, part 10, the composition $\varphi = \gamma^{-1}\bar{\varphi} : G \to \bar{G}$ of homomorphisms $\bar{\varphi}$ and γ^{-1} is a homomorphism. Since $\bar{\varphi}$ maps G onto G/N and γ^{-1} maps G/N onto G, φ maps G onto $\bar{G}$. Finally, $\ker \varphi = \varphi^{-1}(\bar{0}) = \varphi^{-1}(\gamma(\bar{0})) = \bar{\varphi}^{-1}(N) = N$. $\square$

The homomorphism theorem has a number of useful corollaries.

10. Let G be a group, F be its subgroup, and N be its normal subgroup. We will use multiplicative notation and write $AB = \{xy \mid x \in A, y \in B\}$ for $A, B \subseteq G$.

Theorem. $F/(F \cap N) \simeq FN/N$.

Proof. From the definitions, it follows that $F \cap N$ and FN are subgroups of G, $F \subseteq FN$, $N \subseteq FN$, and $F \cap N$ and N are normal subgroups of F and FN. It follows that the factor groups in the assertion of the theorem are well defined. The following diagram illustrates the proof. The natural homomorphism from G onto the factor group G/N is denoted by γ, while α and β are its restrictions to F and FN, respectively. The maps $\imath, \jmath, \bar{\imath}, \bar{\jmath}$ are identity embeddings, which are homomorphisms. From the definition of γ, it follows that $\ker \gamma = N$.

$$
\begin{array}{ccccccc}
G & \xrightarrow{\ \gamma\ } & \bar{G} & = & G/\operatorname{Ker}\gamma & = & G/N \\[2pt]
\jmath \uparrow & & \bar{\jmath} \uparrow & & & & \uparrow \\[2pt]
FN & \xrightarrow{\ \beta\ } & \overline{FN} & \simeq & FN/\operatorname{Ker}\beta & = & FN/N \\[2pt]
\imath \uparrow & & \bar{\imath} \uparrow & & & & \parallel \\[2pt]
F & \xrightarrow{\ \alpha\ } & \bar{F} & \simeq & F/\operatorname{Ker}\alpha & = & F/(F \cap N)
\end{array}
$$

By the homomorphism theorem, $\bar{F} \simeq F/\ker\alpha$, $\overline{FN} \simeq FN/\ker\beta$. We have $\ker\beta = \beta^{-1}(\bar{0}) = (\jmath^{-1}\gamma\jmath)^{-1}(\bar{0}) = \jmath^{-1}\gamma^{-1}\jmath(\bar{0}) = \jmath^{-1}\gamma^{-1}(\bar{0}) = \jmath^{-1}(N) = FN \cap N = N$, $\ker\alpha = \alpha^{-1}(\bar{0}) = (\imath^{-1}\beta\imath)^{-1}(\bar{0}) = \imath^{-1}\beta^{-1}\imath(\bar{0}) = \imath^{-1}\beta^{-1}(\bar{0}) = \imath^{-1}(N) = F \cap N$. This proves the horizontal equalities in the diagram.

From the definitions, it follows that $\gamma^{-1}(\bar{F}) = FN$. Therefore, $\overline{FN} = \gamma(FN) = \gamma(\gamma^{-1}(\bar{F})) = \bar{F}$, which implies that $\bar{\imath} : \bar{F} \to \overline{FN}$ is an identity isomorphism and $F/(F \cap N) \simeq \bar{F} = \overline{FN} \simeq FN/N$. $\square$

This corollary of the homomorphism theorem is called the *first isomorphism theorem*.

Exercise. Prove that FN is a subgroup and $F \cap N$ is a normal subgroup of G.

11. Let G be a group, N and M be its normal subgroups such that $M \subseteq N$.

Theorem. $G/N \simeq (G/M)/(N/M)$.

Proof. It is easy to see that M is a normal subgroup of N. The proof is illustrated by the following diagram:

$$G \xrightarrow{\ \gamma\ } \bar{G} = G/N \simeq (G/M)/(N/M)$$

$$\beta \searrow \qquad \uparrow \alpha$$

$$\bar{\bar{G}} = G/M$$

By γ and β we mean the natural homomorphisms onto the corresponding factor groups, while $\alpha = \gamma\beta^{-1}$ is the composition of the correspondence β^{-1} and the homomorphism γ. We now prove that $\alpha : \bar{\bar{G}} \to \bar{G}$ is a homomorphism. Let $x \in G$ and $\beta(x) = \bar{\bar{x}} \in \bar{\bar{G}}$. Then $\beta^{-1}(\bar{\bar{x}}) = xM$ and $\alpha(\bar{\bar{x}}) = \gamma(\beta^{-1}(\bar{\bar{x}})) = \gamma(xM) = xM \cdot N = x \cdot MN = xN = \gamma(x) = \bar{x} \in \bar{G}$ (since $M \subseteq N$ and $1 \in M$, we have $MN = N$). The correspondence α is injective. Let $x, y \in G$ and $\bar{\bar{x}} = \beta(x)$, $\bar{\bar{y}} = \beta(y)$. We obtain $\bar{\bar{x}}\bar{\bar{y}} = \beta(x)\beta(y) = \beta(xy) = \overline{\overline{xy}}$ and $\alpha(\bar{\bar{x}}\bar{\bar{y}}) = \alpha(\overline{\overline{xy}}) = \gamma(\beta^{-1}(\overline{\overline{xy}})) = \gamma(xyM) = \gamma(xM \cdot yM) = \gamma(xM) \cdot \gamma(yM) = \gamma(\beta^{-1}(x)) \cdot \gamma(\beta^{-1}(y)) = \alpha(\bar{\bar{x}})\alpha(\bar{\bar{y}})$. Hence, α is a homomorphism from $\bar{\bar{G}}$ into $\bar{G}$. By the homomorphism theorem, $\bar{G} \simeq \bar{\bar{G}}/\ker\alpha$. We conclude that $\ker\alpha = \alpha^{-1}(\bar{0}) = (\gamma\beta^{-1})^{-1}(\bar{0}) = \beta(\gamma^{-1}(N)) = \beta(N) = \{\beta(x) \mid x \in N\} = \{x \in M \mid x \in N\} = N/M$. $\square$

The foregoing corollary from the homomorphism theorem is called the *second isomorphism theorem*. It describes the successive factorization of a group. Such factorization can be useful for constructing general solutions to equations.

Example. $\mathbb{Z}/2\mathbb{Z} \simeq (\mathbb{Z}/4\mathbb{Z})/(2\mathbb{Z}/4\mathbb{Z})$.

Exercise. Use the isomorphism theorems to prove that

$$(FM/M)/((F \cap N)/(F \cap M)) \simeq FN/N.$$

Conversely, prove that this equality implies the first and the second isomorphism theorems for $M = \{1\}$ and $F = G$, respectively.

We now formulate a useful proposition that follows from the theorems proved above. Let F, G, and H be groups and let $\beta : G \to F$ and $\gamma : G \to H$ be surjective homomorphisms.

Proposition. *The inclusion* $\ker \beta \subseteq \ker \gamma$ *is true if and only if there exists a unique homomorphism* $\alpha : F \to H$ *such that* $\alpha\beta = \gamma$.

Proof. Let $M = \ker \beta$ and $N = \ker \gamma$. By the homomorphism theorem, there exist isomorphisms $\varphi : F \to G/M$ and $\psi : H \to G/N$. If $M \subseteq N$, then there is a natural homomorphism $G/M \to (G/M)/(N/M)$ and, by the second isomorphism theorem, $(G/M)/(N/M) \simeq G/N$. Then there exists a homomorphism $\bar{\alpha}$ from G/H onto G/N. The required homomorphism is $\alpha = \psi^{-1}\bar{\alpha}\varphi$.

If there is a homomorphism α such that $\alpha\beta = \gamma$, then for $x \in G$ we have $\beta(x) = 0 \Rightarrow \gamma(x) = \alpha(\beta(x)) = \alpha(0) = 0$ and $\ker \beta \subseteq \ker \gamma$. The uniqueness of α follows from the equalities $\bar{\alpha} = \psi\alpha\varphi^{-1}$ and $\alpha = \psi^{-1}\bar{\alpha}\varphi$. $\square$

The proof is illustrated by the following diagram ($\bar{\beta} : G \to G/M$, $\bar{\gamma} : G \to G/N$ are natural homomorphisms).

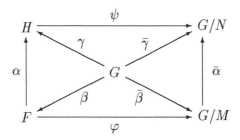

12. Any group G is isomorphic to the group $\mathcal{T} = \mathcal{T}(G)$ of *left-shift operators*. In additive and multiplicative notations, the shift τ_a by an element $a \in G$ is defined as follows:

$$\tau_a(x) = a + x, \quad \tau_a(x) = ax \quad (x \in G).$$

In part 7 of Subsection 2.1.1, we proved that $(\mathcal{T}, +)$ and $(\mathcal{T}, \cdot)$ are subgroups. They are groups whenever G is a group. The shift by the neutral element is the neutral shift. For every shift τ_a, its inverse is the shift by the element inverse to a:

$$\tau_0(x) = 0 + x = x, \quad \tau_1(x) = 1 \cdot x = x;$$
$$\tau_a(\tau_{-a}(x)) = a + (-a + x) = x, \quad \tau_a(\tau_{a^{-1}}(x)) = a(a^{-1}x) = x;$$
$$\tau_a\tau_{-a} = \tau_0 = \mathrm{id}, \quad \tau_a\tau_{a^{-1}} = \mathrm{id}.$$

The same is true for right-shift operators. For Abelian groups, the notions of right shift and left shift coincide.

13. Let E be a set, $(F, +)$ be an Abelian group, and let $(\mathcal{F}, +)$ be an Abelian group such that $\mathcal{F} = \mathcal{F}(E, F)$ is the set of all maps from E to F and the addition operation for $f, g \in \mathcal{F}$ is defined as follows:

$$(f + g)(x) = f(x) + g(x) \quad (x \in E).$$

The neutral element in $\mathcal{F}$ is the constant 0. For $f \in \mathcal{F}$, its inverse is $-f \in \mathcal{F}$ such that $(-f)(x) = -f(x)$ for all $x \in E$. The addition operation on $\mathcal{F}$ is commutative because so is the addition operation on F.

If $(E, +)$ is a semigroup, then every element $a \in E$ determines the *difference operator* $\Delta_a : f \to \Delta_a f$ $(f \in \mathcal{F})$, where a function $\Delta_a f$ is defined by the formula

$$\Delta_a f(x) = f(a + x) - f(x) \quad (x \in E).$$

The operator Δ_a is *additive*, i.e., it is a homomorphism from the group $(\mathcal{F}, +)$ into itself:

$$\Delta_a(f + g)(x) = (f(a + x) + g(a + x)) - (f(x) + g(x))$$
$$= (f(a + x) - f(x)) + (g(a + x) - g(x))$$
$$= \Delta_a f(x) + \Delta_a g(x) \quad (f, g \in \mathcal{F}).$$

It follows from the definitions that

$$\Delta_a f = f\tau_a - f \quad (f \in \mathcal{F}).$$

14. Let $(E, +) = (\mathbb{N}, +)$, $(F, +) = (\mathbb{Z}, +)$, $\mathcal{F} = \mathcal{F}(\mathbb{N}, \mathbb{Z})$, and $a = 1$. Then, for the difference operator $\Delta = \Delta_1$ and a function $f \in \mathcal{F}$, the function $\Delta f \in \mathcal{F}$ is defined by the formula

$$\Delta f(x) = f(x + 1) - f(x) \quad (x \in \mathbb{N}).$$

In particular, $\Delta f(x) = 1$ for $f(x) = x$ and $\Delta f(x) = 2x + 1$ for $f(x) = x^2$.

Let $g \in \mathcal{F}$. We now solve the difference equation

$$\Delta f = g, \quad \Delta f(x) = g(x) \quad (x \in \mathbb{N}).$$

From the definitions, it follows that $f(2) = f(1) + g(1)$ and $f(x+1) = f(x) + g(x)$ for $x \in \mathbb{N}$. By induction, the function f is defined by $f(1)$ and the equality

$$f(x+1) = f(1) + \sum_{1 \leq n \leq x} g(n) \quad (x \in \mathbb{N}).$$

Thus, for any right-hand side $g \in \mathcal{F}$, there exist a solution to the equation $\Delta f = g$. The solution, however, is not unique. Indeed, any constant function $f(x) = f(1) = c \in \mathbb{Z}$ $(x \in \mathbb{N})$ is a solution of the *homogeneous* difference equation $\Delta f = 0$. This means that the problem of solving the equation $\Delta f = g$ is *ill-posed*.

For the problem to be *well-posed*, i.e., to provide the *existence* and *uniqueness* of a solution $f \in \mathcal{F}$ to the equation $\Delta f = g$ for any right-hand side $g \in \mathcal{F}$, it is necessary to add an initial condition $f(1) = c$. Together with the equation for $f(x+1)$, it determines the unique solution of the equation $\Delta f = g$. Let $\mathcal{F}_c$ denote the set of all $f \in \mathcal{F}$ such that $f(1) = c$. It follows that the equation $\Delta f = g$ has a *unique solution* f in $\mathcal{F}_c$ for any $g \in \mathcal{F}$.

For $c = 0$, the set $\mathcal{F}_0 = \{f \in \mathcal{F} \mid f(1) = 0\}$ is a subgroup of the group $\mathcal{F}$. The restriction $\Delta_0 : \mathcal{F}_0 \to \mathcal{F}$ of the operator $\Delta : \mathcal{F} \to \mathcal{F}$ is an isomorphism because $\ker \Delta_0 = \{f_0 \in \mathcal{F}_0 \mid \Delta_0 f_0 = 0\} = 0$. For any $g \in \mathcal{F}$, the equation $\Delta_0 f_0 = g$ has a unique solution $f_0 = \Delta_0^{-1} g \in \mathcal{F}_0$ determined by the equalities

$$f_0(1) = 0, \quad f_0(x+1) = \sum_{1 \leq n \leq x} g(n) \quad (x \in \mathbb{N}).$$

The operator Δ_0^{-1} inverse to the difference operator Δ_0 is the *summation operator*.

Exercise. Solve the equation $\Delta f = g$ for $g(x) = x^m$ $(x \in \mathbb{N})$, $m = 1, 2, 3$.

The group $(\mathcal{C}, +)$ of constant sequences of integers is the kernel of the difference operator Δ. This group is isomorphic to the group of integers $(\mathbb{Z}, +)$. Hence, $\ker \Delta \simeq \mathbb{Z}$. Consider the factor group $(\bar{\mathcal{F}}, +) = (\mathcal{F}/\mathcal{C}, +)$ of $(\mathcal{F}, +)$ over its subgroup $(\mathcal{C}, +)$. Elements of $\bar{\mathcal{F}} = \mathcal{F}/\mathcal{C}$ are sets $\bar{f} = f + \mathcal{C}$ containing sequences $f + c$ of integer numbers $f(n) + c$ $(n \in \mathbb{N})$. The difference operator $\bar{\Delta}$ on $\bar{\mathcal{F}}$ maps $\bar{f}$ to $\bar{\Delta}\bar{f} = \Delta f$. From our analysis of the equation

$\Delta f = g$, it follows that there exists a unique solution $\bar{f} \in \bar{\mathcal{F}}$ to the equation $\bar{\Delta}\bar{f} = g$ for any right-hand side $g \in \mathcal{F}$. The problem is now *well-posed*.

The operator $\bar{\Delta} : \bar{\mathcal{F}} \to \mathcal{F}$ is an isomorphism. Indeed, Δ is a homomorphism, and so is $\bar{\Delta}$. We have ker $\bar{\Delta} = \{\bar{f} \in \bar{\mathcal{F}} \mid \bar{\Delta}\bar{f} = 0\} = \{f + c \mid f \in \mathcal{F}, c \in C, \Delta f = 0\} = C = \bar{0}$.

Remark. We have shown that there are two ways to make a problem well-posed. The first way is to provide additional conditions that restrict the set of admissible solutions. The second way is to consider factor sets. The equation $\Delta f = g$ without the initial condition $f(1) = c$ is sometimes called *underdetermined* or *ill-posed*. The addition of the initial condition makes it *determined* or *well-posed*.

15. Let $(G(i), +)$ $(i \in I)$ be a collection of Abelian semigroups with zero. The Cartesian product $P = \prod G(i)$ with the componentwise addition operation such that

$$(x + y)(i) = x(i) + y(i) \quad (x = (x(i)), \ y = (y(i)) \in P)$$

is called the *direct product* of the collection $(G(i), +)$, written $(P, +)$. It is clear that $(P, +)$ is an Abelian semigroup.

If $G(i)$ are Abelian groups, then their direct product is also an Abelian group because $x = 0 \Leftrightarrow x(i) = 0$ and $-x = (-x(i))$ $(i \in I)$.

The *direct sum* of Abelian semigroups $(G(i), +)$ with zero $(i \in I)$ is the subsemigroup $S = \sum G(i)$ of the semigroup $P = \prod G(i)$ that consists of finite collections $x = x(i) \in P$ $(|\{i : x(i) \neq 0\}| < \infty)$. If $G(i)$ are Abelian groups, then their direct sum $S = \sum G(i)$ is also an Abelian group (a subgroup of P).

Exercise. Verify that the notion of direct sum is well defined and prove the preceding statement.

Example 1. Let E be a set, $(F, +)$ be an Abelian semigroup with zero and $G(x) = F$ for all $x \in E$. Then $\prod G(x) = \mathcal{F} = \mathcal{F}(E, F)$ is the set of all functions $f : E \to F$ and the addition operation is defined by the formula $(f + g)(x) = f(x) + g(x)$ $(x \in E; \ f, g \in \mathcal{F})$. It is conventional to write $\mathcal{F} = F^E$ instead of $\mathcal{F} = \mathcal{F}(E, F)$. The direct sum $\sum G(x)$ consists of finite functions $f : E \to F$ $(|\operatorname{supp} f| < \infty)$.

Example 2. Consider an infinite field $\mathbb{K}$, the nth power function s^n : $\mathbb{K} \to \mathbb{K}$ with values $s^n(\xi) = \xi^n$ ($\xi \in \mathbb{K}$), and the additive group $G(n) = \mathbb{K}s^n = \{\alpha s^n \mid \alpha \in \mathbb{K}\}$ generated by s^n such that $(\alpha + \beta)s^n = \alpha s^n + \beta s^n$ ($\alpha, \beta \in \mathbb{K}$). The additive group of polynomials $G = \mathcal{M}(\mathbb{K})$ is the direct sum of power groups $G(n) = \mathbb{K}s^n$ ($n = 0, 1, 2, \dots$).

The group of polynomials $\mathcal{M}(\mathbb{K})$ is a subgroup of the additive group of functions $\mathcal{F}(\mathbb{K}, \mathbb{K})$. For a polynomial $p : \mathbb{K} \to \mathbb{K}$ with values $p(\xi) = \alpha_0 + \alpha_1\xi + \cdots + \alpha_n\xi_n$ ($\alpha_n \neq 0$), its coefficients α_i ($0 \leq i \leq n$) are uniquely determined if the field $\mathbb{K}$ is infinite.

An additive Abelian group G is the *direct sum of a finite collection of subgroups* $G(i)$, written $G = \sum G(i)$, if for any $\bar{x} \in G$ there exists a unique collection $x = (x(i)) \in \prod G(i)$ such that $\bar{x} = \sum x(i)$.

For the direct sum of groups and subgroups, there following correspondence is an isomorphism:

$$\varphi : \prod G(i) \to G, \quad \varphi(x) = \bar{x} = \sum x(i) \quad (x = (x(i)) \in \prod G(i)).$$

To prove this, we write the following equalities using the rules of addition for the groups $\prod G(i)$ and G:

$$\varphi(x + y) = \varphi(x(i) + y(i)) = \sum(x(i) + y(i))$$
$$= \sum x(i) + \sum y(i) = \varphi(x) + \varphi(y) \quad (x, y \in \prod G(i)).$$

Hence, φ is a homomorphism from $\prod G(i)$ onto G. Since the equality $\bar{0} = \sum x(i)$ holds only for the collection $x = 0$ with $x(i) = 0$, it follows that $\ker \varphi = \{0\}$ and φ is an isomorphism. This justifies the choice of the term *direct sum* and the designation $\sum G(i)$ for finite collections of groups and subgroups $G(i)$.

If a group G is the direct sum of subgroups A and B, then $B = A^{ac}$ is called the (*algebraic*) complement of A with respect to G. Let $L(X)$ be a subgroup of G generated by a set $X \subseteq G$.

Take a finite collection of subgroups $G(i)$ of the additive Abelian group G such that $\cup G(i)$ is a generating set of G, i.e., for every $\bar{x} \in G$ there is a collection $x = (x(i)) \in \prod G(i)$ such that $\bar{x} = \sum x(i)$. The collection $x = (x(i))$ is not assumed to be unique.

Proposition. $G = \sum G(i) \Leftrightarrow G(i) \cap \sum_{j \neq i} G(j) = \{0\}$.

Proof. The assertion is true for the collection consisting of one group, which is the group G itself. Suppose that the assertion is true for collections consisting of n or less subgroups. Consider a collection of $(n+1)$ subgroups $G(1), \ldots, G(n), G(n+1)$. If $\bar{x} = \sum_{1 \leq i \leq n} x(i) + x(n+1) = \sum_{1 \leq i \leq n} y(i) + y(n+1)$, then the element $z = \sum_{1 \leq i \leq n} x(i) - \sum_{1 \leq i \leq n} y(i) = x(n+1) - y(n+1)$ belongs to $Z = \left(\sum_{1 \leq i \leq n} G(i) \right) \cap G(n+1)$. Let $Z = \{0\}$. Then $x(n+1) = y(n+1)$ and $\sum x(i) = \sum y(i)$ $(1 \leq i \leq n)$. The representation $\bar{x} = \sum x(i)$ $(1 \leq i \leq n+1)$ is unique and $G = \sum G(i)$ $(1 \leq i \leq n+1)$. By induction, the implication $\Rightarrow$ is true for any finite collection of subgroups $G(i)$.

The converse implication $\Leftarrow$ follows directly from the uniqueness of the representation. Let $a \in G(j) \cap G(k)$ for $j \neq k$. Then $a = \sum x(i) = \sum y(i)$ for $x(j) = a$, $x(i) = 0$ $(i \neq j)$ and $y(k) = a$, $y(i) = 0$ $(i \neq k)$. For the direct sum, $x(i) = y(i)$ for all i. Hence $a = x(j) = y(j) = 0$. □

Example 3. Let $\mathbb{K}$ be an infinite field. Consider the group of polynomials $G = \mathcal{M}_n(\mathbb{K})$ generated by the power functions s^i $(0 \leq i \leq n)$. Every power function s^i generates the subgroup $G(i) = \mathbb{K}s^i = \{\alpha s^i \mid \alpha \in \mathbb{K}\}$. For every i we have $\mathbb{K}s^i \cap \sum \mathbb{K}s^j = \{0\}$ $(j < i)$; therefore the group of polynomials $G = \mathcal{M}_n(\mathbb{K})$ is the direct sum of subgroups $\mathbb{K}s^i$ $(0 \leq i \leq n)$. The group $G = \mathcal{M}_n(\mathbb{K})$ is a subgroup of the group $G = \mathcal{M}(\mathbb{K})$.

16. Assume that sets X, Y, and Z are isomorphic to each other and consider isomorphisms $j : X \to Y$, $k : X \to Z$, and $l = kj^{-1} : Y \to Z$. Let $S = I(Y, Z)$ and $P = I(X, X)$ be the sets of substitutions $s : Y \to Z$ and $p : X \to X$, respectively. We specify the *generalized symmetric group* $(S, *)$ isomorphic to the *symmetric group* $(P, \cdot)$ of automorphisms of X. The following diagram shows the relations between the above maps.

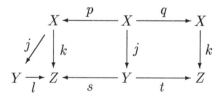

The equality $s = kpj^{-1}$ $(p \in P, s \in S)$ defines the isomorphism $\varphi : P \to S$. The product $s * t$ of substitutions $s = \varphi(p)$ and $t = \varphi(q) \in S$ is defined by the formula $\varphi(p) * \varphi(q) = \varphi(pq)$. Thus, φ is also an isomorphism from the group $(P, \cdot)$ onto the group $(S, *)$.

For the identity permutation $e : X \to X$ and the substitution $l : Y \to Z$, we have $\varphi(e) = kej^{-1} = kj^{-1} = l$. Hence, l is the identity element of the group $(S, *)$. The inverse element for the substitution $s = \varphi(p)$ in this group is the substitution $s^{-1} = \varphi(p^{-1})$ (which is not to be confused with the inverse map $s^{-1} : Z \to Y$).

We emphasize that the choice of isomorphisms j and k is essential for the definition of the multiplication operation $*$, although for any choice of j and k the resulting group is isomorphic to the group of substitutions $(P, \cdot)$.

17. Let $(S(n), \cdot)$ denote the symmetric group on an ordered set X consisting of n elements. Based on the results of the preceding part, we can assume that $X = \{1, \ldots, n\}$. We will describe the group $S(n)$ because of its particular importance.

First, we note that $|S(n)| = n!$, which follows from the combinatorial addition formulas.

Exercise. Prove that $|S(n)| = n!$.

For every permutation $p \in S(n)$, its *signature* is defined as follows:

$$\operatorname{sgn} p = \prod \operatorname{sgn}(p(j) - p(i)) \quad (1 \le i < j \le n).$$

A permutation p is said to be *even* if $\operatorname{sgn} p = 1$ and *odd* if $\operatorname{sgn} p = -1$.

For $i \ne j$, a *transposition* t_{ij} is a permutation such that $t_{ij}(i) = j$, $t_{ij}(j) = i$ and $t_{ij}(k) = k$ for all $k \ne i$, $k \ne j$ ($1 \le i, j, k \le n$). Clearly, $t_{ij}^{-1} = t_{ij} = t_{ji}$. Let $p \in S(n)$. For all $1 \le i \le n$, we set $p^0(i) = i$ and $p^j(i) = p(p^{j-1}(i))$ for $j \ge 1$. A permutation $i \to p(i) \to \cdots \to p^{l-1}(i) \to i$ such that $p^j(i) \ne p^k(i)$ ($0 < j < k < l$) is called a *cycle* of length $l > 1$ beginning with i, written $(i, p(i), \ldots, p^{l-1}(i))$. In particular, $t_{ij} = (ij)$. If $p(i) = i$, then i determines the cycle (i) of length 1. Two cycles are said to be *disjoint* if the elements appearing in of one of them do not appear in the other. Disjoint cycles commute.

Lemma. *Every permutation can be represented as a product of disjoint cycles.*

Proof. By induction, we establish the following representation for every $p \in S(n)$:

$$p = (i(1)p(1) \ldots p^{n(1)-1}(1)) \ldots (i(m)p(m) \ldots p^{n(m)-1}(m)),$$

where $i(1), \ldots, i(m)$ are selected elements, $p^j(k) = p^j(i(k))$ $(1 \le k \le m)$. The cycles in the above representation are disjoint and $n(1) + \cdots + n(m) = n$. $\square$

The representation of every permutation as a product of disjoint cycles is unique up to the order of the cycles. For this reason, every permutation $p \in S(n)$ can be associated with the number $n - m$, where m is the number of disjoint cycles in the representation of p. The number $n - m$ is called the *decrement* of the permutation p.

Note that $(ip(i) \ldots p^{l-1}(i)) = (ip(i)) \ldots (ip^{l-1}(i))$.

Exercise. Give a detailed proof of the above statements concerning disjoint cycles.

Proposition. *A permutation is even if and only if it can be represented as the product of an even number of transpositions.*

Proof. Every transposition $t = (ij)$ is odd. Indeed, we have $\operatorname{sgn} t = (-1)^{2(j-i)-1} = -1$. Hence, $\operatorname{sgn}(tu) = -\operatorname{sgn} u$ for every permutation u. By the lemma of this subsection, any permutation p can be represented as a product of cycles. Consequently, p is the product of transpositions, each of which has negative signature and therefore changes the signature of the entire product. It remains to note that the signature of p is equal to the product of the signatures of its factors. $\square$

Corollary. *A permutation is even if and only if its decrement is even.*

Proof. Let a permutation p be the product of m cycles of lengths $n(1), \ldots, n(m)$. Since a cycle of length l is a product of $l-1$ transpositions, p is a product of $(n(1) - 1) + \cdots + (n(m) - 1) = n - m$ transpositions. $\square$

Exercise. Prove that every permutation can be represented as the product of transpositions that move only adjacent elements.

The propositions proved above imply that $\operatorname{sgn}(pq) = \operatorname{sgn} p \cdot \operatorname{sgn} q$ for any permutations p and q. The product of even permutations is even. Hence, the even permutations form a subgroup of the symmetric group $S(n)$. This subgroup is denoted by $A(n)$ and is called the *alternating group* on an ordered set X of n elements. Since there are as many even permutations as there are odd permutations, we have $|A(n)| = n!/2$.

Exercise. Prove the preceding statement.

Remark. The map sgn : $S(n) \to \{+1, -1\}$ is a homomorphism from the group $(S(n), \cdot)$ onto the multiplicative group $(\{+1, -1\}, \cdot)$, the alternating group $A(n)$ being its kernel.

2.1.3. Rings and fields

These algebraic structures have two operations that are consistent with each other. A ring is formed by a group and a semigroup, while a field is formed by two groups. Rings and fields are the subject of Chapters 3 and 6 of van der Waerden (1991).

1. We first consider a more general algebraic structure with two operations. Two semigroups (H, f) and (H, g) sharing the same set, with operations f and g related to each other by the property of *distributivity*

$$g(f(x,y), z) = f(g(x,z), g(y,z)), \quad g(x, f(y,z)) = f(g(x,y), g(x,z)),$$

form a *semiring* (H,f,g). In this definition, g distributes over f. The distributivity of f over g is not required.

The semigroup (H, f) is assumed to be *Abelian*. If (H, g) is also commutative, then the semiring (H, f, g) is said to be *commutative*.

Examples. 1) Semigroups $(\mathcal{P}, \cup)$ and $(\mathcal{P}, \cap)$ form the semiring $(\mathcal{P}, \cup, \cap)$ whose elements are subsets of a given set.

2) The semigroups $(\mathbb{N}, +)$ and $(\mathbb{N}, \cdot)$ form the semiring $(\mathbb{N}, +, \cdot)$ whose elements are natural numbers.

3) The group $(\mathbb{Z}, +)$ and the semigroup $(\mathbb{Z}, \cdot)$ form the semiring $(\mathbb{Z}, +, \cdot)$ whose elements are integers.

2. The semiring $(R, +, \cdot)$ formed by an *Abelian* group $(R, +)$ and a *semigroup* $(R, \cdot)$ is called a *ring*. In a ring, multiplication distributes over addition:

$$(x + y)z = xz + yx, \quad x(y + z) = xy + xz.$$

Depending on its multiplicative semigroup $(R, \cdot)$, a ring can be *commutative* or *noncommutative*.

Remark. The definition of a semigroup usually postulates the associativity of the operation. Thus, both operations in a ring are assumed to be associative. There is a theory of nonassociative rings, where multiplication is not necessarily associative. The case of nonassociative multiplication is substantially more complicated. In this situation, some weaker requirements are often introduced instead.

There exists an element 0 in the ring $(R, +, \cdot)$, since it exists in its additive group $(R, +)$. The ring may or may not have a multiplicative identity 1, depending on its multiplicative semigroup $(R, \cdot)$. A ring with a multiplicative identity is called a *unital* ring.

Examples. 1) The *trivial ring* $(\{0\}, +, \cdot)$ with operations defined as $0 + 0 = 0 \cdot 0 = 0$. Its only element 0 coincides with 1.

2) The *binary ring* $(\mathbb{B}, +, \cdot)$ with operations defined by the table in the example of Subsection 2.1.1, part 4.

3) The ring of integers $(\mathbb{Z}, +, \cdot)$.

Every additive Abelian group $(R, +)$ is an additive group of some commutative ring $(R, +, \cdot)$. Indeed, one example is the *zero ring* with the multiplication operation $xy = 0$ for all $x, y \in R$. For this reason, Abelian groups can be considered a special case of commutative rings.

All the general notions and statements formulated for groups and semigroups can be extended to rings, provided that extension is consistent with both operations. In particular, the homomorphism $\varphi : R \to S$ from a ring $(R, +, \cdot)$ into a ring $(S, +, \cdot)$ is a homomorphism from the group $(R, +)$ into the group $(S, +)$ and from the semigroup $(R, \cdot)$ into the semigroup $(S, \cdot)$. It has the following properties:

$$\varphi(x + y) = \varphi(x) + \varphi(y), \quad \varphi(xy) = \varphi(x)\varphi(y) \quad (x, y \in R).$$

The composition of homomorphisms (isomorphisms) of rings is a homomorphism (isomorphism).

A ring $(R, +, \cdot)$ is a subring of $(S, +, \cdot)$, written $(R, +, \cdot) \subseteq (S, +, \cdot)$, if the group $(R, +)$ is a subgroup of the group $(S, +)$ and the semigroup $(R, \cdot)$ is a subsemigroup of the semigroup $(S, \cdot)$. Thus, by definition,

$$(R, +, \cdot) \subseteq (S, +, \cdot) \iff (R, +) \subseteq (S, +), \quad (R, \cdot) \subseteq (S, \cdot).$$

Example. The ring of even integers $(2\mathbb{Z}, +, \cdot)$ is a subring of the ring of integers $(\mathbb{Z}, +, \cdot)$.

Remark. Any ring $(R, +, \cdot)$ can be embedded in a unital ring. Consider the ring $(\mathbb{Z} \times R, +, *)$ with operations defined by the formulas $(m, x) + (n, y) = (m + n, x + y)$, $(m, x) * (n, y) = (mn, my + nx + xy)$ $(m, n \in \mathbb{Z},$

$x, y \in R$) and consider its subring $(\{0\} \times R, +, *)$. It is easy to verify that $\varphi : x \to (0, x)$ is an isomorphism from $(R, +, \cdot)$ to $(\{0\} \times R, +, *)$. The ring $(\mathbb{Z} \times R, +, *)$ has an identity element $(1, 0)$. If there exists a multiplicative identity $e \in R$, then $(0, e)$ is an identity in the subring $(\mathbb{Z} \times R, +, *)$, but not in the ring $(\mathbb{Z} \times R, +, *)$. This must be taken into account when constructing the embedding.

3. An important example of a ring is the *Boolean ring of sets* $(\mathcal{P}, +, \cdot)$. This ring is closely related to the Boolean algebra of sets $(\mathcal{P}, \cup, \cap, ')$ described in Subsection 1.1.2, part 10. Elements if these structures are subsets of a given set E. Addition and multiplication in the ring $(\mathcal{P}, +, \cdot)$ is defined as follows:

$$X + Y = (X \cup Y) \cap (X \cap Y)', \quad XY = X \cap Y.$$

Exercises. 1) Verify the associativity of addition and the distributivity rule.

2) Prove the identity $X + Y = (X \cap Y') \cup (Y \cap X')$.

3) Prove the identity $X \cup Y = X + Y + XY$, $X' = E - X$.

In the ring $(\mathcal{P}, +, \cdot)$, the empty set $\varnothing$ is a zero and the set E is an identity.

The opposite of an element $X \in \mathcal{P}$ (its inverse with respect to addition) coincides with X, i.e., $-X = X$. This follows from the identities $X + X = (X \cup X) \cap (X \cap X)' = X \cap X' = \varnothing$. Addition in $(\mathcal{P}, +, \cdot)$ is the same as substraction: $X + Y = X - Y = Y - X = Y + X$. For this reason, the sum of sets is often called the *symmetric difference* of sets . Furthermore, $X + Y = X \cup Y$ whenever $X \cap Y = \varnothing$, i.e., the sum of two disjoint sets is equal to their union.

Every element $X \in \mathcal{P}$ is a multiplicative idempotent: $X^2 = X$. The natural order on $\mathcal{P}$ coincides with the inclusion relation $\subseteq$: $X \subseteq Y \iff XY = X$. The empty set and the set E are the greatest and the least elements of $\mathcal{P}$, respectively: $\varnothing = \min \mathcal{P}$, $E = \max \mathcal{P}$. Any two elements $X, Y \in \mathcal{P}$ have the supremum and the infimum:

$$X \vee Y = \sup\{X, Y\} = X + Y + XY = X \cup Y,$$
$$X \wedge Y = \inf\{X, Y\} = XY = X \cap Y.$$

Exercise. Prove the statements contained in the preceding paragraph.

Consider the class $\mathcal{B} = \{0, E\}$. The subring $(\mathcal{B}, +, \cdot)$ of the ring $(\mathcal{P}, +, \cdot)$ is isomorphic to the binary ring $(\mathbb{B}, +, \cdot)$.

Another important example is the ring of functions.

Let $(\mathcal{F}, +)$ be an additive Abelian group of functions on a set X with values in a ring $(R, +, \cdot)$. Similarly to addition, multiplication of functions is defined by the formula

$$(fg)(x) = f(x) \cdot g(x) \quad (x \in X, \; f, g \in \mathcal{F}).$$

It is associative and consistent with addition:

$$((f + g)h)(x) = (f(x) + g(x)) \cdot h(x)$$
$$= f(x) \cdot h(x) + g(x) \cdot h(x) = (fh)(x) + (fh)(x),$$

$$(f(g + h))(x) = (fg)(x) + (fh)(x) \quad (x \in X, \; f, g, h \in \mathcal{F}).$$

With this multiplication, the group $(\mathcal{F}, +)$ becomes the *ring of functions* $(\mathcal{F}, +, \cdot)$.

Remark. The composition of functions is often called the product. For $X \neq R$, the composition may be undefined, whereas for $X = R$ the composition of functions may be not consistent with addition, i.e., $f \circ (g + h) \neq f \circ g + f \circ h$ in the general case.

If the ring R is unital, then the set of functions $\mathcal{F} = \mathcal{F}(X, R)$ contains indicators f with images $f(X) \subseteq \{0, 1\} \subseteq R$. The set of indicators $I = I(X)$ is isomorphic to the set $\mathcal{P}(X)$ of subsets of X. The set of indicators, however, does not form a subring of $(\mathcal{F}, +, \cdot)$ in the general case.

Let $0 < 1$ and let the *Boolean sum* $f \oplus g$ of indicators $f, g \in I$ be defined as follows:

$$(f \oplus g)(x) = \max\{f(x), g(x)\} - f(x) \cdot g(x) \quad (x \in X).$$

Then $(I, \oplus, \cdot)$ is a Boolean ring isomorphic to $(\mathcal{P}(X), +, \cdot)$.

Exercise. Prove the preceding statement.

4. A subring $(N, +, \cdot)$ of a ring $(R, +, \cdot)$ is called an *ideal* of this ring if the semigroup $(N, \cdot)$ is an ideal of the semigroup $(R, \cdot)$, i.e., $xy, yx \in N$ for $x \in N$, $y \in R$. The definitions of proper and maximal ideals of rings are similar to those for semigroups. The same is true for *left* and *right* ideals of rings.

Example. The subring of even numbers $(2\mathbb{Z}, +, \cdot)$ is a maximal ideal of the ring of integers $(\mathbb{Z}, +, \cdot)$. Indeed, we have $(2m)n = 2(mn) \in 2\mathbb{Z}$ $(2m \in 2\mathbb{Z}, n \in \mathbb{Z})$ and $2\mathbb{Z} \neq \mathbb{Z}$. Therefore, $2\mathbb{Z}$ is a proper ideal of the ring $\mathbb{Z}$. To prove that it is maximal, we note that every ideal of the ring $\mathbb{Z}$ that includes $2\mathbb{Z}$ as a proper subset contains at least one odd element $2m + 1$. Then it also contains the element $1 = (2m + 1) - 2m$ and therefore includes all integers. Hence, it is not a proper ideal.

Every set $A \subseteq R$ *generates* an ideal $(J(A), +, \cdot)$ in the ring $(R, +, \cdot)$. It is defined as the intersection of all ideals of the ring $(R, +, \cdot)$ that include A.

Exercise. Prove that the intersection of any collection of ideals of a ring is an ideal.

The ideal $J(a)$ generated by a single element $a \in R$ is called a *principal ideal*. If $(R, +, \cdot)$ is a commutative ring, then a principal ideal $(J(a), +, \cdot)$ is formed by elements $y = xa + na$ $(x \in R, n \in \mathbb{Z})$. If $(R, +, \cdot)$ is a unital ring, then $(J(a), +, \cdot)$ is formed by elements $y = xa$ $(x \in R)$.

Exercise. Prove the preceding statements.

Example. It is easy to see that $2\mathbb{Z} = J(2)$.

5. Let $(R, +, \cdot)$ be a ring and $(N, +, \cdot)$ be its ideal. By the definition of a ring, the group $(R, +)$ is Abelian, which implies that $(N, +)$ is its normal subgroup and there is a factor group $(\bar{R}, +) = (R/N, +)$ whose elements are factor sets $\bar{x} = x + N$ $(x \in R)$.

In part 7 of Subsection 2.1.2, we showed $\bar{a} + \bar{b} = \overline{a + b}$ $(a, b \in R)$. For $\bar{R}$ multiplication is defined by the formula $\bar{a}\bar{b} = \overline{ab}$. The operation is well defined because the product does not depend on the choice of particular elements a and b of the factor sets under consideration. Indeed, let $x \sim a$ and $y \sim b$, i.e., $x = a + u$ and $y = b + v$ for some $u, v \in N$. Then

$$xy = (a + u)(b + v) = ab + ub + av + uv = ab + w$$

for $w = ub + av + uv \in N$ because $(N, +, \cdot)$ is an ideal of $(R, +, \cdot)$. It follows that $xy \sim ab$. We now verify the distributive rule:

$$(\bar{a} + \bar{b})\bar{c} = \overline{(a + b)c} = \overline{ac + bc} = \bar{a}\bar{c} + \bar{b}\bar{c}.$$

Similarly, $\bar{a}(\bar{b} + \bar{c}) = \bar{a}\bar{b} + \bar{a}\bar{c}$ $(a, b, c \in R)$. Hence, $(\bar{R}, +, \cdot) = (R/N, +, \cdot)$ is a ring. It is called a factor ring of the ring $(R, +, \cdot)$ over the ideal $(N, +, \cdot)$. If $(R, +, \cdot)$ is commutative, then so is $(R/N, +, \cdot)$.

Examples. 1) The factor ring $\mathbb{Z}/2\mathbb{Z}$ consists of the sets $\bar{0} = 2\mathbb{Z}$ and $\bar{1} = 2\mathbb{Z} + 1$ of even and odd integers, respectively. The factor ring $\mathbb{Z}/2\mathbb{Z}$ is isomorphic to the binary ring $\mathbb{B}$.

2) Let $m \in \mathbb{N}$ and $J(m) = \{ml \mid l \in \mathbb{Z}\}$. Then the factor ring $\mathbb{Z}/J(m)$ consists of the sets $\bar{k} = \{n = ml + k \mid l \in \mathbb{Z}\}$ $(k = 0, \ldots, m - 1)$, which are called *residues modulo* m. For example, $\bar{0}$ and $\bar{1}$ are residues modulo 2.

3) Let $\mathcal{P} = \mathcal{P}(E)$ and let $\mathcal{K} = \mathcal{K}(E)$ be the class of all finite subsets of a set E. The class $\mathcal{K}$ forms an ideal in the Boolean ring $(\mathcal{P}, +, \cdot)$. The factor ring $\mathcal{P}/\mathcal{K}$ consists of subsets $X \subset E$ that differ from each other by a finite set, i.e., $\bar{X} = X + \mathcal{K}$ $(X \in \mathcal{P})$. If E is finite, then $\mathcal{P}(E) = \mathcal{K}(E)$ and therefore the factor ring $\mathcal{P}/\mathcal{K} = \mathcal{K}/\mathcal{K}$ is trivial.

6. A *field* is a commutative unital ring in which every nonzero element has a multiplicative inverse. A field $(F, +, \cdot)$ is formed by two Abelian groups: the additive group $(F, +)$ and the multiplicative group $(F \setminus \{0\}, \cdot)$. The operations in these groups are related by the distributivity rule. In addition, it is assumed that $0x = x0 = 0$ $(x \in F)$.

Remark. The general theory also studies noncommutative fields. Such fields are also called *bodies*.

The natural operations, such as addition, subtraction, multiplication, and division (except division by zero), are defined for elements of a field. A field does not contain *zero divisors*, i.e., there are no nonzero elements whose product is zero. This is due to the existence of multiplicative inverses: if $ab = 0$ and $a \neq 0$, then $b = a^{-1} \cdot 0 = 0$.

Some factor rings turn out to be fields. For example, let R be a unital commutative ring and N be its proper ideal.

Theorem. *The factor ring $\bar{R} = R/N$ is a field if and only if the ideal N is maximal.*

Proof. From the maximality of N it follows that for any $a \in R \setminus N$ and $b \in R$ there exist $x \in R$ and $y \in N$ such that $ax + y = b$ and therefore $\bar{a}\bar{x} = \bar{b}$ for the corresponding $\bar{a}, \bar{b}, \bar{x} \in \bar{R}$. This means that for any $\bar{a} \neq 0$ there exists an $\bar{x} = \bar{a}^{-1}$. $\quad\square$

Example. The factor ring $\mathbb{Z}/2\mathbb{Z}$ is a field.

Exercise. Prove that the factor ring $\mathbb{Z}/J(m) = \{\bar{0}, \ldots, \overline{m-1}\}$ in Example 2 of part 5 of this subsection is a field if and only if m is a prime number.

7. Every nonempty unital commutative ring R with no zero divisors can be embedded into a field $Q = Q(R)$ by associating a *fraction* a/b with every pair $a, b \in R$. We now describe this embedding in detail.

Since there are no zero divisors in R, for any $x, y \in R$ and $z \in R$ such that $z \neq 0$ the equality $xz = yz$ implies $(x-y)z = xz - yz = 0$ and $x - y = 0$. Hence $x = y$.

Consider the set $P = R \times (R \setminus \{0\})$ consisting of pairs (x, y) with the first element $x \in R$ and the second element $y \in R \setminus \{0\}$ ($y \in R$, $y \neq 0$). We define an equivalence relation on P as follows: $(x, y) \sim (a, b) \Leftrightarrow ay = bx$. It is reflexive and symmetric because multiplication in R is commutative. The transitivity follows from the relations $(x, y) \sim (a, b)$, $(a, b) \sim (u, v) \Leftrightarrow ay = bx$, $av = bu \Rightarrow abyu = abxv \Rightarrow yu = xv \Leftrightarrow (x, y) \sim (u, v)$, $a, u, x \in R$ and $b, v, y \in R \setminus \{0\}$. (If $a = 0$, then $ay = bx$, $av = bu \Rightarrow x = 0$, $u = 0 \Rightarrow yu = 0 = xv$.) Elements $\overline{(a, b)}$ of the factor set $Q = \bar{P} = P/\sim$ will be called *fractions* and denoted by a/b. The equality $a/b = x/y$ means $ay = bx$. Addition and multiplication for fractions is defined by the usual rules

$$(a/b) + (x/y) = (ay + bx)/by, \quad (a/b)(x/y) = ax/by.$$

Theorem. $(Q, +, \cdot)$ *is a field.*

Proof. 1) It is clear that $(Q, +)$ is an Abelian group. Indeed, addition of fractions is commutative. It is easy to verify that it is also associative. The fraction $0/1$ is the zero element in $(Q, +)$. For a fraction a/b, its additive inverse is $(-a)/b$.

2) $(Q, \cdot)$ is an *Abelian group with identity.* Since multiplication in the ring R is commutative and associative, so is multiplication of fractions. The fraction $1/1$ is an identity element.

3) $(Q, +, \cdot)$ is a *commutative unital ring.* It only remains to verify the distributivity rule, which is trivial.

4) $(Q, +, \cdot)$ *is a field.* Every fraction $a/b \neq 0/1$ has an inverse $(a/b)^{-1} = b/a$. Indeed, $a/b \neq 0/1 \Leftrightarrow a \neq 0$ and $(a/b)(b/a) = ab/ab = 1/1$.   □

Exercise. Give a more detailed proof of the preceding theorem without using $1 \in R$.

The field $Q = Q(R)$ is said to be the *ring of fractions* generated by the ring R. The map $\varphi : R \rightarrow Q$, $\varphi : x \rightarrow x/1$ $(x \in R)$ is an embedding of R into the field Q. Indeed, φ is a one-to-one map because $x/1 = y/1 \Leftrightarrow x = y$. It preserves the operations: $x/1 + y/1 = (x+y)/1$, $(x/1)(y/1) = (xy)/1$. The ring R can be identified with its image $\varphi(R) \subseteq Q$ so that it can be assumed that $R \subseteq Q$. It is conventional to write x instead of $x/1$.

8. The field $\mathbb{Q} = Q(\mathbb{Z})$ is called the *field of rational numbers.* It is assumed that $\mathbb{Z} \subseteq \mathbb{Q}$ and $n/1 = n$ $(n \in \mathbb{Z})$.

Proposition. *The set* $\mathbb{Q}$ *is countable.*

Proof. 1) The set of integers $\mathbb{Z}$ is countable. Indeed, the map $\alpha : \mathbb{N} \rightarrow \mathbb{Z}$ such that $\alpha(2n) = n$ and $\alpha(2n-1) = -n+1$ $(n \in \mathbb{N})$ is an isomorphism.

2) The set $\mathbb{Z} \times \mathbb{Z}$ is countable because so is $\mathbb{N} \times \mathbb{N}$. The map $\beta : \mathbb{N} \times \mathbb{N} \rightarrow \mathbb{N}$ such that $\beta(m, n) = (m+n)(m+n+1)/2 + n$ $(m, n \in \mathbb{N})$ is an isomorphism.

3) Since $\mathbb{N}$ can be embedded into $\mathbb{Q}$, we have $|\mathbb{N}| \leq |\mathbb{Q}|$. At the same time, the map $\gamma : \mathbb{Z} \times (\mathbb{Z} \setminus \{0\}) \rightarrow \mathbb{Q}$ such that $\gamma(a, b) = a/b$ is an onto map and therefore $|\mathbb{N}| = |\mathbb{Z} \times \mathbb{Z}| \geq |\mathbb{Z} \times (\mathbb{Z} \setminus \{0\})| \geq |\mathbb{Q}|$. Hence $|\mathbb{N}| = |\mathbb{Q}|$.   □

Using the order on $\mathbb{Z}$, we can introduce an order on $\mathbb{Q}$. Set $a/b \leq x/y \Leftrightarrow ay \leq bx$ $(a/b, x/y \in \mathbb{Q}, b > 0, y > 0)$. This order is antisymmetric: $a/b \leq x/y, x/y \leq a/b \Leftrightarrow a/b = x/y$. It is also a linear order because any two rational numbers a/b and x/y are comparable $(a/b \leq x/y$ or $x/y \leq a/b)$. The order on $\mathbb{Q}$ is consistent with the addition and multiplication operations. We have

$$p < q \Leftrightarrow p + z < q + z \ (z \in \mathbb{Q}) \Leftrightarrow pz < qz \quad (z \in \mathbb{Q}, \ z > 0).$$

This follows from the fact that the order on $\mathbb{Z}$ are consistent with the operations.

Lemma. *Every rational number is majorized by a natural number.*

Proof. Let $q = a/b \in \mathbb{Q}$. Since $(-a)/(-b) = a/b$, we can assume that $b > 0$. If $a \leq 0$, then $q \leq 1$. Let $a > 0$. Then $1 \leq b \Rightarrow q \leq a$. $\square$

Archimedes' theorem. *For any two rational numbers $p > 0$ and q, there exists a natural number n such that $np \geq q$.*

Proof. From the preceding lemma, it follows that there exists a natural number $n \geq r = q/p$. This inequality is equivalent to the desired one. $\square$

Obviously, the inequality $np \geq q$ in the theorem can be replaced by the strict inequality $np > q$.

In view of Archimedes' theorem and the consistency of the order with the operations, $(\mathbb{Q}, +, \cdot, \leq)$ is called an *Archimedean field*. Since the order is total, $(\mathbb{Q}, \leq)$ can be called the *rational line*. Unlike $(\mathbb{Z}, \leq)$, the line $(\mathbb{Q}, \leq)$ is not discrete, i.e., for any two distinct rational numbers there are rational numbers between them. This can be proved using Archimedes' theorem, but the easiest way is to use the observation that $p < r = (p+q)/2 < q$ for all rational $p < q$.

9. The Archimedean field of rational numbers has a disadvantage of being *incomplete*, i.e., its subsets may have no infimum or supremum.

Example. The set $A = \{x \in \mathbb{Q} \mid x^2 \leq 2\}$ has neither infimum nor supremum in $\mathbb{Q}$.

Suppose that $q = a/b = \sup A \in \mathbb{Q}$. Then $q^2 = 2$. Indeed, if $q^2 < 2$, then $(q + 1/n)^2 < 2$ for some $n \in \mathbb{N}$ because, by Archimedes' theorem, there exists an $n \in \mathbb{N}$ such that $2q + 1 < n(2 - q^2)$ and therefore $(q + 1/n)^2 = q^2 + 2q/n + (1/n)^2 \leq q^2 + 2q/n + 1/n = q^2 + (2q + 1)/n < q^2 + 2 - q^2 = 2$. Hence $q < q + 1/n \in A$, which contradicts the assumption that $q = \sup A$.

We now prove that the equality $q^2 = 2$ cannot hold for any $q = a/b \in \mathbb{Q}$. Since $(-q)^2 = q^2$, we can assume that $a > 0$ and $b > 0$. We can also assume that the fraction a/b is irreducible. The equality $q^2 = 2$ is equivalent to $a^2 = 2b^2$. Therefore, there is an $m \in \mathbb{N}$ such that $a = 2m$ (since the squared odd number is odd). Hence, $4m^2 = 2b^2$, $2m^2 = b^2$ and $b = 2n$ for some $n \in \mathbb{N}$. We have $a/b = (2m)/(2n) = m/n$, which contradicts the assumption that the fraction a/b is irreducible. Consequently, the equality $q^2 = 2$ does not hold and $\mathbb{Q}$ does not contain $\sup A$.

Since A is symmetric with respect to zero, $\inf A$ is equal to $-\sup A$ and therefore $\mathbb{Q}$ does not contain it either.

10. Let $(F, +, \cdot)$ be an infinite field and consider a collection of elements $a_i \in F$ $(0 \le i \le n)$. The function $p : F \to F$ with values

$$p(x) = a_0 + a_1 x + \cdots + a_n x^n \quad (x \in F)$$

is called a *polynomial function* or simply *polynomial*. If $a_n \ne 0$, then n is called the degree of the polynomial p. *Solving the equation* $p(x) = 0$ was long considered the principal algebraic problem.

A first degree equation $a_0 + a_1 x = 0$ $(a_1 \ne 0)$ is easy to solve. Its proper solvability is due to the existence of additive and multiplicative inverses in a field. The unique solution to this solution is $x = -a_1^{-1} a_0$. Indeed, $a_0 + a_1(-a_1^{-1} a_0) = a_0 - a_0 = 0$. The solution is unique because $a_0 + a_1 x = 0 \Rightarrow a_1 x = -a_0 \Rightarrow x = -a_1^{-1} a_0$.

However, there are equations of the second degree that are unsolvable in some fields. For example, so are the equations $-2 + x^2 = 0$ and $1 + x^2 = 0$ in $\mathbb{Q}$. For the first equation, this was proved in part 9 of this subsection. The unsolvability of the second one follows from the *rule of signs*: since the order is consistent with the operations, if $x \in \mathbb{Q}$ and $y \in \mathbb{Q}$ such that either $x > 0$ and $y > 0$ or $x < 0$ and $y < 0$, then $xy > 0$. Therefore, $x^2 > 0$ for $x \ne 0$.

Remark. In what follows, we consider mostly nonzero groups, rings, and fields.

2.1.4. Lattices

Ordered sets with binary bounds are naturally associated with algebraic structures. The general theory of such sets is presented in Birkhoff (1979).

1. An *upper semilattice* $(L, \vee, \le)$ is formed by an *Abelian semigroup of idempotents* $(L, \vee)$ and a partially ordered set $(L, \le)$ such that $x \le y \Leftrightarrow x \vee y = y$. A *lower semilattice* $(L, \wedge, \le)$ is formed by an Abelian semigroup $(L, \wedge)$ and a partially ordered set $(L, \le)$ such that $x \le y \Leftrightarrow x \wedge y = x$.

From the definition, it follows that $x \vee y = \sup\{x, y\}$ and $x \wedge y = \inf\{x, y\}$. This explains the term *semilattice*. We now prove that $x \vee y = \sup\{x, y\}$ for the upper semilattice $(L, \vee, \le)$. We have $x \le (x \vee y)$ since $x \vee (x \vee y) = (x \vee x) \vee y = x \vee y$ and if $x \le z$ and $y \le z$ then $(x \vee y) \le z$ because $(x \vee y) \vee z = x \vee (y \vee z) = x \vee z = z$. It can be proved in a similar way that $x \wedge y = \inf\{x, y\}$ for the lower semilattice $(L, \wedge, \le)$.

We will often write $(L, \vee)$ and $(L, \wedge)$ or simply L instead of $(L, \vee, \le)$ and $(L, \wedge, \le)$.

Examples. 1) $(\mathcal{P},\cup)$ and $(\mathcal{P},\cap)$. 2) $(\mathbb{Z},\max)$, $(\mathbb{Z},\min)$. 3) $(\mathbb{Q},\max)$, $(\mathbb{Q},\min)$. 4) $(\mathbb{Z}\times\mathbb{Z},\sup)$, $(\mathbb{Z}\times\mathbb{Z},\inf)$. 5) $(\mathbb{Q}\times\mathbb{Q},\sup)$, $(\mathbb{Q}\times\mathbb{Q},\inf)$.

In the last two examples, the order is componentwise: $(x,y)\le(a,b)\Leftrightarrow x\le a,\ y\le b$.

Exercise. Let $(H,\cdot)$ be a multiplicative Abelian semigroup. We define the operation $\vee$ and the order relation on H as follows: for any $a,b\in H$ $a\vee b=ab$; $a\le b$ if and only if the equation $ax=b$ is solvable (b is divisible by a). Prove that the operation $\vee$ is consistent with the introduced order and $(H,\vee,\le)$ is an upper semilattice.

2. A *lattice* $(L,\vee,\wedge)$ is formed by an upper semilattice $(L,\vee)$ and a lower semilattice $(L,\wedge)$ satisfying the *absorption* condition

$$x\vee(x\wedge y)=x\wedge(x\vee y)=x.$$

These equalities represent a weaker version of the distributivity rule.

Examples. 1) $(\mathcal{P},\cup,\cap)$. 2) $(\mathbb{Z},\max,\min)$. 3) $(\mathbb{Q},\max,\min)$. 4) $(\mathbb{Z}\times\mathbb{Z},\sup,\inf)$. 5) $(\mathbb{Q}\times\mathbb{Q},\sup,\inf)$.

A lattice is said to be (*relatively*) *complete* if every (*bounded*) subset of it has a least upper bound and a greatest lower bound.

Examples. 1) $(\mathcal{P},\cup,\cap)$ is a complete lattice. 2) $(\mathbb{Z},\max,\min)$ is a relatively complete lattice. 3) $(\mathbb{Q},\max,\min)$ is not a complete lattice because the set $A=\{x\in\mathbb{Q}\mid x^2\le 2\}$ has no supremum or infimum in $\mathbb{Q}$ (see part 9 of Subsection 2.1.3).

If the order for a lattice has to be specified explicitly, then it is included in the designation so that we write $(L,\vee,\wedge,\le)$ instead of $(L,\vee,\wedge)$. On the other hand, if it is not necessary to specify the operations and order, then the lattice is denoted simply by L.

3. An *ortholattice* $(L,\vee,\wedge,\backslash)$ is a lattice $(L,\vee,\wedge)$ with a smallest element $0=\min L$ and the third operation $\backslash$ that is consistent with $\vee$ and $\wedge$, which is provided by the conditions

$$y\backslash x=y\backslash(x\wedge y),\quad (y\backslash x)\vee(x\wedge y)=y,$$
$$(y\backslash x)\wedge(x\wedge y)=0,\quad y\backslash(y\backslash x)=x\wedge y.$$

The operation $\backslash$ is called the *difference* or the *relative complement*.

Example. $(\mathcal{P}, \cup, \cap, \setminus)$ is an ortholattice.

Exercise. For the ortholattice in the preceding example, verify that the operation $\setminus$ is consistent with $\vee$ and $\wedge$.

We now define the orthogonality relation $\perp$ for $(L, \vee, \wedge, \setminus)$ as follows: $x \perp y \Leftrightarrow (x \vee y) \setminus y = x$. (This explains the choice of the term *ortholattice*.) The relation $\perp$ is symmetric because $(y \vee x) \setminus x = (x \vee y) \setminus ((x \vee y) \setminus y) = (x \vee y) \wedge y = y$. Note that $y \setminus x \perp x \wedge y$, which follows from the equalities $((y \setminus x) \vee (x \wedge y)) \setminus (x \wedge y) = y \setminus (x \wedge y) = y \setminus x$.

Example. For the lattice $(\mathcal{P}, \cup, \cap, \setminus)$, orthogonality means disjointness, i.e., $X \perp Y \Longleftrightarrow X \cap Y = \varnothing$.

If an ortholattice $(L, \vee, \wedge, \setminus)$ has not only $0 = \min L$ but also $1 = \max L$, then every $x \in L$ has a *complement* $x' = 1 \setminus x$, which satisfies the following conditions:
$$x \wedge x' = 0, \quad x \vee x' = 1, \quad (x')' = x.$$
Such a lattice is said to be *orthomodular* if the difference operation satisfies to the additional condition
$$y \setminus x = y \wedge (x \wedge y)'.$$

Exercise. Prove that the following conditions hold for elements of an orthomodular lattice: $y \setminus 0 = y$ $0 \setminus x = 0$, $y \setminus y = 0$, $x \perp y \Rightarrow x \wedge y = 0$, $x \perp y \Longleftrightarrow x \leq y'$.

4. A lattice $(L, \vee, \wedge)$ is said to be *distributive*, if the operations $\vee$ and $\wedge$ are distributive over each other:
$$(x \vee y) \wedge z = (x \wedge z) \vee (y \wedge z), \quad (x \wedge y) \vee z = (x \vee z) \wedge (y \vee z).$$

Exercise. Prove that these identities are equivalent.

Examples. 1) The lattice $(\mathcal{P}, \cup, \cap)$ is distributive. 2) The lattice $(\mathcal{L}(G), \vee, \wedge)$ of subgroups of a group G is not distributive. By definition, $F \vee H = G(F \cup H)$ is a subgroup of G generated by the set $F \cup H$ and $F \wedge H = F \cap H$ $(F, Y \in \mathcal{L}(G))$. 3) The lattice $(I(R), \vee, \wedge)$ of ideals of a ring R is not distributive. By definition, $A \vee B = I(A \cup B)$ and $A \wedge B = A \cap B$ $(A, B \in I(R))$.

Exercise. Prove that the lattices $\mathcal{L}(G)$ and $I(R)$ are not distributive.

5. A distributive ortholattice is called *Boolean*. If a Boolean lattice has a greatest element, it is called a *Boolean algebra*.

Example. The lattice $(\mathcal{P}, \cup, \cap, \setminus)$ is a Boolean algebra.

For a Boolean lattice $(B, \vee, \wedge, \setminus)$, addition and multiplication are defined by analogy with the Boolean ring of sets $(\mathcal{P}, +, \cdot)$ described in part 3 of Subsection 2.1.3:

$$x + y = (x \vee y) \setminus (x \wedge y), \quad xy = x \wedge y$$

The set B with addition and multiplication defined this way is a commutative ring $(B, +, \cdot)$ of multiplicative idempotents ($x^2 = x \wedge x = x$ for all $x \in B$). Such rings are called *Boolean algebras*.

This is justified by the fact that for any Boolean ring $(B, +, \cdot)$ the identities

$$x \vee y = x + y + xy, \quad x \wedge y = xy, \quad y \setminus x = y + xy$$

define the Boolean lattice $(B, \vee, \wedge, \setminus)$. We have $0 = \min B$ and $1 = \max B$ if $(B, +, \cdot)$ is a ring with identity. Furthermore, $x + y = (x \vee y) \setminus (x \wedge y)$.

Exercise. Prove the preceding statements.

In Boolean algebras, the duality rules hold:

$$(x \vee y)' = x' \wedge y', \quad (x \wedge y)' = x' \vee y'.$$

They are equivalent to the identities $1 + (x + y + xy) = (1 + x)(1 + y)$ and $1 + xy = (1 + x) + (1 + y) + (1 + x)(1 + y)$.

Example. Let $\mathcal{K} = \mathcal{K}(E)$ be the class of all finite subsets of a set E. This class forms the Boolean lattice $(\mathcal{K}, \cup, \cap)$ and the ring $(\mathcal{K}, +, \cdot)$, which are Boolean algebras if and only if E is finite.

The additive inverse of any element x in a Boolean ring coincides with x, i.e., $-x = x$. Indeed, $x + x = (x + x)(x + x) = x + x + x + x$ and therefore $x + x = 0$. Addition in a Boolean ring is equivalent to subtraction: $x + y = x - y = y - x = y + x$. For this reason, the sum of elements in a Boolean ring is often called the *symmetric difference* of these elements.

Remark. The commutativity of a Boolean ring follows from the idempotency of its elements: $x+y = (x+y)(x+y) = x+xy+yx+y \Rightarrow xy+yx = 0 \Rightarrow xy = -yx = yx$. Therefore, a Boolean ring can be defined as a ring of multiplicative idempotents, the commutativity requirement being omitted.

Since $xy = -xy$, $x \lor y = x+y-xy = x \star y$ is a *star product* of elements of a Boolean ring (see Section 97 in van der Waerden, 1991). The star product can be also defined by the equality $(1 - x)(1 - y) = 1 - (x \lor y)$.

6. Let $(A, \lor, \land, \backslash)$ and $(B, \lor, \land, \backslash)$ be Boolean lattices and let $(A, +, \cdot)$ and $(B, +, \cdot)$ be the corresponding Boolean rings. The homomorphism $\varphi : A \to B$ from the ring $(A, +, \cdot)$ into the ring $(B, +, \cdot)$ is a homomorphism from the lattice $(A, \lor, \land, \backslash)$ into the lattice $(B, \lor, \land, \backslash)$. This follows from the identities

$$\varphi(x \lor y) = \varphi(x + y + xy) = \varphi(x) + \varphi(y) + \varphi(xy) = \varphi(x) \lor \varphi(y),$$
$$\varphi(x \land y) = \varphi(xy) = \varphi(x)\varphi(y) = \varphi(x) \land \varphi(y),$$
$$\varphi(y \backslash x) = \varphi(y + xy) = \varphi(y) + \varphi(x)\varphi(y) = \varphi(y) \backslash \varphi(x).$$

The homomorphism $\varphi : A \to B$ is a monotonic map from the lattice A into the lattice B:

$$x \leq y \Rightarrow xy = x \Rightarrow \varphi(x)\varphi(y) = \varphi(x) \Rightarrow \varphi(x) \leq \varphi(y).$$

Example. Consider the class $\mathcal{P} = \mathcal{P}(E)$ of all subsets of a set E and the set $\mathcal{F} = \mathcal{F}(E, \mathbb{B})$ of all maps from E to $\mathbb{B} = \{0, 1\}$. We define addition and multiplication on $\mathcal{F}$ based on addition and multiplication in $(\mathbb{B}, +, \cdot)$:

$$(f + g)(x) = f(x) + g(x), \quad (fg)(x) = f(x)g(x) \quad (x \in E).$$

As a result, $(\mathcal{F}, +, \cdot)$ is a Boolean algebra. Its zero and identity elements are the zero map and the constant map with constant value 1.

The image of any $A \in \mathcal{P}$ under the map $\varphi : \mathcal{P} \to \mathcal{F}$ is the function $\varphi_A \in \mathcal{F}$ such that $\varphi_A(x) = 1$ for $x \in A$ and $\varphi_A(x) = 0$ for $x \notin A$. The map φ is an isomorphism from the Boolean algebra $(\mathcal{P}, +, \cdot)$ onto the Boolean algebra $(\mathcal{F}, +, \cdot)$.

Exercise. Prove the statements in the preceding example.

7. The correspondences between the sets A and B were defined as subsets of the Cartesian product $A \times B$. They form the Boolean algebra $\mathcal{P}(A \times B)$. The order relation of inclusion for correspondences determines their restrictions and extensions. Specific types of correspondences form lattices, although they are not necessarily Boolean. Most common types of lattices are formed by equivalences. The reason is that equivalences define partitions and vice versa.

Let $\mathcal{S} = \mathcal{S}(E) \subseteq \mathcal{P}(E \times E)$ be the class of all equivalence relations on E. Let $R, S \in \mathcal{S}$. Then $R \wedge S = R \cap S$. If R and S define partitions R_i and S_j of E, then $R \cap S$ defines the partition $R_i \cap S_j$ of E. Furthermore, $R \vee S$ is an equivalence relation on $E \times E$ generated by $R \cup S$.

The least equivalence is the *equality* $\Delta = \{(x, x) \mid x \in E\}$, and the greatest one is the *trivial equivalence* $E \times E$.

The class $\mathcal{S}$ and the operations $\vee$ and $\wedge$ form the lattice $(\mathcal{S}, \vee, \wedge)$. This lattice is not Boolean.

Exercise. Prove the above statements.

Remark. The lattice $(\mathcal{S}(E), \vee, \wedge)$ is not a sublattice of $\mathcal{P}(E \times E), \cup, \cap)$ because $\vee \neq \cup$ although $\wedge = \cap$.

2.1.5. Numbers

The real and complex type fields include the field of real numbers and the field of complex numbers. They are described in van der Waerden (1991), Chapter 11.

1. Any relatively complete Archimedean field is called the *field of real numbers*. All such fields are isomorphic in the sense of order and algebraic structure and are therefore considered identical and denoted by $\mathbb{R}$. The field of rational numbers $\mathbb{Q}$ is identified with the subfield of $\mathbb{R}$ isomorphic to it, which is generated by integer multiples of unity, and is considered a subset of the field of real numbers: $\mathbb{Q} \subset \mathbb{R}$.

There are many representations of the field $\mathbb{R}$. Cantor's model, which is based on the notion of convergence, is the most convenient representation for calculus. Let $\mathcal{R} = \mathcal{F}(\mathbb{N}, \mathbb{Q})$ be the ring of sequences of rational numbers. The *absolute value* $|x|$ of a number $x \in \mathbb{Q}$ is defined as follows: $|x| = x$ for $x \geq 0$ and $|x| = -x$ for $x < 0$. A sequence $f \in \mathcal{R}$ *converges* (*in itself*) if for any rational number $\varepsilon > 0$ there is a number $n(\varepsilon)$ such that $|f(p) - f(n)| \leq \varepsilon$

for $p, n \geq n(\varepsilon)$. The convergent sequences form a subring $\mathcal{K}$ of the ring $\mathcal{R}$. A sequence $\alpha \in \mathcal{R}$ is called *infinitesimal* if for any rational number $\varepsilon > 0$ there is a number $n(\varepsilon)$ such that $|\alpha(n)| \leq \varepsilon$ for $n \geq n(\varepsilon)$. The infinitesimal sequences form a maximal ideal $\mathcal{V}$ of the ring $\mathcal{K}$ of convergent sequences. It is maximal since for any $f \in \mathcal{K} \setminus \mathcal{V}$ there exists a $g \in \mathcal{K}$ such that $fg = (1) + \alpha$, where $(1) \in \mathcal{K}$ is the *constant unity* and $\alpha \in \mathcal{V}$ is infinitesimal. Therefore, if we add f to $\mathcal{V}$ then $(1) \in \mathcal{I}$ and $\mathcal{I} = \mathcal{K}$ for any ideal $\mathcal{I}$ generated by the set $\{f\} \cup \mathcal{V}$. By the theorem of Subsection 2.1.3, part 6, the factor ring $\mathbb{R} = \mathcal{K}/\mathcal{V}$ is a field. In this representation, real numbers are sets $\bar{f} = f + \mathcal{V}$ of *equivalent* convergent sequences of rational numbers (which differ from each other by an infinitesimal sequence). A rational number $q \in \mathbb{Q}$ is identified with the real number $\bar{q} = (q) + \mathcal{V}$, where (q) is the constant sequence with constant value q.

Based on the order on $\mathbb{Q}$, we can define an order on $\mathbb{R}$. Let $\bar{f} = f + \mathcal{V}$ and $\bar{g} = g + \mathcal{V} \in \mathbb{R}$. Then $\bar{g} > \bar{f}$ means that there exists a rational number $\varepsilon > 0$ and a number $n(\varepsilon)$ such that $g(n) - f(n) > \varepsilon$ $(n \geq n(\varepsilon))$. This relation is well defined since it does not depend on the choice of the representative sequences $f, g \in \mathcal{K}$. The order on $\mathbb{R}$ defined above is consistent with the operations. The field $\mathbb{R}$ with such an order is a relatively complete Archimedean field.

Exercise. Prove the preceding statements.

The equation $x^m = b$ has a solution in the set of real numbers for all $m \in \mathbb{N}$ and $b \in \mathbb{R}$ such that $b \geq 0$.

Example. Let $b > 0$, $b \in \mathbb{Q}$. Then a solution $x = b^{1/m} > 0$ to the equation $x^m = b$ is determined by the recurrent sequence of rational numbers

$$f(n + 1) = f(n) + (b - f^m(n))/mf^{m-1}(n) \quad (n \geq 0)$$

with an arbitrary initial term $x(0) = a \in \mathbb{Q}$. Using Newton's binomial formula, we can show that

$$|f(p) - f(n)| \leq \varepsilon \ (p, n \geq m^2 b^{m-1}|a^m - b|\varepsilon^{-1} + 1)$$

for every rational $\varepsilon > 0$. The sequence $f(n)$ converges. Moreover, $f(n + 1) \leq f(n)$ and $0 \leq f^m(n) - b \leq (1 - 1/m)^n|a^m - b|$ $(n \geq 1)$. Hence $x = b^{1/m} = f + \mathcal{V}$. The rule that determines the sequence f is called the *Newton's algorithm*. This algorithm is convenient for calculating specific roots (such as $\sqrt{2}$). It is also applicable in the case of real number $b > 0$.

2. *The field of complex numbers* $\mathbb{C}$ is obtained from the field of real numbers $\mathbb{R}$ by adding the imaginary unit i, which is a solution to the equation $x^2 = -1$. Complex numbers can be identified with the points on the real plane $\mathbb{R}^2 = \mathbb{R} \times \mathbb{R}$. In this representation, $\mathbb{R} \times \{0\}$ is identified with the real line $\mathbb{R}$ and it is assumed that $\mathbb{R} \subset \mathbb{C}$. The *imaginary unit* i is identified with the point $(0, 1)$, and $\{0\} \times \mathbb{R}$ is called the *imaginary axis*, written $\mathbb{R}i$. The representation of $\mathbb{C}$ as $\mathbb{R} \times \mathbb{R}$ leads to the formula $\mathbb{C} = \mathbb{R} + \mathbb{R}i$ and provides a convenient notation for complex numbers: $c = (a, b) = a + bi$ $(a, b \in \mathbb{R})$. The numbers $bi = (0, b)$ are called *imaginary*.

The sum of complex numbers $c = a + bi$ and $z = x + yi$ is equal to $c + z = (a + x) + (b + y)i$. The product of complex numbers is defined using the distributivity rule and the equality $i^2 = -1$:

$$cz = (a + bi)(x + yi) = (ax - by) + (ay + bx)i.$$

The equality $i^2 = -1$ shows that the rule of signs does not hold in the field of complex numbers and therefore there is no linear order consistent with the operations.

The *absolute value* of a complex number $z = x + yi$ is the positive number $|z| = (x^2 + y^2)^{1/2}$. The following relations hold:

$$|z| > 0 \; (z \neq 0) \text{ and } |0| = 0, \quad |c + z| \leq |c| + |z|, \quad |cz| = |c||z|.$$

The middle relation is called the *triangle inequality*. It implies the inequality

$$||c| - |z|| \leq |c - z|.$$

The *conjugate* of a complex number $z = x + yi$ is defined to be the number $z^* = x - yi$. The following equalities hold:

$$(c + z)^* = c^* + z^*, \quad (cz)^* = z^* c^*, \quad c^* c = |c|^2, \quad c^{**} = c.$$

Conjugation has a clear geometrical interpretation: it represents the *reflection* of the plane with respect to the real axis $\mathbb{R}$, the points of $\mathbb{R}$ remaining invariant. The equality $c^* = c$ implies that c is a real number.

In view of the representation of a complex number z as a point (x, y) of the real plane, the expression $z = x + iy$ is called the *Cartesian representation* of the number z. The numbers $\operatorname{Re} z = x$ and $\operatorname{Im} z = y$ are called the *real* and *imaginary* parts of z, respectively. Another representation of complex numbers $z \neq 0$ is the *polar representation* $z = |z|w$, where $w = u + iv = z/|z|$ so that $|w| = 1$. This representation is unique. Indeed, let $z = rc$, where

$c = a + bi$, $|c| = 1$, and $r > 0$. Then $|z|u = ra$, $|z|v = rb$, and $|z| = r|c| = r$, which implies $u = a$ and $v = b$. The Cartesian and polar representations have clear geometrical interpretations. The former is the decomposition of a vector z into components parallel to the coordinate axes, whereas the latter specifies the length of the vector z and the angle between the vector and the horizontal axis.

Remark. Let $c_0, c_1, \ldots, c_m$ $(m \geq 1)$ be complex numbers. Consider a polynomial $p : \mathbb{C} \to \mathbb{C}$ with values $p(z) = c_0 + c_1 z + \cdots + c_m z^m$ $(z \in \mathbb{C})$. We assume that $c_m \neq 0$. The principal theorem of algebra of complex numbers states that the algebraic equation $p(z) = 0$ has a solution $z = z_0 \in \mathbb{C}$. In particular, the equation $z^2 + 1 = 0$ has a solution $z = i$.

3. The field $\mathbb{Q}$ of rational numbers is an example of an Archimedean field that is not relatively complete. It is associated with the field $\mathbb{G} = \mathbb{Q} + \mathbb{Q}i$ of *Gaussian numbers* obtained from Q by adding the imaginary unit i. By definition, $\mathbb{G}$ is an algebra of vectors with scalar field $\mathbb{Q}$, basis $\{1, i\}$, and multiplication table $1 \cdot 1 = 1$, $1 \cdot i = i = i \cdot 1$, $i \cdot i = -1$. Since the multiplication table is symmetric and it has unity 1, $(\mathbb{G}, +, \cdot)$ is a unital commutative ring. This ring is also a field. Indeed, if $a + bi \neq 0$ $(a, b \in \mathbb{Q})$, then $a^2 + b^2 \neq 0$ and $(a + bi)^{-1} = (a^2 + b^2)^{-1}a - (a^2 + b^2)^{-1}bi$. It follows from the definitions that $1^2 + i^2 = 1 - 1 = 0$.

The fields $\mathbb{Q}$ and $\mathbb{G}$ have characteristic 0, i.e., $n \cdot 1 \neq 0$ for $n \in \mathbb{N}$. An example of a field with nonzero characteristic is the field $\mathbb{Z}/I(5) = \{\bar{0}, \bar{1}, \bar{2}, \bar{3}, \bar{4}\}$ of residues modulo 5. The residues satisfy the equalities $5 \cdot \bar{1} = \bar{0}$ and $\bar{1}^2 + \bar{2}^2 = \bar{0}$.

Operator equations for functions with values in $\mathbb{Q}$, $\mathbb{G}$, and various finite fields are used in solving some theoretical and applied problems.

4. Nonstandard analysis uses a different kind of real numbers. This theory is constructed as a logical model, although its numbers can also be understood from the standard algebraic point of view.

The set $\mathcal{N} = \mathcal{F}_0(\mathbb{N}, \mathbb{Q})$ of finite sequences forms an ideal of the ring $\mathcal{R} = \mathcal{F}(\mathbb{N}, \mathbb{Q})$. This ideal is not maximal since it is included in the ideal $\mathcal{M} = \{h \in \mathcal{R} \mid h(2n) = 0 \ (n \geq n(0))\}$ of sequences whose even component is finite. Therefore, the factor ring $\bar{\mathbb{Q}} = \mathcal{R}/\mathcal{N}$ is not a field. Its elements are sets $\bar{f} = f + \mathcal{N}$ of sequences that differ from f by only a finite number of terms, i.e., $g \in f + \mathcal{N}$ means that $g(n) - f(n) = 0$ for $n \geq n(0)$. The elements of $\bar{\mathbb{Q}}$ will be called *generalized rational numbers*. After identifying $q \in \mathcal{R}$ with $\bar{q} = (q) + \mathcal{N} \in \bar{\mathbb{Q}}$, we can assume that $\mathcal{R} \subset \bar{\mathbb{Q}}$.

A generalized rational number can be a zero divisor. For example, let $a(2n) = 0$, $a(2n-1) = 1$, $b(2n) = 1$, and $b(2n-1) = 0$ $(n \geq 1)$. Then $\bar{a} \neq \bar{0}$, $\bar{b} \neq \bar{0}$ and $\bar{a}\bar{b} = \bar{0}$. The inverse $\bar{g} = 1/\bar{f}$ exists if and only if the sequences f that make up the number $\bar{f}$ have no zero subsequences, i.e., $f(n) \neq 0$ for $n \geq n(0)$. In particular, the numbers $1/\bar{a}$ and $1/\bar{b}$ do not exist.

Based on the order on $\mathbb{Q}$, we can define an order on $\bar{\mathbb{Q}}$ as follows: $\bar{g} \leq \bar{f}$ means that $f(n) \leq g(n)$ for $n \geq n(0)$. The introduced order is not total since two arbitrary elements are not necessarily comparable. Furthermore, we have $\bar{a} > 0$ and $\bar{b} > 0$, but $\bar{a}\bar{b} = \bar{0}$, which means that the order is not consistent with multiplication.

Note that $\mathcal{N} \subseteq \mathcal{V}$, which means that every finite sequence is infinitesimal. From the second isomorphism theorem (Subsection 2.1.2, part 11), it follows that $\mathbb{R} = \mathcal{K}/\mathcal{V} = \bar{\mathcal{K}}/\bar{\mathcal{V}}$ for $\bar{\mathcal{K}} = \mathcal{K}/\mathcal{N}$, $\bar{\mathcal{V}} = \mathcal{V}/\mathcal{N}$. Standard real numbers can be identified with sets $\bar{f} + \bar{\mathcal{V}}$, where $\bar{f} = f + \mathcal{N}$ $(f \in \mathcal{K})$ and $\bar{\mathcal{V}} = \{\alpha + \mathcal{N} \mid \alpha \in \mathcal{V}\}$. Elements $\bar{f} \in \bar{\mathcal{K}}$ are called *generalized real numbers*, and elements $\bar{\alpha} \in \bar{\mathcal{V}}$ are called *infinitesimal numbers*. Identifying $\mathbb{R}$ with $\bar{\mathcal{K}}/\bar{\mathcal{V}}$ makes it possible to represent every real number as a set of generalized numbers. This provides more detailed distinction between convergent sequences of rational numbers.

Depending on the representative sequences, the generalized rationals are subdivided into bounded and unbounded. A number $\bar{f} = f + \mathcal{N} \in \bar{\mathbb{Q}}$ is said to be bounded if there exists an $r \in \mathbb{Q}$, $r > 0$, such that $|f(n)| \leq r$ for all $n \in \mathbb{N}$. All generalized real numbers are bounded. They include rational and infinitesimal ones. There are infinitely large unbounded numbers, which are the inverses of infinitesimal ones. Infinitesimal numbers are less than any rational number, while infinitely large ones are greater than any rational number.

Remark. Furthermore, it is possible to consider an arbitrary maximal ideal $\mathcal{M}$ that contains $\mathcal{N}$ and the field $\bar{\bar{\mathbb{Q}}} = \mathcal{R}/\mathcal{M} \simeq \bar{\mathbb{Q}}/\bar{\mathbb{M}}$ for $\bar{\mathbb{Q}} = \mathcal{R}/\mathcal{N}$ and $\bar{\mathbb{M}} = \mathcal{M}/\mathcal{N}$. If the continuum hypothesis is adopted as an axiom, then all such fields $\bar{\bar{\mathbb{Q}}}$ can be shown to be isomorphic to each other. A disadvantage of this model is that the set $\mathcal{M}$ of infinitesimal numbers is undetermined. For example, since the ideal $\mathcal{M}$ is maximal, it follows that only one of the numbers $\bar{\bar{a}} = a + \mathcal{M}$ and $\bar{\bar{b}} = b + \mathcal{M}$ with a and b defined above is infinitesimal, but it has to be specified which one.

A similar representation of nonstandard real numbers can be constructed by replacing the field $\mathbb{Q}$ of rational numbers by the field $\mathbb{R}$ of standard real numbers and the ring $\mathbb{Q}$ by the ring $\mathbb{R}$.

5. Instead of real numbers, non-Archimedean analysis deals with p-adic numbers obtained by completion of the field $\mathbb{Q}$, the absolute value being replaced by the p-adic norm (where p is a prime natural number). Such a completion is described in Chapter 18 of van der Waerden (1991).

Let $m \in \mathbb{Z}$, $n \in \mathbb{N}$, $x = m/n \in \mathbb{Q}$, $m = p^a u$, and $n = p^b v$, where $u \in \mathbb{Z}$, $v \in \mathbb{N}$ is not divisible by p, and $a, b \in \{0\} \cup \mathbb{N}$. Then $x = p^{a-b} \cdot u/v$, where $a - b = c \in \mathbb{Z}$. The numbers $a = \operatorname{ord} m$, $b = \operatorname{ord} n$, and $c = \operatorname{ord} x$ are called the *orders* a, b, x. Whenever p needs to be specified explicitly, we say p-*order* and write ord_p. For convenience, it is assumed that $\operatorname{ord} 0 = \infty$. It is easy to verify that $\operatorname{ord}(lm) = \operatorname{ord} l + \operatorname{ord} m$ and, consequently, $\operatorname{ord}(lm) - \operatorname{ord}(ln) = \operatorname{ord} m - \operatorname{ord} n$ for every $l, m, n \in \mathbb{Z}$. The order of a fraction is well defined since it does not depend on its representation.

The p-*norm* of a number $x \in \mathbb{Q}$ is defined by the formula

$$|x|_p = p^{-\operatorname{ord} x}.$$

The following relations hold for any $x, y \in \mathbb{Q}$:

$$|x|_p > 0 \ (x \neq 0) \text{ and } |0|_p = 0, \quad |x + y|_p \leq |x|_p \vee |y|_p, \quad |xy|_p = |x|_p |y|_p.$$

The middle relation is called the *isosceles triangle inequality*. It follows that

$$|x - y|_p = |y|_p \quad (|x|_p < |y|_p).$$

The norms that satisfy the isosceles triangle inequality are called *non-Archimedean*. The other norms are called *Archimedean*.

Exercise. Prove that $|x|_p \leq 1$ for $x \in \mathbb{Z}$.

Similar to the definition of $\mathcal{K}$ and $\mathcal{V}$, with the absolute value replaced by the p-norm, we define the ring $\mathcal{K}_p$ of p-*convergent* sequences and its maximal ideal $\mathcal{V}_p$ of *infinitely p-small* sequences of rational numbers. The field $\mathbb{Q}_p = \mathcal{K}_p / \mathcal{V}_p$ is called the *field of p-adic numbers*. These numbers are sets $f + \mathcal{V}_p$ of p-equivalent p-convergent sequences of rational numbers (i.e., which differ from each other only by a p-infinitesimal sequence).

We emphasize that p-convergence is essentially different from the conventional convergence. It is related to the divisibility by p.

Examples. 1) Let $f(n) = p^n$. Then $\operatorname{ord} f(n) = n$, $|f(n)|_p = p^{-n} \to 0$. In particular, $2^n \to 0$ for $p = 2$.

2) Let $g(n) = p^{-n}$. Then $\operatorname{ord} g(n) = -n$ and $|g(n)|_p = p^n \to \infty$. In particular, $2^{-n} \to \infty$ for $p = 2$.

Remark. We do not use nonstandard and non-Archimedean analysis in this book. Nonstandard real and p-adic numbers are introduced here only for the purpose of demonstrating that the standard model is not unique. Nonstandard and non-Archimedean analysis is extensively covered in the literature (see Monna, 1970; Robinson, 1966).

6. Chapter 2 of Maslov (1987), introduces a commutative ordered ring $(A, \oplus, \odot, \leq)$ with neutral elements $\bar{0}$ and $\bar{1}$ such that $x \odot \bar{0} = \bar{0}$, $x \odot \bar{1} = x$, and $x \oplus \bar{0} = x$ for all $x \in A$. The consistency of the order with the operations is provided by the following relations for comparable elements: $x \odot x \geq \bar{0}$, $x \oplus z \leq y \oplus z \ (x \leq y)$, and $x \odot z \leq y \odot z \ (x \leq y, z \geq \bar{0})$.

Examples provided there include semirings whose elements are numbers of the extended real line $[-\infty, \infty]$ and whose algebraic operations are consistent with order relations. The order in every example is induced by the order on $[-\infty, \infty]$.

Example 1. $A = [-\infty, \infty]$, $\oplus = \vee$, $\odot = +$, $\bar{0} = -\infty$, $\bar{1} = 0$.

Example 2. $A =]-\infty, \infty]$, $\oplus = \wedge$, $\odot = +$, $\bar{0} = \infty$, $\bar{1} = 0$.

Example 3. $A = [0, \infty]$, $\oplus = \vee$, $\odot = \wedge$, $\bar{0} = 0$, $\bar{1} = \infty$.

Example 4. $A = [0, \infty]$, $\oplus = \wedge$, $\odot = \vee$, $\bar{0} = \infty$, $\bar{1} = 0$.

Example 5. $A = [0, \infty[$, $\oplus = \vee$, $\odot = \cdot$, $\bar{0} = 0$, $\bar{1} = 1$.

Exercise. Verify that the operations in these examples are consistent with the order.

Note that all elements in Examples 1–5 are additive idempotents, the elements in Examples 3 and 4 also being multiplicative idempotents.

2.2. LINEAR ALGEBRA

Linear algebra The main objects studied in linear algebra are vector spaces, vector algebras, and their homomorphisms. Vector spaces and vector algebras are the subject of Chapters 12 and 13 of van der Waerden (1991). Various applications of linear algebra in mathematics and physics are described in the book Kostrikin and Manin (1997), which contains many interesting problems. The connection between linear algebra and geometry is demonstrated in detail in Postnikov (1986b).

2.2.1. Vector spaces

Vector spaces are Abelian groups with multiplicative field operators.

1. Consider a set $\mathbb{K}$, an Abelian group (E, g), and the *multiplication*
$\cdot : \mathbb{K} \times E \to E$ satisfying the *distributivity rule*

$$\alpha g(x, y) = g(\alpha x, \alpha y) \quad (\alpha \in \mathbb{K}, \ x, y \in E).$$

These components form the *group with left operators* $(\mathbb{K}, E) = (\mathbb{K}, E, g, \cdot)$.
Operators are elements of the set $\mathbb{K}$. Elements of the group (E, g) are called
vectors. In a similar way we define the *group with right operators* $(E, \mathbb{K}, g, \cdot)$
with outer multiplication $\cdot : E \times \mathbb{K} \to E$ and the rule

$$g(x, y)\alpha = g(x\alpha, y\alpha) \quad (\alpha \in \mathbb{K}, \ x, y \in E).$$

The distributivity rules in left additive and multiplicative notations are
written as follows:

$$\alpha(x + y) = \alpha x + \alpha y, \quad \alpha(xy) = (\alpha x)(\alpha y) \quad (\alpha \in \mathbb{K}, \ x, y \in E).$$

In what follows, we will usually use left additive notation and left operators
will be called simply *operators*.

Note that $\alpha 0 = 0$ ($\alpha \in \mathbb{K}$, $0 \in E$.) Indeed, $\alpha 0 + \alpha 0 = \alpha(0 + 0) = \alpha 0$.
Therefore, $\alpha(-x) = -\alpha(x)$ because $\alpha(x) + \alpha(-x) = \alpha(x - x) = \alpha 0 = 0$
($\alpha \in \mathbb{K}$, $x, y \in E$).

Example. The set of integers $\mathbb{Z}$, the additive Abelian group $(E, +)$,
and outer multiplication $nx = x + \cdots + x$ (n times) such that $(-n)x = -nx$
($n \in \mathbb{N}$) and $0x = 0$ ($x \in E$) form the group with operators $(\mathbb{Z}, E)$. Since
outer multiplication for $(\mathbb{Z}, E)$ is defined in terms of addition in E, the
group E can be identified with $(\mathbb{Z}, E)$. Every group can be considered as a
group with integer operators.

In multiplicative notation, instead of nx, we write $x^n = x \cdots x$ (n times),
$x^{-n} = (x^n)^{-1}$ ($n \in \mathbb{N}$), $x^0 = 1$ ($x \in E$). Integer multiples of nx and
powers x^n were considered in part 3 of Subsection 2.1.2.

All general definitions and propositions for groups can be extended to
groups with operators. A subgroup $(\mathbb{K}, A)$ of the group $(\mathbb{K}, E)$ is made
up of the group $\mathbb{K}$ and a subgroup A of E that is *invariant* under outer
multiplication ($\mathbb{K}A \subseteq A$). A factor group $(\mathbb{K}, E/N)$ is made up of the $\mathbb{K}$
and the factor group E/N over an invariant subgroup N of the group E.

Since N is normal, the factor group over N exists. Since N is invariant, it follows that $(\mathbb{K}, N)$ and $(\mathbb{K}, E/N)$ are groups with operators.

If $\mathbb{K} = \mathbb{Z}$, then $\mathbb{Z}N = N$ for every normal subgroup N and the factor group $(\mathbb{Z}, E/N)$ can be identified with E/N.

2. Consider a multiplicative group $(\mathbb{K}, \cdot)$, an additive group $(E, +)$ and outer multiplication $\mathbb{K} \times E \to E$, which will be also denoted by $\cdot$. We obtain the *group with multiplicative operators* $(\mathbb{K}, E) = (\mathbb{K}, E, +, \cdot)$ if the outer multiplication is distributive and *associative*:

$$\alpha(\beta x) = (\alpha\beta)x \quad (\alpha, \beta \in \mathbb{K}, \ x \in E).$$

Multiplicative operators are called *multipliers* for short.

Example. The multiplicative semigroup $(\mathbb{Z}, \cdot)$ of integers, the additive group $(E, +)$, and the outer multiplication $(n, x) \to nx$ $(n \in \mathbb{Z}, x \in N)$ make up the group with multipliers $(\mathbb{Z}, E) = (\mathbb{Z}, E, +, \cdot)$.

3. Let $(\mathbb{K}, +, \cdot)$ be a ring, $(E, +)$ be an additive *Abelian* group, and let $\cdot \colon \mathbb{K} \times E \to E$ be outer multiplication. They are said to form the *left* $\mathbb{K}$*-module* $(\mathbb{K}, E) = (\mathbb{K}, E, +, +, \cdot, \cdot)$ if outer multiplication, in addition to being associative and distributive over addition on E, is also *distributive over addition on* $\mathbb{K}$, i.e.,

$$(\alpha + \beta)x = \alpha x + \beta x \quad (\alpha, \beta \in \mathbb{K}, \ x \in E).$$

If the ring $\mathbb{K}$ is unital and outer multiplication *preserves the neutrality of unity*, i.e.,

$$1x = x \quad (1 \in \mathbb{K}, \ x \in E),$$

then the $\mathbb{K}$-module $(\mathbb{K}, E)$ is said to be *unital*. A left $\mathbb{K}$-module will be called a $\mathbb{K}$-*module* or simply a *module* if it is clear from the context what ring $\mathbb{K}$ is meant. The right $\mathbb{K}$-module $(E, \mathbb{K}) = (E, \mathbb{K}, +, +, \cdot, \cdot)$ is defined in a similar way. It uses right operators.

Example. The ring of integers $(\mathbb{Z}, +, \cdot)$, the additive Abelian group $(E, +)$, and the outer multiplication $(n, x) \to nx$ form the module $(\mathbb{Z}, E)$. It follows from the definition of integer multiples that $(m + n)x = mx + nx$ for $m, n \in \mathbb{Z}$ and $x \in E$. The module $(\mathbb{Z}, E)$ is unital. Thus, every additive Abelian group can be considered to be a $\mathbb{Z}$-module.

A *submodule* $(\mathbb{K}, N)$ of a module $(\mathbb{K}, E)$ is formed by the ring $\mathbb{K}$ and an invariant subgroup N of the group E (which means that $\alpha N \subseteq N$ for $\alpha \in \mathbb{K}$).

A *factor module* $(\mathbb{K}, E/N)$ of the module $(\mathbb{K}, E)$ over the submodule $(\mathbb{K}, N)$ is formed by the ring $\mathbb{K}$ and the factor group E/N. Outer multiplication is defined by the formula $\alpha(x + N) = \alpha x + N$ ($\alpha \in \mathbb{K}$, $x \in E$).

Exercise. Prove that the outer multiplication defined above is consistent with the other operations.

Remark. Every module can be embedded in a unital module since every ring can be embedded in a unital ring. This allows one to focus on studying unital modules.

Exercise. Describe in detail how an arbitrary module can be embedded into a unital module.

4. If an operator ring $(\mathbb{K}, +, \cdot)$ is a *field*, then the unital module $(\mathbb{K}, E)$ is called a *vector space*. Elements of the field $\mathbb{K}$ are called *scalar operators* or *scalars*.

A vector space *has no zero divisors*: if $\alpha\beta = 0$ or $\alpha x = 0$, where α and β are scalars and x is a vector, then $\alpha = 0$ or $\beta = 0$, and $\alpha = 0$ or $x = 0$, respectively. Indeed, $\alpha\beta = 0$, $\alpha \neq 0 \Rightarrow \beta = \alpha^{-1}0 = 0$ and $\alpha \neq 0$, $\alpha x = 0 \Rightarrow \alpha^{-1}(\alpha x) = (\alpha^{-1}\alpha)x = 1x = x = 0$. A module may have zero divisors.

Example. The module $(\mathbb{K}, \mathbb{K})$, where $\mathbb{K} = \mathbb{Z}/4\mathbb{Z} = \{\bar{0}, \bar{1}, \bar{2}, \bar{3}\}$ is a residue class ring (see part 5 of Subsection 2.1.3), contains zero divisors: $\bar{2} \cdot \bar{2} = \bar{0}$.

Remark. In the general theory, vector spaces include modules with operators from a noncommutative field. In some problems, it is necessary to consider nonassociative rings of operators and assume that distributivity does not hold in some cases. Operations with vectors are much more complicated in such situations.

There are special features of operations in vector spaces with finite scalar fields (in particular, the Boolean field $\mathbb{B} = \{0, 1\}$) In what follows, we will usually consider infinite scalar fields. Some of the propositions below are true only for such fields.

Example 1. Let $(\mathbb{K}, +, \cdot)$ be a field, X be a set, and let $\mathcal{F} = \mathcal{F}(X, \mathbb{K})$ be the set of functions on X with values in $\mathbb{K}$ (*scalar functions on X*). Addition and multiplication on $\mathbb{K}$ determine addition and multiplication on $\mathcal{F}$:

$$(f + g)(x) = f(x) + g(x), \quad (fg)(x) = f(x)g(x) \quad (f, g \in \mathcal{F}, \ x \in X).$$

The following statements ate true: 1) $(\mathcal{F}, +, \cdot)$ is a commutative unital ring. 2) $(\mathbb{K}, \mathcal{F}) = (\mathbb{K}, \mathcal{F}, +, +, \cdot, \cdot)$ is a vector space. 3) $\varphi : \mathbb{K} \to \mathcal{F}$ such that $\varphi(\alpha) = \bar{\alpha}, \bar{\alpha}(x) = \alpha \ (x \in X)$ is an embedding of the field $\mathbb{K}$ into the ring $\mathcal{F}$. 4) The constant $\bar{0}$ is the zero of the ring $\mathcal{F}$. 5) The constant $\bar{1}$ is the identity of the ring $\mathcal{F}$. 6) If $|X| > 1$, then the ring $\mathcal{F}$ has zero divisors.

Exercise. Verify the statements 1)–6).

Example 2. Let $(\mathbb{K}, E)$ be a vector space, X be a set, and let $\mathcal{G} = \mathcal{F}(X, E)$ be the set of functions on X with values in E (*vector functions on X*). Addition and multiplication on $\mathcal{G}$ are determined by addition and outer multiplication on E:

$$(f + g)(x) = f(x) + g(x), \quad (\alpha \cdot f)(x) = \alpha \cdot f(x) \quad (f \in \mathcal{F}, \ \alpha \in \mathbb{K}).$$

The field $\mathbb{K}$ and the set $\mathcal{G}$ of vector functions with these operations form the vector space $(\mathbb{K}, \mathcal{G}) = (\mathbb{K}, \mathcal{G}, +, +, \cdot, \cdot)$.

Exercise. Prove the assertion in the preceding example.

5. For the geometry of a vector space is the fundamental concept of linear independence of vectors is very essential.

Let $(\mathbb{K}, E)$ be a vector space. A collection of scalars $\alpha_i \ (i \in I)$ is called *finite*, if $\alpha_i = 0$ for al $i \in I \setminus K$, where K is a finite subset of the set of indices I. In particular, it is possible that $K = \varnothing$ and $\alpha_i = 0$ for all $i \in I$. This collection is called the *zero collection*. If $\alpha_i \neq 0$ for some $i \in I$, then the collection α_i is said to be *nonzero*. For every finite collection of scalars α_i and every collection of vectors x_i, their *linear combination* $\sum \alpha_i x_i$ is the sum of the finite set of nonzero vectors $\alpha_i x_i$. The scalars α_i are called the *coefficients* of the linear combination $\sum \alpha_i x_i$. A linear combination with zero or nonzero collection of coefficients is said to be *zero* or *nonzero*, respectively.

If a vector x is equal to a linear combination of vectors $x_i \ (i \in I)$, then x is said to be *expressed linearly* in terms of x_i. In particular, the vector $x = 0$ is expressed linearly in terms of any vectors x_i. A collection of vectors x_i is said to be *linearly independent*, if every linear combination of vectors of this collection is nonzero. The vectors of a linearly independent collection are called *linearly independent*. It is clear all vectors in a linearly independent collection are distinct. Therefore, linear independent *sets* of vectors can be considered instead of linearly independent *collections*.

Examples. The empty set of vectors and any singleton whose element is a nonzero vector are linearly independent. Every set of vectors that contains the vector $x = 0$ is not linearly independent.

We will often write *independent* instead of *linearly independent* and *space* instead of *vector space*.

Lemma. *The vectors of a collection (x_i) $(i \in I, |I| > 1)$ are linearly independent if and only if none of them can be expressed linearly in terms of the other.*

Proof. The assertion of the lemma follows from the equivalence of the equalities $\sum \alpha_i x_i = 0$ and $x_{i(0)} = -\sum \alpha_{i(0)}^{-1} \alpha_i x_i$ for some index $i(0)$, where $i \neq i(0)$ and $\alpha_{i(0)} \neq 0$. $\square$

Remark. The lemma does not hold for modules, since $\sum \alpha_i x_i = 0$ does not necessarily imply $x_{i(0)} = -\sum \alpha_{i(0)}^{-1} \alpha_i x_i$ $(i \neq i(0))$ if none of the nonzero coefficients have inverses.

Example. In the module $(\mathbb{Z}, \mathbb{Z})$, the equality $(-3) \cdot 2 + 2 \cdot 3 = 0$ is true, but none of the vectors 2 and 3 can be expressed linearly in terms of the other.

6. Maximal independent collections of vectors (which are not contained in any other independent collections) are called *bases* of a given vector space. The vectors that make up a basis are called *basis vectors*.

Let $(\mathbb{K}, E)$ be a vector space and (e_i) be a collection of vectors $e_i \in E$ $(i \in I)$.

Lemma. *Let the vectors e_i be linearly independent. The collection (e_i) is a basis of $(\mathbb{K}, E)$ if and only if every vector $x \in E$ is equal to a linear combination of vectors e_i.*

Proof. 1) Let (e_i) be a basis. Then, if we add x to (e_i), the resulting collection is not linearly independent. By the lemma of part 5 of this subsection, $\alpha(x)x + \sum \alpha_i e_i = 0$ for some nonzero finite collection of scalars $\alpha(x)$, α_i. Since the vectors e_i are independent, $\alpha(x) = 0$ does not hold. Consequently, $x = \sum \alpha_i(x)e_i$ for $\alpha_i(x) = -\alpha^{-1}(x)\alpha_i$.

2) If every vector $x \in E$ is linearly dependent on (e_i) and the collection (e_i) is linearly independent, then it is maximal and therefore it is a basis. $\square$

Theorem. *Let* (e_i) *be a basis. Then for every vector* $x \in E$ *there exists a unique finite collection of scalars* $\alpha_i(x)$ *such that* $x = \sum \alpha_i(x)e_i$.

Proof. By the preceding lemma, such a collection $\alpha_i(x)$ does exist. Consider an arbitrary collection of scalars β_i such that $x = \sum \beta_i e_i$. The equality $\sum \alpha_i(x)e_i = \sum \beta_i e_i$ implies that $\sum(\alpha_i(x) - \beta_i)e_i = 0$. By the preceding lemma, $\alpha_i(x) - \beta_i = 0$ and $\alpha_i(x) = \beta_i$ for all indices i. Hence, the collection $(\alpha_i(x))$ is unique. □

The scalar $\alpha_i(x)$ is called the *ith coordinate* of the vector x with respect to the basis (e_i). In particular, $\alpha_i(0) = 0$ for all indices i and $\alpha_i(e_j) = \delta_{ij}$, where δ_{ij} is the *Kronecker delta* ($\delta_{ii} = 1$ and $\delta_{ij} = 0$ for $i \neq j$).

Remark. In the case of modules, the definition of a maximal linearly independent collection of vectors is also meaningful. However, a theorem analogous to the preceding one does not hold for modules. In the definition of a basis, the possibility for every vector to be represented as a linear combination of the basis vectors has to be postulated explicitly.

A *subspace* $(\mathbb{K}, N)$ of a vector space $(\mathbb{K}, E)$ is formed by the field $\mathbb{K}$ and a subgroup $(N, +)$ of the group $(E, +)$ that is invariant under outer multiplication, i.e., $(\mathbb{K}, N)$ is a subgroup of the group with operators $(\mathbb{K}, E)$. By the definition of a vector space, the group E is Abelian and therefore its every subgroup N is normal. Thus, there exists a factor space $(\mathbb{K}, E/N)$ of the space $(\mathbb{K}, E)$ over the subspace $(\mathbb{K}, N)$. Vectors of $\bar{E} = E/N$ are factor sets $\bar{x} = x + N$ $(x \in E)$.

Example. Let $\mathcal{F} = \mathcal{F}(X, \mathbb{K})$ be the space of all scalar functions on a set X, and let $\mathcal{F}_0 = \mathcal{F}_0(X, \mathbb{K})$ be its subspace consisting of all finite scalar functions on X. For any $f \in \mathcal{F}_0$, let Supp f denote the finite set on which f assumes nonzero values, i.e., $f(x) \neq 0$ for $x \in$ Supp f and $f(x) = 0$ for $x \notin$ Supp f. The set $S(f) = $ Supp f is called the *support* of the function f. A *finite function* is also called a function with *finite support*.

For any set X and any additive Abelian group Y, we denote the set of all functions $f : X \to Y$ by $Y^X = \mathcal{F}(X, Y)$ and the set of all finite functions $f_0 : X \to Y$ by $Y_0^X = \mathcal{F}_0(X, Y)$.

Every $i \in X$ is associated with a scalar finite function $\chi_i \in \mathcal{F}$ which is called the *indicator of i*: $\chi_i(x) = 1$ for $x = i$ and $\chi_i(x) = 0$ for $x \neq i$. Every scalar finite function f can be expressed linearly in terms of indicators:

$f = \sum f(i)\chi_i$ $(i \in X)$. The sum in the right-hand side of the preceding equality is finite because so is the support of f: if $i \notin S(f)$, then $f(i) = 0$. It follows from the definition of indicators χ_i that

$$\sum f(i)\chi_i(x) = f(x)\chi_x(x) = f(x) \quad (x \in X).$$

The indicators χ_i are linearly independent: the equality $\sum \alpha_i \chi_i = 0$ for a finite collection of scalars α_i means that $\alpha_i = 0$ for every $i \in X$. By the preceding lemma, the collection (χ_i) is a basis of the vector space $(\mathbb{K}, \mathcal{F}_0)$. If X is infinite, then so is the basis.

7. We now prove that any vector space E has a basis and all bases in E are equipollent.

The set $L(C) = \operatorname{Lin} C$ of all linear combinations of vectors of a set $C \subseteq E$ is called the *linear span* of C in E. If B is a basis, then $L(B) = E$. Obviously, $L(E) = E$.

Exercise. Prove that the linear span $\operatorname{Lin} C$ of a set $C \subseteq E$ is equal to the intersection of all vector subspaces of E that include C.

If $L(C) = E$, then we say that C *generates* the space E or is a *generating system* of E.

Let $A \subseteq E$ be a linearly independent set and $C \supseteq A$ be a set that generates the space E. (For example, we may take $C = E$.)

Lemma. *There is a basis B in E such that $A \subseteq B \subseteq C$.*

Proof. Let $\mathcal{A}$ be the class of of linearly independent sets $X \subseteq E$ such that $A \subseteq X \subseteq C$. If C is linearly independent, then $B = C$ and the proof is complete. Assume that C is not linearly independent.

Consider the order on $\mathcal{A}$ introduced by the inclusion relation Let $\mathcal{L}$ be a chain in $\mathcal{A}$. Then the union $L = \cup \mathcal{L}$ of all linearly independent sets $X \in \mathcal{L}$ is linearly independent, belongs to $\mathcal{A}$, and majorizes $\mathcal{L}$. By Zorn's lemma, the class $\mathcal{A}$ contains a maximal set B. This set is the required basis. We have $A \subseteq B \subseteq C$, B is linearly independent, and $L(B) = E$. Indeed, if $L(B) \neq E = L(C)$, then there exists a $c \in C \setminus L(B)$ ($C \subseteq L(B)$ implies $E = L(C) = L(B)$). Therefore, $X = B \cup \{c\} \in \mathcal{A}$, which contradicts the maximality of B in $\mathcal{A}$. □

Theorem. *Every vector space has a basis. All the bases in a vector space are equipollent.*

Proof. If $A = \varnothing$ and $C = E$, the existence of a basis B in E follows from the preceding lemma.

1) Assume that a space E has a finite basis A such that $|A| = m$. Take an arbitrary basis $B \subseteq E$. If $m < |B|$, then there is an embedding $\varphi : A \to B$ with range $\varphi(A) \neq B$. Since $\varphi(A) \subset B$ and B is linearly independent, it follows that $\varphi(A)$ is also linearly independent.

Rearrange the vectors in A and B so that $\varphi(a_i) = b_i$ $(1 \leq i \leq m)$. The equalities

$$\bar\varphi(x) = \sum \xi_i \varphi(a_i) = \sum \xi_i b_i \quad (x = \sum \xi_i a_i)$$

determine the extension $\bar\varphi : E \to E$ of the embedding $\varphi : A \to B$ with range $\bar\varphi(E) = L(\varphi(A))$.

We now prove that $\bar\varphi(E) = E$. To this end, we solve the equation $\bar\varphi(x) = y$ for every $y = \sum \eta_j a_j$ $(1 \leq j \leq m)$. Let $b_i = \varphi(a_i) = \sum \alpha_{ij} a_j$. Then $\bar\varphi(x) = \sum_i \xi_i \sum_j \alpha_{ij} a_j = \sum_j \left(\sum_i \xi_i \alpha_{ij} \right) a_j$ and the equation $\bar\varphi(x) = y$ is equivalent to the system of linear equations

$$\sum_i \xi_i \alpha_{ij} = \eta_j \quad (1 \leq i, j \leq m).$$

Since A is a basis, this system is determined and its unique solution $\xi = (\xi_j)$ is obtained by Gaussian elimination (see Section 22 in van der Waerden, 1991).

Thus, $L(\varphi(A)) = \bar\varphi(E) = E$ and $\varphi(A)$ is a basis in E. Since $\varphi(A) \subset B$, we have arrived at a contradiction. We conclude that our assumption $m < |B|$ leads to a contradiction. The same is true for the assumption that $m > |B|$. It follows that $m = |B|$.

2) Assume that E has no finite bases. Let A and B be bases in E. For every $a \in A$ there is a finite set $B(a) \subseteq B$ of vectors of the basis B such that a is expressed linearly in terms of these vectors. Consider the union $B(A)$ of all such sets $B(a)$. Since A is infinite and $B(a)$ are finite, it follows that $|A| = |B(A)|$. Since $B(A) \subseteq B$, we have $|B(A)| \leq |B|$. Consequently, $|A| \leq |B|$. By symmetry, we also have $|B| \leq |A|$. Therefore $|A| = |B|$. $\square$

This theorem makes it possible to define the *dimension* of a vector space to be the power (cardinal number) of any of its bases. The dimension of a space E is denoted by $\dim E$. Depending on whether the dimension of E is finite ($\dim E < \infty$) or infinite ($\dim E = \infty$), E is said to be *finite-dimensional* or *infinite-dimensional*. If $\dim E = n \in \mathbb{P} = \{0\} \cup \mathbb{N}$, then E is called *n-dimensional*. Infinite-dimensional spaces include *countably infinite dimensional*. The properties of cardinal numbers are described in detail in Chapter 3 of Bourbaki (2004a).

Exercise. Prove that $|\mathcal{K}(I)| = |I|$ for an infinite I and the class $\mathcal{K}(I)$ of its infinite subsets.

Example. Let $\mathbb{K}$ be an infinite field and let $\mathcal{M} = \mathcal{M}(\mathbb{K})$ be the subspace of polynomials in the space of functions $\mathcal{F} = \mathcal{F}(\mathbb{K})$. Vectors in $\mathcal{M}$ are functions $p : \mathbb{K} \to \mathbb{K}$ with values $p(\xi) = \alpha_0 + \alpha_1\xi + \cdots + \alpha_n\xi^n$ ($\xi \in \mathbb{K}$), determined by collections of coefficients $\alpha_i \in \mathbb{K}$ ($0 \le i \le n$; $n = 0, 1, 2, \dots$). The sequence of powers $s^n : \mathbb{K} \to \mathbb{K}$ ($n = 0, 1, 2, \dots$) with values $s^n(\xi) = \xi^n$ ($\xi \in \mathbb{K}$) is a basis in the space $\mathcal{M}$. The powers s^n are linearly independent, which follows from the theorem on the number of roots of a polynomial (van der Waerden, 1991; Section 28), and every polynomial is a linear combination of these powers.

The polynomials make up the linear span of the set $\mathcal{S}$ of powers in the space $\mathcal{F}$: $\mathcal{M} = \operatorname{Lin}\mathcal{S} \subseteq \mathcal{F}$. For every $n \in \mathbb{P} = \{0\} \cup \mathbb{N}$, the powers s^i ($0 \le i \le n$) generate the $(n + 1)$-dimensional subspace $\mathcal{M}_n = \mathcal{M}_n(\mathbb{K})$ of the space $\mathcal{M} = \mathcal{M}(\mathbb{K})$. In particular, for $n = 0$ the subspace $\mathcal{M}_0 = \mathbb{K}s^0$ is isomorphic to the field $\mathbb{K}$.

8. We now formulate two corollaries of the theorem on the dimension of a vector space.

Corollary 1. *If X and Y are subspaces of a space F such that $X \subseteq Y$, then $\dim X \le \dim Y$.*

Proof. Let $E = Y$ and let A be a basis of the subspace X. By the preceding lemma, the subspace Y has a basis B that includes A. Since $A \subseteq B$, we have $\dim X = |A| \le |B| = \dim Y$. $\quad\square$

Examples. 1) $\dim \mathcal{F}_0(U, \mathbb{K}) \le \dim \mathcal{F}(U, \mathbb{K})$. 2) $\dim \mathcal{M}_n(\mathbb{K}) = n + 1 \le \infty = \dim \mathcal{M}(\mathbb{K})$ ($n \in \mathbb{N}$).

Let A be a linearly independent set and C be a generating system of a space E.

Corollary 2. *There exists a set $D \subseteq C$ such that D and A are disjoint and D complements A with respect to the basis $B = A \cup D$ of E.*

The second corollary is equivalent to the lemma of part 7 in the present subsection. It is often called the *change of basis theorem*, since it makes it possible to replace some or all of the vectors in a basis.

Example. Let I be a set and $\mathbb{K}$ be a field. Consider the vector spaces $\mathbb{K}^I$ and $\mathbb{K}_0^I$. The basis $B^* = \{\chi_i \mid i \in I\}$ of the space $\mathbb{K}_0^I$ is included in $\mathbb{K}^I$ and so is $\mathbb{K}_0^I$ itself. By Corollary 2, there exists a linearly independent set $D^* \subseteq \mathbb{K}^I$ such that D^* and B^* are disjoint and D^* is the complement of B^* with respect to the basis $G^* = B^* \cup D^*$ of $\mathbb{K}^I$. The basis G^* is called a *Hamel basis*. In the general case, it is only known that a Hamel basis exists (even with $\mathbb{K} = \mathbb{Q}$ and $I = \mathbb{N}$).

Exercise. Prove that $\left| \mathbb{K}_0^{\mathbb{N}} \right| = |\mathbb{K}|$ if $\mathbb{K}$ is infinite (see Chapter 3, Section 6, part 4 in Bourbaki, 2004).

9. Rings $\mathbb{K} = (\mathbb{K}, +, \cdot)$ and $E = (E, +, \cdot)$ form a *left $\mathbb{K}$-algebra* $(\mathbb{K}, E) = (\mathbb{K}, E, +, +, \cdot, \cdot, \cdot)$ if the ring $\mathbb{K}$ and the group $(E, +)$ form a left $\mathbb{K}$-module and vector multiplication is consistent with the other operations, which is provided by the *associativity conditions*

$$\alpha(xy) = (\alpha x)y \quad (\alpha \in \mathbb{K}, \ x, y \in E).$$

In addition, it is usually required that $(\alpha x)y = x(\alpha y)$, i.e., the following identities hold in the algebra $(\mathbb{K}, E)$:

$$\alpha(xy) = (\alpha x)y = x(\alpha y) \quad (\alpha \in \mathbb{K}, \ x, y \in E).$$

Unless specified explicitly, this *bilinear* property of multiplication will be assumed to hold.

A *right $\mathbb{K}$-algebra* $(E, \mathbb{K}) = (E, \mathbb{K}, +, +, \cdot, \cdot, \cdot)$ is defined analogously. The above algebras are obtained from modules by adding the operation of vector multiplication. All general definitions and propositions for modules can be extended to algebras. If the vector ring E is commutative, then the algebra $(\mathbb{K}, E)$ is said to be *commutative*. If the scalar ring $\mathbb{K}$ is a field, then the algebra $(\mathbb{K}, E)$ is called a *vector algebra*. It is usually assumed that $1 \in \mathbb{K}$ and $1 \cdot x = x$ for $x \in E$.

If $(\mathbb{K}, E)$ is an algebra, the ring $\mathbb{K}$ and a subring A of E that is *invariant* under outer multiplication ($\mathbb{K}A \subseteq A$) form the *subalgebra* $(\mathbb{K}, A)$ of $(\mathbb{K}, E)$. The ring $\mathbb{K}$ and the factor ring E/N over an invariant ideal N of the ring E form the *factor algebra* $(\mathbb{K}, E/N)$.

If $\mathbb{K} = \mathbb{Z}$, then $\mathbb{Z}N = N$ for every ideal N and factor algebras are identified with the factor rings.

Remark. Any Abelian group $(E, +)$ can be turned into a zero ring $(E, +, \cdot)$ by introducing zero multiplication ($xy = 0$ for all $x, y \in E$). Thus, the module $(\mathbb{K}, E)$ can be considered to be an algebra and all general propositions for algebras can be applied to modules. Similarly, any vector space can be identified with a zero vector algebra.

Example 1. Any ring $\mathbb{K}$ defines an algebra $(\mathbb{K}, \mathbb{K})$ with vector multiplication coinciding with scalar multiplication. The ring $\mathbb{K}$ is identified with the algebra $(\mathbb{K}, \mathbb{K})$. If the ring $\mathbb{K}$ is noncommutative, then the identity $(\alpha x)y = x(\alpha y)$ does not hold.

Example 2. A field $\mathbb{K}$ and a ring $\mathcal{F} = \mathcal{F}(X, \mathbb{K})$ form the vector algebra $(\mathbb{K}, \mathcal{F})$.

Example 3. A field $\mathbb{K}$ and the ring $\mathcal{F}_0 = \mathcal{F}_0(X, \mathbb{K})$ of finite scalar functions form the vector algebra $(\mathbb{K}, \mathcal{F}_0)$. It is a subalgebra of $(\mathbb{K}, \mathcal{F})$.

Example 4. Let $\mathbb{K}$ be a field and $A \subseteq X$. Consider the ring $A\mathcal{F} = A\mathcal{F}(X, \mathbb{K})$ of scalar functions on X whose supports are subsets of A. The vector algebra $(\mathbb{K}, A\mathcal{F})$ is a subalgebra of $(\mathbb{K}, \mathcal{F})$.

Example 5. A field $\mathbb{K}$ and the ring $\mathcal{M} = \mathcal{M}(\mathbb{K})$ of polynomials form the vector algebra $(\mathbb{K}, \mathcal{M})$. The product pq of polynomials $p = \sum \alpha_i s^i$, $q = \sum \beta_j s^j$ is defined by the formulas $pq = \sum \gamma_k s^k$, $\gamma_k = \sum\limits_{i+j=k} \alpha_i \beta_j$. This product is often called a *convolution*. The vector space of polynomials is a subspace of the vector space $\mathcal{F} = \mathcal{F}(\mathbb{K})$ of scalar functions. The algebra of polynomials, however, is not a subalgebra of $\mathcal{F}$ because its multiplication is different from that in $\mathcal{F}$.

10. Bases of a vector algebra $(\mathbb{K}, E)$ are bases of the corresponding vector space. Vector multiplication can be defined using the multiplication tables for the basis vectors.

Let $B = \{e_i \mid i \in I\}$ be a basis of an algebra $(\mathbb{K}, E)$. Consider two arbitrary vectors $x = \sum \alpha_i e_i$ and $y = \sum \beta_j e_j$ $(i, j \in I)$. The product

$$xy = \left(\sum \alpha_i e_i \right) \left(\sum \beta_j e_j \right) = \sum_{ij} (\alpha_i \beta_j)(e_i e_j)$$

is expressed linearly in terms of the products $e_i e_j$ of basis vectors. The collection $\{e_i e_j \mid i, j \in I\}$ represents the *multiplication table* of the algebra $(\mathbb{K}, E)$ with basis B. It is clear that commutative algebras have *symmetric* multiplication tables ($e_i e_j = e_j e_i$).

Examples. 1) The multiplication table for the algebra $\mathcal{F}_0$ of finite functions with the basis consisting of indicators e_i is diagonal: $e_i e_i = e_i$, $e_i e_j = 0$ for $i \neq j$.

2) The multiplication table for the algebra of polynomials $\mathcal{M}$ with the basis consisting of powers s^n has the form $s^m s^n = s^{m+n}$ $(m, n \in \mathbb{P} = \{0\} \cup \mathbb{N})$.

Both of the above tables are symmetric.

11. The set $[a, b] = \{x = (1 - t)a + tb \mid 0 \leq t \leq 1\}$ is called the *segment with endpoints* a *and* b of a real or complex space E. If for any two points of a set $C \subset E$ the segment with these endpoints is a subset of C, then C is said to be *convex*.

Example. The empty set and the set of all points of a space E are convex.

Proposition. *The only convex sets in* $\mathbb{R}$ *are intervals.*

Proof. 1) Let $A \subseteq \mathbb{R}$ be convex and $a = \inf A$, $b = \sup A \in \bar{\mathbb{R}}$. If A is the empty set or a singleton $\{a\} \subseteq \mathbb{R}$, then A is an interval. Assume that $a < b$. Then for every $x \in \,]a, b[$ there exist $u, v \in A$ such that $a \leq u \leq x \leq v \leq b$. Since A is convex, we have $[u, v] \subseteq A$ and therefore $x \in A$. Then $]a, b[\subseteq A$ and A is an interval with left endpoint a and right endpoint b.

2) Let A be an interval in $\mathbb{R}$ with left endpoint a and right endpoint b, where $a, b \in \bar{\mathbb{R}}$. If A is the empty set or a singleton $\{a\} \subseteq \mathbb{R}$, then A is convex. Assume that $a < b$. By transitivity, $a \leq u \leq v \leq b$ implies $[u, v] \subseteq A$, which means that A is convex. $\square$

Examples. 1) The balls $B(c, r) = \{z : |z - c| < r\}$ and $\bar{B}(c, r) = \{z : |z - c| \leq r\}$ in the complex plane $\mathbb{C}$ are convex. 2) The rectangles in the complex plane $\mathbb{C}$ are convex. 3) The circles $S(c, r) = \{z : |z - c| = r\}$ in the complex plane $\mathbb{C}$ are not convex.

Lemma. *The intersection of any collection of convex sets is convex.*

Proof. Assume that sets $A_i \subseteq E$ are convex and $A = \cap A_i$. Then, for any $a, b \in A$, if $[a, b] \subseteq A_i$ for all i then $[a, b] \subseteq A$. $\square$

Exercise. Give examples of convex sets whose union is not convex.

The intersection Con X of all convex sets in E that include the set $X \subseteq E$ is called the *convex hull* of X.

Corollary. *A set is convex if and only if it coincides with its convex hull.*

Example. The convex hull of the set $X = \{(0,0), (0,1), (1,0), (1,1)\} \subseteq \mathbb{C}$ is the square with elements of X as vertices: Con $X = [0,1] \times [0,1]$.

12. The union $L(a,b) = \cup[a(i), b(i)]$ of segments $[a(i), b(i)] \subseteq E$ $(1 \leq i \leq n)$ such that $a(1) = a$, $b(i) = a(i+1)$ $(1 \leq i < n)$, $b(n) = b$, is called a *broken line connecting the points* a and b in the space E. If for any two points of a set $C \subset E$ there exists a broken line connecting these points that is a subset of C, then C is said to be *polygonally connected*.

Example. Every convex set is polygonally connected.

Lemma. *The union of a collection of polygonally connected sets whose intersection is nonempty is polygonally connected.*

Proof. Assume that sets $C_i \subseteq E$ are polygonally connected $C = \cup C_i$, and $D = \cap C_i \neq 0$ $(i \in I)$. Take arbitrary points $a, b \in C$, $c \in D$, and choose indices $i(a), i(b) \in I$ such that $a \in C_{i(a)}$ and $b \in C_{i(b)}$. Since $c \in C_{i(a)}$ and $c \in C_{i(b)}$, there is a broken line in $C_{i(a)}$ connecting a and c and a broken line in $C_{i(b)}$ connecting c and b. Therefore, $C_{i(a)} \cup C_{i(b)}$ includes a broken line $L(a,b)$, connecting a and b. $\square$

Example. The union of a collection of broken lines connecting a point $a \in E$ with every point of a set $B \subseteq E$ is polygonally connected.

Exercise. Give examples of polygonally connected sets whose intersection is not polygonally connected.

Let $A \subseteq E$ and $x \in A$. The set $A(x)$ of all $y \in A$ that are connected with x by broken lines in A is called the *polygonally connected component of* x *in the set* A. The sets $A(x)$ are called the *polygonally connected components of the set* A. It follows from the preceding lemma that they are polygonally connected.

Proposition. *The polygonally connected components of a set make up a partition of this set.*

Proof. Every $x \in A$ belongs to its connected component $A(x)$ in A. For the intersection of $A(x)$ and $A(y)$ for x and y in A is nonempty, then $A(x) = A(y)$. Indeed, if $c \in A(x) \cap A(y)$, then the union of broken lines in A connecting x with c and c with y connects x and y. Hence, $y \in A(x)$ and $x \in A(y)$, which implies $A(y) \subseteq A(x)$ and $A(x) \subseteq A(y)$. This means that distinct polygonally connected components of A are pairwise disjoint and their union is equal to A. $\square$

Example. Intervals $[0, 1]$ and $[2, 3]$ are polygonally connected components of the set $A = [0, 1] \cup [2, 3]$.

2.2.2.   Linear operators

Linear operators are additive homogeneous maps of modules and vector spaces. They play an important role in mathematics. In what follows, we usually consider unital modules.

1. Let $\mathbb{L}$ be a ring, $\mathbb{K}$ be its subring, $(\mathbb{K}, E)$ and $(\mathbb{L}, F)$ be modules, and $T : (E, +) \to (F, +)$ be a homomorphism. By definition,

$$T(x + y) = Tx + Ty \quad (x, y \in E).$$

Such homomorphisms will be called *additive maps*. The consistency of T : $E \to F$ with outer multiplication is expressed by the equality $T(\alpha x) = \alpha \cdot Tx$ ($\alpha \in \mathbb{K}$, $x \in E$). Such maps are called *homogeneous*. Since $\mathbb{K} \subseteq \mathbb{L}$, the product $\alpha \cdot Tx$ is well defined. Maps that are additive and homogeneous are called *linear maps* or *linear operators*. If $\mathbb{K} = \mathbb{Z}$ and the modules are unital, then an operator is linear whenever it is additive.

The property of an operator $T : E \to F$ to be linear can be expressed as follows:

$$T(\alpha x + \beta y) = \alpha \cdot Tx + \beta \cdot Ty \quad (\alpha, \beta \in \mathbb{K}; \ x, y \in E).$$

This means that a linear operator maps a linear combination of vectors to the linear combination of its images, which explains the choice of the term.

From the definitions, it follows that $T0 = 0$, i.e., the image of zero of the module E is the zero of the module F.

The set of all linear operators $T : E \to F$ is denoted by $\mathcal{L} = \mathcal{L}(E, F)$. Whenever the scalar ring $\mathbb{K}$ of the module $(\mathbb{K}, E)$ must be explicitly specified, we say that an operator is $\mathbb{K}$-*linear*. If only scalars of $\mathbb{K}$ are used, then the ring of scalars $\mathbb{L}$ of the module $(\mathbb{L}, F)$ that includes $\mathbb{K}$ can be replaced with $\mathbb{K}$ so that the module $(\mathbb{K}, F)$ is considered instead of $(\mathbb{L}, F)$.

We now extend the notion of a submodule. A module $(\mathbb{K}, E)$ will be called a submodule of the module $(\mathbb{L}, F)$ if $\mathbb{K}$ and E are the subring and the subgroup of $\mathbb{L}$ and F, respectively, while the outer multiplication on $(\mathbb{L}, F)$ is an extension of the outer multiplication on $(\mathbb{K}, E)$. We will also say "a $\mathbb{K}$-submodule E" instead of "a submodule $(\mathbb{K}, E)$". It should be noted that a $\mathbb{K}$-submodule is not necessarily an $\mathbb{L}$-module if $\mathbb{K} \neq \mathbb{L}$. We will write $(\mathbb{K}, E) \subseteq (\mathbb{L}, F)$.

Let $\mathbb{K} = E = \mathbb{Z}$ and $\mathbb{L} = F = \mathbb{Q}$. Then $(\mathbb{K}, E) = (\mathbb{Z}, \mathbb{Z}) \subseteq (\mathbb{Q}, \mathbb{Q}) = (\mathbb{L}, F)$.

Similarly, we extend the notion of a subalgebra and define a $\mathbb{K}$-subalgebra of an $\mathbb{L}$-algebra. By introducing zero vector multiplication, $\mathbb{K}$-subalgebras can be identified with $\mathbb{K}$-submodules of the $\mathbb{L}$-module.

Example 1. Let $(\mathbb{K}, E)$ be a submodule of the module $(\mathbb{L}, F)$: $(\mathbb{K}, E) \subseteq (\mathbb{L}, F)$. Every scalar $\gamma \in \mathbb{L}$ defines a linear operator $T : E \to F$ by the equality $Tx = \gamma x$ $(x \in E)$. This operator is called a *homothety* with coefficient γ.

If $\mathbb{K} = E$, $\mathbb{L} = F$, and the ring $\mathbb{K}$ is unital, then $Tx = T(1x) = T(1)x$ $(x \in \mathbb{K})$ and $\gamma = T(1)$.

Example 2. Let $(\mathbb{K}, E)$ be a subalgebra of the algebra $(\mathbb{L}, F)$: $(\mathbb{K}, E) \subseteq (\mathbb{L}, F)$. Every scalar $c \in \mathbb{L}$ defines an operator $T : E \to F$ by the equality $Tx = cx$ $(x \in E)$. The operator T is called the *multiplier* with coefficient c. If $c = 0$, then $Tx = 0$ $(x \in E)$. If the ring E is unital and $c = 1$, then T is the *identity embedding*, i.e., $Tx = x$ for all $x \in E$.

Example 3. Let $E = \mathcal{F} = \mathcal{F}(X, \mathbb{K})$, $a \in X$, and $\mathbb{K} = \mathbb{L} = F$. We define a functional $\delta_a : \mathcal{F} \to \mathbb{K}$ by the formula $\delta_a f = f(a)$ $(f \in \mathcal{F})$. This functional is linear:

$$\delta_a(f + g) = (f + g)(a) = f(a) + g(a) = \delta_a f + \delta_a g,$$
$$\delta_a(\alpha f) = (\alpha f)(a) = \alpha \cdot f(a) = \alpha \cdot \delta_a f \quad (\alpha \in \mathbb{K}, \; f, g \in \mathcal{F}).$$

The functional δ_a is called the *delta function* at the point a. Delta functions are also called *point functionals*. They play an important role in calculus.

Remark. Every homomorphism $T : E \to F$ from the algebra $(\mathbb{K}, E)$ into the algebra $(\mathbb{L}, F)$ is a linear operator. The converse is not true because a linear operator is not necessarily multiplicative and the identity $T(xy) = Tx \cdot Ty$ may not hold. In particular, the linear operator in Example 1 is a homomorphism if and only if its coefficient γ is idempotent. In Example 2, in addition to being idempotent, the coefficient c must commute with every vector $x \in E$.

2. An important characteristic of a linear operator $T \in \mathcal{L}(E, F)$ that maps the $\mathbb{K}$-module E to the $\mathbb{L}$-module F is its *range* $R(T) = \operatorname{Ran} T = T(E) \subseteq F$ and *kernel* $K(T) = \operatorname{Ker} T = T^{-1}(0) \subseteq E$. (These characteristics are important for every group homomorphism.)

Lemma. *The range of $T(E)$ is a $\mathbb{K}$-module.*

Proof. Let $x, y \in T(E)$. Then there exist $u, v \in E$ such that $x = Tu$ and $y = Tv$. Therefore, $\alpha x + \beta y = \alpha \cdot Tu + \beta \cdot Tv = T(\alpha u + \beta v) \in T(E)$ for all $\alpha, \beta \in \mathbb{K}$. $\quad\square$

Proposition. *A linear operator is a one-to-one map if and only if its kernel is zero.*

Proof. By definition, a linear operator is a homomorphism of Abelian groups. Thus, the proposition follows from the lemma of Subsection 2.1.2, part 5. $\quad\square$

Example 1. Let $\mathbb{K}$ be a field and $(\mathbb{K}, E)$ be a vector space. For the homothety $T : E \to F$ with coefficient γ, the conditions $K(T) = \{0\}$ and $\gamma \neq 0$ are equivalent.

Corollary. *Let $T \in \mathcal{L}(E, F)$ and $K(T) = \{0\}$. Then $T^{-1} \in \mathcal{L}(R(T), E)$.*

Proof. From the assumption, it follows that the inverse correspondence $T^{-1} : R(T) \to E$ is one-to-one. We now show that it is linear. Take an arbitrary linear combination $z = \alpha x + \beta y$ $(\alpha, \beta \in \mathbb{K}, x, y \in R(T))$. By the preceding lemma, $z \in R(T)$. Let $u = T^{-1}x$, $v = T^{-1}y$, and $w = T^{-1}z$. Then, since T is a one-to-one linear operator, we have $T^{-1}(\alpha x + \beta y) = T^{-1}z = w \Rightarrow \alpha x + \beta y = z = Tw \Rightarrow T(\alpha u + \beta v) = \alpha \cdot Tu + \beta \cdot Tv = \alpha x + \beta y = Tw \Rightarrow \alpha \cdot T^{-1}x + \beta \cdot T^{-1}y = \alpha u + \beta v = w$. Consequently, $T^{-1}(\alpha x + \beta y) = w = \alpha \cdot T^{-1}x + \beta \cdot T^{-1}y$. $\quad\square$

Example 2. The inverse of a homothety with coefficient γ is a homothety with coefficient γ^{-1}.

Example 3. Let N be a subspace of a vector space E and $\bar{\varphi} : E \to \bar{E}$ be the natural homomorphism from the group $(E, +)$ onto the factor group $\bar{E} = E/N$. The homomorphism $\bar{\varphi}$ is a linear operator: $\bar{\varphi}(x+y) = \bar{\varphi}(x) + \bar{\varphi}(y)$ $(x, y \in E)$ and $\bar{\varphi}(\alpha x) = \alpha(x + N) = \alpha\bar{\varphi}(x)$ $(\alpha \in \mathbb{K}, x \in E)$.

3. Let $(\mathbb{K}, E) \subseteq (\mathbb{L}, F) \subseteq (\mathbb{M}, G)$ be modules and let $S : E \to F$ and $T : F \to G$ be linear operators. The composition $TS : E \to G$ is called the *product* of linear operators S and T.

Proposition. *The product TS of linear operators S and T is a linear operator.*

Proof. Since S and T are homomorphisms of additive groups, their product TS is additive. Its homogeneity follows from the homogeneity of S and T: $TS(\alpha x) = T(S(\alpha x)) = T(\alpha \cdot Sx) = \alpha \cdot TSx$ $(\alpha \in \mathbb{K}, x \in E)$. $\quad\square$

Corollary. *The product of linear isomorphisms is a linear isomorphism.*

Lemma. *The sum $R + S$ of linear operators $R : E \to F$ and $S : E \to F$ is a linear operator.*

Proof. We have $(R+S)(\alpha x + \beta y) = R(\alpha x + \beta y) + S(\alpha x + \beta y) = \alpha \cdot Rx + \beta \cdot Ry + \alpha \cdot Sx + \beta \cdot Sy = \alpha(Rx + Ry) + \beta(Rx + Ry) = \alpha \cdot (R+S)x + \beta \cdot (R+S)y$ $(\alpha, \beta \in \mathbb{K}; x, y \in E)$. $\quad\square$

Let $(\mathbb{K}, E)$ be a space and $\mathcal{L} = \mathcal{L}(E)$ be the set of its linear transformations. They make up a subspace of the vector space $\mathcal{F} = \mathcal{F}(E)$ of all transformations of the space $(\mathbb{K}, E)$. Multiplication of linear operators is consistent with the other operations:

$$(R + S)T = RT + ST, \quad R(S + T) = RS + RT,$$
$$\alpha(ST) = (\alpha S)T = S(\alpha T) \quad (\alpha \in \mathbb{K}, \ R, S, T \in \mathcal{F}).$$

The first of these equalities holds for the product (composition) of any transformations $R, S, T \in \mathcal{F}$. For the other equalities, the linearity is essential:

$$R(S + T)x = R(Sx + Tx) = RSx + RTx,$$
$$((\alpha S)T)x = (\alpha S)(Tx) = S(\alpha Tx) = (S(\alpha T))x$$

for all $\alpha \in \mathbb{K}, x \in E$.

Exercise. Give examples of functions $f, g, h \in \mathcal{F}$ such that $f \circ (g + h) \neq f \circ g + f \circ h$ and $(\alpha f) \circ g \neq f \circ (\alpha g)$.

Thus, the set $\mathcal{L} = \mathcal{L}(E)$ of linear operators on the space $(\mathbb{K}, E)$ with the specified operations is an algebra. The zero of this algebra is the *zero operator* O: $Ox = 0$ $(x \in E)$. The identity of the algebra $\mathcal{L}$ is the *identity operator* I $(Ix = x$ for all $x \in E)$. There are *scalar operators* αI, which are often identified with scalars $\alpha \in \mathbb{K}$ and are also written as α.

Since operators of $\mathcal{L}$ are defined on the entire space E, isomorphisms are the only invertible operators. Indeed, if T is a linear isomorphism from E onto E, then so is T^{-1} and $TT^{-1} = T^{-1}T = I$.

The algebra $\mathcal{L}$ is *noncommutative* since $ST \neq TS$ in the general case.

Example. Let $E = \mathbb{K}$ and $f, b \in \mathbb{K}$ $(a \neq 0,\ b \neq 1)$. Consider the operators S and T of additive and multiplicative shift defined by the formulas $Sx = a+x$ and $Tx = bx$ $(x \in \mathbb{K})$. We have $STx = S(bx) = a+bx \neq ba+bx = T(a+x) = TSx$ $(x \in \mathbb{K})$.

Note that the scalar operator αI commutes with any operator $S \in \mathcal{L}$:

$$(\alpha I)Sx = \alpha \cdot Sx = S(\alpha x) = S(\alpha I x) \quad (x \in E),$$
$$\alpha I \cdot S = S \cdot \alpha I \quad (\alpha S = S\alpha).$$

4. For vector spaces, the dimension $r(T)$ of the range $R(T)$ of a linear operator T is called the *rank* of T $(r(T) = \dim R(T))$. The dimension $d(T)$ of the kernel $K(T)$ of a linear operator T is called the *nullity* of T $(d(T) = \dim K(T))$. These dimensions may be finite or infinite.

In the space $\mathcal{L}(E, F)$ of linear operators $T : E \to F$, there is a subspace $\mathcal{K}(E, F)$ of operators of *finite rank*: $T \in \mathcal{K}(E, F) \Leftrightarrow r(T) < \infty$. If $\dim F < \infty$, then $\mathcal{K}(E, F) = \mathcal{L}(E, F)$. In the algebra $\mathcal{L} = \mathcal{L}(E)$ of linear operators $T : E \to F$, there is a subalgebra $\mathcal{K} = \mathcal{K}(E)$ of operators of finite rank. Let $S, T \in \mathcal{K}(E)$. Then $(S+T)(E) \subseteq \mathcal{L}(S(E) \cup T(E))$ and therefore $r(S+T) = \dim(S+T)(E) \leq \dim \mathcal{L}(S(E) \cup T(E)) \leq \dim S(E) + \dim T(E) = r(S) + r(T) < \infty$. Furthermore, $r(ST) \leq r(S) < \infty$ and $r(\alpha S) \leq r(S)$ $(\alpha \in \mathbb{K})$. The inequalities for the dimensions follow from the Corollary 1 of Subsection 2.2.1, part 8.

Example. Let $E = \mathbb{K}^\infty = \mathcal{F}(\mathbb{N}, \mathbb{K})$ be the space of sequences of elements of a field $\mathbb{K}$. Let $\mathcal{K}(n, E)$ denote the set of all operators $T \in \mathcal{L}(E)$ whose ranges are subsets of $\mathbb{K}^n \times \{0\}^\infty$ (the set of sequences $(x_i) \in \mathbb{K}^\infty$ such that $x_i = 0$ for $i > n$). It is clear that $\cup \mathcal{K}(n, E) \subseteq \mathcal{K}(E)$.

In particular, the operators in $\mathcal{K}(E)$ include *projections* P_n defined by the formulas $P_n(x_i) = (\delta_{ni}x_i)$, $\delta_{nn} = 1$, and $\delta_{ni} = 0$ for $i \neq n$. These projections are *idempotent* and *pairwise orthogonal*: $P_n^2 = P_n$ and $P_m P_n = P_n P_m = 0$ for $m \neq n$. Indeed, $P_n(P_n(x_i)) = P_n(\delta_{ni} x_i) = (\delta_{ni}^2 x_i) = (\delta_{ni} x_i) = P_n(x_i)$ and $P_m(P_n(x_i)) = P_m(\delta_{ni} x_i) = (\delta_{mi} \delta_{ni} x_i) = (0) = O(x_i) = P_n(P_m(x_i))$ $((x_i) \in \mathbb{K}^\infty)$.

Obviously, for any finite sequence $(\alpha_n) \in \mathbb{K}^\infty$, the linear combination $T = \sum \alpha_n P_n$ has finite rank.

Exercise. Find out whether there exist any other operators of finite rank in $\mathcal{L}(\mathbb{K}^\infty)$.

5. We now prove a few more propositions related to dimensions.

For a space $(\mathbb{K}, E)$ with basis $B = \{e_i \mid i \in I\}$, consider the space $(\mathbb{K}, \mathbb{K}_0^I)$ of finite functions on I.

Lemma. $E \simeq \mathbb{K}_0^I$

Proof. The set $\{\chi_i \mid i \in I\}$ of indicators on I makes up the basis $\mathbb{K}_0^I$. The map $\varphi : E \to \mathbb{K}_0^I$ defined by the formula

$$\varphi(x) = \sum \alpha_i \chi_i \quad (x = \sum \alpha_i e_i, \ (\alpha_i) \in \mathbb{K}_0^I)$$

is a linear isomorphism. Indeed, we have $\varphi(x + y) = \varphi\left(\sum (\alpha_i + \beta_i)e_i\right) = \sum (\alpha_i + \beta_i)\chi_i = \sum \alpha_i \chi_i + \sum \beta_i \chi_i = \varphi(x) + \varphi(y)$, $\varphi(\alpha x) = \varphi\left(\sum (\alpha \alpha_i)e_i\right) = \sum (\alpha \alpha_i)\chi_i = \alpha \sum \alpha_i \chi_i = \alpha \varphi(x)$ for all $\alpha \in \mathbb{K}$ and $x = \sum \alpha_i e_i$, $y = \sum \beta_i e_i \in E$. Since the indicators χ_i are linearly independent, $\varphi(x) = \sum \alpha_i \chi_i = 0$ implies $\alpha_i = 0$ and $x = \sum \alpha_i e_i = 0$. Hence, $\operatorname{Ker} \varphi = \{0\}$. Moreover, every vector $f = \sum \alpha_i \chi_i \in \mathbb{K}_0^I$ is an image of $x = \sum \alpha_i e_i$. $\quad\square$

Thus, instead of the vector space $(\mathbb{K}, E)$ with basis $B = \{e_i \mid i \in I\}$, we can consider its *coordinate space* $(\mathbb{K}, \mathbb{K}_0^I)$ with basis $\{\chi_i \mid i \in I\}$. By definition, $\dim E = |I|$. If the specific form of the index set is of no importance, then we write $\mathbb{K}^{|I|}$ or $\mathbb{K}^{\dim E}$. instead of $\mathbb{K}^I$. In particular, if $\dim E = n \in \mathbb{N}$, then $E \simeq \mathbb{K}^n$ $(I = \{1, \ldots, n\})$.

Consider spaces $(\mathbb{K}, E)$ and $(\mathbb{K}, F)$.

Proposition. $E \simeq F \Leftrightarrow \dim E = \dim F$.

Proof. Let $\varphi : E \to F$ be a linear isomorphism and $A = \{a_i \mid i \in I\}$ be a basis of the space E. Then the image $\varphi(A) = B = \{b_i \mid i \in I\}$ is a basis of the space F. Indeed, B is a linearly independent set because

$0 = \sum \alpha_i b_i = \sum \alpha_i \varphi(a_i) = \varphi\left(\sum \alpha_i a_i\right) \Rightarrow \sum \alpha_i a_i = 0 \Rightarrow \alpha_i = 0$ for any finite collection $\alpha_i \in \mathbb{K}$. The set B generates F:

$$y \in F \Rightarrow y = \varphi\left(\sum \alpha_i a_i\right) = \sum \alpha_i(b_i) = \sum \alpha_i b_i$$

for some finite collection $\alpha_i \in \mathbb{K}$. Therefore, $\dim E = |A| = |B| = \dim F$.

Conversely, let $\dim E = \dim F$ and let A and B be bases of the spaces E and F. Since $|A| = \dim E = \dim F = |B|$, we have $\mathbb{K}_0^A \simeq \mathbb{K}_0^B$ and, by the preceding lemma, $E \simeq \mathbb{K}_0^A \simeq \mathbb{K}_0^B \simeq F$. $\quad\square$

Let $(\mathbb{K}, E)$ and $(\mathbb{K}, G)$ be linear spaces and $T \in \mathcal{L}(E, G)$.

Corollary. *If* $T : E \to G$ *is a linear embedding, then* $\dim T(E) = \dim E$.

Proof. It was proved in part 2 of this subsection that $F = T(E)$ is a subspace of the space G. If T is a linear isomorphism from E onto F, then the equality $\dim E = \dim F$ follows from the preceding proposition. $\quad\square$

Remark. For any linear operator $T : E \to G$, we have $\dim T(E) \leq \dim E$. Indeed, let A be a basis of the space E and $B = T(A)$ be its image in G. It is obvious that $T(E) = \mathcal{L}(B)$ and tehrefore $\dim T(E) \leq |B| \leq |A| = \dim E$. The images of linearly independent vectors under a linear map may be linearly dependent: $\sum \beta_i T e_i = 0$ may hold for some $\beta_i \neq 0$, even if $\sum \alpha_i e_i = 0$ only for all $\alpha_i = 0$ (for example, if $T e_i = 0$ for some i). However, the images of linearly dependent vectors cannot be linearly independent because $\sum \alpha_i e_i = 0 \Rightarrow \sum \beta_i T e_i = 0$.

6. Let X and Y be subspaces of a space E. In the context of definitions for additive Abelian groups, Y is called the (*linear*) *complement* of X if E is the direct sum of X and Y, i.e., $E = X + Y$ and $X \cap Y = \{0\}$.

Example. The subspace $Y = L\{s^i \mid i > n\}$ is the complement of the subspace $X = L\{s^i \mid i \leq n\}$ in the space $E = \mathcal{M}(\mathbb{K})$.

Remark. For a given space E and its subspace X, there may exist more than one subspace Y such that $X \cap Y = \{0\}$ and $X + Y = E$. If we choose a basis B in Y, then we can speak of the *complement relative to the basis B* and write $Y = X^c = X^c(B)$.

Example. Let $E = \mathbb{K}^3$, $X = \mathbb{K}^2 \times \{0\}$, $b = (0,0,1)$, and $c = (1,1,1)$. Then $Y = \mathbb{K}b$ and $Z = \mathbb{K}c$ are the complements of X relative to the bases $B = \{b\}$ and $C = \{c\}$.

Lemma. $E = X + Y$, $X \cap Y = \{0\} \Rightarrow \dim E = \dim X + \dim Y$.

Proof. Let $A = (a_i)$ and $B = (b_j)$ be bases of X and Y, respectively. Then $A \cup B = C = (c_k)$ is a basis of E. Indeed, C is linearly independent: $0 = \sum \gamma_k c_k = \sum \alpha_i a_i + \sum \beta_j b_j \Rightarrow \sum \alpha_i a_i = -\sum \beta_j b_j \in X \cap Y \Rightarrow \sum \alpha_i a_i = \sum \beta_j b_j = 0 \Rightarrow \alpha_i = 0$, $\beta_j = 0 \Rightarrow \gamma_k = 0$ for any finite collection $\gamma_k = (\alpha_i, \beta_j)$. Moreover, C generates E. If $z \in E$, then there exist $x = \sum \alpha_i a_i$ and $y = \sum \beta_j b_j$ such that $z = x + y = \sum \gamma_k c_k = \sum \alpha_i a_i + \sum \beta_j b_j$. $\quad\square$

Since $X \cap Y = \{0\}$ and $0 \notin A \cap B$, we have $A \cap B = \varnothing$. Hence, $\dim E = |C| = |A| + |B| = \dim X + \dim Y$ (see Kuratowski and Mostowski, 1976; Chapter V, Section 4).

Example. If $\dim X < \infty$ and $\dim Y = \infty$, then $\dim E = \dim Y$ (see part 5 of Subsection 1.3.3). In particular, we have $\dim \mathcal{M}(\mathbb{K}) = \dim L\{s^i \mid i \geq 0\} = \dim L\{s^i \mid i > n\}$ for any $n \geq 0$. Furthermore, $\dim L\{s^i \mid 1 \leq i \leq m + n\} = m + n = \dim L\{s^i \mid 1 \leq i \leq m\} + \dim L\{s^i \mid m + 1 \leq i \leq m + n\}$.

7. For spaces $(\mathbb{K}, E)$ and $(\mathbb{K}, F)$, consider an operator $T \in \mathcal{L}(E, F)$ with range $R(T) = T(E)$, kernel $K(T) = T^{-1}\{0\}$, rank $r(T) = \dim R(T)$, and nullity $d(T) = \dim K(T)$.

Theorem. $\dim E = r(T) + d(T)$.

Proof. The space E is the direct sum of the subspace $X = K(T)$ and its complement $Y = X^c$ relative to a selected basis. Therefore $\dim E = \dim X + \dim Y$, where $\dim X = d(T)$. We now prove that $\dim Y = r(T)$. Let $A = (a_i)$ and $B = (b_j)$ be bases of X and Y, respectively, and let $C = A \cup B = (c_k)$ be a basis of E. The image $T(B)$ is a basis of $R(T) = T(E)$. Indeed, $T(B)$ is linearly independent: $0 = \sum \beta_j T b_j = T(\sum \beta_j b_j) \Rightarrow \sum \beta_j b_j \in X \cap Y = \{0\} \Rightarrow \sum \beta_j b_j = 0 \Rightarrow \beta_j = 0$. Moreover, $T(B)$ generates $R(T)$: $y \in R(T) \Rightarrow y = Tx = T(\sum \gamma_k c_k) = T(\sum \alpha_i a_i + \sum \beta_j b_j) = 0 + \sum \beta_j T b_j$. We conclude that $\dim R(T) = |T(B)| = |B| = \dim Y$. $\quad\square$

Corollary. If $\dim E < \infty$, then $T \in \mathcal{L}(E, F)$ is a one-to-one operator if and only if $\dim E = r(T)$.

Proof. By the preposition of part 2 of this subsection, T is a one-to-one operator if and only if its nullity is zero. By the preceding theorem, the equality $d(T) = 0$ is equivalent to $\dim E = r(T)$ for $\dim E < \infty$. $\quad\square$

In particular, for $\dim E < \infty$, $T : E \to F$ is an isomorphism if and only if $\dim E = r(T) = \dim F$.

Remark. If $\dim E > \dim F$, then $T \in \mathcal{L}(E, F)$ cannot be a one-to-one operator. Under this condition, the image $T(A)$ of the basis A of E cannot be a linearly independent set because otherwise we would have $\dim E = |A| = |T(A)| \leq \dim F$.

Bases satisfy the *Dirichlet principle* (Kuratowski and Mostowski, 1976, Chapter III, Section 4): If the cardinality of a set A is strictly less than that of a set B, any map from A onto B is not one-to-one. (The combinatorial analog of this statement is as follows: If $m > n$, then m items cannot be distributed over n cells so that each cell contains one item.)

8. Consider the factor space $\bar{E} = E/K(T)$.

Proposition. $r(T) = \dim E/K(T)$.

Proof. In the proof of the theorem on group homomorphisms (see Subsection 2.1.2, part 9), it was shown that the composition $\bar{\psi} = \bar{\varphi} T^{-1} : R(T) \to \bar{E}$ of the inverse correspondence $T^{-1} : R(T) \to E$ and the natural homomorphism $\bar{\varphi} : E \to \bar{E}$ is an isomorphism from the group $(R(T), +)$ onto $(\bar{E}, +)$. The map $\bar{\varphi}$ is linear, and so is $\bar{\psi}$. Indeed, let $a \in E$, $b = Ta$, and $\alpha \in \mathbb{K}$. Then $\bar{\varphi}(u) = \bar{\varphi}(b)$, $\bar{\varphi}(v) = \bar{\varphi}(\alpha b)$ for every $u \in T^{-1}(b)$ and $v \in T^{-1}(\alpha b)$. Therefore, $\bar{\psi}(\alpha b) = \bar{\varphi}(T^{-1}(\alpha b)) = \bar{\varphi}(\alpha b) = \alpha \cdot \bar{\varphi}(b) = \alpha \cdot \bar{\varphi}(T^{-1}(b)) = \alpha \cdot \bar{\psi}(b)$. Hence $R(T) \simeq E/K(T)$, which means that $\dim R(T) = \dim E/K(T)$. $\quad\square$

Let X be a subspace of E and X^c be the complement of X relative to some basis. Then $X = K(T)$ and $X^c = R(T)$ for the operator $T \in \mathcal{L}(E, E)$ defined by the formula $T(x + y) = y$ ($x \in X$, $y \in X^c$).

Corollary 1. $X^c \simeq E/X$.

The next corollary follows from the above proposition and the preceding theorem

Corollary 2. $\dim E = \dim E/X + \dim X$.

This equality also holds for the kernel $X = K(T)$ of every operator $T \in \mathcal{L}(E, F)$. To emphasize the isomorphism $E/K(T) \simeq R(T)$, the factor space $CR(T) = E/K(T)$ is called *coimage*. From the statements proved above, it follows that the dimension of the coimage $CR(T)$ is equal to the dimension of the image $R(T)$, i.e., the rank $r(T)$ of the operator T. The dimension of E/X is called the *codimension* of X, written $\operatorname{codim} X$. Corollary 1 implies that

$$\operatorname{codim} X = \dim E/X = \dim X^c.$$

It is natural to consider the equality

$$\dim E = \dim E/K(T) + \dim K(T)$$

together with its *dual* equality

$$\dim F = \dim F/R(T) + \dim R(T).$$

The factor space $CK(T) = F/R(T)$ is called the *cokernel*, of T, and its dimension $c(T) = \dim F/R(T)$ is called the *conullity* of the operator T.

The rank $r(T)$, nullity $d(T)$, and conullity $c(T)$ of an operator T can be finite or infinite.

Lemma. *A linear operator is an onto operator if and only if it has zero cokernel.*

Proof. Let $T \in \mathcal{L}(E, F)$, $CK(T) = F/R(T) = \{0\}$. Then $R(T) = F$. Conversely, if T is an onto operator, then $CK(T) = F/F = \{0\}$. $\square$

Thus, the condition $c(T) = 0$ is a criterion for T to be an onto operator, as well as $d(T) = 0$ is a criterion for T to be a one-to-one operator.

Example. Let $E = F = \mathbb{K}_0^\infty$ $(= \mathbb{K}_0^\mathbb{N})$ and $e_n = (\delta_{nj})$ $(j = 1, \ldots)$ be standard basis sequences, i.e., $x = (\alpha_n) \in \mathbb{K}_0^\infty \Leftrightarrow x = \sum \alpha_n e_n$.

We introduce operators $S, T, U, V \in \mathcal{L}(E)$ such that $S(e_n) = e_n$, $Te_n = e_{2n}$, $Ue_{2n-1} = Ue_{2n} = e_n$, $Ve_{2n-1} = Ve_{2n} = e_{2n}$ $(n \in \mathbb{N})$. These operators have infinite rank: $r(S) = r(T) = r(U) = r(V) = \infty$. The identity operator $S = I$ is an onto and one-to-one operator. We have $R(I) = F$, $c(I) = 0$, and $d(I) = 0$. The operator T is a one-to-one operator, but not an onto operator. We have $R(T) = L\{e_{2n} \mid n \in \mathbb{N}\}$, $CK(T) = L\{e_{2n-1} \mid n \in \mathbb{N}\}$, $c(T) = \infty$, and $d(T) = 0$. The operator U is an onto operator, but it is not one-to-one. We have $R(U) = F$, $K(U) = \{x = (\alpha_n) \mid \alpha_{2n-1} + \alpha_{2n} = 0, n \in \mathbb{N}\}$, $c(U) = 0$, and $d(U) = \infty$. The operator V is neither onto nor one-to-one. We have $R(V) = R(T)$, $CK(V) = CR(T)$, $K(V) = K(U)$, $c(V) = \infty$, and $d(V) = \infty$.

9. For spaces E and F, consider an operator $T \in \mathcal{L}(E, F)$ which is not necessarily an onto or one-to-one operator. The following theorem follows from the proposition of part 2 and the lemma of part 7 of this subsection.

Theorem. *The following statements hold:*

(1) *T is an onto and one-to-one operator if and only if $c(T) = 0$ and $d(T) = 0$.*

(2) *T is an onto operator, but not a one-to-one operator if and only if $c(T) = 0$ and $d(T) \neq 0$.*

(3) *T is a one-to-one operator, but not an onto operator if and only if $c(T) \neq 0$ and $d(T) = 0$.*

(4) *T is neither onto nor one-to-one if and only if $c(T) \neq 0$ and $d(T) \neq 0$.*

The theorem covers all the possibilities for the operator T. For the linear equation $Tx = y$, these possibilities are formulated as follows:

(1) *For any $y \in F$, there is a unique solution $x \in E$.*

(2) *For any $y \in F$ there is more than one solution $x \in E$.*

(3) *For some (but not all) $y \in F$ there is a unique solution $x \in E$, whereas for the other $y \in F$ there are no solutions $x \in E$.*

(4) *For some (but not all) $y \in F$ there is more than one solution $x \in E$, whereas for the other $y \in F$ there are no solutions $x \in E$.*

In the case (1), we say that the equation $Tx = y$ is *well posed* or *uniquely solvable*. In the other cases, it is called *ill posed*.

Remark. For a nonlinear operator $T \in \mathcal{F}(E, F)$, each of the cases (2)–(4) should have included the possibility that the nonlinear equation $Tx = y$ has a unique solution for some y and many solutions for some other y. This is not possible for linear equations $Tx = y$. If $y = 0$, a linear equation $Tx = y$ is said to be *homogeneous*. If $y \neq 0$, it is said to be *nonhomogeneous*. The inverse image $T^{-1}(y)$ is called the *general solution*, and every $x(y) \in T^{-1}(y)$ is called the *special solution* of the equation. The general solution of the homogeneous equation $Tx = 0$ is the kernel $K(T) = T^{-1}(0)$. The *zero solution* $x(0) = 0$ is a special solution of any homogeneous equation $Tx = 0$. The relationship between the above solutions is illustrated by the following proposition.

Proposition. *The general solution of a linear equation $Tx = y$ is equal to the sum of any special solution $x(y)$ and the general solution $K(T)$ of the homogeneous equation $Tx = 0$: $T^{-1}(y) = x(y) + K(T)$.*

Proof. If $Tx(y) = y$ and $z \in K(T)$, then $T(x(y) + z) = Tx(y) + Tz = y + 0 = y$. Hence $x(y) + K(T) \subseteq T^{-1}(y)$. Conversely, for every $x \in T^{-1}(y)$, the equalities $Tx = y$ and $Tx(y) = y$ imply that $T(x - x(y)) = Tx - Tx(y) = y - y = 0$. Consequently, $x - x(y) = z \in K(T)$ and $x = x(y) + z \in x(y) + K(T)$. As a result, $T^{-1}(y) \subseteq x(y) + K(T)$. □

Remark. For any vector a and any subspace N of a space E, $M = a + N$ is called a *linear manifold* in E. The *dimension* of M is assumed to be the same as the dimension of N. The general solution $T^{-1}(y) = x(y) + K(T)$ of a solvable linear equation $Tx = y$ is a linear manifold. All such manifolds have the same dimension $d(T) = \dim K(T)$.

10. A linear operator $T \in \mathcal{L}(E, F)$ such that $c(T) < \infty$ and $d(T) < \infty$ is called a *Fredholm* operator. The difference $i(T) = c(T) - d(T)$ is called the *index* of the Fredholm operator T. The equality $i(T) = 0$ implies $c(T) = d(T)$.

Examples. 1) If $\dim E < \infty$ and $\dim F < \infty$, then all $T \in \mathcal{L}(E, F)$ are Fredholm operators.

2) Let $E = F$ and M be subspaces of dimension $\dim M = m < \infty$. Consider the identity operator I and an operator $S \in \mathcal{L}(E)$ such that $R(S) \subseteq M$ and $K(S) \supseteq E - M$ ($Sx = 0$ for $x \notin M$). The operator $T = I - S$ is a Fredholm operator. Indeed, we have $0 = Tx = x - Sx \Rightarrow x = Sx \in R(S)$ and therefore $K(T) \subseteq R(S) \subseteq M$ and $d(T) \leq m < \infty$. If $y \notin M$, then $Ty = y - Sy = y - 0 = y$ and $y \in R(T)$. It follows that $E - R(T) \subseteq M$ and $c(T) = \dim E/R(T) = \dim(E - R(T)) \leq m < \infty$. The complements $E - X$ are taken relative to the selected basis.

Question. Are there any Fredholm operators in $\mathcal{L}(E)$ other than those described in the preceding example?

For spaces E and F, let $\dim E < \infty$, $\dim F < \infty$, and $T \in \mathcal{L}(E, F)$. Since E and F are finite-dimensional, T is a Fredholm operator. The following lemma holds in the case of finite-dimensional spaces.

Lemma. $i(T) = \dim F - \dim E$.

Proof. The assertion of the lemma follows immediately from the equalities $\dim E = r(T) + d(T)$ and $\dim F = r(T) + c(T)$. □

Corollary. $c(T) = d(T) \Leftrightarrow \dim F = \dim E$

Let E and F be spaces and let $T \in \mathcal{L}(E, F)$. The operator may or may not be an isomorphism. Under the conditions (2)–(4) of the theorem of part 9 of this subsection, T is not an isomorphism. For a Fredholm operator T with $i(T) = 0$, only case (4) is possible.

The Fredholm alternative. *For any Fredholm operator with zero index, either 1) it is an isomorphism, or 2) it is neither an onto operator nor a one-to-one operator.*

Proof. If $i(T) = 0$, then $c(T) = d(T) = a$. For $a = 0$, the operator T is an isomorphism of the space E onto the space F. For $a \neq 0$, the operator T is neither an onto operator nor a one-to-one operator. $\square$

Proposition. *If* $\dim E = \dim F < \infty$, *then, for every* $T \in \mathcal{L}(E, F)$, *either 1)* T *is an isomorphism, or 2)* T *is neither an onto operator nor a one-to-one operator.*

Proof. As was proved above, from the assumption it follows that T is a Fredholm operator and $i(T) = 0$. $\square$

Example. Consider the spaces $E = F = \mathbb{K}_0^\infty$, the standard basis $\{e_i \mid i \geq 1\}$, a natural number m, the subspace $M = L\{e_i \mid 1 \leq i \leq m\}$, and a collection (α_{ij}) $(1 \leq i, j \leq m)$ of elements of the field $\mathbb{K}$. We introduce an operator $S \in \mathcal{L}(\mathbb{K}_0^\infty)$ as follows: $Se_i = \sum_{1 \leq j \leq m} \alpha_{ij} e_j$ $(1 \leq i \leq m)$, $Se_i = 0$ $(i > m)$.

Since $R(S) \subseteq M$ and $K(S) \supseteq E - M$, it follows that $T = I - S$ is a Fredholm operator. If $\det(\delta_{ij} - \alpha_{ij}) \neq 0$, then T an isomorphism. Otherwise, if $\det(\delta_{ij} - \alpha_{ij}) = 0$, then T is not an onto or a one-to-one operator. In particular, if $\alpha_{ii} = \alpha$ and $\alpha_{ij} = 0$ $(i \neq j)$ then $\alpha = 1$ implies $K(T) = M$ and $R(T) = M^c$ (the complement of M relative to the standard basis), whereas $\alpha \neq 1$ implies $K(T) = \{0\}$ and $R(T) = E$.

Exercise. Calculate the index of the operator T in the above example.

A linear equation $Tx = y$ such that T is a Fredholm operator is said to be a *Fredholm equation*. For such equations, we have the following alternative: *If* T *has zero index, then either 1) the equation* $Tx = y$ *is well-posed, or 2) the equation* $Tx = y$ *has no solutions for some* y *and the equation* $Tx = 0$ *has nonzero solutions.*

Remark. Thus, for an operator of zero index, the *uniqueness* of solutions to the homogeneous Fredholm equation *ensures the existence* of unique solutions of the nonhomogeneous equation for any right-hand side.

2.2.3.   Linear functionals

Linear functionals are linear operators with scalar values. They play the role of coordinates in vector spaces.

1. For a space $(\mathbb{K}, E)$, the space $(\mathbb{K}, E^*)$ whose vectors are linear functionals $f : E \to \mathbb{K}$ $(E^* = \mathcal{L}(E, \mathbb{K}))$ is called the (*algebraic*) *conjugate* of $(\mathbb{K}, E)$.

We now prove several propositions for conjugate spaces.

Lemma 1. *For any spaces* $(\mathbb{K}, E)$ *and* $(\mathbb{K}, F)$, $E \simeq F \Rightarrow E^* \simeq F^*$.

Proof. Let $T : E \to F$ be a linear isomorphism. We define $T^* : E^* \to F^*$ to be a map such that $T^* g = gT$ $(g \in F^*)$ or, equivalently, $(T^* g)x = g(Tx)$ $(x \in E)$.

The map T^* is linear because $T^*(g + h) = (g + h)T = gT + hT = T^* g + T^* h$, $T^*(\alpha g) = (\alpha g)T = \alpha(T^* g)$ $(\alpha \in \mathbb{K}; g, h \in F^*)$. It is also a one-to-one map: $T^* g = 0 \Leftrightarrow gT = 0 \Leftrightarrow g(Tx) = 0$ $(x \in E) \Leftrightarrow g(y) = 0$ $(y \in F)$ because $T(E) = F$. Moreover, T^* is a surjection, since for any $f \in E^*$ we have $T^*(fT^{-1}) = (fT^{-1})T = f(T^{-1}T) = f$.   □

Lemma 2. *Let* $\mathbb{K}_0^I$ *be the space of finite collections of elements* $\alpha_i \in \mathbb{K}$ $(i \in I)$. *Then* $\left(\mathbb{K}_0^I\right)^* \simeq \mathbb{K}^I$.

Proof. Every collection $a = (\alpha_i) \in \mathbb{K}^I$ defines the functional $f_a \in \left(\mathbb{K}_0^I\right)^*$ with values $f_a(x) = \sum \alpha_i \xi_i$, where $x = \sum \xi_i e_i$ $((\xi_i) \in \mathbb{K}_0^I)$. Since $x = (\xi_i)$ is a finite collection, the set of nonzero terms $\alpha_i \xi_i$ is finite. We have $f_a \in \left(\mathbb{K}_0^I\right)^*$:

$$f_a(\alpha x + \beta y) = \sum \alpha_i(\alpha \xi_i + \beta \eta_i) = \alpha \sum \alpha_i \xi_i + \beta \sum \alpha_i \eta_i$$
$$= \alpha f_a(x) + \beta f_a(y) \quad (\alpha, \beta \in \mathbb{K}; \ x = (\xi_i), \ y = (\eta_i) \in \mathbb{K}_0^I).$$

A map $S : \mathbb{K}^I \to \left(\mathbb{K}_0^I\right)^*$ such that $Sa = f_a$ $(a = (\alpha_i) \in \mathbb{K}^I)$ is linear:

$$S(\alpha a + \beta b)x = \sum (\alpha \alpha_i + \beta \beta_i)\xi_i = (\alpha Sa + \beta Sb)x$$
$$(\alpha, \beta \in \mathbb{K}; \quad a = (\xi_i), b = (\eta_i) \in \mathbb{K}^I, \quad x = (\xi_i) \in \mathbb{K}_0^I).$$

Moreover, S has zero kernel: $Sa = 0 \Leftrightarrow \sum \alpha_i \xi_i = 0$ $(x = (\xi_i) \in \mathbb{K}_0^I) \Leftrightarrow \alpha_j = \sum \alpha_i \delta_{ij}$ $(j \in I$; summation is with respect to the repeating index i). The map S is a one-to-one map whose range is the whole $\left(\mathbb{K}_0^I\right)^*$. Indeed, every functional $f \in \left(\mathbb{K}_0^I\right)^*$ is defined by its values $\alpha_i = f(e_i)$ for the basis collections $e_i = (\delta_{ij})$:

$$f(x) = \sum \alpha_i \xi_i \quad \left(x = (\xi_i) = \sum \alpha_i e_i \in \mathbb{K}_0^I\right).$$

Hence, $f = Sa$ for $a = (\alpha_i) \in \mathbb{K}^I$. Therefore, S is an isomorphism from $\mathbb{K}^I$ onto $\left(\mathbb{K}_0^I\right)^*$. $\square$

Corollary. If $\dim E < \infty$, then $E^* \simeq E$.

Proof. Let $\dim E = m < \infty$. Since $\mathbb{K}_0^m = \mathbb{K}^m$, we have $E \simeq \mathbb{K}^m$ (see Subsection 2.2.2, part 5). By Lemma 1 and Lemma 2, $E^* \simeq (\mathbb{K}^m)^* \simeq \left(\mathbb{K}_0^m\right)^* \simeq E$. $\square$

Proposition. $\dim E \le \dim E^*$.

Proof. Let $\{e_i \mid i \in I\}$ be a basis of E and let $e_i^* \in E^*$ be defined by the equality $e_i^*(x) = \xi_i$ $(x = \sum \xi_i e_i \in E)$. The collection $\{e_i^* \mid i \in I\}$ is linearly independent. By the lemma of Subsection 2.2.1, part 7, there exists a basis B^* of the space E^* that includes the above collection. Hence $\dim E = |I| \le |B^*| = \dim E^*$. $\square$

Example. Let $\mathbb{K} = \mathbb{Q}$ and $I = \mathbb{N}$. Then $\left|\mathbb{K}_0^I\right| = \left|\mathbb{Q}_0^\mathbb{N}\right| = \sum |\mathbb{Q}^n| = |\mathbb{Q}| \cdot |\mathbb{N}| = |\mathbb{N}| \cdot |\mathbb{N}| = |\mathbb{N}|$ and the set of finite rational sequences is countable. By the Cantor theorem, $|\mathbb{N}| < |\mathbb{Q}^\mathbb{N}| = |\mathbb{K}^I|$ and the set of all rational sequences is uncountable. There is no isomorphism from the set $\mathbb{Q}_0^\mathbb{N}$ to the set $\mathbb{Q}^\mathbb{N}$, and neither does it exist for the corresponding spaces.

Exercise. Prove that $\dim E^* = |\mathbb{K}^{\dim E}|$ and therefore $\dim E < \dim E^*$ if $\dim E = \infty$.

Remark. Based on the rules for operations with *cardinal numbers* (see Theorem 7, Section 8, Chapter V in Kuratowski and Mostowski, 1976) and the Cantor theorem, we obtain the following relations, which hold for any infinite field $\mathbb{K}$ and any set I:

$$\left|\mathbb{K}_0^I\right| = \sum |\mathbb{K}^K| = |\mathbb{K}| \cdot |I| \quad (K \in \mathcal{K}(I)), \quad |I| < |\mathbb{K}^I|.$$

Here $\mathcal{K}(I)$ is the class of all finite subsets of I. We have used the equality $|\mathbb{K}^K| = |\mathbb{K}|$ for $|\mathbb{K}| = \infty$ (see Theorem 2 and Corollary 1, Section 6, Chapter III in Bourbaki, 2004).

Remark. The set E of points of the space $(\mathbb{K}, E)$ is embedded into the vector space $\mathcal{F} = \mathcal{F}(E, \mathbb{K})$ of scalar functions on E by means of the map $a \rightarrow \chi_a$ ($\chi_a(a) = 1$, $\chi_a(x) = 0$ for $x \neq a$, $a \in X$, $x \in X$). The image of E is included in the subspace $\mathcal{F}_0 = \mathcal{F}_0(E, \mathbb{K})$ of finite functions on E. It is possible to consider the conjugate spaces $\mathcal{F}_0^*$ and $\mathcal{F}^*$. It should be emphasized that indicators χ_a are not linear functionals on E ($\chi_a \notin \mathcal{L}(E, \mathbb{K})$).

2. The conjugation of spaces can be extended, i.e., for E one can consider the *second conjugate space* $E^{**} = (E^*)^*$, which is the conjugate of (E^*).

For any $x \in E$, the functional $\delta_x \in E^{**}$ is defined as follows: $\delta_x(x^*) = x^*(x)$ ($x^* \in E^*$). Consider the map $\delta : E \rightarrow E^{**}$ that takes every point $x \in E$ to the functional $\delta_x \in E^{**}$. This map is called the *natural embedding* of E into E^{**}. The map δ is a linear one-to-one map. The range $\delta(E)$ is a subspace of E^{**}. The range of δ is often identified with E and it is therefore assumed that $E \subseteq E^{**}$.

The following inequalities hold: $\dim E \leq \dim E^* \leq \dim E^{**}$. If $\dim E < \infty$, then $\dim E = \dim E^* = \dim E^{**}$ and $E \simeq E^* \simeq E^{**}$.

Functionals of nonzero E^* are said to *separate* points of E: if $a, b \in E$ and $a \neq b$, then there exists an $x^* \in E^*$ such that $x^*a \neq x^*b$ (for which $x^*(a - b) \neq 0$). This makes it possible to use the values of functionals of E^* as *coordinates* of points in E: every point a is associated with a collection of (linear) coordinates $x^*(a) \in \mathbb{K}$ ($x^* \in E^*$). From the property of functionals to separate points, it follows that different points correspond to different collections of coordinates, i.e., $a \neq b$ implies $E^*a \neq E^*b$.

Remark. Since there are linearly dependent functionals among $x^* \in E^*$, the collection E^*a could be replaced by a smaller collection B^*a of basis coordinates for any basis B^* of the space E^*. However, in the general case, if $\dim E = \infty$ then it is only known that such bases exist. If $m = \dim E < \infty$, then it is natural to use basis coordinates e_i^* ($1 \leq i \leq m$).

Any subspace $E' \subseteq E^*$ that separates points of E will be called a *coordinate system* for E. Since functionals $x' \in E'$ are linear, the subspace E' is a coordinate system for E if and only if for any $z \in E$ such that $z \neq 0$ there exists an $x' \in E'$ such that $x'(z) \neq 0$ (x' separates the point z from 0).

Let $E' \subseteq E^*$ be a coordinate system for E. Consider a coordinate system $E'' \subseteq (E')^*$ for E'. For example, we can choose $E'' = \delta(E)$: if $z \neq 0$, then there is a point $a \in E$ such that $\delta_a(z) = z'(a) \neq 0$. Since E is identified with $\delta(E)$, we can consider E to be a coordinate system for E'. Thus, transition to the coordinate system turns out to be reflexive since performing it twice brings us back to the original point. We will call the space E with coordinate system E' for E and coordinate system $E'' = \delta(E)$ for E' a *reflexive space* and denote it by (E, E') or simply E, specifying the scalar field $\mathbb{K}$ whenever necessary.

Example. If $\dim E < \infty$, then $E \simeq E^*$ and $E' = E^*$ with $E'' = E^{**}$ form a reflexive space.

Remark. Every functional $x'' \in \mathcal{L}(E', \mathbb{K})$ can be extended to the functional $x^{**} \in \mathcal{L}(E^*, \mathbb{K})$ using the equalities $x^{**}(x' + z^*) = x''(x')$ for $x' \in E'$ and $z^* \in Z^*$, where $Z^* = (E')^c$ is the complement of E' with respect to E^* relative to the selected basis.

3. For spaces $(\mathbb{K}, E)$ and $(\mathbb{K}, F)$, consider a linear operator $T \in \mathcal{L}(E, F)$. Transition from the points of these spaces to their coordinates yields the conjugate spaces $(\mathbb{K}, E^*)$, $(\mathbb{K}, F^*)$, and the operator $T^* \in \mathcal{L}(F^*, E^*)$, which maps the coordinates of a point $Tx = y \in F$ to the coordinates of the point $x \in E$. We introduce the correspondence $T^* : F^* \to E^*$ defined by the equality $T^*y^* = y^*T$ $(y^* \in F^*)$. Note that $y^*T = x^* \in E^*$.

Lemma. $T^* \in \mathcal{L}(F^*, E^*)$.

Proof. The correspondence T^* is many-to-one. We have $T^*y_1^* \ni x_1^* \neq x_2^* \in T^*y_2^* \Rightarrow y_1^*(Ta) = x_1^*a \neq x_2^*a = y_2^*(Ta)$ for some $a \in E \Rightarrow y_1^* \neq y_2^*$. The map T^* is linear because $T^*(\alpha_1 y_1^* + \alpha_2 y_2^*) = (\alpha_1 y_1^* + \alpha_2 y_2^*)T = \alpha_1(y_1^*T) + \alpha_2(y_2^*T) = \alpha_1 \cdot T^*y_1^* + \alpha_2 \cdot T^*y_2^*$ $(\alpha_1, \alpha_2 \in \mathbb{K}; y_1^*, y_2^* \in F^*)$. $\quad\square$

The linear operator T^* is said to be the *conjugate* of T. The equalities $x^*a = (T^*y^*)a = y^*(Ta) = y^*b$ show that the operator T^* maps the coordinate y^*b of the image $Ta = b \in F$ of a point $a \in E$ to the coordinate x^*a of this point. If $R(T^*) \neq E^*$, then the result does not include all coordinates of a.

Examples. 1) O^* is the zero operator such that $O^*y^* = y^*O = o^*$ $(y^* \in F^*)$.

2) Let $E = F$ and $T = I$. Then I^* is the identity operator such that $I^*y^* = y^*I = y^*$ $(y^* \in F^*)$.

Let $(\mathbb{K}, E)$, $(\mathbb{K}, F)$, and $(\mathbb{K}, G)$ be spaces, $T, U \in \mathcal{L}(E, F)$ and $V \in \mathcal{L}(F, G)$ be operators, and $\alpha, \beta \in \mathbb{K}$ be scalars.

Theorem. $(\alpha T + \beta U)^* = \alpha T^* + \beta U^*$ and $(VU)^* = U^* V^*$.

Proof. Indeed, we have $(\alpha T + \beta U)^* y^* = y^*(\alpha T + \beta U) = \alpha(y^* T) + \beta(y^* U) = \alpha \cdot T^* y^* + \beta \cdot U^* y^*$ $(y^* \in F^*)$ and $(VU)^* z^* = z^*(VU) = (z^* V) U = (V^* z^*) U = U^*(V^* z^*) = (U^* V^*) z^*$ $(z^* \in G^*)$. $\square$

Corollary. If T is an isomorphism from E onto F, then T^* is an isomorphism from F^* onto E^* and $(T^*)^{-1} = (T^{-1})^*$.

Proof. Applying the theorem to $F = R(T)$, $G = E$, $U = T$, and $V = T^{-1}$, we obtain $T^*(T^{-1})^* = I^* : E^* \to E^*$. Similarly, applying the theorem to the spaces F, $E = R(T^{-1})$, and $G = F$ and the operators $V = T^{-1} : F \to E$ and $U = T = V^{-1} : E \to F$, we obtain $(T^{-1})^* T^* = (TT^{-1})^* = I^* : F^* \to F^*$. $\square$

Remark. If $F \neq R(T)$, then $T^{-1} : R(T) \to E$ if $T : E \to F$ and $T^* : F^* \to E^*$ are injective; $(T^{-1})^* : E^* \to R(T)^*$, and the correspondence $(T^*)^{-1} : R(T^*) \to F^*$ is defined on the subspace $R(T^*)$ of the space E^*. To compare the operator $(T^{-1})^*$ with the correspondence $(T^*)^{-1}$, we need to consider the restriction of $(T^{-1})^*$ to $R(T^*)$ and an embedding of $R(T^*)$ into F^*, extending the functionals on $R(T)$ to the functionals on F.

Conjugation of operators can be extended so that we can consider the second conjugate $T^{**} = (T^*)^* \in \mathcal{L}(E^{**}, F^{**})$. Using the natural embedding of the spaces E and F into E^{**} and F^{**}, respectively, and identifying x with x^{**} and y with y^{**}, the restriction of the operator T^{**} can be identified with T:

$$Tx = y \Rightarrow x^* = T^* y^* \Rightarrow T^{**} x^{**} = y^{**}.$$

If $\dim E = \dim F < \infty$, then the above identification implies $T^{**} = T$.

4. Let $A = \{a_i \mid i \in I\}$ and $B = \{b_j \mid j \in J\}$ be bases in spaces $(\mathbb{K}, E)$ and $(\mathbb{K}, F)$, respectively. Consider vectors $x = \sum \xi_i a_i \in E$ and $y = \sum \eta_j b_j \in F$ and an operator $T \in \mathcal{L}(E, F)$. Since T is linear, the equality $Tx = y$ is equivalent to the equalities

$$Tx = \sum \xi_i T a_i = \sum_j \left(\sum_i \xi_i \alpha_{ij} \right) b_j,$$

$$T a_i = \sum_j \alpha_{ij} b_j, \quad \eta_j = \sum_i \xi_i \alpha_{ij}.$$

The collection $M = (\alpha_{ij})$ is called a *matrix* of the operator T in the bases A and B. By definition, $M : I \times J \to \mathbb{K}$. Elements of a matrix M will be also denoted by $\alpha(i,j)$. The maps $\alpha(i,\cdot) : J \to \mathbb{K}$ and $\alpha(\cdot, j) : I \to \mathbb{K}$ are called the *ith row* and the *jth column* of M, respectively. These terms are associated with the traditional representation of a matrix in the form of a table. The rows $\alpha(i, \cdot)$ are finite collections, since they are collections of coordinates of the images Ta_i of the basis vectors a_i. The collection of coordinates (ξ_i) of the vector x is also finite. Therefore, there are finitely many nonzero terms in all the sums being considered.

Finite collections of coordinates

$$\xi = \{\xi(1, i) = \xi_i \mid i \in I\}, \quad \eta = \{\eta(1, j) = \eta_j \mid j \in J\}$$

will be called *coordinate rows*. By definition, $\xi \in \mathbb{K}_0^I$, $\eta \in \mathbb{K}_0^J$ and $\alpha(i, \cdot) \in \mathbb{K}_0^J$, $\alpha(\cdot, j) \in \mathbb{K}^I$ (columns of a matrix M are not necessarily finite).

The operations with matrices are determined by those with operators. Addition and multiplication by a scalar are componentwise:

$$\alpha(\alpha_{ij}) + \beta(\beta_{ij}) = (\alpha\alpha_{ij} + \beta\beta_{ij})$$

for $\alpha, \beta \in \mathbb{K}$ and $M = (\alpha_{ij})$, $N = (\beta_{ij})$. A row is multiplied by a matrix using the *row-by-column method*:

$$\eta = \xi M \iff \eta_j = \xi \cdot \alpha(\cdot, j) \Leftrightarrow \eta_j = \sum_i \xi_i \alpha_{ij}.$$

The same method is used when multiplying two matrices. Consider sets I, J, K, and matrices $M = (\alpha_{ij})$ and $N = (\beta_{jk})$, where $i \in I$, $j \in J$, $k \in K$. The *product of matrices* M and N with finite rows is the matrix $MN = (\gamma_{ik})$ with elements

$$\gamma(i, k) = \alpha(i, \cdot) \cdot \beta(\cdot, k) \Leftrightarrow \gamma_{ik} = \sum_j \alpha_{ij} \beta_{jk}.$$

Since $\alpha(i, \cdot)$ are finite, summation is well-defined. Since $\beta(j, \cdot)$ are finite, so are the rows $\gamma(i, \cdot)$. Note that $\gamma(i, \cdot) = \alpha(i, \cdot) \cdot N$, $\gamma(\cdot, k) = M \cdot \beta(\cdot, k)$.

A matrix $M : I \times J \to \mathbb{K}$ is called a $I \times J$- or $|I| \times |J|$-*matrix* whenever it is necessary to explicitly specify the index sets or their power. In some cases, matrices with nonscalar elements are considered. These elements, in their turn, are often represented by matrices.

5. Consider an example with Gaussian rationals. Let $I = J = \{1,2\}$ and $\mathbb{K} = \mathbb{Q}$. Let $\mathcal{M}(2,2)$ denote the set of all 2×2-matrices whose elements are rational numbers. We select the matrices

$$\mathbf{1} = \begin{pmatrix} 1 & 0 \\ 0 & 1 \end{pmatrix}, \quad \mathbf{i} = \begin{pmatrix} 0 & 1 \\ -1 & 0 \end{pmatrix}$$

and consider the set $\mathbb{C}(\mathbb{Q})$ of all linear combinations of these matrices

$$\mathbf{c} = \alpha\mathbf{1} + \beta\mathbf{i} = \begin{pmatrix} \alpha & \beta \\ -\beta & \alpha \end{pmatrix} \quad (\alpha, \beta \in \mathbb{Q}).$$

A scalar matrix $\alpha\mathbf{1}$ will be identified with its coefficient α and also denoted by α. Then $\mathbf{c} = \alpha + \beta\mathbf{i}$. The matrix $\mathbf{i}$ is called the *imaginary unit*.

From the properties of matrix operations, it follows that $(\mathbb{C}(\mathbb{Q}), +, \cdot)$ is a field:

$$\mathbf{c}_1 + \mathbf{c}_2 = (\alpha_1 + \alpha_2) + (\beta_1 + \beta_2)\mathbf{i},$$
$$\mathbf{c}_1\mathbf{c}_2 = (\alpha_1\alpha_2 - \beta_1\beta_2) + (\alpha_1\beta_2 + \alpha_2\beta_1)\mathbf{i},$$
$$\mathbf{c}^{-1} = (\alpha^2 + \beta^2)^{-1}(\alpha - \beta\mathbf{i}) \quad (\mathbf{c} \neq 0 \Leftrightarrow \alpha^2 + \beta^2 \neq 0),$$

and the commutativity, associativity, and distributivity rules hold. The field $\mathbb{Q}$ can be embedded into $\mathbb{C}(\mathbb{Q})$ using the map $\alpha \to \alpha\mathbf{1}$ ($\alpha \in \mathbb{Q}$). Identifying $\mathbb{Q}$ with $\mathbb{Q}\mathbf{1}$, we can say that $\mathbb{C}(\mathbb{Q}) = \mathbb{Q} + \mathbb{Q}\mathbf{i}$. The numbers $\mathbb{Q}\mathbf{i}$ are called *imaginary*.

Every 2×2-matrix with elements in $\mathbb{Q}$ represents a linear transformation of the plane $\mathbb{Q} \times \mathbb{Q}$. The vectors $e_1 = (1,0)$ and $e_2 = (0,1)$ make up a basis of $\mathbb{Q} \times \mathbb{Q}$. Matrices of linear transformations with respect to this basis can be considered. The scalar matrices $\alpha\mathbf{1}$ represent homotheties of the plane $\mathbb{Q} \times \mathbb{Q}$. The matrix $\mathbf{i}$ represents the rotation of the plane $\mathbb{Q} \times \mathbb{Q}$ such that $e_1 \to e_2$ and $e_2 \to -e_1$:

$$e_1\mathbf{i} = (1,0)\begin{pmatrix} 1 & 0 \\ 0 & 1 \end{pmatrix} = (0,1) = e_2,$$

$$e_2\mathbf{i} = (0,1)\begin{pmatrix} 0 & 1 \\ -1 & 0 \end{pmatrix} = (-1,0) = -e_1$$

(*counterclockwise rotation by 90 degrees*). Gaussian rationals represent transformations of the plane $\mathbb{Q} \times \mathbb{Q}$ that are linear combinations of the identity transformation and the above rotation. Note that

$$\mathbf{i}^2 = \begin{pmatrix} 0 & 1 \\ -1 & 0 \end{pmatrix} \begin{pmatrix} 0 & 1 \\ -1 & 0 \end{pmatrix} = \begin{pmatrix} -1 & 0 \\ 0 & -1 \end{pmatrix} = -\begin{pmatrix} 1 & 0 \\ 0 & 1 \end{pmatrix} = -\mathbf{1}.$$

The imaginary unit $\mathbf{i}$ is a solution to the equation $\mathbf{i}^2 + 1 = 0$.

A field $\mathbb{K}$ in which $\sum \alpha_i^2 = 0$ if and only if $\alpha_i = 0$ for any finite nonempty collection α_i is said to be a *real type field*. The condition that $\sum \beta_i^2 \neq -1$ for any finite collection $\beta_i \in \mathbb{K}$ is equivalent (see Section 81 in Vladimirov, 1979). The field of rationals $\mathbb{Q}$ is a real type field. This follows from the fact that the field $\mathbb{Q}$ is ordered and the order relation is consistent with the operations. If a finite collection of rational numbers α_i contains a number $\alpha_{i(0)} = \alpha \neq 0$, then $\sum \alpha_i^2 \geq \alpha^2 > 0$.

The field $\mathbb{C}(\mathbb{Q})$ is not a real type field. Similar to $\mathbb{C}(\mathbb{Q})$, any real type field $\mathbb{K}$ defines the field $\mathbb{C}(\mathbb{K})$ formed by 2×2-matrices with elements in $\mathbb{K}$. This field is also not a real type field and the equation $x^2 + 1 = 0$ is also solvable in it. The property of $\mathbb{K}$ to be a real type field is required for finding inverses in $\mathbb{C}(\mathbb{Q})$ using $(\alpha^2 + \beta^2)^{-1}$ for $\alpha, \beta \in \mathbb{K}$. The field $\mathbb{C}(\mathbb{Q})$ can be called a *complex type field*.

6. Consider spaces $(\mathbb{K}, E)$, $(\mathbb{K}, F)$, $(\mathbb{K}, G)$, operator spaces $\mathcal{L}(E, F)$, $\mathcal{L}(F, G)$, and the spaces of matrices with finite rows $\mathcal{M}(I, J)$, $\mathcal{M}(J, K)$, where I, J, K are index sets of bases $A = \{a_i \mid i \in I\}$, $B = \{b_j \mid j \in J\}$, $C = \{c_k \mid k \in K\}$ of the spaces E, F, and G, respectively.

For selected bases, let the matrices of operators $T, U \in \mathcal{L}(E, F)$ and $V \in \mathcal{L}(F, G)$ be $\operatorname{mat} T = M = (\alpha_{ij})$, $\operatorname{mat} U = N = (\beta_{ij})$, and $\operatorname{mat} V = P = (\gamma_{jk})$. Multiplication and addition for matrices and operators, as well as multiplication by a scalar, were defined in part 4 of this subsection.

Lemma. *The correspondence $\mathcal{L}(E, F) \to \mathcal{M}(I, J)$ is an isomorphism of vector spaces that preserves the operations.*

Proof. Consider the equalities $Ta_i = \sum_j \alpha_{ij} b_j$, $Ua_i = \sum_j \beta_{ij} b_j$, $Vb_j = \sum_k \gamma_{jk}$, and $Tx = \sum_j (\sum_i \xi_i \alpha_{ij}) b_j$, $Ux = \sum_j (\sum_i \xi_i \beta_{ij}) b_j$, $Vy = \sum_k (\sum_j \eta_j \gamma_{jk}) c_k$ for $x = \sum \xi_i a_i$, $y = \sum \eta_j b_j$. The first three equalities determine the matrices M, N, P for the operators T, U, V. The existence of the elements α_{ij}, β_{ij}, γ_{jk} is due to the bases A, B, C. Different operators correspond to different matrices. Indeed, if $T \neq U$, then there is an index $i(0) \in I$ such that $Ta_{i(0)} \neq Ua_{i(0)}$ and therefore $\alpha_{i(0)j(0)} \neq \beta_{i(0)j(0)}$ for some index $j(0) \in J$, which means that $M \neq N$.

The equalities for Tx and Ux are used to determine the operators $T, U \in \mathcal{L}(E, F)$ from the corresponding matrices. Thus, mat is an injective map from $\mathcal{L}(E, F)$ onto $\mathcal{M}(I, J)$. The equalities

$$\mathrm{mat}(\alpha T + \beta U) = \alpha \cdot \mathrm{mat}\, T + \beta \cdot \mathrm{mat}\, U,$$
$$\mathrm{mat}(VU) = (\mathrm{mat}\, U)(\mathrm{mat}\, V)$$

follow from the definition of matrix operations.   □

Remark. When representing the composition of operators as the product of their matrices, the order of the terms is reversed. This is illustrated by the equalities $(VU)x = V(Ux) = (xN)P = x(NP)$ $(x \in E)$.

Let $E = F = G$, $\mathcal{L}(E, F) = \mathcal{L}(F, G) = \mathcal{L}$, $A = B = C$, $I = J = K$, and $\mathcal{M}(I, J) = \mathcal{M}(J, K) = \mathcal{M}$.

Corollary. *The correspondence* mat $: \mathcal{L} \to \mathcal{M}$ *is an antisymmetric isomorphism from the operator algebra* $\mathcal{L}$ *onto the matrix algebra* $\mathcal{M}$.

The antisymmetry of the above isomorphism means that for any product the order of terms in the corresponding product is reversed. Antisymmetric isomorphisms are called *anti-isomorphisms*. A matrix is said to be *square* if its index sets are equal.

7. It is convenient to generalize the notion of a basis for the conjugate space E^*.

Choose a basis $A = \{e_i \mid i \in I\}$ in the space $(\mathbb{K}, E)$. Every vector e_i of the basis defines the functional $e_i^* \in E^*$ such that $e_i^*(x) = \xi_i$ $(x = \sum \xi_i e_i \in E)$. The conjugate space E^* is isomorphic to the space $\mathbb{K}^I$ of collections $\alpha_i \in \mathbb{K}$ $(i \in I)$. Every such collection defines the generalized linear combination $f = \sum \alpha_i e_i^*$ of functionals e_i^*:

$$fx = \sum \alpha_i \left(e_i^* x\right) = \sum \alpha_i \xi_i \quad \left(x = \sum \xi_i e_i\right).$$

Since the collection of coordinates ξ_i of the vector x is finite, there are finitely many nonzero terms in the above sums, although the collection of coefficients α_i is not necessarily finite.

The functionals e_i^* make up a *generalized basis* $A^* = \{e_i^* \mid i \in I\}$ of the space E^* and are called *basis functionals*. The generalized basis A^* is said to be *conjugate* to A. If $\dim E < \infty$, then A^* is a conventional basis of the space E^*. The quantity $\dim^* E^* = |A^*| = \dim E$ is called the *generalized dimension* of the conjugate space E^*. If $\dim E < \infty$, then $\dim^* E^* = \dim E^*$. If $\dim E = \infty$, then $\dim^* E^* < \dim E^*$. The uncountable dimension makes it difficult to deal with conventional bases of E^*. If a basis of E is countable, it is also more convenient to use a countable generalized basis of the conjugate space E^*.

Since $E^* = \mathcal{L}(E, F)$ for $F = \mathbb{K}$ and $\dim F = 1$, the matrix of the functional $f = \sum \alpha_i e_i^*$ with respect to the bases $A = \{e_i \mid i \in I\}$ and $B = \{1\}$ is the $|I| \times 1$-matrix (column) $\mathrm{mat}\, f = f^t$ with elements $\alpha(i, 1) = \alpha_i$ $(i \in I)$. In view of the isomorphism $E^* \simeq \mathbb{K}^I$, the functional $f = \sum \alpha_i e_i^*$ can be identified with its coordinate row $(1 \times |I|$-matrix) $\mathrm{mat}\, f = \{\alpha(i, 1) = \alpha_i \mid i \in I\}$. The column f^t is obtained by *transposing* this row. Similarly, the vector $x = \sum \xi_i e_i$ can be identified with its coordinate row $\mathrm{mat}\, x = \{\xi(i, 1) = \xi_i \mid i \in I\}$. A scalar $\alpha \in \mathbb{K}$ can be identified with the 1×1-matrix (α): $\mathrm{mat}\, \alpha = \alpha$.

With these conventions, the following equalities hold for any $f \in E^*$ and $x \in E$:

$$f x = \mathrm{mat}(f x) = \mathrm{mat}\, x \cdot \mathrm{mat}\, f = \xi f^t.$$

For an operator $T \in \mathcal{L}(E, F)$, its matrix M with respect to the bases $A \subseteq E$ and $B \subseteq F$, vectors $x \in E$, $T x = y \in F$, and their coordinate collections $\xi = (\xi_i)$ and $\eta = (\eta_j)$ with respect to the bases A and B, the following equalities hold:

$$\eta = \mathrm{mat}(T x) = \mathrm{mat}\, x \cdot \mathrm{mat}\, T = \xi M.$$

Adding an operator $V \in \mathcal{L}(F, G)$, its matrix P with respect to the bases B and C, the vector $V y = z \in G$ and its coordinate row $\mathrm{mat}\, z = \zeta$ in with respect to the basis C, we obtain the equalities $\zeta = \mathrm{mat}((VT)x) = \mathrm{mat}\, x \cdot \mathrm{mat}\, T \cdot \mathrm{mat}\, V = \xi(MP)$.

Remark. The matrices M and P have finite rows and arbitrary rows The finiteness of rows makes it possible to multiply matrices by the *row-by-column* rule. An appropriately definition of sums $\sum \gamma_j \delta_j$ for some (not necessarily finite) collections makes it possible to define the product $MN = (\gamma_{ik})$ with elements $\gamma_{ik} = \sum \alpha_{ij} \beta_{jk}$ for matrices $M = (\alpha_{ij})$ and $N = (\beta_{jk})$ of a more general form. This may be useful for generalizing the notion of a basis of a vector space and the corresponding matrix representations of operators. The present subsection, however, deals only with matrices whose rows or columns are finite.

8. Consider an operator $T \in \mathcal{L}(E, F)$, its matrix $M = (\alpha_{ij})$ with respect to bases $A = (a_i) \subseteq E$ and $B = (b_j) \subseteq F$, the conjugate operator $T^* \in \mathcal{L}(F^*, E^*)$, its matrix $M^* = (\alpha_{ji})$ with respect to the conjugate bases $B^* = (b_j^*)$ and $A^* = (a_i^*)$, and the transposed matrix $M^t = (\alpha_{ji}^t)$ with elements $\alpha_{ji}^t = \alpha_{ij}$. The rows of the matrix M are the columns of M^t and the columns of M are the rows of M^t: $\alpha^t(\cdot, i) = \alpha(i, \cdot)$, $\alpha^t(j, \cdot) = \alpha(\cdot, j)$.

Proposition. $M^* = M^t$.

Proof. By definition, $T^* y^* = y^* T$ $(y^* \in F^*)$. As a result, we have

$$\alpha_{ji}^* = \sum_k \alpha_{jk}^* (a_k^* a_i) = \left(\sum_k \alpha_{jk}^* a_k^* \right) a_i = (Tb_j^*) a_i = (b_j^* T) a_i = b_j^* (Ta_i) =$$

$$b_j^* \left(\sum_l \alpha_{il} b_l \right) = \sum_l \alpha_{il} b_j^* b_l = \alpha_{ij} = \alpha_{ji}^t. \quad \square$$

The coordinate rows η^* and ξ^* of functionals $y^* \in F^*$ and $T^* y^* = x^* \in E^*$ satisfy the equality $\xi^* = \eta^* M^*$. As follows from the above proof, it is equivalent to $\xi^* = \eta^* M^t$.

Remark. Since the rows of M are finite, the columns of M^t are finite and the product $\eta^* M^* = \eta^* M^t$ is defined for arbitrary collections $\eta^* \in \mathbb{K}^J$. Being the columns of M, the rows of M^t are not necessarily finite, the collections $\xi^* = \eta^* M^* = \eta^* M^t \in \mathbb{K}^I$ are also not necessarily finite.

The matrix notation can also be used for the bases A, B, B^*, A^*, considering them the columns with elements $a(i, 1) = a_i$, $b(j, 1) = b_j$, $b^*(j, 1) = b_j^*$, $a^*(i, 1) = a_i^*$. Using the *row-by-column* multiplication rule, the vectors $x = \sum \xi_i a_i \in E$, $y = \sum \eta_j b_j \in F$ and the functionals $y^* = \sum \eta_j^* b_j^* \in F^*$, $x^* = \sum \xi_i^* a_i^* \in E^*$ can be expressed in terms of their coordinate rows $\xi = (\xi(i, 1) = \xi_i)$, $\eta = (\eta(j, 1) = \eta_j)$ and $\eta^* = (\eta^*(j, 1) = \eta_j^*)$, $\xi^* = (\xi^*(i, 1) = \xi_i^*)$ in the matrix form: $x = \xi A$, $y = \eta B$ and $y^* = \eta^* B^*$, $x^* = \xi^* A^*$. The images of the bases can also be conveniently written in the same form (in the right side notation for operators, $xT = y$ and $y^* T^* = x^*$): $AT = MB$, $B^* T^* = M^* B^*$. After transposing the result, in the left side notation $Tx = y$, we obtain $TA^t = B^t M^t$.

Exercise. Verify that $M^* = T^* B^* A^t = B^* T A^t = B^* B^t M^t = M^t$.

9. The equality $\xi M = \eta$ makes it possible to identify the matrix $M \in \mathcal{M}(I, J)$ with the operator $M \in \mathcal{L}(\mathbb{K}_0^I, \mathbb{K}_0^J)$. We will use the left side notation of operators. Consider the isomorphisms $K \in \mathcal{L}(E, \mathbb{K}_0^I)$ and $L \in \mathcal{L}(F, \mathbb{K}_0^J)$ defined by the formulas $Kx = \xi$ and $Ly = \eta$. The relationship between the operators K, L, M, and T is shown in the first diagram.

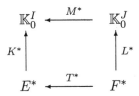

It follows from the definitions that $T = L^{-1}MK$, $M = LTK^{-1}$. After transposing the matrices and passing to the right side notation, we obtain the equalities $M^t\xi^t = \eta^t$ and $T = KM^tL^{-1}$, $M^t = K^{-1}TL$.

Exercise. Verify the above equalities for T, M, and M^t.

Due to the equalities $\eta^*M^* = \xi^*$, the matrix $M^* \in \mathcal{M}(J, I)$ can be identified with the operator $M^* \in \mathcal{L}(\mathbb{K}^J, \mathbb{K}^I)$. As before, we will use the left side notation for operators. Let $L^* \in \mathcal{L}(F^*, \mathbb{K}^J)$ and $K^* \in \mathcal{L}(E^*, \mathbb{K}^I)$ be isomorphisms defined by the equalities $L^*y^* = \eta^*$ and $K^*x^* = \xi^*$. The relationship between the operators M^*, L^*, K^*, and T^* is illustrated by the second diagram.

$$
\begin{array}{ccc}
\mathbb{K}_0^I & \xleftarrow{\quad M^* \quad} & \mathbb{K}_0^J \\
\Big\uparrow{\scriptstyle K^*} & & \Big\downarrow{\scriptstyle L^*} \\
E^* & \xleftarrow{\quad T^* \quad} & F^*
\end{array}
$$

It follows from the definitions that $T^* = (K^*)^{-1}M^*L^*$ and $M^* = K^*T^*(L^*)^{-1}$, where M^* can be replaced by M^t.

Exercise. Use the identity embeddings $\varphi : \mathbb{K}_0^I \to \mathbb{K}^I$ and $\psi : \mathbb{K}_0^J \to \mathbb{K}^J$ to derive the appropriate equalities for T and M from the above equalities for T^* and M^*.

Remark. Although the use of several types of notations for operations with matrix and operators may sometimes be confusing, each of these types has its advantages.

2.2.4. Scalar products

A scalar product defines the geometry of a vector space.

1. For vector spaces $(\mathbb{K}, X)$, $(\mathbb{K}, Y)$, and $(\mathbb{K}, Z)$, consider a map p : $X \times Y \to Z$ and the maps $p(\cdot, b) : X \to Z$ and $p(a, \cdot) : Y \to Z$ $(a \in X,$ $b \in Y)$. If the maps $p(\cdot, b)$ and $p(a, \cdot)$ are linear, then the map $p = p(\cdot, \cdot)$ is said to be *bilinear*. Bilinear functions will be called *products*. Multiplicative notation will be used for their values: $p(x, y) = xy \in Z$ $(x \in X,\ y \in Y)$. Sometimes yx is written as xy. Let $\mathcal{B} = \mathcal{B}(X \times Y, Z)$ be the set of bilinear maps $p : X \times Y \to Z$. It is a subset of the set $\mathcal{F} = \mathcal{F}(X \times Y, Z)$ of all maps $f : X \times Y \to Z$.

Lemma. $(\mathbb{K}, \mathcal{B})$ *is a subspace of the vector space* $(\mathbb{K}, \mathcal{F})$.

Proof. Let $\alpha, \beta \in \mathbb{K}$ and $p, q \in \mathcal{B}$. Then $(\alpha p + \beta q)(\cdot, b) = \alpha p(\cdot, b) + \beta q(\cdot, b) \in \mathcal{L}(X, Z)$ and $(\alpha p + \beta q)(a, \cdot) = \alpha p(a, \cdot) + \beta q$ $(a, \cdot) \in \mathcal{L}(Y, Z)$ $(a \in X,\ b \in Y)$. Hence, $\alpha p + \beta q \in \mathcal{B}$. □

If $Z = \mathbb{K}$, then products $p : X \times Y \to \mathbb{K}$ in $\mathcal{B}(X \times Y, \mathbb{K})$ are called *scalar*. A product $p : X \times X \to Z$ such that $p(x, y) = p(y, x)$ $(x \in X,\ y \in Y)$, is said to be *symmetric*. If $p(x, x) \neq 0$ whenever $x \neq 0$, then the product $p : X \times X \to Z$ is said to be *nondegenerate*. In multiplicative form, the relations in the above definitions are written as $xy = yx$ and $x^2 = xx \neq 0$.

Examples. 1) $X = Y = Z = \mathbb{K}$ and the product of scalars $p(\xi, \eta) = \xi\eta$. 2) Vector algebras $X = Y = Z = E$ and the product of scalars $p(x, y) = xy$. 3) $X = E$, $Y = E^*$, $Z = \mathbb{K}$, and $p(x, f) = f(x) = xf$ (the value of a linear functional $f \in E^*$ at the point $x \in E$). 4) $X = E$, $Y = \mathcal{L}(E, F)$, $Z = F$, and $p(x, T) = Tx = xT$ (the value of a linear operator $T \in \mathcal{L}(E, F)$ at the point $x \in E$. 5) $X = \mathcal{L}(E, F)$, $Y = \mathcal{L}(F, G)$, $Z = \mathcal{L}(E, G)$ and $p(T, U) = UT$ (the composition of operators, where $UTx = U(Tx) \in G$ for all $x \in E$).

Together with the space $\mathcal{B} = \mathcal{B}(X \times Y, Z)$ of bilinear maps $p : X \times Y \to Z$, we consider the space $\mathcal{L} = \mathcal{L}(X, \mathcal{L}(Y, Z))$ of linear operators $T : X \to \mathcal{L}(Y, Z)$. Let $\varphi : \mathcal{L} \to \mathcal{B}$ be a map such that $\varphi(T)(x, y) = (Tx)y$ $(x \in E,$ $Tx \in \mathcal{L}(Y, Z),\ y \in Y,\ (Tx)y \in Z)$.

Proposition. *The map* φ *is an isomorphism from* $(\mathbb{K}, \mathcal{L})$ *onto* $(\mathbb{K}, \mathcal{B})$.

Proof. 1) Let $p \in \mathcal{B}$. We set $Tx = p(x, \cdot) \in \mathcal{L}(Y, Z)$. Since $p(\cdot, y) \in \mathcal{L}(X, Z)$, we have $T \in \mathcal{L}$. Moreover, $(Tx)y = p(x, y)$. Therefore, $\varphi(T) = p$ and φ maps $\mathcal{L}$ onto $\mathcal{B}$.

If $S, T \in \mathcal{L}$ and $S \neq T$, then there is an $a \in X$ such that $Sa \neq Ta$. Consequently, there is a $b \in Y$ such that $(Sa)b \neq (Ta)b$ and $\varphi(S) \neq \varphi(T)$. Hence, φ is injective.

2) For any $\alpha, \beta \in \mathbb{K}$ and $S, T \in \mathcal{L}$,

$$\varphi(\alpha S + \beta T)(x, y) = ((\alpha S + \beta T)x)y$$
$$= \alpha(Sx)y + \beta(Tx)y = \alpha\varphi(S)(x, y) + \beta\varphi(T)(x, y).$$

Therefore, $\varphi(\alpha S + \beta T) = \alpha\varphi(S) + \beta\varphi(T)$ and the map φ is linear. $\square$

Remark. The isomorphism $\mathcal{L} \simeq \mathcal{B}$ makes it possible to replace repeating linear maps $T \in \mathcal{L}$ with bilinear maps $p \in \mathcal{B}$. This appears to be convenient in many situations.

2. Consider spaces $(\mathbb{K}, E)$, $(\mathbb{K}, E^*)$, and the scalar product $p(x, x^*) = x^*(x) = xx^*$ $(x \in E,\ x^* \in E^*)$. If $xx^* = 0$, the functional x^* is said to be *orthogonal* to the vector x, written $x \perp x^*$. A set $X^* \subseteq E^*$ is said to be orthogonal to a set $X \subseteq E$ if $x \perp x^*$ for all $x^* \in X^*$ and $x \in X$. In this case, we write $X \perp X^*$. If $X = \{x\}$ or $X^* = \{x^*\}$, we write $x \perp X^*$ or $X \perp x^*$.

Example. Choose a basis $A = (a_i) \subseteq E$ and take the conjugate basis $A^* = (a_i^*) \subseteq E^*$ $(i \in I)$. By definition, $a_i a_k^* = \delta_{ik}$ and therefore $a_i \perp a_k^*$ for $i \neq k$ $(i, j \in I)$.

Let $A = (a_i)$ and $B = (b_j)$ be bases in spaces $(\mathbb{K}, E)$ and $(\mathbb{K}, F)$, respectively. Let $x = \sum \xi_i a_i \in E$ and $y = \sum \eta_j b_j \in F$ $(i \in I,\ j \in J)$. Set $p(x, y) = \sum_{ij} \xi_i \gamma_{ij} \eta_j$. Then $p \in \mathcal{B}(E \times F, \mathbb{K})$ is a scalar product with matrix $C = (\gamma_{ij})$ with respect to the bases $A = (a_i)$ and $B = (b_j)$. Elements of the matrix C are scalar products of the basis vectors: $\gamma_{ij} = a_i b_j$ $(i \in I,\ j \in J)$. The matrix C defines the multiplication table for the scalar product p. In particular, if $E = F$, $A = B$, and $\gamma_{ij} = \delta_{ij}$, then $xy = \xi \eta^t = \sum \xi_i \eta_i$. The matrix C is usually assumed to have specific properties. If $E = F$, then the scalar product is called an *inner product*.

For $E = F$, any basis $A = (a_i)$ such that $a_i a_j = \delta_{ij}$, is said to be *orthonormal*.

Exercise. Answer the question whether there exists an orthonormal basis in every vector space with a scalar product.

3. Consider a reflexive space (E, E'). For any sets $X \subseteq E$ and $X' \subseteq E'$ their *orthogonal complements* are defined as follows:

$$X^{\perp} = \{x' \mid xx' = 0 \ (x \in X)\} \subseteq E',$$
$$X'^{\perp} = \{x \mid x'x = 0 \ (x' \in X')\} \subseteq E.$$

In the second equality, we used the identification $x'' = \delta_x = x$ and $E'' = \delta(E) = E$. In addition, consider a reflexive space (F, F') with the same scalars and identification $y'' = y$ and $F'' = F$. Choose an operator $T \in \mathcal{L}(E, F)$ such that $y'T = x' \in E'$ for every $y' \in F'$. Let $T' \in \mathcal{L}(F', E')$ be an operator such that $T'y' = y'T$ $(y' \in F')$. The operator T' is said to be the *conjugate* of T (analogous to T^*, whose restriction to F' is equal to T').

Exercise. Verify that the operator T' is linear.

Note that $x''T' = y'' \in F''$ for every $x'' \in E''$. This allows us to define an operator $T'' = (T')' \in \mathcal{L}(E'', F'')$ by the formula $T''x'' = x''T'$ $(x'' \in E'')$. The operator T'' is called the *second conjugate* of T (analogous to T^{**}). Identifying E'' with E and F'' with F allows us to assume that $T'' = T$ and write $Tx = xT'$ instead of $T''x'' = x''T'$.

Consider the range $R(T) \subseteq F$, its orthogonal complement $R(T)^{\perp} \subseteq F'$ and the kernel $K(T') \subseteq F'$.

Lemma. $R(T)^{\perp} = K(T'), \ R(T')^{\perp} = K(T)$.

Proof. It follows from the definitions that $y' \in R(T)^{\perp} \Leftrightarrow yy' = 0$ $(y \in R(T)) \Leftrightarrow x(y'T') = x(Ty') = (xT)y' = 0 \ (x \in E) \Leftrightarrow T'(y') = 0 \Leftrightarrow y' \in K(T')$.

Since $T'' = T$, the second equality follows from the first one. $\square$

Exercise. Prove the equality $R(T')^{\perp} = K(T)$ directly, without using the equality $R(T)^{\perp} = K(T')$.

4. For the sake of clarifying the algebraic aspect of the matter, let the definition of a *Fredholm operator* $T \in \mathcal{L}(E, F)$ include the following equalities in addition to the inequalities $c(T) < \infty$ and $d(T) < \infty$:

$$\dim F/R(T) = \dim R(T)^{\perp}, \quad \dim E'/R(T') = \dim R(T')^{\perp}.$$

Exercise. For a subspace N of the space F, find the conditions under which $F/N \simeq N^{\perp}$.

Remark. If $\dim E < \infty$ and $\dim F < \infty$, then every operator $T \in \mathcal{L}(E, F)$ is a Fredholm operator for $E' = E^*$ and $F' = F^*$. There exist other important classes of spaces and operators for which the equalities included in the above definition hold true.

Proposition. Let $c(T) = \dim F/R(T)$, $d(T) = \dim K(T)$, $c(T') = \dim E'/R(T')$, $d(T') = \dim K(T')$. If T is a Fredholm operator, then $c(T') = d(T)$ and $d(T') = c(T)$.

Proof. These properties follow from the equalities in the lemma of part 3 of this subsection and the above equalities included in the definition of the Fredholm operator.   □

Corollary. If T is a Fredholm operator, then the conjugate operator T' is also a Fredholm operator and $i(T') = -i(T)$.

In particular, we have $i(T') = 0 \Leftrightarrow i(T) = 0$.

5. Let $T \in \mathcal{L}(E, F)$. The operators T and $T' \in \mathcal{L}(F', E')$ define the equations

$$Tx = y, \tag{1}$$
$$Tx = 0, \tag{2}$$
$$T'y' = x', \tag{3}$$
$$T'y' = 0' \tag{4}$$

where $x \in E$, $y, 0 \in F$, $y' \in F'$, $x', 0' \in E'$. Equations (1) and (3) are well posed if they have a unique solution for any right-hand side. Equations (2) and (4) are well posed if they have only zero solutions (the well-posedness is reduced to the uniqueness of solutions in this case). The relationship between the above equations for Fredholm operators T and T' is described by the three classical Fredholm's theorems.

Theorem 1 (The well-posedness theorem). *Let a Fredholm operator T have index $i(T) = 0$. Then either all of the equations (1)–(4) are well-posed or none of them are well-posed.*

Proof. If $i(T) = 0$, then $i(T') = 0$ and $c(T) = d(T)$, $d(T') = c(T')$. These equalities and the equalities in the proposition of part 4 of this subsection imply that $c(T') = d(T) = d(T') = c(T) = a$. If $a = 0$, then all of the equations (1)–(4) are well-posed (see Subsection 2.2.3, part 10). If $a \neq 0$, then none of them are well-posed. □

Let $R(T) = R(T)^{\perp\perp}$. Then the lemma of part 3 of this subsection implies the following theorem.

Theorem 2 (The existence theorem). *Equation* (1) *is solvable if and only if its right-hand side is orthogonal to every solution of equation* (4).

Proof. The assertion of the theorem is equivalent to the equality $R(T) = K(T')^{\perp}$, which follows from the equalities $R(T) = R(T)^{\perp\perp}$ and $R(T)^{\perp} = K(T')$. □

Exercise. Prove that the condition $R(T) = R(T)^{\perp\perp}$ is necessary.

Remark. A similar statement holds for the dual equations (3) and (2).

Theorem 3 (Theorem on the number of solutions). *Assume that a Fredholm operator T has index $i(T) = 0$. Then equations* (2) *and* (4) *have the same finite number of linearly independent solutions.*

Proof. The assertion of the theorem is equivalent to the relation $d(T) = d(T') < \infty$. It was shown in the proof of Theorem 1 that $d(T) = d(T')$ if $i(T) = 0$. Since T and T' are Fredholm operators, the dimensions $d(T)$ and $d(T')$ are finite. □

Remark. If $\dim E = \dim F < \infty$, then $E \simeq E' \simeq F' \simeq F$ for the natural embeddings with $E' = E^*$ and $F' = F^*$. In this case, every $T \in \mathcal{L}(E, F)$ together with its conjugate $T' \in \mathcal{L}(F', E')$ are Fredholm operators and $i(T) = i(T') = \dim F - \dim E = 0$. All three Fredholm's theorems hold for these operators.

6. Consider a reflexive space (E, E'). For every subspace $M \subseteq E$, its *closure* is defined as follows: $\bar{M} = M^{\perp\perp} = (M^{\perp})^{\perp} \subseteq E$. If $M = \bar{M}$, then the subspace M is said to be *closed*.

Examples. 1) Let $M = \{0\}$. Then $M^{\perp} = E'$ and $\bar{M} = (E')^{\perp} = \{0\} = M$. Indeed, $x'x = xx' = 0$ ($x' \in E'$) $\Leftrightarrow x = 0$ because points of E are separated (if $x \neq 0$, then there exists an $x'_0 \in E'$ such that $x'_0(x) \neq 0$). The zero subspace $M = \{0\}$ of the reflexive space E is closed.

2) Let $M = E$. Then $M^\perp = \{0'\}$ and $\bar{M}^\perp = \{0'\}^\perp = E = M$ because $0'x = x0' = 0$ $(x \in E)$. The whole reflexive space $M = E$ is closed.

Exercise. Prove that all subspaces of a finite-dimensional space are closed.

7. Consider a reflexive space (E, E') and subspaces $M, X \subseteq E$. It follows from the definitions that $M \subseteq \bar{M}$. Indeed, we have $x \in M \Rightarrow xx' = x'x = 0$ $(x' \in M^\perp) \Rightarrow x \in M^{\perp\perp} = \bar{M}$. Since $X \subseteq M \Rightarrow X^\perp \supseteq M^\perp \Rightarrow X^{\perp\perp} \subseteq M^{\perp\perp}$, it follows that $X \subseteq M \Rightarrow \bar{X} \subseteq \bar{M}$. Furthermore, $\bar{M}^\perp = M^\perp$. Indeed, we have $M \subseteq \bar{M} \Rightarrow \bar{M}^\perp \subseteq M^\perp$. At the same time, $M^\perp \subseteq (M^\perp)^{\perp\perp} = (M^{\perp\perp})^\perp = \bar{M}^\perp$.

Exercises. 1) Prove that the intersection $M = \cap M_i$ of closed subspaces $M_i \subseteq E$ is closed. 2) Prove that the closure $\bar{M}$ is equal to the intersection of all closed subspaces of E that contain the subspaces M. 3) Prove that $\bar{\bar{M}} = \bar{M}$.

Note that $M = \bar{M}$ if and only if $M = X'^\perp$ for some $X' \subseteq E'$. Indeed, if $M = \bar{M} = M^{\perp\perp}$ then $X' = M^\perp$. Conversely, if $M = X'^\perp$ then $\bar{M} = M^{\perp\perp} = X'^{\perp\perp\perp} = X'^\perp = M$.

Lemma. $(X + M)^\perp = X^\perp \cap M^\perp$.
Proof. Let $x' \in X^\perp \cap M^\perp$. Then $xx' = 0$ for $x \in X$ and $yx' = 0$ for $y \in M$. Therefore, $(x+y)x' = xx' + yx' = 0$ $(x+y \in X+M)$, $x' \in (X+M)^\perp$, $X^\perp \cap M^\perp \subseteq (X+M)^\perp$. Conversely, let $x' \in (X+M)^\perp$. Since $X \subseteq X + M$ and $M \subseteq X + Y$, it follows that $x' \in X^\perp$, $x' \in M^\perp$, $x' \in X^\perp \cap M^\perp$, and $(X+M)^\perp \subseteq X^\perp \cap M^\perp$. $\square$

8. Assume that the subspace M is finite dimensional. Choose a basis $A = (a_i)$ in M and the complement of A with respect to the basis $B = (b_j)$ of the space $(\mathbb{K}, E)$. The collection $A^* \cup B^* = (a_i^*, b_j^*)$ is a generalized basis of the conjugate space E^*.

Lemma. If $\dim M < \infty$, then $\operatorname{codim} M^\perp = \dim M$.
Proof. The proof can be reduced to solving a system of linear equations. Since $a_i b_j^* = 0$ for all i and j, we have $M^\perp = \{x' \in E' \mid a_i x' = 0 \ (a_i \in A)\}$ $(x' \perp M \Leftrightarrow x' \perp A)$. Consider an operator $T \in \mathcal{L}(E', \mathbb{K}^A)$ such that $Tx' = Ax' = (a_i x')$. From the proposition of part 8 of Subsection 2.2.2, the equality $M^\perp = K(T)$, the linear dependence of the vectors a_i, and the separability of (E, E'), it follows that $R(T) = \mathbb{K}^A$ and $\operatorname{codim} M^\perp = \operatorname{codim} K(T) = \dim R(T) = |A| = \dim M$. $\square$

Remark. As far as matrices are concerned, the equality $\dim R(T) = |A|$ means that the number of linearly independent rows of the corresponding matrix is equal to the number of its linearly independent columns.

Exercise. Give a detailed proof of the equality $R(T) = \mathbb{K}^A$.

Proposition. If $M = \bar{M}$ and $\operatorname{codim} M < \infty$ then $\operatorname{codim} M = \dim M^{\perp}$.

Proof. In part 8 of Subsection 2.2.2 it was proved that $\dim M + \operatorname{codim} M = \dim E$ and there is a subspace X isomorphic to E/M such that $M + X = E$ and $M \cap X = \{0\}$. We have $\dim X = \operatorname{codim} M < \infty$ by assumption. From the lemma of part 7 of this subsection, it follows that $X^{\perp} \cap M^{\perp} = (X + M)^{\perp} = E^{\perp} = \{0'\}$. Therefore, the restriction $\varphi : M^{\perp} \to E'/X^{\perp}$ is injective: $\varphi(x') = x' + X^{\perp} = y' + X^{\perp} = \varphi(y') \Rightarrow x' - y' \in X^{\perp} \cap M^{\perp} = \{0'\} \Rightarrow x' = y'$ $(x', y' \in M^{\perp})$. Consequently, $M^{\perp} \simeq \varphi(M^{\perp})$ and $\dim M^{\perp} = \dim \varphi(M^{\perp}) \le \operatorname{codim} X^{\perp}$. Since $\dim X < \infty$, the preceding lemma implies that $\operatorname{codim} X^{\perp} = \dim X$. It follows that $\dim M^{\perp} < \infty$ and $\operatorname{codim}(M^{\perp})^{\perp} = \dim M^{\perp}$. By assumption $M = \bar{M} = (M^{\perp})^{\perp}$. As a result, $\operatorname{codim} M = \dim M^{\perp}$. $\square$

9. Consider reflexive spaces (E, E') and (F, F'). We point out the class of operators $T \in \mathcal{L}(E, F)$ with *closed ranges* $R(T)$ and the class of operators $T' \in \mathcal{L}(F', E')$ with closed ranges $R(T')$.

Example. If $\dim E < \infty$ and $\dim F < \infty$, then all operators $T \in \mathcal{L}(E, F)$ and $T^* \in \mathcal{L}(F^*, E^*)$ have closed ranges $R(T)$ and $R(T^*)$.

Remark. The equalities $K(T) = R(T')^{\perp}$ and $K(T') = R(T)^{\perp}$, which were proved in part 3 of this subsection, imply that the kernels $K(T)$ and $K(T')$ of the operators T and T' are closed.

Proposition. If T and T' have closed ranges $R(T)$ and $R(T')$ of finite codimensions $c(T) = \operatorname{codim} R(T)$ and $c(T') = \operatorname{codim} R(T')$, then T and T' are Fredholm operators.

Proof. If $R(T) = \overline{R(T)}$ and $\operatorname{codim} R(T) < \infty$, then, as follows from part 8 of this subsection, $\dim R(T)^{\perp} = \operatorname{codim} R(T) < \infty$. Similarly, we have $\dim R(T')^{\perp} = \operatorname{codim} R(T') < \infty$. From these relations and the equalities $K(T) = R(T')^{\perp}$ and $K(T') = R(T)^{\perp}$ proved in part 3, it follows that $d(T) = \dim K(T) < \infty$ and $d(T') = \dim K(T') < \infty$. Hence, T and T' are Fredholm operators. $\square$

Remark. The operators with closed ranges form an important class of linear operators. There exist alternative definitions of the property of being closed (for example, see Section 21 in Spivak, 1965). Another class of operators being studied consists of operators $T \in \mathcal{L}(E, F)$ whose graph $G(T) = \{(x, Tx) \mid x \in E\}$ is closed in the space $E \times F$.

Exercise. Determine the conditions under which an operator with closed graph becomes a Fredholm operator.

10. Consider a reflexive space (E, E') such that E is identified with a subspace of the conjugate space E' that is isomorphic to E. Under this identification Under this identification, the operator $T \in \mathcal{L}(E, F)$ is equal to the restriction of its conjugate T' to E and is called *symmetric*: $T \subseteq T'$. If $T = T'$, then the operator T is said to be *self-adjoint*.

Choose the appropriate conjugate bases in E, F, E', and F'. Let M and M' denote the matrices of the operators T and T' with respect to these bases. It was proved in Subsection 2.2.3, part 8, that $M' = M^t$. The operator T is symmetric if and only if its matrix is symmetric, i.e., $M = M^t$. The following statement follows from the proposition in part 9 of this subsection.

Corollary. *Any self-adjoint operator with closed range of finite codimension is a Fredholm operator.*

Exercise. Prove the analogous statement for a symmetric operator.

2.2.5. Normed spaces

A normed space is a vector space equipped with a norm. A norm is used to measure the length of vectors.

1. Let $\mathbb{M}$ be an *ordered field* and let $\mathbb{P}$ be the set of *strictly positive* elements $p > 0$. The following properties hold: (1) *trichotomy* (for any $a \in \mathbb{M}$, either $a > 0$, or $a = 0$, or $-a > 0$); (2) *the rule of signs* (if $a > 0$ and $b > 0$, then $a+b > 0$ and $ab > 0$). Elements $-a < 0$ that are additive inverses of strictly positive elements $a > 0$ are called *strictly negative*, making up the set $-\mathbb{P}$. Strictly positive and strictly negative elements are often called simply *positive* and *negative*, assuming that zero is not included. Since the order is consistent with the operations, we have $-\mathbb{P} < \{0\} < \mathbb{P}$ and $\mathbb{M}$ is the order sum of these sets, i.e., $\mathbb{M} = (-\mathbb{P}) \cup \{0\} \cup \mathbb{P}$.

The *absolute value* of an element $a \in \mathbb{M}$, written $|a|$, is defined as follows: $|a| = a$ for $a \geq 0$ and $|a| = -a$ for $a < 0$. The absolute value is positive, nondegenerate, semi-additive, and multiplicative:

$$|a| > 0 \ (a \geq 0), \quad |a + b| \leq |a| + |b|, \quad |ab| = |a||b|$$

for all $a, b \in \mathbb{M}$. Furthermore, $a^2 = (-a)^2 = |a|^2 \geq 0$ and $\sum a_i^2 \geq 0$ for every finite collection of elements $a_i \in \mathbb{M}$. In particular, $1 = 1^2 > 0$ and $n \cdot 1 > 0$ for any $n \in \mathbb{N}$. Consequently, the ordered field $\mathbb{M}$ has characteristic zero.

Examples of ordered fields are $\mathbb{Z}$, $\mathbb{Q}$, and $\mathbb{R}$.

2. Let $\mathbb{K}$ be a field and $\mathbb{M}$ be an ordered field. Assume that there exists a *positive nondegenerate semi-additive and multiplicative* function $|\cdot| : \mathbb{K} \to \mathbb{M}$. It follows that $|\alpha| \in \mathbb{M}$, $|0| = 0$ and

$$|\alpha| > 0 \ (\alpha \neq 0), \quad |\alpha + \beta| \leq |\alpha| + |\beta|, \quad |\alpha\beta| = |\alpha||\beta|$$

for all $\alpha, \beta \in \mathbb{K}$. A function with these properties is called a *norm* or an *absolute value* on $\mathbb{K}$. If a norm is defined on $\mathbb{K}$, it is called a *normed field*.

From the properties of an absolute value, it follows that

$$|1| = 1, \quad |-\alpha| = |\alpha| \quad ||\alpha| - |\beta|| \leq |\alpha - \beta|.$$

In the last inequality, $||\alpha| - |\beta||$ denotes the absolute value of the element $|\alpha| - |\beta| \in \mathbb{M}$.

Examples. 1) $\mathbb{K} = \mathbb{M}$. 2) $\mathbb{K} = \mathbb{C}$, $\mathbb{M} = \mathbb{R}$. 3) $\mathbb{K} = \mathbb{Q}$, $\mathbb{M} = \mathbb{R}$.

The example with the field of complex numbers $\mathbb{C}$ (see Subsection 2.1.5, part 2) shows that unordered fields also can be normed. The example of p-norm on the field of rationals $\mathbb{Q}$ (Subsection 2.1.5, part 4) shows that the above norms can have substantially different properties on the same field. The absolute value of a number in $\mathbb{Q}$ is an Archimedean norm, whereas the p-norm is non-Archimedean because it satisfies the *isosceles* triangle inequality.

Remark. In operator theory, for a normed scalar field we usually take the field of real or complex numbers, the norming field being the field of real numbers. For this reason, in what follows we assume that $\mathbb{K} = \mathbb{R}$ or $\mathbb{K} = \mathbb{C}$ and $\mathbb{M} = \mathbb{R}$. The norm is assumed to be Archimedean. Whenever we need to drop this assumption, this will be clear from the context or mentioned explicitly. The main advantage of the field of real numbers $\mathbb{R}$ as normirujuschee is in its properties of being complete and Archimedean. The completeness ensures the existence of the supremum and infimum of sets of real numbers bounded from above and from below, respectively.

3. Consider a vector space $(\mathbb{K}, E)$ with normed scalar field $\mathbb{K}$. A *semi-additive* and *absolutely homogeneous* function $\| \ \| : E \to \mathbb{R}$ is called a *norm* on E:

$$\|x + y\| \leq \|x\| + \|y\|, \quad \|\alpha x\| = |\alpha| \|x\| \quad (x, y \in E, \ \alpha \in \mathbb{K}).$$

These relations imply that $\|x\| \geq 0$ and $\|0\| = 0$ (the vector zero is not to be confused with the scalar zero). Indeed, we have

$$\|0\| = \|0 \cdot x\| = 0 \cdot \|x\| = 0, \quad 0 = \|x - x\| \leq \|x\| + \|x\| = 2\|x\|.$$

A vector space with a norm make up a *normed space*. A norm is said to be *nondegenerate* and *separating the points* of the vector space if $\|x\| > 0$ whenever $x \neq 0$. A normed space with nondegenerate norm is said to be *separated*.

If E is a normed vector space, a function $\rho : E \times E \to \mathbb{R}$ with values $\rho(x, y) = \|x - y\|$ is called a *metric* on E and the quantity $\rho(x, y)$ is called the *distance* between the points x and y. In particular, $\rho(0, x) = \|x\|$ is the distance from the point 0 to x.

Remark. Definitions of a norm usually include the requirement that the norm be nondegenerate and the normed space be separated. A degenerate norm is called a *seminorm*. A space equipped with a seminorm is said to be *seminormed*. We will follow these conventions unless otherwise specified.

If a norm is not assumed to be nondegenerate, every vector space E can be normed using the zero norm $\|x\| = 0$ $(x \in E)$. Thus, every vector space can formally be considered a normed space.

Example 1. Consider the subspace $\mathcal{F}_0 = \mathcal{F}_0(X, \mathbb{K})$ of finite scalar functions on a set X (with real or complex values). Let $A \subseteq X$. Then the equality

$$\|f\| = \max\{|f(x)| : x \in A\} \quad (f \in \mathcal{F}_0)$$

defines a norm on $\mathcal{F}_0$. The norm is nondegenerate if $A = X$ and degenerate otherwise.

Example 2. Let $E = \prod E_i$ be the product of a finite collection of vector spaces E_i with norms $\| \ \|_i$. The equalities

$$\|x\| = \vee \|x_i\|_i, \quad \|x\|_1 = \sum \|x_i\|_i, \quad \|x_i\|_2 = \left(\sum \|x_i\|_i^2 \right)^{1/2}$$

$(x = (x_i), \ x_i \in E_i)$ define norms on E. If $\| \ \|_i$ are nondegenerate, then so are $\| \ \|$, $\| \ \|_1$, and $\| \ \|_2$.

Example 3. Consider the factor space $\bar{E} = E/N$ of a vector space E over a subspace N. Let $\|\ \|_0$ be a norm in E and let φ be the natural map from E to $\bar{E}$. Then the equality

$$\|\bar{x}\| = \inf\{\|x\|_0 : \varphi(x) = \bar{x}\} \quad (x \in E)$$

defines a norm on $\bar{E}$. This norm is a measure of the distance from the point 0 to linear manifolds $x + N$ in the normed space E.

Exercises. 1) Verify that the norms in examples 1–3 satisfy the definition of a norm. 2) Establish whether $\|x\|_{1/2} = \left(\sum_i \|x_i\|_i^{1/2}\right)^2$ is a norm.

4. Let E be a real or complex normed space and E^* be the conjugate of E. A linear functional $x' : E \to \mathbb{K}$ is said to be *bounded*, if there exists a $\gamma \geq 0$ such that $|x'(x)| \leq \gamma\|x\|$ $(x \in E)$. The bounded functionals form a subspace E' of E^*:

$$\left|(\alpha_1 x_1' + \alpha_2 x_2')(x)\right| \leq (|\alpha_1|\gamma_1 + |\alpha_2|\gamma_2)\|x\|$$

for $|x_1'(x)| \leq \gamma_1\|x\|$ and $|x_2'(x)| \leq \gamma_2\|x\|$.

If a linear functional is bounded, this does not necessarily imply that its range is bounded in $\mathbb{K}$. Indeed, since $x'(\alpha x_0) = \alpha x'(x_0)$, we have $x'(E) = \mathbb{K}$ for $x'(x_0) \neq 0$ for some $x_0 \in E$. It is the set of ratios $x'(x)/\|x\|$ $(\|x\| \neq 0)$ that is bounded. For a bounded linear functional x', the equality

$$\|x'\| = \sup\{|x'(x)|/\|x\| : \|x\| \neq 0\}$$

defines a norm, which turns E' into a normed space. It follows from the definition that

$$\|x'(x)\| \leq \|x'\| \cdot \|x\| \quad (x \in E).$$

Exercise. Prove the preceding statement.

The spaces E^* and E' are called the *algebraic conjugate* and *geometric conjugate* of the normed space E.

Example. If $\dim E < \infty$, then $E' = E^*$. Every linear functional on a finite-dimensional normed space is bounded. In particular, if $E = \mathbb{K}$ and $\dim E = 1$, then $E' \simeq \mathbb{K}$, $x'(x) = \alpha x$ $(x \in \mathbb{K})$ and $\|x'\| = |\alpha| = |x'(1)|$, where α is the coefficient of x'. If $\mathbb{K} = \mathbb{R}$, the coefficient is a measure of the angle between the straight line x' and the constant zero line $0'$.

The following two questions arise in view of the definition of the geometric conjugate space E': *Do nonzero bounded linear functionals exist? If they exist, do they separate the points of E?* The Hahn-Banach theorem gives positive answers to both of these questions.

5. The Hahn-Banach theorem states that any bounded linear functional on a subspace of a normed space can be extended to the whole space so that the norm of the functional is preserved.

First, consider a simple special case where Y is a real normed space and a subspace $X \subseteq Y$ is a *hyperplane*, i.e., there exists a $b \in Y \setminus X$ such that $Y = X + \mathbb{R}b = \{y = x + tb \mid x \in X, t \in \mathbb{R}\}$. Note that $b \neq 0$ since $0 \in X$.

Lemma. *Let $x' : X \to \mathbb{R}$ be a bounded linear functional on a hyperplane X of a real normed space Y. Then there exists a bounded linear functional $y' : Y \to \mathbb{R}$ that is an extension of x' whose norm is equal to the norm of x'.*

Proof. Any linear functional $y^* \in Y^*$ that is an extension of a functional $x' \in X'$ has values

$$y^*(x + tb) = y^*(x) + ty^*(b) = x'(x) + t\beta \quad (x \in X, \ t \in \mathbb{R})$$

and is defined by a real number $y^*(b) = \beta$. The uniqueness of y^* follows from the equivalence of the equalities $x + tb = y + ub$, $(t - u)b = y - x$, $(t - u)b = 0 = y - x$, and $t = u$ for $x, y \in X$ and $t, u \in \mathbb{R}$ for $b \notin X$, $X \cap \{\mathbb{R}b\} = \{0\}$. Since x' is linear, so is y^*. We have

$$y^*((x + tb) + (y + ub)) = y^*((x + y) + (t + u)b) = x'(x + y) + (t + u)\beta$$
$$= x'(x) + x'(y) + t\beta + u\beta = y^*(x + tb) + y^*(y + ub),$$

$$y^*(\alpha(x + tb)) = y^*(\alpha x + \alpha tb)$$
$$= x'(\alpha x) + \alpha t\beta = \alpha x'(x) + \alpha t\beta = \alpha y^*(x + tb) \quad (\alpha \in \mathbb{R}).$$

It is required to choose a real number β such that the corresponding linear functional y^* is bounded and $\|y^*\| = \|x'\|$. It is necessary that the condition $|y^*(y)| \leq \|x'\|\|y\|$ $(y \in Y)$ is satisfied. Note that this condition is equivalent to the inequality $y^*(y) \leq \|x'\|\|y\|$ $(y \in Y)$. Indeed, since $-y^*(y) = y^*(-y) \leq \|x'\|\| - y\| = \|x'\|\|y\|$ $(y \in Y)$, it follows that $|y^*(y)| \leq \|x'\|\|y\|$ $(y \in Y)$. Therefore, the desired number $\beta \in \mathbb{R}$ must satisfy the inequality

$$x'(x) + t\beta \leq \|x'\|\|x + tb\| \quad (x \in X, \ t \in \mathbb{R}). \tag{1}$$

For $t = 0$, inequality (1) is written as $x'(x) \leq \|x'\|\|x\|$ $(x \in X)$, which follows from the definition of the norm of a bounded linear functional x'.

For $t < 0$ and $-t > 0$, we have

$$t^{-1}\|x + tb\| = -\|(-t^{-1})(x + tb)\| = -\|(t^{-1}(x + tb)\| = -\|t^{-1}x + b\|.$$

Therefore, for $t > 0$ inequality (1) is equivalent to the inequality $\beta \leq \|x'\|\|t^{-1}x + b\| - x'(t^{-1}x)$, while for $t < 0$ it is equivalent to the inequality $\beta \geq -\|x'\|\|t^{-1}x + b\| - x'(t^{-1}x)$. Let $u, v \in X$. We can assume that $u = t^{-1}x$ for $t < 0$ and $x = tu \in X$, and $v = t^{-1}x$ for $t > 0$ and $x = tv \in X$. Therefore, the obtained inequalities for β are equivalent:

$$-\|x'\|\|u + b\| - x'(u) \leq \beta \leq \|x'\|\|v + b\| - x'(v) \quad (u, v \in X). \tag{2}$$

The inequalities (2) are equivalent to (1).

A number $\beta \in \mathbb{R}$ satisfying inequalities (2) exists if and only if

$$-\|x'\|\|u + b\| - x'(u) \leq \|x'\|\|v + b\| - x'(v) \quad (u, v \in X). \tag{3}$$

Inequality (3) follows from the relations $x'(v) - x'(u) = x'(v - u) \leq \|x'\| \cdot \|v - u\| = \|x'\| \cdot \|(v + b) - (u + b)\| \leq \|x'\| \cdot \|v + b\| + \|x'\| \cdot \|u + b\|$. Consider the nonempty sets $A = \{-\|x'\|\|u + b\| - x'(u) \mid u \in X\} \subseteq \mathbb{R}$, $C = \{\|x'\|\|v + b\| - x'(v) \mid v \in X\} \subseteq \mathbb{R}$. Inequality (3) implies that A is bounded from above by any number in C (for example, $\|x'\|\|b\|$), while C is bounded from below by any number in A (for example, $-\|x'\|\|b\|$). Since the set of real numbers $\mathbb{R}$ is complete, there exist $\alpha = \sup A$ and $\gamma = \inf C$. From inequality (3), it follows that $\alpha \leq \gamma$. Therefore, inequalities (2) are equivalent to the inequality $\alpha \leq \beta \leq \gamma$. Hence, inequality (1) holds for any β such that $\alpha \leq \beta \leq \gamma$. In particular, (1) holds for $\beta = \alpha$ and for $\beta = \gamma$.

Take a number $\beta \in [\alpha, \gamma]$ and the corresponding linear functional y^* with values $y^*(x + tb) = x'(x) + t\beta$ ($t \in \mathbb{R}$). By inequality (1) y^* is bounded and $\|y^*\| \leq \|x'\|$. Since y^* is an extension of x', the reverse inequality $\|y^*\| \geq \|x'\|$ is also true. Indeed, we have $\|y^*\| = \sup\{|y^*(y)|/\|y\| \mid y \in Y, y \neq 0\} \geq \sup\{|y^*(x)|/\|x\| \mid x \in X, x \neq 0\} = \sup\{|x'(x)|/\|x\| \mid x \in X\} = \|x'\|$. Hence $\|y^*\| = \|x'\|$ and $y' = y^*$ is the desired functional .   □

6. We now consider a more general case where the normed space Y is real or complex and X is an arbitrary subspace of Y.

The Hahn-Banach Theorem. *Let $x' : X \to \mathbb{K}$ be a bounded linear functional on a subspace X of a normed space Y with scalar field $\mathbb{K}$. Then there exists a bounded linear functional $y' : Y \to \mathbb{K}$ that is an extension of x', its norm being equal to the norm of x'.*

Proof. 1) If $\mathbb{K} = \mathbb{R}$, then the Hahn-Banach theorem easily follows from the preceding lemma and Zorn's lemma.

Let $\mathcal{M}$ be a set of all bounded linear functionals $z' : Z \to \mathbb{R}$ on subspaces $Z \subseteq Y$ that are extensions of $x' : X \to \mathbb{R}$ and such that $\|z'\| = \|x'\|$. We introduce an inclusion order on $\mathcal{M}$ as follows: $z'_2 \supseteq z'_1$ means that $z'_2 \in \mathcal{M}$ is an extension of $z'_1 \in \mathcal{M}$. The ordered set $(\mathcal{M}, \supseteq)$ is inductive since any chain $\mathcal{L} \in \mathcal{M}$ has a majorant $l' = \cup\mathcal{L} \in \mathcal{M}$.

The correspondence l' is many-to-one. Indeed, let $z \in L = \text{Dom}(l') = \cup \text{Dom}(z')$ $(z' \in \mathcal{L})$. Then there is a $z'_1 \in \mathcal{L}$ such that $z \in Z_1 = \text{Dom}(z'_1)$ and $l'(z) = z'_1(z)$. If $z \in Z_2 = \text{Dom}(z'_2)$ for $z'_2 \in \mathcal{L}$, then $l'(z) = z'_2(z)$. Since $\mathcal{L}$ is a chain, it follows that $z'_2 \supseteq z'_1$ or $z'_1 \supseteq z'_2$. In any case, $z'_1(z) = z'_2(z)$ for $z \in Z_1 \cap Z_2$.

The domain of definition L of the functional l' is a subspace of Y. Indeed, let $\alpha_1, \alpha_2 \in \mathbb{R}$ and $z_1, z_2 \in L$. Then there are $z'_i \in \mathcal{L}$ $(i = 1, 2)$ such that $z_i \in Z_i = \text{Dom}(z'_i)$. Since $z'_1 \supseteq z'_2$ or $z'_2 \supseteq z'_1$, we have $Z_1 \cup Z_2 = Z_1 \subseteq L$ or $Z_1 \cup Z_2 = Z_2 \subseteq L$ and, in any case, $\alpha_1 z_1 + \alpha_2 z_2 \in L$. Furthermore, $l'(\alpha_1 z_1 + \alpha_2 z_2) = z'_i(\alpha_1 z_1 + \alpha_2 z_2) = \alpha_1 z'_i(z_1) + \alpha_2 z'_i(z_2) = \alpha_1 l'(z_1) + \alpha_2 l'(z_2)$, where i is the index of the extension functional: $i = 1$ for $z'_1 \supseteq z'_2$ and $i = 2$ for $z'_2 \supseteq z'_1$. Therefore, the functional $l' : L \to \mathbb{R}$ is linear.

The functional $l' : L \to \mathbb{R}$ is bounded, it is an extension of $x' : X \to \mathbb{R}$, and $\|l'\| = \|x'\|$. Indeed, $l'(x) = z'(x) = x'(x)$ $(x \in X, z' \in \mathcal{L})$. Let $z \in L$. Then there is a $z' \in \mathcal{L}$ such that $z \in Z = \text{Dom}(z')$ and $|l'(z)| = |z'(z)| \leq \|x'\|\|z\|$. Hence, l' is bounded and $\|l'\| \leq \|x'\|$. Since l' is an extension of x', it follows that $\|l'\| \geq \|x'\|$ (which was shown in the proof of the preceding lemma). Therefore, $\|l'\| = \|x'\|$ and $l' \in \mathcal{M}$. Clearly, l' is a majorant of $\mathcal{L}$, since the definition of l' implies that l' is an extension of every functional $z' \in \mathcal{L}$. We have proved that $(\mathcal{M}, \supseteq)$ is an inductive ordered set.

By Zorn's lemma, there is a maximal element $m' : M \to \mathbb{R}$ in $(\mathcal{M}, \supseteq)$. This element is the desired extension. Indeed, since $m' \in \mathcal{M}$, m' is a bounded linear functional on the subspace $M \subseteq Y$ such that m' is an extension of x' and $\|m'\| = \|x'\|$. It remains to show that $M = Y$. Let $M \neq Y$ and $b \in Y \setminus M$. By the lemma of part 5 of this subsection, there exists a bounded linear functional $n' : N \to \mathbb{R}$ on the subspace $N = M + \mathbb{R}b \subseteq Y$ such that n' is an extension of $m' : M \to \mathbb{R}$ and $\|n'\| = \|m'\|$. Since $m' \supseteq x'$ and $\|m'\| = \|x'\|$, we have $n' \supseteq x'$ and $\|n'\| = \|x'\|$. Hence, $n' \in \mathcal{M}$ and $n' \supset m'$, which contradicts the maximality of m'. Therefore, $M = Y$.

Thus, there exists a bounded linear functional $y' = m' : Y \to \mathbb{R}$ such that y' is an extension of $x' : X \to \mathbb{R}$ and $\|y'\| = \|x'\|$. The theorem is proved in the case of real spaces.

2) Let $\mathbb{K} = \mathbb{C}$. Any complex bounded linear functional $x' : X \to \mathbb{C}$ on the subspace $X \subseteq Y$ defines a real linear functional $a' : X \to \mathbb{R}$ with values $a'(x) = \text{Re}(x'(x))$ $(x \in X)$ and norm $\|a'\| \leq \|x'\|$. Indeed, a' is linear because so is x' and $\text{Re} : \mathbb{C} \to \mathbb{R}$, while the inequality $\|a'\| \leq \|x'\|$ follows from the inequality $|\text{Re}(x'(x))| \leq |x'(x)|$ $(x \in X)$.

As was proved before, for $\mathbb{K} = \mathbb{R}$ there exists a bounded linear functional $b' : Y \to \mathbb{R}$ on the real space Y such that b' is an extension of a' and $\|b'\| = \|a'\| \leq \|x'\|$. Consider a functional $y' : Y \to \mathbb{C}$ with values $y'(y) = b'(y) - ib'(iy)$ $(y \in Y)$. It is a complex linear functional on the complex subspace Y. Indeed, for all $\alpha \in \mathbb{R}$ and $y, z \in Y$ we have

$$y'(y + z) = b'(y + z) - ib'(iy + iz) = b'(y) + b'(z) - i(b'(iy) + b'(iz))$$
$$= (b'(y) - ib'(iy)) + (b'(z) - ib'(iz)) = y'(y) + y'(z),$$

$$y'(\alpha y) = b'(\alpha y) - ib'(i(\alpha y)) = \alpha b'(y) - i(\alpha b'(iy)) = \alpha y'(y),$$

$$y'(iy) = b'(iy) - ib'(i(iy)) = b'(iy) - ib'(-y) = b'(iy) + ib'(y) = iy'(y).$$

Hence $y'((\alpha + \beta i)y) = y'(\alpha y + \beta iy) = \alpha y'(y) + \beta iy'(y) = (\alpha + \beta i)y'(y)$ for all $\alpha, \beta \in \mathbb{R}$ and $y \in Y$. Since b' is an extension of $a' = \text{Re}(x')$, it follows that y' is an extension of x': $y'(x) = b'(x) - ib'(ix) = a'(x) - ia'(ix) = \text{Re}(x'(x)) - i\,\text{Re}(x'(ix)) = \text{Re}(x'(x)) - i\,\text{Re}(ix'(x)) = \text{Re}(x'(x)) - i(-\text{Im}\,x'(x))) = \text{Re}(x'(x)) + (\text{Im}\,x'(x))i = x'(x)$ for all $x \in X$.

We now prove that $|y'(y)| \leq \|x'\|\|y\|$. Assume the contrary. Then there is a $y_0 \in Y$ such that $|y'(y_0)| > \|x'\|\|y_0\|$. Since $c = y'(y_0) \neq 0$, there is a unique polar representation $c = \rho u$ with $\rho = |c| > 0$ and $u = \sigma + \tau i$ $(\sigma, \tau \in \mathbb{R})$, $|u| = \sigma^2 + \tau^2 = 1$. Let $z_0 = u^{-1}y_0$. Then $\|y_0\| = \|uz_0\| = |u|\|z_0\| = \|z_0\|$ and $b'(z_0) = \text{Re}(y'(z_0)) = \text{Re}(y'(u^{-1}y_0)) = \text{Re}(u^{-1}y'(y_0)) = \text{Re}(u^{-1}c) = \rho = |y'(y_0)| > \|x'\|\|y_0\| = \|x'\|\|z_0\|$. This contradicts the inequality $\|b'\| \leq \|x'\|$. Therefore, $|y'(y)| \leq \|x'\|\|y\|$ for all $y \in Y$. As a result, the linear functional y' is bounded and $\|y'\| \leq \|x'\|$. Since y' is an extension of x', it follows that $\|y'\| \geq \|x'\|$ and $\|y'\| = \|x'\|$.

Thus, there exists a bounded linear functional $y' : Y \to \mathbb{C}$ such that y' is an extension of $x' : X \to \mathbb{C}$ and $\|y'\| = \|x'\|$. The assertion is now proved for $\mathbb{K} = \mathbb{C}$, which completes the proof of the theorem. $\square$

7. The Hahn-Banach theorem has many useful corollaries. In particular, it provides the answers to the questions formulated in part 4 of this subsection. Throughout this part, we will assume the normed space Y under consideration to be separated.

Proposition. *Let $a \in Y$ and $a \neq 0$. Then there exists a $y' \in Y'$ such that $\|y'\| = 1$ and $y'(a) = \|a\|$.*

Proof. Let $x' : X \to \mathbb{K}$ be a functional on the subspace $X = \mathbb{K}a \subseteq Y$ with values $x'(ta) = t\|a\|$ ($t \in \mathbb{K}$). Obviously, x' is linear, bounded, and $\|x'\| \leq 1$. Since $x'(\|a\|^{-1}a) = \|a\|^{-1}\|a\| = 1$, we have $\|x'\| = 1$. By the Hahn-Banach theorem, there exists a functional $y' \in Y'$ that is an extension of x' whose norm is equal to the norm of x'. Consequently, $y'(a) = x'(a) = \|a\|$ and $\|y'\| = \|x'\| = 1$. $\square$

Example. Assume that $\mathbb{K} = \mathbb{R}$, $Y = \mathbb{R} \times \mathbb{R} \times \{0\}$ is a horizontal plane in the space $\mathbb{R}^3 = \mathbb{R} \times \mathbb{R} \times \mathbb{R}$, $a = (\alpha, 0, 0)$, and $\alpha > 0$. Then $X = X_1 = \mathbb{R} \times \{0\} \times \{0\}$ is the first horizontal axis, the graph of x' is a bisector of the angle between X and the vertical axis $X_3 = \{0\} \times \{0\} \times \mathbb{R}$, and the graph of y' — is the plane passing through this bisector and the second horizontal axis $X_2 = \{0\} \times \mathbb{R} \times \{0\}$.

This example also helps to explain the meaning of the Hahn-Banach theorem. We now give three corollaries of the preceding proposition.

Corollary 1. *There exist nonzero bounded linear functionals on any nonzero normed vector space.*

Proof. Let $Y \neq \{0\}$ be a normed vector space. Let $a \in Y$ and $a \neq 0$. By the preceding proposition, there exists a functional $y' \in Y'$ such that $y'(a) = \|a\| \neq 0$. $\square$

Corollary 2. *Bounded linear functionals on a normed space separate its points.*

Proof. Assume that Y is a normed vector space, $a, b \in Y$, $a \neq b$, and $a - b = c \neq 0$. By the preceding proposition, there exists a functional $z' \in Y'$ such that $z'(c) = \|c\| \neq 0$. Then $z'(a) - z'(b) = z'(c) = \|c\| \neq 0$ and $z'(a) \neq z'(b)$. $\square$

Thus, the geometric conjugate space Y' is a coordinate system of the real or complex separated normed space Y.

Corollary 3. *For any subspace Z of a space Y and any point $b \notin Z$, there exists a bounded linear functional y' on Y such that $y'(z) = 0$ for all $z \in Z$ and $y'(b) = 1$.*

Proof. Consider the subspace X of Y that consists of vectors of the form $x = z + tb$, where $z \in Z$ and t is a number. The equality $x'(z + tb) = t$ defines a bounded linear functional x' on X. Clearly, $x'(z) = 0$ for all $z \in Z$ and $x'(b) = 1$. By the Hahn-Banach theorem, there exists a bounded linear functional y' on Y that is an extension of x' and has all the desired properties.   □

The above statements have a simple geometric interpretation.

8. The set $N = \{z \in E \mid \|z\| = 0\}$ is a subspace of a normed space $(\mathbb{K}, E)$:

$$\|\alpha_1 z_1 + \alpha_2 z_2\| \le |\alpha_1| \|z_1\| + |\alpha_2| \|z_2\| = 0 \quad (\alpha_1, \alpha_2 \in \mathbb{K}; \ z_1, z_2 \in E).$$

Consider the factor space $\bar{E} = E/N$. The equality

$$\|\bar{x}\| = \|x\| \quad (\bar{x} = x + N, \ x \in E)$$

defines a norm on $\bar{E}$. The norm is well defined:

$$\|x\| \le \|(x + z) - z\| \le \|x + z\| + \|z\| = \|x + z\| \le \|x\| + \|z\| = \|x\|;$$

therefore, $\|x + z\| = \|x\|$ for all $z \in N$. This definition is equivalent to the definition in Example 3 of part 3 in this subsection.

Thus, factorization provides a way to turn a nonseparated normed space into a separated one. For this reason, usually only separated normed spaces are studied in pure mathematics. In applications, transition to a more complex form of vectors is not always convenient.

If $x - y = z \in N$, then the vectors x and y will be called *equivalent*, written $x \approx y$. By definition, such proximity means that $\|x - y\| = 0$. In particular, $z \approx 0$ implies that $z \in N$. The proposition of part 7 and Corollaries 1 and 2 can be applied to the separated factor space $\bar{E}$ and its conjugate $\bar{E}'$. The results can also be used for the nonseparated space E after replacing the equality relation with the above equivalence relation. This process requires special care to be taken.

2.2.6. Euclidean spaces

Euclidean spaces are vector spaces with scalar product of a special form. The scalar product defines a norm and serves as a measure of linear and angular characteristics of vectors.

1. Let $(\mathbb{K}, E)$ be a vector space with normed scalar field $\mathbb{K}$. Assume that the operation of conjugation $* : \mathbb{K} \to \mathbb{K}$ is defined, which is an *involutive isomorphism*, i.e.,

$$\alpha^{**} = \alpha, \quad (\alpha + \beta)^* = \alpha^* + \beta^*, \quad (\alpha\beta)^* = \beta^*\alpha^*, \quad |\alpha^*| = |\alpha|$$

$(\alpha, \beta \in \mathbb{K})$. Elements $\alpha \in \mathbb{K}$ such that $\alpha^* = \alpha$ are called *self-adjoint*. In particular, $0^* = 0$, $1^* = 1$.

Examples. 1) The identity map $\alpha^* = \alpha$ ($\alpha \in \mathbb{R}$) is a conjugation in $\mathbb{K} = \mathbb{R}$. 2) The equality $\gamma^* = \alpha - \beta i$ ($\gamma = \alpha + \beta i \in \mathbb{C}$) defines a conjugation in $\mathbb{K} = \mathbb{C}$. The set of self-adjoint elements under this conjugation coincides with $\mathbb{R}$.

A function $p : E \times E \to \mathbb{K}$ is called a *Hermitian form* if it is linear in the second argument and commutes with respect to conjugation. We will use multiplicative notation for the values of p: $p(x, y) = xy$ ($x, y \in E$). The following equalities hold:

$$x(\beta y) = \beta(xy), \quad x(y + z) = xy + xz, \quad (xy)^* = yx,$$
$$(\alpha x)y = \alpha^*(xy), \quad (x + y)z = xz + yz \quad (x, y, z \in E; \ \alpha, \beta \in \mathbb{K}).$$

Exercise. Prove these inequalities.

If $\mathbb{K} = \mathbb{R}$, then $(xy)^* = yx$ means that the Hermitian form p is symmetric, i.e., $xy = yx$ for all $x, y \in E$.

Every Hermitian form $p : E \times E \to \mathbb{K}$ defines a quadratic form $q : E \to \mathbb{K}$ such that $q(x) = p(x, x) = xx = x^2$ ($x \in E$). Note that $q(x) = xx = (xx)^* = q(x)^*$ ($x \in E$). The values of the quadratic form q are self-adjoint elements of the field $\mathbb{K}$. If $\mathbb{K} = \mathbb{C}$, then $q(x) \in \mathbb{R}$, i.e., q is real-valued.

2. Consider a vector space $(\mathbb{K}, E)$ with $\mathbb{K} = \mathbb{R}$ or $\mathbb{K} = \mathbb{C}$. The Hermitian form $p : E \times E \to \mathbb{K}$ that defines the positive quadratic form $q(x) = x^2$ for all $x \in E$ is called the *(Euclidean) scalar product* on E. The quadratic form q and the corresponding scalar product p are said to be *nondegenerate* if $q(x) = x^2 > 0$ for all nonzero vectors $x \in E$. In this case $x^2 = 0 \Leftrightarrow x = 0$.

A vector space with Euclidean scalar product is called a *Euclidean* space. It is usually assumed that the scalar product is nondegenerate. The equality $\|x\| = (xx)^{1/2}$ ($x \in E$) defines a norm on the Euclidean space E. This norm is called the *Euclidean norm*.

The connection between the scalar product and the Euclidean norm is partially illustrated by the following inequality.

The Cauchy inequality. $|xy| \leq \|x\| \cdot \|y\|$ $(x, y \in E)$.

Proof. We have

$$0 \leq (\lambda x + y)(\lambda x + y) = |\lambda|^2 \|x\|^2 + \|y\|^2 + \lambda^*(xy) + \lambda(xy)^*$$

for all $x, y \in E$ and $\lambda \in \mathbb{K}$. Let $\|x\| \neq 0$. Then for $\lambda = -xy/\|x\|^2$ we obtain $0 \leq \|y\|^2 - |xy|/\|x\|^2$, which is equivalent to the desired inequality. By symmetry, the same is true for $\|y\| \neq 0$.

Let $\|x\| = \|y\| = 0$. Then for $\lambda = -xy$ we obtain $0 \leq -2|xy|^2$, which implies $|xy| = 0 = \|x\|\|y\|$. $\square$

Corollary. For all $x, y \in E$, $\|x + y\| \leq \|x\| + \|y\|$.

Proof. We have

$$\|x + y\|^2 = \|x\|^2 + \|y\|^2 + xy + yx \leq \|x\|^2 + \|y\|^2 + 2|xy| \leq$$
$$\leq \|x\|^2 + \|y\|^2 + 2\|x\| \cdot \|y\| = (\|x\| + \|y\|)^2,$$

which immediately yields the desired inequality. $\square$

Thus, the function $\| \ \| : E \to \mathbb{K}$ with values $\|x\| = (xx)^{1/2}$ $(x \in E)$ is semi-additive. It is also absolutely homogeneous, which follows from the equalities $\|\alpha x\|^2 = \alpha x \cdot \alpha x = \alpha^* \alpha \cdot xx = |\alpha|^2 \|x\|^2$ $(x \in E, \alpha \in \mathbb{K})$. This proves that the function $\| \ \|$ is, indeed, a norm on the Euclidean space E, and its name has been justified.

The Euclidean norm separates the points of the space E if and only if the corresponding scalar product is nondegenerate.

3. The geometry of a Euclidean space is determined by the properties of the scalar product. Throughout this part of the present subsection, x, y, and z are vectors of a Euclidean space E.

The triangle identity:

$$\|x + y\|^2 = \|x\|^2 + \|y\|^2 + (xy + yx).$$

The parallelogram law:

$$\|x + y\|^2 + \|x - y\|^2 = 2(\|x\|^2 + \|y\|^2).$$

The polarization identity:

$$\|x+y\|^2 - \|x-y\|^2 = 2(xy+yx).$$

The triangle identity follows from the equality $\|x+y\|^2 = (x+y)(x+y)$. The parallelogram law is obtained by adding the triangle identity for x and y to the same identity written for x and $-y$. The polarization identity is obtained by subtracting the former from the latter. These identities have clear geometric interpretation and express the relations between the lengths of the sides, diagonals, and projections in a parallelogram.

If $\mathbb{K} = \mathbb{R}$, we have $xy + yx = 2xy$ and the polarization identity becomes

$$xy = 4^{-1}(\|x+y\|^2 - \|x-y\|^2).$$

If $\mathbb{K} = \mathbb{C}$, we have $xy + yx = 2\,\mathrm{Re}(xy)$ and $\mathrm{Im}(xy) = -\,\mathrm{Re}(x \cdot iy)$, which implies

$$xy = 4^{-1}[(\|x+y\|^2 - \|x-y\|^2) - i(\|x+iy\|^2 - \|x-iy\|^2)].$$

If $xy = 0$, then the vectors x and y are said to be *orthogonal*, which is written as $x \perp y$. Since $yx = (xy)^*$, $xy = 0$ implies $yx = 0$ and the orthogonality relation is symmetric: $x \perp y \Leftrightarrow y \perp x$. The zero vector is orthogonal to all vectors in the space. It is the only vector with such a property in separated spaces.

The Pythagorean theorem. *If $x \perp y$, then $\|x+y\|^2 = \|x\|^2 + \|y\|^2$.*

If $\mathbb{K} = \mathbb{R}$, then the triangle inequality also implies the converse statement, i.e., if $\|x+y\|^2 = \|x\|^2 + \|y\|^2$ then $x \perp y$.

Exercises. 1) Find out whether the preceding statement true if $\mathbb{K} = \mathbb{C}$.
2) Prove the equivalent *tetrahedron identity* and *Apollonius' identity*:

$$\|(z-x)+(z-y)\|^2 + \|(z-x)-(z-y)\|^2 = 2(\|z-x\|^2 + \|z-y\|^2),$$
$$2^{-1}\|x-y\|^2 + 2\|z - 2^{-1}(x+y)\|^2 = \|z-x\|^2 + \|z-y\|^2.$$

4. The parallelogram law is a distinctive feature of Euclidean spaces as compared to other normed spaces.

Proposition. *A scalar product that defines a norm exists if and only if the parallelogram law holds for the norm.*

Proof. Let $(\mathbb{K}, E)$ be a normed vector space with norm $\| \ \|$.
1) Let $\mathbb{K} = \mathbb{R}$ and

$$p(x, y) = xy = 4^{-1}(\|x + y\|^2 - \|x - y\|^2) \quad (x, y \in E).$$

We now prove that this equality defines the required scalar product p : $E \times E \rightarrow \mathbb{R}$. It follows from the definition that $xx \geq 0$, $xy = yx$, and $\|x\| = (xx)^{1/2}$ for all $x, y \in E$. We need to prove that the function $p(x, \cdot) : E \rightarrow \mathbb{R}$ is linear. The additivity of $p(x, \cdot)$ follows from the equality

$$\|a + b + c\|^2 + \|a + b - c\|^2 = 2(\|a + b\|^2 + \|c\|^2) \quad (a, b, c \in E).$$

We now apply this equality to $a = x$, $b = y$, $c = z$ and $a = x$, $b = -z$, $c = y$. and subtract the second of the two obtained identities from the first one. As a result, we have

$$\|x + y + z\|^2 - \|x - y - z\|^2 = 2(\|x + y\|^2 - \|x - z\|^2 + \|z\|^2 - \|y\|^2),$$
$$\|x + z + y\|^2 - \|x - z - y\|^2 = 2(\|x + z\|^2 - \|x - y\|^2 + \|y\|^2 - \|z\|^2),$$

$$\|x + z + y\|^2 - \|x - y - z\|^2$$
$$= \|x + y\|^2 - \|x - y\|^2 + \|x + z\|^2 - \|x - z\|^2$$

and $x(y + z) = xy + xz$.

We first establish the homogeneity of $p(x, \cdot)$ in the case of rational numbers. Obviously, $x(0y) = 0(xy)$ and $x(1y) = 1(xy)$. Since $p(x, \cdot)$ is additive, the equality $x(ny) = n(xy)$ for all $n \in \mathbb{N}$ follows by induction. It follows from the definition that $x(-y) = -(xy)$ and therefore $x(my) = m(xy)$ for all $m \in \mathbb{Z}$. Hence $x \cdot (m/n)y = m(x \cdot (1/n)y) = (m/n)(x \cdot n(1/n)y) = (m/n)(xy)$ for all $m \in \mathbb{Z}$ and $n \in \mathbb{N}$, i.e., $x \cdot qy = q \cdot xy$ for all $q \in \mathbb{Q}$.

Let $\alpha \in \mathbb{R}$. Since $\mathbb{R}$ is an Archimedean field, for any $k \in \mathbb{N}$ there exists a $q = q(k) \in \mathbb{Q}$ such that $|\alpha - q| \leq 1/k$. Consequently,

$$x \cdot \alpha y = x \cdot qy + x \cdot (\alpha - q)y = q \cdot xy + x \cdot (\alpha - q)y = \alpha \cdot xy + \Delta(x, y),$$
$$\Delta(x, y) = x \cdot (\alpha - q)y - (\alpha - q) \cdot xy.$$

Using the definition of xy and the properties of $\| \ \|$, we obtain $|(\alpha - q) \cdot xy| = 4^{-1}|\alpha - q| \cdot |\|x+y\|^2 - \|x-y\|^2| = 4^{-1}|\alpha - q| \cdot |\|x+y\| + \|x-y\|| \cdot |\|x+y\| - \|x-y\|| \leq 4^{-1}|\alpha - q| \cdot 2(\|x\| + \|y\|) \cdot \|(x+y) - (x-y)\| \leq (1/k)(\|x\| + \|y\|)\|y\| = \gamma/k$, $|x \cdot (\alpha - q)y| = 4^{-1}|\|x + (\alpha - q)y\|^2 - \|x - (\alpha - q)y\|^2| \leq (\|x\| + |\alpha - q|\|y\|) \cdot$

$|\alpha - q|\|y\| \le \gamma/k, |\Delta(x, y)| \le 2\gamma/k$ for $\gamma = (\|x\| + \|y\|)\|y\|$ and for all $k \in \mathbb{N}$. Hence $\Delta(x, y) = 0$ and $x \cdot \alpha y = \alpha \cdot xy$.

2) Let $\mathbb{K} = \mathbb{C}$ and $x \times y = xy - i(x \cdot iy)$ $(x, y \in E)$, where, as before, xy and $x \cdot iy$ are defined by the equality

$$ab = 4^{-1}(\|a + b\|^2 - \|a - b\|^2) \quad (a, b \in E)$$

(with $a = x$ and $b = y$ for the former and $a = x$ and $b = iy$ for the latter). We have $x \times x = \|x\|^2 - 4i(\|x + ix\|^2 - \|x - ix\|^2) = \|x\|^2 - 4^{-1}i(|1 + i|^2 - |1 - i|^2)\|x\|^2 = \|x\|^2$ because $|1 + i|^2 = 2 = |1 - i|^2$. Then $x \times x \ge 0$ and $\|x\| = (x \times x)^{1/2}$ $(x \in E)$. From the equalities proved in part 1, it follows that

$$x \times (y + z) = x(y + z) - i(x \cdot i(y + z))$$
$$= xy + xz - i(x \cdot iy) - i(x \cdot iz) = x \times y + x \times z,$$

$$x \times (iy) = x \cdot iy - i(x \cdot i^2 y) = x \cdot iy + i \cdot xy = i(xy - i(x \cdot iy)) = i(x \times y)$$

$(x, y, z \in E)$. Then, using the statements proved for the case of real numbers, we obtain $x \times (\beta iy) = \beta(x \times (iy))$. By additivity, together with $x \times (\alpha y) = \alpha(x \times y)$ this yields $x \times (\gamma y) = \gamma(x \times y)$ for $\gamma = \alpha + \beta i$. It follows that the function $y \to x \times y$ is linear.

Furthermore, we have $iy \cdot x = 4^{-1}(\|iy + x\|^2 - \|iy - x\|^2) = 4^{-1}(\|iy - i^2 x\|^2 - \|iy + i^2 x\|^2) = 4^{-1}(\|y - ix\|^2 - \|iy + ix\|^2) = -(y \cdot ix)$. As a result,

$$(x \times y)^* = (xy)^* - i^*(x \cdot iy)^* = yx + i(iy \cdot x) = yx - i(y \cdot ix) = y \times x,$$

and $(x, y) \to x \times y$ is the required scalar product. $\square$

Remark. Different designations for the scalar product in the cases $\mathbb{K} = \mathbb{C}$ and $\mathbb{K} = \mathbb{R}$ were only necessary within the proof for the sake of convenience. In what follows, as a rule, we will use the general designation xy. In some situations, we will also write $\langle x, y \rangle$ or $\langle \cdot, \cdot \rangle$.

5. Let $(\mathbb{K}, E)$ be a Euclidean space. A collection of vectors $x(j) \in E$ is said to be *orthogonal*, if $x(j) \perp x(k)$ whenever $j \ne k$. In particular, any collection whose index set is empty or consists of one element is orthogonal because there are no distinct indices. An orthogonal collection $(x(j))$ is said to be *orthonormal*, if $\|x(j)\| = 1$ for every j.

Proposition. *Any orthogonal collection of nonzero vectors in a separated Euclidean space is linearly independent.*

Proof. Let a collection $x(j) \in E$ be orthogonal and $\sum \alpha(j)x(j) = 0$ for a finite collection $\alpha(j) \in \mathbb{K}$. Then $\alpha(k) \cdot x(k) = \sum \alpha(j)x(j) \cdot x(k) = 0 \cdot x(k) = 0$ for all k. If $x(k) \neq 0$ and the scalar product is nondegenerate, then $x(k)x(k) \neq 0$ and $\alpha(k) = 0$. □

The converse is not true.

Example 1. Let $\mathbb{K} = \mathbb{R}$, $E = \mathbb{R} \times \mathbb{R}$, and $cz = \alpha\xi + \beta\eta$ for any $c = (\alpha, \beta)$ and $z = (\xi, \eta) \in E$. Then the vectors $c = (0, 1)$ and $z = (1, 1)$ are linearly independent, but not orthogonal.

In the general case, a vector space with a scalar product may contain vectors that are orthogonal and linearly independent at the same time.

Example 2. Let $\mathbb{K} = \mathbb{C}$, $E = \mathbb{C} \times \mathbb{C}$, and $(a, b) \cdot (x, y) = ax + by$ for any (a, b) and (x, y) in E. Then the vectors $(1, i)$ and $(2, 2i)$ are orthogonal and linearly dependent: $(1, i) \cdot (2, 2i) = 2 + 2i^2 = 2 - 2 = 0$ and $2(1, i) - (2, 2i) = 0$. The scalar product in this example is non-Euclidean and degenerate, which follows from the equalities $(i, 0) \cdot (i, 0) = i^2 + 0 = -1$, $(1, i) \cdot (1, i) = 1 + i^2 = 1 - 1 = 0$.

6. Let $(\mathbb{K}, E)$ be a real or complex vector space and let a collection $B = \{e_j \mid j \in J\}$ be its basis. We define a scalar product in E by the formula

$$xy = \sum \xi_j^* \eta_j \quad \left(x = \sum \xi_j e_j, \, y = \sum \eta_j e_j \in E \right).$$

The coordinate collections ξ_j and η_j are finite and thus there are finitely many nonzero terms in the sums. If $\mathbb{K} = \mathbb{R}$, then $\xi_j^* = \xi_j$ and $\xi_j^* \eta_j = \xi_j \eta_j$.

This scalar product is well defined, which follows from the equalities $x(y + z) = \sum \xi_j^* (\eta_j + \zeta_j) = \sum \xi_j^* \eta_j + \sum \xi_j^* \zeta_j = xy + xz$, $x(\beta y) = \sum \xi_j^* (\beta \eta_j) = \beta \sum \xi_j^* \eta_j = \beta \cdot xy$, $xx = \sum \xi_j^* \xi_j = \sum |\xi_j|^2 > 0$, $yx = \sum \eta_j^* \xi_j = \sum (\eta_j \xi_j^*)^* = \left(\sum \xi_j^* \eta_j \right)^* = (xy)^*$ $(x = \sum \xi_j e_j, \, y = \sum \eta_j e_j, \, z = \sum \zeta_j e_j \in E, \, \beta \in \mathbb{K})$. The following equality defines a norm on E:

$$\|x\| = (xx)^{1/2} = \left(\sum |\xi_j|^2 \right)^{1/2}.$$

If $\mathbb{K} = \mathbb{R}$, then $|\xi_j|^2 = \xi_j^2$. The following statement is a generalization of the Pythagorean theorem.

Parseval's identity. $\|x\|^2 = \sum |\xi_j|^2$.

We emphasize that $e_i e_j = \delta_{ij}$. The coordinates $\xi_j = e_j x$ of a vector $x = \sum \xi_j e_j$ with respect to the orthonormal basis (e_j) are called the *Fourier coefficients* with respect to this basis. Let $K \subseteq J$ be an index set and $P_K : E \to E$ be a linear operator such that $P_K e_k = e_k$ $(k \in K)$ and $P_K e_j = 0$ $(j \notin K)$. By definition,

$$P_K x = \sum_{k \in K} \xi_k e_k \quad \left(x = \sum \xi_j e_j \right).$$

The operator P_K is a projection of the space E onto the subspace $E_K = \{x = \sum \xi_j e_j \mid \xi_j = 0 \ (j \notin K)\}$, since $P_K^2 = P_K$ and $P_K x = x$ $(x \in E_K)$. Parseval's identity implies that

$$\|P_K x\|^2 = \sum_{k \in K} |\xi_k|^2 \quad (x = \sum \xi_j e_j \in E).$$

With the above notation, the following equality holds.

The Fourier identity. $\|x\|^2 = \|P_K x\|^2 + \|x - P_K x\|^2$.

Proof. Let $y = P_K x$ and $z = x - P_K x$. Then $x = y + z$ and $y \perp z$. Indeed, we have $yz = \left(\sum_{k \in K} \xi_k e_k \right) \left(\sum_{j \notin K} \xi_j e_j \right) = \sum_{k \in K, j \notin K} \xi_k^* \xi_j e_k e_j = 0$. Applying the Pythagorean theorem yields the required identity. $\square$

Bessel's inequality. $\|P_K x\|^2 \leq \|x\|^2$.
This inequality follows from the Fourier identity.

Remark. Let $J = K \cup L$ and $L = J \setminus K$. In addition to P_K, we introduce the linear operator $P_L : E \to E$ that projects E onto the subspace $E_L = \{x = \sum \xi_j e_j \mid \xi_j = 0 \ (j \notin L)\}$, that complements E_K with respect to E relative to the basis B. Let $I : E \to E$ denote the identity transformation of E, and let $O : E \to E$ be the map with constant value 0. It follows from the definitions that

$$I = P_K + P_L, \quad P_K P_L = O.$$

The Fourier identity is equivalent to the equality

$$\|x\|^2 = \|P_K x\|^2 + \|P_L x\|^2 \quad (x \in E, \ K \subseteq J, \ L = J \setminus K).$$

In addition, we have $I = P_J$ and $O = P_O$ (if the index set is empty, then the range of projection is the zero subspace). These projections are said to be *orthogonal*. The convenience of application of projections makes them a widely used instrument.

7. We now describe orthogonal projections in detail. Let E be a real or complex Euclidean space and let A be its subspace. We will denote by

$$\rho(z, A) = \inf\{\|z - x\| \mid x \in A\}$$

the distance from a point $z \in E$ to the subspace A. A point $a = a(z) \in A$ such that $\|z - a\| = \rho(z, A)$ is a point of A that is *closest* to z. If such a point exists for every $z \in E$, then the subspace A is said to be *closed* in E.

A point $p = p(z) \in A$ such that $z - p \perp x - p$ for all $x \in A$ is called an *orthogonal projection* (or *orthoprojection*, for short) of z onto A. If $z \in A$, then $a(z) = p(z) = z$. If the space is not separated, closest points and orthogonal projections are not necessarily unique.

Lemma. *1) Any point $a \in A$ that is closest to z is an orthogonal projection of z onto A.*

2) Any orthogonal projection of z onto A is closest to z.

Proof. 1) Let $\|z - a\| = \alpha = \rho(z, A)$. Consider a point $x \in A$ and any point $y = a + \lambda(x - a)$ of the line that passes through a and x. ($\lambda \in \mathbb{R}$). Since A is a subspace, we have $y \in A$. Therefore,

$$\alpha^2 \leq \|z - y\|^2 = \|z - a\|^2 + \lambda^2\|x - a\|^2 - 2\lambda \cdot \mathrm{Re}(z - a)(x - a).$$

Using the equality for α and assuming $\lambda > 0$ and then $\lambda < 0$, we obtain

$$\pm\,\mathrm{Re}(z - a)(x - a) \leq 2^{-1}(\pm\lambda)\|x - a\|^2,$$

$$|\mathrm{Re}(z - a)(x - a)| \leq 2^{-1}|\lambda|\|x - a\|^2$$

for all $\lambda \neq 0$. Hence $\mathrm{Re}(z - a)(x - a) = 0$. If E is a real space, then $(z - a)(x - a) = re(z - a)(x - a)$, $z - a \perp x - a$ and $a = p(z)$. If E is a complex space, then for $y = a + \lambda i(x - a)$ we also have $\mathrm{Im}(z - a)(x - a) = -\mathrm{Re}(z - a) \cdot i(x - a) = 0$. As a result, $(z - a)(x - a) = 0$, $z - a \perp x - a$ and $a = p(z)$. We conclude that a is an orthogonal projection of z onto A.

2) Let $p = p(z)$ be an orthogonal projection of z onto A. By the Pythagorean theorem,

$$\|z - x\|^2 = \|(z - p) - (x - p)\|^2 = \|z - p\|^2 + \|x - p\|^2 \geq \|z - p\|^2$$

for any $x \in A$. Then $p = a(z)$. The point $p \in A$ is closest to z. $\square$

Let $A(z)$ denote the set of all points of A that are closest to z, and let $P(z)$ denote the set of all orthogonal projections of z onto A.

Proposition. *1) Let $a \in A(z)$. Then $b \in A(z)$ for $b \in A$ if and only if $\|a - b\| = 0$.*

2) Let $p \in P(z)$. Then $q \in P(z)$ for $q \in A$ if and only if $\|p - q\| = 0$.

Proof. 1) If $a \in A(z)$, $b \in A$, and $\|a - b\| = 0$, then

$$\|z - a\| \le \|z - b\| \le \|z - a\| + \|a - b\| = \|z - a\|, \quad \|z - a\| = \|z - b\|.$$

If $b, a \in A(z)$, then from $2^{-1}(a + b) \in A$ and the parallelogram law it follows that $\|a - b\|^2 = 2[\|z - a\|^2 + \|z - b\|^2] - \|(z - a) + (z - b)\|^2 \le 4\alpha^2 - 4\|z - 2^{-1}(a + b)\|^2 \le 0$ for $\alpha = \|z - a\| = \|z - b\|$.

2) By the preceding lemma, the second assertion follows from the first one. $\square$

If $\|u - v\| = 0$, the points $u, v \in E$ are said to be *equivalent*, which is written as $u \approx v$. The preceding proposition implies that the closest points and the orthogonal projections are *defined up to equivalence*. Suppose that E is a separated Euclidean space. If a point $a(z) \in A$ is closest to z and a point $p(z) \in A$ is an orthogonal projection of z onto A, then $p(z)$ and $a(z)$ are unique and coincide.

The lemma proved above has a clear geometric interpretation. It states that the perpendicular is the shortest segment from a point to a subspace.

Exercise. Verify that the conditions $b \in A$ and $q \in A$ in the above proposition are true.

8. Let X be a closed subspace of a Euclidean space E, i.e., $a(z) \in X$ $(z \in E)$. We denote by $X^{\perp}$ the *orthogonal complement* (*orthocomplement*, for short) of X consisting of all vectors $y \in E$ that are orthogonal to every $x \in X$.

The orthogonal projection theorem. *1) Every vector $z \in E$ can be represented as the sum $x + y$ of the orthogonal projection $x = p(z)$ of z onto a closed subspace X and the orthogonal projection $y = q(z)$ of z onto the orthocomplement $Y = X^{\perp}$.*

2) The decomposition $z = x + y$ is unique up to equivalence.

Proof. 1) Since X is closed, there exists an orthogonal projection $x = p(z)$, which is a point of X that is closest to z (as follows from the lemma of part 7). Let $y = z - x$. Then for all $u \in X$ we have $uy = u(z - x) = ((u + x) - x)(z - x) = (u + x)(z - x) + (0 - x)(z - x) = 0 + 0 = 0$. Hence $y \in Y = X^{\perp}$.

We now prove that y is an orthogonal projection of z onto Y. Choose a vector $v \in Y$. By definition, $uv = 0$ for all $u \in X$. In particular, $xv = 0$ and $xy = 0$. Consequently, $(z - y)(v - y) = x(v - y) = xv - xy = 0 - 0 = 0$. We conclude that $y = q(z)$ is an orthogonal projection of z onto Y.

2) Let $z = x_1 + y_1 = x_2 + y_2$ ($x_1, x_2 \in X$; $y_1, y_2 \in Y$). Then $u = x_1 - x_2 = y_1 - y_2 \in X \cap Y$, $\|u\|^2 = (x_1 - x_2)(y_1 - y_2) = 0$ and $\|u\| = \|x_1 - x_2\| = \|y_1 - y_2\| = 0$. Therefore, $x_1 \approx x_2$ and $y_1 \approx y_2$. $\square$

The equality $z = x + y$ above is said to be an *orthogonal decomposition* of z with respect to the subspaces X and $Y = X^{\perp}$. If the space E is separated, then such a decomposition is unique. The Cartesian coordinates method is based on the orthogonal decomposition. This relationship is illustrated by the following classic picture.

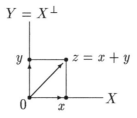

Remark. The set $N = \{z \in E \mid \|z\| = 0\}$ is a subspace of the Euclidean space E. The Cauchy inequality $|vz| \leq \|v\| \cdot \|z\|$ implies that every vector $z \in N$ is orthogonal to all $v \in E$. In particular, $z \perp z$. Consider the factor space $\bar{E} = E/N$. The equality $\bar{u}\bar{v} = uv$ ($\bar{u} = u + N$, $\bar{v} = v + N$) defines a scalar product on $\bar{E}$. The scalar product is well defined because $(u + w)(v + z) = uv + wv + uz + wz = uv$ ($w, z \in N$).

Thus, a nonseparated space can be turned into a separated space by means of factorization. For this reason, in pure theory, usually only separated Euclidean spaces are studied. In applications, transition to a more complex form of vectors is not always convenient.

2.3. MULTILINEAR ALGEBRA

Multilinear functions are vector functions of many vector variables that are linear in every argument. They are the subject of Part 4 of Kostrikin and Manin (1997) and Chapter 3 of Bourbaki (2003).

2.3.1. Tensor product

Elements of the tensor product of vector spaces are specific factor sets of multilinear functions.

1. Consider a finite collection of vector spaces $(\mathbb{K}, E_i)$, their product $(\mathbb{K}, E)$ $(E = \prod E_i)$, and a vector space $(\mathbb{K}, F)$. Elements of E are collections $x = (x_i)$ of vectors $x_i \in E_i$. These collections will be called *multivectors*. It is assumed that the index set I is linearly ordered. Usually $I = \{1, \ldots, n\}$.

For any function $f : E \to F$, point $a \in E$, and index i, the corresponding *partial function* is defined as $f_{ai} : E_i \to F$, $f_{ai}(x_i) = f(a - a_i^0 + x_i^0)$ $(x_i \in E_i)$, where x_i^0 denotes the collection $z = (z_j) \in E$ such that $z_i = x_i$ and $z_j = 0$ for $j \neq i$. A function f is called *multilinear*, if all its partial functions f_{ai} are linear. The term *multilinear map* is sometimes used as an equivalent for *multilinear function*.

Example. Bilinear functions described in part 1 of Subsection 2.2.4 are an important example of multilinear functions. The properties of multilinear functions can be derived by induction from the corresponding properties of bilinear functions.

We denote by $\mathcal{M} = \mathcal{ML}(E, F)$ the set of all multilinear maps $f : E \to F$. The set $\mathcal{M}$ with the conventional addition and multiplication by a scalar constitutes a vector space.

2. We now define the tensor product of vector spaces. In Subsections 2.2.1 (part 6), 2.2.1 (part 9), and 2.2.2 (part 1) we described the algebra $\mathcal{F}_0(X, \mathbb{K})$ of finite scalar functions on a set X. If $X = E$, the algebra $\Phi(E) = \Phi(E) = \mathcal{F}_0(E, \mathbb{K})$ has a basis consisting of delta functions δ_x $(x \in E)$. We will identify δ_x with the multivector x and write $\varphi = \sum \alpha(x)x$ instead of $\varphi = \sum \alpha_x \delta_x$ for $\varphi \in \Phi(E)$ and a finite collection $\alpha_x = \alpha(x) \in \mathbb{K}$ $(x \in E)$.

Let N be a subspace of the vector space Φ generated by a set of special linear combinations of delta functions of two kinds:

$$b = (a - a_i^0 + x_i^0 + y_i^0) - (a - a_i^0 + x_i^0) - (a - a_i^0 + y_i^0),$$
$$c = (a - a_i^0 + \alpha x_i^0) - \alpha(a - a_i^0 + x_i^0) \quad (a, x, y \in E; \ \alpha \in \mathbb{K}).$$

Note that $f(b) = f(c) = 0$ for all $f \in \mathcal{M}$.

The factor space $\bar{E} = \Phi(E)/N$ is called the *tensor product* of the spaces E_i, written $\otimes E_i$. Its elements $\bar{\varphi} = \varphi + N$ ($\varphi \in \Phi(E)$) are called *tensors* Tensors of the form $\bar{x} = \otimes x_i = x + N$ ($x = (x_i) \in E$) are called *decomposable*. The decomposable tensors form a generating system of the tensor product $\bar{E} = \otimes E_i$, but they do not form a basis because of linear dependence: $\bar{\varphi} = \sum \alpha(x)\bar{x}$ for a finite collection $\alpha(x) \in \mathbb{K}$ ($x \in E$), which is *not unique*.

Exercise. Give examples of linearly dependent decomposable tensors.

If $E_i = X$ and $I = \{1, \ldots, n\}$, then $\prod E_i = X^n$ and $\otimes E_i = X^{\otimes n}$.

3. We will now find out the relationship between the conventional product $E = \prod E_i$ and the tensor product $\bar{E} = \otimes E_i$ of vector spaces E_i. Let $u : E \to \bar{E}$ be defined as $u(x) = \bar{x}$ ($x = (x_i) \in E$, $\bar{x} = \otimes x_i \in \bar{E}$). The function u is multilinear. Indeed, from the definition of the factorizing subspace N it easily follows that any partial function u_{ai} satisfies the equalities $u_{ai}(\alpha x_i + \beta y_i) = u(a - a^0 + (\alpha x_i + \beta y_i)^0) = (a - a^0 + (\alpha x_i + \beta y_i)^0) + N = (a - a^0 + \alpha x_i^0) + (a - a^0 + \beta y_i^0) + N = \alpha(a - a^0 + x_i^0) + \beta(a - a^0 + y_i^0) + N = \alpha u_{ai}(x_i) + \beta u_{ai}(y_i)$ ($\alpha, \beta \in \mathbb{K}$; $a \in E$; $x_i, y_i \in E_i$). The function $u \in \mathcal{ML}(E, \bar{E})$ will be called the *natural map* from $E = \prod E_i$ to $\bar{E} = \otimes E_i$.

Exercises. 1) Prove that $E_1 \otimes (E_2 \otimes E_3) \simeq (E_1 \otimes E_2) \otimes E_3 \simeq E_1 \otimes E_2 \otimes E_3$. Formulate and prove the analogous associativity rule for $\otimes E_i$.

2) Any permutation $\sigma : I \to I$ of an ordered index set I, defines the linear map $l_\sigma : \bar{E} \to \bar{E}_\sigma$, $l_\sigma(\bar{x}) = \bar{x}_\sigma$ ($\bar{x} = \otimes x_i \in \bar{E} = \otimes E_i$, $\bar{x}_\sigma = \otimes x_{\sigma(i)} \in \bar{E}_\sigma = \otimes E_{\sigma(i)}$) and the map $f_\sigma : E \to \bar{E}_\sigma$, $f_\sigma(x) = \otimes x_{\sigma(i)}$ ($x = (x_i) \in E = \prod E_i$). Prove that f_σ is multilinear, $f_\sigma = l_\sigma u$, l_σ is an isomorphism, and $l_{\sigma\tau} = l_\sigma \circ l_\tau$ for any permutations σ and τ of the set I.

3) Every decomposable tensor $\overline{x^*} = \otimes x_i^* \in \otimes(E_i^*)$ defines the multilinear function $x^* \in \mathcal{ML}(\prod E_i, \mathbb{K})$ such that $x^*(x) = \prod x_i^*(x_i)$ ($x = (x_i) \in \prod E_i$). This function is associated with the linear functional $l^* = l^*(x^*) \in (\otimes(E_i^*))^*$. Prove that the composition $\overline{x^*} \to l^*$ defines the isomorphism $\otimes(E_i^*) \simeq (\otimes(E_i^*))^*$.

4. Consider a multilinear function $f : E \to F$, a linear function $m :$ $\Phi \to F$ such that $m(x) = f(x)$ for $x = \delta_x$, the natural map $k : \Phi \to \bar{E}$ from the space Φ onto the factor space $\bar{E} = \Phi/N$, and the natural map $u : E \to \bar{E}$ from the product $E = \prod E_i$ to the product $\bar{E} = \otimes E_i$. The relationship between these maps is shown in the diagram.

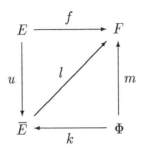

Lemma. *There exists a unique linear function* $l : \bar{E} \to F$ *such that* $f = lu$.

Proof. We have $f = mk^{-1}u$, where the correspondence $k^{-1} : \bar{E} \to \Phi$ is the inverse of k. Since $f(b) = f(c) = 0$ for all functions b and c generating N, we have $\operatorname{Ker} k = N \subseteq \operatorname{Ker} m$, and from the proposition of Subsection 2.1.2 (part 11) with $G = \Phi$, $F = \bar{E}$, $H = f(E)$, $\beta = k$, and $\gamma = m$ it follows that there exists a linear function $l : \bar{E} \to F$ such that $m = lk$. Then $f = lkk^{-1}u = lu$. We have proved the existence of the required function.

It remains to prove that it is unique. Let $f = l_1 u = l_2 u$ for $l_1, l_2 \in \mathcal{L}(\bar{E}, F)$. Then $f(x) = l_1(u(x)) = l_2(u(x)) = l_1(\bar{x}) = l_2(\bar{x})$ and $l_1 = l_2$. $\square$

We denote by $l(f)$ the linear function whose existence and uniqueness is asserted by the lemma. The function $l(f)$ is called the *linear representation* of the multilinear function f. By definition, $f = l(f)u$. Let $L : \mathcal{ML}(E, F) \to \mathcal{L}(\bar{E}, F)$ be a map such that $L(f) = l(f)$ $(f \in \mathcal{ML}(E, F))$.

Theorem. *The map L is a linear isomorphism.*

Proof. The map L is linear. Indeed, we have $L(\alpha f + \beta g)\bar{x} = l(\alpha f + \beta g)\bar{x} = (\alpha f + \beta g)(x) = \alpha f(x) + \beta g(x) = \alpha l(f)\bar{x} + \beta l(g)\bar{x} = (\alpha Lf + \beta Lg)\bar{x}$ $(\bar{x} \in \bar{E})$ and therefore $L(\alpha f + \beta g) = \alpha Lf + \beta Lg$ $(\alpha, \beta \in \mathbb{K}; f, g \in \mathcal{ML}(E, F))$.

The map L is *surjective*. Indeed, if $l \in \mathcal{L}(\bar{E}, F)$, then $f = lu \in \mathcal{ML}(E, F)$ and $l = l(f) = Lf$, which follows from the preceding lemma.

The map L is *injective* because $f = l(f)u \neq 0$ implies $l(f) \neq 0$ and $\operatorname{Ker} L = \{0\}$. $\square$

The isomorphism L makes it possible to identify the space $\mathcal{ML}(E, F)$ with the space $\mathcal{L}(\bar{E}, F)$ and the multilinear function $f \in \mathcal{ML}(E, F)$ with the linear function $l(f) \in \mathcal{L}(\bar{E}, F)$.

Exercise. Prove by induction that $\mathcal{ML}(E, F) \simeq \mathcal{L}(\bar{E}, F)$ using the proposition of Subsection 2.2.4 (part 1), which states that $\mathcal{ML}(X \times Y, Z) \simeq \mathcal{L}(X, \mathcal{L}(Y, Z))$.

5. If any of the subspaces E_i in the product $E = \prod E_i$ is zero, then the space $\mathcal{ML}(E, F)$ is also zero because the only multilinear function is the function identically equal to zero. Indeed, let $E_i = \{0\}$ for some index i. Then $x_i = 0 = 0 \cdot 0$ for the scalar and vector zero. Therefore, $f(x) = 0 \cdot f(x) = 0$ for all $f \in \mathcal{ML}(E, F)$ and $x \in E$. This corresponds to the well-known rule for the conventional products: the product is equal to zero whenever any of the factors is equal to zero.

Exercise. Verify whether the converse holds, i.e., if $E = \prod E_i$ and $\mathcal{ML}(E, F) = \{0\}$, then there exists an index i such that $E_i = \{0\}$.

Similarly, $\bar{E} = \otimes E_i = \{\bar{0}\}$, if there is an index i such that $E_i = \{0\}$.

We choose a basis $A_i = (a_j^i)$ in every space E_i and denote by ξ_i^j the coordinates of a vector $x_i \in E_i$ with respect to this basis, i.e., $x_i = \sum_j \xi_i^j a_j^i$.

We introduce *multi-indices* $\mu = (\mu(i))$ formed by the indices $\mu(i)$ of the basis vectors a_j^i. The *multidegree* ξ^μ and the *basis multivector* a_μ are defined as follows:

$$\xi^\mu = \prod \xi_i^{\mu(i)} \in \mathbb{K}, \quad a_\mu = \left(a_{\mu(i)}^i\right) \in E.$$

For any multilinear function $f \in \mathcal{ML}(E, F)$,

$$f(x) = \sum \xi^\mu f(a_\mu) \quad (x = (x_i),\ x_i = \Pi_j \xi_i^j a_j^i).$$

The values of f are linear combinations of the values it assumes at the basis multivectors. The multidegrees ξ^μ, which are coefficients in these linear combinations, can be any elements of the scalar field $\mathbb{K}$.

The decomposable tensors $\bar{a}_\mu = \otimes a^i_{\mu(i)}$, which are natural images of the basis multivectors a_μ, comprise the basis of the tensor product $\bar{E} = \otimes E_i$ since they are linearly independent and generate $\bar{E}$. Therefore,

$$\dim(\otimes E_i) = \prod(\dim E_i).$$

Exercise. Verify that the basis tensors $\bar{a}_\mu$ are linearly independent.

Let $E_i = \mathbb{K}$ ($1 \leq i \leq n$). Then $\dim E_i = 1$ and $\dim(\mathbb{K}^{\otimes n}) = 1^n = 1 = \dim \mathbb{K}$. Hence, $\mathbb{K}^{\otimes n} \simeq \mathbb{K}$.

6. We now consider two examples of tensor products.

Example 1. Let $E_i = \mathbb{K}_0^{S_i}$ be the spaces of finite scalar functions on sets S_i. Then $\bar{E} = \otimes \mathbb{K}_0^{S_i} \simeq \mathbb{K}_0^{\Pi S_i} = \mathbb{K}_0^S$, where $S = \prod S_i$. After the appropriate identification, we have

$$\bar{x}(s) = \otimes \left(\sum x_i(s_i) s_i \right) = \prod x_i(s_i) \quad (s = (s_i) \in S, \ x = (x_i) \in E).$$

Example 2. Let (E_i) and (F_i) be finite collections of vector spaces with common scalar field and index set. Consider the product $E = \prod E_i$, the tensor products $\bar{E} = \otimes E_i$ and $\bar{F} = \otimes F_i$, and the collection of operators $T_i \in \mathcal{L}(E_i, F_i)$. The equality $T\bar{x} = \otimes(T_i x_i)$ ($\bar{x} = \otimes x_i \in \bar{E}$) defines the linear operator $T = \otimes T_i \in \mathcal{L}(\bar{E}, \bar{F})$, which is called the *tensor product of linear operators* T_i.

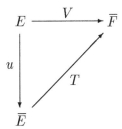

The operator $T = \otimes T_i$ exists since the natural map $u : E \to \bar{E}$ and the map $V : E \to \bar{F}$, $Vx = \otimes(T_i x_i)$ ($x = (x_i) \in E$) are multilinear. The equality that defines $T = \otimes T_i$ can also be written as $(\otimes T)(\otimes x_i) = \otimes(T_i x_i)$.

Exercises. 1) Prove the existence of the operator $T = \otimes T_i$.

2) Verify the following properties of tensor products of operators: $T \otimes 0 = 0 \otimes T = 0$, $I \otimes I = I$, $(S + T) \otimes U = S \otimes U + T \otimes U$, $S \otimes (T + U) = S \otimes T + S \otimes U$, $(\alpha S) \otimes (\beta T) = \alpha\beta(S \otimes T)$, $(S \otimes T)^{-1} = S^{-1} \otimes T^{-1}$, $(ST) \otimes (UV) = (S \otimes U)(T \otimes V)$.

7. For any integer $p, q \geq 0$ and a vector space $(\mathbb{K}, X)$, it is natural to define the tensor space $T_p^q(X) = (X^*)^{\otimes p} \otimes X^{\otimes q}$. Its elements are called *tensors of type* (p, q) *on* X or pq-*tensors*, for short. The number $p + q$ is called the *rank* of such tensors.

Examples. 1) Scalars are tensors of zero rank: $T_0^0(X) = \mathbb{K}$.

2) Vectors are 01-tensors: $T_0^1(X) = X$.

3) Linear functionals are 10-tensors: $T_1^0(X) = X^*$.

4) Linear operators are identified with 11-tensors: $T_1^1(X) = X^* \otimes X \simeq \mathcal{L}(X, X)$ (see Kostrikin and Manin, 1997, Part 4, Section 1(5)).

5) Bilinear forms are identified with 20-tensors: $T_2^0(X) = X^* \otimes X^* \simeq (X \otimes X)^* \simeq \mathcal{B}(X, \mathbb{K})$.

Note that pq-tensors on X can be identified with scalar multilinear functions (*forms*) on $X^p \times (X^*)^q$, while, as before, multilinear functions can be identified with their linear representations.

Exercise. Give a detailed description of the above identifications.

Consider the direct product $T(X) = \sum T_p^q(X)$ of vector spaces $T_p^q(X)$ $(p, q \geq 1)$. The equality

$$(f \otimes g)(x, y) = f(x) \cdot g(y) \quad (x \in X^p, \, y \in (X^*)^q)$$

defines the tensor product $f \otimes g \in T_{p+r}^{q+s}(X)$ of functions $f \in T_p^q(X)$ and $g \in T_r^s(X)$. The tensor product is bilinear and associative, but not commutative.

2.3.2. Exterior product

The definition of an exterior product uses the notion of the tensor product for antisymmetric multilinear functions.

1. Let $\mathcal{M} = \mathcal{ML}(E, F)$ be the set of multilinear maps $f : E \to F$ from the product $E = X^q$ of a collection of vector spaces $E_i = X$ $(i \in I = \{1, \ldots, q\})$ to the vector space F with the same scalar field $\mathbb{K}$. By definition, $\mathcal{ML}(X^0, F) = F$. Assume that $\mathbb{K}$ has characteristic zero, i.e., $p \cdot 1 = 0$ if and only if $p = 0$. Then $q! \neq 0$ for all $q = q \cdot 1$ $(1 \in \mathbb{K})$.

A multivector $x = (x(i))$, $x : I \to E$, and a permutation $\sigma : I \to I$ determine the multivector $x\sigma = x(\sigma(i))$. Thus, for σ, x and $f : E \to F$ we can define the composition $fx\sigma : I \to E$ together with the function $\sigma f : E \to F$ such that $\sigma f(x) = f(x\sigma)$ $(x \in E)$. If f is multilinear, then so is σf. A multilinear map f is said to be *symmetric* if $\sigma f = f$ for all σ. and *antisymmetric* if $\sigma f = \operatorname{sgn} \sigma \cdot f$. The symmetric and antisymmetric maps form the subspaces $S = S(E, F)$ and $A = A(E, F)$ of the space $ML(E, F)$, respectively.

Since every permutation is equal to the product of transpositions, f is antisymmetric if and only if $\tau f = -f$ for any transposition $\tau \in P(I)$. (It suffices to consider the transpositions of adjacent indices only.) If the scalar field has characteristic zero, then $2f(x) \neq 0$ for $f(x) \neq 0$ and the antisymmetry of f means that there are $i \neq j$ such that $f(x) = 0$ for $x_i = x_j$. (It suffices to assume that $j = i + 1$.)

Exercise. Give a detailed proof of the preceding statements.

2. Every multilinear map $f \in M = ML(X^q, \mathbb{K})$ can be associated with the symmetric map

$$Sf = \frac{1}{q!} \sum \sigma f \in S \qquad (\sigma \in P(I), \quad I = \{1, \dots, q\}).$$

The operator $S : M \to S$ is called the *symmetrization* operator. It is a projection that maps M onto S:

$$S^2 f = S(Sf) = Sf \quad (f \in M) \quad \text{and} \quad Sf = f \quad (f \in S).$$

The operator S can be applied, in particular, to the natural map $u :$ $E \to \bar{E}$ from the Cartesian power $E = X^q$ ti the tensor power $\bar{E} = X^{\otimes q}$:

$$Su(x) = \frac{1}{q!} \sum \overline{x\sigma} \qquad (x = (x(i)), \; \overline{x\sigma} = \otimes x(\sigma(i))).$$

As was done in Subsection 2.3.1 (part 7), we will identify $ML(X^q, \mathbb{K})$ with $T_0^q(X)$ and consider multilinear forms to be $0q$-tensors. We will write $T^q(X)$ instead of $T_0^q(X)$ and denote by $S^q(X)$ the subspace of $T^q(X)$ that consists of all symmetric tensors.

Exercise. Prove that $\dim S^q(X) = \binom{n+q-1}{q}$ for $\dim X = n$.

Consider the direct sum $S(X) = \sum S^q(X)$ of vector spaces $S^q(X)$ $(q \geq 1)$. Multiplication on $S(X)$ is defined as follows:

$$fg = S(f \otimes g) \in S^{q+r}(X) \quad (f \in S^q(X), \; g \in S^r(X)).$$

We call it the *symmetrized* multiplication. It is commutative and associative.

Exercise. Prove that the symmetrized multiplication is commutative and associative.

The direct sum $S(X) = \sum S^q(X)$ with symmetrized multiplication form the *symmetric algebra of the space* X.

3. Every multilinear map $f \in \mathcal{M} = \mathcal{ML}(X^{*p}, F)$ can be associated with the antisymmetric map

$$Af = (1/p!) \cdot \sum \operatorname{sgn} \sigma \cdot \sigma f \quad (\sigma \in P(I), \; I = \{1, \ldots, p\}).$$

The operator $A : \mathcal{M} \to \mathcal{A}$ is called the *antisymmetrization* or *alternation* operator. It is a projection that maps $\mathcal{M}$ onto $\mathcal{A}$:

$$A^2 f = A(Af) = Af \; (f \in \mathcal{M}) \quad \text{and} \quad Af = f \; (f \in \mathcal{A}).$$

Exercise. Verify the preceding statement.

As in Subsection 2.3.1 (part 7), we identify $\mathcal{ML}(X^{*p}, \mathbb{K})$ with $T_p(X) = T_p^0(X)$ and consider the subspace $A_p(X)$ of $T_p(X)$ that consists of all antisymmetric tensors.

Exercise. Prove that $\dim A_p(X) = \binom{n}{p}$ for $\dim X = n$.

Consider the direct product $A(X) = \sum A_p(X)$ of the vector spaces $A_p(X)$ $(p \geq 1)$. Multiplication on $A(X)$ is defined as follows:

$$f \wedge g = A(f \otimes g) \in A_{p+r}(X) \quad (f \in A_p(X), \; g \in A_r(X)).$$

It is called the *exterior product*. It is associative and anticommutative:

$$g \wedge f = (-1)^{pr} f \wedge g \quad (f \in A_p(X), \; g \in A_r(X)).$$

Exercise. Prove the preceding equality.

The direct sum $A(X) = \sum A_p(X)$ $(p \geq 1)$ with the exterior product form the *exterior algebra* or the *Grassmann algebra of the space* X.

4. If $p = 1$, then $A_1(X) \simeq X^*$. Let $x^* = (x_i^*)$ $(x_i^* \in X^*)$ and $x = (x_i)$ $(x_i \in X)$ be finite collections with index set $I = \{1, \ldots, n\}$. It follows from the definitions that

$$\wedge x^*(x) = A(\otimes x^*)(x) = (1/n!) \sum_\sigma \mathrm{sgn}\, \sigma \left(\prod_i x_i^*(x_{\sigma(i)}) \right)$$

$(\sigma \in P(I),\ i \in I)$. The scalars $\alpha_{ij} = x_i^*(x_j)$ form an $n \times n$-matrix $(1 \le i, j \le n)$. The number

$$\det M = (n!) \wedge x^*(x)$$

is called the *determinant* of the matrix M. It is equal to the sum of the products of the elements of M that include one element from each row and column, the sign of each term being equal to the signature of the corresponding permutation of indices.

Exercise. Prove that $x_i^* \in X^*$ are linearly independent if and only if $\wedge x^* \ne 0$ for $x^* = (x_i^*)$ $(1 \le i \le n)$.

5. Suppose that $\dim E = n < \infty$, (e_i) is a basis of E, (e_i^*) is the conjugate basis of E^* $(i \in I = \{1, \ldots, n\})$, F is a vector space, and $\mathbb{K}$ is a scalar field for the above spaces. Note that $E \simeq \mathbb{K}^n$. Let $\mu = \mu(k)$ $(1 \le k \le p \le n)$ denote a strictly increasing permutation of p indices selected from I.

Consider the subspace $A_p(E, F)$ of $\mathcal{ML}(E^p, F)$ formed by all antisymmetric multilinear functions. For any $f \in A_p(E, F)$, there exists a unique collection of vectors $c_\mu(f) \in F$ such that

$$f = \sum c_\mu(f) \wedge e_\mu^* \qquad \left(\wedge e_\mu^* = \bigwedge_i e_{\mu(i)}^* \right).$$

This representation is obtained from the general equality in part 5 of Subsection 2.3.1 and the equality $f(x\sigma) = \mathrm{sgn}\, \sigma \cdot f(x)$ $(x = (x_i))$, which is equivalent to $c_\sigma = \mathrm{sgn}\, \sigma \cdot c_\mu$, for every permutation σ of the selected indices (the terms with coinciding sets of indices involved in the permutation are grouped together).

Exercise. Present a detailed derivation of the above representation.

If $p = n$, then the representation for f contains only one term $c(f) \wedge e_i^*$ with coefficient $c(f) \in F$. For $F = \mathbb{K}$ and $p \le n$, the exterior products $\wedge e_i^*$ form the basis of the space $A_p(E, \mathbb{K})$.

Exercise. Prove the preceding assertion.

Chapter 3.

Calculus

In this chapter, we describe the three main concepts of calculus: the limit, the differential, and the integral. We use the lectures in Savel'ev (1969a, 1969b, 1973, 1974, 1975) and Kelley (1975).

3.1. LIMIT

The notion of a limit formalizes the idea of approximating functions by constant functions. For an exposition of the general theory of limits, the reader is referred to Bourbaki (1998a, 1998b) and Kelley (1975).

3.1.1. Topological spaces

Topological spaces serve as domains of definitions of functions for which limits are defined. In the theory of topological spaces, a convenient and illustrative geometric language is adopted. Topology gives a formal description of neighborhoods of elements in a set.

1. Let E be a set. A *topology* for E is a class $\mathcal{U}$ of sets $U \subseteq E$ with the following two properties: (1) The intersection of *any finite* collection of sets in $\mathcal{U}$ belongs to $\mathcal{U}$. (2) The union of *any* collection of sets in $\mathcal{U}$ belongs to $\mathcal{U}$.

The union of an empty collection of subsets of E is equal to the empty set $\varnothing$, and its intersection is E. Therefore, $\varnothing$ and E belong to any topology $\mathcal{U}$ for E. If $\mathcal{T}$ and $\mathcal{U}$ are topologies for E and $\mathcal{T} \subseteq \mathcal{U}$, then $\mathcal{T}$ is said to be *weaker* than $\mathcal{U}$.

Examples. 1) The *weakest* topology is $\mathcal{A} = \{\varnothing, E\}$. It is included in any topology $\mathcal{U}$ for E. 2) The *strongest* topology is $\mathcal{D} = \mathcal{P}(E)$. It includes every topology $\mathcal{U}$ for E.

A set E and a topology $\mathcal{U}$ for E form a *topological space* $(E, \mathcal{U})$, denoted by E for short. Elements of E are called *points* of this topological space. Sets belonging to the topology are said to be *open*. Any set $V \subseteq E$ including a set $U \in \mathcal{U}$ that contains a point $x \in E$ is called a *neighborhood* of the point x in the topological space $(E, \mathcal{U})$. We denote the class of all neighborhoods of a point x by $\mathcal{V}(x)$. Neighborhoods describe the proximity of points of the space to a given point. The definition implies that an open set is a neighborhood of each of its points.

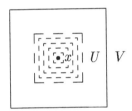

Examples. 1) In the weakest topology $\mathcal{A} = \{\varnothing, E\}$, all points in E have a unique neighborhood, the set E: $\mathcal{V}(x) = \{E\}$ $(x \in E)$. 2) In the strongest topology $\mathcal{D} = \mathcal{P}(E)$, every set $X \subseteq E$ containing a point $x \in E$ is a neighborhood of x: $\mathcal{V}(x) = \{X \subseteq E \mid x \in X\}$. In particular, the singleton $V = \{x\}$ is a neighborhood of x. Therefore, the strongest topology is also called *discrete*. The topological space with such a topology is called *discrete*.

Exercise. Prove that the intersection of a finite collection of neighborhoods of a point is a neighborhood of the point.

A topological space and its topology are called *separated* (or *Hausdorff*) if every two distinct points in the space have disjoint neighborhoods.

Examples. 1) Every discrete space is separated. 2) A two-point set $E = \{a, b\}$ $(a \neq b)$ admits four topologies: $\mathcal{A} = \{\varnothing, E\}$, $\mathcal{B} = \{\varnothing, \{b\}, E\}$, $\mathcal{C} = \{\varnothing, \{a\}, E\}$, and $\mathcal{D} = \{\varnothing, \{a\}, \{b\}, E\}$, $\mathcal{D}$ being the only Hausdorff topology. The space $(E, \mathcal{A})$ is called an *indiscrete space* whose points are "lumped together". The spaces $(E, \mathcal{B})$ and $(E, \mathcal{C})$ are called *connected* two-point spaces, and $(E, \mathcal{D})$ is called a *discrete two-point space*.

Sets that are complements to open sets are said to be *closed*. If every neighborhood of any point in the space includes a closed neighborhood of the point, then the space is called *regular*. The notion of regularity often includes the assumption that the space is separated.

Examples. 1) In any topological space $(E,\mathcal{U})$, the sets $\varnothing$ and E are open and closed. 2) A discrete space is regular. Every set therein is open and closed.

Suppose that $A \subseteq B \subseteq E$. If each neighborhood of any point $x \in B$ contains a point $a \in A$, then A is said to be *dense in* B. If $B = E$, then A is also said to be *everywhere dense*. A topological space that includes a countable everywhere dense set is called *separable*. Such a space can be regarded as *approximately countable* (up to a chosen collection of neighborhoods).

A class $\mathcal{B}(x) \subseteq \mathcal{V}(x)$ such that, for any neighborhood $V \in \mathcal{V}(x)$, there exists a *basic neighborhood* $B \in \mathcal{B}(x)$ included in V, is called a *local base* at the point $x \in E$. A space having a local base at every point is called a *space with countable local bases*.

Every class $\mathcal{B} \subseteq \mathcal{U}$ such that any set $U \in \mathcal{U}$ is the union of a collection of sets $B \in \mathcal{B}$ is called a *base for the topology* $\mathcal{U}$.

Spaces whose topologies have countable bases form an important class of topological spaces. These spaces are called *spaces with countable base*. Clearly, these are spaces with countable local base.

Example. In a discrete space $(E, \mathcal{D})$, the class $\mathcal{B}(x)$ consisting of the only singleton $\{x\}$ is a base for $\mathcal{V}(x)$. If E is countable, then $(E, \mathcal{D})$ has a countable base.

A set $A \subseteq E$ with the *induced topology* $A \cap \mathcal{U} = \{A \cap U \mid U \in \mathcal{U}\}$ forms a *subspace* of the topological space $(E,\mathcal{U})$. If $A \in \mathcal{U}$, then $A \cap \mathcal{U} \subseteq \mathcal{U}$ and $(A, A \cap \mathcal{U}) \subseteq (E,\mathcal{U})$. The intersection $A \cap U$ is called the *trace of* U on A.

2. Every set $A \subseteq E$ generates a natural partition of points in E into three types: *interior*, *exterior*, and *boundary* points of A. It suffices to define interior points. Exterior and boundary points are defined using the notion of interior points. A point $a \in A$ is called *interior* for a set A if there exists a neighborhood U of a included in A. A point $b \notin A$ is called *exterior* for A if it is interior for the complement $B = E \setminus A$.

A point $c \in E$ is called a *boundary point* of A if it is neither interior nor exterior for A. For an exterior point b, there exists a neighborhood V such that V and A are disjoint. For every neighborhood W of a boundary point of A, the intersection of W and A is nonempty, and so is the intersection of W and the complement of A.

We call the sets of interior, exterior, and boundary points of A the *interior*, the *exterior*, and the *boundary* of A and denote them by $I(A) = \operatorname{Int} A$,

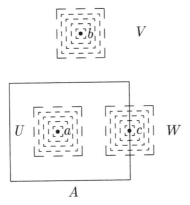

$E(A) = \text{Ext } A$, and $F(A) = \text{Fr } A$, respectively. These sets are pairwise disjoint and their union is the entire space: $I(A) + E(A) + F(A) = E$. The definitions imply that the sets $I(A)$ and $E(A)$ are open and $F(A)$ is closed. Open sets have no boundary points.

The union of a set A and its boundary is called the *closure* of A, written $\bar{A} = A \cup F(A)$. The definitions imply that a set A is closed if and only if it coincides with its closure. As above, we use $+$ instead of $\cup$ for disjoint sets.

Exercise. Prove that $\bar{A}$ is equal to the intersection of all closed sets that include A.

Points of the closure $\bar{A}$ are also called *adherent points* for A: by definition, these are points of A that are interior and exterior at the same time. A point $x \in E$ is an adherent point for A if and only if every neighborhood of x intersects A.

3. An important class of topological spaces is the class of metric spaces, in which the proximity of points can be measured by means of real numbers.

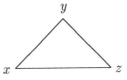

A *positive symmetric* function $\rho : E \times E \to \mathbb{R}$ *vanishing on the diagonal* and satisfying the triangle inequality

$$\rho(x, z) \le \rho(x, y) + \rho(y, z) \quad (x, y, z \in E)$$

is called a *metric* on E. The number $\rho(x,y)$ is called the *distance* between x and y. The set E and the function ρ form the *metric space* (E,ρ). The sets

$$B(c,r) = \{x \mid \rho(x,c) < r\}, \quad \overline{B}(c,r) = \{x \mid \rho(x,c) \le r\}$$

are called the *open ball* and the *closed ball* with center c and radius $r \ge 0$. If $r = 0$, then the ball $B(c,r) = \varnothing$ is closed and the ball $\overline{B}(c,r)$ consists of the points such that the distance from them to the center c is zero. The balls of zero radius are called *degenerate*. They can consist of a single point.

The topology of a metric space is generated by balls: for any point in a set of this topology, this set includes an open ball centered at this point. By definition, a set $U \subseteq E$ is *metrically open* if, for any point $u \in U$, there exists a number $r = r(u) > 0$ such that $B(u, r(u)) \subseteq U$. Metrically open sets constitute the *metric topology* for E: the intersections of finite collections of metrically open sets and the unions of any collections of such sets are metrically open. Every open ball is an open set in this topology, and every closed ball is a closed set.

Exercise. Prove these assertions.

A metric $\rho : E \times E \to \mathbb{R}$ is called *nondegenerate* if $\rho(x,y) > 0$ for $x \ne y$ $(x,y \in E)$. The metric topology generated by ρ is a Hausdorff topology: every two distinct points x and y in E are separated by the open balls $B(x,r)$ and $B(y,r)$ of radius $r = \rho(x,y)/2$ and by the closed balls $\overline{B}(x,s)$ and $\overline{B}(y,s)$ of radius $s = \rho(x,y)/3$. The assumption of nondegeneracy is usually included in the definition of a metric. As a rule, in the sequel we consider nondegenerate metrics, which will not always be specified explicitly. A nondegenerate metric defines a regular topological space.

Exercise. Prove that the equality $\rho(x,y) = 0$ defines an equivalence for the set E of points of a metric space (E,ρ). The factorization of E with respect to this equivalence relation yields a nondegenerate metric $\bar{\rho}$ for an equivalence class $\overline{E}$ defined by the equality $\bar{\rho}(\bar{x},\bar{y}) = \rho(x,y)$ ($\bar{x}$ and $\bar{y}$ are the equivalence classes of points in E equivalent to x and y).

4. Consider several examples.

Example 1. The *discrete metric* $\rho : E \times E \to \mathbb{R}$ with $\rho(x,x) = 0$ and $\rho(x,y) = 1$ if $x \ne y$ $(x,y \in E)$ defines a discrete topology for E.

Example 2. The metric topology of a normed space $(E, \|\cdot\|)$ is defined by the distance $\rho(x,y) = \|x-y\|$ $(x,y \in E)$. A nondegenerate norm defines a nondegenerate distance. The equalities $\rho(x+z, y+z) = \rho(x,y)$, $\rho(\alpha x, \alpha y) = |\alpha|\rho(x,y)$ hold for any scalar α and vectors x, y, z. If a distance ρ for a vector space E possesses these properties, then the equality $\|x\| = \rho(x, 0)$ ($x \in E$) defines a norm for E.

Example 3. Let E be a vector space and let $(\|\cdot\|_i)$ be a collection of norms for E, called a *multinorm*. They constitute a *multinormed space* $(E, (\|\ \|_i))$.

We denote by $B_i(c, r) = \{x : \|x - c\|_i < r\}$ the metrically open ball in E with center c and radius $r > 0$ in the norm $\|\cdot\|_i$. Assume that, for each i, j, there exists k such that $\|x\|_i \vee \|x\|_j \leq \|x\|_k$ ($x \in E$). The topology for E is defined similarly to the topology of a metric space: a set $U \subseteq E$ is open if for any $u \in U$ there exist an index $i = i(u)$ and a number $r_i = r_i(u) > 0$ such that $B_i(u, r_i) \subseteq U$. As in a metric space, for any point in an open set, this set includes a nondegenerate ball centered at this point. In this case, however, these balls may be defined in distinct norms for distinct points. In this topology, every point has a local base consisting of convex sets (balls); therefore, the topology is called *locally convex*. A vector space with a locally convex topology is called *locally convex*.

Note that different multinorms can define the same topology. The connection between locally convex spaces and multinormed spaces is described in Bourbaki (1987) (Chapter 2, Section 5) and Kirillov and Gvishiani (1988) (Chapter 3, Section 2).

Exercises. 1) Prove that, with the above definition, the intersection of a finite collection of open sets is open. 2) Prove that a locally convex space is separated if and only if there is a norm among those defining the topology that does not vanish at any nonzero point.

Example 4. Consider the vector space $\mathcal{F} = \mathcal{F}(X, \mathbb{R})$ of real functions on a set X. For any finite $K \subseteq X$, we define the norm $\|\ \|_K$ by the equality $\|f\|_K = \max\{|f(x)| : x \in K\}$. Clearly, $\|f\|_K \vee \|f\|_L \leq \|f\|_M$ if $K \cup L = M$ for any function $f : X \to \mathbb{R}$. The ball with center f and radius $r \geq 0$ is defined by the equality $B_K(f, r) = \{g : |f(x) - g(x)| < r, x \in K\}$. It consists of all functions $g : X \to \mathbb{R}$ whose values are close to the values of f on K and are arbitrary at the remaining points. If X is uncountable then the locally convex topology defined by the collection of norms $\|\ \|_K$ is *nonmetrizable*, i.e., it cannot be defined with the use of a single metric.

Exercise. Prove the preceding assertion.

Example 5. Consider a vector space E, its algebraic conjugate E^*, and a set $S^* \subseteq E^*$ that is a coordinate system for E: if $x \neq 0$, $x \in E$, then $s^*(x) \neq 0$ for some functional $s^* \in S^*$. Every finite set $K^* \subseteq S^*$ defines a norm $\|\ \|_{K^*}$ for E: $\|x\|_{K^*} = \max\{|s^*(x)| : s^* \in K^*\}$ ($x \in E$).

Clearly, $\|x\|_{K^*} \vee \|x\|_{L^*} \leq \|x\|_{M^*}$ if $K^* \cup L^* = M^*$ for any x. The ball with center x and radius $r > 0$ is defined by the equalities $B_{K^*}(x,r) = \{y : |s^*(x) - s^*(y)| < r, \ s^* \in K^*\}$. It consists of all points $y \in E$ whose coordinates $s^*(y)$ are close to $s^*(x)$. The obtained topology for the vector space E is called the *weak topology* defined by the coordinate system S^*, or the S^*-*topology*.

If E is a normed space, then we can take its topological conjugate E', consisting of bounded linear functionals in E, for the coordinate system defining the weak topology. Then $|x'(x) - x'(y)| = |x'(x-y)| \leq \|x'\|\|x-y\| \leq \gamma\|x-y\|$ if $\gamma = \max\{\|x'\| : x' \in K'\}$ for any finite $K' \subseteq E'$: if the points are close in the norm, then their corresponding coordinates are close too. The converse is not necessarily true. In this sense, the coordinate proximity is weaker than the norm proximity. By contrast, the topology on E defined by the norm is often called *strong*.

Exercise. Prove that if E is a finite-dimensional normed space then the strong topology and the weak E'-topology coincide.

Weak topologies for the algebraic conjugate E^* and its subspaces are called *-weak*. Important examples are the E^{**}-*topology* and the E-*topology* defined by the second dual space E^{**} and by its subspace identified with E. For the geometric conjugate E', apart from the strong topology, we have the E'^*-*topology*, the E''-*topology*, and the E-*topology*.

Exercises. 1) Define the E-topology for $E^* = \mathcal{L}(E, \mathbb{R})$ by analogy with the topology for $\mathcal{F} = \mathcal{F}(X, \mathbb{R})$ described in Example 4. 2) Prove that the E-topology for E^* is weaker than the E^{**}-topology and the E-topology for E' is weaker than the E''-topology, which is, in its turn, weaker than the E'^*-topology.

5. We define multimetric spaces by analogy with multinormed spaces. Let E be a set and let (ρ_i) be a *multimetric*, i.e., a collection of metrics on E. They constitute the multimetric space $(E, (\rho_i))$. Every multinormed space is a multimetric space.

We denote by $B_i(c, r) = \{x : \rho_i(x, c) < r\}$ the metrically open ball in E with center c and radius $r > 0$ with respect to the metric ρ_i. Assume that for any i and j there exists a k such that $\rho_i(x, y) \vee \rho_j(x, y) \leq \rho_k(x, y)$ $(x, y \in E)$. The topology for E will be defined similarly to the topology of a metric space: a set $U \subseteq E$ is open if for any $u \in U$ there is an index $i = i(u)$ and a number $r_i = r_i(u) > 0$ such that $B_i(u, r) \subseteq U$. This topology is called *uniform*.

Exercises. 1) Prove that, with the above definition, the intersection of a finite collection of open sets is open. 2) Prove that a multimetric space is separated if and only if there is a metric among those defining the topology that separates any two distinct points.

The notion of a metric may be generalized by introducing infinite distances. Consider the additive semigroup $[0, \infty] = [0, \infty[\cup \{\infty\}$ obtained by adding the *infinite number* ∞ to the ordered additive semigroup $[0, \infty[$ of finite positive real numbers and extending the addition by the rule $x + \infty = \infty + x = \infty$ for all $x \in [0, \infty]$. The order is extended by the relation $x \leq \infty$ ($x \in [0, \infty]$). By a *generalized metric* for a set E we mean a function $\Delta : E \times E \to [0, \infty]$ that is *symmetric*, *vanishes on the diagonal*, and satisfies the *triangle inequality*. It is obvious that

$$\Delta(x, x) = 0, \quad \Delta(x, y) = \Delta(y, x), \quad \Delta(x, z) \leq \Delta(x, y) + \Delta(y, z)$$

for all $x, y, z \in E$. The number $\Delta(x, y)$, which may be finite or infinite, is called the *generalized distance* between x and y. A *generalized metric space* (E, Δ) includes open and closed balls of infinite radius. When it does not lead to a confusion, we will use the same terms for conventional and generalized metric spaces. Then almost all that was said about metric spaces can be applied to generalized metric spaces.

Similar to a collection of metrics ρ_i, a collection of generalized metrics Δ_i for E defines a *generalized multimetric space* $(E, (\Delta_i))$.

Remark. The equalities $\rho_i = \Delta_i/(1+\Delta_i)$ or $\rho_i = \inf(1, \Delta_i)$ make it possible to replace generalized metrics by conventional metrics (see Chapter 9, Section 1 in Bourbaki, 1998b)

Example 1. Consider the set $\mathcal{F} = \mathcal{F}(X, \mathbb{R})$ of real functions on a set X. Put $\Delta(f, g) = \sup\{|f(x) - g(x)| : x \in X\}$ ($f, g \in \mathcal{F}$). The function $\Delta : \mathcal{F} \times \mathcal{F} \to [0, \infty]$ is symmetric and $\Delta(f, f) = 0$ ($f \in \mathcal{F}$). It satisfies the triangle inequality: $|f(x) - g(x)| + |g(x) - h(x)| \geq |f(x) - h(x)|$ implies $\Delta(f, g) + \Delta(g, h) \geq |f(x) - h(x)|$ ($f, g, h \in \mathcal{F}$, $x \in X$). Hence, $\Delta(f, g) + \Delta(g, h) \geq \Delta(f, h)$. Consequently, Δ is a (generalized) metric space.

Example 2. Consider the set $\mathcal{F} = \mathcal{F}(X, E)$ of vector functions mapping a set X into a multinormed space $(E, (\| \ \|_i))$. Put

$$\Delta_i(f, g) = \sup\{\|f(x) - g(x)\|_i : x \in E\} \quad (f, g \in \mathcal{F}).$$

The functions $\Delta_i : \mathcal{F} \times \mathcal{F} \to [0, \infty]$ are metrics for $\mathcal{F}$ and, together with $\mathcal{F}$, they constitute a multimetric space $(\mathcal{F}, (\Delta_i))$.

Remark. A uniform topology for E defined by a multimetric may be defined with the help of a *uniform structure*, which is a special class of subsets of the Cartesian product $E \times E$ (see Chapter 2 in Bourbaki, 1998a). The connection between uniform structures and topologies is described in detail in Savel'ev (1969a). For the definition of uniform structures in terms of multimetrics, the reader is referred to Chapter 9, Section 1 in Bourbaki (1998a).

A set with a uniform structure is called a *uniform space*. We will also use this term for multimetric spaces and spaces with uniform topology.

The general theory also includes the study of *proximity spaces* (see Section 8.4 in Engelking, 1989). Proximity for E is defined as a special relation between subsets of the set E. It generates a topology for E.

6. A nonempty class $\mathcal{B}$ of nonempty subsets of a set E such that the intersection of any finite collection of these subsets includes a set in $\mathcal{B}$ is said to be *strongly centered*. A class of subsets of E that contains all supersets of each of its elements will be called *hereditary*. A hereditary strongly centered class $\mathcal{F}$ of subsets of E is called a *filter* for E. It follows from the definitions that the intersection of any finite collection of sets in a filter $\mathcal{F}$ is nonempty and belongs to $\mathcal{F}$. The set E belongs to any filter $\mathcal{F}$ in E. Every strongly centered class $\mathcal{B}$ of sets in E *generates* the filter $\mathcal{F}(\mathcal{B})$ for E which consists of all subsets of E that include sets of $\mathcal{B}$. The class $\mathcal{B}$ is called a *filter base* for $\mathcal{F}(\mathcal{B})$.

Examples. 1) The class $\mathcal{V}(x)$ of all neighborhoods of a point x in a topological space E is a filter for E. The class $\mathcal{U}(x)$ of open neighborhoods of x is a filter base for $\mathcal{V}(x)$. 2) The class $\mathcal{F}(x)$ of all subsets of E that contain an element x is a filter for E. The class consisting of a single set $\{x\}$ is a filter base for $\mathcal{F}(x)$. 3) Let $(E, (\Delta_i))$ be a multimetric space. Given an index i and a number $r > 0$, consider the (i, r)-entourage

$$W(i, r) = \{(x, y) \mid \Delta_i(x, y) < r\} \subseteq E \times E.$$

By the definition of a multimetric space, for any i and j there exists a k such that $\Delta_i(x, y) \vee \Delta_j(x, y) \le \Delta_k(x, y)$ $(x, y \in E)$. Consequently, the class of sets $W(i, r)$ is strongly centered and generates the *filter of entourages* $\mathcal{W}$ for $E \times E$. A set $X \subseteq E$ with $X \times X \subseteq W \in \mathcal{W}$ is said to be W-small. The filter $\mathcal{W}$ is a *uniform structure* for E.

Note that $B_i(x, r) = \{y \mid \Delta_i(x, y) < r\} = \{y \mid (x, y) \in W(i, r)\}$. Therefore, every neighborhood $V \in \mathcal{V}(x)$ of a point $x \in E$ in the uniform topology for E is defined by some entourage $W \in \mathcal{W}$: $V = \{y \mid (x, y) \in W\}$.

Exercises. 1) Prove that the image $f(\mathcal{A}) = \{f(A) \mid A \in \mathcal{A}\}$ of a strongly centered class $\mathcal{A} \subseteq \mathcal{P}(E)$ under a map $f : E \to F$ is a strongly centered class of subsets of F. 2) Suppose that $\mathcal{A} \subseteq \mathcal{P}(E)$, $\mathcal{B} \subseteq \mathcal{P}(E)$, and $\mathcal{A} \times \mathcal{B} = \{A \times B \mid A \in \mathcal{A}, B \in \mathcal{B}\} \subseteq \mathcal{P}(E \times F)$. Prove that $\mathcal{A} \times \mathcal{B}$ is strongly centered if $\mathcal{A}$ and $\mathcal{B}$ are strongly centered.

The set $\mathbb{F} = \mathbb{F}(E)$ of all filters for E is ordered by inclusion.

Lemma. *The set $\mathbb{F}$ is linearly majorized.*

Proof. Any nonempty chain $\mathbb{L} \subseteq \mathbb{F}$ is majorized by its union $\mathcal{U} = \cup \mathbb{L}$ composed of sets belonging to some filter $\mathcal{L} \in \mathbb{L}$. Clearly, $\mathcal{L} \subseteq \mathcal{U}$ ($\mathcal{L} \in \mathbb{L}$). However, $\mathcal{U} \in \mathbb{F}$. Indeed, the class $\mathcal{U}$ is nonempty and does not contain the empty set $\varnothing$, since the classes $\mathcal{L}$ are nonempty and $\varnothing \notin \mathcal{L}$. Suppose that $X, Y \in \mathcal{U}$. Then $X \in \mathcal{L}$, $Y \in \mathcal{M}$ for some $\mathcal{L}, \mathcal{M} \in \mathbb{L}$. Since $\mathcal{L} \subseteq \mathcal{M}$ or $\mathcal{M} \subseteq \mathcal{L}$, it follows that $X, Y \in \mathcal{L}$ or $X, Y \in \mathcal{M}$ and, consequently, $X \cap Y \in \mathcal{L}$ or $X \cap Y \in \mathcal{M}$. In any case, $X \cap Y \in \mathcal{U}$. Similarly, if $X \in \mathcal{U}$ and $X \subseteq Y$, then $X \in \mathcal{L}$ and $Y \in \mathcal{L}$ for some $\mathcal{L} \in \mathbb{L}$. Hence, $\mathcal{U}$ is a filter for E. $\quad\square$

Filters for E that are maximal elements of the ordered set $\mathbb{F}(E)$ are called *maximal filters* or *ultrafilters*. By definition, a maximal filter for E is not included in any other filter for E.

Example. The filter $\mathcal{F}(x)$ in Example 2 is maximal.

The above lemma and Zorn's lemma imply the following proposition.

Proposition. *Every filter for a set E is included in a maximal filter for E.*

Example. The filter $\mathcal{V}(x)$ of neighborhoods of a point x in a topological space E is included in the filter $\mathcal{F}(x)$ generated by $\{x\}$.

A filter $\mathcal{F}$ for a set E is said to be *prime* if, for any $X \subseteq E$ and $Y \subseteq E$, from $X \cup Y \in \mathcal{F}$ it follows that $X \in \mathcal{F}$ or $Y \in \mathcal{F}$. This property is analogous to the property of prime numbers: if the product mn of naturals m and n is divisible by a prime p, then m or n is divisible by p.

Theorem. *A maximal filter is the same as a prime filter.*

Proof. 1) Suppose that a filter $\mathcal{F}$ for E is prime and $\mathcal{G}$ is a filter such that $\mathcal{G} \supseteq \mathcal{F}$ and $Y \in \mathcal{G}$. Since $Y \cup Y^c = E \in \mathcal{F}$, we have $Y \in \mathcal{F}$ or $Y^c \in \mathcal{F}$. Since $\mathcal{F} \subseteq \mathcal{G}$, it follows that $Y^c \notin \mathcal{G}$ implies $Y^c \notin \mathcal{F}$. Hence, $Y \in \mathcal{F}$ and $\mathcal{F} = \mathcal{G}$. The filter $\mathcal{F}$ is maximal.

2) Suppose that a filter $\mathcal{F}$ for E is maximal, $X \subseteq E$, $Y \subseteq E$, and $X \cup Y \in \mathcal{F}$. Assume that $X \notin \mathcal{F}$ and $Y \notin \mathcal{F}$. Consider the class $\mathcal{G}$ of all $Z \subseteq E$ such that $X \cup Z \in \mathcal{F}$. The class $\mathcal{G}$ is nonempty, since $Y \in \mathcal{G}$. It does not contain the empty set $Z = \varnothing$ because $X \notin \mathcal{F}$. If $A \in \mathcal{G}$ and $B \in \mathcal{G}$, then $X \cup (A \cap B) = (X \cup A) \cap (X \cup B) \in \mathcal{F}$ and therefore $A \cap B \in \mathcal{G}$. Finally, if $A \in \mathcal{G}$ and $A \subseteq B \subseteq E$, from $X \cup A \in \mathcal{F}$ and $X \cup A \subseteq X \cup B$ it follows that $X \cup B \in \mathcal{F}$ and $B \in \mathcal{G}$. Therefore, $\mathcal{G}$ is a filter for E. However, if $Z \in \mathcal{F}$, then $X \cup Z \in \mathcal{F}$ and $Z \in \mathcal{G}$. Consequently, $\mathcal{F} \subseteq \mathcal{G}$. But $\mathcal{F} \neq \mathcal{G}$, since $Y \notin \mathcal{F}$ and $Y \in \mathcal{G}$. This contradicts the maximality of $\mathcal{F}$. Hence, $X \in \mathcal{F}$ or $Y \in \mathcal{F}$ and the filter $\mathcal{F}$ is prime. $\square$

Examples. 1) The filter $\mathcal{F}(a)$ generated by $\{a\} \subseteq E$ is prime. Indeed, if $a \in X \cup Y$ then $a \in X$ or $a \in Y$ for all $X \subseteq E$ and $Y \subseteq E$. 2) The filter $\mathcal{F}(a,b)$ for E generated by $\{a,b\} \subseteq E$ ($a \neq b$) is not prime: if $A = \{a\}$ and $B = A^c = E \setminus A$, then $A \cup B = E \in \mathcal{F}(a,b)$, but $A \notin \mathcal{F}(a,b)$ and $B \notin \mathcal{F}(a,b)$. Consequently, $\mathcal{F}(a,b)$ is not maximal either.

7. A set C in a topological space $(E, \mathcal{U})$ is called *compact* if every cover of C consisting of open sets $U(i) \in \mathcal{U}$ ($i \in I$) includes a finite cover $U(k)$ ($k \in K(0) \in \mathcal{K}(I)$). Clearly, any finite subset of E is finite. Compact sets can be considered *approximately finite* (up to a chosen collection of neighborhoods). It is conventional to say "a *compactum*" instead of "a *compact set*". If the set $C = E$ is compact, then the space $(E, \mathcal{U})$ is called a *compact space* or simply a *compactum*.

The following negative statement is often convenient: a set C is *noncompact* if there exists a cover of C consisting of open sets $U(i)$ ($i \in I(0)$) that does not include any finite cover $U(k)$ ($k \in K \in \mathcal{K}(I(0))$).

A set with compact closure is called *relatively compact*.

Examples. 1) An interval $[a,b] \subseteq \mathbb{R}$ is compact in the metric topology for $\mathbb{R}$. Intervals $[a,b[$, $]a,b]$, and $]a,b[$ are relatively compact. Unbounded sets in $\mathbb{R}$ are not relatively compact. 2) A rectangle $[a,b] \times [c,d] \subseteq \mathbb{R}^2$ is compact in the Euclidean plane $\mathbb{R}^2$. Rectangles whose sides are bounded intervals are relatively compact. Unbounded sets in $\mathbb{R}^2$ are not relatively compact. 3) A closed ball and its boundary sphere are compact in the Euclidean space $\mathbb{R}^3$.

Proposition 1. *A closed subset of a compactum is a compactum.*

Proof. Suppose that $C \subseteq E$ is a compactum, $F \subseteq C$ is a closed set, a collection of open sets $V(i)$ covers F, and $U = F^c$. Then the collection of open sets $U(i) = V(i) \cup U$ covers C. Indeed, $\cup U(i) = \cup V(i) \cup U = F \cup F^c = E \supseteq C$. Since C is compact, $\cup U(k) \supseteq C$ for some compact set of indices k. Consequently, $\cup V(k) \supseteq F$: if $\cup V(k) = V \subset F$, then the set $F \setminus V \subseteq C$ (as well as C) is not covered by the sets $U(k) = V(k) \cup F^c$. Hence, F is compact. $\square$

Proposition 2. *Any compactum in a separated space is closed.*

Proof. Suppose that E is a separated space and $C \subseteq E$ is a compact set. Assume that $C \neq \overline{C}$ and take $y \in \overline{C} \setminus C$. Since E is separated, for any $x \in C$ there exist open disjoint neighborhoods $U(x)$ and $V(x, y)$ of x and y. Clearly, $U(x)$ covers C. Therefore, $U = \cup U(k) \supseteq C$ for some finite set of points $k \in C$. The set $V(y) = \cap V(k, y)$ is an open neighborhood of y and does not intersect U, which implies that $V(y)$ and C are disjoint. Consequently, $y \in \operatorname{Ext} C$ in contradiction with $y \in \overline{C}$. Hence, $C = \overline{C}$. $\square$

Corollary. *In a separated space, compactness and closedness are equivalent for subsets of a compactum.*

8. A collection of sets $A(i) \subseteq E$ ($i \in I$) is called *centered* if the intersection $A(K) = \cap A(k)$ of any finite collection $A(k)$ ($k \in K \in \mathcal{K}(I)$) is nonempty. The following negative form of this definition is often convenient: a collection of sets $A(i) \subseteq E$ ($i \in I$) is *not centered* if the intersection $A(K_0) = \cap A(k)$ of some finite collection $A(k)$ ($k \in K_0 \in \mathcal{K}(I)$) is empty.

We denote by $\mathcal{A}$ the class of all distinct sets in $(A(i))$ (the image of this collection). The class $\mathcal{A}$ is identified with the collection of its sets, which also serve as indices for themselves. Clearly, $(A(i))$ is centered if and only if $\mathcal{A}$ is centered. Therefore, the inclusion order for classes defines an order for centered collections. Consider the set $C(\mathcal{A})$ of all centered collections $\mathcal{B}$ of subsets of E that contain $\mathcal{A}$ (with the above order). Every chain $L(\mathcal{A}) \subseteq C(\mathcal{A})$ is majorized by $\mathcal{B} = \cup L(\mathcal{A}) \subseteq C(\mathcal{A})$. Indeed, since $L(\mathcal{A})$ is linearly ordered, every finite collection of sets $B(k) \in \mathcal{B}$ is included in some collection $\mathcal{L} \in L(\mathcal{A})$ and therefore has nonempty intersection. By Zorn's lemma, $C(\mathcal{A})$ contains a maximal centered collection $\mathcal{M} \supseteq \mathcal{A}$.

Exercise. Find out whether $\mathcal{M}$ is an ultrafilter.

Example. Let $a \in E$ and $\mathcal{A} = \{a\}$. Then the collection $\mathcal{M}(a)$ of all subsets of E containing a is a maximal centered collection that includes $\mathcal{A}$.

Suppose that a centered collection $\mathcal{M} \supseteq (\mathcal{A})$ is maximal and $X \subseteq E$.

Lemma. (1) If $X = \cap M(k)$ $(k \in K_0)$ for some $K_0 \in \mathcal{K}(I)$, then $X \in \mathcal{M}$.
(2) If $X \cap M(i) \neq \varnothing$ for all $i \in I$, then $X \in \mathcal{M}$.
(3) If $X \supseteq M(i_0)$ for some $i_0 \in I$, then $X \in \mathcal{M}$.

Proof. If $X = \cap M(k) \notin \mathcal{M}$, then $\{\mathcal{M}, X\} \in C(\mathcal{A})$ and $\{\mathcal{M}, X\} \supset \mathcal{M}$; a contradiction to the maximality of $\mathcal{M}$.

If $X \cap M(i) \neq \varnothing$ $(i \in I)$, then, as follows from the statements proved above, $X \cap M(K) \neq \varnothing$ for $M(K) = \cap M(k)$ $(k \in K \in \mathcal{K}(I))$. Hence, $\{\mathcal{M}, X\} \in (\mathcal{A})$ and $X \in \mathcal{M}$.

If $X \supseteq M(i_0)$, then $X \cap M(i) \supseteq M(i_0) \cap M(i) \neq \varnothing$ $(i \in I)$ and we conclude that $X \in \mathcal{M}$. $\square$

The following compactness criterion for a space is often convenient.

Proposition. *A space E is compact if and only if every centered collection of closed sets $F(i) \subseteq E$ has nonempty intersection.*

Proof. Let E be a compact space and let $(F(i))$ be a collection of closed subsets of E. If $\cap F(i) = \varnothing$, then $\cup U(i) = E$ for a collection of open sets $U(i) = F^c(i)$. Since E is compact, $\cup U(k) = E$ and $\cap F(k) = \varnothing$ for some finite set of indices k. The collection $(F(i))$ is not centered.

Suppose now that a space E satisfies the condition in the hypothesis. Let $(U(i))$ be a collection of open subsets of E. If $\cup U(i) = E$, then $\cap F(i) = \varnothing$ for the collection of closed sets $F(i) = U(i)^c$. By hypothesis, $\cap F(k) = \varnothing$ and therefore $\cup U(k) = E$ for some finite set of indices k. The space E is compact. $\square$

Corollary 1. *A space E is compact if and only if for any centered collection of sets $A(i) \subseteq E$ the collection of the closures $\bar{A}(i)$ has nonempty intersection.*

Proof. Let E be compact, and let $(A(i))$ be a centered collection of subsets of E. Then the collection $(\bar{A}(i))$ is also centered and, in view of the above, has nonempty intersection. Conversely, if $A(i) = \bar{A}(i)$, then the condition in the corollary guarantees the compactness of E. $\square$

Corollary 2. *A closed set $C \subseteq E$ is compact if and only if every centered collection of closed sets $B(i) \subseteq C$ has nonempty intersection.*

Proof. Since C is closed, the class of subsets of C closed in the topology $\mathcal{U} \cap C = \{U \cap C \mid U \in \mathcal{U}\}$ for C coincides with the class of subsets of C closed in the topology $\mathcal{U}$ for E. Therefore, the assertion of the corollary follows from the proposition applied to the space $(C, \mathcal{U} \cap C)$. ☐

Exercise. Prove that a closed set $C \subseteq E$ is compact if and only if the space $(C, \mathcal{U} \cap C)$ is compact.

9. Consider the Cartesian product $E = \prod E_i$ of a collection of sets E_i with topologies $\mathcal{U}_i$ $(i \in I)$. By *open rectangles* or *rectangles with open sides* we mean products $U(K) = \prod U_i$ such that $U_i \in \mathcal{U}_i$ for all $i \in I$ and $U_i = E_i$ if $i \notin K$, where $K \in \mathcal{K}(I)$. We have $U(K) \cap U(L) = U(K \cup L)$ for all $K, L \in \mathcal{K}(I)$. Therefore, open rectangles form a base of some topology $\mathcal{U}$ for E. This topology is referred to as the *Tikhonov topology*, and the space $(E, \mathcal{U})$ is called the *Tikhonov product* of the topological spaces $(E_i, \mathcal{U}_i)$.

Examples. 1) In $\mathbb{R}^m$, a base of the Tikhonov topology is composed of the m-dimensional rectangles with sides open in $\mathbb{R}$. Since every open set in $\mathbb{R}$ is a union of bounded open intervals, a base can be composed of usual m-dimensional open rectangles. The Tikhonov topology for $\mathbb{R}^m$ coincides with the Euclidean topology.

2) In the space $\mathcal{F} = \mathcal{F}(X, \mathbb{R}) = \mathbb{R}^X$ of real functions on a set X, a topology base is composed of the sets $U(K) = \{f \mid f(x) \in U(x), x \in K\}$. Here $K \subseteq X$ is finite, $U(x) \subseteq \mathbb{R}$ is open, and the values $f(x)$ of functions $f : X \to \mathbb{R}$ are arbitrary for $x \notin K$. The Tikhonov topology for $\mathcal{F} = \mathbb{R}^X$ coincides with the weak topology defined by the functionals δ_x.

3) A base of the *generalized Cantor's discontinuum* $\mathbb{B}^{\mathbb{R}}$ with discrete two-point set $\mathbb{B} = \{0, 1\}$ is formed by the sets $U(K, L) = \{f \mid f(x) = 1 \ (x \in K), f(x) = 0 \ (x \in L)\}$. Here K and L are finite, and the values $f(x)$ of the functions $f : \mathbb{R} \to \mathbb{B}$ are arbitrary if $x \notin K \cup L$.

Exercise. Prove that the product of a collection of separated spaces is separated.

The Tikhonov theorem. *The product $(E, \mathcal{U})$ of any collection of compact spaces $(E_i, \mathcal{U}_i)$ is compact.*

Proof. Let $\mathcal{A}$ be a centered collection of subsets of E. Consider a maximal centered collection $\mathcal{M}$ of subsets of E that includes $\mathcal{A}$ and the collection $\overline{\mathcal{M}}$ of the closures of these sets. Since $\mathcal{M} \supseteq \mathcal{A}$, it follows that $\overline{\mathcal{M}} \supseteq \overline{\mathcal{A}}$ and $\cap\overline{\mathcal{M}} \subseteq \cap\overline{\mathcal{A}}$. Therefore, to prove the theorem, it suffices to show that $\cap\overline{\mathcal{M}} \neq \varnothing$.

We denote by $p_i : E \rightarrow E_i$ the projection of E onto E_i: $p_i(x) = x_i \in E_i$ for $x = (x_i) \in E$. Furthermore, $M_i = p_i(M) \subseteq E_i$ is the projection of a set $M \in \mathcal{M}$ onto E_i and $M(i) = p_i^{-1}(M_i) \subseteq E$ is the rectangle with one side M_i and the other sides equal to E_j for the remaining $j \neq i$. Since the collection $\mathcal{M}$ is centered, for any i the collection $\mathcal{M}_i$ of the projections $M_i \subseteq E_i$ of the sets $M \in \mathcal{M}$ is also centered. Since E_i is compact, the collection $\overline{\mathcal{M}}_i$ of the closures $\overline{M}_i \subseteq E_i$ has nonempty intersection. Take an x_i in this intersection for each i. We now prove that $x = (x_i) \in \overline{M}$. Let V be an arbitrary neighborhood of a point x. By the definition of the Tikhonov topology, there exists an open rectangle $U = U(K) \subseteq V$ containing x. Since $x_i \in \overline{M}_i$, it follows that $U(i)\cap M_i \neq \varnothing$ for any $M \in \mathcal{M}$. Consequently, $U(i)\cap M \neq \varnothing$ for $U(i) = p_i^{-1}(U_i)$ and every $M \in \mathcal{M}$. Indeed, if $u_i = m_i \in U_i \cap M_i$, then $u = (u_j) \in U(i)$ for any $u_j \in E_j$ for all $j \neq i$ and, in particular, $u_j = m_j$ if $m = (m_j) \in M$. Then $u = m \in U(i)\cap M$. By the lemma of part 8, we obtain $U(i) \in \mathcal{M}$ and $U = U(K) = \cap U(k) \in \mathcal{M}$ ($k \in K$) for any index set K. Since $\mathcal{M}$ is centered, $U \cap M \neq \varnothing$ and therefore $V \cap M \neq \varnothing$ for all $M \in \mathcal{M}$. As a result, $x \in \overline{M}$ for any $M \in \mathcal{M}$ and $\cap\overline{\mathcal{M}} \neq \varnothing$. $\square$

Example. The Tikhonov cube $E = \mathcal{F}(\mathbb{R}, [0,1]) = [0,1]^{\mathbb{R}}$, whose points are functions $f : \mathbb{R} \rightarrow [0,1]$, is compact in the topology for $[0,1] \subseteq \mathbb{R}$ induced from $\mathbb{R}$.

10. The Tikhonov theorem implies Lebesgue's classical theorem describing compact sets in $\mathbb{R}^m$.

Lemma. *Every interval $[a,b] \subseteq \mathbb{R}$ is compact.*
Proof. Consider a collection of open sets $U(i) \subseteq \mathbb{R}$ that covers $[a,b]$ and the set X of the endpoints x of the intervals $[a,x]$ ($a \leq x \leq b$) that are covered by sets $U(k)$ for a finite set of indices k. Since $a \in U(i(a))$ for some index $i(a)$, it follows that $a \in X$ and $X \neq \varnothing$. Since $x \leq b$ by definition, X is bounded from above. Consequently, there exists $c = \sup X$ and $c \in U(i(c))$ for some index $i(c)$. If $c < b$, then there exists $u \in U(i(c))\cap]c,b]$. Obviously, $u \in X$. Since $c < u$, this is impossible. Hence, $c = b$ and $[a,b]$ is covered by a finite collection $U(k)$. $\square$

Open balls constitute a base of the Euclidean topology for $\mathbb{R}^m$ and open rectangles form a base of the Tikhonov topology. Since every ball includes some rectangle containing the center of the ball and vice versa, it follows that the Euclidean and Tikhonov topologies for $\mathbb{R}^m$ coincide. The class of closed sets for them is the same.

Rectangles in $\mathbb{R}^m$ with equal sides (*cubes*) are balls in the norm $\|x\|_{\max} = \max|x_i|$ ($x = (x_i)$, $x_i \in \mathbb{R}$, $1 \le i \le m$). This norm is *equivalent* to the Euclidean norm $\|x\|_2 = \left(\sum x_i^2\right)^{1/2}$: $\|x\|_{\max} \le \|x\|_2 \le m^{1/2}\|x\|_{\max}$. The norms $\|\ \|_{\max}$ and $\|\ \|_2$ define the same class of bounded sets.

Lebesgue's theorem. *A subset of $\mathbb{R}^m$ is compact if and only if it is bounded and closed.*

Proof. 1) Let $C \subseteq \mathbb{R}^m$ be a compact set. Since $\mathbb{R}^m$ is separated, from Proposition 2 of part 7 it follows that C is closed. Consider the cover of C consisting of balls of radius 1 and choose a finite subcollection that covers C. Clearly, the chosen balls are included in a ball of a sufficiently large radius. Hence, C is bounded.

2) Suppose that $X \subseteq \mathbb{R}^m$ is bounded and closed. Then X is a closed subset of a rectangle that is a product of intervals. By the above lemma and the Tikhonov Theorem, this rectangle is compact. By Proposition 1 of part 7, it follows that X is compact. $\square$

Corollary. *A subset of $\mathbb{R}^m$ is relatively compact if and only if it is bounded.*

This corollary is equivalent to Lebesgue's theorem. In the general case, a metric space can include bounded sets that are not relatively compact.

Example. The discrete space $(\mathbb{N}, d)$ with metric d such that $d(n, n) = 0$ and $d(m, n) = 1$ for $m \ne n$ is a ball of radius 1 with an arbitrary center. The set $\mathbb{N}$ is bounded and closed but not compact, since a set in a discrete space is compact if and only if it is finite. Any infinite set $X \subseteq \mathbb{N}$ is also bounded and closed, but not compact.

11. Consider a topological space $(E, \mathcal{U})$, a set $C \subseteq E$, open sets $U \in \mathcal{U}$ and $V \in \mathcal{U}$, and their traces $A = UC$ and $B = VC$ on C. We say that U and V *partition* C if $A \ne \varnothing$, $B \ne \varnothing$, $A + B = C$, and $AB = \varnothing$. A set that is not partitioned by any open sets is called *topologically connected* or just *connected*. If the set $C = E$ is connected, then $(E, \mathcal{U})$ is called a *connected space*. If a space E is connected, then $\varnothing$ and E are the only sets that are closed and open at the same time.

Examples. 1) The empty set is connected. 2) A set in $(\mathbb{R}, |\,|)$ is connected if and only if it is an interval. 3) Balls in $(\mathbb{R}, \|\,\|)$ for $\|\,\| = \|\,\|_1, \|\,\|_2, \|\,\|_{\max}$ are connected.

Exercises. Prove the following assertions: 1) A set is connected if and only if it is not partitioned by closed sets. 2) A set $C \subseteq E$ is connected if and only if $UVC \neq \varnothing$ for any open $U \subseteq E$ and $V \subseteq E$ intersecting C. 3) If $C \subseteq E$ is connected and included in the union of *disjoint* open sets $U \subseteq E$ and $V \subseteq E$, then $C \subseteq U$ or $C \subseteq V$.

Proposition. *The union of a collection of connected sets with nonempty intersection is connected.*

Proof. Suppose that sets $C_i \subseteq E$ are connected, $C = \cup C_i$, and $D = \cap C_i \neq \varnothing$ $(i \in I)$. Assume that some $U \subseteq E$ and $V \subseteq E$ form a partition of C. Let $a \in D$. Since $UC + VC = C$, we have $UD + VD = D$ and $a \in UD$ or $a \in VD$. Suppose that $a \in UD$ and hence $a \in UC_i$ for any $i \in I$. By hypothesis, $VC_{i(0)} \neq \varnothing$ and therefore $C_{i(0)} \neq \varnothing$ for some $i(0) \in I$. Consequently, U and V form a partition of $C_{i(0)}$, which contradicts its connectedness. Therefore, U and V cannot partition C. □

Exercise. Give examples of collections of connected sets with disconnected intersections.

We say that a connected set *joins* its points.

Connectedness criterion. *A set is connected if and only if every two points in it are joined by a connected subset of this set.*

Proof. If a set is connected then it joins every two of its points.

Suppose that a set $X \subseteq E$ is disconnected and open sets $U \subseteq E$ and $V \subseteq E$ make up a partition of X. Take $a \in UX$, $b \in VX$, and a set $Y \subseteq X$ that contains a and b. Clearly, $a \in UXY = UY$, $b \in VXY = VY$. Since $UX + VX = X$ and $(UX)(VX) = \varnothing$, we have $UY + VY = Y$ and $(UY)(VY) = \varnothing$. The sets U and V make up a partition of Y and therefore it cannot be connected. □

Exercise. Prove that the product of connected spaces is connected.

Lemma. *If C is connected and $x \in \overline{C}$, then so is $C(x) = C \cup \{x\}$.*

Proof. Suppose that open sets U and V partition $C(x)$ and let $x \in V$. Since $x \in \overline{C}$, it follows that $VC \neq \varnothing$. Since $x \notin UC(x)$, we have $UC(x) = UC \cup U\{x\} = UC$ and $UC \neq \varnothing$. At the same time, from $C \subseteq C(x)U \cup V$ and $UVC(x) = \varnothing$ it follows that $C \subseteq U \cup V$ and $UVC = \varnothing$. This contradicts the connectedness of C. $\quad\square$

Corollary. *If C is connected and $C \subseteq X \subseteq \overline{C}$, then X is connected.*

Proof. If $C = \varnothing$, then $\overline{C} = X = C = \varnothing$ and therefore X is connected. Assume that $C \neq \varnothing$ and $x, y \in X$. Then, as follows from the lemma, $C(x)$ and $C(y)$ are connected. Since $C \subseteq C(x) \cap C(y)$, the proposition implies that $C(x, y) = C(x) \cup C(y)$ is connected. By the connectedness criterion, X is connected. $\quad\square$

Remark. By the proposition, the connectedness of $X \neq \varnothing$ also follows from that of $C(x)$ and the relations $\varnothing \neq C \subseteq \cap C(x)$, $X = \cup C(x)$ $(x \in X)$.

12. The union of all connected sets of a topological space $(E, \mathcal{U})$ that contain a point $x \in E$ is called a *connected component of the point x.*

Proposition 1. *The connected component of a point is a connected closed set.*

Proof. A point x belongs to each of the sets that form the connected component of this point, and this component is equal to the union of a collection of connected sets with nonempty intersection. By the proposition of part 11, it is connected. By the corollary of the lemma of part 11, the closure $\overline{C}$ of the connected component C of x is also connected. Since $x \in C \subseteq \overline{C}$, it follows that $\overline{C} \subseteq C$. Hence, $\overline{C} = C$. $\quad\square$

Examples. 1) A connected space is the connected component of each of its points. 2) In a discrete space, the connected component of a point consists of this point. 3) In the metric space (A, Δ) with $A = [0, 1] \cup [2, 3]$ and $\Delta(x, y) = |x - y|$, the connected components of $x \in [0, 1]$ and $y \in [2, 3]$ are $[0, 1]$ and $[2, 3]$.

Suppose that $A \subseteq E$. The connected component of a point $x \in A$ in the subspace $(A, A \cap \mathcal{U})$ is equal to the union of all connected subsets of A containing x. The connected component $A(x)$ of x in A may differ from the trace $AE(x)$ on A of the connected component $E(x)$ of x in E.

Example. The connected component of the point $x = 0$ in the metric space $E = \mathbb{R}$ is equal to $\mathbb{R}$ while in the subspace $A = [0,1] \cup [2,3]$ it is equal to $[0,1]$, whereas $A\mathbb{R} = A$.

The connected components of points $x \in A$ in a subspace $(A, A\cap\mathcal{U})$ are briefly called the *connected components of the set* A.

Proposition 2. *The connected components of a set form a partition of the set.*

Proof. Every $x \in A$ belongs to its connected component $C(x)$ in A. If the connected components $C(x)$ and $C(y)$ of points $x, y \in A$ in A intersect, then they are equal. Indeed, if $x \in C(x) \cap C(y)$, then $C = C(x) \cup C(y)$ is connected in A by the proposition of part 11 of this subsection. Since $x \in C$ and $y \in C$, we have $C \subseteq C(x)$ and $C \subseteq C(y)$; hence, $C(x) = C = C(y)$. Therefore, different connected components of A are pairwise disjoint and their union is A.   □

Example. The intervals $[0,1]$ and $[2,3]$ are the connected components of the set $A = [0,1] \cup [2,3]$ in $\mathbb{R}$.

13. We defined connectedness of two kinds for a multinormed space $(E, (\| \ \|_i))$, namely polygonal connectedness and topological connectedness.

Lemma. *Every broken line in a multinormed space is topologically connected.*

Proof. Consider a broken line $L(a,b) = \cup[a(i), b(i)]$ $(1 \le i \le n)$ with endpoints $a, b \in E$. If the intervals $[a(i), b(i)]$ are topologically connected, then, by the proposition of part 11, the equalities $b(i) = a(i+1)$ imply that the unions $[a(i), b(i)] \cup [a(i+1), b(i+1)]$ are topologically connected. It can be easily proved by induction that the entire broken line $L(a,b)$ is topologically connected.

We now prove that any interval $[a,b] = \{x = (1-t)a + tb \mid 0 \le t \le 1\} \subseteq E$ is topologically connected. Assume the contrary and take open sets $U, V \subseteq E$ making up a partition of $[a,b]$.

Consider the map $\varphi : t \to (1-t)a + tb$ $(0 \le t \le 1)$, a number $s = \sup\{t : \varphi[0,1] \subseteq U\} \in [0,1]$, and a point $x = \varphi(s) \in [a,b]$. Let $x \in U$. Since U is open, we have $B(x, r) \subseteq U$ for some norm and radius $r > 0$. Consequently, $y = \varphi(t) \in U$ for all $t \in [s - \delta, s + \delta]$ for some $\delta > 0$. This contradicts the definition of s. The assumption that $x \in V$ also yields a contradiction to the definition of s. In any case, the assumption that $[a,b]$ is disconnected leads to a contradiction.   □

Corollary. *Every polygonally connected set is topologically connected.*

Proof. Any two points in a polygonally connected set are joined by a broken line included in this set. By the lemma, this broken line is topologically connected. By the criterion of part 11, this implies that the entire set is topologically connected. □

Proposition. *An open set in a multinormed space is topologically connected if and only if it is polygonally connected.*

Proof. The necessity part follows from the preceding corollary. We now prove the sufficiency part. Consider an open topologically connected set $C \subseteq E$. Suppose that C is not polygonally connected, and take points $a, b \in C$ that are not joined by any broken line in C. Let A be the polygonally connected component of a and let $B = C \setminus A$ be its complement in C. Since $a \in A$ and $b \in B$, it follows that $A \neq \varnothing$ and $B \neq \varnothing$. Moreover, $A \cup B = C$ and $AB = \varnothing$. At the same time, A and B are open. Indeed, let $x \in A$. Then $B(x, r) \subseteq C$ for some norm and radius $r > 0$ because C is open. Since a and x are joined by an open line in C, every point in $B(x, r)$ is joined with a through x by a broken line in C. Hence, $B(x, r) \subseteq A$, and A is open. Similarly, if $x \in B$, then $B(x, r) \subseteq B$ because $y \in AB(x, r)$ implies $x \in A$. The open sets A and B partition C, which contradicts its topological connectedness. □

Examples. 1) The open annulus $C = \{(x, y) : 0 < \alpha < x^2 + y^2 < \beta\}$ in the Euclidean plane $\mathbb{R}^2$ is polygonally and topologically connected. 2) The circle $S(0, r) = \{(x, y) : x^2 + y^2 = r^2\}$ in $\mathbb{R}^2$ is topologically connected but not polygonally connected.

14. A topological space in which every point has a compact neighborhood is called *locally compact*. Every compact space is locally compact. The class of locally compact spaces is much larger than the class of compact spaces.

Example. The Euclidean space $\mathbb{R}^n$ is locally compact but not compact.

It is conventional to consider separated compact and locally compact spaces.

Lemma. *A separated compact space is regular.*

Proof. Let $(E,\mathcal{U})$ be a separated compact space. In a separated space, the intersection of all closed neighborhoods of a point $x \in E$ is equal to $\{x\}$, since every point $y \in E$, $y \neq x$, belongs to the exterior of an open neighborhood U of x and therefore to the exterior of a closed neighborhood $\overline{U}$ of x. Suppose that the space $(E,\mathcal{U})$ is not regular and there exists a point $a \in E$ and its neighborhood $A \in \mathcal{U}$ such that $A^c\overline{U} \neq \varnothing$ for any neighborhood $U \in \mathcal{U}$ of a. Since the intersection of any finite collection of closed neighborhoods of a point a is a closed neighborhood of a, the collection of sets $A^c\overline{U}$ is centered. Since A is open, these sets are closed. At the same time, their intersection is empty because the intersection of all $\overline{U}$ is equal to $a \in A$. A contradiction to the compactness of E. $\square$

Proposition. *A separated locally compact space is regular.*

Proof. Consider a separated locally compact space $(E,\mathcal{U})$, an arbitrary point $x \in E$, and its compact neighborhood $C \subseteq E$, whose existence is assumed. The space $(C, C\mathcal{U})$ is compact and regular, which follows from the preceding lemma. Since $(E,\mathcal{U})$ is separated, it follows that $C = \overline{C}$ and the sets closed in $(C, C\mathcal{U})$ are closed in $(E,\mathcal{U})$. Therefore, the regularity of $(C, C\mathcal{U})$ implies the regularity of $(E,\mathcal{U})$. $\square$

Corollary. *In a separated locally compact space, the filter of neighborhoods of a point has a filter base of compact neighborhoods of the point.*

Proof. Since the intersection $C\overline{U}$ of any closed neighborhood $\overline{U}$ of a point x with its compact neighborhood C is compact and, in view of the proposition, closed neighborhoods are a filter base for the filter of neighborhoods of x, it follows that compact neighborhoods also form its filter base. $\square$

Remark. For any separated locally compact space, there exist multimetrics generating its topology (see Chapter II, Section 4, Corollary 2 to Theorem 1 in Bourbaki, 1998a).

3.1.2.   Directed sets

Directed sets serve as domains of definitions of functions for which the limit is defined. A direction is described in terms of a special order.

1. By a *direction* for a set X we mean an order relation on X with the following property: *any* $x, y \in X$ *have an immediate successor* $z \in X$. An order with this property is also called a *lattice order* or a *filtering order*. All these terms give an intuitive idea about the properties of a direction.

A set X and a direction $\succeq$ on it constitute a *directed set* $(X, \succeq)$. It is conventional to write X instead of $(X, \succeq)$. A directed set with a last element and the corresponding direction are called *degenerate*.

2. We give examples of directed sets.

Example 1. Consider the set $(\mathbb{N}, \geq)$ of natural numbers with the conventional order. The direction $\geq$ for $\mathbb{N}$ is denoted by $n \to \infty$. Adding ∞ to natural numbers, we obtain the degenerate directed set $(\overline{\mathbb{N}}, \geq)$, where $\overline{\mathbb{N}} = \mathbb{N} \cup \{\infty\}$ and $\infty > n$ for all $n \in \mathbb{N}$.

Example 2. Consider the set $(\mathbb{R}, \geq)$ of real numbers with the natural order. The direction $\geq$ for $\mathbb{R}$ is denoted by $x \to \infty$, and the opposite direction $\leq$ for $\mathbb{R}$ is denoted by $x \to -\infty$. These directions are degenerate for the extended real line $\overline{\mathbb{R}}$, where $-\infty < x < \infty$ for all $x \in \mathbb{R}$ and $\overline{\mathbb{R}} = \{-\infty\} \cup \mathbb{R} \cup \{\infty\}$.

Example 3. Assume that $a \in \mathbb{R}$ and $y \succeq x$ means $|y - a| \leq |x - a|$, i.e., a point *succeeds* another point if the first point is *closer* to a. This direction for $\mathbb{R}$ is written as $x \to a$. To make this direction nondegenerate, the point a is usually excluded. In this case, it is conventional to write $x \to a$, $x \neq a$. The expression $x \neq a$ is often implied but omitted in writing.

Example 4. Suppose that $A \subseteq \mathbb{R}$ and $a \in \bar{A}$. As in Example 3, the relation $y \succeq x$ means that $|y - a| \leq |x - a|$ but is defined only for points $x, y \in A$. Such direction is designated as $x \to a$, $x \in A$, and is described as follows: x *tends to* a *along the set* A. It may be regarded as a direction for $\mathbb{R}$ and as a direction for A. If $a \notin A$, then the direction $x \to a$, $x \in A$, is nondegenerate.

If $A = \mathbb{R} \setminus \{a\}$, then the direction $x \to a$, $x \in A$, becomes $x \to a$, $x \neq a$. If $A = \,]-\infty, a[$ and $A = \,]a, +\infty[$, then we obtain the directions $x \to a-$ and $x \to a+$, or $x \uparrow a$ and $x \downarrow a$, respectively. These directions are described as follows: x *tends to* a *from the left* and x *tends to* a *from the right*.

Example 5. Suppose that (X, d) is a metric space, $A \subseteq X$, $a \in \bar{A}$, and $y \succeq x$ means that $d(y, a) \leq d(x, a)$, i.e., a point *succeeds* another point if it is *closer* to a. This direction for X is designated as $x \to a$, $x \in A$ and is described as follows: x *tends to a along the set A*. If $a \notin A$, then the direction $x \to a$, $x \in A$ is nondegenerate. In a normed space $(X, \| \; \|)$, the proximity to a point a is measured with the help of the norm: $y \succeq x$ means that $\|y - a\| \leq \|x - a\|$.

Example 6. Consider the set $\mathbb{C}$ of complex numbers. Suppose that $a = (a_1, a_2)$, $x = (x_1, x_2)$, $y = (y_1, y_2) \in \mathbb{C}$. Then $y \succeq x$ means that $|y - a| \leq |x - a|$. In the geometric interpretation, greater proximity to a means being inside a disk of smaller radius with center at a. Choosing different $A \subseteq \mathbb{C}$ yields different directions $x \to a$, $x \in A$, in particular, along a given ray or a sector with vertex a. One can also choose more complicated directions (for example, along a given curve).

Example 7. Any linear order is a direction.

Example 8. Consider a set X and the class $\mathcal{K} = \mathcal{K}(X)$ of all finite subsets of X. The inclusion order $\supseteq$ is a direction for $\mathcal{K}$: given two sets $K, L \in \mathcal{K}$, their union $K \cup L = M \in \mathcal{K}$ succeeds both of the sets. We denote this direction by $K \to X$. If X is finite, then this direction is degenerate.

Example 9. Consider the real line $\mathbb{R}$, a finite set $X \subseteq \mathbb{R}$, and a disjoint collection $\mathcal{A}$ of intervals $A(x) = [a(x), b(x)[$, making an interval and such that $x \in A(x)$ for any $x \in X$. We call $\mathcal{X} = (X, \mathcal{A})$ a *Riemannian pair* for $\mathbb{R}$. The direction $X \to \mathbb{R}$ for the class $\mathcal{K} = \mathcal{K}(\mathbb{R})$ defines a direction for the set $\mathcal{R} = \mathcal{R}(\mathbb{R})$ of Riemannian pairs: $\mathcal{Y} \succeq \mathcal{X}$ means that $Y \supseteq X$ for the pair $\mathcal{Y} = (Y, \mathcal{B})$. The collections $\mathcal{A}$ and $\mathcal{B}$ of intervals are not taken into account here. Any pairs $\mathcal{X} = (X, \mathcal{A})$ and $\mathcal{Y} = (Y, \mathcal{B})$ are succeeded by the pair $\mathcal{Z} = (X \cup Y, \mathcal{C})$ with any admissible collection of intervals. Such a direction for $\mathcal{R}$ is also denoted by $X \to \mathbb{R}$.

Example 10. The set $\mathcal{R}^m = \mathcal{R}(\mathbb{R}^m)$ of Riemannian pairs for $\mathbb{R}^m$ and the direction $X \to \mathbb{R}^m$ for $\mathbb{R}^m$ can be defined in a similar way. In this case, m-dimensional intervals $[a, b[= [a_1, b_1[\times \cdots \times [a_m, b_m[\subseteq \mathbb{R}^m$ are considered instead of one-dimensional intervals.

3. The Cartesian product $(X_1 \times X_2, \succeq)$ of directed sets $(X_1, \succeq)$ and $(X_2, \succeq)$ with the coordinate order is a directed set. It is easy to verify that the order for $X_1 \times X_2$ is a lattice order: any pairs (x_1, x_2), $(y_1, y_2) \in X_1 \times X_2$ are succeeded by a pair $(z_1, z_2) \in X_1 \times X_2$ composed of elements $z_1 \in X_1$ and $z_2 \in X_2$ succeeding $x_1, y_1 \in X_1$ and $x_2, y_2 \in X_2$.

The Cartesian product of an arbitrary collection of directed sets $(X_i, \succeq)$ is defined in the same manner: collections $x = (x_i)$, $y = (y_i)$ are succeeded by a pair $z = (z_i)$ composed of elements $z_i \in X_i$ that succeed $x_i, y_i \in X_i$.

Examples. 1) $(\mathbb{N} \times \mathbb{N}, \geq) = (\mathbb{N}, \geq) \times (\mathbb{N}, \geq)$ with the direction $(m, n) \to (\infty, \infty)$ or $m \to \infty$, $n \to \infty$. 2) $(\mathbb{R} \times \mathbb{R}, \geq) = (\mathbb{R}, \geq) \times (\mathbb{R}, \geq)$, $(\mathbb{R}, \geq) \times (\mathbb{R}, \leq)$, $(\mathbb{R}, \leq) \times (\mathbb{R}, \geq)$, and $(\mathbb{R}, \leq) \times (\mathbb{R}, \leq)$ with the directions $(x, y) \to (+\infty, +\infty)$, $(+\infty, -\infty)$, $(-\infty, +\infty)$, and $(-\infty, -\infty)$, respectively.

4. Consider a directed set $(X, \succeq)$ and a set F. A function $f : X \to F$ on $(X, \succeq)$ is called a *net* in F. If $(X, \succeq) = (\mathbb{N}, \geq)$, then a net is a sequence. A net is also called a *generalized sequence*.

Consider directed sets $(T, \succeq)$, $(X, \succeq)$, and a map $r : T \to X$. We say that r *agrees with the directions* for T and X if for any $x \in X$ there exists a $t(x) \in T$ such that $r(t) \succeq x$ for all $t \succeq t(x)$.

Examples. 1) $(T, \succeq) = (2\mathbb{N}, \geq)$, $(X, \succeq) = (\mathbb{N}, \geq)$, and $r(2n) = 2n$.
2) $(T, \succeq) = (]0, 1[, \leq)$, $(X, \succeq) = (]1, \infty[, \geq)$, and $r(t) = 1/t$.

The composition $g = f \circ r : T \to F$ of a net $f : X \to F$ and a map $r : T \to X$ that agrees with the directions for T and X is called the *subnet* of f generated by r. We may say that a subnet is obtained from a given net by means of a change of variables that agrees with the directions.

Examples. 1) Let $(T, \succeq) = (2\mathbb{N}, \geq)$, $(X, \succeq) = (\mathbb{N}, \geq)$, $r(2n) = 2n$, $F = \mathbb{R}$, and $f(n) = 1/n$. Then $g(2n) = f(r(2n)) = 1/2n$. If $r(2n) = n$, then $g(2n) = 1/n$. If $T = \mathbb{N}$ and $r(n) = 2n$, then $g(n) = 1/2n$.
2) Assume that $(T, \succeq) = (X, \succeq) = (\mathbb{N}, \geq)$, $r : \mathbb{N} \to \mathbb{N}$ strictly increases and $f : X \to F$. Then $g = f \circ r : \mathbb{N} \to F$ is called a *subsequence* of f. In the first example, we obtain a subsequence if $r(n) = 2n$.
3) We say that a set $T \subseteq X$ contains *arbitrarily remote elements* if the identity embedding $r : T \to X$ agrees with the direction for X, i.e., for any $x \in X$, there exists a $t(x) \in T$ that succeeds x (by transitivity, from $t \succeq t(x)$ and $t(x) \succeq x$ it follows that $t \succeq x$).

Any restriction of a net $f : X \to F$ to a set $T \subset X$ with arbitrarily remote elements is a subnet of f. If $X = \mathbb{N}$, then such restrictions are also called subsequences. In the first example, we obtain such a subsequence by setting $T = 2\mathbb{N}$ and $r(2n) = 2n$.

4) Let $(T, \succeq) = (]0, 1[, \leq)$, $(X, \succeq) = (]1, \infty[, \geq)$, $r(t) = 1/t$, $F = \mathbb{R}$, and $f(x) = x^2$. Then $g = f \circ r :]0, 1[\to \mathbb{R}$, $g(t) = 1/t^2$ $(0 < t < 1)$, where the domains of definition of g and f are disjoint.

5) Let $(T, \succeq) = (\mathbb{R}, \geq)$, $(X, \succeq) = (]-1, 1[, \geq)$, $r(t) = t/(1 + |t|)$, $F = \mathbb{R}$, and $f(x) = x^2$. Then $g = f \circ r : \mathbb{R} \to \mathbb{R}$, $g(t) = t^2/(1 + |t|)^2$ $(-\infty < t < \infty)$. The domain of definition of the subnet g is much larger than that of f.

5. The examples show that the notion of a subnet is very general. Therefore, it is logical to distinguish special classes of subnets. By a *subsequence* of a net $f : X \to F$ we mean any of its subnets $g = f \circ r : T \to F$ defined on the set $T = \mathbb{N}$ of natural numbers with the conventional order $\geq$. Not all nets have subsequences. Nets that have subnets, sets admitting such nets, and their directions are called *sequentiable*. If a directed set $(X, \succeq)$ is sequentiable, this means that there is a map $r : \mathbb{N} \to X$ such that for any $x \in X$ there exists a number $n(x) \in \mathbb{N}$ such that $r(n) \succeq x$ for all $n \geq n(x)$. That is, X must include a countable set D with arbitrarily remote elements. The property of being sequentiable is analogous to the property of being Archimedean.

Examples. 1) Every sequence is a sequentiable net. 2) The directed sets in Examples 2–6 of part 2 are sequentiable. 3) The linearly ordered set $(\mathbb{R}, \geq)$ is sequentiable because of the Archimedean principle, which is equivalent to the assertion that any real number is less than some natural number. 4) In Examples 9 and 10 for Riemannian pairs (in part 2 of this subsection), the direction $X \to \mathbb{R}$ is not sequentiable.

Exercise. Prove that the inclusion direction for Riemannian pairs is not sequentiable.

6. We call the set $S(a) = [a, \to [= \{x \mid x \succeq a\} \subseteq X$ the *section of a directed set* $(X, \succeq)$ defined by an element $a \in X$, or simply the *a-section of X*. (The set X is assumed to be nonempty.) If X is linearly ordered, then the a-section of X is an interval with left boundary a. The class $\{S(a) \mid a \in X\}$ of sections of a directed set X is strongly centered. Indeed, since $X \neq \varnothing$, this class is nonempty and does not contain the empty set. Since for any a and b in X there is an element c that succeeds them, it follows that $S(c) \subseteq S(a) \cap S(b)$. The filter $\mathcal{S} = \mathcal{S}(X)$ for X generated by the class of sections is called the *filter of sections for the directed set X*. By definition, any set $S \in \mathcal{S}$ includes a section $S(a) \subseteq X$.

Example. The class $\{[n, \to [: n \in \mathbb{N}\}$ generates the *Fréchet filter* for $\mathbb{N}$. It consists of the subsets of $\mathbb{N}$ that have finite complements.

We call the set $fS(a) = f[a, \to[= \{f(x) \mid x \succeq a\} \subseteq F$ the *section of the net $f : X \to F$* defined by an element $a \in X$, or simply the *a-section of the net f*. The class $\{fS(a) \mid a \in X\}$ of sections of a net f is strongly centered together with $\{S(a) \mid a \in X\}$. The filter $f\mathcal{S} = f\mathcal{S}(X)$ for F generated by the class $\{fS(a) \mid a \in X\}$ is called the *filter of sections of the net f*.

Let $g = f \circ r : T \to F$ be the subnet defined on a set $(T, \succeq)$ using a change of variables $r : T \to X$ that agrees with the directions.

Lemma 1. *The filter of sections of a net is included in the filter of sections of each of its subnets.*

Proof. It follows from the definitions that for any $a \in X$ there exists a $t(a) \in T$ such that $rS(t(a)) \subseteq fS(a)$. Hence, $gS(t(a)) \subseteq fS(a)$ and $fS(a) \in g\mathcal{S}(T)$. Since the class $\{fS(a) \mid a \in X\}$ generates the filter $f\mathcal{S}(X)$, it follows that $f\mathcal{S}(X) \subseteq g\mathcal{S}(T)$. $\square$

Exercise. Prove that the filter of sections of a sequentiable net has countable base.

A net $f : X \to F$ is said to be *elementary* if the filter of its sections $f\mathcal{S}(X)$ is prime, i.e., for any finite covering $(Y(k))$ of the image fX of an elementary net $f : X \to F$, there exist an index $k(0)$ and $x(0) \in X$ such that $fS(x(0)) \subseteq Y(k(0))$. In particular, for any $Y \subseteq F$ there exists an $x(0) \in X$ such that $fS(x(0)) \subseteq Y$ or $fS(x(0)) \subseteq Y^c = F \setminus Y$.

Example. Every constant net is elementary.

By the theorem of part 6 of Subsection 3.1.1, a filter is prime if and only if it is maximal. Consequently, if a net f is elementary, then the preceding lemma implies that the filter of sections of each of its subnets g is equal to $f\mathcal{S}(X)$ and is prime too. Hence, subnets of an elementary net are also elementary.

Exercise. Deduce this statement directly from the definitions.

Lemma 2. *Any filter containing all sections of a net is the filter of sections of some of its subnets.*

Proof. Consider the filter $\mathcal{G}$ for F that contains all sections $fS(a)$ of a net $f : X \to F$, and the set $T = \{(x, Y) \mid Y \in \mathcal{G}, \, f(x) \in Y\} \subseteq X \times \mathcal{G}$. The order $(x, Y) \succeq (u, Z) \Leftrightarrow x \succeq u, \, Y \subseteq Z$ defines a direction for T. Indeed, suppose that $t(a) = (a, U)$ and $t(b) = (b, V) \in T$ and $x \in X$ succeeds a and b. By hypothesis, $W = fS(x) \cap U \cap V \in \mathcal{G}$. Take $c \in X$ such that $c \succeq x$ and $f(c) \in W$. Then $t(c) = (c, W) \in T$. Since $c \succeq x \succeq a$, $c \succeq x \succeq b$, and $W \subseteq U$, $W \subseteq V$, it follows that $t(c) \succeq t(a)$ and $t(c) \succeq t(b)$.

The projection map $r : T \to X$ that takes a pair $t = (x, Y) \in T$ to the element $r(t) = x \in X$ agrees with the directions for T and X. Indeed, for any $u \in X$ there exists a pair $t(u) = (u, S(u)) \in T$ such that $r(t) = x \succeq u$ for all $t = (x, Y) \in T$, $t \succeq t(u)$: by the definition of the direction T, from $(x, Y) \succeq (u, S(u))$ it follows that $x \succeq u$. Hence, $g = f \circ r : T \to F$ is a subnet of f.

Consider arbitrary $a \in X$ and $Y \in \mathcal{G}$. By hypothesis, $Z = fS(a) \cap Y \in \mathcal{G}$. Let $u \in X$ be such that $u \succeq a$, $f(u) \in Z$, and $t(u) = (u, Z) \in T$. It follows from the definitions that $g(t) = f(r(t)) = f(x) \in Y$ for any $t = (x, W) \in T$, $t \succeq t(u)$, since $x \succeq u \succeq a$ and $f(x) \in W \subseteq Z \subseteq Y$. Hence, $gS(t(u)) \subseteq Y$ and $\mathcal{G}$ is the filter of sections of g. $\square$

Lemma 2 implies the following proposition.

Proposition. *Every net has an elementary subnet.*

Proof. The filter of sections of any subnet of a net f is included in a maximal filter $\mathcal{G}$. A filter $\mathcal{G}$ is prime and it is the filter of sections of a subnet g of f. By definition, g is an elementary subnet. $\square$

Example. The sequence $f(n) = (-1)^n$ has elementary subsequences $g(n) = 1$ and $h(n) = -1$ $(n \in \mathbb{N})$.

7. The statement that the class $\mathcal{B}$ of subsets of a set E is strongly centered is equivalent to the statement that there is a direction $\subseteq$ defined on $\mathcal{B}$, i.e., for any $X, Y \in \mathcal{B}$ there exists a $Z \in \mathcal{B}$ such that $Z \subseteq X$ and $Z \subseteq Y$. Any choice function $\xi : \mathcal{B} \to E$ with $\xi(X) \in X$ defines a direction $\succeq (\xi)$ for $\xi(\mathcal{B}) \subseteq E$: the relation $x \succeq (\xi)y$ for $x = \xi(X)$ and $y = \xi(Y)$ means that $X \subseteq Y$. We call the collection $(\succeq (\xi))$ the *multidirection* generated by the class $\mathcal{B}$, or simply the *direction* $\mathcal{B}$ for E. Like $(E, \succeq)$, the pair $(E, \mathcal{B})$ will be referred to as a *directed set*.

Example 1. The class $\mathcal{B} = \{S(x) \mid x \in E\}$ of sections of a directed set $(E, \succeq)$ generates a multidirection for E. The initial direction $\succeq$ is defined by the choice function with $\xi(S(x)) = x$. In this case, the directed set $(E, \succeq)$ is identified with $(E, \mathcal{B})$.

Example 2. Consider a topological space $(E, \mathcal{U})$ and a set $A \subseteq E$. If $A \cap U \neq \emptyset$ for any neighborhood U of a point a, then the class $A \cap \mathcal{U}(a) = \{A \cap U \mid U \in \mathcal{U}(a)\}$ is strongly centered and generates a direction $x \to a$, $x \in A$ for E. In particular, if $A = E \setminus \{a\}$, then we obtain the direction $x \to a$, $x \neq a$. These multidirections are frequently used.

3.1.3. Convergence

Convergence for nets is defined by analogy with the standard definition of convergence for sequences.

1. Consider a directed set $(X, \succeq)$, a topological space $(F, \mathcal{V})$, a function $f : X \to F$, and a point $c \in F$. The function f is said to *converge to c in the direction* $\succeq$, written $f \to c$, if for any neighborhood $V \in \mathcal{V}(c)$ there exists an element $x(V) \in X$ such that $f(x) \in V$ for all $x \succeq x(V)$. Informally, the convergence of a function f to a point c in a given direction means that the values $f(x)$ at all *sufficiently advanced* x are as close to c as desired. Clearly, the property of being *advanced* depends on the choice of the direction. For some directions, whenever it is intuitively appropriate, advanced values will be called *remote* or, on the contrary, *close*.

It is conventional to write $f(x) \to c$ instead of $f \to c$. Sometimes the direction is added. For example, $f(n) \to c$ $(n \to \infty)$ or $f(x) \to c$ $(x \to a)$. A point c to which a function f converges along a given direction is called its *limit* in this direction, which is written as $\lim f = c$ or $\lim f(x) = c$. Also, the direction is often specified, for example, $\lim_{n \to \infty} f(n) = c$ or $\lim_{x \to a} f(x) = c$. Sometimes it is convenient to use the approximate equality $\approx$ instead of $\to$ and write, for example, $f(n) \approx c$ $(n \approx \infty)$ or $f(x) \approx c$ $(x \approx a)$. The abbreviated designations are convenient in theoretical arguments, whereas the more detailed notation is good for calculations.

If a direction $x \to a$ is degenerate, then $f(x) \to f(a)$ for any $f : X \to F$. Therefore, the point a is usually excluded and the direction $x \to a$, $x \neq a$ is considered.

For a given direction, all functions $f : X \to F$ are subdivided into those *convergent* to some point and those that are *divergent* (not convergent in this direction to any point in F).

If F is *separated*, then a net $f : X \to F$ cannot converge to more than one point in F.

Exercise. Prove the preceding assertion. Formulate and verify the converse.

Example 1. A net $f : X \to F$ in a multimetric space $(F, (\Delta_i))$ converges to a point $c \in F$ if and only if $\Delta_i(f, c) \to 0$ for each i. Convergence in a multimetric means convergence in each of its metrics. In particular, convergence in a multinorm means convergence in each of its norms.

Example 2. Consider the vector space $\mathcal{F} = \mathcal{F}(X, E)$ of functions $f : X \to E$ that map a set X into a separated multimetric space $(\mathbb{K}, E)$ and the class $\mathcal{K} = \mathcal{K}(X)$ of finite subsets of X with the direction $\supseteq$. Every $f \in \mathcal{F}$ defines the net $S(f) : \mathcal{K} \to E$ of finite sums $S(f, K) = \sum_{x \in K} f(x)$. Its limit $\sum_{x \in X} f(x) = \lim_{K \to X} \sum_{x \in K} f(x)$ is called the *sum* (of the collection) f. If the set X is known, this sum is denoted by $\sum f$. The collections f for which the sum $\sum f$ exists are called *summable*. They constitute a subspace $\mathcal{S} = S(X, E)$ of the space $\mathcal{F} = \mathcal{F}(X, E)$. Obviously, the operator $\sum : f \to \sum f$ ($f \in \mathcal{S}$) is linear. It is the composition of the operators $S : f \to S(f)$ and $\lim : S(f) \to \sum f$.

Exercise. Prove that the operators S, $\lim$, and $\sum$ are linear.

Remark. If a set X is finite, then the above definition of the sum coincides with the conventional one because the direction $K \to X$ is degenerate.

Example 3. Consider a separated multinormed space $(\mathbb{K}, E)$, a vector-function $f : \mathbb{R}^m \to E$, and a scalar function $g : \mathbb{R}^m \to \mathbb{K}$. For any Riemannian pair $\mathcal{X} = (X, \mathcal{A}) \in \mathbb{R}^m$, the scalars $\Delta g(x) = g(b(x)) - g(a(x))$ and the vector

$$S(f, g, X) = \sum f(x) \cdot \Delta g(x) \quad (x \in X)$$

are defined. The net $S(f, g) : X \to S(f, g, X)$ is called the *integral sum* for f and g, and the limit

$$\int f \, dg = \lim S(f, g, X) \quad (X \to \mathbb{R}^m)$$

is called the *integral* of f with respect to g.

If such a limit exists for a function f, then f is said to be *integrable* with respect to g. The set $\mathcal{I}(g)$ of such functions forms a subspace in the vector space $\mathcal{F} = \mathcal{F}(\mathbb{R}^m, E)$. It is easy to verify that the operator $\int : \mathcal{I}(g) \to E$ is linear, i.e., $\int \in \mathcal{L}(\mathcal{I}(g), E)$. It is the composition of the operators $S : f \to S(f, g)$ and $\lim : S(f, g) \to \int f \, dg$.

Exercise. Prove that the operators S, lim, and $\int$ are linear.

Remark. In the same manner, we can define the integral of a scalar function $f : \mathbb{R}^m \to \mathbb{K}$ with respect to a vector-function $g : \mathbb{R}^m \to E$. We can also define the integral of a vector-function $f : \mathbb{R}^m \to F$ with respect to a vector-function $g : \mathbb{R}^m \to G$, provided that the product $F \times G \to H$ for vector spaces F and G and a separated multinormed space H with common scalar field $\mathbb{K}$ is defined. In any case, integral sums are sums of values of f weighted by g, and the integral is the limit of the integral sums as the number of the summands increases infinitely. The Riemann integrability of a continuous map from an interval in $\mathbb{R}$ into a Banach space is proved in Chapter 5, Section 3 in Zaidman (1999).

Informative theories are obtained by various restrictions of the classes of functions under consideration. Elements of the theory of the Riemann–Stieltjes integrals for $m = 1$ are presented in Hille and Phillips (1974), Chapter 3, Section 1. The book McLeod (1980) is devoted to the generalized Riemann integral. The Stieltjes integral is also considered there.

2. Consider the directed set $(E, \mathcal{B})$ constructed in part 7 of Subsection 3.1.2, a topological space $(F, \mathcal{V})$, a function $f : E \to F$, and a point $c \in F$. We call the function f a *net* in F. We say that f *converges to* c and write $f \to c$ (along $\mathcal{B}$) if for any $V \in \mathcal{V}$ there exists $B \in \mathcal{B}$ such that $f(B) \subseteq V$. The terms and the notation of part 1 of this subsection can be extended to this convergence. In this case, the degeneracy of the direction $x \to a$ means that $\{a\} \in \mathcal{B}$.

A collection of choice functions $\xi : \mathcal{B} \to E$ with $\xi(X) \in X$ $(X \in \mathcal{B})$ and a function $f : E \to F$ generate the collection of nets $f \circ \xi : \mathcal{B} \to F$. If $f \to c$, then $f \circ \xi \to c$ for any ξ; moreover, this convergence is uniform: for any $V \in \mathcal{V}(c)$, there exists a $B \in \mathcal{B}$ such that $f(\xi(X)) \in V$ for all $X \subseteq B$ in $\mathcal{B}$ and all ξ. Since for any $x \in B$ there exists a ξ such that $\xi(B) = x$, the condition $f(\xi(B)) \in V$ for all ξ is equivalent to $f(B) \subseteq V$. Therefore, the above uniform convergence $f \circ \xi \to c$ is equivalent to the convergence $f \to c$ along $\mathcal{B}$.

Example 1. If $\mathcal{B} = \{S(x) \mid x \in E\}$ is the class of sections of a directed set $(E, \succeq)$, then the convergence in $\mathcal{B}$ is equivalent to the convergence in $\succeq$.

Example 2. Consider topological spaces $(E, \mathcal{U})$ and $(F, \mathcal{V})$, points $a \in E$ and $c \in F$, a set $A \subseteq E$ such that the intersection of A and every neighborhood of a is nonempty, and a map $f : E \to F$. The convergence $f \to c$ ($x \to a$, $x \in A$) means that for any $V \in \mathcal{V}(c)$ there exists a $U \in \mathcal{U}(a)$ such that $f(A \cap U) \subseteq V$.

Exercise. Prove that $f \to c(\mathcal{B})$ if and only if the filter generated by the class $f(\mathcal{B}) = \{f(X) \mid X \in \mathcal{B}\}$ includes the filter of neighborhoods of c.

Remark. A topological space $(F, \mathcal{V})$ can be associated with the topological space $(\widehat{F}, \widehat{\mathcal{V}})$, where $\widehat{F} = \mathcal{P}(F)$ and $\widehat{\mathcal{V}}$ is the topology with the base formed by the classes $\widehat{V} = \mathcal{P}(V)$ ($V \in \mathcal{V}$). (Since $\mathcal{P}(U) \cap \mathcal{P}(V) = \mathcal{P}(U \cap V)$ and $U \cap V \in \mathcal{V}$ if $U, V \in \mathcal{V}$ and $F \in \mathcal{V}$, it follows that $\{\widehat{V} \mid V \in \mathcal{V}\}$ is a topology base for $\widehat{F}$; see Section 11, Chapter 1 in Kelley, 1975). A function $f : E \to F$ on a directed set $(E, \mathcal{B})$ generates the function $\widehat{f} : \widehat{E} \to \widehat{F}$ defined on $\widehat{E} = \mathcal{P}(E)$ such that $\widehat{f}(X) = f(X)$ ($X \subseteq E$). Its restriction $\widehat{f} : \mathcal{B} \to \widehat{F}$ is defined on the directed set $(\mathcal{B}, \subseteq)$. By definition, the convergence $\widehat{f} \to \{c\}$ ($c \in F$) means that for any $\widehat{V} \in \widehat{\mathcal{V}}(\{c\})$ there exists a $B \in \mathcal{B}$ such that $\widehat{f}(X) \in \widehat{V}$ ($x \subseteq B$, $X \in \mathcal{B}$). That is, for any open $V \in \mathcal{V}(c)$, there exists a $B \in \mathcal{B}$ such that $f(B) \subseteq V$. Therefore, $\widehat{f} \to \{c\}$ is equivalent to $f \to c$. The convergence in the direction $\mathcal{B}$ for f is reduced to the convergence of $\widehat{f}$ in the direction $\subseteq$.

Convergence in a filter base (multidirection) is the most general, although it is defined in a rather complicated way. The definition of directional convergence is simpler, but it does not cover all important cases. To cover them, one has to consider uniform convergence or pass to images of the base sets. This makes it possible to reduce convergence in a filter base to directional convergence. The reverse reduction is obtained by using the filter of sections.

3. If a net $f : X \to F$ converges to a point $c \in F$, then each of its subnets $g = f \circ r : T \to F$ also converges to c. Therefore, if a net f in a separated space F has two subnets $g = f \circ r$ and $h = f \circ s$ converging to two different points, then f does not converge to any point (diverges).

Example. In Example 2 of part 2, assume that $\mathbb{R}^m = \mathbb{K} = E = \mathbb{R}$, $f = \operatorname{ind} \mathbb{Q}$, and let $g(x) = 0$ for $x < 0$, $g(x) = x$ for $0 \le x \le 1$, and $g(x) = 1$ for $x > 1$. Then $S(f, g, U, \eta) = 0$ if all $\eta_j \in \mathbb{R} \setminus \mathbb{Q}$ and $S(f, g, T, \xi) = 1$ if $\xi_i \in \mathbb{Q}$ for all i. Thus, f is not integrable with respect to g.

Heine's criterion. *A sequentiable net in a separated space converges if and only if each of its subsequences converges.*

Proof. Necessity follows from the definitions. We now establish sufficiency.

Let $(F, \mathcal{V})$ be a separated space, and let $f : X \to F$ be a sequentiable net. Note that if all subsequences $f \circ r : \mathbb{N} \to F$ converge, then they converge to the same point of F. Indeed, if $f \circ r \to a$, $f \circ s \to b$, and $a \neq b$, then the subsequence $f \circ t$ such that $f(t(2m-1)) = f(r(m))$ and $f(t(2m)) = f(s(m))$ diverges. We now consider the common limit $c \in F$ of all subsequences f and show that $f \to c$.

Suppose the contrary, i.e., $f \nrightarrow c$. Then there exists a neighborhood $V \in \mathcal{V}(c)$ of c such that for any $x \in X$ there is a $y(x) \succeq x$ such that $f(y(x)) \notin V$. Since the direction $\succeq$ for X is sequentiable, we can take a sequence with elements $x(n) \in X$ as remote as desired. Put $r(n) = y(x(n))$ $(n \in \mathbb{N})$. It follows from the definitions that $f \circ r : \mathbb{N} \to F$ is a subsequence of f. At the same time, $f(r(n)) = f(y(x(n))) \notin V$ $(n \in \mathbb{N})$, and therefore $f \circ r \nrightarrow c$, which contradicts the assumption. $\square$

The limits of subnets are called the *partial limits* or the *limit points* of the net. The set of all limit points is called the *limit set*. If a net in a separated space converges, then its limit set consists of a single point, namely, the limit of the net.

Exercise. Prove that an elementary net that has a limit point converges to this point.

The limits of subsequences are called the *sequential partial limits* or the *sequential limit points* of the net. The set of all sequential limit points is called the *sequential limit set*.

Exercise. Give examples of sequentiable nets for which the sequential limit set is not equal to the limit set.

Consider the limit set $\mathrm{Lim}\, f$ of a net $f : X \to F$ and the intersection $\cap \overline{fS(x)}$ of the closures of its sections $fS(x) = \{f(y) \mid y \succeq x\}$.

Lemma. $\mathrm{Lim}\, f = \cap \overline{fS(x)}$.

Proof. Let $c \in \cap \overline{fS(x)}$. Then $U \cap fS(x) \neq \varnothing$ for all $U \in \mathcal{V}(c)$, $x \in X$; therefore, there exists an element $r(U, x) \succeq x$ such that $f(r(U, x)) \in V$ if $U \subseteq V \in \mathcal{V}(c)$, $x \in X$. Hence, $f \circ r \to c$.

Assume that $c \notin \cap \overline{fS(x)}$. Then $U_0 \cap fS(x_0) = \varnothing$ for some $U_0 \in \mathcal{V}(c)$ and $x_0 \in X$ and therefore f does not have a subnet converging to c. $\square$

4. An important class of uniform spaces is the class of complete uniform spaces, in which the convergence of nets is described by the Cauchy criterion.

Consider a directed set $(X, \succeq)$, the product $(X, \succeq) \times (X, \succeq) = (X \times X, \succeq)$ with the coordinatewise order, and a multimetric space $(F, (\Delta_i))$. Every net $f : X \to F$ defines a collection of real *double nets* $\Delta_i f : X \times X \to \mathbb{R}$, $\Delta_i f(x, y) = \Delta_i(f(x), f(y))$ $(x, y \in X)$.

The net f is said to *converge in itself* (written $f \to$) if $\Delta_i f \to 0$ for each i.

Lemma. *Every net converging to a point converges in itself.*

Proof. The lemma follows from the inequalities

$$0 \le \Delta_i(f(x), f(y)) \le \Delta_i(f(x), c) + \Delta_i(c, f(y)) \le \varepsilon/2 + \varepsilon/2 = \varepsilon,$$

which hold for all $\varepsilon > 0$, indices i, and $x, y \succeq x(\varepsilon)$ for some $x(\varepsilon) \in X$. □

In the general case, the converse does not hold.

Example. Assume $(X, \succeq) = (]0, 1[, \ge)$, $F = X =]0, 1[$ with the conventional metric, and $f(x) = x$ $(0 < x < 1)$. Then $\Delta(f(x), f(y)) = |x - y| \to 0$ as $x \to 1$, $y \to 1$, and $f \nrightarrow c$ for any $c \in F$.

Exercises. 1) Prove that for a net converging in itself in a separated uniform space, the existence of a unique limit point is a sufficient condition for convergence to this point. 2) Prove that a net in a uniform space converges in itself if and only if the square of its filter of sections includes the filters of entourages: $f \to \Leftrightarrow f\mathcal{S} \times f\mathcal{S} \supseteq \mathcal{W}$. This means that the filter of sections $f\mathcal{S}$ for f contains arbitrarily *small* sets, i.e., for any entourage $W \in \mathcal{W}$, there exists a section $S \in \mathcal{S}$ with $f\mathcal{S} \times f\mathcal{S} \subseteq W$.

A uniform space in which every net converging in itself converges to some point is said to be *complete*. A uniform space in which every sequence converging in itself converges to some point is said to be *sequentially complete*.

Exercises. 1) Prove that a sequentially complete metric space is complete. 2) Give an example of an incomplete sequentially complete uniform space. 3) Prove that a finite-dimensional complex normed space is complete.

The following assertion often helps establish the existence of a limit for a net.

The Cauchy criterion. *A net in a complete metric space converges to some point if and only if the net converges in itself.*

Proof. Necessity follows from the lemma, and sufficiency follows from the definition of a complete space. □

Example 1. A collection of vectors $f(x) \in E$ ($x \in X$) in a complete multimetric space $(E, \| \ \|_i)$ is summable if and only if for any number $\varepsilon > 0$ and any index i there exists a finite set $K = K(\varepsilon, i) \subseteq X$ such that $\left\| \sum_{x \in M} f(x) \right\|_i \leq \varepsilon$ for all finite sets $M \subseteq X$ such that M and K are disjoint.

Example 2. In Example 3 of part 1, a function $f : \mathbb{R}^m \to E$ with values in a complete metric space $(\mathbb{K}, E, \| \ \|_i)$ is integrable with respect to a function $g : \mathbb{R}^m \to \mathbb{K}$ if and only if, for any number $\varepsilon > 0$ and any index i, there exists a Riemannian pair $(X_0, \mathcal{A}_0)$ such that $\|S(f, g, X) - S(f, g, X_0)\|_i \leq \varepsilon$ for all pairs $(X, \mathcal{A}) \succeq (X_0, \mathcal{A}_0)$.

Exercises. 1) Deduce the above summability criterion from the Cauchy criterion. 2) Transform the inequality for integral sums in the above integrability criterion so that the left-hand side be the sum of the products of the norms of values of the vector-function f and the moduli of the differences of the scalar function g.

Remark. Every uniform space can be completed with nets converging in themselves (Chapter II, Section 3, Theorem 3 in Bourbaki, 1998a). In particular, it is thus possible to complete normed fields, multinormed spaces, and many other algebraic systems while preserving the operations (see Chapters 18 and 20 in van der Waerden, 1991).

5. Consider a metric space $E \neq \varnothing$ with metric d. A sequence of closed balls $\overline{B}_n = \overline{B}(c_n, r_n) = \{x \mid d(x, c_n) \leq r_n\} \subseteq E$ is called *shrinking* if the balls are embedded in each other and their radii vanish: $\overline{B}_{n+1} \subseteq \overline{B}_n$, $r_n \to 0$. (For example, if $c_n = c \in E$ and $r_n = 1/n$, then we obtain a sequence of concentric balls centered at c.) We say that a sequence of balls *shrinks into a point* $c \in E$ if $c \in \cap \overline{B}_n$.

Lemma. *In a complete metric space, every shrinking sequence of closed balls shrinks into some point.*

Proof. If a sequence of balls $\overline{B}_n = \overline{B}(c_n, r_n)$ shrinks, the sequence of their centers c_n converges in itself: $d(c_{n+m}, c_n) \leq 2r_n$ for all m, n. If the space E is complete, then the sequence c_n converges to a point $c \in E$. Since $c_{n+m} \in \overline{B}_n$ for all m and n and the ball $\overline{B}_n$ is closed, it follows that $c \in \overline{B}_n$ for all n. $\square$

Remark. The existence of a common point for any shrinking sequence of closed balls is a criterion of completeness for a metric space (see Subsection 4.3.8 in Postnikov, 1987).

Exercise. Prove that an incomplete metric space has a shrinking sequence of closed balls that does not shrink into any point.

The shrinking balls lemma implies the following important assertion.

The Baire theorem. *In a complete metric space, the union of any sequence of closed sets without interior points has no interior points.*

Proof. Consider a sequence of closed sets $F_n \subseteq E$ without interior points and let $F = \cup F_n$. Assume that the theorem fails and there exists a closed ball $\overline{B} = \overline{B}(c, r) \subseteq F$ ($c \in E$, $r > 0$). Note that the intersections $E_n = \overline{B} \cap F_n$ are closed, have no interior points, and $\overline{B} = \cup E_n$. The closed ball $\overline{B}$ with the metric d is a complete metric space. Therefore, it suffices to prove the theorem for $E = F$. This is technically easier. Assume now that $E = \cup F_n$.

We define a shrinking sequence of closed balls in E that will lead us to a contradiction to the preceding lemma. Put $U_n = F'_n = E \setminus F_n$. Since $E = \cup F_n$, it follows that $\cap U_n = \varnothing$. By hypothesis, F_n is closed and has no interior points, and E does have interior points. Therefore, $F_n \neq E$ and U_n is a nonempty open set for each n. All its points are interior. Hence, there is a closed ball $\overline{B}_1 = \overline{B}(c_1 r_1) \subseteq U_1$ ($c_1 \in E$, $0 < r_1 < 1$). Let $n \geq 1$ and consider a closed ball $\overline{B}_n = \overline{B}(c_n, r_n) \subseteq U_n$ with center $c_n \in E$ and radius $0 < r_n < n^{-1}$. Since F_{n+1} has no interior points, it follows that the open ball $B_n = B_n(c_n, r_n)$ cannot be entirely in F_{n+1} and therefore B_n and its complement U_{n+1} intersect. Since B_n and U_{n+1} are open, so is $B_n \cap U_{n+1}$, and all its points are interior. Consequently, there is a closed ball $\overline{B}_{n+1} = \overline{B}(c_{n+1}, r_{n+1}) \subseteq B_n \cap U_{n+1}$ with center $c_{n+1} \in E$ and radius

$0 < r_{n+1} < (n+1)^{-1}$. Note that $\overline{B}_{n+1} \subseteq B_n \subseteq \overline{B}_n$ and $\overline{B}_{n+1} \subseteq U_{n+1}$. By induction, this implies that there exists a shrinking sequence of closed balls $\overline{B}_{n+1} \subseteq U_n$. By the preceding lemma, $\cap \overline{B}_n \neq \varnothing$. Since $\cap \overline{B}_n \subseteq \cap U_n$, this contradicts the condition $\cap U_n = \varnothing$. ☐

The Baire theorem can be reformulated as follows: *If the union of a sequence of closed sets in a complete metric space has an interior point, then at least one of these sets has an interior point.*

For applications of the Baire theorem, the reader is referred to Chapter 2 in Kantorovich and Akilov (1982).

Exercise. Give an example of an incomplete space in $\mathbb{R}^2$ for which the Baire theorem holds (see Section 4.3.C in Postnikov, 1987).

6. Consider directed sets $(X, \succeq)$, $(Y, \succeq)$, their product $(X \times Y, \succeq)$, and a separated regular topological space $(F, \mathcal{V})$. Every *double net* $h : X \times Y \to F$ defines *simple nets* $h(x, \cdot) : Y \to F$ and $h(\cdot, y) : X \to F$ $(x \in X, y \in Y)$. We call the limit $c \in F$ of h the *double limit* and write $\lim_{x,y} h(x, y) = c$. The limits $\lim_y h(x, y) = f(x)$ and $\lim_x h(x, y) = g(y)$ of the nets $h(x, \cdot)$ and $h(\cdot, y)$ are said to be *simple*, and the limits $\lim_x \lim_y h(x, y) = \lim_x f(x) = a$ and $\lim_y \lim_x h(x, y) = \lim_y g(y) = b$ are called *iterated*.

Theorem. *Suppose that the double limit $\lim_{x,y} h(x, y) = c$ exists.*

(1) If the limit $\lim_y h(x, y) = f(x)$ exists for all x, then there exists a limit

$$\lim_x \lim_y h(x, y) = \lim_x f(x) = a \text{ and } a = c.$$

(2) If the limit $\lim_x h(x, y) = g(y)$ exists for all y, then there exists a limit

$$\lim_y \lim_x h(x, y) = \lim_y g(y) = b \text{ and } b = c.$$

Proof. We now prove (1). Suppose the contrary: $h \to c$, $h(x, \cdot) \to f(x)$ $(x \in X)$, and $f \not\to c$. Then the regularity of F implies the existence of a closed neighborhood $\overline{V}$ of c such that for any x we have $f(r(x)) \notin \overline{V}$ for some $r(x) \succeq x$. At the same time, there exist $\bar{x} \in X$ and $\bar{y} \in Y$ such that $h(x, y) \in \overline{V}$ for all $x \succeq \bar{x}$, $y \succeq \bar{y}$. In particular, $h(r(\bar{x}), y) \in \overline{V}$ if $y \succeq \bar{y}$. The complement $U = F \setminus \overline{V}$ is an open neighborhood of $f(r(\bar{x}))$. Since $h(r(\bar{x}), y) \notin U$ for all $y \succeq \bar{y}$, it follows that $h(r(\bar{x}), y) \not\to f(r(\bar{x}))$. A contradiction. Assertion (2) follows from (1) by symmetry. ☐

The theorem implies a sufficient condition for the equality of iterated limits.

Corollary. *Suppose that the double limit and all simple limits exist. Then both iterated limits exist and are equal to the double limit.*

Exercise. Give an example showing that the existence of a double limit is not necessary for the existence of equal iterated limits.

7. Replace a topological space $(F, \mathcal{V})$ by a uniform space $(F, (\Delta_i))$. We say that a convergence $h(x, \cdot) \to f(x)$ is *uniform in x* if, for any number $\varepsilon > 0$ and index i, there exists an element $y_i(\varepsilon) \in Y$ such that $\Delta_i(h(x, y), f(x)) \le \varepsilon$ for all $y \succeq y_i(\varepsilon)$ and all $x \in X$. Similarly, a convergence $h(\cdot, y) \to g(y)$ is *uniform in y* if, for any $\varepsilon > 0$ and i, there exists $x_i(\varepsilon) \in X$ such that $\Delta_i(h(x, y), g(y)) \le \varepsilon$ for all $x \succeq x_i(\varepsilon)$ and all $y \in Y$.

Exercise. Give examples of a uniform and a nonuniform convergence.

Theorem. (1) *If a convergence $h(x, y) \to f(x)$ with respect to y is uniform in x and the iterated limit $\lim_{x} \lim_{y} h(x, y) = \lim_{x} f(x) = a$ exists, then there exists a double limit $\lim_{x,y} h(x, y) = c$ and $c = a$.*

(2) *If a convergence $h(x, y) \to g(y)$ with respect to x is uniform in y and the iterated limit $\lim_{y} \lim_{x} h(x, y) = \lim_{y} g(y) = b$ exists, then there exists a double limit $\lim_{x,y} h(x, y) = c$ and $c = b$.*

Proof. We now prove (1). Take $\varepsilon > 0$, i, and $y_i(\varepsilon) \in Y$ such that $\Delta_i(h(x, y), f(x)) \le \varepsilon/2$ for all $y \succeq y_i(\varepsilon)$ and $x \in X$. Such $y_i(\varepsilon)$ exists because of the uniform convergence $h(x, y) \to f(x)$. Since $f(x) \to a$, there is an $x_i(\varepsilon) \in X$ such that $\Delta_i(f(x), a) \le \varepsilon/2$ for all $x \succeq x_i(\varepsilon)$. Consequently, $\Delta_i(h(x, y), a) \le \varepsilon$ for all $x \succeq x_i(\varepsilon)$, $y \succeq y_i(\varepsilon)$. Hence, $\lim_{x,y} h(x, y) = a$. Item (2) follows from (1) by symmetry. $\square$

Corollary. *If all simple limits exist, the convergence with respect to one of the variables is uniform, and the corresponding iterated limit exists, then the double limit and the other repeated limit exists and the three limits are equal.*

Proof. Assume that all simple limits $f(x)$ and $g(y)$ exist, the convergence $h(x, \cdot) \to f(x)$ with respect to y is uniform in x, and there is a double limit $\lim\limits_{x} f(x) = a$. Then, by the preceding theorem, $\lim\limits_{x,y} h(x, y) = c = a$. Since the simple limits $g(y)$ exist, using Theorem 1 of part 6, we conclude that $\lim\limits_{y} g(y) = b = c$. $\square$

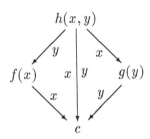

The connection between the simple, iterated, and double limits is shown in the diagram. This connection is also expressed by the equalities

$$\lim_{x} \lim_{y} h(x, y) = \lim_{y} \lim_{x} h(x, y) = \lim_{x,y} h(x, y),$$

which hold under the above conditions.

Exercise. Verify that the condition for uniform convergence is necessary for the existence of a double limit.

8. The closedness and compactness of sets in a space $(E, \mathcal{U})$ are closely connected with the convergence of nets therein.

Birkhoff's theorem. *A set $F \subseteq E$ is closed if and only if the limits of any converging net in F belong to F.*

Proof. Suppose that $F \subseteq E$ is closed, $x(i) \in F$, and $x(i) \to x \in E$. Assume that $x \in U = F^c$. Since F is closed, U is open and $x(i) \to x$ implies $x(i(0)) \in U$ for some index $i(0)$, which contradicts the condition $x(i) \in F$ for all i. Hence, $x \in F$.

Suppose now that $F \subseteq E$ is not closed and $x \in \overline{F} \setminus F$. The class $\mathcal{V}(x)$ of neighborhoods of a point x is directed by the inclusion $\subseteq$ (a smaller neighborhood is a successor). Since $x \in \overline{F}$, for any $V \in \mathcal{V}(x)$ there exists an $x(V) \in F \cap V$. Clearly, $x(V) \to x \notin F$. This completes the proof. $\square$

Remark. In a nonseparated space, some of the limits of a convergent net in a given set may not belong to this set. A closed set contains all limits of any convergent net whose points belong to this set.

Consider a set C in a topological space E. We say that a net of points in a set C *partially converges to* C if one of its subnets converges to a point in C (that is, if this convergence has a limit point in C).

Weierstrass's partial convergence theorem. *A set $C \subseteq E$ is compact if and only if every net of points in C partially converges in C.*

Proof. 1) Suppose that a set $C \subseteq E$ is compact. Consider a net $f : X \to C$ and a collection of its sections $fS(x) = \{f(y) \mid y \succeq x\}$. By the lemma of part 3, $\operatorname{Lim} f = \cap \overline{fS(x)}$. Since the collection of sections is centered and C is compact, it follows that $\cap \overline{fS(x)} \neq \varnothing$.

2) Suppose that a set $C \subseteq E$ satisfies the assumption of the theorem. Consider a collection of open sets $U(i) \subseteq E$ $(i \in I)$ covering C. Assume that $C \not\subseteq U(K) = \cup U(k)$ $(k \in K)$ for any finite $K \subseteq I$. Then there exists a collection of points $x(K) \in C \setminus U(K)$ $(K \in \mathcal{K})$. The class $\mathcal{K} = \mathcal{K}(I)$ of all finite subsets of a set I is directed by inclusion, and the points $x(K)$ are the values of a net $x : \mathcal{K} \to C$. By hypothesis, it has a subnet $x \circ r : T \to C$ converging to a point $a \in C$. Since $U(i)$ cover C, there exists an $i(a) \in I$ such that $a \in U(i(a))$. Therefore, $x(K) \in U(i(a))$ if $K = r(t)$ for all $t \in T$ that succeed some $t(a) \in T$. Put $K(a) = r(t(a)) \cup \{i(a)\}$. Since $r : T \to \mathcal{K}$ agrees with the directions for T and $\mathcal{K}$, there exists an element $t(0) \in T$ such that $K = r(t) \supseteq K(a)$ for any $t \in T$ succeeding $t(0)$. We can choose one of these elements that succeeds $t(a)$. If $t \succeq t(0)$ and $t \succeq t(a)$, then $x(K) \in U(i(a))$ and $x(K) \notin U(K)$ if $K = r(t)$. At the same time, from $i(a) \in K(a) \subseteq K$ and $U(K) = \cup U(k)$ $(k \in K)$ it follows that $x(K) \notin U(i(a))$. The obtained contradiction shows that $C \subseteq U(K)$ for a finite $K \subseteq I$. We conclude that the set C is compact. $\square$

Remark. If $C = E$, then Weierstrass's theorem states that the compactness of a space E is equivalent to the partial convergence of every net in E.

Exercise. Prove that a set $C \subseteq E$ is compact if and only if every simple net of points in C converges in C.

Corollary. *If a set $B \subseteq E$ is relatively compact, then every net of points in B partially converges in E.*

Proof. Every net $(x(i))$ in B is a net in $\overline{B}$. If $\overline{B}$ is compact, then, by Weierstrass's theorem, $(x(i))$ partially converges in $\overline{B}$. $\square$

Remark. The converse fails if the topological space E is arbitrary (see Chapter 1, Section 9, Exercise 23 in Bourbaki, 1998a).

Counterexample (P. E. Alaev). Suppose that $E = \mathbb{N}$ and the topology $\mathcal{U}$ for E is defined by the base $\mathcal{B} = \{1\} \cup \{\{1, n\} \mid n \geq 2\}$. The closure of the set $B = \{1\}$ is noncompact $\overline{B} = E$, although all nets in B converge to 1.

Exercise. Verify the converse for a regular space E.

9. Consider a uniform space $(E, (\Delta_i))$ with metrics Δ_i $(i \in I)$ and a set $X \subseteq E$. A collection of balls $B_i(k, \varepsilon) = \{x \mid \Delta_i(k, x) < \varepsilon\}$ $(k \in K \subseteq E)$ that covers X is called an (i, ε)-*net* for X. If there exists a finite (i, ε)-net in X for all $i \in I$ and $\varepsilon > 0$, then the set X is said to be *precompact* or *totally bounded*. A net *partially converges in itself* if one of its subnets converges in itself.

Theorem. *A set $X \subseteq E$ is precompact if and only if every net of points in X partially converges in itself.*

Proof. 1) Suppose that X is precompact and $(x(t))$ is a net in X. By the proposition in part 6 of Subsection 3.1.2, there is a simple subnet $(y(u))$ of points $y(u) = x(r(u)) \in X$. Since X is precompact, for any $i \in I$ and $\varepsilon > 0$ there exists a finite collection of balls $B_i(k, \varepsilon)$ covering X. Since the subnet $(y(u))$ is simple, from part 6 of Subsection 3.1.2 it follows that there exists a ball $B_i(k(0), \varepsilon)$ and an index $u(0)$ such that $\{y(u) \mid u \succeq u(0)\} \subseteq B_i(k(0), \varepsilon)$. Hence, $(y(u))$ converges in itself.

2) Suppose that X is not precompact and there are $i = i(0) \in I$ and $\varepsilon = \varepsilon(0) > 0$ such that there is no finite (i, ε)-net, i.e., there is no finite collection of balls $B_i(k, \varepsilon)$ covering X. Let $x(1) \in X$. By induction, we can prove that there exists a sequence of points $x(n) \in X$ such that $x(n) \notin B(n) = \bigcup_{m<n} B_i(x(m), \varepsilon)$. Since $\Delta_i(x(n), x(p)) \geq \varepsilon$ if $n \neq p$, it follows that $x(n)$ cannot converge in itself even partially. This completes the proof. $\square$

Corollary. *In a complete separated space, a set is compact if and only if it is precompact and closed.*

Proof. Consider a set X in a complete separated space E. If X is compact, then it is closed (Subsection 3.1.1, part 7). By Weierstrass' theorem, any net of points in X has a limit point and therefore converges in itself. By the preceding theorem, X is precompact. Assume now that X is precompact and closed. Since E is complete and nets of points in X partially converge in themselves, it follows that they partially converge to some points of E. By Birkhoff's theorem, these points belong to X. By Weierstrass's theorem, this implies that X is compact.   □

Exercise. 1) Prove that the relative compactness of a set in a complete metric space is equivalent to its precompactness. 2) Verify this assertion for a multimetric space using its regularity.

In a uniform space $(E, (\Delta_i))$, any open cover $\mathcal{C} = (U(j))$ of a compact set $C \subseteq E$ can be associated with a metric characteristic $(i \in I, j \in J)$.

Lemma. There exist an index $i(0) \in I$ and a number $\varepsilon(0) > 0$ such that the ball $B_{i(0)}(x, \varepsilon(0))$ is included in the set $U(j(x))$ for any $x \in C$.

Proof. For any $x \in C$, take $j(x) \in J$, $i(x) \in I$, and $\varepsilon(x) > 0$ such that $B_{i(x)}(x, 2\varepsilon(x)) \subseteq U(j(x))$. The balls $B_{i(x)}(x, \varepsilon(x))$ are open and make up a cover of C. The set C is compact. Therefore, there exists a finite set $K \subseteq C$ such that the balls $B_{i(k)}(k, \varepsilon(k))$ $(k \in K)$ make up a cover of C. Suppose that $\Delta_{i(0)}(x, y) \geq \max \Delta_{i(k)}(x, y)$ and $\varepsilon(0) = \min \varepsilon(k)$ $(k \in K)$, $x \in B_{i(k)}(k, \varepsilon(k))$ and $y \in B_{i(k)}(k, \varepsilon(k))$. It follows from the definitions that $B_{i(0)}(x, \varepsilon(0)) \subseteq B_{i(k)}(x, \varepsilon(0)) \subseteq B_{i(k)}(k, \varepsilon(k))$. Since $\Delta_{i(k)}(k, y) \leq \Delta_{i(k)}(k, x) + \Delta_{i(k)}(x, y)$, we have $B_{i(k)}(x, \varepsilon(k)) \subseteq B_{i(k)}(k, 2\varepsilon(k))$. Consequently, for any $x \in C$ there exists a point $k = k(x) \in K$ and an index $j(x)$ such that $B_{i(0)}(x, \varepsilon(0)) \subseteq B_{i(k)}(k, 2\varepsilon(k)) \subseteq U(j(x))$.   □

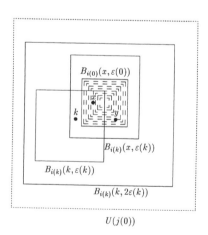

$$U(j(0))$$

Consider the metric space (E, Δ). In this case, by the preceding lemma, for an open cover $\mathcal{C} = (U(j))$ of a compact set $C \subseteq E$, there exists a number

$\varepsilon(0) > 0$ such that for $x \in C$ the ball $B(x, \varepsilon(0))$ is covered by one set $U(j(x))$. The supremum $\lambda = \lambda(C)$ of the set of such numbers $\varepsilon(0)$ is called *Lebesgue's number* of C. This number can be infinite.

10. Consider topological spaces $(E, \mathcal{U})$ and $(F, \mathcal{V})$, a function $f : E \to F$, a point $a \in E$, and a direction $x \to a$, $x \neq a$, for E. If $\lim\limits_{x \to a, x \neq a} f(x) = f(a)$, then the function f is said to be *continuous at a*. This means that for any neighborhood $V \in \mathcal{V}$ of $f(a)$ there exists a neighborhood $U \in \mathcal{U}$ of $a \in E$ such that $f(U) \subseteq V$, i.e., for all points x that are U-close to a, their images $f(x)$ are V-close to $f(a)$. A function $f : E \to F$ that is continuous at any point a of a set $A \subseteq E$ is said to be *continuous on the set A*. A function that is continuous on the entire space is called simply *continuous*. The continuity of restrictions means their continuity on the corresponding subspaces.

Examples. 1) Constant functions are continuous in all topologies. 2) If $(E, \mathcal{U})$ is discrete $(\mathcal{U} = \mathcal{P}(E))$ or $(F, \mathcal{V})$ is indiscrete $(\mathcal{V} = \{\varnothing, F\})$, then all functions $f : E \to F$ are continuous. 3) If $(F, \mathcal{V})$ is discrete, then a function $f : E \to F$ is continuous at a point $a \in E$ if and only if f is constant in a neighborhood $U \subseteq E$ of a. The continuity of f at any point $a \in E$ means that f is *locally constant*.

Exercise. Prove that arbitrary neighborhoods can be replaced by open neighborhoods in the definition of continuity.

Proposition. *A function $f : E \to F$ is continuous if and only if the inverse image $X = f^{-1}(Y)$ of any open set Y in F is an open set in E.*

Proof. Suppose that f is continuous, $Y \subseteq F$ is open, $x \in X = f^{-1}(Y)$, and $y = f(x)$. Then $V = Y$ is a neighborhood of y, and there is a neighborhood $U \subseteq E$ of x such that $f(U) \subseteq Y$ and $U \subseteq f^{-1}(f(U)) \subseteq f^{-1}(Y) = X$. Hence, X is open. Conversely, assume that the inverse image $X = f^{-1}(Y)$ of any open set Y in F is open in E. Let $x \in E$ and $y = f(x) \in F$. Then $X = f^{-1}(Y)$ is an open neighborhood of x for any open neighborhood Y of y and $f(X) \subseteq Y$. Every neighborhood V of y includes its open neighborhood Y and, consequently, $f(X) \subseteq Y \subseteq V$. We conclude that f is continuous. □

Exercise. Prove that the proposition still holds if open sets are replaced by closed sets.

Remark. The image of an open set under a continuous map can fail to be open. For example, suppose that $E = F = \mathbb{R}$ and $f(x) = x^2$ $(x \in \mathbb{R})$. Then $f(]-1, 1[) = [0, 1[$.

Theorem. *A composition of continuous functions is continuous.*

Proof. Apart from topological spaces $(E, \mathcal{U})$, $(F, \mathcal{V})$ and a function $f : E \to F$, consider a topological space $(G, \mathcal{W})$, a function $g : F \to G$, and the composition $h = g \circ f : E \to G$. Suppose that f and g are continuous and the inverse images $f^{-1}(V)$ and $g^{-1}(W)$ of open sets $V \subseteq F$ and $W \subseteq G$ are open. Then the inverse image $h^{-1}(W) = f^{-1}(g^{-1}(W))$ of any open set $W \subseteq G$ is open and h is continuous. $\square$

Exercise. Prove that the continuity of a function $f : E \to F$ is equivalent to each of the following properties: (1) $f(\overline{X}) \subseteq \overline{f(X)}$ $(X \subseteq E)$, (2) $\overline{f^{-1}(Y)} \subseteq f^{-1}(\overline{Y})$ $(Y \subseteq F)$.

11. Compactness is preserved under continuous maps.

Weierstrass's compactness theorem. *The image of a compact set under a continuous map is compact.*

Proof. Assume that $f : E \to F$ is continuous, $X \subseteq E$ is compact, and $f(X) = Y \subseteq F$. Consider a net of points $y(i) \in Y$ and a net of points $x(i) \in X$ such that $f(x(i)) = y(i)$. Since X is compact, there exists a subnet $x(r(i))$ of $x(i)$ that converges to a point $x \in X$. Since f is continuous, it follows that $y(r(i)) = f(x(r(i))) \to f(x) = y \in Y$. We conclude that Y is compact. $\square$

Weierstrass's compactness theorem is widely used.

Corollary. *A product* $(E, \mathcal{U}) = \prod(E_i, \mathcal{U}_i)$ *is compact if and only if each* $(E_i, \mathcal{U}_i)$ *is compact.*

Proof. The *if* part follows from the Tikhonov theorem. Prove the *only if* part. It follows from the definitions that the projection $p_i : E \to E_i$ that maps a point $x = (x_i) \in E$ to $p_i(x) = x_i \in E_i$ is continuous. If E is compact, then, by Weierstrass's theorem, $E_i = p_i(E)$ is also compact. $\square$

Consider topological spaces $(E, \mathcal{U})$ and $(F, \mathcal{V})$, sets $X \subseteq E$ and $Y \subseteq F$, a bijective map $f : X \to Y$, and the inverse map $g = f^{-1} : Y \to X$. Clearly, g is also a bijective map. If f and g are continuous, then they are called *homeomorphisms* from X onto Y and from Y onto X, respectively, and the sets X and Y are called *homeomorphic*, as well as the subspaces X and Y.

The Homeomorphism Theorem. *A one-to-one continuous map from a compactum into a separated space is a homeomorphism.*

Proof. We use the above notation and assume that X is compact and f is continuous. Then, by Weierstrass's theorem, $Y = f(X)$ is a compactum. If $g = f^{-1}$ has a discontinuity at a point $b \in Y$, then there exists a net of points $y(i) \in Y$ such that $y(i) \to b$ but $x(i) = g(y(i)) \nrightarrow g(b) = a$. Since $x(i) \nrightarrow a$, it follows that $x(r(i)) \notin U(a)$ for some subnet $(x(r(i)))$ and open neighborhood $U(a)$ of a. The closed subset $C = X \setminus U(a)$ of the compactum X is compact; therefore, there exists a subnet $(x(sr(i)))$ of the net of points $(x(r(i))) \in C$ that converges to a point $c \in C$. Since f is continuous, $y(sr(i))) = f(x(sr(i))) \to f(c)$. Since $y(i) \to b$, it follows that $y(sr(i))) \to b = f(a)$. Since F is separated, we have $f(a) = f(c)$, though $a \neq c$. This contradicts the injectivity of f. We conclude that $g = f^{-1}$ is continuous. $\quad\square$

Exercise. Find out whether the homeomorphism theorem still hold without the requirement of separatedness in its assumption.

Remark. M. A. Lavrent'ev proved an important theorem on the extension of homeomorphisms in complete metric spaces to sets of a special kind (see Subsection 4.3.21 in Engelking, 1989).

12. Continuous maps preserve connectedness.

Bolzano's theorem. *The image of a connected set under a continuous map is connected.*

Proof. Suppose that $f : E \to F$ is continuous, $X \subseteq E$ is connected, and $f(X) = Y \subseteq F$. Let Y be disconnected and let $V, W \subseteq F$ make up a partition of Y. Since f is continuous, the inverse images $T = f^{-1}(V)$ and $U = f^{-1}(W) \subseteq E$ are open. Since $YV \neq \varnothing$, $YW \neq \varnothing$, $Y \subseteq V \cup W$, and $YVW = \varnothing$, it follows that $XT \neq \varnothing$, $XU \neq \varnothing$ $X \subseteq T \cup U$, and $XTU = \varnothing$. Indeed, if $y \in YV$, then $y = f(x) \in V$ for some $x \in X$, and $f(x) \in V$ means that $x \in T = f^{-1}(V)$. Consequently, $x \in XT$ and $XT \neq \varnothing$. Similarly, $XU \neq \varnothing$. Moreover, $X \subseteq f^{-1}(Y) \subseteq f^{-1}(V \cup W) = f^{-1}(V) \cup f^{-1}(W) = T \cup U$ and $XTU \subseteq f^{-1}(Y)f^{-1}(V)f^{-1}(W) = f^{-1}(YVW) = f^{-1}(\varnothing) = \varnothing$. Hence, the open sets T and U make up a partition of X. This contradicts the assumption that X is connected. $\quad\square$

These general theorems imply classical theorems for real functions of one variable.

Exercises. Prove the following assertions for a continuous function $f : \mathbb{R} \to \mathbb{R}$: 1) The image $Y = f(X)$ of any interval $X \subseteq \mathbb{R}$ is an interval in $\mathbb{R}$; therefore, f assumes all intermediate values. 2) The image $Y = f([a, b])$ of any closed interval $[a, b] \subseteq \mathbb{R}$ is a closed interval in $\mathbb{R}$; therefore, f assumes its smallest and greatest values in $[a, b]$.

13. A map $f : E \to F$ from a multimetric space $(E, (\rho_i))$ to a multimetric space $(F, (\Delta_j))$ is said to be *uniformly continuous* if for any index j and number $\varepsilon > 0$ there exists an index i and a number $\delta > 0$ such that $\Delta_j(f(x), f(y)) < \varepsilon$ for all $x, y \in E$ with $\rho_i(x, y) < \delta$. Let $\mathcal{V}$ and $\mathcal{W}$ be the filters of entourages for E and F (part 6 of Subsection 3.1.1). It follows from the definitions that the above condition is equivalent to the following condition: for any $W \in \mathcal{W}$, there exists a $V \in \mathcal{V}$ such that $(f(x), f(y)) \in W$ for all $x, y \in E$ that belong to V.

Proposition. *A uniformly continuous map is continuous in uniform topologies.*

Proof. The assertion follows from the definitions: if $y = a \in E$, then the inequalities $\Delta_j(f(x), f(y)) < \varepsilon$ and $\rho_i(x, a) < \delta$ mean that $f(x) \in B_j(f(a), \varepsilon)$ and $x \in B_i(a, \delta)$.   □

The continuity of $f : E \to F$ means that, for any point $a \in E$, index j, and number $\varepsilon > 0$, there exists an index $i(a)$ and a number $\delta(a) > 0$ such that $\Delta_j(f(x), f(a)) < \varepsilon$ for $x \in E$ satisfying the condition $\rho_{i(a)}(x, a) < \delta(a)$. If f is uniformly continuous, then there exist an index $i = i(a)$ and a number $\delta = \delta(a) > 0$, common for all points $a \in E$, that provide the inequality $\Delta_j(f(x), f(a)) < \varepsilon$ if $\rho_i(x, a) < \delta$.

Example. The function $f(x) = 1/x$ is uniformly continuous in $[1, \infty[$ and nonuniformly continuous in $]0, \infty[$.

For maps of compact spaces, continuity is equivalent to uniform continuity.

Theorem. *A continuous map from a compact multimetric space to a multimetric space is uniformly continuous.*

Proof. If $f : E \to F$ is continuous, then, for any point $a \in E$, index j, and number $\varepsilon > 0$, there exists an index $i(a)$ and a number $\delta(a) > 0$ such that $\Delta_j(f(x), f(a)) < \varepsilon/2$ if $\rho_{i(a)}(x, a) < \delta(a)$. If E is compact, then, by the lemma of part 8, there exists an index i and a number $\delta > 0$ such that for any $x \in E$ the ball $B_i(x, \delta)$ is included in a ball $B_{i(a)}(x, \delta(a))$. Suppose that $\Delta_i(x, y) < \delta$. Then $y \in B_i(x, \delta(a)) \subseteq B_{i(a)}(x, \delta(a))$ and $\Delta_{i(a)}(x, a) < \delta(a)$, $\Delta_{i(a)}(y, a) < \delta(a)$. Therefore, $\Delta_j(f(x), f(a)) < \varepsilon/2$, $\Delta_j(f(y), f(a)) < \varepsilon/2$, and $\Delta_j(f(x), f(y)) < \varepsilon$. Hence, f is uniformly continuous. $\quad\square$

Remark. The topology of any compact space is defined by a multimetric (see Chapter II, Section 4, Theorem 1 in Bourbaki, 1998a). Therefore, we can speak of a uniformly continuous map from a compact space into a multimetric space.

14. Consider topological spaces $(E, \mathcal{U})$ and $(F, \mathcal{V})$, a set $A \subseteq E$ and its closure $\bar{A}$, and a function $f : A \to F$. Suppose that $(F, \mathcal{V})$ is separated and regular and there exists a limit $\bar{f}(\bar{a}) = \lim f(x)$ $(x \to \bar{a}, x \in A)$ for any point $\bar{a} \in \bar{A}$. We denote by $\bar{f}$ the function on $\bar{A}$ with $\bar{f}(\bar{a}) \in F$ $(\bar{a} \in \bar{A})$. Since F is separated, $\bar{f}$ is a many-to-one correspondence.

Theorem. *The function $\bar{f}$ is the only continuous function on $\bar{A}$ with values in F that is an extension of f.*

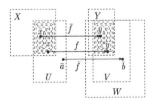

Proof. If $\bar{a} = a \in A$, then $\bar{f}(a) = f(a)$. The function $\bar{f}$ is an extension of f. We now prove that $\bar{f}$ is continuous. The regularity of F and the existence of a limit $\bar{b} = \bar{f}(\bar{a})$ imply that, for any neighborhood $W \subseteq F$ of $\bar{b}$, there exists a closed neighborhood $V \subseteq W$ of $\bar{b}$ and an open neighborhood $U \subseteq E$ of $\bar{a}$ such that $f(AU) \subseteq V$. Suppose that $\bar{x} \in \bar{A}U$. Since $\bar{y} = \bar{f}(\bar{x}) = \lim f(x)$ $(x \to \bar{x}, x \in A)$, it follows that, for any neighborhood $Y \subseteq F$ of $\bar{y}$, there exists a neighborhood $X \subseteq E$ of $\bar{x}$ such that $f(AX) \subseteq Y$. Since $\bar{x} \in \bar{A}$, $\bar{x} \in U$, and U is open, we conclude that UX is a neighborhood of $\bar{x}$ and $AUX \neq \varnothing$. Consequently, the intersection $VY \supseteq f(AU)f(AX) \supseteq$

$f(AUX)$ is nonempty ($f(x) \in f(AU) \subseteq V$ and $f(x) \in f(AX) \subseteq Y$ if $x \in AUX$). Therefore, $\bar{y} = \bar{f}(\bar{x}) \in \overline{V} = V \subseteq W$ and $\bar{f}(\bar{A}U) \subseteq W$. Hence, $\bar{f}$ is continuous.

We now prove the uniqueness of $\bar{f}$. Suppose that $\bar{g} : \bar{A} \to F$ is continuous and $\bar{g}(x) = f(x)$ ($x \in A$). Then $\bar{g}(\bar{a}) = \lim f(x) = \bar{f}(\bar{a})$ as $x \to \bar{a}$, $x \in A$ for any $\bar{a} \in \bar{A}$, and $\bar{g} = \bar{f}$. $\square$

Remark. A function f is continuous together with its extension $\bar{f}$. The equalities $\bar{f}(a) = \lim f(x) = f(a)$ ($x \to a$, $x \in A$) for $a \in A$ follow from the definitions.

Since a separated locally compact space is regular (Proposition of part 14 of Subsection 3.1.1), the continuous extension theorem holds for a separated locally compact space $(F, \mathcal{V})$.

The function $\bar{f} : \bar{A} \to F$ such that $\bar{f}(\bar{a}) = \lim f(x)$ ($x \to \bar{a}$, $x \in A$) for $\bar{a} \in \bar{A}$ is called the *continuous extension of a function* $f : A \to F$ *to* $\bar{A}$.

Example. Suppose that $E = F = \mathbb{R}$, $A = \mathbb{R} \backslash \{0\}$, and $f(x) = x \sin(1/x)$ ($x \in A$). Then $\bar{f}(0) = 0$. If $f(x) = \sin(1/x)$ ($x \in A$), then the extension $\bar{f} : \mathbb{R} \to \mathbb{R}$ of the function $f : \mathbb{R} \to \mathbb{R}$ is discontinuous for any value $\bar{f}(0) \in \mathbb{R}$.

15. Consider multimetric spaces $(E, (\rho_i))$, $(F, (\Delta_j))$, a set $A \subseteq E$ and its closure $\bar{A}$, and a uniformly continuous function $f : A \to F$. Assume that $(F, (\Delta_j))$ is separated and complete.

Theorem. *The continuous extension $\bar{f}$ of a uniformly continuous function f exists and is uniformly continuous.*

Proof. The definition implies that every multinormed space is regular. Since f is uniformly continuous and the space $(F, (\Delta_j))$ is complete, there exists a limit $\bar{f}(\bar{a}) = \lim f(x)$ ($x \to \bar{a}$, $x \in A$). Hence, by the theorem of part 14, there exists a continuous extension $\bar{f} : \bar{A} \to F$ for $f : A \to F$. Since f is uniformly continuous and $\bar{f}$ is continuous at $\bar{x}, \bar{y} \in \bar{A}$, it follows that for any index j and any $\varepsilon > 0$ there are indices $i(1), i(2), i(3)$, and numbers $\delta(1), \delta(2), \delta(3) > 0$ such that $\Delta_j(f(x), f(y)) < \varepsilon/3$, $\Delta_j(f(x), \bar{f}(\bar{x})) < \varepsilon/3$, and $\Delta_j(f(y), \bar{f}(\bar{y})) < \varepsilon/3$ whenever $\rho_{i(1)}(x, y) < \delta(1)$, $\rho_{i(2)}(x, \bar{x}) < \delta(2)$, and $\rho_{i(3)}(y, \bar{y}) < \delta(3)$, respectively. By the definition of a multimetric, there exists an index i and a number δ that make it possible to replace the last inequalities by the inequalities $\rho_i(x, y) < 3\delta$, $\rho_i(x, \bar{x}) < 3\delta$, and $\rho_i(y, \bar{y}) < 3\delta$.

Suppose that $\rho_i(\bar{x}, \bar{y}) < \delta$. Since $\bar{x}, \bar{y} \in \bar{A}$, there exist $x, y \in A$ such that $\rho_i(x, y) < \delta$, $\rho_i(y, \bar{y}) < \delta$, and therefore $\rho_i(x, y) < 3\delta$. Consequently, $\Delta_j(\bar{f}(\bar{x}), \bar{f}(\bar{y})) < \Delta_j(f(x), \bar{f}(\bar{x})) + \Delta_j(f(x), f(y)) + \Delta_j(f(y), \bar{f}(\bar{y})) < \varepsilon$. We conclude that $\bar{f}$ is uniformly continuous. $\quad\square$

Example. Let $E = \mathbb{R} \times \mathbb{R}$, $A = \mathbb{Q} \times \mathbb{Q}$, $F = \mathbb{R}$, $f(x, y) = x + y$, and $g(x, y) = xy$ $(x, y \in \mathbb{Q})$. Then $\bar{f}(\bar{x}, \bar{y}) = \bar{x} + \bar{y}$ and $\bar{g}(\bar{x}, \bar{y}) = \bar{x}\bar{y}$ $(\bar{x}, \bar{y} \in \mathbb{R})$.

Exercise. Let $E = \mathbb{R} \times \mathbb{R}$, $A = \mathbb{R} \times \mathbb{R}$, $F = \bar{\mathbb{R}}$, $f(x, y) = x + y$, and $g(x, y) = xy$ $(x, y \in \mathbb{R})$. Prove that there are no continuous extensions $\bar{f} : \bar{\mathbb{R}} \times \bar{\mathbb{R}} \to \bar{\mathbb{R}}$ and $\bar{g} : \bar{\mathbb{R}} \times \bar{\mathbb{R}} \to \bar{\mathbb{R}}$ for $f : \mathbb{R} \to \mathbb{R}$ and $g : \mathbb{R} \to \mathbb{R}$.

Remark. Since a separated compact space is multimetrizable and complete (see Chapter II, Section 4, Theorem 1 in Bourbaki, 1998a), the uniformly continuous extension theorem holds for a separated compact space $(F, \mathcal{V})$ for the multimetric (Δ_j) defined by the topology $\mathcal{V}$. It should be noted that, as opposed to this subsection, the main notion in the theory of limit in Bourbaki (1998a) is not a net, but a filter. The connection between them is described in Subsection 3.1.3, part 2.

3.2. DIFFERENTIAL

The notion of the differential formalizes the idea of linear approximation. The theory of differentiation in normed spaces is described in detail in Cartan (1970) and Lavrent'ev and Savel'ev (1995). Differentiation on manifolds is studied in the second part of Cartan (1970). Tensor differentiation is described in Chapter 4 of Sulanke and Wintgen (1972). The book Frölicher and Bucher (1966) is devoted to the general theory of differentiation in vector spaces with convergence.

The theory of differentiation in Banach spaces is the most developed one. It has extensive applications. The section is mainly devoted to this theory and contains lectures from Savel'ev (1975).

3.2.1. The definition of the differential

The differential of a function is defined as the main linear part of the increment of the function.

1. Consider two vector spaces $(\mathbb{K}, E)$ and $(\mathbb{L}, F)$ with real or complex scalar fields $\mathbb{K}$ and $\mathbb{L}$. We assume that $\mathbb{K} \subseteq \mathbb{L}$. This means that $\mathbb{K} = \mathbb{L} = \mathbb{R}$ or $\mathbb{K} = \mathbb{L} = \mathbb{C}$, or $\mathbb{K} = \mathbb{R}$ and $\mathbb{L} = \mathbb{C}$. The case $\mathbb{K} = \mathbb{C}$ and $\mathbb{L} = \mathbb{R}$ is excluded. The inclusion $\mathbb{K} \subseteq \mathbb{L}$ is assumed throughout the section.

Suppose that $U \subseteq E$, $u \in U$, $V = U - u = \{v = t - u \mid t \in U\}$. Then, given a function $f : U \to F$, we can define the function $\Delta f u : V \to F$ such that

$$\Delta f u(v) = f(u + v) - f(u) \quad (v \in V)$$

Since $u \in U$, we have $0 = u - u \in V$. The function $\Delta f u$ is called the *increment of f at the point u*. The increment $\Delta f u$ can be associated with the operators

$$\Delta_u : \mathcal{F}(U, F) \to \mathcal{F}(V, F), \quad \Delta f : U \to \mathcal{F}(V, F),$$

defined by the equalities $\Delta_u(f) = \Delta f u$ and $\Delta f(u) = \Delta f u$. The operator Δ_u is linear:

$$\Delta_u(af + bg) = a \cdot \Delta_u(f) + b \cdot \Delta_u(g) \quad (a, b \in \mathbb{K}, \ f, g \in \mathcal{F}(U, F)).$$

As a rule, Δf is nonlinear.

Example. Let $U = E = F = \mathbb{K} = \mathbb{L} = \mathbb{C}$ and $f(x) = x^2$ ($x \in \mathbb{C}$). Then $V = \mathbb{C}$ and $\Delta f u(v) = (u + v)^2 - u^2 = 2uv + v^2$ ($u, v \in \mathbb{C}$).

Exercise. Prove that the operator $\Delta : \mathcal{F}(U, F) \to \mathcal{F}(U, \mathcal{F}(V, F)$ defined by the equality $\Delta(f) = \Delta f$ ($f \in \mathcal{F}(U, F)$) is continuous.

2. Assume that the spaces E and F are normed, and consider another normed space $(\mathbb{M}, G)$ with real or complex scalar field $\mathbb{M} \supseteq \mathbb{L}$. For simplicity, we denote the norms $\| \ \|_E$, $\| \ \|_F$, and $\| \ \|_G$ by the same symbol $\| \ \|$. We say that $h : V \to F$ is *small relative to* $g : V \to G$ in a neighborhood of 0 and write $h = o(g)$ if for any $\varepsilon > 0$ there exists a $\delta > 0$ such that

$$\|h(v)\| \le \varepsilon \|g(v)\| \quad (\|v\| \le \delta, \ v \in V).$$

We say that h is *bounded relative to* g in a neighborhood of 0 and write $h = O(g)$ if there exist $\alpha > 0$ and $\delta > 0$ such that

$$\|h(v)\| \le \alpha \|g(v)\| \quad (\|v\| \le \delta, \ v \in V).$$

Clearly, $h = O(g)$ if $h = o(g)$.

Examples. 1) Let g be constant: $g(v) = c \in G$ ($v \in V$). Then $h = o(c)$ means that $h(v) \to 0$ ($v \to 0$). It is conventional to write $o(1)$ instead of $h = o(c)$. The expression $h = O(c) = O(1)$ means that h is bounded in a neighborhood of 0. 2) Suppose that $E = G$ and $g(v) = v$ ($v \in V$). Then $h = o(v)$ means that $h(0) = 0$ and $\|v\|^{-1}h(v) \to 0$ ($v \to 0$, $v \neq 0$). If $h = O(v)$, then $h = o(1)$. 3) Suppose that $E = F = G = V = \mathbb{C}$ and (s^n) is the sequence of power functions $s^n(v) = v^n$ ($v \in \mathbb{C}$). Then $s^1 = o(1)$, $s^2 = o(v)$, and $s^{n+1} = o(s^n)$.

3. Consider the vector spaces $\mathcal{L}(E,F)$, $\mathcal{L}(F,G)$, $\mathcal{L}(E,G)$, and the vector algebra $\mathcal{L}(E,E)$ of linear operators described in Subsection 2.2.2. A linear operator $A \in \mathcal{L}(E,F)$ is said to be *bounded* if there exists an $\alpha \geq 0$ such that $\|Ax\| \leq \alpha\|x\|$ ($x \in E$).

Bounded linear operators form subspaces $\mathcal{B}(E,F)$, $\mathcal{B}(F,G)$, and $\mathcal{B}(E,G)$ in the corresponding spaces and a subalgebra $\mathcal{B}(E,E)$ in $\mathcal{L}(E,E)$. This follows from the equalities

$$\|Ax + Bx\| \leq (\alpha + \beta)\|x\|, \quad \|\gamma Ax\| \leq |\gamma|\alpha\|x\|,$$

for $A, B \in \mathcal{L}(E,F)$ and $x \in E$, where $\|Ax\| \leq \alpha\|x\|$ and $\|Bx\| \leq \beta\|x\|$. Moreover, if $\|Ax\| \leq \alpha\|x\|$ and $\|By\| \leq \beta\|y\|$ for $A \in \mathcal{B}(E,F)$, $B \in \mathcal{B}(F,G)$, and $x \in E$, then $\|BAx\| \leq \beta\|Ax\| \leq \beta\alpha\|x\|$. In particular, the latter inequalities hold for $E = F = G$. The equality

$$\|A\| = \sup\{\|x\|^{-1}\|Ax\| : x \in E, \|x\| \neq 0\}$$

defines the *norm* of an operator $A \in \mathcal{B}(E,F)$. Since $\|x\|^{-1}\|Ax\| \leq \alpha$, we have $\|A\| < \infty$. The inequality $\|x\|^{-1}\|Ax\| \leq \|A\|$ implies

$$\|Ax\| \leq \|A\|\,\|x\| \quad (x \in E).$$

Applying this inequality to the operators $A \in \mathcal{B}(E,F)$ and $B \in \mathcal{B}(F,G)$ consecutively, we obtain

$$\|BA\| \leq \|B\|\,\|A\|.$$

This inequality can be strict. This is a distinguishing feature of the operator norm, as opposed to the absolute value of a number.

Exercise. Prove that $\|A\| = \sup\{\|Au\| : u \in E, \|u\| \leq 1\} = \inf\{\alpha \geq 0 : \|Ax\| \leq \alpha\|x\| \ (x \in E)\}$.

4. Consider the product $E = \prod E_j$ of normed spaces E_j $(1 \leq j \leq n)$, the projection $p_j : E \to E_j$ such that $p_j(x) = x_j \in E_j$ for all $x = (x_l) \in E$, and the embedding $r_j : E_j \to E$, $r_j(x_j) = x_j^0 \in E$ $(x_j^0 = (z_l)$, $z_j = x_j$, and $z_l = 0$ for $l \neq j)$. The operator r_j is an isomorphism from E_j onto $E_j^0 = r_j(E_j) \subseteq E$. Clearly, the operators p_j, r_j, and $r_j p_j$ are linear, and the map $r_j p_j : E_j \to E_j$ is the identity map.

Proposition. *The linear operators p_j, r_j, and $r_j p_j$ are bounded and have norm 1.*

Proof. It follows from the definitions that $\|p_j(x)\| = \|x_j\| \leq \|x\| = \max \|x_l\|$ and $\|r_j(x_j)\| = \|x_j\|$ for all $x = (x_l) \in E$ and $x_j \in E_j$. Therefore, $\|p_j\| \leq 1$ and $\|r_j\| = 1$. Since $\|p_j(x_j^0)\| = \|x_j^0\|$, it follows that $\|p_j\| \geq 1$. Consequently, the composition $r_j p_j$ is also bounded and $\|r_j p_j\| \leq 1$. The reverse inequality $\|r_j p_j\| \geq 1$ follows from the equality $r_j p_j(x_j^0) = x_j^0$. $\square$

The projection p_j singles out the jth component $x_j \in E_j$ of a vector $x = (x_l) \in E$, and the embedding identifies it with the vector $x_j^0 \in E_j^0 \subseteq E$. The composition $r_j p_j$ projects the space E onto its subspace E_j^0: $(r_j p_j)(r_j p_j) = r_j(p_j r_j)p_j = r_j p_j$.

5. Complete separated normed spaces are called *Banach spaces*.

Proposition. *If F is a Banach space, then $\mathcal{B} = \mathcal{B}(E, F)$ is a Banach space.*

Proof. Let $T \in \mathcal{B}$ and $\|T\| = 0$. Then $\|Tx\| = 0$ and $Tx = 0$ $(x \in E)$, $T = 0$ if F is separated. Hence, $\mathcal{B}$ is separated.

Consider a sequence of operators $T_n \in \mathcal{B}$ converging in itself. Since $\|T_p x - T_n x\| = \|(T_p - T_n)x\| \leq \|T_p - T_n\| \cdot \|x\|$, the sequence of points $T_n x \in F$ converges in itself for any $x \in E$. If F is complete, then $T_n x$ converges to a point $y = \lim T_n x \in F$. Therefore, if F is separated, the map $T : E \to F$ such that $Tx = \lim T_n x$ $(x \in E)$ is defined. Since the operators T_n and the limit are linear, we conclude that T is linear.

We now prove that $T \in \mathcal{B}$ and $T_n \to T$. Suppose that $\|T_p - T_n\| \leq \varepsilon$ $(p, n \geq n(\varepsilon))$. Then $\|T_p x - T_n x\| \leq \varepsilon \|x\|$, $\|Tx - T_n x\| \leq \varepsilon \|x\|$ $(n \geq n(\varepsilon))$, $\|Tx\| \leq \|T_{n(1)} x\| + \|x\| \leq (\|T_{n(1)}\| + 1)\|x\|$ $(x \in E)$. Hence, $T \in \mathcal{B}$ and the inequality $\|Tx - T_n x\| \leq \varepsilon \|x\|$ $(n \geq n(\varepsilon), x \in E)$ implies $\|T - T_n\| \leq \varepsilon$ for $n \geq n(\varepsilon)$. As a result, $T_n \to T$ and and the normed space $\mathcal{B}$ is sequentially complete. By Heine's criterion of convergence, this implies that $\mathcal{B}$ is complete. $\square$

Exercise. Prove this proposition for a net of operators.

Remark. If $E = F$, then $\mathcal{B}$ becomes a *normed algebra* with multiplication represented by the composition of operators. If E is a Banach space, then $\mathcal{B}$ is a *Banach algebra*. If $E \neq F$, then $\mathcal{B}$ is a Banach space if F is a Banach space and E is a normed space. In what follows, as a rule, we consider separated spaces.

6. The boundedness of a linear operator is equivalent to its continuity.

Lemma. *If an operator $T \in \mathcal{L}(E, F)$ is continuous at a point $0 \in E$, then T is uniformly continuous.*

Proof. Suppose that $\varepsilon > 0$, $\delta > 0$, and $\|Tz\| \leq \varepsilon$ for $\|z\| \leq \delta$. Then $\|Tx - Ty\| = \|T(x - y)\| \leq \varepsilon \ (\|x - y\| \leq \delta)$. $\square$

Proposition. *An operator in $\mathcal{L}(E, F)$ is bounded if and only if it is continuous.*

Proof. Suppose that an operator $T \in \mathcal{L}(E, F)$ is continuous. We now prove that $T \in \mathcal{B}(E, F)$. If $\|Tz\| \leq 1 \ (\|z\| \leq \delta, \ z \in E)$ and $x \in E \ (\|x\| \neq 0)$, then $\delta \|x\|^{-1} \|Tx\| = \|Tz\| \leq 1$ and $\|Tx\| \leq \delta^{-1} \|x\|$ for $z = \delta \|x\|^{-1} x$. Hence, T is bounded.

The continuity of an operator $T \in \mathcal{B}(E, F)$ follows from the inequality $\|Tx\| \leq \alpha \|x\| \ (x \in E)$. By the preceding lemma, T is continuous because it is continuous at zero. $\square$

Remark. Since any operator $T \in \mathcal{L}(E, F)$ is linear, its uniform convergence is equivalent to its continuity at any point $a \in E$.

7. Consider an open set $U \subseteq E$, a function $f : U \to F$, and a point $u \in U$. Assume that the spaces E and F are separated. Note that the set $V = U - u \subseteq E$ is open because so is U. The function f is said to be *differentiable at the point u* if there exists a bounded linear operator $df u : E \to F$ and a function $r f u : V \to F$ that is small relative to the identity function $\mathrm{id} : V \to E$ and satisfies the inequality

$$\Delta f u(v) = df u(v) + r f u(v) \quad (v \in V).$$

In this case, the operator $dfu \in \mathcal{B}(E, F)$ is called the *differential of f at u*, and the function rfu is called the *remainder of differentiation*. The relative smallness of the remainder means that

$$rfu(0) = 0, \quad \|v\|^{-1}\|rfu(v)\| \to 0 \quad (v \to 0, \ v \neq 0).$$

The phrase *the differential dfu exists* means that f is differentiable at u.

Theorem. *If the differential dfu exists, then it is unique.*

Proof. Let $T \in \mathcal{B}(E, F)$ and $\Delta fu(v) = dfu(v) + rfu(v) = T(v) + o(v)$ $(v \in V)$. Then, for any $\varepsilon > 0$, there exists a $\delta > 0$ such that

$$\|(T - dfu)(v)\| = \|T(v) - dfu(v)\| \leq \|rfu(v)\| + \|o(v)\| \leq 2\varepsilon \cdot \|v\|,$$

where $\|v\| < \delta$. Consequently, $\|T - dfu\| \leq 2\varepsilon$ for any $\varepsilon > 0$. Hence, $\|T - dfu\| = 0$ and $T = dfu$. □

Remark. The remainder $rfu = \Delta fu - dfu$ is also defined uniquely. However, it plays a secondary role and usually only its estimate is used.

Proposition. *If a function f is differentiable at a point u, then f is continuous at u.*

Proof. We have $f(u + v) = f(u) + dfu(v) + rfu(v)$ $(v \in V)$ and $dfu(v) \to 0$, $rfu(v) \to 0$ as $v \to 0$, $v \neq 0$. Therefore, $f(u + v) \to f(u)$ as $v \to 0$, $v \neq 0$. □

Examples. 1) Let $U = E = F = \mathbb{C}$, $u = 1$, and $f(x) = x^2$ $(x \in \mathbb{C})$. Then $V = U - 1 = \mathbb{C}$, $\Delta fu(v) = (1 + v)^2 - 1 = 2v - v^2$, $dfu(v) = 2v$, and $rfu(v) = v^2$ $(v \in \mathbb{C})$.

2) If $f \in \mathcal{B}(E, F)$, then $dfu = f$ $(u \in U)$.

If a function $f : U \to F$ is differentiable at any $u \in A \subseteq U$, then f is said to be *differentiable on A*. If $A = U$, then f is said to be *differentiable*.

8. Consider the following objects: the products $E = \prod E_j$ and $F = \prod F_i$ of normed spaces E_j, F_i $(1 \leq i \leq m, 1 \leq j \leq n)$; a rectangle $U = \prod U_j \subseteq E$ with open sides $U_j \subseteq E$; a point $u = (u_j) \in U$; a set $V = U - u$ and sets $V_j = U_j - u_j$; the projections $p_j : E \to E_j$, $q_i : F \to F_i$ and the embeddings $r_j : E_j \to E$, $s_i : F_i \to F_i$; the translation $Tu : E \to E$ such that $Tu(v) = u + v$ $(v \in E)$; a map $f : U \to F$. The compositions

$$f_{ij}u = q_i \cdot f \cdot Tu \cdot r_j : V_j \to F_i$$

are called the *partial functions* for f at a point u. By definition,

$$f_{ij}u(v_j) = q_i\left(f\left(u + v_j^0\right)\right) \quad (v_j \in V_j).$$

The compositions

$$f_{ij}^0 u = s_i \cdot f_{ij}u \cdot p_j = s_i q_i \cdot f \cdot Tu \cdot r_j p_j : V \to F$$

are also called the partial functions for f at u. By definition,

$$f_{ij}^0 u(v) = f_i^0(u + v_j^0) \quad (v \in V).$$

Since the maps $p_j r_j$ and $q_i s_i$ are identities,

$$f_{ij}u = q_i \cdot f_{ij}^0 u \cdot r_j.$$

For the increments of the partial functions, we have

$$\Delta f_{ij}u = q_i \cdot \Delta f_{ij}^0 u \cdot r_j = q_i \cdot \Delta f u \cdot r_j : V_j \to F_i,$$
$$\Delta f_{ij}^0 u = s_i \cdot \Delta f_{ij}u \cdot p_j = s_i q_i \cdot \Delta f u \cdot r_j p_j : V \to F.$$

Since we used the translation Tu in the definition of the partial functions $f_{ij}u$ and $f_{ij}^0 u$, the increment and the differential of $f_{ij}u$ and $f_{ij}^0 u$ at $0_j \in V_j$ and $0 \in V$ are connected with the increment and the differential of f at u. To avoid cumbersome calculations, similarly to the increments $\Delta f_{ij}u$ and $\Delta f_{ij}^0 u$, we will denote the corresponding differentials by $df_{ij}u$ and $df_{ij}^0 u$. Moreover, we put

$$\Delta_{ij}fu = \Delta f_{ij}^0 u, \quad d_{ij}fu = df_{ij}^0 u.$$

Thus, by definition,

$$\Delta f_{ij}u = df_{ij}u + r f_{ij}u, \quad \Delta_{ij}fu = d_{ij}fu + r_{ij}fu,$$

where $df_{ij}u \in \mathcal{B}(E_j, F_i)$, $d_{ij}fu \in \mathcal{B}(E, F)$, and $r f_{ij}u = o(v_j)$, $r_{ij}fu = o(v)$. We call $\Delta_{ij}fu$ and $d_{ij}fu$ the *partial increments* and the *partial differentials* of f at u, respectively. These terms are also used for $\Delta f_{ij}u$ and $df_{ij}u$.

9. Consider an open set $U \subseteq E = \mathbb{K}$, a point $u \in U$, a set $V = U - u$, and a normed space $(F, \| \ \|)$. Every linear function $l : \mathbb{K} \to F$ is defined by its coefficient $l(1)$: $l(t) = l(1) \cdot t$ $(t \in \mathbb{K})$. A linear function l is bounded on the unit ball by $\|l(1)\|$ and is therefore continuous. Let $f : U \to F$ be a function differentiable at u. The coefficient $f'(u) = dfu(1)$ of the differential dfu is called the *derivative* of the function f at the point u. The differential and the derivative of a function f are related by the inequality

$$dfu(v) = f'(u) \cdot v \quad (v \in \mathbb{K}).$$

Proposition. *If a function f is differentiable at a point u, then*

$$f'(u) = \lim(v^{-1}\Delta fu(v)) \quad (v \to 0, \ v \neq 0, \ v \in V).$$

Proof. Since f is differentiable at u, we have

$$v^{-1}\Delta fu(v) = v^{-1}dfu(v) + v^{-1}rfu(v) = f'(u) + v^{-1}rfu(v)$$

$(v \to 0, \ v \neq 0, \ v \in V)$. Since the remainder rfu is relatively small, we obtain the desired equality. ☐

Corollary. *A function $f : U \to F$ is differentiable at a point u if and only if there exists a limit of the quotient $|v^{-1}\Delta fu(v)|$ $(v \to 0, \ v \neq 0, \ v \in V)$.*

Proof. The *only if* part follows from the proposition. The *if* part is proved as follows. Assume that $\lim(v^{-1}\Delta fu(v)) = c$ $(v \to 0, \ v \neq 0, \ v \in V)$. Then, for any $\varepsilon > 0$ there exists a $\delta > 0$ such that $\|v^{-1}\Delta fu(v) - c\| \leq \varepsilon$ $(0 < |v| \leq \delta, \ v \in V)$ and $\|\Delta fu(v) - cv\| \leq v\varepsilon$ $(|v| \leq \delta, \ v \in V)$. Consequently, the function $rfu : V \to F$ such that $rfu(v) = \Delta fu(v) - cv$ $(v \in V)$ is relatively small, and f is differentiable at u. ☐

Example. Consider the Banach space with $F = \mathbb{K}^m$ as the set of points, the Euclidean norm $\| \ \|$, and the standard basis. A function $f : U \to F$ defines the coordinate functions $f_i : U \to F$ such that $f(t) = (f_i(t))$ for $i \in [1, m]$, $t \in U$. It is easy to check that $f'(t) = (f_i'(t))$ $(i \in [1, m], u \in U)$. In particular, if $U = \mathbb{R}$, $F = \mathbb{R}^2$, and $u \in \mathbb{R}$, then $f'(u) = (-\sin u, \cos u)^t$, $f_1'(u) = -\sin u$, and $f_2'(u) = \cos u$ for $f(u) = (\cos u, \sin u)^t$, and

$$df(u) = f'(u) \cdot v = \begin{pmatrix} -\sin u \\ \cos u \end{pmatrix} \cdot v \quad (v \in \mathbb{R}).$$

Remark. Many different symbols are used to designate derivatives. In particular, it is conventional to write dfu or Dfu instead of $f'(u)$. This operator notation is useful in all respects.

10. Suppose that $E_j = \mathbb{K}$, $F_i = \mathbb{L}$, and $E = \mathbb{K}^n$, $F = \mathbb{L}^m$. Then $f_{ij}u : V_j \to \mathbb{L}$ ($V_j \subseteq \mathbb{K}$), i.e., the partial functions are scalar functions of a scalar argument. If the partial differentials $df_{ij}u$ exist, then the *partial derivatives* $f'_{ij}(u)$ exist and

$$df_{ij}u = f'_{ij}u \cdot v_j \quad (v_j \in V_j).$$

The partial differential df_{ij} defines the derivative f'_{ij}. The continuity of df_{ij} is equivalent to the continuity of f'_{ij}. Indeed,

$$|(df_{ij}(u+z) - df_{ij}u)v_j| \cdot |v_j|^{-1} = |f'_{ij}(u+z) - f'_{ij}(u)| \quad (v_j \neq 0)$$

and therefore

$$\|df_{ij}(u+z) - df_{ij}u\| = |f'_{ij}(u+z) - f'_{ij}(u)| \quad (u, u+z \in U).$$

Consider the matrices $f'(u) = (f'_{ij}(u))$ and $f' = (f'_{ij})$ composed of the numbers $f_{ij}(u) \in \mathbb{L}$ and the functions $f'_{ij} : U \to \mathbb{L}$. The scalar matrix $f'(u)$ is the value of the matrix function f'. The equality

$$dfu(v) = f'(u) \cdot v,$$

which is similar to the equality for scalar functions, holds for the vectors $dfu(v) = (q_i \cdot dfu(v)) \in \mathbb{L}^m$, $v = (v_j) \in \mathbb{K}^n$ and the $m \times n$-matrix $f'(u) = (f'_{ij}(u))$. The matrix function f' is called the *derivative* of f. The matrix $f'(u)$ is called the *Jacobi matrix*. If f is differentiable at a point u, then $f'(u)$ exists. The continuity of f' means the continuity of all partial derivatives f'_{ij}.

Example. Set $E = U = \mathbb{R}^2$, $F = \mathbb{R}^3$, $f(x) = (x_1 \sin x_2, x_2 cos x_1, x_1 x_2)^t$. Then

$$f'(u) = \begin{pmatrix} \sin u_2 & u_1 \cos u_2 \\ -u_2 \sin u_1 & \cos u_1 \\ u_2 & u_1 \end{pmatrix} \quad (x = (x_1, x_2),\ u = (u_1, u_2) \in \mathbb{R}^2).$$

3.2.2.   Differentiation rules

An Appropriate application of some simple differentiation rules makes it possible to find the differentials of many functions.

1. Consider normed spaces $(E, \| \, \|)$ and $(F, \| \, \|)$ with scalar fields $\mathbb{K}$ and $\mathbb{L}$ ($\mathbb{K} \subseteq \mathbb{L}$), an open set $U \subseteq E$, a point $u \in U$, a set $V = U - u$, functions $f : U \to F$ and $g : U \to F$ differentiable at u, and numbers $a, b, c \in \mathbb{L}$.

Proposition. $d(f + g)u = dfu + dgu$, $d(cf)u = c \cdot dfu$.

Proof. 1) We have $\Delta(f+g)u(v) = \Delta fu(v) + \Delta gu(v) = (dfu + dgu)(v) + (rfu + rgu)(v)$ $(v \in V)$. The function $dfu + dgu$ is continuous and linear, while $rfu + rgu$ is relatively small. Therefore, $f + g$ is differentiable at u and the first equality of the proposition holds.

2) We have $\Delta(cf)u(v) = c\Delta fu(v) = (c \cdot dfu)(v) + (c \cdot rfu)(v)$. The function $c \cdot dfu$ is continuous and linear, while $c \cdot rfu$ is relatively small. Therefore, cf is differentiable at u and the second equality holds.   □

This proposition implies the rule of differentiating a linear combination.

Corollary. $d(af + bg)u = f \cdot dfu + b \cdot dgu$.

2. Consider normed spaces E, F, G, and H, whose norms are designated by $\| \, \|$, an open set $U \subseteq E$, a point $u \in U$, and the set $V = U - u$. Let functions $f : U \to F$ and $g : U \to F$ be differentiable at u.

In addition, consider the normed space $F \times G$ with the norm

$$\|(x, y)\| = \max\{\|x\|, \|y\|\} \quad (x \in F, \ y \in G).$$

Continuous bilinear functions will be called *products*. Take a product $(x, y) \to xy$ on $F \times G$ with values in H. Together with f and g, this product defines the function $fg : U \to H$ such that $(fg)(t) = f(t)g(t)$ $(t \in U)$. This function is called the *product of the functions f and g*. It is easy to prove that fg is differentiable at u and its differential $d(fg)u$ is obtained by successive differentiation of the factors. The proof will involve the use of the following continuity criterion for bilinear maps.

Lemma. *A bilinear function is continuous if and only if it is bounded on the unit ball.*

Proof. Consider the bilinear function $(x, y) \to xy$.

1) If this function is continuous, then for any $\varepsilon > 0$ there exists a $\delta > 0$ such that $\|pq\| \leq \varepsilon$ whenever $\|(p, q)\| \leq \delta$ for $p \in F$ and $q \in G$. Suppose that $\|(x, y)\| \leq 1$, $x \in F$, $y \in G$. Then $\|(p, q)\| = \|\delta(x, y)\| \leq \delta$ for $(p, q) = \delta(x, y)$, $(x, y) = \delta^{-1}(p, q) = (\delta^{-1}p, \delta^{-1}q)$, and $\|xy\| = \|\delta^{-1}p \cdot \delta^{-1}q\| = \delta^{-2}\|pq\| \leq \delta^{-2}\varepsilon = r$. Hence, the function $(x, y) \to xy$ is bounded on the unit ball $\overline{B}(0, 1) \subseteq F \times G$.

2) If the function under consideration is bounded on the unit ball, then $\|xy\| \leq c$ ($\|(x, y)\| \leq 1$, $x \in F$, $y \in G$) for some finite $c > 0$. Suppose that $\delta > 0$, $a, p \in F$, $b, q \in G$, $\|(p, q)\| \leq \delta$. Then $\|(x, y)\| = \|\delta^{-1}(p, q)\| \leq 1$ as $(x, y) = \delta^{-1}(p, q)$, $(p, q) = \delta(x, y) = (\delta x, \delta y)$, and $\|aq\| = \|a\|\|a\|^{-1}\|a \cdot \delta y\| = (\|a\|\delta)\|\|a\|^{-1}a \cdot y\| \leq \|a\|\delta c$, $\|pb\| = \|\delta x \cdot \|b\|\|b\|^{-1}b\| = (\delta\|b\|)\|x \cdot \|b\|^{-1}b\| \leq \delta\|b\|c$, $\|pq\| = \|\delta x \cdot \delta y\| = \delta^2\|xy\| \leq \delta^2 c$. Therefore, $\|(a + p)(b + q) - ab\| \leq \|aq\| + \|pb\| + \|pq\| \leq (\|a\| + \|b\| + \delta)c\delta$ and, consequently, the function $(x, y) \to xy$ is continuous. □

Proposition. $d(fg)u(v) = dfu(v) \cdot g(u) + f(u) \cdot dgu(v)$.

Proof. We have $\Delta(fg)u(v) = f(u + v)g(u + v) - f(u)g(u) = (f(u) + dfu(v) + rfu(v))(g(u) + dgu(v) + rgu(v)) - f(u)g(u) = l(v) + r(v)$ $(v \in V)$, $l(v) = dfu(v) \cdot g(u) + f(u) \cdot dgu(v)$ $(v \in E)$, $r(v) = f(u) \cdot rgu(v) + dfu(v) \cdot (dgu(v) + rgu(v)) + rfu(v) \cdot (g(u) + dgu(v) + rgu(v))$ $(v \in V)$. The function $l = dfu \cdot g(u) + f(u) \cdot dgu$ is continuous and linear. The function $r(v)$ is relatively small. Indeed, using the preceding lemma, the definition of relative smallness, and the inequality for the norm of a continuous linear function, we conclude that for any $\varepsilon \in]0, 1[$ there exist $\delta \in]0, 1[$ and $c > 0$ such that $\|f(u) \cdot rgu(v)\| \leq c\|f(u)\| \cdot \|rgu(v)\| \leq c\|f(u)\| \cdot \varepsilon\|v\|\|dfu(v) \cdot (dgu(v) + rgu(v))\| \leq c\|dfu(v)\|\|dgu(v) + rgu(v)\| \leq c\|dfu\|\|v\|(\|dgu\|\|v\| + \varepsilon\|v\|) \leq c\|dfu\|(\|dgu\| + \varepsilon)\|v\|^2$, $\|rfu(v) \cdot (g(u) + dgu(v) + rgu(v))\| \leq c\|rfu(v)\|\|g(u) + dgu(v) + rgu(v)\| \leq c(\|g(u)\| + \|dgu\|\|v\| + \varepsilon\|v\|) \cdot \varepsilon\|v\|\|r(v)\| \leq c(\|f(u)\| + \|g(u)\| + (\|dfu\| + 1)(\|dgu\| + 1)) \cdot \varepsilon\|v\|$ $(\|v\| \leq \delta, v \in V)$. As a result, fg is differentiable at u and the desired inequality holds. □

3. Consider normed spaces E, F, and G, whose norms are designated by $\| \|$, an open set $U \subseteq E$, a point $u \in U$ and the set $V = U - u$, an open set $Y \subseteq F$, a point $y \in Y$ and the set $Z = Y - y$, functions $f : U \to F$ and $g : Y \to G$ differentiable at u and y. Let $f(U) \subseteq Y$ and $f(u) = y$. Then the composition $g \circ f : U \to G$ is differentiable at u and its differential $d(g \circ f)u$ is the composition of the differentials dfu and dgy.

Theorem. $d(g \circ f)u = dgy \circ dfu$.

Proof. For $v \in V$, we have

$$
\begin{aligned}
\Delta(g \circ f)u(v) &= \Delta gy(\Delta fu(v)) = dgy(\Delta fu(v)) + rgy(\Delta fu(v)) \\
&= dgy(dfu(v) + rfu(v)) + rgy(\Delta fu(v)) \\
&= dgy(dfu(v)) + dgy(rfu(v)) + rgy(\Delta fu(v)) = l(v) + r(v),
\end{aligned}
$$

where $l(v) = dgy(dfu(v))$ $(v \in E)$ and $r(v) = dgy(rfu(v)) + rgy(\Delta fu(v))$ $(v \in V)$. The function $l = dgy \circ dfu$ is continuous and linear. The function r is relatively small. Indeed, involving the definition of relative smallness and the inequality for the norm of a continuous linear function, we conclude that for any $\varepsilon \in \,]0, 1[$ there exists an $\alpha > 0$ such that

$$
\|dgy(rfu(v))\| \leq \|dgy\| \|rfu(v)\| \leq \|dgy\| \cdot \varepsilon \|v\|,
$$

$$
\|\Delta fu(v)\| \leq \|dfu(v)\| + \|rfu(v)\| \leq \|dfu\| \|v\| + \varepsilon \|v\| \leq (\|dfu\| + 1)\|v\|,
$$

where $\|v\| \leq \alpha$, $v \in V$. At the same time, there exists a $\beta > 0$ such that $\|rgy(z)\| \leq \varepsilon \|z\|$ ($\|z\| \leq \beta$, $z \in Z$). Then $\|rgy(\Delta fu(v))\| \leq \varepsilon \|\Delta fu(v)\| \leq \varepsilon(\|dfu\| + 1)\|v\|$, $\|r(v)\| \leq (\|dfu\| + \|dgy\| + 1)\varepsilon \|v\|$, $\|v\| \leq \delta = \min\{\alpha, (1 + |dfu\|)^{-1}\beta\}$, $v \in V$. Consequently, the function $g \circ f$ is differentiable at u and the desired inequality holds.   □

Consider the standard Euclidean spaces $E = \mathbb{R}^n$, $F = \mathbb{R}^m$, and $G = \mathbb{R}^l$. In this case, the differentials dfu, dgy, and $d(g \circ f)u$ are defined by their matrices $f'(u)$, $g'(y)$, and $(g \circ f)'(u)$ with respect to the standard bases. These matrices are called the derivatives. They satisfy the equalities

$$
dfu(v) = f'(u) \cdot v, \quad dgy(z) = g'(y) \cdot z,
$$
$$
d(g \circ f)u(v) = (g \circ f)'(u) \cdot v \quad (v \in E, \ z \in F).
$$

Since the matrix of a composition of linear maps is equal to the product of the matrices of these linear maps, the above theorem implies the following corollary.

Corollary. $(g \circ f)'(u) = g'(f(u)) \cdot f'(u)$.

4. We now consider several examples.

Example 1. $E = \mathbb{R}^2$, $F = \mathbb{R}^3$, $G = \mathbb{R}^2$, $U = E$, $Y = F$; $t = (t_1, t_2)$, $u = (u_1, u_2) \in \mathbb{R}^2$, $x = (x_1, x_2, x_3)$, $y = (y_1, y_2, y_3) \in \mathbb{R}^3$; $f(t) = (t_1 \cdot \sin t_2, t_2 \cdot \cos t_1, t_1 t_2)$, $g(x) = (x_1 x_2 x_3, x_3^2)$, $h(t) = (g \circ f)(t) = (t_1^2 t_2^2 \cos t_1 \cdot \sin t_2, t_1^2 t_2^2)$,

$$f'(u) = \begin{pmatrix} \sin u_2 & u_1 \cdot \cos u_2 \\ -u_2 \cdot \sin u_1 & \cos u_1 \\ u_2 & u_1 \end{pmatrix}, \quad g'(y) = \begin{pmatrix} y_2 y_3 & y_1 y_3 & y_1 y_2 \\ 0 & 0 & 2y_3 \end{pmatrix},$$

$$h'(u) = (g \circ f)'(u) = g'(f(u)) \cdot f'(u) =$$

$$\begin{pmatrix} u_1 u_2^2 \cdot \cos u_1 & u_1^2 u_2 \cdot \sin u_2 & u_1 u_2 \cdot \cos u_1 \cdot \sin u_2 \\ 0 & 0 & 2u_1 u_2 \end{pmatrix} \begin{pmatrix} \sin u_2 & u_1 \cdot \cos u_2 \\ -u_2 \cdot \sin u_1 & \cos u_1 \\ u_2 & u_1 \end{pmatrix}.$$

Example 2. $E = \mathbb{R}$, $F = \mathbb{R}^2$, $G = \mathbb{R}$, $U = \,]-2^{-1}, 2^{-1}[$, $Y = B(0, 1)$; $t, u \in U$, $x = (x_1, x_2)$, $y = (y_1, y_2) \in Y$; $f(t) = (t, (2^{-1} - t^2)^{1/2})$, $g(x) = (1 - x_1^2 - x_2^2)^{1/2})$, $(g \circ f)(t) = 2^{-1/2}$;

$$f'(u) = \begin{pmatrix} 1 \\ -u(2^{-1} - u^2)^{-1/2} \end{pmatrix}, \quad g'(y) = -(1 - y_1^2 - y_2^2)^{-1/2}(y_1, y_2),$$

$$(g \circ f)'(u) = g'(f(u)) \cdot f'(u)$$

$$= -2^{-1}(u, (2^{-1} - u^2)^{1/2}) \begin{pmatrix} 1 \\ -u(2^{-1} - u^2)^{-1/2} \end{pmatrix} = -2^{-1}(u - u) = 0.$$

Example 3. $E = F = G = \mathbb{R}^2$, $U = E$, $Y = F$; $t = (t_1, t_2)$, $u = (u_1, u_2)$, $x = (x_1, x_2)$, $y = (y_1, y_2) \in \mathbb{R}^2$; $f(t) = (t_1 \cdot \cos t_2, t_1 \cdot \sin t_2)$, $g(x) = (x_1^2 + x_2^2, x_1 - x_2)$, $(g \circ f)(t) = (t_1^2, t_1 \cdot (\cos t_2 - \sin t_2))$,

$$f'(u) = \begin{pmatrix} \cos u_2 & -u_1 \cdot \sin u_2 \\ \sin u_2 & u_1 \cdot \cos u_2 \end{pmatrix}, \quad g'(y) = \begin{pmatrix} 2y_1 & 2y_2 \\ 1 & -1 \end{pmatrix},$$

$$(g \circ f)'(u) = g'(f(u)) \cdot f'(u)$$
$$= \begin{pmatrix} 2u_1 \cos u_2 & 2u_1 \cdot \sin u_2 \\ 1 & -1 \end{pmatrix} \begin{pmatrix} \cos u_2 & -u_1 \cdot \sin u_2 \\ \sin u_2 & u_1 \cdot \cos u_2 \end{pmatrix}$$
$$= \begin{pmatrix} 2u_1 & 0 \\ \cos u_2 - \sin u_2 & -u_1(\cdot \sin u_2 + \cos u_2) \end{pmatrix}.$$

3.2.3. Lagrange's theorem

Lagrange's theorem can be considered to be the main theorem of differential calculus. It establishes a connection between the differential and the variation of a function.

1.   Consider an open interval $U \subseteq \mathbb{R}$, a normed space $(F, \| \; \|)$, and differentiable functions $h : U \to F$ and $\varphi : U \to \mathbb{R}$.

Lagrange's theorem. *If* $\|h'(u)\| < \varphi'(u)$ $(u \in U)$, *then*

$$\|h(t) - h(u)\| < \varphi(t) - \varphi(u) \quad (t > u; \; t, u \in U).$$

Proof. Suppose that the theorem is false, which means that there exist points $t_0 > u_0$ in U such that $\|h(t_0) - h(u_0)\| \geq \varphi(t_0) - \varphi(u_0)$. For a natural number n, suppose that

$$\|h(t_n) - h(u_n)\| \geq \varphi(t_n) - \varphi(u_n) \tag{1}$$

for points $t_n > u_n$ in U. Then $c_n = 2^{-1}(u_n + t_n)$ satisfies at least one of the inequalities

$$\|h(t_n) - h(c_n)\| \geq \varphi(t_n) - \varphi(c_n), \quad \|h(c_n) - h(u_n)\| \geq \varphi(c_n) - \varphi(u_n).$$

If the first inequality holds, we take $u_{n+1} = c_n$, $t_{n+1} = t_n$. Otherwise, we take $u_{n+1} = u_n$, $t_{n+1} = c_n$. In any case, $\|h(t_{n+1}) - h(u_{n+1})\| \geq \varphi(t_{n+1}) - \varphi(u_{n+1})$. By induction, this implies that there exists a shrinking sequence of intervals $[u_n, t_n]$ $(t_n - u_n \to 0)$ with endpoints that satisfy (1). This sequence shrinks into a point $c \in U$: $t_n \downarrow c \uparrow u_n$, $\cap[u_n, t_n] = \{c\}$. We now prove that there exists a sequence of points $s_n \in U$ converging to c such that

$$\|(s_n - c)^{-1}(h(s_n) - h(c))\| \geq (s_n - c)^{-1}(\varphi(s_n) - \varphi(c)). \tag{2}$$

By inequality (1), since h and φ are continuous, it follows that at least one of the following inequalities holds for any n:

$$\|h(t_n) - h(c)\| \geq \varphi(t_n) - \varphi(c), \quad \|h(c) - h(u_n)\| \geq \varphi(c) - \varphi(u_n). \quad (3)$$

Moreover, there are three alternatives: $u_n < c < t_n$, $u_n = c < t_n$, and $u_n < c = t_n$. If $u_n < c < t_n$ and the first inequality in (3) holds, then we take $s_n = t_n$. If $u_n < c < t_n$ and the first inequality in (3) fails, then we take $s_n = u_n$. If $u_n = c < t_n$, then (1) implies the first inequality in (3) and we take $s_n = t_n$. Finally, if $u_n < c = t_n$, then the second inequality in (3) holds, and we may take $s_n = u_n$. In any case, inequality (2) holds. Since $u_n \to c$ and $t_n \to c$, it follows that $s_n \to c$. Inequalities (2) imply that $\|h'(c)\| \geq \varphi'(u)$. This contradicts the hypothesis. $\quad\square$

Remark. The theorem does not hold if the open set $U \in \mathbb{R}$ is assumed to be arbitrary.

Exercise. Provide the corresponding counterexamples.

2. Consider several corollaries to Lagrange's theorem. We preserve the notation of part 1 of this subsection, where U is an *open interval* in $\mathbb{R}$.

Corollary 1. *If* $\|h'(u)\| \leq \varphi'(u)$ $(u \in U)$, *then*

$$\|h(t) - h(u)\| \leq \varphi(t) - \varphi(u) \quad (t \geq u;\ t, u \in U).$$

Proof. If $u = t$, then the desired inequality is equivalent to the equality $0 = 0$. For any number $n > 0$, take the function $\varphi_n : U \to \mathbb{R}$ such that $\varphi_n(t) = \varphi(t) + n^{-1}t$ $(t \in U)$. This function is differentiable and $\varphi_n'(u) = \varphi'(u) + n^{-1} > \varphi'(u) \geq \|h'(u)\|$ $(u \in U)$. Then, by Lagrange's theorem,

$$\|h(t) - h(u)\| < \varphi_n(t) - \varphi_n(u) = \varphi(t) - \varphi(u) + n^{-1}(t - u) \quad (t > u).$$

This proves the desired inequality. $\quad\square$

Let $c \geq 0$.

Corollary 2. *If* $\|h'(u)\| \leq c$ $(u \in U)$, *then*

$$\|h(t) - h(u)\| \leq c(t - u) \quad (t \geq u;\ t, u \in U).$$

Proof. Let $\varphi : U \to \mathbb{R}$, $\varphi(t) = ct$ $(t \in U)$. This function is differentiable and $\varphi'(u) = c$ $(u \in U)$. The desired inequality follows from Corollary 1.   □

3. Consider normed spaces $(F, \| \, \|)$ and $(G, \| \, \|)$, an open set $Y \subseteq F$, a function $g : Y \to G$, points $a, b \in F$ $(a \neq b)$, a number $c \geq 0$, $[a, b] = \{x = a + t(b - a) : t \in [0, 1]\} \subseteq Y$, $]a, b[= \{x = a + t(b - a) : t \in]0, 1[\} \subseteq [a, b]$. Suppose that g is continuous at any point of $[a, b]$ and differentiable at any point in $]a, b[$.

Corollary 3. *If* $\|dgy\| \leq c$ $(y \in]a, b[)$, *then* $\|g(b) - g(a)\| \leq c|b - a|$.

Proof. Consider the function $f :]0, 1[\to F$ such that $f(t) = a + t(b - a)$ $(t \in]0, 1[)$ and the function $h = g \circ f$. The function f is differentiable and $f'(u) = b - a$ $(u \in]0, 1[)$. Consequently, h is differentiable and $dhu(v) = dgy(dfu(v)) = dgy((b - a)v) = dgy(b - a) \cdot v$, $h'(u) = dgy(b - a)$, $\|h'(u)\| \leq \|dgy\|\|b - a\| \leq c\|b - a\|$ for $u \in]0, 1[$, $v \in \mathbb{R}$, and $y = f(u)$. Hence $\|h(x) - h(y)\| \leq c\|b - a\|$ if $x = f(t)$, $y = f(u)$ for any $t > u$ in $]0, 1[$. Since $f(t) \to b$ $((t \to 1)$, $f(u) \to a$ $(u \to 0)$, and g is continuous at a and b, it follows that $h(x) = g(f(t)) \to g(b)$ $(t \to 1)$, $h(y) = g(f(t)) \to g(a)$ $(t \to 0)$, and the desired inequality follows.   □

Given $y \in]a, b[$, choose a number $c(y) \geq 0$. As before, assume that g is continuous on $[a, b]$ and differentiable on $]a, b[$.

Corollary 4. *If* $\|dg(x) - dg(y)\| \leq c(y)$ *for* $x \in]a, b[$, *then*

$$\|g(b) - g(a) - dgy(b - a)\| \leq c(y)\|b - a\|.$$

Proof. Consider the function $h : Y \to G$ such that $h(x) = g(x) - dgy(x)$ for $x \in Y$. This function is continuous on $[a, b]$ and differentiable on $]a, b[$ together with g, dgy, and $dhx = dgx - dgy$ for $x \in]a, b[$. By hypothesis, $\|dhx\| = \|dgx - dgy\| \leq c(y)$ $(x \in]a, b[)$. Since $h(b) - h(a) = (g(b) - dgy(b)) - (g(a) - dgy(a)) = g(b) - g(a) - (dgy(b) - dgy(a)) = g(b) - g(a) - dgy(b - a)$, the desired inequality follows from Corollary 3.   □

Remark. Corollary 4 yields an estimate for the mixed remainder $\Delta ga(b - a) - dgy(b - a)$ $(y \neq a)$.

4. In addition, assume that the open set $Y \subseteq F$ is convex and $g : Y \to G$ is differentiable. Corollary 3 readily implies the following assertion.

Corollary 5. *If* $\|dgy\| \leq c$ $(y \in Y)$, *then* $\|g(x) - g(y)\| \leq c\|x - y\|$ $(x, y \in Y)$.

Remark. In this case, g is said to satisfy the *the Lipschitz condition with Lipschitz constant c*. Corollary 5 states that if a function on an open convex set has a bounded differential, then it satisfies the Lipschitz condition.

Assume that the open set Y is connected and the function $g : Y \to G$ is differentiable.

Corollary 6. *The function* g *is constant if and only if* $dgy = 0$ *for all* $y \in Y$.

Proof. 1) If $g(y) = q \in G$ $(y \in Y)$, then $\Delta gy(z) = g(y + z) - g(y) = q - q = 0$ $(z \in Y - y)$, the function $g : Y \to G$ is differentiable, and $dgy = 0$ $(y \in Y)$.

2) For any $x \in Y$, consider a ball $B(x, r) \subseteq Y$. By Corollary 5, from $dgy = 0$ $(y \in Y)$ it follows that $g(x) = g(y)$ for $y \in B(x, r)$. The function g is constant (locally constant) on any ball in Y. Since Y is connected, g is constant on the entire set Y. Indeed, let $a \in Y$, $p = g(a) \in G$, $A = g^{-1}(p)$, and $B = Y - A$. The set A is open. Indeed, if $x \in A$, then $g(y) = g(x) = p$, $y \in A$ for $y \in B(x, r)$ and $B(x, r) \subseteq A$. Since g is continuous and the set $F \backslash \{p\}$ is open, the set $B = g^{-1}(F \backslash \{p\})$ is open. The open sets A and B are disjoint and their union is Y. Since Y is connected, at least one of the sets A and B must be empty. Since A contains a, it is nonempty. Consequently, $B = \varnothing$, $A = Y$, and $g(x) = g(a)$ $(x \in Y)$. $\square$

Remark. If $Y = U \subseteq \mathbb{R}$ is an interval, Corollary 6 follows directly from Corollary 2 for $c = 0$.

5. Consider normed spaces E and F, an open set $U \subseteq E$, a point $u \in U$, and a map $f : U \to F$. Assume that for any $v \in E$ there exists a limit

$$d_w fu(v) = \lim_{t \to 0, t \neq 0} \frac{f(u + tv) - f(u)}{t} \quad (t \in \mathbb{R}, \ u + tv \in U).$$

If the function $d_w fu : E \to F$ is a bounded linear operator, then $d_w fu$ is called the *weak differential* of the *Gâteaux differential* of the function f at u. By contrast, the usual differential dfu is called the *strong differential* or the *Fréchet differential*.

Clearly, strong differentiability implies weak differentiability and the equality $df u = d_w f u(v)$. The converse fails even if E is finite-dimensional.

Counterexample. (Kolmogorov and Fomin, 1989, Chapter 10, Section 1). Let $U = E = \mathbb{R}^2$, $F = \mathbb{R}$, $f(x_1, x_2) = x_1^3 x_2 (x_1^4 + x_2^2)^{-1}$ $((x_1, x_2) \neq (0,0))$, $f(0,0) = 0$. Then $d_w f u(v) = 0$ for $u = (0,0)$ and $df u$ does not exist.

The chain rule fails for weak differentials.

Exercise. Provide the corresponding counterexamples.

In addition, let $(E, \| \ \|)$ and $(F, \| \ \|)$ be Banach spaces and let the function f be weakly differentiable at any $[a, b] \subseteq U$. From Lagrange's theorem and the Hahn–Banach theorem, it follows that

$$\| f(b) - f(a) - d_w f a \cdot v \| \leq \sup_{0 \leq \theta \leq 1} \| d_w f(a + \theta(b - a)) - d_w f a \| \cdot \| b - a \| \quad (1)$$

(see Kolmogorov and Fomin, 1989, Chapter 10, Section 1).

A sufficient condition for strong differentiability is given by the following theorem.

Theorem. *If $d_w f$ is defined in a neighborhood of a point u and is continuous at u, then df exists and $df u = d_w f u$.*

Proof. By assumption, for any $\varepsilon > 0$ there exists a $\delta > 0$ such that

$$\| d_w f(u + v) - d_w f u \| \leq \varepsilon \quad (\|v\| \leq \delta). \quad (2)$$

Inequalities (1) and (2) yield

$$\| f(u + v) - f(u) - d_w f u \cdot v \| \leq \varepsilon \|v\| \quad (\|v\| \leq \delta). \quad (3)$$

Inequality (3) is equivalent to the assertion of the theorem.   □

Example. Consider a Hilbert space $E = H$. Assume that $U = E$, $F = \mathbb{R}$, and f is a functional such that $f(x) = \|x\|^2$ $(x \in H)$. Then $df u \cdot v = d_w f u \cdot v = 2u \cdot v$ $(u, v \in H.)$

Exercise. Find the points at which the function $g(x) = \|x\|$ $(x \in H)$ is differentiable.

3.2.4. Termwise differentiation

Consider normed spaces $(F, \| \ \|)$ and $(G, \| \ \|)$, an open set $Y \subseteq F$, a point $a \in Y$, and a sequence of differentiable functions $g_n : Y \to G$. Assume that G is a Banach space and Y is connected.

1. If for any $y \in Y$ there exists a neighborhood $B(y, r) \subseteq Y$ on which a sequence of functions $f_n : Y \to G$ uniformly converges to $f : Y \to G$, then the sequence f_n is said to converge *locally uniformly* to f. A sequence of differentiable functions $g_n : Y \to G$ is said to be *termwise differentiable* if the sequence of functions $g_n : Y \to G$ converges locally uniformly to a function $g : Y \to G$ and the sequence of differentials $dg_n : Y \to \mathcal{L}(E, F)$ converges locally uniformly to the differential $dg : Y \to \mathcal{L}(E, F)$.

Suppose that the open set $Y = B$ is convex. We now prove a lemma on uniform convergence.

Lemma. *If the sequence of differentials dg_n converges uniformly and the sequence of points $g_n(a)$ converges, then the sequence of functions g_n is termwise differentiable and converges uniformly on any bounded subset of B.*

Proof. By hypothesis, for any $\varepsilon > 0$ there exists a number l such that $\|g_n(a) - g_m(a)\| \leq \varepsilon$, $\|d(g_n - g_m)y\| = \|dg_n y - dg_m y\| \leq \varepsilon$ $(y \in B)$ if $n, m \geq l$. Since B is convex, we have $\|(g_n - g_m)(x) - (g_n - g_m)(a)\| \leq \varepsilon \|x - a\|$ $(x \in B)$, $\|g_n(x) - g_m(x)\| \leq \|g_n(a) - g_m(a)\| + \varepsilon \|x - a\| \leq (1 + r)\varepsilon$ $(\|x - a\| \leq r$, $x \in B)$ for any $r > 0$ and $n, m \geq l$. Since G is complete, the last inequality implies the convergence of g_n to a function $g : B \to G$ that is uniform on any bounded subset of B. The function g is differentiable and its differential dgy is equal to the limit of the sequence of differentials dg_n at any $y \in B$. To prove this, we suppose that $y \in B$ and $hy = \lim dg_n y \in \mathcal{L}(E, F)$. Then

$$
\begin{aligned}
\|g(x) - g(y) - hy(x - y)\| &\leq \|(g(x) - g(y)) - (g_n(x) - g_n(y))\| \\
&+ \|g_n(x) - g_n(y) - dg_n(x - y)\| + \|dg_n(x - y) - hy(x - y)\| \quad (x \in B).
\end{aligned}
$$

We have proved that g_n converges at any point of B. Therefore, the inequality obtained for the difference of the values of the function $g_n - g_m$ at x and a holds if we replace a by y. Therefore, for any $\varepsilon > 0$ there exists a number m such that $\|dg_m y(x - y) - hy(x - y)\| = \|(dg_m y - hy)(x - y)\| \leq \|dg_m y - hy\| \|x - y\| \leq \varepsilon \|x - y\|$, $\|(g_n(x) - g_n(y)) - (g_m(x) - g_m(y))\| = \|(g_n - g_m)(x) - (g_n - g_m)(y)\| \leq \varepsilon \|x - y\|$ if $n \geq m$ and $x \in B$. Consequently, $\|(g(x) - g(y)) - (g_m(x) - g_m(y))\| \leq \varepsilon \|x - y\|$ $(x \in B)$. At the same time, there is a $\delta > 0$ such that $\|g_m(x) - g_m(y) - dg_m y(x - y)\| = \|r g_m y(x - y)\| \leq \varepsilon \|x - y\|$ $(\|x - y\| \leq \delta$, $x \in B)$. Hence, g is differentiable at y and $dgy = hy$. $\quad\square$

2. We now assume that the open set Y is connected.

Theorem. *If the sequence dg_n converges locally uniformly and the sequence $g_n(a)$ converges, then g_n is termwise differentiable.*

Proof. By hypothesis, for any $y \in Y$ there exists a neighborhood $B(y, r) \subseteq Y$ ($r \in \,]0, \infty[$) on which the sequence dg_n converges uniformly. If the sequence of the points $g_n(y)$ converges, then, by the lemma of part 1 of this subsection, the sequence $g_n(x)$ converges for all $x \in B(y, r)$. Hence, the set A of all points $y \in Y$ such that $g_n(y)$ converges is open. The set $B = Y \setminus A$ is also open. Indeed, take a point $y \in B$ and its neighborhood $B(y, r) \subseteq Y$ ($r \in \,]0, \infty[$) on which the sequence dg_n converges uniformly. If the sequence of the points $g_n(x)$ converges for some $x \in B(y, r)$, then, by Lemma 1, the sequence $g_n(y)$ converges and $y \in A$, which is impossible. Hence, $B(y, r) \subseteq B$. By assumption, $a \in A$ and $A \neq \varnothing$. Since Y is connected, this implies that $B = \varnothing$ and $A = Y$. Therefore, there exists a function $g : Y \to G$ such that $g(x) = \lim g_n(x)$ ($x \in Y$).

Take a point $y \in Y$ and its neighborhood $B(y, r) \subseteq Y$ on which the sequence of differentials dg_n converges uniformly. By the preceding lemma, this implies that the function g is differentiable at any $x \in B(y, r)$, g_n converges to g uniformly on $B(y, r)$, and the sequence dg_n converges to the differential of g uniformly on $B(y, r)$. □

3. Let $F = \mathbb{R}$. Then, every differentiable function $g_n : Y \to G$ has a derivative $g_n' : Y \to G$.

Lemma. *A sequence $g_n : Y \to G$ is termwise differentiable if and only if it converges locally uniformly to a differentiable function $g : Y \to G$ and the sequence $g_n' : Y \to G$ converges locally uniformly to $g' : Y \to G$.*

Proof. The assertion of the lemma follows from the definitions and the equalities $\|g'(y) - g'(y)\| = \|dg y(1) - dg_n y(1)\| = \|d(g - g_n)y(1)\| = \|d(g - g_n)y\| = \|dg y - dg_n y\|$, valid for all $n \in \mathbb{N}$ and $y \in Y$. □

A set in $\mathbb{R}$ is open and connected if and only if it is an open interval. Every interval is convex. Let $Y = B \subseteq \mathbb{R}$ be an open interval. The lemma and the termwise differentiation theorem imply the following statements.

Corollary 1. *If the sequence of derivatives g_n' converges uniformly and the sequence of points $g_n(a)$ converges, then the sequence of functions g_n is termwise differentiable and converges uniformly on any bounded subset of B.*

Corollary 2. *If the sequence of the derivatives g'_n converges locally uniformly and $g_n(a)$ converges, then the sequence g_n is termwise differentiable.*

Remark. From the assertions on termwise differentiation of sequences, it is possible to obtain similar assertions on termwise differentiation for sums.

Exercise. Formulate and prove the assertions mentioned in the above remark.

3.2.5.   Total differentials

We continue the study of connections between total and partial differentials started in part 8 of Subsection 3.2.1.

1. If the differentials $df_{ij}u$, $d_{ij}fu$, and dfu exist at any point $u \in U$, then the maps $df_{ij} : U \to \mathcal{B}(E_j, F_i)$, $d_{ij}f : U \to \mathcal{B}(E, F)$, and $df : U \to \mathcal{B}(E, F)$ that map a point u to the linear operators $df_{ij}u$, $d_{ij}fu$, and dfu, respectively, are defined. The maps df_{ij}, $d_{ij}f$, and df are called the *partial* and *total differentials* of f. If the total differential df exists and is continuous, then f is said to be *continuously differentiable*. Continuously differentiable functions are often called *smooth*.

Remark. The continuous differentiability of the partial functions is not equivalent to the continuity of partial differentials.

Counterexample. Consider the function $f : \mathbb{R}^2 \to \mathbb{R}$ defined by the equalities $f(x) = x_1 x_2 (x_1^2 + x_2^2)$ $(x = (x_1, x_2) \neq (0, 0))$, $f(0, 0) = 0$. Its partial functions and their derivatives are as follows: $f_1 u(v_1) = (u_1 + v_1)u_2((u_1 + v_1)^2 + u_2^2)$, $f_2 = u_1(u_2 + v_2)(u_1^2 + (u_2 + v_2)^2)$ at a point $u = (u_1, u_2) \neq (0, 0)$,

$$f'_1 u(0) = -u_2(u_1^2 - u_2^2)(u_1^2 + u_2^2)^{-2}, \quad f'_2 u(0) = -u_1(u_1^2 - u_2^2)(u_1^2 + u_2^2)^{-2}.$$

Suppose that $\delta > 0$ and $u_1 = \delta$, $u_2 = 2\delta$. Then the partial derivatives of f with respect to the first and the second variables are $f'_1 u(0) = (6.5^{-2}\delta^{-1}, 0)$ and $f'_2 u(0) = (0, 6.5^{-2}\delta^{-1})$. Consequently, if $v_j = \delta$ $(j = 1, 2)$, then $\|df_{1j}u(v)\| = 6.5^{-2}\|\delta^{-1}v_j\|$, $\|df_{1j}u\| \geq 6.5^{-2}$ for $\|u\| = \max\{\|u_1\|, \|u_2\|\} = 2\delta$. It follows that the partial differentials df_j are discontinuous at $u = (0, 0)$. At the same time, the partial functions $f_j u : \mathbb{R} \to \mathbb{R}$ vanish at this point and are therefore continuously differentiable.

2. The differentials of the bounded linear operators p_j, q_i, r_j, s_i are equal to these operators. Differentiating the compositions that relate the functions $f_{ij}u$, $f_{ij}^0 u$, and f, we obtain similar equalities for their differentials: $df_{ij}u = d(q_i \cdot f \cdot Tu \cdot r_j)0_j = q_i \cdot d(f \cdot Tu \cdot r_j)0_j = q_i \cdot dfu \cdot r_j$, $d_{ij}fu = df_{ij}^0 u = s_i q_i \cdot dfu \cdot r_j p_j$, $df_{ij}u = q_i \cdot d_{ij}fu \cdot r_j$, $d_{ij}fu = s_i \cdot df_{ij}u \cdot p_j$. In these equalities, the differentials on the right-hand sides are assumed to exist. By the chain rule, this guarantees the existence of the differentials on the left-hand sides and provides the validity of the equalities themselves.

Proposition. *If the total differential dfu exists, then all partial differentials $df_{ij}u$ and $d_{ij}fu$ exist and $dfu = \sum d_{ij}fu$.*

Proof. The existence of the partial differentials follows from the chain rule. The desired equality follows from the equalities $\sum s_i q_i = id_F$, $\sum r_j p_j = id_E$, and $d_{ij}fu = s_i q_i \cdot dfu \cdot r_j p_j$, and the linearity of the operators under consideration: $dfu = id_F \cdot dfu \cdot id_E = \sum s_i q_i \cdot dfu \cdot \sum_{ij} r_j p_j = \sum (s_i q_i \cdot dfu \cdot r_j p_j) = \sum_{ij} d_{ij}fu$. □

Remark. The partial differentials of a function can exist even if the function is not differentiable. This is shown in Counterexample 1. The function f therein is discontinuous and therefore is not differentiable at u: if $x_1 = x_2 \neq 0$, then $f(x_1, x_2) = 2^{-1} \nrightarrow 0 = f(0,0)$ as $x_1 = x_2 \to 0$. Nevertheless, all partial differentials of f exist.

3. If a function of two variables is continuously differentiable with respect to one variable and is differentiable with respect to the other variable, then it is differentiable. This is expressed more precisely by the existence theorem for the total differential. For clarity, consider the case of $m = 1$, $n = 2$ and a rectangular domain of definition. We now simplify the notation.

Put $E = E_1 \times E_2$, $F = F_1$, $f_1 = f(\cdot, u_2)$, $f_2 = f(u_1, \cdot)$, $d_j fu = df_j u \cdot p_j$, $d_j f : u \to d_j fu$ ($u = (u_1, u_2)$, $j = 1, 2$). We preserve the remaining notation of part 8 of Subsection 3.2.1: $U = U_1 \times U_2 \subseteq E$, $f : U \to F$.

Theorem. *If the partial differential $d_1 f$ exists on an open set U and is continuous at a point $u \in U$ and if the partial differential $d_2 f$ exists at u, then the total differential df exists at u.*

Proof. The idea of the proof is simple. If dfa exists, then $d_1 fu$ and $d_2 fu$ exist and $dfu = df_1 u + df_2 u$. Therefore, in order to prove the existence of dfu, it suffices to prove that the function $r : V \to F$ ($v \in V = U - u$) with values

$$r(v) = \Delta fu(v) - (d_1 fu + d_2 fu)(v) \quad (v \in V)$$

is relatively small. We divide the proof into several steps.

(1) REPRESENTATION OF THE REMAINDER. We have $r(v) = f(u_1 + v_1, u_2 + v_2) - f(u_1, u_2) - d_1 fu(v) - d_2 fu(v) = h_1(v) + h_2(v)$, $h_1(v) = f(u_1 + v_1, u_2 + v_2) - f(u_1, u_2 + v_2) - d_1 fu(v)$, $h_2(v) = f(u_1, u_2 + v_2) - f(u_1, u_2) - d_2 fu(v)$ ($v = (v_1, v_2) \in V$).

(2) ESTIMATION OF THE PURE REMAINDER h_2. Since $d_2 fu$ exists, it follows that $df_2 u_2$ exists and, for any $\varepsilon > 0$ and some $\delta > 0$, we have

$$\|h_2(v)\| = \|\Delta fu_2(v_2) - df_2 u_2(v_2)\| = \|r f_2 u_2(v_2)\| \le \varepsilon \|v_2\| \le \varepsilon \|v\|$$

if $\|v\| = \max\{\|v_1\|, \|v_2\|\} \le \delta$, $v \in V$, and h_2 is relatively small.

(3) ESTIMATION OF THE MIXED REMAINDER h_1. The value $h_1(v)$ is not equal to $r f_1 u_1(v_1)$ because the increment is expressed in terms of the function $f_2(\cdot, u_2 + v_2)$ for $t_2 = u_2 + v_2$, i.e., $f(u_1 + v_1, t_2) - f(u_1, t_2) = \Delta f_2(\cdot, t_2) u_1(v_1)$, and the differential is expressed in terms of the differential of the partial function $f(\cdot, u_2)$ for u_2, i.e., $d_1 fu(v) = df(\cdot, u_2) u_1(v_1)$. Therefore, we cannot estimate h_1 as easily as h_2. We have to use the continuity of $d_1 f$ at u and Lagrange's theorem.

(1.3) DEFINITION OF AN AUXILIARY FUNCTION g_1. Assume that $t_2 \in U_2$, $t = (t_1, t_2) \in U$, and $v = (v_1, v_2) = (t_1 - u_1, t_2 - u_2) = t - u \in V$. Let $g_1 : U_1 \to F$ be defined by the equality $g_1(t_1) = h_1(t - u) = h_1(v)$ ($t_1 \in U_1$). We have

$$g_1(u_1) = h_1(0, v_2) = f(u_1, t_2) - f(u_1, t_2) - df(\cdot, u_2) u_1(0) = 0.$$

Therefore, $g_1(t_1) - g_1(u_1) = h_1(v)$ and the estimation of h_1 is reduced to the estimation of the increment of g_1.

(2.3) VERIFYING THAT THE ASSUMPTIONS OF LAGRANGE'S THEOREM ARE SATISFIED. By assumption, U_1 is open. Then there exists a $\rho_1 > 0$ such that $B(u_1, \rho_1) \subseteq U_1$. Consequently, $[u_1, t_1] = \{u_1 + \xi(t_1 - u_1) \colon \xi \in [0, 1]\} \subseteq B(u_1, \rho_1) \subseteq U_1$. The function g_1 is equal to the difference of the functions $\Delta_1 : U_1 \to F$ and $\varphi_1 : U_1 \to F$ such that

$$\Delta_1(t_1) = f(t_1, t_2) - f(u_1, t_2), \quad \varphi_1(t_1) = df(\cdot, u_2) u_1(t_1 - u_1) \ (t_1 \in U_1).$$

By hypothesis, the function $f(\cdot, t_2)$ is differentiable at any $t_1 \in U_1$. Consequently, the function Δ_1 is differentiable and $d\Delta_1 t_1 = df(\cdot, t_2)t_1$ $(t_1 \in U_1)$. The function $\varphi_1(t_1) = df(\cdot, u_2)u_1 \circ \tau_1$ is composed of the continuous linear function $df(\cdot, u_2)u_1 : E_1 \to F$ and the differentiable function $\tau_1 : U_1 \to V_1$ $(V_1 = U_1 - u_1)$ such that $\tau(t_1) = t_1 - u_1$ $(t_1 \in U_1)$. The differential $d\tau_1 t_1$ is equal to the identity map from E_1 to itself. From the chain rule it follows that φ_1 is differentiable and $d\varphi_1 t_1 = df(\cdot, u_2)u_1$ $(t_1 \in U_1)$. Consequently, the function $g_1 = \Delta_1 - \varphi_1$ is differentiable and $dg_1 t_1 = df(\cdot, t_2)t_1 - df(\cdot, u_2)u_1$ $(t_1 \in U_1)$. Expressing the differentials of the partial functions in terms of the partial differentials with the use of the embedding $r_1 : E_1 \to E$, $r_1(v_1) = (v_1, 0)$ $(v_1 \in E_1)$, we obtain $dg_1 t_1 = d_1 ft \circ r_1 - d_1 fu \circ r_1 = (d_1 ft - d_1 fu) \circ r_1$ $(t_1 \in U_1)$.

By assumption, $d_1 f$ is continuous at u: for any $\varepsilon > 0$, there exists a $\gamma > 0$ such that $\|d_1 ft - d_1 fu\| \le \varepsilon$ when $\|t - u\| \le \gamma$, $t \in U$. Consequently, $\|dg_1 t_1(v_1)\| = \|(d_1 ft - d_1 fu)(v_1, 0)\| \le \|dg_1 t - d_1 fu\| \|(v_1, 0)\| \le \varepsilon \|v_1\|$ $(v_1 \in V_1)$ and $\|dg_1 t_1(v_1)\| \le \varepsilon$ when $\|t_1 - u_1\| \le \gamma$, $t_1 \in U_1$. Therefore, the function $g = g_1$ satisfies the assumption of Corollary 3 of Lagrange's theorem for $c = \varepsilon$, $a = u_1$, $b = t_1$ $(t_1 \in B(u_1, \delta)$, $\delta = \min\{\gamma, \rho_1\})$.

(3.3) APPLICATION OF LAGRANGE'S THEOREM. From (1.3) and (2.3), the proof of Lagrange's theorem, and Corollary 3, it follows that for any $\varepsilon > 0$ there exists a $\delta > 0$ such that

$$\|h_1(v)\| = \|g_1(t_1) - g_1(u_1)\| \le \varepsilon \|t_1 - u_1\| \le \varepsilon \|v\| \quad (\|v\| \le \delta, \ v \in V).$$

Hence, h_1 is relatively small.

(4) RELATIVE SMALLNESS OF THE REMAINDER r. Since h_1 and h_2 are relatively small, so is $r = h_1 + h_2$ $\square$

4. We now return to the general case with arbitrary m, n. The corresponding assertion is deduced from the special case of $m = 1$, $n = 2$ by induction.

Theorem. *The total differential df exists and is continuous if and only if all partial differentials df_{ij} exist and are continuous.*

Proof. (1) If df_{ij} exist and are continuous, then there exist continuous differentials $df_{ij} = df_{ij}^0$. Hence, the differentials df_i and $d_i f = df_i^0$ of the projections $f_i = q_i \cdot f : U \to F_i$ and their embeddings $f_i^0 = s_i \cdot f_i = s_i q_i \cdot f : U \to F$ also exist. The existence of the differentials df_i and $d_i f$ follows from the theorem of part 3 of this subsection and the principle of induction, while their continuity follows from the equalities $df_i = \sum_j df_{ij}$, $d_i f = \sum_j d_{ij} f$. The existence and continuity of df follows from the equality $df = \sum d_i f$.

(2) If the differential df exists and is continuous, then all partial differentials df_{ij} exist, are continuous, and the equalities $df_{ij} = q_i \cdot df u \cdot r_j$ hold. The continuity of df_{ij} follows from the relations $\|(df_{ij}(u+z) - df_{ij}u)(v_j)\| = \|q_i(df(u + z) - df u)(v_j^0)\| \leq \|(df(u + z) - df u)(v_j^0)\| \leq \|df_{ij}(u + z) - df u\| \|v_j^0\| \leq \|df(u+z) - df u\| \|v\|$, $\|df_{ij}(u+z) - df_{ij}u\| \leq \|df(u+z) - df u\|$ for all $u, z, u + z \in U$ and $v \in V$. $\square$

Remark. The existence and continuity of all partial differentials of a function guarantee that it is not only differentiable, but also continuously differentiable (smooth).

For $E = \mathbb{K}^n$ and $F = \mathbb{F}^n$, part 10 of Subsection 3.2.1 and the preceding theorem imply the following statement.

Corollary. *A function f is continuously differentiable if and only all its partial derivatives exist and are continuous.*

Exercise. Give a detailed proof of this assertion.

5. We now consider several examples.

Example 1. Polar coordinates. Let $Y = \,]0,\infty[\, \times \,]0, \pi/2[$, $X = \,]0,\infty[\, \times \,]0,\infty[$, and let functions $g : Y \to \mathbb{R}^2$ and $f : X \to \mathbb{R}^2$ be defined by the equalities $x = g(y)$, $y = f(x)$,

$$\begin{pmatrix} x_1 \\ x_2 \end{pmatrix} = \begin{pmatrix} y_1 \cdot \cos y_2 \\ y_1 \cdot \sin y_2 \end{pmatrix}, \quad \begin{pmatrix} y_1 \\ y_2 \end{pmatrix} = \begin{pmatrix} (x_1^2 + x_2^2)^{1/2} \\ \arctan(x_1^{-1} x_2) \end{pmatrix}.$$

The functions g and f express Cartesian coordinates in terms of polar coordinates and polar coordinates in terms of Cartesian coordinates, respectively. These functions are inverse to each other. Applying appropriate differentiation rules, we have

$$g'(y) = \begin{pmatrix} \cos y_2 & -y_1 \sin y_2 \\ \sin y_2 & y_1 \cos y_2 \end{pmatrix}, \quad f'(x) = \begin{pmatrix} x_1(x_1^2 + x_2^2)^{-1/2} & x_2(x_1^2 + x_2^2)^{-1/2} \\ -x_2(x_1^2 + x_2^2)^{-1} & x_1(x_1^2 + x_2^2)^{-1} \end{pmatrix}.$$

Like the functions g and f, the derivatives g' and f' at the corresponding points are inverse to each other: $g'(y) \cdot f'(x) = I$.

Example 2. Spherical coordinates. Let $X =]0, \infty[^3$, $Y =]0, \infty[\times]0, \pi/2[\times]0, \pi/2[$ and let $g : Y \to \mathbb{R}^3$ and $f : X \to \mathbb{R}^3$ be defined by the equalities $x = g(y)$, $y = f(x)$,

$$
\begin{pmatrix} x_1 \\ x_2 \\ x_3 \end{pmatrix} = \begin{pmatrix} y_1 \cdot \cos y_2 \cdot \sin y_3 \\ y_1 \cdot \sin y_2 \cdot \sin y_3 \\ y_1 \cdot \cos y_3 \end{pmatrix}, \quad \begin{pmatrix} y_1 \\ y_2 \\ y_3 \end{pmatrix} = \begin{pmatrix} (x_1^2 + x_2^2 + x_3^2)^{1/2} \\ \arctan(x_1^{-1} x_2) \\ \arccos(x_3 (x_1^2 + x_2^2 + x_3^2)^{-1/2}) \end{pmatrix}.
$$

The functions g and f represent Cartesian coordinates in terms of spherical coordinates and vice versa. Applying appropriate differentiation rules, we obtain

$$
g'(y) = \begin{pmatrix} \cos y_2 \cdot \sin y_3 & -y_1 \sin y_2 \cdot \sin y_3 & i_1 \cos y_2 \cdot \cos y_3 \\ sin y_2 \cdot \sin y_3 & y_1 \cos y_2 \cdot \sin y_3 & y_1 \sin y_2 \cdot \cos y_3 \\ \cos y_3 & 0 & -y_1 \cdot \sin y_3 \end{pmatrix},
$$

$$
f'(x) = \begin{pmatrix} x_1 r^{-1} & x_2 r^{-1} & x_3 r^{-1} \\ -x_2(x_1^2 + x_2^2)^{-1} & x_1(x_1^2 + x_2^2)^{-1} & 0 \\ x_1 x_3(x_1^2 + x_2^2)^{-1/2} r^{-2} & x_2 x_3(x_1^2 + x_2^2)^{-1/2} r^{-2} & -(x_1^2 + x_2^2)^{1/2} r^{-2} \end{pmatrix},
$$

$r = \|x\|_2 = (x_1^2 + x_2^2 + x_3^2)^{1/2}$. Like the functions g and f, the derivatives g' and f' at the corresponding points are inverse to each other: $g'(y) \cdot f'(x) = I$.

Example 3. Divergence. Let $f : \mathbb{R}^n \to \mathbb{R}^n$ be a differentiable function with derivative $f'(x) = (f'_{ij}(x))$ at $x \in \mathbb{R}^n$. The trace

$$
\mathrm{tr} f'(x) = f'_{11}(x) + \cdots + f'_{nn}(x)
$$

of the matrix $A = f'(x)$ is called the *divergence* of the function f at x and is denoted by $\mathrm{div}\, f(x)$. We will prove that

$$
\mathrm{div}(L^{-1} f L)(t) = \mathrm{div}\, f(x) \quad (x = L(t))
$$

for any invertible linear operator $L : \mathbb{R}^n \to \mathbb{R}^n$. The desired equality is equivalent to the equality $\mathrm{tr}(L^{-1} f L) = \mathrm{tr} f'(x)$. Applying the chain rule and using the property of L to be linear and bounded, we obtain

$$
B = (L^{-1} f L)'(t) = (L^{-1})'(f(L(t))) \cdot f'(L(t)) \cdot L'(t) = C^{-1} A C.
$$

Here $A = f'(x)$ and C is the matrix of the operator L with respect to the standard bases.

Let $\lambda \in \mathbb{C}$ and let I be the identity $n \times n$-matrix. Then

$$\det(\lambda I - B) = \det(\lambda I - C^{-1}AC) = \det(C^{-1}(\lambda I - A)C) = \det(\lambda I - A),$$

$$\det(\lambda I - A) = \lambda^n - \operatorname{tr} A \cdot \lambda^{n-1} + \cdots + (-1)^n \det A$$
$$= \prod(\lambda - \lambda_i) = \lambda^n - (\lambda_1 + \cdots + \lambda_n) \cdot \lambda^{n-1} + \cdots + (-1)^n \lambda_1 \ldots \lambda_n,$$

$\operatorname{tr} A = \lambda_1 + \cdots + \lambda_n$, $\det A = \lambda_1 \ldots \lambda_n$, where $\lambda_1, \ldots, \lambda_n$ are the eigenvalues of the matrix A. Consequently, $\operatorname{tr} B = \operatorname{tr} A$.

Therefore, the divergence does not change under invertible linear transformations.

Example 4. Determinants. Consider an open set $X \subseteq \mathbb{R}$, a collection of differentiable functions $y_{ij} : X \to \mathbb{R}$ ($1 \le i, j \le n$), and a function $h : X \to \mathbb{R}$ defined by the formula $h(x) = \det(y_{ij}(x))$ ($x \in X$). We will find the derivative of h. Arrange the pairs ij in the lexicographic order: $m(11) = 1$, $m(12) = 2, \ldots, m(1n) = n$, $m(21) = n+1, m(22) = n + 2, \ldots, m(2n) = n+n, \ldots, m(nn) = n \times n$. The equalities $y_{m(ij)} = y_{ij}$ define a linear transformation $L : M \to \mathbb{R}^{n \times n}$ from the space of real $n \times n$-matrices $\bar{y} = (y_{ij})$ into the space of real $(n \times n) \times 1$-matrices $y = L\bar{y} = (y_1, \ldots, y_{n \times n})^t$ (columns). Let $\bar{y}(x) = (y_{ij}(x))$ be an $n \times n$-matrix. Put

$$f(x) = L\bar{y}(x) = (y_1, \ldots, y_{n \times n})^t.$$

This equality defines a function $f : X \to \mathbb{R}^{n \times n}$. It is differentiable together with the functions y_{ij} and $f'(x) = (y_1', \ldots, y_{n \times n}')^t$ ($x \in X$). a function Let a function $g : \mathbb{R}^{n \times n} \to \mathbb{R}$ be defined by the equality

$$g(y) = \det(L^{-1}y) \quad (y \in \mathbb{R}^{n \times n}).$$

Let $z_{ij}(\bar{y})$ be the algebraic complement of the element y_{ij} of the matrix $\bar{y} = L^{-1}Y$. Then $g(y) = \sum_{1 \le p \le n} z_{pq} y_{pq}$ ($1 \le q \le n$). Therefore, $g_{m(ij)}'(y) = z_{ij}(\bar{y})$. The partial products $g_1', \ldots, g_{n \times n}'$ are continuous. Hence, g is continuously differentiable and $g'(y) = (g_1'(y), \ldots, g_{n \times n}'(y))$. Consequently, the composition $h = g \circ f$ is differentiable and

$$h'(x) = g'(f(x)) \cdot f'(x) = \sum_i \left(\sum_j z_{ij}(f(x)) y_{ij}'(x) \right).$$

Remark. The interior sum in this inequality is equal to the determinant obtained from $h(x)$ by replacing the jth column $(y_{1j}(x), \ldots, y_{nj}(x))^t$ by the column $(y'_{1j}(x), \ldots, y'_{nj}(x))^t$. Thus, $h'(x)$ is equal to the sum of the n determinants that are obtained by consecutively differentiating the columns of $h(x)$.

Example 5. The Cauchy–Riemann equations. Every complex number is identified with a pair of real numbers: $\mathbb{C} = \mathbb{R}^2$. The field of complex numbers is a one-dimensional complex vector space. The standard basis of $\mathbb{C}$ is represented by the vector $e_1 = (1,0)^t$. The pair of vectors $e_1 = (1,0)^t$, $e_2 = (0,1)^t$ forms the standard basis of $\mathbb{R}^2$. The equalities

$$L(ax + by) = aL(x) + bL(y) \quad (a, b \in \mathbb{C}; \; x, y \in \mathbb{C}), \tag{1}$$

$$M(\alpha x + \beta y) = \alpha M(x) + \beta M(y) \quad (\alpha, \beta \in \mathbb{R}, \; x, y \in \mathbb{R}^2) \tag{2}$$

define linear transformations of $\mathbb{C}$ and $\mathbb{R}^2$, respectively. These equalities are equivalent to the matrix equalities

$$L(v) = \begin{pmatrix} L_1(e_1) & -L_2(e_1) \\ L_2(e_1) & L_1(e_1) \end{pmatrix} \begin{pmatrix} v_1 \\ v_2 \end{pmatrix}, \quad M(v) = \begin{pmatrix} M_1(e_1) & M_1(e_2) \\ M_2(e_1) & M_2(e_2) \end{pmatrix} \begin{pmatrix} v_1 \\ v_2 \end{pmatrix}$$

$(v = (v_1, v_2)^t \in \mathbb{R}^2)$. Indeed, equality (1) implies $L(v) = v \cdot L(e_1)$ $(v \in \mathbb{C})$ and vice versa: $L(ax + by) = (ax + by) \cdot L(e_1) = ax \cdot L(e_1) + by \cdot L(e_1) = aL(x) + bL(y)$. At the same time, $L(v) = L_1(v)e_1 + L_2(v)e_2$, $vL(e_1) = (v_1 e_1 + v_2 e_2)(L_1(e_1)e_1 + L_2(e_1)e_2) = (v_1 L_1(e_1) - v_2 L_2(e_1))e_1 + (v_1 L_2(e_1) + v_2 L_1(e_1))e_2$ if $e_2^2 = -e_1$ and

$$L_1(v) = v_1 L_1(e_1) - v_2 L_2(e_1), \quad L_2(v) = v_1 L_2(e_1) + v_2 L_1(e_1).$$

Hence, (1) and (2) are equivalent to the above matrix equalities. The matrix $M = (\mu_{ij})$ with real elements $\mu_{ij} = M_i(e_j)$ can be arbitrary. The matrix $L = (\lambda_{ij})$ has a special form: $\lambda_{11} = \lambda_{22}$, $\lambda_{12} = -\lambda_{21}$.

Remark. Every linear transformation of $\mathbb{C}$ is equivalent to the combination of a homothety and a rotation in $\mathbb{R}^2$. All other transformations of $\mathbb{R}^2$ are not linear transformations of $\mathbb{C}$.

Choose the Euclidean norm $\| \ \|_2$ for $\mathbb{R}^2$, which corresponds to the absolute value for $\mathbb{C}$. We take a function $f : \mathbb{C} \to \mathbb{C}$, regarding it as a map $f : \mathbb{R}^2 \to \mathbb{R}^2$, and consider the coordinate functions $f_1 : \mathbb{R}^2 \to \mathbb{R}$ and

$f_2 : \mathbb{R}^2 \to \mathbb{R}$. We will prove that f is differentiable if and only if f_1 and f_2 are differentiable and

$$f'_{11} = f'_{22}, \quad f'_{12} = -f'_{21}.$$

The differentiability of $f : \mathbb{C} \to \mathbb{C}$ at a point $u \in \mathbb{C}$ means that

$$\Delta f u(v) = df u(v) + r f u(v) \quad (v \in \mathbb{C})$$

for a linear function $df u : \mathbb{C} \to \mathbb{C}$ and a relatively small $r f u : \mathbb{C} \to \mathbb{C}$. The differentiability of f_1 and f_2 at a point $u \in \mathbb{R}^2$ is equivalent to the equality

$$\Delta f u(v) = \begin{pmatrix} f'_{11}(u) & f'_{12}(u) \\ f'_{21}(u) & f'_{22}(u) \end{pmatrix} \begin{pmatrix} v_1 \\ v_2 \end{pmatrix} + r f u(v), \qquad v = \begin{pmatrix} v_1 \\ v_2 \end{pmatrix} \in \mathbb{R}^2$$

for the matrix $f'(u) = (f'_{ij}(u))$ and a relatively small $r f u : \mathbb{R}^2 \to \mathbb{R}^2$. The differentiability of $f : \mathbb{R}^2 \to \mathbb{R}^2$ is equivalent to the differentiability of $f_1 : \mathbb{R}^2 \to \mathbb{R}$ and $f_2 : \mathbb{R}^2 \to \mathbb{R}$ in any of the norms $\|\ \|_1$, $\|\ \|_2$. This follows from the equivalence of these norms, i.e., $\|x\|_1 \le \|x\|_2 \le \sqrt{2}\,\|x\|_1$ ($x \in \mathbb{R}^2$). The differentiability of $f : \mathbb{C} \to \mathbb{C}$ is equivalent to that of $f : \mathbb{R}^2 \to \mathbb{R}^2$ and the condition that $df u$ is a linear transformation of $\mathbb{C}$. As was demonstrated, the equality $df u(v) = f'(u) \cdot v$ defines a linear transformation of $\mathbb{C}$ if and only if $f'_{11} = f'_{22}$ and $f'_{12} = -f'_{21}$. These equalities are called the *Cauchy–Riemann equations*.

In particular, suppose that $x = (x_1, x_2)^t$ and $f(x) = e^{x_1} \cdot e^{i x_2}$. The Euler formula yields $f_1(x) = e^{x_1} \cdot \cos x_2$, $f_2(x) = e^{x_1} \cdot \sin x_2$, and

$$\begin{pmatrix} f'_{11}(u) & f'_{12}(u) \\ f'_{21}(u) & f'_{22}(u) \end{pmatrix} = \begin{pmatrix} e^{u_1} \cdot \cos u_2 & -e^{u_1} \cdot \sin u_2 \\ e^{u_1} \cdot \sin u_2 & e^{u_1} \cdot \cos u_2 \end{pmatrix}$$

for each $u = (u_1, u_2)^t$. The Cauchy–Riemann equations are satisfied. The function f is differentiable and

$$df u(v) = e^u \cdot v = e_1^u \cdot \begin{pmatrix} \cos u_2 & -\sin u_2 \\ \sin u_2 & \cos u_2 \end{pmatrix} \begin{pmatrix} v_1 \\ v_2 \end{pmatrix} = e_1^u \cdot \begin{pmatrix} \cos u_2 \cdot v_1 - \sin u_2 \cdot v_2 \\ \sin u_2 \cdot v_1 + \cos u_2 \cdot v_2 \end{pmatrix}.$$

Let $x = x_1 + x_2 i$ and $\bar{f}(x) = \bar{x} = x_1 - x_2 i$. Then $\bar{f}_1(x) = x_1$, $\bar{f}_2(x) = -x_2$ and $\bar{f}'_{11}(x) = 1$, $\bar{f}'_{22}(x) = -1$, $\bar{f}'_{12}(x) = 0$, $\bar{f}'_{21}(x) = 0$. The Cauchy–Riemann equations are not satisfied. The function $\bar{f} : \mathbb{C} \to \mathbb{C}$ is not differentiable at any point $u \in \mathbb{C}$.

3.2.6.   Solution of functional equations

Under certain conditions, an equation $g(x, y) = 0$ for a smooth function g can be reduced to the linear equation $P(x, y) \cdot u + Q(x, y) \cdot v = 0$ for the partial differentials $P(x, y) = dg(\cdot, y)x$ and $Q(x, y) = dg(x, \cdot)y$ of g. This makes it possible to establish the existence and uniqueness of local solutions $y = f(x)$ to the equation $g(x, y) = 0$ and calculate their derivatives.

We identify functions with their graphs and write $f = g^{-1}(0) \cap A \times B$ whenever such geometric language is convenient.

1. Consider normed spaces E, F, and G with norms denoted by $\| \ \|$, open sets $X \subseteq E$ and $Y \subseteq F$, a function $g : X \times Y \to G$, and a point $(a, b) \in X \times Y$. We now introduce the terms and the notation that will be necessary to formulate the implicit function theorem. Let $A \subseteq X$ and $B \subseteq Y$ be open neighborhoods of a and b, respectively. They form the open rectangular neighborhood $A \times B$ of (a, b). Any function $f : A \to F$ such that $f(A) \subseteq B$ and $g(x, f(x)) = 0$ for $x \in A$ is called a *solution* to the equation

$$g(x, y) = 0 \tag{1}$$

defined on A and included in $A \times B$. Such a solution is a subset of the intersection of the sets $g^{-1}(0)$ and $A \times B$: $f \subseteq g^{-1}(0) \cap A \times B$. Solutions that satisfy the condition $f(a) = b$ are said to *pass through the point* (a, b). For the existence of such a solution, it is necessary that equation (1) *hold at* (a, b), i.e., $g(a, b) = 0$. This solution is unique if and only if the set $f = g^{-1}(0) \cap A \times B$ is the graph of a function on A. In this case, the equalities $f(a) = b$ and $g(a, b) = 0$ are equivalent.

Equation (1) for a smooth function g is called *nondegenerate (in* y) at the point (a, b) if $Q(a, b) = dg(a, \cdot)b$ is a homeomorphism from F onto G. If $F = G = \mathbb{R}^m$, then this is equivalent to the condition $\det Q(a, b) \neq 0$.

The equation $g(x, y) = 0$ is said to *have a unique solution* f *in a neighborhood of* (a, b) if and only if there exists a rectangular neighborhood $A \times B$ of (a, b) such that the set $g^{-1}(0) \cap A \times B$ is the graph of a function f on A.

Remark. If a solution h with graph $g^{-1}(0) \cap C \times D$ is defined on $C \supseteq A$ and takes values in $D \supseteq B$, then it is an extension of the solution f. The solution f is a unique function on A such that $f(A) \subseteq B$, $g(x, f(x)) = 0$ for any $x \in A$. We can say that the *local solution* f to the equation $g(x, y) = 0$ is unique up to the choice of a neighborhood $A \times B$ of (a, b). The solution f is also referred to as the *implicit function*.

Let g be continuously differentiable on $X \times Y$ for any $(x, y) \in X \times Y$. Consider the functions $g(\cdot, y) : X \to G$, $g(x, \cdot) : Y \to G$, the differentials

$$P(x, y) = dg(\cdot, y)x \in \mathcal{B}(E, G), \quad Q(x, y) = dg(x, \cdot)y \in \mathcal{B}(F, G)$$

of these functions at x, y, and the maps $P : X \times Y \to \mathcal{B}(E, G)$ and $Q : X \times Y \to \mathcal{B}(F, G)$.

Lemma. *The functions P and Q are continuous.*

Proof. Any points $(x, y) \in X \times Y$, $(\Delta x, \Delta y) \in (X - x) \times (Y - y)$, $(u, v) \in E \times F$ satisfy the relations $\|(P(x + \Delta x, y + \Delta y) - P(x, y))(u)\| = \|(dg(\cdot, y + \Delta y)(x + \Delta x) - dg(\cdot, y)x)(u)\| = \|(dg(x + \Delta x, y + \Delta y) - dg(x, y))(u, 0)\| \leq \|dg(x + \Delta x, y + \Delta y) - dg(x, y)]\|\|(u, v)\|$. Therefore, $\|P(x + \Delta x, y + \Delta y) - P(x, y)\| \leq \|dg(x + \Delta x, y + \Delta y) - dg(x, y)\|$ and P is continuous because so is dg. The continuity of Q is proved in the same way. $\square$

The differential $dg(x, y)$ and the functions $P(x, y)$ and $Q(x, y)$ are related by the equality

$$dg(x, y)(u, v) = P(x, y)u + Q(x, y)v \quad (u \in E, \ v \in F).$$

Thus, it is natural to associate the equation $g(x, y) = 0$ with the equation

$$P(x, y) \cdot u + Q(x, y) \cdot v = 0. \tag{2}$$

Equations (1) and (2) are said to be *nondegenerate at a point* (a, b) if the differential $Q(a, b)$ is a homeomorphism from F onto G.

Remark. If $F = G = \mathbb{R}^m$, then the fact that $Q(a, b)$ is a homeomorphism means that its matrix is nondegenerate and its determinant $\det Q(a, b)$ is nonzero. If $F = \mathbb{R}^m$, $G = \mathbb{R}^l$, and $m \neq l$, then there are no linear homeomorphisms from F onto G.

The implicit function theorems will be proved later in this subsection.

2. Consider a metric space (M, ρ), a set $C \subseteq M$, a transformation $h : C \to C$, and a number $\alpha > 0$. A transformation h is called a *Lipschitz transformation with constant* α if

$$\rho(h(x), h(y)) \leq \alpha \rho(x, y) \quad (x, y \in C).$$

If $\alpha < 1$, then h is called a *contraction*; if $\alpha = 1$, then h is said to be *nondilating*. The transformation h is called a *dilation* if

$$\rho(h(x), h(y)) \geq \alpha \rho(x, y) \quad (x, y \in C).$$

An isometry is nondilating and dilating at the same time.

A point $c \in C$ such that $h(c) = c$ is said to be *fixed*. Fixed points are solutions to the equation $h(x) = x$ $(x \in C)$. Clearly, any dilation is uniformly continuous.

Theorem. *Every contraction of a closed set in a complete separated metric space has a unique fixed point.*

Proof. Take $x_0 \in C$ and consider the sequence of points $x_n \in C$ defined by the equality $x_{n+1} = h(x_n)$ $(n \geq 0)$. This sequence converges because h is a contraction. Indeed, we have $\rho(x_2, x_1) = \rho(h(x_1), h(x_0)) \leq \alpha\rho(x_1, x_0)$, $\rho(x_{n+1}, x_n) = \rho(h(x_n), h(x_{n-1})) \leq \alpha\rho(x_n, x_{n-1})$, $(n \geq 1)$. It follows that $\rho(x_{n+1}, x_n) \leq \alpha^n \rho(x_1, x_0)$, $\rho(x_{n+m}, x_n) \leq \sum\limits_{0 \leq k < m} \rho(x_{n+k+1}, x_{n+k}) \leq \sum\limits_{0 \leq k < m} \alpha^{n+k}\rho(x_1, x_0) \leq \alpha^n(1 - \alpha)^{-1}\rho(x_1, x_0)$ $(n \geq 0)$ for any m. If $\alpha \in {]0, 1[}$, then $\alpha^n \to 0$ $(n \to \infty)$. The sequence (x_n) converges in itself. Since M is complete, it follows that (x_n) converges to a point $c \in M$. Since $x_n \in C$ and C is closed, we conclude that $c \in C$. The equality $x_{n+1} = h(x_n)$ and the continuity of h imply that $c = h(c)$.

Suppose that there is another fixed point $b \in C$ $(b \neq c)$ for h. Since h is a contraction, we have $\rho(c, b) = \rho(h(c), h(b)) \leq \alpha\rho(c, b)$. Since $\alpha < 1$, it follows that $\rho(c, b) = 0$, which is impossible for $b \neq c$ because ρ separates points in M. We have arrived at a contradiction. $\square$

Remark. In the general case, the theorem does not hold for nondilating transformations: if h is the identity, i.e., $h(x) = x$ $(x \in \mathbb{R})$, then all points are fixed, whereas the translation $h(x) = x + 1$ has no fixed points. The theorem also fails for closed sets: if $h(x) = 2^{-1}x$, then there are no fixed points in the set $C = \{x : x \neq 0\} \subseteq \mathbb{R}$.

At the same time, every nondilating transformation of a bounded closed convex set in a Hilbert space has a nonempty closed convex set of fixed points (see Subsection 5.1.1 in Aubin and Ekeland, 1984, and Section 2, Chapter 5 in Nirenberg, 2001). A fixed point theorem for multivalued contractions is proved in Aubin and Ekeland (1984).

3. Consider a Banach space $(F, \|\ \|)$ and the space $(\mathcal{B}, \|\ \|)$ of all continuous linear transformations of a set F. The space $(\mathcal{B}, \|\ \|)$ is also a Banach space. Consider a linear operator $T \in \mathcal{B}$, the identity operator $I \in \mathcal{B}$, the difference $I - T \in \mathcal{B}$, and the sequence of powers $T^n \in \mathcal{B}$. By definition, $T^0 = I$.

Theorem. *If* $\|T\| < 1$, *then* $I - T$ *is a linear homeomorphism from* F *onto* F *and* $(I - T)^{-1} = \sum_{n \geq 0} T^n$.

Proof. Since $\|T^n\| \leq \|T\|^n$ for all $n \geq 0$, the sum of the series of powers $T^n \in \mathcal{B}$ is $S = \sum_{n \geq 0} T^n \in \mathcal{B}$ if $\|T\| < 1$. This assertion follows from the Cauchy criterion and the inequalities $\left\| \sum_{0 \leq k < m} T^{n+k} \right\| \leq \| \sum_{0 \leq k < m} \|T^{n+k}\| \leq \sum_{0 \leq k < m} \|T\|^{n+k}$ $(m > 0)$. Since $(I - T) \cdot \sum_{k < n} T^k = \sum_{k < n} T^k \cdot (I - T) = 1 - T^n$ $(n > 0)$ and the sequence T^n converges to the zero operator $O \in \mathcal{B}$, we have $(I - T)S = S(I - T) = I$. Hence, $(I - T)$ and S are one-to-one maps from F onto F that are inverse to each other. They are linear and continuous. $\square$

The theorem implies an auxiliary assertion that will be necessary in the next part of this subsection.

Lemma. $\|(I - T)^{-1} - I - T\| \leq \|T\|^2(1 - \|T\|)^{-1}$ $(\|T\| < 1)$.

Proof. By the preceding theorem, $\|(I - T)^{-1} - I - T\| = \| \sum_{n \geq 2} T^n \| \leq \sum_{n \geq 2} \|T\|^n = \|T\|^2(1 - \|T\|)^{-1}$ if $T \in \mathcal{B}$ and $\|T\| < 1$. $\square$

The operator progression (T^n) is called the *Neumann series*.

4. Consider Banach spaces F, G, $\mathcal{B} = \mathcal{B}(F, G)$ and the set $\mathcal{H}(F, G) \subseteq \mathcal{B}(F, G)$ of all linear homomorphisms from F onto G. Assume that $\mathcal{H}(F, G)$ is nonempty. Also, consider the Banach space $\mathcal{B}(G, F)$ of all continuous linear maps from G to F and the set $\mathcal{H}^{-1} = \mathcal{H}(G, F) \subseteq \mathcal{B}(G, F)$ of all linear homeomorphisms from G onto F. If $\mathcal{H} \neq \varnothing$, then $\mathcal{H}^{-1} \neq \varnothing$.

Let $\varphi : \mathcal{H} \to \mathcal{H}^{-1}$ be the map that takes any homeomorphism $H \in \mathcal{H}$ to its inverse: $\varphi(H) = H^{-1}$ $(H \in \mathcal{H})$. We call φ the *natural map* from $\mathcal{H}$ onto $\mathcal{H}^{-1}$. This map appears in the proof of the implicit function theorem. It is described by the following assertion, which is a consequence of Theorem 3.

Corollary. *The set* $\mathcal{H}$ *is open,* φ *is differentiable, and* $d\varphi H(T) = -H^{-1}TH^{-1}$ $(H \in \mathcal{H}, T \in \mathcal{B})$.

Proof. Let $H \in \mathcal{H}$. We will prove that there exists an $r > 0$ such that $H + T \in \mathcal{H}$ for any $T \in \mathcal{B}$ such that $\|T\| < r$. This will mean that $B(H, r) \subseteq \mathcal{H}$ and $\mathcal{H}$ is open. The theorem of part 2 implies that $H + T = H(I + H^{-1}T) \in \mathcal{H}$ if $\|H^{-1}T\| \leq \|H^{-1}\|\|T\| < 1$, $\|T\| < r = \|H^{-1}\|^{-1}$. Note that

$(H + T)^{-1} = (I + H^{-1}T)^{-1}H^{-1}$. Consider the increment of the function φ at a point $H \in \mathcal{H}$ for $T \in \mathcal{B}$, $\|T\| < \|H^{-1}\|^{-1}$. We have $\Delta\varphi H(T) = \varphi(H + T) - \varphi(H) = (H+T)^{-1} - H^{-1} = ((I+H^{-1}T)^{-1} - I)H^{-1} = ((1+H^{-1}T)^{-1} - I + H^{-1}T - H^{-1}T)H^{-1} = ((I+H^{-1}T)^{-1} - I + H^{-1}T) \cdot H^{-1} - H^{-1}TH^{-1}$.
Applying the lemma of part 3, we obtain $\|\Delta\varphi H(T) + H^{-1}TH^{-1}\| \leq \|(I + H^{-1}T)^{-1} - I + H^{-1}T\| \cdot \|H^{-1}\| \leq \|H^{-1}T\|^2(I - \|H^{-1}T\|)^{-1} \cdot \|H^{-1}\| \leq c\|T\|^2$
if $c = 2\|H^{-1}\|^3$ and $\|T\| \leq 2^{-1}\|H^{-1}\|$. Therefore, φ is differentiable and its differential $d\varphi H(T)$ is equal to $-H^{-1}TH^{-1}$.   □

Remark. The differentiability of the natural map φ implies its continuity.

Example. If $F = G = \mathbb{R}$, then the map $T \to T(1)$ $(T \in \mathcal{B}(\mathbb{R}, \mathbb{R}))$ from $\mathcal{B} = \mathcal{B}(\mathbb{R}, \mathbb{R})$ onto $\mathbb{R}$ is an isomorphism. This map takes the set $\mathcal{H} = \mathcal{B} \setminus \{O\}$ to the set $\mathbb{R} \setminus \{0\}$, while the map φ corresponds to the function $f : x \to x^{-1}$. It is differentiable and $f'(x) = -x^{-2}$.

5.   In what follows, we formulate the main theorem in this section, namely, the smooth (continuously differentiable) implicit function theorem in the form of a theorem on the solution of a functional equation.

Consider functional spaces E, F, and G with norms denoted by $\| \, \|$, open sets $X \subseteq E$ and $Y \subseteq F$, a function $g : X \times Y \to G$, and a point $(a, b) \in X \times Y$. Assume that F and G are Banach spaces and the function g is smooth. We take the functions $g(\cdot, y)$, $g(x, \cdot)$ and their differentials $P(x, y) = dg(\cdot, y)x$, $Q(x, y) = dg(x, \cdot)y$ at any $(x, y) \in X \times Y$.

Theorem.   An equation $g(x, y) = 0$ with smooth g that holds at a point (a, b) and is nondegenerate at (a, b) has a unique solution f in a neighborhood of (a, b). This solution passes through (a, b), is smooth, and $df x = -Q^{-1}(x, f(x)) \cdot P(x, f(x))$ $(x \in \text{Dom } f)$.

6.   The proof of the theorem is lengthy and is carried out in several steps. We consecutively prove the following: 1) the existence, 2) the continuity, 3) the differentiability, 4) the continuous differentiability of the implicit function.

1) **Existence.** The equation

$$g(x, y) = 0 \qquad (1)$$

is equivalent to the equation

$$h(x, y) = y \tag{2}$$

for the function $h(x, y) = y - Q^{-1}(a, b) \cdot g(x, y)$ $(x \in X, y \in Y)$. For any $x \in X$, the solution $y = f(x)$ to (2) is a fixed point for the function $h(x, \cdot) : Y \to F$:

$$h(x, \cdot)(y) = h(x, y) = y \quad (y \in Y).$$

Thus, in order to prove the existence of a solution f to (1) in a neighborhood of (a, b), it suffices to prove that for any point x in a neighborhood of a the function $h(x, \cdot)$ is a contraction and it satisfies the assumption of the fixed point theorem. Before doing this, we prove an important auxiliary assertion.

Lemma. *There exists an open rectangular neighborhood $A_0 \times B_0 \subseteq X \times Y$ of (a, b) such that $Q(x, y)$ is a homeomorphism from F onto G for any point (x, y) in this neighborhood.*

Proof. By the corollary of the theorem on the Neumann series, the set $\mathcal{H}$ of all linear homeomorphisms from F onto G is open. Therefore, the condition $Q(a, b) \in \mathcal{H}$ implies that there exists a number $\varepsilon > 0$ such that any linear transformation $T \in \mathcal{B}$ with the property $\|T - Q(a, b)\| < \varepsilon$ belongs to $\mathcal{H}$. By the lemma of part 1, the map $Q : X \times Y \to \mathcal{B}$ is continuous. Hence, there exists a number $\delta > 0$ such that $\|Q(x, y) - Q(a, b)\| < \varepsilon$ $(\|(x, y) - (a, b)\| < \delta)$. Therefore, $Q(x, y) \in \mathcal{H}$ if $x \in A_0 = B(a, \delta) \subseteq F$ and $y \in B_0 = B(b, \delta) \subseteq G$. $\square$

Proposition. *The equation $g(x, y) = 0$ has a unique solution f in a neighborhood of (a, b).*

Proof. It is required to prove that there exists an open rectangular neighborhood $A \times B \subseteq A_0 \times B_0$ such that $f = g^{-1}(0) \cap A \times B$ is the graph of a function on A.

We now prove that there exists an open neighborhood $A \subseteq A_0$ of a and a ball $B = B(b, r) \subseteq B_0$ such that the restriction of $h(x, \cdot)$ is a contraction of $\overline{B}$ and $h(x, \cdot)(\overline{B}) \subseteq B$ for any $x \in A$.

To this end, we estimate the differential $dh(x, \cdot)y$, apply Lagrange's theorem, specify the ranges for the restrictions $h(x, \cdot)$, and apply the fixed point theorem.

ESTIMATION OF THE DIFFERENTIAL $dh(x, \cdot)y$. The following equalities hold:

$$h(x, \cdot)(y) = h(x, y) = y - Q^{-1}(a, b) \cdot g(x, y)$$
$$= y - Q^{-1}(a, b) \cdot g(x, \cdot)(y) \quad (x \in A_0, \ y \in B_0). \quad (3)$$

The differentiability of g implies that the function $g(x, \cdot)$ is differentiable for any $x \in A_0$, and so is the function $h(x, \cdot)$. Applying the chain rule and taking into account the continuity of the linear map $Q^{-1}(a, b)$, we find that

$$dh(x, \cdot)y = I - Q^{-1}(a, b) \cdot Q(x, y) \quad (x \in A_0, \ y \in B_0). \quad (4)$$

In particular,

$$dh(a, \cdot)b = I - Q^{-1}(a, b) \cdot Q(a, b) = I - I = O. \quad (5)$$

(Here the symbol I stands for the identity transformation of F.)

By the lemma of part 1, Q is continuous. Therefore, from (4) and (5) it follows that there exists a neighborhood $A_1 \times B_1 \subseteq A_0 \times B_0$ of (a, b) such that

$$\|dh(x, \cdot)y\| \leq 2^{-1} \quad (x \in A_1, \ y \in B_1). \quad (6)$$

APPLICATION OF LAGRANGE'S THEOREM. The neighborhood B_1 of b includes a closed ball with center b and radius $r > 0$:

$$\overline{B} = \overline{B}(b, r) = \{y \in F : \|y - b\| \leq r\} \subseteq B_1.$$

By Corollary 3 of Lagrange's theorem, inequality (6) implies that

$$\|h(x, \cdot)(y) - h(x, \cdot)(y_0)\| \leq 2^{-1}\|y - y_0\| \quad (x \in A_1; \ y, y_0 \in \overline{B}). \quad (7)$$

Hence, for any $x \in A_1$, the restriction of the function $h(x, \cdot)$ to the closed ball $\overline{B}$ is a contraction of this ball if $h(x, \cdot)(\overline{B}) \subseteq \overline{B}$.

SPECIFICATION OF THE RANGE OF $h(x, \cdot)$.

We now consider the open ball $B = B(b, r) = \{y \in F : \|y - b\| < r\} \subset \overline{B}$ and prove that there exists an open neighborhood $A \subseteq A_1$ of a such that

$$h(x, \cdot)(\overline{B}) \subseteq B \quad (x \in A). \quad (8)$$

Since $h(a, \cdot)(b) = h(a, b) = b - Q^{-1}(a, b) \cdot g(a, b) = b - Q^{-1}(a, b) \cdot 0 = b - 0 = b$, it follows that

$$\|h(x, \cdot)(y) - b\| \leq \|h(x, \cdot)(y) - h(x, \cdot)(b)\| + \|h(x, \cdot)(b) - b\|$$
$$= \|h(x, \cdot)(y) - h(x, \cdot)(b)\| + \|h(x, \cdot)(b) - h(a, \cdot)(b)\| \quad (9)$$

for all $x \in A_0$ and $y \in B_0$. The continuity of g implies the continuity of h at (a, b) and the existence of an open neighborhood $A \subseteq A_1$ such that

$$\|h(x, \cdot)(b) - h(a, \cdot)(b)\| = \|h(x, b) - h(a, b)\| < 2^{-1}r \quad (x \in A). \quad (10)$$

Inequalities (7), (9), and (10) imply that

$$\|h(x, \cdot)(y) - b\| < 2^{-1}\|y - b\| + 2^{-1}r \leq r \quad (x \in A, \ y \in \overline{B}). \quad (11)$$

We conclude that inclusion (8) holds.

APPLICATION OF THE FIXED POINT THEOREM. As follows from the above, the restriction of $h(x, \cdot)$ to the closed ball $\overline{B}$ is its contraction for any $x \in A$. By the fixed point theorem, for any $x \in A$ there exists a unique point $y = f(x) \in \overline{B}$ with $h(x, \cdot)(f(x)) = f(x)$ and therefore

$$g(x, f(x)) = Q(a, b)(f(x) - h(x, f(x)))$$
$$= Q(a, b)(f(x) - h(x, \cdot)(f(x))) = Q(a, b)(f(x) - f(x)) = 0 \quad (x \in A).$$

Inclusion (8) shows that $f(x) \in B$.

Thus, $f = g^{-1}(0) \cap A \times B$ is the graph of a unique solution f to the equation $g(x, y) = 0$ in the open rectangular neighborhood $A \times B$ of (a, b) that is defined on A. The proposition is proved. $\square$

2) Continuity. Let $f = g^{-1}(0) \cap A \times B$.

Proposition. *The function f is continuous.*

Proof. Suppose that $x \in A$, $\Delta x \in A - x$. Then $\|f(x + \Delta x) - f(x)\| = \|h(x + \Delta x, f(x + \Delta x)) - h(x, f(x))\| \leq \|h(x + \Delta x, f(x + \Delta x)) - h(x + \Delta x, f(x))\| + \|h(x + \Delta x, f(x)) - h(x, f(x))\|$. From (7) it follows that $\|h(x + \Delta x, f(x + \Delta x)) - h(x + \Delta x, f(x))\| \leq 2^{-1}\|f(x + \Delta x) - f(x))\|$. From this and (11) we obtain $\|f(x + \Delta x) - f(x)\| \leq 2\|h(x + \Delta x, f(x)) - h(x, f(x))\|$. Consequently, the continuity of h at $(x, f(x))$ implies the continuity of f at x: for any $\varepsilon > 0$, there is a $\delta > 0$ such that $\|h(x + \Delta x, f(x)) - h(x, f(x))\| \leq 2^{-1}\varepsilon$, $\|f(x + \Delta x) - f(x)\| \leq \varepsilon$ if $\|(x + \Delta x, f(x)) - (x, f(x))\| = \|(\Delta x, 0)\| = \|\Delta x\| \leq \delta$. $\square$

3) Differentiability. We use the continuity of f to prove that it is differentiable.

Proposition. *The function f is differentiable and*

$$df x = -Q^{-1}(x, f(x)) \cdot P(x, f(x)) \quad (x \in A).$$

Proof. The proof goes as follows: we represent the increment of f as the sum of the linear part and the remainder, provide a preliminary estimate for the remainder, then provide an estimate of the increment itself, and establish the relative smallness of the remainder.

REPRESENTATION OF THE INCREMENT. Let $x \in A$ and $\Delta x \in A - x$. Since g is differentiable and $g(x + \Delta x, f(x + \Delta x)) = g(x, f(x)) = 0$, we have

$$
\begin{aligned}
0 = q(x + \Delta x, f(x + \Delta x)) &- q(x, f(x)) \\
&= dq(x, f(x))(\Delta x, f(x + \Delta x) - f(x)) \\
&+ rg(x, f(x))(\Delta x, f(x + \Delta x) - f(x)) \\
&= P(x, f(x)) \cdot \Delta x + Q(x, f(x)) \cdot (f(x + \Delta x) - f(x)) \\
&+ rg(x, f(x))(\Delta x, f(x + \Delta x) - f(x)). \quad (12)
\end{aligned}
$$

Since $(x, f(x)) \in A \times B \subseteq A_0 \times B_0$, the preceding lemma implies that $Q(x, f(x))$ is a homeomorphism from F onto G. By equalities (12),

$$
\begin{aligned}
f(x + \Delta x)) - f(x)) \\
= -Q^{-1}(x, f(x)) \cdot P(x, f(x)) \cdot \Delta x + r(x, \Delta x), \quad (13)
\end{aligned}
$$

$$
\begin{aligned}
r(x, \Delta x) \\
= -Q^{-1}(x, f(x)) \cdot rg(x, f(x))(\Delta x, f(x + \Delta x) - f(x)). \quad (14)
\end{aligned}
$$

PRELIMINARY ESTIMATE OF THE REMAINDER. Let $x \in A$, $\Delta x \in A - x$, $\varepsilon_1 > 0$. Since g is differentiable, the inequality

$$
\begin{aligned}
\|rg(x, f(x))(\Delta x, f(x + \Delta x) - f(x))\| \\
\leq \varepsilon_1 \|(\Delta x, f(x + \Delta x) - f(x))\|, \quad (15)
\end{aligned}
$$

holds for some $\delta_0 > 0$ if

$$
\|(\Delta x, f(x + \Delta x) - f(x))\| \leq \delta_0. \quad (16)
$$

Since f is continuous, it follows that

$$
\|f(x + \Delta x) - f(x)\| \leq \delta_0 \quad (\|\Delta x\| \leq \delta_1) \quad (17)
$$

for some $\delta_1 \in]0, \delta_0[$. Since $\|(\Delta x, f(x + \Delta x) - f(x))\| = \max\{\|\Delta x\|, \|f(x + \Delta x) - f(x)\|\} \leq \max\{\delta_0, \delta_1\} = \delta_0$, from (17) we have (16). Hence, (15) holds if $\|\Delta x\| \leq \delta_1$. Using the inequality

$$
\|(\Delta x, f(x + \Delta x) - f(x))\| \leq \|\Delta x\| + \|f(x + \Delta x) - f(x)\|,
$$

equality (14), the properties of the norm of a linear operator, and inequality (15), we obtain

$$\|r(x, \Delta x)\| \leq \varepsilon_1 \|Q^{-1}(x, f(x))\| (\|\Delta x\| + \|f(x + \Delta x) - f(x)\|)$$

$$(18)$$

$$(\|\Delta x\| \leq \Delta_1).$$

ESTIMATION OF THE INCREMENT. If $x \in A$, $\Delta \in A - x$, and $\varepsilon_1 \in$ $]0, \|Q^{-1}(x, f(x))\|^{-1}[$, then (13), (18), and the inequality for the norm of the product of linear operators imply that

$$\|f(x + \Delta x) - f(x)\| \leq \|Q^{-1}(x, f(x))\| \|P(x, f(x))\| \|\Delta x\|$$
$$+ \varepsilon_1 \|Q^{-1}(x, f(x))\| (\|\Delta x\| + \|f(x + \Delta x) - f(x)\|),$$

$$(19)$$

$$\|f(x + \Delta x) - f(x)\| \leq \gamma(x) \|\Delta x\| \quad (\|\Delta x\| \leq \delta_1),$$
$$\gamma(x) = \|Q^{-1}(x, f(x))\| (\|P(x, f(x))\| + \varepsilon_1)(1 - \|Q^{-1}(x, f(x))\| \cdot \varepsilon_1)^{-1}$$

for some $\delta_1 > 0$.

RELATIVE SMALLNESS OF THE REMAINDER. Let $x \in A$, $\Delta x \in A - x$, $\varepsilon > 0$, $\varepsilon_1 \in]0$, $Q^{-1}(x, f(x))\|^{-1}[$, and $\varepsilon_1 \cdot \|Q^{-1}(x, f(x))\| \cdot (1 + \gamma(x)) \leq \varepsilon$. Take a $\delta = \delta_1 > 0$ that satisfies (19). Using (18) and (19), we have

$$\|r(x, \Delta x)\| \varepsilon_1 \|Q^{-1}(x, f(x))\| (1 + \gamma(x)) \|\Delta x\| \leq \varepsilon \|\Delta x\| \quad (\|\Delta x\| \leq \delta).$$

Hence, the remainder $r(x, \cdot)$ defined by (14) is relatively small. In view of (13), this implies that f is differentiable, and its differential satisfies the desired equality. □

4) **Continuous differentiability.** Consider the differential $df : A \to \mathcal{B}(E, F)$ of f. Its values are the differentials $df\,x = -Q^{-1}(x, f(x)) \cdot P(x, f(x))$ $(x \in A)$.

Proposition. *The function f is continuously differentiable.*

Proof. The differential df is equal to the composition of the maps described by the following diagram.

$$
\begin{array}{ccc}
x & \xrightarrow{\;\;df\;\;} & -Q^{-1}(x, f(x)) \cdot P(x, f(x)) \\
\Big\downarrow \varphi_1 & & \Big\uparrow \varphi_5 \\
(x, f(x)) & & Q^{-1}(x, f(x)) \cdot P(x, f(x)) \\
\Big\downarrow \varphi_2 & & \Big\uparrow \varphi_4 \\
(P(x, f(x)), Q(x, f(x))) & \xrightarrow{\;\;\varphi_3\;\;} & (P(x, f(x)), Q^{-1}(x, f(x)))
\end{array}
$$

It means that $df = \varphi_5 \circ \varphi_4 \circ \varphi_3 \circ \varphi_2 \circ \varphi_1$, where φ_1, φ_2, φ_3, φ_4, φ_5 are defined as follows: $\varphi_1(x) = (x, f(x))$, $\varphi_2(x, y) = (P(x, y), Q(x, y))$, $\varphi_3(T, H) = (T, H^{-1})$, $\varphi_4(T, S) = S \cdot T$, $\varphi_5(T) = -T$. In order to prove that df is continuous, it suffices to prove that each of the maps φ_1, φ_2, φ_3, φ_4, φ_5 is continuous.

The continuity of φ_1 follows from the continuity of f:

$$
\|\varphi_1(x + \Delta x) - \varphi_1(x)\| = \|(\Delta x, f(x + \Delta x) - f(x))\|
$$
$$
\leq \|\Delta x\| + \|f(x + \Delta x) - f(x)\| \quad (x \in A, \ \Delta x \in A - x).
$$

The continuity of φ_2 follows from the continuity of P and Q:

$$
\|\varphi_2(x + \Delta x, y + \Delta y) - \varphi_2(x, y)\|
$$
$$
= \|(P(x + \Delta x, y + \Delta y) - P(x, y), Q(x + \Delta x, y + \Delta y) - Q(x, y))\|
$$
$$
\leq \|(P(x + \Delta x, y + \Delta y) - P(x, y)\| + \|Q(x + \Delta x, y + \Delta y) - Q(x, y))\|
$$
$$
(x \in A, \quad \Delta x \in A - x, \quad y \in B, \quad \Delta y \in B - y).
$$

The continuity of φ_3 follows from the continuity of the natural map $H \to H^{-1}$:

$$
\|\varphi_3(T + \Delta T, h + \Delta h) - \varphi_3(T, h)\|
$$
$$
= \|(\Delta T, (H + \Delta H)^{-1} - H^{-1})\| \leq \|\Delta T\| + \|(H + \Delta H)^{-1} - H^{-1}\|
$$
$$
(T, \Delta T \in \mathcal{B}(E, G), \quad H, \Delta H, H + \Delta H \in \mathcal{H}(F, G)).
$$

The continuity of φ_4 follows form the properties of the norm of the product of linear operators:

$$
\|\varphi_4(T + \Delta T, S + \Delta S) - \varphi_4(T, S)\|
$$
$$
= \|(S + \Delta S)(T + \Delta T) - ST\| \leq \|\Delta S\|\|T\| + \|S\|\|\Delta T\| + \|\Delta S\|\|\Delta T\|
$$
$$
(T, \Delta T \in \mathcal{B}(E, G), \quad S, \Delta S \in \mathcal{B}(G, F)).
$$

The continuity of φ_5 is obvious:

$$\|\varphi_5(T + \Delta T) - \varphi_5(T)\| = \|\Delta T\| \quad (T, \Delta T \in \mathcal{B}(E, F)).$$

The proposition is proved. $\square$

The above propositions imply the smooth implicit function theorem formulated in part 5 of this subsection.

7. In its turn, the theorem proved above implies the smooth inverse function theorem, which states that a continuously differentiable function is invertible in the neighborhood of any point at which its differential is invertible. Moreover, the inverse function is also continuously differentiable, and its differential at the corresponding points is equal to the inverse differential of the direct function.

Consider the following objects: Banach spaces E, F; open sets $X \subseteq E$, $Y \subseteq F$; points $a \in X$, $b \in Y$, and open sets $A \subseteq X$, $B \subseteq Y$; a function $h : X \to F$ such that $h(X) \subseteq Y$, and a function $f : A \to F$ with $f(A) = B$. If f is one-to-one, then the inverse function $f^{-1} : B \to E$ with $f^{-1}(B) = A$ is defined. A function f is called a *diffeomorphism* from A onto B if f maps A onto B and both f and f^{-1} are continuously differentiable.

Theorem. *If a function h is continuously differentiable and the differential dha is a homeomorphism from E onto F, then there exists an open neighborhood A of a and an open neighborhood B of $b = h(a)$ such that the restriction f of h to A is a diffeomorpism from A onto B and*

$$df^{-1}(f(x)) = (dfx)^{-1} \quad (x \in A).$$

Proof. Let $g : Y \times X \to F$, $g(y, x) = y - h(x)$ $(y \in Y, x \in X)$. Then $y = h(x)$ is equivalent to $g(y, x) = 0$. We have $h^{-1} = \{(y, x): y = h(x)\} = \{(y, x) : g(y, x) = 0\} = g^{-1}(0)$. To prove the theorem, it suffices to demonstrate that:

(1) g satisfies the assumption of the implicit function theorem and, consequently, there exists an open rectangular neighborhood $B \times A_0 \subseteq Y \times X$ of (b, a) such that $\varphi = g^{-1}(0) \cap B \times A_0$ is a function on B;

(2) $A = \varphi(B)$ is open and $f = \varphi^{-1}$ is the restriction of h to A;

(3) the functions f and $\varphi = f^{-1}$ are smooth and $d\varphi y = (dfx)^{-1}$ $(x \in A, y = f(x))$.

We now prove these assertions.

(1) THE FUNCTION g. It is required to prove that g is continuously differentiable, $g(b, a) = 0$, and the differential $Q(b, a) = dg(b, \cdot)a$ is a homeomorphism from E onto F. We use the theorem on the continuity of the total differential. Consider the functions

$$g(\cdot, x)(u) = u - h(x), \quad g(y, \cdot)(t) = y - h(t) \quad (u, y \in Y; \ t, x \in X).$$

Clearly, $g(\cdot, x)$ is differentiable. The diferentiability of $g(y, \cdot)$ follows form the diferentiability of h. Let I denote the identity transformation of F and let q and p be the projections of $F \times E$ onto F and E, respectively. We have

$$P(y, x) = dg(\cdot, x)y = I, \quad Q(y, x) = dg(y, \cdot)x = -dhx;$$
$$d_1g(y, x) = P(y, x) \cdot q = I \cdot q = q, \quad d_2g(y, x) = Q(y, x) \cdot p = -dhx \cdot p.$$

The first partial differential d_1g of g is a constant map; therefore, it is continuous. The second partial differential d_2g of g is equal to the composition of the maps $\psi_1(y, x) = (p, dhx)$, $\psi_2(T, S) = S \cdot T$, $\psi_3(T) = -T$. By assumption, dh is continuous. Since

$$\|\psi_1(y + \Delta y, x + \Delta x) - \psi_1(y, x)\| = \|(0, dh(x + \Delta x) - dhx)\|$$
$$= \|dh(x + \Delta x) - dhx\| \quad (y \in Y, \ \Delta y \in Y - y, \ x \in X, \ \Delta x \in X - x),$$

it follows that the continuity of dh implies the continuity of ψ_1.

The continuity of ψ_2 follows form the inequalities for the norm of the product of linear maps. The continuity of ψ_3 is obvious. The second partial differential $d_2g = \psi_3 \circ \psi_2 \circ \psi_1$ of g is continuous. By the total differential theorem, the continuity of the partial differentials d_1g and d_2g implies the continuity of the differential dg of g.

Since $b = h(a)$, it follows that $g(b, a) = b - h(a) = 0$ ($b = h(a)$). By hypothesis, the differential dha is a homeomorphism from E into F. Consequently, the differential $Q(b, a) = -dha$ is a homeomorphism form E onto F.

(2) THE FUNCTION φ. From the preceding arguments, it follows that g satisfies the assumption of the implicit function theorem. Therefore, there exists an open rectangular neighborhood $B \times A_0 \subseteq Y \times X$ of (b, a) such that $\varphi = g^{-1}(0) \cap B \times A_0$ is a smooth function on B and

$$d\varphi y = -Q^{-1}(y, x) \cdot P(y, x) = -(-dhx)^{-1} \cdot I = (dhx)^{-1}$$

$(y \in B, \ x = \varphi(y))$. It follows from the definition of φ that

$$A = \varphi(B) = \{x = \varphi(y) : y \in B\}$$
$$= \{x \in A_0 : (y, x) \in B \times A_0, \ q(y, x) = y - h(x) = 0\}$$
$$= \{x \in A_0 : y = h(x) \in B\} = A_0 \cap h^{-1}(B).$$

Since B is open and h is continuous, $h^{-1}(B)$ is open. Hence, since A_0 is also open, it follows that $A = A_0 \cap h^{-1}(B)$ is open.

(3) THE FUNCTION f. We have established that φ maps the open set B onto the open set A. Consider the restriction f of h onto A.

Suppose that $y \in B$ and $x = \varphi(y)$. Then the definition of φ implies that $y - f(x) = g(y, x) = g(y, \varphi(y)) = 0$. Hence, $y = f(x)$. Conversely, suppose that $x \in A$ and $y = f(x)$. Then the definition of f implies that $0 = y - f(x) = g(y, x)$. Since there exists a unique solution φ to the equation $g(y, x) = 0$ in $B \times A$, the obtained inequality implies that $x = \varphi(y)$. Hence, f is a one-to-one map from A onto B.

Being the restriction of the smooth function h, the function f is smooth. The smoothness of the function $f^{-1} = \varphi$ follows from the implicit function theorem. Thus, f is a diffeomorphism from A onto B and the differential of the inverse function $f^{-1} = \varphi$ is expressed by the equality in the assertion of the theorem. □

8. Analyzing the proofs of the smooth implicit and inverse function theorems, we can obtain similar theorems on continuous implicit and inverse functions. We use the notation of part 5. For any $x \in X$, let the function $g(x, \cdot) : Y \to G$ be differentiable at any $y \in Y$ and the map $Q : X \times Y \to \mathcal{B}(F, G)$, $Q(x, y) = dg(x, \cdot)y$, be continuous. In this case, g is said to be *continuously differentiable with respect to the second variable*.

Theorem. *If $g(x, y)$ is a continuous function that is continuously differentiable with respect to the second variable and the equation $g(x, y) = 0$ holds for a point (a, b) and is nondegenerate at (a, b), then this equation has a unique solution in a neighborhood of (a, b). The solution is continuous and passes through (a, b).*

Proof. In the proof of the existence and uniqueness of a solution f for a smooth function g, we used only the nondegeneracy of the equation and the continuity and continuous differentiability of g with respect to the second variable. The same properties of f were sufficient for f to be differentiable. Therefore, the assertion of the theorem holds. □

Remark. The continuous implicit function theorem is not a generalization of the implicit function theorem. Its assumption is weaker, but so is its assertion.

Some theorems on the solutions of nondegenerate equations are presented in Nirenberg (2001), Chapter 6. In particular, the Kolmogorov – Arnold – Moser theorem is proved. An implicit function theorem associated with the abstract form of the nonlinear Cauchy – Kovalevskaya theorem is proved in Appendix.

9. Consider Banach spaces $(E, \| \|)$ and $(F, \| \|)$, points $a \in E$ and $b \in F$, their neighborhoods A and B, an isomorphism f from A onto B, and the inverse isomorphism $\varphi = f^{-1}$.

Theorem. *If the function f is differentiable at a, the differential $df\,a$ is a homeomorphism from E onto F, and the inverse function $\varphi = f^{-1}$ is continuous at $b = f(a)$, then the function $\varphi = f^{-1}$ is differentiable at $b = f(a)$ and $d\varphi b = (df\,a)^{-1}$.*

Proof. The proof of this theorem repeats the proof of the differentiability of an implicit function and is reduced to estimating the corresponding remainder. The difference is in notation and assumptions.

Since f is an isomorphism from A onto B, the increment $\Delta f a$ is an isomorphism from $A - a$ onto $B - b$. At the same time,

$$\Delta\varphi b(v) = \varphi(b+v) - \varphi(b) = \varphi(f(a) + f(a+u) - f(a)) - \varphi(f(a))$$
$$= \varphi((f(a+u)) - \varphi(f(a)) = (a+u) - a = u$$

if $u \in A - a$, $v = \Delta f a(u) = f(a+u) - f(a) \in B - b$. Therefore,

$$\Delta\varphi b(v) = (\Delta f a(u))^{-1}. \tag{1}$$

Let $x \in A$, $y = f(x)$, $u = x - a$, $v = y - b$. In view of (1) and since f is differentiable at a, it follows that

$$0 = (y - b) - (f(a+u) - f(a))$$
$$= v - df a(u) - rf a(u) = v - df a(\Delta\varphi b(v)) - rf a(\Delta\varphi. \tag{2}$$

By hypothesis $L = (df a)^{-1}$ exists. By equalities (2),

$$\Delta\varphi b(v) = L(v) + \rho(v), \tag{3}$$
$$\rho(v) = -L(rf a(\Delta\varphi b(v))). \tag{4}$$

We now give a preliminary estimate of the remainder ρ. Let $\varepsilon_1 > 0$. Since the function $rf a$ is relatively small, there exists a $\delta_0 > 0$ such that

$$\|rf a(\Delta\varphi b(v))\| \leq \varepsilon_1 \|\Delta\varphi b(v)\| \tag{5}$$

whenever

$$\|\Delta \varphi b(v)\| \le \delta_0. \tag{6}$$

By hypothesis, φ is continuous. Consequently, there exists a $\delta_1 > 0$ such that

$$\|\Delta \varphi b(v)\| \le \delta_0 \quad (\|v\| \le \delta_1). \tag{7}$$

Since $L = (df a)^{-1}$ is a continuous map, from (4) and (5)–(7) it follows that

$$\|\rho(v)\| = \|L(rf a(\Delta \varphi b(v)))\|$$
$$\le \|L\| \cdot \|rf a(\Delta \varphi b(v))\| \le \le \varepsilon_1 \|L\| \cdot \|\Delta \varphi b(v)\| \quad (\|v\| \le \delta_1). \tag{8}$$

Using this inequality, we estimate the increment $\Delta \varphi b(v)$. Let $\varepsilon_1 \in]0, \|L\|^{-1}[$. Then (3) and (8) imply

$$\|\Delta \varphi b(v)\| \le \|L(v)\| + \|\rho(v)\| \le \|L\| \cdot \|v\| + \varepsilon_1 \|L\| \cdot \|\Delta \varphi b(v)\|,$$
$$\|\Delta \varphi b(v)\| \le \gamma \cdot \|v\| \quad (\|v\| \le \delta_1, \ \gamma = \|L\|(1 - \varepsilon_1 \|L\|)^{-1}). \tag{9}$$

It is now easy to establish that ρ is relatively small. Let $\varepsilon > 0$, $\varepsilon_1 \in]0, \|L\|^{-1}[$, and $\varepsilon_1 \|L\| \gamma \le \varepsilon$. Take a $\delta = \delta_1 > 0$ that satisfies (9). Using (8) and (9), we obtain

$$\|\rho(v)\| \le \varepsilon_1 \|L\| \cdot \|\Delta \varphi b(v)\| \le \varepsilon_1 \|L\| \gamma \cdot \|v\| \le \varepsilon \|v\| \quad (\|v\| \le \delta).$$

Hence, ρ is relatively small. By (3), this implies that φ is differentiable at b and its differential satisfies the equality in the assertion of the theorem. $\square$

Remark. The continuous inverse function theorem is not a generalization of the continuously differentiable implicit function theorem. It assumes the existence of the inverse function that is continuous at the point under consideration.

Exercise. Formulate corollaries for the matrix derivatives and deduce them from the implicit and inverse function theorems.

3.2.7. Taylor's formula

Taylor's formula expresses the idea of approximating smooth functions by polynomials in a general form.

1. Consider normed spaces $(E, \| \ \|)$, $(F, \| \ \|)$, an open set $U \subseteq E$, and a function $f : U \to F$. If f is continuous, then f is said to be 0 *times* differentiable. Put $\mathcal{B}^{(0)} = F$. If f is a continuous function, then it is considered to be the 0th differential $d^{(0)}f : U \to \mathcal{B}^{(0)}$ of itself.

Suppose that $n \in \mathbb{N}$ and the nth differential $d^{(n)}f : U \to \mathcal{B}^{(n)}$ of f is defined. If it is differentiable, then f is said to be $n+1$ times differentiable. Its $(n+1)$th differential $d^{(n+1)}f : U \to \mathcal{B}^{(n+1)}$ is defined as the *differential* $d(d^{(n)}f)$ of $d^{(n)}f$. By definition, $\mathcal{B}^{(n+1)} = \mathcal{B}(E, \mathcal{B}^{(n)})$. The notions of an *$n$ times differentiable function* $f : U \to F$ and the *nth differential* $d^{(n)}f : U \to \mathcal{B}^{(n)}$ are, by induction, defined for all natural n:

$$d^{(0)}f = f, \quad d^{(n+1)}f = d(d^{(n)}f).$$

Functions that are n times differentiable for any n are called *infinitely differentiable*. The values $d^{(n)}fu$ of the differential $d^{(n)}$ of a function f at a point $u \in U$ for different n lie in different spaces $\mathcal{B}^{(n)}$: $d^{(n)}fu \in \mathcal{B}(E, \mathcal{B}^{(n-1)})$ ($n \geq 1$). Therefore, $d^{(n)}fu \cdot v_1 \in \mathcal{B}^{(n-1)}$, $d^{(n)}fu \cdot v_1 \cdot v_2 \in \mathcal{B}^{(n-2)}, \ldots, d^{(n)}fu \cdot v_1 \cdot v_2 \cdots v_n \in \mathcal{B}^{(0)}$ ($v_1, \ldots, v_n \in E$). These values of the differential $d^{(n)}fu$, determined successively, are points of the same space F. If $v_1 = \cdots = v_n = v$, then we also write v^n instead of $v \cdots v$:

$$d^{(n)}fu \cdot v^n = d^{(n)}fu \cdot v \cdots v \in F \quad (v \in E).$$

In particular, $d^{(0)}fu \cdot v^0 = f(u)$ and $d^{(1)}fu \cdot v^1 = dfu \cdot v$.

Remark. As a rule, F is assumed to be a Banach space. In this case, all $\mathcal{B}^{(n)}$ are Banach spaces. This follows from the proposition of part 5 of Subsection 3.2.1.

2. Let $u \in U$, $B(u, \delta) \subseteq U$ and $u + v \in B(u, \delta)$. Choose an open interval $X \in \mathbb{R}$ such that $[0, 1] \subseteq X$ and $u + xv \in B(u, \delta)$ ($x \in X$). Consider the function $\varphi : X \to U$, $\varphi(x) = u + xv$, an $n + 1$ times differentiable function $f : U \to F$, and the composition $g = f \circ \varphi : X \to F$, $g(x) = f(u + xv)$.

Lemma. *The function g is $n + 1$ times differentiable and*

$$g^{(k)}(x) = d^{(k)}f(u + xv) \cdot v^k$$

$(0 \leq r \leq n + 1)$.

Proof. Since f is differentiable, it is continuous. The function φ is also continuous. Consequently, so is $g = f \circ \varphi$. Hence, g is 0 times differentiable and $g^{(0)}(x) = d^{(0)}f(u + xv) \cdot v^0$. Take a natural number $k \leq n$. If g is k times differentiable and $g^{(k)}(x) = d^{(k)}f(u + xv) \cdot v^k$, then g is $k + 1$ times differentiable and $g^{(k+1)}(x) = d^{(k+1)}f(u+xv) \cdot v^{k+1}$. Indeed, let $x \in X$ and $z \in X - x$. Then

$$g^{(k)}(z+x) - g^{(k)}(x) = d^{(k)}f(u+xv+zv) \cdot v^k - d^{(k)}f(u+xv) \cdot v^k$$
$$= (d^{(k)}f(u+xv+zv) - d^{(k)}f(u+xv)) \cdot v^k$$
$$= (d(d^{(k)}f)(u+xv) \cdot (zv) + r(d^{(k)}f)(u+xv)(zv)) \cdot v^k$$
$$= d^{(k+1)}f(u+xv) \cdot v^{k+1} \cdot z + r(d^{(k)}f)(u+xv)(zv) \cdot v^k.$$

The first equality follows from the equality for $g^{(k)}(x)$, the second is proved by induction, the third holds because $d^{(k)}f$ is differentiable, and the fourth follows from the equality $d^{(k+1)}f = d(d^{(k)}f)$ and the fact that $d(d^{(k)}f)(u+xv)$ is linear. For any $\varepsilon > 0$, there exists a $\gamma > 0$ such that $\|r(d^{(k)}f)(u+xv)(zv) \cdot v^k\| \le \|r(d^{(k)}f)(u+xv)(zv)\| \cdot \|v^k\| \le \varepsilon\|v^k\| \cdot |z|$ ($|z| \le \gamma$). The first inequality is proved by induction, and the second follows from the differentiability of $d^{(k)}f$. Therefore, the above relations imply that $g^{(k)}(x)$ is differentiable and the desired equality for $g^{(k+1)}(x)$ holds. By the principle of induction, the assertion of the lemma follows. $\square$

3. Let $(F, \|\,\|)$ be a normed space and $I \subseteq \mathbb{R}$ be an open interval. By analogy with the definition in part 1, a continuous function $f : I \to F$ is said to be 0 *times differentiable*. Its 0th derivative $f^{(0)}$ is, by definition, the function f itself. The property of being n times differentiable and the nth derivative are defined by induction: $f^{(0)} = f$, $f^{(n+1)} = (f^{(n)})'$. Functions that are n times differentiable for any natural n are called *infinitely differentiable*.

We will prove two lemmas. Suppose that $u \in I$, $u + v \in I$. Choose an open interval $X \subseteq \mathbb{R}$ such that $[0, 1] \subseteq X$ and $u + xv \in I$ for any $x \in X$. Consider the function $\varphi : X \to I$, $\varphi(x) = u+xv$, an $n+1$ times differentiable function $f : I \to F$, and the composition $g = f \circ \varphi : X \to F$, $g = f(u+xv)$.

Lemma 1. *The function g is $n + 1$ times differentiable and*

$$g^{(k)}(x) = f^{(k)}(u+xv) \cdot v^k$$

for $0 \le k \le n+1$.

Proof. Since f is differentiable, it is continuous. The function φ is continuous too. Hence, it is 0 times differentiable and $g^{(0)}(x) = f^{(0)}(u+xv) \cdot v^0$. Take a natural $k \le n$. If g is k times differentiable and $g^{(k)}(x) = f^{(k)}(u+xv) \cdot v^k$, then g is $k+1$ times differentiable and $g^{(k+1)}(x) = f^{(k+1)}(u+$

$xv) \cdot v^{k+1}$. Indeed, by the chain rule, since φ and $f^{(k)}$ are differentiable, so is $f^{(k)} \circ \varphi$, and

$$(f^{(k)} \circ \varphi)'(x) = f^{(k+1)}(\varphi(x)) \cdot \varphi'(x) = f^{(k+1)}(u + xv) \cdot v.$$

This equality together with the equality for $g^{(k)}(x)$ imply the desired equality for $g^{(k+1)}(x) = (g^{(k)}(x))'$. By induction, the assertion of the lemma follows. $\square$

Let $\psi : X \to \mathbb{R}$, $\psi(x) = (n!)^{-1}(1 - x)^n$.

Lemma 2. *The function ψ is infinitely differentiable. Moreover, $\psi^{(l)} = (-1)^l((n - l)!)^{-1}(1 - x)^{n-l}$ for $0 \le l \le n$ and $\psi^{(m)} = 0$ for $m > n$.*

Proof. It follows from the definition of ψ that it is continuous and $\psi^{(0)} = (-1)^0((n - 0)!)^{-1}(1 - x)^{n-0}$. Take an $n \in \mathbb{N}$. If ψ is l times differentiable and $\psi^{(l)} = (-1)^l((n - l)!)^{-1}(1 - x)^{n-l}$, then it is $l + 1$ times differentiable and $\psi^{(l+1)} = (-1)^{l+1}((n - l - 1)!)^{-1}(1 - x)^{n-l-1}$. By induction, this implies that ψ is n times differentiable and the lth derivative satisfies the equality in the assertion of the lemma if $0 \le l \le n$. In particular, $\psi^{(n)} = (-1)^n$. This implies that ψ is infinitely differentiable and its mth derivative is identically equal to zero for each $m > n$. $\square$

4. Consider an open interval $X \subseteq \mathbb{R}$ that includes $[0, 1]$, a function $g : X \to F$, a number $n \in \mathbb{N}$, and a number $c > 0$. Suppose that g is $n + 1$ times differentiable and $\|g^{(n+1)}(x)\| \le c$ $(x \in X)$.

Lemma. $g(1) = \sum\limits_{0 \le k \le n} (1/k!)g^{(k)}(0) + r_n$ $(\|r_n\| \le c/(n + 1)!)$.

Proof. Let $h : X \to F$, $h(x) = \sum\limits_{0 \le k \le n} (1/k!)g^{(k)}(x)(1 - x)^k$ $(x \in X)$. We now calculate the derivative $h'(x)$ and apply Lagrange's theorem.

Since h is equal to the sum of the products of the differentiable functions $g^{(k)}$ and $(-1)^{n-k}\psi^{(n-k)}$ $(0 \le k \le n)$, it follows that it is differentiable and

$$h'(x) = \sum_{0 \le k \le n} (-1)^{n-k}(g^{(k)}\psi^{(n-k)})'(x)$$

$$= g^{(n+1)}(x) \cdot \psi(x) + (-1)^n g(x) \cdot \psi^{(n+1)}(x) = g^{(n+1)}(x) \cdot \psi(x).$$

Let $q : X \to \mathbb{R}$, $q(x) = -(c/(n + 1)!)(1 - x)^{n+1}$. The function q is differentiable and $q'(x) = (c/n!)(1 - x)^n$. Consequently, $\|h'(x)\| \le q'(x)$ $(x \in X)$. Then, by Lagrange's theorem, $\|h(1) - h(0)\| \le q(1) - q(0) = c/(n + 1)!$. Since $h(1) - h(0) = g(1) - \sum\limits_{0 \le k \le n} (1/k!)g^{(k)}(0) = r_n$, the desired assertion follows. $\square$

The equality in the assertion of the lemma is called *Maclaurin's formula*.

Consider a normed space $(F, \| \; \|)$, an open interval $I \subseteq \mathbb{R}$, a function $f : I \to F$, points $u \in I$ and $v \in I - u$, numbers $n \in \mathbb{N}$ and $c > 0$. Suppose that f is $n + 1$ times differentiable and $\|f^{(n+1)}(t)\| \leq c \; (t \in I)$.

Theorem. $f(u+v) = \sum\limits_{0 \leq k \leq n} (1/k!) f^{(k)}(u) v^k + r f u(v)$, where $\|r f u(v)\| \leq (c/(n+1)!) \|v\|^{n+1}$.

Proof. Assume that $X \subseteq \mathbb{R}$ be an open interval such that $[0, 1] \subseteq X$ and $u + xv \in I \; (x \in X)$. The function $g : X \to F$, $g(x) = f(u + xv)$, satisfies the assumption of Lemma 1 of part 3 of this subsection. Therefore, g is $n+1$ times differentiable and $g^{(k)}(x) = f^{(k)}(u + xv) \cdot v^k \; (0 \leq k \leq n+1)$. Hence $\|g^{(n+1)}(x)\| \leq c \|v^k\|$. Consequently, the above lemma holds for g. Since $g(1) = f(u + v)$ and $g^{(k)}(x) = f^{(k)}(u) \cdot v^k$, the theorem follows. $\qquad \square$

The equality in the assertion of the theorem is called *Taylor's formula* for derivatives.

5. Consider normed spaces $(E, \| \; \|)$ and $(F, \| \; \|)$, an open set $U \subseteq E$, a point $u \in U$, a ball $B(u, \delta) \subseteq U$, numbers $n \in \mathbb{N}$ and $c > 0$, and a function $f : U \to F$. Let f be $n + 1$ times differentiable and let $\|d^{(n+1)} ft\| \leq c$ for $t \in B(u, \delta)$, i.e., the values of the $(n+1)$th differential $d^{(n+1)} f : U \to \mathcal{B}^{(n+1)}$ are bounded in the norm by the number c in a neighborhood of u.

Theorem. $f(u+v) = \sum\limits_{0 \leq k \leq n} (1/k!) d^{(k)} fu \cdot v^k + r_n fu(v)$, where $\|r_n fu(v)\| \leq (c/(n+1)!) \cdot \|v\|^{n+1} \; (\|v\| \leq \delta)$.

Proof. Let $X \subseteq \mathbb{R}$ be an interval such that $[0, 1] \subseteq X$, $u + xv \in B(u, \delta)$ $(x \in X)$. By the lemma of part 2, the function $g(x) = f(u + xv)$ is $n + 1$ times differentiable on X and $g^{(k+1)}(x) = d^{(k+1)} f(u+xv) \cdot v^{k+1} \; (0 \leq k \leq n)$. Hence, by induction, we obtain

$$\|g^{(n+1)}(x)\| = \|d^{(n+1)} f(u + xv) \cdot v^{n+1}\| \leq c \cdot \|v\|^{n+1}.$$

Thus, the lemma of part 4 holds for g. Since $g(1) = f(u + v)$ and $g^{(k)}(0) = d^{(k)} f(u) \cdot v^k$, the assertion of the theorem follows from the lemma. $\qquad \square$

The equality in the assertion of the theorem is called *Taylor's formula* for differentials.

6. Let f be twice differentiable. The differential $d^{(2)} fu : E \to \mathcal{B}(E, F)$ defines a continuous bilinear map $d^2 fu : E \times E \to F$, $d^2 fu(x, y) = d^{(2)} fuxy$ $(x, y \in E)$.

Theorem. *The bilinear map $d^2 fu$ is symmetric.*

Proof. We need to prove that

$$d(df)uxy = d(df)uyx \quad (x, y \in E). \tag{1}$$

This equation for the iterated differentials follows from the equality

$$\Delta(\Delta f)uvw = \Delta(\Delta f)uwv \quad (u + v, u + w, u + v + w \in U) \tag{2}$$

for the iterated increments, which is readily verified:

$$\Delta(\Delta f)uvw = (\Delta f(u + v) - \Delta fu)w = \Delta f(u + v)w - \Delta fuw$$
$$= f(u + v + w) - f(u + v) - f(u + w) + f(u) = \Delta(\Delta f)uwv.$$

We first prove that the difference between the iterated increment and the iterated differential

$$t(u, w) = \Delta(\Delta f)uvw - d(df)uvw \tag{3}$$

is relatively small. We have

$$t(v, w) = s(u, w) + r(df)uvw,$$
$$s(v, w) = \Delta(\Delta f)uvw - \Delta(df)uvw, \tag{4}$$
$$r(df)uvw = \Delta(df)uvw - d(df)uvw.$$

Since df is differentiable, the remainder $r(df)u$ is relatively small and for any $\varepsilon > 0$ there exists a $\delta > 0$ such that $\|r(df)uv\| \le \varepsilon\|v\|$ ($\|v\| \le \delta$), $r(df)uv \in \mathcal{B}(E, F)$. Therefore,

$$\|r(df)uvw\| \le \|r(df)uv\| \cdot \|w\| \le \varepsilon\|v\| \cdot \|w\| \quad (\|v\| \le \delta). \tag{5}$$

To estimate s, consider the function $h(w) = f(u + v + w) - f(u + w) - df(u + v)w + dfuw$ for any v. The function s is equal to the increment $\Delta h0$ of h: $s(v, w) = f(u+v+w) - f(u+w) - f(u+v) + f(u) - df(u+v)w + dfuw = h(w) - h(0)$. Therefore, for estimating s it suffices to estimate h and apply Lagrange's theorem. Applying the chain rule to $f(u + v + \cdot)$ and $f(u + \cdot)$ and using the continuity and linearity of $df(u + v)$ and dfu, we obtain

$$dhw = df(u + v + w) - df(u + w) - df(u + v) + dfu.$$

Since df is differentiable, we have

$$df(u+v+w) = dfu + d(df)u(v+w) + r(df)u(v+w),$$
$$df(u+w) = dfu + d(df)uw + r(df)uw,$$
$$df(u+v) = dfu + d(df)uv + r(df)uv.$$

Therefore, since $d(df)u$ is linear, it follows that

$$dhw = r(df)u(v+w) - r(df)uw - r(df)uv.$$

Hence, for any $\varepsilon > 0$ there exists a $\gamma \in]0, \delta[$ such that $\|dhw\| \le 3\varepsilon(\|v\|+\|w\|)$ ($\|v\| + \|w\| \le \gamma$). By Corollary 5 to Lagrange's theorem,

$$\|s(v,w)\| = \|h(w) - h(0)\| \le 3\varepsilon(\|v\| + \|w\|)\|w\|. \tag{6}$$

Equality (4) and inequalities (5) and (6) imply

$$\|t(v,w)\| \le 4\varepsilon(\|v\| + \|w\|)^2. \tag{7}$$

Equalities (2) and (3) and inequality (7) yield

$$\|d(df)uwv - d(df)uvw\| = \|t(v,w) - t(w,v)\|$$
$$\le \|t(v,w)\| + \|t(w,v)\| \le 8\varepsilon(\|v\| + \|w\|)^2. \tag{8}$$

Let $x, y \in E$, $\alpha > 0$, and $\|\alpha x\| + \|\alpha y\| \le \gamma$. From (8) it follows that

$$\alpha^2 \cdot \|d(df)uyx - d(df)uxy\|$$
$$= \|d(df)u(\alpha y)(\alpha x) - d(df)u(\alpha x)(\alpha y)\| \le \alpha^2 \cdot 8\varepsilon(\|x\| + \|y\|)^2,$$

$$\|d(df)uyx - d(df)uxy\| \le 8\varepsilon(\|x\| + \|y\|)^2$$

for any $\varepsilon > 0$. Consequently, $\|d(df)uyx - d(df)uxy\| = 0$ for $x, y \in E$, which yields (1). $\square$

7. Assume that $E = E_1 \times E_2$ is a product of normed spaces, $U = U_1 \times U_2$ is a rectangle with open sides $U_1 \subseteq E_1$ and $U_2 \subseteq E_2$, $u = (u_1, u_2) \in U$, and $f : U \to F$ is a three times differentiable function with values in a normed space F.

Take the partial derivatives $d_j f$ of f at u: $d_j f u = df u \cdot r_j p_j$ ($j = 1, 2$). Since df and the constant map $u \to r_j p_j$ are differentiable, from the differentiation rule for a product it follows that the maps $d_j f$ are differentiable and $d(d_j)fu = d(df)u \cdot r_j p_j$.

Lemma. $d(df) = d_1(d_1f) + d_2(d_1f) + d_1(d_2f) + d_2(d_2f)$.

Proof. Using the relationship between the total and partial differentials, by linearity we obtain $d(df) = d(d_1f + d_2f) = d(d_1f) + d(d_2f) = d_1(d_1f) + d_2(d_1f) + d_1(d_2f) + d_2(d_2f)$. □

Let $U_1 = E_1 = U_2 = E_2 = \mathbb{R}$. Consider the partial functions f_j for f, their derivatives f'_j, the partial derivatives f'_{ij} for f'_j, and their derivatives f''_{ij} ($i = 1, 2$; $j = 1, 2$). Note that

$$d_i(d_jf)uyx = f''_{ij}(u)x_iy_j, \quad d_1(d_2f)uyx = d(df)u(0, y_2)(x_1, 0),$$
$$d_2(d_1f)uyx = d(df)(x_1, 0)u(0, y_2),$$

where $x = (x_1, x_2) \in \mathbb{R}^2$, $y = (y_1, y_2) \in \mathbb{R}^2$. The theorem of part 6 and the preceding lemma imply the following statement.

Corollary. $f''_{12} = f''_{21}$.

Proof. Indeed, we have $f''_{12}(u)x_1y_2 = d_1(d_2f)uyx = d(df)u(0, y_2)(x_1, 0)$ $= d(df)(x_1, 0)u(0, y_2) = d_2(d_1f)uyx = f''_{21}(u)y_2x_1$. Putting $x_1=y_2=1$, we obtain the desired equality. □

Remark. The equality $f''_{12}(u) = f''_{21}(u)$ for the derivatives is equivalent to the equality $d_1(d_2f)u = d_2(d_1f)u$ for the differentials. If these equalities fail, then f is not twice differentiable.

Exercise. Using the principle of induction, prove the theorem on the symmetry of the derivatives for functions of several variables.

8. Consider normed spaces E and F, an open set $U \subseteq E$, a function $f : U \to F$, a point $u \in U$, the set $V = U - u$, and a ball $B(u, \delta) \subseteq E$ ($\delta > 0$).

Suppose that f is twice differentiable.

Proposition. *There exists a $\gamma \in \,]0, \delta[$ such that*

$$f(u + v) = f(u) + dfu(v) + 2^{-1}d^2fu(v, v) + r_2fu(v),$$

where $\|v\|^{-2} \cdot r_2fu(v) \to 0$ ($v \to 0$, $0 < \|v\| < \gamma$).
Proof. Consider the function

$$\bar{g}(x, y) = f(u + x) - f(u) - dfu(x) - 2^{-1}d^2fu(x, y) \quad (x, y \in B(0, \delta)).$$

Since f is differentiable, so is $h_1(x) = f(u+x)$ $(x \in B(0,\delta))$ and we have $dh_1s = df(u+s)$ $(s \in B(0,\delta))$. The continuous linear function $h_2 = dfu$ is differentiable and $dh_2s = dfu$ $(s \in B(0,\delta))$. Let $h(x,y) = h_1(x) - f(u) - h_2(x)$ $(x,y \in B(0,\delta))$. The partial differentials d_1h and d_2h of h are as follows:

$$d_1h(s,t) = (df(u+s) - dfu) \cdot p_1, \quad d_2h(s,t) = 0$$
$$(p_1(x,y) = x; \quad x,y \in E; \quad s,t \in B(0,\delta)).$$

Since df is differentiable, it is continuous and so is d_1h. The existence and continuity of the partial differentials d_1h and d_2h guarantees the existence of the total differential dh and the equality

$$dh(s,t)(x,y) = (df(u+s) - dfu)(x) \quad (s,t \in B(0,\delta); \; x,y \in E).$$

The map $b = d^2fu$ is differentiable and

$$db(s,t)(x,y) = d^2fu(x,t) + d^2fu(s,y) \quad (s,t \in B(0,\delta); \; x,y \in E).$$

Consequently, the function $\bar{g} = h - 2^{-1}b$ is differentiable and

$$d\bar{g}(s,t)(x,y) = (df(u+s) - dfu)(x) - 2^{-1}(d^2fu(x,t) + d^2fu(s,y)).$$

Consider the function $g(v) = \bar{g}(v,v) = r_2fu(v)$ $(v \in B(0,\delta))$. It is equal to the composition of $\bar{g}$ and the function φ such that $\varphi(v) = (v,v)$ $(v \in B(0,\delta))$. Using the chain rule, the equality for $d\bar{g}(s,t)(x,y)$, and the symmetry theorem, we have $dgs(x) = d\bar{g}(s,s)(x,x) = (df(u+s) - dfu)(x) - 2^{-1}(d^2fu(x,s) + d^2fu(s,x)) = (df(u+s) - dfu)(x) - d(df)usx = (df(u+s) - dfu - d(df)us)x$, $dgs = r(df)us$ $(s \in B(0,\delta), \; x \in E)$.

Since df is differentiable, it follows that for any $\varepsilon > 0$ there exists a $\gamma \in]0,\delta[$ such that $\|dgv\| = \|r(df)uv\| \le \varepsilon\|v\|$ $(\|v\| \le \gamma)$. By Corollary 5 of Lagrange's theorem, $\|r_2fu(v)\| = \|g(v)\| = \|g(v) - g(0)\| \le \varepsilon\|v\|^2$ $(\|v\| \le \gamma)$.
□

Remark. The above equality does not follow from the theorem of part 5 because f is assumed to be twice (but not three times) differentiable.

Exercise. Using induction, prove a similar equality for an n times differentiable function.

3.2.8.   Local minima

The parabolic approximation given by Taylor's formula for $n = 2$ makes it possible to formulate general conditions for local minima and maxima of real functions.

1. Consider a normed space $(E, \| \, \|)$, an open set $U \subseteq E$, a point $u \in U$, and a function $f : U \to \mathbb{R}$. The function f is said to have a *local minimum* at u if there exists a neighborhood $V \subseteq U - u$ of 0 such that $\Delta fu(v) \geq 0$ $(v \in V)$. The function f is said to have a *strict local minimum* at u if there exists a neighborhood $V \subseteq U - u$ of 0 such that $\Delta fu(v) > 0$ $(v \in V, v \neq 0)$. We can formulate similar definitions for a local maximum and a strict local maximum by reversing the inequalities. However, it is easier to observe that f has a (strict) local maximum at u if and only if $-f$ has a (strict) local minimum at u. Therefore, we can confine ourselves to considering only local minima.

Example. Let $E = U = \mathbb{R}$, $u = 0$, $f(t) = t^2$ $(t \in \mathbb{R})$. The parabola f has a strict local minimum at 0.

This example is typical. It explains everything below.

Take a $\delta > 0$ such that $B(u, \delta) \subseteq U$. Let f be twice differentiable. Then, by the proposition of part 8 of Subsection 3.2.7, there is a $\gamma \in \,]0, \delta[$ such that

$$f(u + v) = f(u) + dfu(v) + 2^{-1}d^2fu(v, v) + r_2fu(v),$$

where $\|v\|^{-2}|r_2fu(v)| \to 0$ $(v \to 0, 0 < \|v\| < \gamma)$.

The form d^2fu is said to be *positive*, written $d^2fu \geq 0$, if $d^2fu(v, v) \geq 0$ $(v \in E)$. We say that d^2fu is *strictly positive* and write $d^2fu > 0$ if $d^2fu(v, v) \geq \alpha\|v\|^2$ $(v \in E)$ for some $\alpha > 0$. In this case, d^2fu is positive definite and nondegenerate.

Remark. If $E = \mathbb{R}^k$, then the strict positivity of the form d^2fu is equivalent to its positive definiteness and nondegeneracy. Indeed, the unit sphere $S = \{t : \|t\| = 1\} \subseteq \mathbb{R}^k$ is compact. The continuity of d^2fu implies the continuity of the function $\varphi : E \to \mathbb{R}$ such that $\varphi(v) = d^2fu(v, v)$ $(v \in E)$. If d^2fu is positive definite and nondegenerate, then $\varphi(v) > 0$ $(v \in S)$. Since S is compact and φ is continuous, there exists a point $a \in S$ such that $\alpha = \varphi(a) = \min \varphi(S)$. As a result, $d^2fu(v, v) = \|v\|^2 d^2fu(v\|v\|^{-1}, v\|v\|^{-1}) = \|v\|^2\varphi(v\|v\|^{-1}) \geq \alpha\|v\|^2$ $(v \in E, v \neq 0)$. If $v = 0$, then the desired inequality also holds.

2. The following assertion yields necessary conditions for a local minimum.

Proposition. *If f has a local minimum at u, then $df u = 0$ and $d^2 f u \geq 0$.*
Proof. 1) If f has a local minimum at u, then there exists a $\gamma \in \,]0, \delta[$ such that for $\|v\| \leq \gamma$ we have

$$\Delta f u(v) + 2^{-1} d^2 f u(v, v) + r_2 f u(v) \geq 0.$$

Since $df u$ is linear and the $d^2 f u$ is bilinear, we obtain

$$\xi(df u(v) + 2^{-1} \xi d^2 f u(v, v) + \xi^{-1} r_2 f u(\xi v)) \geq 0 \qquad (1)$$

for $v \neq 0$, $v \in E$, and $\xi \neq 0$, $|\xi| \leq \|v\|^{-1} \gamma$. Assume that $df u \neq 0$ and take an $a \neq 0$ such that $df u(a) \neq 0$. Since $|\xi|^{-2} |r_2 f u(\xi a)| \to 0$ ($\xi \to 0$, $0 < |\xi| < \|a\|^{-1} \gamma$), it follows that there exists an $\alpha \in \,]0, \|a\|^{-1} \gamma[$ such that the sign of the quantity in the parentheses on the left-hand side of (1) is equal to the sign of $df u(a)$ if $0 < |\xi| < \alpha$. Then the left-hand side is strictly negative for some ξ and (1) fails. The assumption that $df u \neq 0$ leads us to a contradiction.

2) Thus, $df u = 0$, $2^{-1} \xi^2 d^2 f u(v, v) + r_2 f u(\xi v) \geq 0$, and $d^2 f u(v, v) + 2\xi^{-2} r_2 f u(\xi v) \geq 0$ for any $v \neq 0$, $v \in E$, and $\xi \neq 0$, $|\xi| \leq \|v\|^{-1} \gamma$. Since $|\xi|^{-2} |r_2 f u(\xi v)| \to 0$ ($\xi \to 0$), $v \neq 0$ and $d^2 f u \geq 0$, we have $d^2 f u(v, v) \geq 0$.
$\square$

Remark. Let $E = U = \mathbb{R}$, $u = 0$, $f(t) = t^3$ ($t \in \mathbb{R}$). The function f has a local minimum at u, although $df u = 0$ and $d^2 f u \geq 0$. Therefore, the above conditions are necessary but not sufficient.

3. The following assertion provides sufficient conditions for a strict local minimum:

Proposition. *If $df u = 0$ and $d^2 f u > 0$, then f has a strict local minimum at u.*
Proof. Using Taylor's formula of the proposition of part 8 of Subsection 3.2.7 and the condition $df u = 0$, for some $\gamma \in \,]0, \delta[$ we have

$$\Delta f u(v) = 2^{-1} d^2 f u(v, v) + r_2 f u(v) \quad (\|v\| < \gamma).$$

Since $d^2 f u(v, v) \geq \alpha \|v\|^2$ ($v \in E$) for some $\alpha > 0$ and $\|v\|^{-2} |r_2 f u(v)| \to 0$ ($v \to 0$), there exists a $\beta \in \,]0, \gamma[$ such that

$$\Delta f u(v) \geq (\alpha - 2^{-1} \alpha) \|v\|^2 = 2^{-1} \alpha \|v\|^2 \quad (\|v\| < \beta).$$

As a result, $\Delta f u(v) > 0$ if $0 < \|v\| < \beta$. $\square$

Remark. Assume that $E = U = \mathbb{R}$, $u = 0$, and $f(t) = t^4$ ($t \in \mathbb{R}$). The function f has a strict local minimum at $u = 0$, although $df\,u = 0$ and $d^2 f\,u = 0$. Therefore, the conditions in the proposition are sufficient but not necessary.

4. Consider normed spaces E, F, and G with norms denoted by $\|\ \|$, open sets $X \subseteq E$ and $Y \subseteq F$, points $a \in X$, $b \in Y$, and a function $g : X \times Y \to G$. Assume that F and G are Banach spaces, g is continuously differentiable, and the equation $g(x, y) = 0$ holds and is nondegenerate at (a, b). The function g satisfies the assumption of the implicit function theorem. Its partial differentials are denoted by

$$P(x, y) = dg(\cdot, y)x, \quad Q(x, y) = dg(x, \cdot)y.$$

Furthermore, consider a function $\varphi : X \times Y \to \mathbb{R}$ and the restriction of φ to the set $g^{-1}(0) = \{(x, y) : g(x, y) = 0\}$. The local minima of this restriction are called the *conditional local minima* of φ on $g^{-1}(0)$. This means that φ has a *conditional local minimum* at (a, b) on $g^{-1}(0)$ if and only if there exists an open neighborhood $A \times B$ of (a, b) such that $\varphi(x, y) \geq \varphi(a, b)$ $((x, y) \in g^{-1}(0) \cap A \times B)$. Moreover, φ has a *conditional strict local minimum* at (a, b) on $g^{-1}(0)$ if and only if there exists an open neighborhood $A \times B$ of (a, b) such that $\varphi(x, y) > \varphi(a, b)$ $((x, y) \neq (a, b), (x, y) \in g^{-1}(0) \cap A \times B)$. It is conventional to say "*under the condition $g(x, y) = 0$*" instead of "on $g^{-1}(0)$".

By the implicit function theorem, there exists an open neighborhood $A \times B$ of (a, b) such that $f = g^{-1}(0) \cap A \times B$ is a function on A. Consider the function ψ on A such that

$$\psi(x) = \varphi(x, f(x)) \quad (x \in A).$$

It follows from the definitions that that φ has a conditional (strict) local minimum at (a, b) on $g^{-1}(0)$ if and only if ψ has a (strict) local minimum at a. Therefore, instead of conditional local minima for φ, we can consider local minima for ψ and apply the propositions of parts 2 and 3.

Remark. This approach usually leads to cumbersome calculations connected with differentiating compositions and implicit functions. Iterated differentiation is especially tedious. In practice, only the necessary conditions are verified and then special techniques or computer programs are applied.

The investigation of conditional local maxima for a function φ is reduced to the study of local minima for $-\varphi$.

5. Suppose that φ is differentiable. Consider the identity map e on E and the function (e, f) on A, $(e, f)(x) = (x, f(x))$ $(x \in A)$. The function ψ is equal to the composition $\varphi \circ (e, f)$. The identity e is differentiable. So is the implicit function f. Therefore, (e, f) and ψ are differentiable. The following lemma provides simple conditions that are necessary for a differentiable function to have a local minimum at a given point of an open set.

Lemma. *If a function* $h : A \to \mathbb{R}$ *is differentiable at a point* $a \in A$ *and has a local minimum at* a, *then* $dha = 0$.

Proof. Let a number $\delta > 0$ be such that $B(a, \delta) \subseteq A$. If a function h satisfies the hypothesis, then $dha(v) + rha(v) \geq 0$ $(\|v\| < \delta)$. Therefore,

$$t(dha(v) + t^{-1}rha(tv)) \geq 0 \qquad (2)$$

for any $v \neq 0$, $v \in E$, and $t \neq 0$, $|t| < \|v\|^{-1}\delta$.

Suppose that $dha \neq 0$ and take a $v \neq 0$ in E such that $dha(v) \neq 0$. Since the remainder $rha(v)$ is relatively small, there exists a number $\gamma \in$ $]0, \|v\|^{-1}\delta[$ such that $|t^{-1}rha(tv)| \leq 2^{-1}\|dha\| \cdot \|v\|$ $(0 < |t| < \gamma)$ and therefore $\operatorname{sgn}(dha(v) + t^{-1}rha(tv)) = \operatorname{sgn} dha(v)$ $(0 < |t| < \gamma)$. The left-hand side of (2) is negative for some t, and (2) is false. A contradiction. $\square$

Remark. This lemma is referred to as *Fermat's theorem*.

6. Consider the functions $\varphi(\cdot, y)$ and $\varphi(x, \cdot)$ and their differentials

$$p(x, y) = d\varphi(\cdot, y)x, \quad q(x, y) = d\varphi(x, \cdot) \quad (x \in A, \ y \in B).$$

Using the differentiation rules and the equality $dex = e$ $(x \in A)$, we obtain

$$d\psi x(\Delta x) = d\varphi(x, f(x))(\Delta x, dfx(\Delta x))$$
$$= p(x, f(x)) \cdot \Delta x + q(x, f(x)) \cdot dfx(\Delta x) \quad (x \in A, \ \Delta x \in E),$$
$$d\psi x = p(x, f(x)) + q(x, f(x)) \cdot dfx, \quad (x \in A).$$

Then, by the implicit function theorem,

$$d\psi x = p(x, f(x)) - q(x, f(x)) \cdot Q^{-1}(x, f(x)) \cdot P(x, f(x))$$
$$= p(x, f(x)) + l(x, f(x)) \cdot P(x, f(x)), \quad (3)$$

where $l(x, f(x)) = -q(x, f(x)) \cdot Q^{-1}(x, f(x))$. Equivalently,

$$q(x, f(x)) + l(x, f(x)) \cdot Q(x, f(x)) = 0.$$

Proposition. *If φ has a conditional local minimum at (a, b) on $g^{-1}(0)$, then*

$$p(a, b) - q(a, b) \cdot Q^{-1}(a, b) \cdot P(a, b) = 0.$$

Proof. If φ has a conditional local minimum at (a, b) on $g^{-1}(0)$, then ψ has a local minimum at (a, b). The assumptions on φ and g imply that ψ is differentiable and its differential satisfies (3). Since ψ has a local minimum at a, from the preceding lemma it follows that $d\psi a = 0$. As a result, equality (3) with $x = a$ yields the desired equality.   □

7. In addition, consider a bounded linear functional $l \in \mathcal{B}(G, \mathbb{R})$. By the *Lagrange function for l, φ, and g* we mean the function

$$\lambda = \varphi + l \circ g, \quad \lambda(x, y) = \varphi(x, y) + l(g(x, y))$$

$(x \in A, \, y \in B)$. The function λ is differentiable because so are φ and g. The differentials of the functions $\lambda(\cdot, y)$, $\lambda(x, \cdot)$ $(x \in A, \, y \in B)$ satisfy the equalities

$$d\lambda(\cdot, y)x = p(x, y) + l \cdot P(x, y), \quad d\lambda(x, \cdot)y = q(x, y) + l \cdot Q(x, y).$$

If $x = a$, $y = b$, and $l = -q(a, b) \cdot Q^{-1}(a, b)$, then the first differential is equal to $d\psi a$ and the second is zero. In this case, the total differential of λ at (a, b) satisfies the equality

$$d\lambda(a, b)(\Delta x, \Delta y) = d\psi a(\Delta x) \quad (\Delta x \in E, \, \Delta y \in F)$$

and the equalities $d\lambda(a, b) = 0$, $d\psi a = 0$ are equivalent to $(\psi = \varphi \circ (e, f))$.
The preceding proposition implies the following statement.

Corollary. *If φ has a conditional local minimum at (a, b) on $g^{-1}(0)$, then there exists a bounded linear functional l such that $d\varphi(a, b) + l \cdot dg(a, b) = 0$.*

Remark. The restrictions of φ and the function $\lambda = \varphi + l \circ g$ to $g^{-1}(0)$ are equal for any $l \in \mathcal{B}(G, \mathbb{R})$. Therefore, φ and λ have conditional local minima at the same points. These points can sometimes be determined together with the corresponding functional l from the equations

$$g(x, y) = 0, \quad d\lambda(x, y) = 0$$

Example. Let $X = E = \mathbb{R}^2$, $Y = F = G = \mathbb{R}$, $g(x, y) = x_1^2 + x_2^2 + y^2 - 1$, $\varphi(x, y) = x_1 + x_2 + y$ $(x_1, x_2, y \in \mathbb{R})$. Let c be an arbitrary number and let $\lambda = \varphi + cg$, i.e.,

$$\lambda(x, y) = x_1 + x_2 + y + c(x_1^2 + x_2^2 + y^2 - 1) \quad (x_1, x_2, y \in \mathbb{R}).$$

Since $\lambda'(x, y) = (1 + 2cx_1, 1 + 2cx_2, 1 + 2cy)$, it follows that $\lambda'(x, y) = (0, 0, 0)$ and $g(x, y) = 0$ if and only if $c = -\sqrt{3}/2$ and $x_1 = x_2 = y = 1/\sqrt{3}$ or $c = \sqrt{3}/2$ and $x_1 = x_2 = y = -1/\sqrt{3}$. It is easy to verify that φ has a conditional local minimum at $(a, b) = -(1/3, 1/3, 1/3)$ equal to -1 if $g = 0$, and a maximum at $(1/3, 1/3, 1/3)$ equal to 1.

8. Return to the notation and assumptions of part 4. If a function g is nondegenerate at any point (a, b) in $L \subseteq X \times Y$, then it is said to be nondegenerate on L. If $g(a, b) = 0$, then, by the implicit function theorem, there exists a neighborhood $A \times B \subseteq X \times Y$ of (a, b) such that $f = g^{-1}(0) \cap A \times B$ is a continuously differentiable function on A. The subset L of the set $X \times Y$ is called a *smooth surface piece* in $X \times Y$ if there exists a smooth function $g : X \times Y \to G$ that is nondegenerate on L and such that $L = g^{-1}(0)$.

A closed subset S of $X \times Y$ is called a *smooth surface* in $X \times Y$ if for any $(a, b) \in S$ there exists a neighborhood $A \times B \subseteq X \times Y$ such that $L = S \cap (A \times B)$ is a smooth surface piece in $X \times Y$. This definition describes the simplest smooth surfaces.

A smooth surface piece $L = g^{-1}(0)$ is said to be *defined by the equation* $g(x, y) = 0$. Smooth surfaces in $E \times F$ defined by linear equations are called *planes*.

Example 1. Let $X = E = \mathbb{R}$, $Y = F = G = \mathbb{R}$. The equation $g(x, y) = x^2 + y^2 - 1 = 0$ $(x, y \in \mathbb{R})$ defines the circle S^1 in $\mathbb{R}^2$.

Example 2. Let $X = E = \mathbb{R}^2$, $Y = F = G = \mathbb{R}$. The equation $g(x, y) = x_1^2 + x_2^2 + y^2 - 1 = 0$ $(x = (x_1, x_2) \in \mathbb{R}^2, y \in \mathbb{R})$ defines the sphere S^2 in $\mathbb{R}^3$.

Any smooth surface S in a neighborhood of each of its points (a, b) can be approximated by a plane $TS(a, b)$. Consider a neighborhood $A \times B \subseteq X \times Y$ of a point (a, b) on a smooth surface S in $X \times Y$ whose piece $L = S \cap (A \times B)$

is defined by an equation $g(x, y) = 0$. The *tangent* to S at (a, b) is the plane $TS(a, b)$ defined by the equation

$$g(a, b)(x - a, y - b) = 0. \tag{1}$$

Equation (1) is equivalent to

$$P(a, b) \cdot (x - a) + Q(a, b)(y - b) = 0. \tag{2}$$

Since $Q(a, b)$ is a homeomorphism, it follows that (2) is equivalent to the equation

$$y = b + Q^{-1}(a, b) \cdot P(a, b) \cdot (x - a). \tag{3}$$

By the implicit function theorem, (3) is equivalent to the equation

$$y = f(a) + df a(x - a). \tag{4}$$

All these equations are called *equations for the tangent to S at the point* (a, b).

Example 1. Equations (2)–(4) for the tangent to the circle S^1 at a point $(a, b) \in S^1$ are written as $2a \cdot (x - a) + 2b \cdot (y - b) = 0$, $y = b - b^{-1} a \cdot (x - a)$.

Example 2. Equations (2)–(4) for the tangent to the sphere S^2 at a point $(a, b) \in S^2$ are written as $2a_1(x_1 - a_1) + 2a_2(x_2 - a_2) + 2b(y - b) = 0$ and $y = b - b^{-1} \cdot (a_1(x_1 - a_1) + a_2(x_2 - a_2))$. (We assume that $b \neq 0$.)

Using the geometric language, we can formulate yet another corollary to the proposition of part 6 for a conditional local minimum of a differentiable function $\varphi : X \times Y \to \mathbb{R}$.

Corollary. *If a function φ has a conditional local minimum at (a, b) on a smooth surface S, then*

$$d\varphi(a, b)(x - a, y - b) = 0 \quad ((x, y) \in TS(a, b)).$$

Proof. Take the piece $L = g^{-1}(0)$ of S, which contains the point (a, b). We have $y - b = df a(x - a)$ for any point (x, y) of the tangent $TS(a, b)$ to S at (a, b). Using the expression for the differential of an implicit function and the equality of the proposition of part 6, we obtain $d\varphi(a, b)(x - a, y - b) = p(a, b)(x - a) + q(a, b)(y - b) = p(a, b)(x - a) + q(a, b)df a(x - a) = p(a, b)(x - a) + q(a, b)Q^{-1}(a, b)P(a, b)(x - a) = 0$ for $(x, y) \in TS(a, b)$.   □

In particular, if φ has a conditional local minimum at $(0,0)$ on a smooth surface S, then the differential $d\varphi(0,0)$ vanishes on the tangent $TS(0,0)$. In general, the corollary claims that $d\varphi(a,b)$ vanishes on the plane $TS(a,b) - (a,b) = \{(x-a,y-b):(x,y)\in TS(a,b)\}$ passing through $(0,0)$ and parallel to the $TS(a,b)$ (which is an image of $TS(a,b)$ under the parallel translation $(x,y)\to(x,y)-(a,b)$).

9. Consider normed spaces $(G,\|\ \|)$ and $(H,\|\ \|)$, an open set $Z\subseteq G$, a point $c\in Z$, and a differentiable function $h:Z\to H$. Take a point $u\in G$ and an open interval $I\subseteq\mathbb{R}$ such that $0\in I$ and $\gamma(\xi)=c+\xi u\in Z\ (\xi\in I)$. The function $\gamma:I\to G$ is differentiable and $\gamma(0)=c$, $\gamma'(0)=u$. The composition $h\circ\gamma:I\to H$ is also differentiable and

$$(h\circ\gamma)(0)=h(c),\quad (h\circ\gamma)'(0)=dhc(u).$$

The derivative $(h\circ\gamma)'(0)$ of $h\circ\gamma$ is called the *derivative $h'_u(c)$ of h at c in the direction u*

$$h'_u(c)=(h\circ\gamma)'(0)=dhc(u).$$

Put $B=\{u:\|u\|=1\}\subseteq G$. If $H=\mathbb{R}$, the set of numbers $h'_u(c)$ for $u\in B$ can contain the greatest element. It is called the *gradient of h at c* and denoted by ∇hc:

$$\nabla hc=\max\{h'_u(c):u\in B\}.$$

In particular, if $G=\mathbb{R}^n$ and h is continuously differentiable, then, since the sphere $B\subseteq\mathbb{R}^n$ is compact and the differential dhc is continuous, in view of the equality $h'_u(c)=(h\circ\gamma)'(0)=dhc(u)$ there exists a gradient ∇hc of h at c. The direction of every vector $u\in B$ such that $h'_u(c)=\nabla hc>0$ is the direction of the maximal growth of h at c.

10. We now return to the notation and assumptions of part 8 and consider a neighborhood $A\times B\subseteq X\times Y$ of a point (a,b) on a smooth surface S in $X\times Y$ that has a piece $L=S\cap A\times B$ defined by the equation $g(x,y)=0$. Using the notion of a directional derivative, we can reformulate the corollary of part 8. Let $\varphi:X\times Y\to\mathbb{R}$ be a differentiable function.

Corollary. *If φ has a conditional local minimum at (a,b) on a smooth surface S, then*

$$\varphi'_{(x-a)(y-b)}(a,b)=0\quad((x,y)\in TS(a,b)).$$

Proof. This is equivalent to the corollary of part 8 of this subsection, since $\varphi'_{(x-a)(y-b)}(a, b) = d\varphi(a, b)(x - a)(y - b)$ for $(x, y) \in X \times Y$. $\square$

In particular, this corollary means that if φ has a conditional local minimum at $(0, 0)$ on a smooth surface S, then its derivative $\varphi'_{(x,y)}(0, 0)$ vanishes in each direction (x, y) in the tangent plane $TS(0, 0)$.

In general, the corollary states that the derivative $\varphi'_{(x-a)(y-b)}(a, b)$ vanishes in each direction $(x - a, y - b)$ in the plane $C = TS(a, b) - (a, b)$. If the gradient of φ at (a, b) exists on C, then the corollary states that it is equal to zero, i.e., $\nabla\varphi(a, b) = 0$.

Remark. It is conventional to write $\operatorname{grad} h(c)$ instead of ∇hc. Suppose that $S = \{u : \|u\| = 1\} \subseteq \mathbb{R}^n$, $h'(c) = (h'_1(c), \ldots, h'_n(c))$, $u = (u_1, \ldots, u_n)^t$: $\operatorname{grad} h(c) = \nabla hc = (h'(c))^t$. Since $h'(c) \cdot u = \sum h'_j(c)u_j$, it follows that if $h'(c) \neq 0$ and $\|u\| = 1$, the greatest value is achieved for

$$u_j = \|h'(c)\|^{-1}h'_j(c) \quad (j = 1, \ldots, n).$$

Comparing the definitions of the gradient and the tangent plane, we see that $\nabla g(a, b)$ is orthogonal to the tangent $TS(a, b)$ to the surface S defined by the equation $g(x, y) = 0$. It defines the normal to S at (a, b). It follows from the definitions that

$$h'_u(c) = (h \circ \gamma)'(0) = \lim_{t \to 0, t \neq 0} [h(c + tu) - h(c)].$$

The partial derivatives $h'_j(c)$ are obtained for $u = e_j$ for the standard basis of $\mathbb{R}^n$. Using the coordinate functions and the coordinate directions, we can calculate the partial derivatives of functions with domains in $\mathbb{R}^n$ and ranges in $\mathbb{R}^m$ as limits.

In the case where $G = \mathbb{R}^n$, the existence of the gradient ∇hx for any x and any differentiable function h makes it possible to consider the operator $\nabla : (h, x) \to \nabla hx$. It is called the *Hamilton operator* and is often written in the form $\nabla = (\partial_1, \ldots, \partial_n)$.

3.2.9.   Smooth curves

In what follows, by a smooth curve we mean a smooth vector function of a real variable.

1. Consider a Banach space $(E, \|\ \|)$, an open interval $I \subseteq \mathbb{R}$ that includes the interval $\overline{X} = [\bar{a}, \bar{b}]$, and an interval $[a, b]$ included in the open interval $X =]\bar{a}, \bar{b}[\ ([a, b] \subset X \subset \overline{X} \subset I)$. The restriction $y : \overline{X} \to E$ of a continuously differentiable function $y : I \to E$ to $\overline{X}$ is called a *smooth curve* in E defined on $\overline{X}$. The inclusions $[a, b] \subset [\bar{a}, \bar{b}] \subset I$ make it possible to consider the derivatives of y at the endpoints a, b, $\bar{a}$, $\bar{b}$.

Smooth curves in E defined on $\overline{X}$ form a vector space H. The equality

$$\|y\| = \sup \|y(x)\| + \sup \|y'(x)\| \quad (x \in \overline{X})$$

defines a norm $\|\ \|$ on H. The existence of the suprema is guaranteed by the compactness of $\overline{X}$ and the continuity of y and y'. The convergence $y \to c$ in H in this norm is equivalent to the uniform convergence of $y(x)$ to $c(x)$ and $y'(x)$ to $c'(x)$ in E $(x \in \overline{X})$. Therefore, $(H, \|\ \|)$ is a Banach space.

Exercise. Prove the above statements.

2. Consider open sets $A \subseteq E$, $B \subseteq E$, and the set

$$Y = \{y \mid y(x) \in A, \ y'(x) \in B \ (x \in \overline{X})\} \subseteq H.$$

Lemma. *The set Y is open.*
Proof. Let $c \in Y$, $U = c(\overline{X}) \subseteq A$, and $V = c'(\overline{X}) \subseteq B$. Since $\overline{X}$ is compact and c and c' are continuous, U and V are compact. Therefore, there exists a number $\delta > 0$ such that for any $x \in \overline{X}$ the closed balls $\overline{B}(c(x), \delta)$ and $\overline{B}(c'(x), \delta)$ are included in A and B, respectively. Choose a closed ball $\overline{B}(c, \delta) \subseteq H$ and a point $y \in \overline{B}(c, \delta)$. Since

$$\|y(x) - c(x)\| \leq \|y - c\| \leq \delta, \quad \|y'(x) - c'(x)\| \leq \|y' - c'\| \leq \delta \ (x \in \overline{X}),$$

it follows that $y(x) \in A$, $y'(x) \in B$ $(x \in \overline{X})$. Hence, $y \in Y$ and $\overline{B}(c, \delta) \subseteq Y$.
□

Consider a smooth function $f : X \times A \times B \to \mathbb{R}$ and the function $g : X \times Y \to \mathbb{R} \times E \times E$, $g(x, y) = (x, y(x), y'(x))$. For their composition $f \circ g = h : X \times Y \to \mathbb{R}$, we have

$$h(x, y) = f(x, y(x), y'(x)) \quad (x \in X, \ y \in Y).$$

Since f and g are continuous, the composition $f \circ g = h$ is continuous. Hence, all functions $h(\cdot, y) : X \to \mathbb{R}$ $(y \in Y)$ are continuous. Therefore, $h(\cdot, y)$ is Riemann integrable on $[a, b]$.

Exercise. Prove that g is continuous and $h(\cdot, y)$ is integrable on $[a, b]$.

A smooth function f and a smooth curve y define the real number

$$\varphi(y) = \int_a^b f(x, y(x), y'(x)) \, dx. \tag{1}$$

The obtained functional $\varphi : Y \to \mathbb{R}$ is defined on an open set in the Banach space H. Therefore, the statements on local minima and maxima in Subsection 3.2.8 are applicable to φ.

3. To formulate the necessary conditions for the existence of a minimum or a maximum of a functional φ on a given curve $y \in Y$, we need its differential $d\varphi y = \varphi'(y) \in \mathcal{B}(Y, \mathbb{R})$. (To simplify the notation, we identify the differentials with the derivatives.)

The vector space $C = C[a, b]$ of continuous functions $u : X \to \mathbb{R}$ on $[a, b]$ with norm $\|u\| = \sup |u(x)|$ $(x \in [a, b])$ is a Banach space. The function $\psi : C \to \mathbb{R}$ with values

$$\psi(u) = \int_a^b u(x) \, dx \quad (u \in C)$$

is linear and $|\psi(u)| \le (b - a)\|u\|$. Hence, ψ is smooth.

Exercise. Prove the preceding assertions.

Note that $g(x, y) = (x, \delta_x(y), d_x(y))$, where $\delta_x, d_x : Y \to E$ are continuous linear maps such that $\delta_x(y) = y(x)$, $d_x(y) = y'(x) \in E$ for any $x \in X$. Therefore, $g_2'(x, y) = (0, \delta_x, d_x)$ at any $y \in Y$:

$$g_2'(x, y) \cdot z = (0, \delta_x, d_x) \cdot z = (0, z(x), z'(x)) \quad (z \in Y).$$

Being a constant function, the differential $g_2'(x, \cdot)$ is continuous. It is now easy to calculate the differential of φ.

Proposition. The functional φ is continuously differentiable and

$$\varphi'(y)z = \int_a^b \left[f_2'(x, y(x), y'(x)) \cdot z(x) + f_3'(x, y(x), y'(x)) \cdot z'(x) \right] dx \tag{2}$$

for any $y \in Y$, $z \in Y$.

Proof. It follows from the definitions that $\varphi(y) = \psi(f(g(\cdot, y)))$ $(y \in Y)$. Applying the chain rule and using the equalities for the differentials $\psi'(u)$ and $g_2'(x, y)$, we have

$$\varphi'(y)z = \psi\big(f'(g(\cdot, y))g_2'(\cdot, y)z\big)$$
$$= \psi\big(f_1'(g(\cdot, y)) \cdot 0 + f_2'(g(\cdot, y)) \cdot z + f_3'(g(\cdot, y)) \cdot z'\big). \quad (3)$$

The function φ' is continuous because so are g, the differentials of f, and the functional ψ. $\quad \square$

3.2.10. A simplest variational problem

We formulate necessary conditions for a local minimum of the special functional φ of Subsection 3.2.9.

1. We use the definitions and the notation of Subsection 3.2.9. Consider the subspace $H_0 = \{y \in H : y(a) = y(b) = 0\}$ of the space H of smooth curves and the set $Y_0 = Y \cap H_0$. Clearly, H_0 is closed in H. Therefore, $(H_0, \|\ \|)$ is a Banach space. The set Y_0 is open in $(H_0, \|\ \|)$ and the restriction $\varphi_0 : Y_0 \to \mathbb{R}$ of the functional $\varphi : Y \to \mathbb{R}$ is a continuously differentiable functional. Its values are defined by (1) and the values of its differential are defined by equalities (2) and (3) in Subsection 3.2.9.

Given a curve $y \in Y_0$, consider the functions p, q:

$$p(x) = f_2'(x, y(x), y'(x)), \quad q(x) = f_3'(x, y(x), y'(x)) \quad (x \in X).$$

A curve $y \in Y_0$ is called an *extremal* if

$$\int_a^b [p(x) \cdot z(x) + q(x) \cdot z'(x)]\, dx = 0 \quad (1)$$

for all $z \in Y_0$.

2. Let $r : X \to \mathcal{B}(E, \mathbb{R})$ be a continuous function and $z \in Y_0$ be a smooth curve. The function $s = r \cdot z : X \to \mathbb{R}$ such that $s(x) = r(x) \cdot z(x)$ $(x \in X)$ is continuous and therefore integrable on $[a, b] \subset X$.

Lemma. If $\int_a^b r(x) \cdot z(x)\, dx = 0$ for any $z \in Y_0$, then $r(x) = 0$ $(x \in [a, b])$.

262   M. Lavrent'ev and L. Savel'ev. Operator theory and ill-posed ...

Proof. Suppose that $r(u) \neq 0$ for some $u \in]a, b[$. Then $r(u) \cdot v > 0$ for some $v \in E$. Since r is continuous, it follows that $r(u) \cdot v > 0$ if $x \in [u - \delta, u + \delta] \subset]a, b[$ for some $\delta > 0$. Take a smooth function $\gamma : X \to \mathbb{R}$ such that $\gamma(x) > 0$ for $x \in [u - \delta, u + \delta]$ and $\gamma(x) = 0$ for $x \notin [u - \delta, u + \delta]$. Let $c \in Y_0$ be a smooth curve such that $c(x) = \gamma(x)v$ $(x \in X)$. We have $r(x) \cdot c(x) = r(x) \cdot \gamma(x)v = r(x)\gamma(x) \cdot v = 0$ $(x \in X)$ and $r(x) \cdot c(x) > 0$ $(x \in [u - \delta, u + \delta])$. Hence, $\int_a^b r(x) \cdot c(x)\, dx \geq \int_{u-\delta}^{u+\delta} r(x) \cdot c(x)\, dx \geq \alpha\delta > 0$, where $\alpha = \min\{r(x) \cdot c(x) : x \in [u - \delta/2, u + \delta/2]\} > 0$. The obtained inequalities for the integrals contradict the assumption. □

3. In addition, suppose that q is smooth.

Lemma. *A curve $y \in Y_0$ is an extremal if and only if the Euler equation holds:*

$$q'(x) = p(x) \quad (x \in [a, b]). \tag{2}$$

Proof. 1) If a curve $y \in Y_0$ satisfies (2), then, by the differentiation rule for a product, for $z \in Y_0$ and $x \in X$ we obtain

$$(qz)'(x) = q'(x) \cdot z(x) + q(x) \cdot z'(x) = p(x) \cdot z(x) + q(x) \cdot z'(x).$$

Since $z(a) = z(b) = 0$, we have $\int_a^b (qz)'(x)\, dx = q(b) \cdot z(b) - q(a) \cdot z(a) = 0$. Consequently, equality (1) holds and y is an extremal.

2) If a curve $y \in Y_0$ is an extremal, then the following equalities hold for $z \in Y_0$:

$$\int_a^b (p(x) - q'(x))z(x)\, dx + q(b)z(b) - q(a)z(a)$$

$$= \int_a^b [p(x)z(x) - q'(x)z(x) + q'(x)z(x) + q(x)z'(x)]\, dx,$$

$$\int_a^b (p(x) - q'(x))z(x)\, dx = \int_a^b [p(x)z(x) + q(x)z'(x)]\, dx = 0.$$

By Lemma 2, this implies $p(x) - q'(x) = 0$ $(x \in [a, b])$. Hence, y satisfies (2). □

Proposition. *If a functional φ_0 has a local minimum or maximum on a curve $y \in Y_0$, then the Euler equation holds for y.*

Proof. If φ_0 has a local minimum or maximum on $y \in Y_0$, we have $\varphi_0'(y) \cdot z = 0$ ($z \in Y_0$) and, by the proposition of part 3 of Subsection 3.2.9, y is an extremal. It remains to apply the preceding lemma. □

4. Consider a simple example of the Euler equation.

Let $E = \mathbb{R}$, $[a, b] = [0, 1]$, $f(x, u, v) = v^2 - \pi^2 u^2$ $(x, u, v \in \mathbb{R})$. Then $\varphi(y) = \int_0^1 [(y'(x))^2 - \pi^2 (y(x))^2] \, dx$ for any smooth function $y : X \to \mathbb{R}$ on an open interval $X \subseteq \mathbb{R}$, $X \supset [0, 1]$ such that $y(0) = y(1) = 0$.

In this case, $p(x) = -2\pi^2 y(x)$, $q(x) = 2y'(x)$ and the Euler equation is equivalent to $y''(x) + \pi^2 y(x) = 0$. The solutions to this equation are the functions $y(x) = a \cos(\pi x) + b \sin(\pi x)$.

Since $y(0) = 0$, it follows that $a = 0$ and the extremals are of the form $y(x) = b \sin(\pi x)$. The functional φ vanishes at these curves.

Exercise. Study the local minima and maxima of φ.

5. The assumption of differentiability in part 3 of this subsection was made to simplify the arguments. The lemma on extremals can be proved without the *a priori* assumption that q is differentiable. A proof can be found in Section 1.4 of Alekseev, Tikhomirov, and Fomin (1987). Chapter 2 of this monograph is devoted to elements of differential calculus in normed spaces. The application of the general theory to a simplest variational problem is the subject of Section 4.4 of the above book. Various sufficient conditions for the extremum of a functional φ are given.

3.3.   INTEGRAL

The abstract integral is a generalization of the concept of area, volume, center of mass, and mean. It is defined to be a functional or operator on a space of integrable functions. Variants of the general theory of the integral can be found in Kolmogorov and Fomin (1989), Halmos (1950), Bourbaki (2004b), Federer (1969), Loève (1977, 1978), Segal and Kunze (1978). The concept of the integral presented in the present book, is described in detail in the lectures of Savel'ev (1988, 1989). The main definitions are introduced in Savel'ev (1982, 1997) and Savel'ev (1983).

3.3.1. Measures

A measure is an additive function. The concept of the integral is based on the notion of a measure.

1. Let $(\mathbb{K}, E)$ and $(\mathbb{L}, F)$ be vector spaces with associative and commutative scalar rings with identity. It is assumed that $\mathbb{K}$ is a subring of the ring $\mathbb{L}$.

Choose a nonempty hereditary class $\mathcal{B}$ of linearly independent sets $B \subseteq E$. The property of $\mathcal{B}$ to be hereditary means that for any $B \in \mathcal{B}$ the class $\mathcal{B}$ contains all subsets of B, i.e., $B \in \mathcal{B} \Rightarrow \mathcal{P}(B) \subseteq \mathcal{B}$. The class $\mathcal{B}$ will be called a *base class* or a *multibase*, while sets $B \in \mathcal{B}$ will be called *base sets* or simply *bases*.

Choose a set $C \subseteq E$ such that for any finite $X \subseteq C$ there exists a base $B \subseteq C$ whose additive span $A(B)$ includes X. This means that every vector in X is equal to a sum of vectors of B, i.e., by definition, $A(B)$ is the set of all finite sums of elements of B. The set C is said to be *spectral* with respect to $\mathcal{B}$.

Let $m : C \to F$ be a map that is additive on every finite base $B \subseteq C$ such that the sum of its vectors belongs to C. The map m is called a *measure*. The triple $(\mathcal{B}, C, m)$ is called a *measure space*.

Let $(\mathbb{K}, E) = \sum(\mathbb{K}, E_i)$ be the direct sum of vector spaces $(\mathbb{K}, E_i) \subseteq (\mathbb{K}, E)$, $(\mathcal{B}_i, C_i, m_i)$ be a collection of measure spaces, and $C = \cup C_i$. Let $\mathcal{B}$ be the class of all unions $B = \cup B_i$ of sets $B_i \in \mathcal{B}_i$, $p_i : E \to E_i$ be the direct projection of E onto E_i, and $m = \sum m_i p_i : C \to F$.

Lemma. *The class $\mathcal{B}$ is a base class, C is a spectral set, and m is a measure.*

Proof. 1) Let $B \in \mathcal{B}$, $B = \cup B_i$, $B_i \in \mathcal{B}_i$, and $X \subseteq B$. Then $X_i = X B_i \in \mathcal{B}_i$ and $X = \cup X_i \in \mathcal{B}$. The class $\mathcal{B}$ is nonempty because so are $\mathcal{B}_i$.

2) Let $X \in \mathcal{K}(C)$, $X = \cup X_i$, $X_i \in \mathcal{K}(C_i)$. Then $X_i \subseteq A(B_i)$ for some $B_i \in \mathcal{B}_i$ and $X \subseteq A(B)$, $B = \cup B_i \in \mathcal{B}$.

3) Let $B \in \mathcal{B}$, $B \in \mathcal{K}(C)$, $\sum b \in C$ $(b \in B)$. Then $m\left(\sum b\right) = \sum_i m_i p_i \left(\sum b\right) = \sum_i m_i \left(\sum p_i b\right) = \sum \sum_i m_i(p_i b) = \sum \sum_i m_i p_i(b) = \sum m(b)$.
$\square$

The space $(\mathcal{B}, C, m) = \sum(\mathcal{B}_i, C_i, m_i)$ is called *the direct sum of the spaces* $(\mathcal{B}_i, C_i, m_i)$, and the measure m is called the *direct sum of the measures m_i*.

2. Consider a field $\mathbb{P}$, the ring $\mathcal{F} = \mathcal{F}(\mathbb{N}, \mathbb{P})$ of all sequences of elements of $\mathbb{P}$, its subring $\mathcal{A}$, an ideal $\mathcal{I}$ of $\mathcal{A}$, the factor ring $\mathbb{K} = \mathcal{A}/\mathcal{I}$, a set U, the ring $E = \mathcal{F}(U, \mathbb{K})$ of all functions $x : U \to \mathbb{K}$, and the algebra $(\mathbb{K}, E)$. The vectors of algebra $(\mathbb{K}, E)$ are scalar functions x, while its scalars are factor sets of the set of sequences in the field $\mathbb{P}$.

Let P denote the set of all indicators defined on U. The subsets of U will be identified with with their indicators, so that $\mathcal{P}(U) = P \subseteq E$. Two indicators are orthogonal whenever the corresponding sets are disjoint.

Collections of pairwise orthogonal vectors are called *orthocollections*. The order relation for the sets defines the order for indicators. We say that a set Q of indicators forms a *subring* if

$$(1)\ 0 \in Q, \qquad (2)\ xy \in Q\ (x, y \in Q), \qquad (3)\ y - x = \sum z_k$$

for a finite orthocollection of vectors $z_k \in Q$ $(x, y \in Q,\ x \leq y)$. This is a term of set theory. Since $x \leq y \Leftrightarrow xy = x$ and $z_j z_k = 0$ $(j \neq k)$, we have $0 = x - x = x(y - x) = \sum x z_k$, $x z_j = 0$, and, consequently, (3) is equivalent to the equality $y = x + \sum z_k$, $x z_j = z_j z_k = 0$ $(j \neq k)$. This equality means that any indicator $y \in Q$ that is greater than $x \in Q$ is equal to the sum of pairwise orthogonal indicators in Q one of which is x, where it is possible that $y - x \notin Q$.

Consider the class $\mathcal{B} = \mathcal{R}$ of all linearly independent orthosets in E, the semiring $C = Q$ of indicators of subsets of U, a vector space $(\mathbb{K}, F)$, and the orthoadditive function $m : Q \to F$. The *Orthoadditivity* of m means that $m\left(\sum b\right) = \sum m(b)$ $(b \in B)$ for any base $B \subseteq Q$ such that $\sum b \in Q$. If a field $\mathbb{P}$ is infinite, then, for any finite collection of indicators $x_i \in P$, $\sum x_i \in P$ if and only if $x_i \perp x_j$ $(i \neq j)$. In this case, the orthoadditivity of the function m is equivalent to the conventional additivity. In what follows, the fields will be usually assumed to be infinite. If the mixed product of a vector and a scalar has no zero divisors, then the orthoset is linearly independent if and only if it does not contain the zero vector:

$$\sum \alpha_i x_i = 0 \Rightarrow \alpha_i x_i^2 = \alpha_i x_i = 0 \Rightarrow \alpha_i = 0 \quad (\alpha_i \in \mathbb{K},\ x_i \in P),$$

if $x_i x_j = 0$ for $i \neq j$ and $x_i \neq 0$.

It is often useful to apply an orthogonalization algorithm for a sequence of indicators. Consider a sequence of indicators $x_i \in Q$ and the sequence of indicators $y_j = x_j - (V_{i<j} x_i) x_j \in P$.

Lemma. *The following relations hold for all j, k, and n:*

$$y_j \leq x_j, \quad y_j \perp y_k\ (j \neq k), \quad \sum_{j \leq n} y_j = V_{j \leq n} x_j.$$

Proof. The inequality $y_j \leq x_j$ follows from the definitions.

By definition, $y_1 = x_1$, $y_2 = x_2 - x_1 x_2$ and $y_1 y_2 = x_1 x_2 - x_1 x_2 = 0$. Let $y_j \perp y_k$ $(j, k < n,\ j \neq k)$, $\sum_{i<n} y_i = V_{i<n} x_i$. Then $y_j y_n = y_j \left(x_n - \left(\sum_{i<n} y_i \right) x_n \right) = y_j x_n - y_j x_n = 0$ $(j < n)$. Hence, $y_j \perp y_k$ $(j, k \leq n,\ j \neq k)$ and therefore $\sum_{j \leq n} y_j = \sum_{j < n} y_j + y_n = \left(\bigvee_{i<n} x_i \right) + x_n - \left(\bigvee_{i<n} x_i \right) x_n = \left(\bigvee_{i<n} x_i \right) \vee x_n = \bigvee_{j \leq n} x_j$. The desired relations follow by induction. $\square$

Remark. The lemma on orthogonalization can also be applied to finite sequences $x_i \in Q$ $(1 \leq i \leq n)$, putting $x_j = 0$ $(j > n)$.

The following proposition provides a generalization of condition (3) in the definition of a semiring.

Proposition. *For any indicator $y \in Q$ and any orthocollection of indicators $u_i \in Q$ such that $u_i \leq y$ $(1 \leq i \leq m)$, there exists an orthocollection of indicators $v_j \in Q$ $(1 \leq j \leq n)$ such that $u_i \perp v_j$ and $y = \sum u_i + \sum v_j$.*

Proof. For $m = 1$, the assertion is equivalent to condition (3) in the definition of a semiring.

Suppose that the proposition holds for $m = l$. We will prove that it also holds for $m = l + 1$. This follows from the equalities $y - \sum_{i \leq l+1} u_i =$

$$\left(y - \sum_{i \leq l} u_i \right) - u_{l+1} = \sum_{j \leq n} (v_j - v_j u_{l+1}) = \sum_{j \leq n} \sum_k v_{jk}, \quad v_j - v_j u_{l+1} = \sum_k v_{jk},$$

where $u_i \perp v_{jk}$, $v_{jk} \perp v_{pq}$ $(j \neq p$ or $k \neq q)$. In addition to the assumption and the definition of a semiring, we have used the equalities $y u_{l+1} = u_{l+1}$, $u_i u_{l+1} = 0$ $(i \leq l)$, $u_{l+1} = \left(y - \sum_{i \leq l} u_i \right) u_{l+1} = \sum_{j \leq n} v_j u_{l+1}$. Since $u_i \perp v_j$ and $v_{jk} \leq v_j$, it follows that $u_i \perp v_{jk}$: $u_i v_{jk} = u_i (v_j v_{jk}) = (u_i v_j) v_{jk} = 0$. Similarly, $v_{jk} \perp v_{pq}$ follows from $v_j \perp v_p$ for $j \neq p$ and $v_{jk} \leq v_j$, $v_{pq} \leq v_p$. We conclude that the proposition holds for all $m \geq 1$. $\square$

The following theorem is the main result of part 2 of this subsection.

Theorem. *The semiring of indicators Q is a spectral set with respect to the orthogonal multibase $\mathcal{R}$.*

Proof. Consider a set of indicators $x_i \in Q$ $(1 \le i \le m)$. The assertion is true for $m = 1$, since the set $X = \{x_1\}$ is the required base in this case.

Suppose that the assertion is true for $m = l$. We will prove that it is also true for $m = l + 1$. Let $x_i = \sum_j \alpha_{ij} b_j$ $(1 \le i \le l)$ with some coefficients $\alpha_{ij} \in \{0, 1\}$ for indicators $b_j \in Q$ in the orthoset B, i.e., $X = \{x_1, \dots, x_l\} \subseteq A(B)$. Let $c_j = b_j x_{l+1}$. From the definition of a semiring, it follows that $c_j \in Q$ and $b_j = c_j + \sum_s z_{js}$ for indicators $z_{js} \in Q$ from the orthosets B_j. Since $b_j \perp b_k$, $c_j \le b_j$, and $c_k \le b_k$, we have $c_j \perp c_k$ $(j \ne k)$. For the same reason, we have $c_j \perp z_{kt}$, $z_{js} \perp z_{kt}$ $(j \ne k)$. In addition, $c_j \perp z_{js}$. Thus, the indicators c_j, z_{js} are pairwise orthogonal and $x_i = \sum_j \alpha_{ij} c_j + \sum_{js} \alpha_{ij} z_{js}$ $(1 \le i \le l)$. At the same time, by the preceding proposition, $x_{l+1} = \sum_j c_j + \sum_p z_p$ for $z_p \in Q$ such that $c_j \perp z_p$ and $z_p \perp z_q$ $(p \ne q)$. Since $z_p \le x_{l+1}$ and $z_{js} \le b_j - c_j = b_j(1 - x_{l+1}) \le 1 - x_{l+1}$, it follows that $z_p \perp z_{js}$. We conclude that the set $\bar{B} = \{c_j, z_{js}, z_p\}$ belongs to the orthogonal multibase $\mathcal{R}$ and the set $\bar{X} = \{x_1, \dots, x_l, x_{l+1}\}$ is included in the additive span $A(\bar{B})$ of the set $\bar{B}$.

By induction, the assertion of the theorem holds for all $m \ge 1$. $\square$

Corollary. *The orthogonal multibase* $\mathcal{R}$, *the semiring of indicators* Q, *and the orthoadditive function* $m : Q \to F$ *constitute a measure space* $(\mathcal{R}, Q, m)$.

Remark. The formal difficulties associated with the additivity of a measure are caused by the fact that a sum of elements of the semiring of indicators does not necessarily belong to it.

Example. We assume the following: $\mathbb{P} = \mathbb{R}$, $\mathcal{F} = \mathcal{F}(\mathbb{N}, \mathbb{R})$; $\mathcal{A} = \mathbb{R}$ is the field of constant real sequences, which are identified with the values of its elements; $\mathcal{I} = \{0\}$, $\mathbb{K} = \mathcal{A}/\mathcal{I} = \mathbb{R}$; $U = \mathbb{R}$, $E = \mathcal{F}(\mathbb{R}, \mathbb{R})$ is the algebra of real functions of a real variable; $\mathcal{P}(\mathbb{R}) = P$ is the class of all subsets of the real line, which are identified with their indicators; $F = \mathbb{R}$.

It is easy to verify that the set S of indicators of bounded intervals in $\mathbb{R}$ forms a semiring. By definition, the number $l(x) = \beta - \alpha$ is the length of an interval x with left endpoint α and right endpoint β. Clearly, $l : S \to \mathbb{R}$ is a measure. The space $(\mathbb{R}, S, l)$ is a prototype of the general case of a measure space.

By choosing an appropriate field $\mathbb{P}$, ring $\mathcal{A}$, ideal $\mathcal{I}$, and group F, it is possible to obtain measures with real, nonstandard, p-adic, vector, and operator values defined on classes of sets.

3. Consider vector spaces $(\mathbb{K}, E)$ and $(\mathbb{L}, F)$ described in part 1 and a set S of linearly independent vectors in E. The class $\mathcal{B} = \mathcal{K}(S)$ of all finite subsets of S is a multibase in E.

Example. A Hamel basis of E can be taken for the set S.

The union $A = A(S) = \cup A(B)$ $(B \in \mathcal{K}(S))$ is called the *additive span* of the set S. It is included in the linear span $L = L(S) = \cup L(B)$ $(B \in \mathcal{K}(S))$ of S.

Lemma. *The additive span $A = A(S)$ is a spectral set with respect to $\mathcal{K}(S)$.*

Proof. Let $X = \{x_i : 1 \le i \le m\} \subseteq A$. Then $x_i \in A(B_i)$ for a $B_i \in \mathcal{B}$. Therefore, $X \subseteq A(B)$ for $B = \cup B_i \in \mathcal{K}(S)$. $\square$

The restriction $m = T|A$ of a linear operator $T : L \to F$ to A is a measure. Indeed, m is additive on $B \in \mathcal{K}(S)$ because T is linear. The class $\mathcal{B} = \mathcal{K}(S)$, the set $A = A(S)$, and the restriction $m = T|A$ constitute the measure space $(\mathcal{K}(S), A(S), T|A)$.

Remark. If $T \in \mathcal{L}(E, F)$, then the restriction $m = T|A$ is a measure on $A = A(S)$ for any linear independent set $S \subseteq E$. Any linear operator generates a collection of measures. Integration is associated with the converse process: the generation of a collection of measures of a linear operator.

Let $\mathbb{K} = \mathbb{C}$ and let H be a complex Hilbert space. Consider the algebra $E = \mathcal{L}(H)$ of bounded linear operators $x : H \to H$. Identifying every subset $X \subseteq H$ with an Hermitian projection $p : H \to X$ makes it possible to identify the class $\mathcal{P} = \mathcal{P}(H)$ of subspaces of H with the set $P = P(H) \subseteq E$ of Hermitian projections onto these subspaces. For Hermitian projections, the orthogonality relation is defined: $p \perp q \Leftrightarrow pq = qp = 0$. If nonzero projections are orthogonal, they are linearly independent: $\sum \alpha_i p_i = 0 \Rightarrow \alpha_j p_j = (\sum \alpha_i p_i) p_j = 0 \Rightarrow \alpha_j = 0$. Therefore, the class $\mathcal{R}$ of all orthosets $B \subseteq P$ of nonzero projections is a base class. We call $\mathcal{R}$ an *orthogonal multibase of projections*.

Take an orthoset $R \in \mathcal{R}$ consisting of projections of finite rank and its additive span $A = A(R) = \cup A(B)$ $(B \in \mathcal{K}(R))$. It is a spectral set with respect to $\mathcal{B} = \mathcal{K}(R)$ and certainly with respect to $\mathcal{R}$. The dimension $d(p) = \dim p(H)$ of the range $p(H)$ of a projection $p \in R$ defines a measure $d : A \to \mathbb{R}$. The class $\mathcal{K}(R)$, the set $A(R)$, and the measure d form a measure space $(\mathcal{K}(R), A(R), d)$.

Remark. Indicators and projectors share a common property of idempotency. We can consider the general model with the algebra E and the set P of its idempotents. If involution is defined for E, then Hermitian idempotents can be considered. We will not deal with this general model in this book. Everything that has been said in this subsection only illustrates the general definition of a measure space. In what follows, we present the theory mainly for the case of the numeric measures defined on classes of sets.

4. Under the definitions of part 1, the spectral additivity of the measure $m : C \to F$ is equivalent to its linearity. Linear operators are usually defined on subspaces of a given space. The spectral set C on which the measure m is defined is not necessarily a subspace of E.

Lemma. *A measure is a linear map.*

Proof. The linearity of a measure $m : C \to F$ means that

$$m(\alpha x + \beta y) = \alpha m(x) + \beta m(y) \quad (\alpha, \beta \in \mathbb{K}; \ x, y, \alpha x + \beta y = z \in C).$$

Since $\mathbb{K} \subseteq \mathbb{L}$, we have $\alpha m(x), \beta m(y) \in F$. Let

$$x = \sum \xi(e)e, \quad y = \sum \eta(e)e, \quad z = \sum \xi(e)e \quad (e \in B)$$

for a base $B \subseteq C$ and finite indicators $\xi, \eta, \zeta : B \to \mathbb{K}$. Then $\alpha\xi(e) + \beta\eta(e) = \zeta(e)$, and since $\zeta(e) \in \{0, 1\} \subseteq \mathbb{K}$ and m is additive on B, it follows that

$$m(z) = \sum \zeta(e)m(e) = \sum (\alpha\xi(e) + \beta\eta(e))m(e)$$
$$= \alpha \sum \xi(e)m(e) + \beta \sum \eta(e)m(e) = \alpha m(x) + \beta m(y).$$

We conclude that the measure m is linear on C. □

Since a measure is linear, it can be extended to its linear span $S = L(C)$, which consists of linear combinations $u = \sum \alpha(x)x \ (x \in C)$ with finite

$\alpha : C \to \mathbb{K}$. The vectors $u \in S$ will be called *simple*. Let a correspondence $l : S \to F$ be defined by the equality

$$l(u) = \left\{ \sum \alpha(x)m(x) : u = \sum \alpha(x)x \right\}.$$

The correspondence l is called the *simple integral*, or *integral sum*, or *simple mean* with respect to the measure m.

Theorem. *The simple integral l is the only linear map from S to F that is an extension of m.*

Proof. 1) The correspondence l is many-to-one. Indeed, the equality $\sum \alpha(x)x = \sum \beta(x)x$ $(x \in C)$ for finite $\alpha, \beta : C \to \mathbb{K}$ is equivalent to the equality $\sum \gamma(x)x = 0$ for $\alpha(x) - \beta(x) = \gamma(x)$ $(x \in C)$. Furthermore, the equality $\sum \alpha(x)m(x) = \sum \beta(x)m(x)$ is equivalent to $\sum \gamma(x)m(x) = 0$. Therefore, l is a many-to-one correspondence if $\sum \gamma(x)x = 0$ implies $\sum \gamma(x)m(x) = 0$. Consider the finite set $X = \{x : \gamma(x) \neq 0\} \subseteq C$ and a base $B \subseteq C$ whose linear span includes X. Let $x = \sum \xi(x,e)e$ $(e \in B)$ for any $x \in X$ and a finite indicator $\xi(x, \cdot) : B \to \mathbb{K}$. Then, since the base vectors $e \in B$ are linearly independent, the equalities

$$0 = \sum_x \gamma(x)x = \sum_x \gamma(x)\left(\sum_e \xi(x,e)e\right) = \sum_e \left(\sum_x \gamma(x)\xi(x,e)\right)e$$

imply $\sigma(e) = \sum_x \gamma(x)\xi(x,e) = 0$ $(e \in B)$. Consequently, since the measure m is linear on C, we have

$$\sum_x \gamma(x)m(x) = \sum_x \sum_e \xi(x,e)m(e) = \sum_e \sigma(e)m(e) = 0.$$

2) The map l is linear. Indeed, let $\alpha, \beta \in \mathbb{K}$ and $u = \sum \alpha(x)x$, $v = \sum \beta(x)x \in S$ for finite collections of scalars $\alpha(x), \beta(x) \in \mathbb{K}$ $(x \in C)$. Then $\alpha u + \beta v = \sum(\alpha\alpha(x) + \beta\beta(x))x$ and, since l is many-to-one, we obtain $l(\alpha u + \beta v) = \sum(\alpha\alpha(x) + \beta\beta(x))m(x) = \alpha \sum \alpha(x)m(x) + \beta \sum \beta(x)m(x) = \alpha l(u) + \beta l(v)$.

3) The map l is an extension of the measure m. This readily follows from the definitions and the property of l to be many-to-one: since $u = 1 \cdot u$ $(u \in S)$ and, by assumption, the scalar rings $\mathbb{K}$ and $\mathcal{L}$ have common identity 1, it follows that $l(u) = 1 \cdot m(u) = m(u)$.

4) If a linear operator $n : S \to F$ is an extension of the measure m, it follows that $n = l$. Indeed, $n(u) = \sum \alpha(x)n(x) = \sum \alpha(x)m(x) = l(u)$ for any $u = \sum \alpha(x)x \in S$ $(x \in C)$. □

Remark. The theorem on the simple integral is a modification of the general principle of linear extension (Bourbaki, 2003, Section 2.1, Chapter II).

Consider the direct sum $(\mathcal{B}, C, m)$ of measure spaces $(\mathcal{B}_i, C_i, m_i)$ described in part 1. Assume that $S_i = L(C_i)$, $l_i : S_i \to F$ is a simple integral with respect to the measure m_i, and $S = \sum S_i$ is the direct sum of spaces S_i. The map $l = \sum l_i p_i : S \to F$ is called the *direct sum* of linear operators l_i. The preceding theorem and the general rules for the operations defined for linear operators (Bourbaki, 2003, Section 2.4, Chapter II) implies the following statement.

Corollary. *The simple integral with respect to the direct sum of measures is equal to the direct sum of simple integrals with respect to these measures.*

Exercise. Provide a detailed proof of the corollary.

5. The most difficult step in the definition of the integral is the continuous extension of the integral sum to the integral. The main problem is in choosing the appropriate conditions for the measure to ensure such an extension. We give only a general scheme for this step.

Assume that the scalar rings $\mathbb{K}$ and $\mathbb{L}$ are fields and there are topologies for $\mathbb{K}$, $\mathbb{L}$, E, and F such that the operations on scalars and vectors are continuous. In topological vector spaces $(\mathbb{K}, E)$ and $(\mathbb{L}, F)$, the closure of any subspace is a subspace, since translations are continuous. A linear operator is continuous if and only if it is continuous at zero. This implies the following general criterion for the continuity of an integral sum.

Lemma. *The integral sum l is continuous if and only if for any neighborhood V of $0 \in F$ there exists a neighborhood U of $0 \in E$ such that $\sum \alpha(e) m(e) \in V$ for $\sum \alpha(e) e \in U$ $(e \in B,\ B \in \mathcal{B},\ B \subseteq C)$.*

In some cases, this criterion makes it possible to find measures such that simple integrals are continuous with respect to these measures. Instead of topologies for the spaces E and F, we can consider topologies for their subspaces that include S and $l(S)$, respectively.

The continuity of linear operators in topological vector spaces is uniform. Under certain assumptions, a uniformly continuous map can be extended to the closure of its domain of definition. We will make an additional assumption that the space F is separated and complete. Then F is regular and there exists a unique continuous extension $\bar{l} : \bar{S} \to F$ of the continuous integral sum $l : S \to F$ to the closure $\bar{S}$ of its domain of definition S. The values of $\bar{l}$ are defined by the equality $\bar{l}(\bar{u}) = \lim l(u)$ ($u \to \bar{u}$, $u \in S$). The operator l is linear, and so is $\bar{l}$.

Theorem. *Under the assumptions of the preceding lemma and the above additional assumptions, there exists a unique continuous map $\bar{l} : \bar{S} \to F$ that is an extension of the integral sum $l : S \to F$.*

Vectors of $\bar{S}$ will be called *regularly integrable*, and the operator $\bar{l}$ will be called a *regular integral* with respect to the measure m. This term is introduced to help us to distinguish $\bar{l}$ from other continuations of the integral sum l.

Remark. In Savel'ev (1976, 1982, 1983, 1984, 1997), some specific implementations of the continuous extension of a measure to an integral are described. In Segal and Kunze (1978), an integral is considered that is defined as the trace of a linear operator in a Hilbert space. A special norm is used to construct a continuous extension.

In the article Savel'ev (1983), nonadditive outer measures are defined whose sums of, in a certain sense, sufficiently small elements are as small as desired. Outer measures are defined on Boolean algebras and assume their values in uniform spaces. There are Jordan and Lebesgue outer measures that are continuous in the corresponding topologies. Corollaries are deduced for measures with abstract values. It would be logical to study continuous extensions of outer measures to the appropriate spaces of functions in the form of nonlinear integrals (such as the upper and the lower integrals in the numeric case; see Federer, 1969, Chapter 2).

3.3.2. Classical definition of the integral

In what follows, we describe the integration of numeric functions with respect to numeric measures. By the integral we mean the limit of integral sums, which is in accordance with the classical definition.

1. Consider the measure space $(\mathcal{R}, Q, \mu)$, which is a special case of the measure space $(\mathcal{R}, Q, m)$ described in Subsection 3.3.1, part 2. By Q we mean the semiring of subsets of U, which is identified with the semiring Q

of their indicators, and $\mu : Q \to \mathbb{K}$ is a numeric measure (real or complex) that satisfies some continuity conditions. ($\mathbb{K} \in \{\mathbb{R}, \mathbb{C}\}$). The algebra $\mathcal{F} = \mathcal{F}(U, \mathbb{L})$ consists of numeric functions on U ($\mathbb{L} \in \{\mathbb{R}, \mathbb{C}\}$). We will assume that $\mathbb{K} \subseteq \mathbb{L}$. The situation where the fields $\mathbb{K}$ and $\mathbb{L}$ are different causes difficulties. On the other hand, in the case $\mathbb{K} = \mathbb{L} = \mathbb{C}$ it is often necessary to distinguish $\mathbb{R} \subseteq \mathbb{C}$, whereas in the case $\mathbb{K} = \mathbb{L} = \mathbb{R}$ it is necessary to extend the results to the complex case, which is not convenient either.

The sets in Q and their indicators in Q will be called *primary*, while the functions in the linear span $\mathcal{S} = L(Q)$ of Q in the algebra $\mathcal{F}$ will be called *simple*. The sets whose indicators are simple will be also called *simple*. Thus, simple sets are sums of disjoint primary sets, while simple functions are are linear combinations of simple indicators. Identifying sets with their indicators makes it possible to view simple functions as linear combinations of simple sets, i.e., $Q \subseteq \mathcal{F}$, $\mathcal{S} = L(Q)$.

Example. If Q is a semiring of intervals in $\mathbb{R}$, then simple sets are unions of finite collections of disjoint intervals, while simple functions are represented by step functions.

The equalities $f = \sum a_j X_j = \sum b_k Y_k$ can hold for a simple function $f \in \mathcal{S}$ for different collections of numbers $a_j, b_k \in \mathbb{L}$ and primary sets $X_j, Y_k \in Q$. Representations of f such that the sets X_j are disjoint are said to be *disjoint*. A disjoint representation can be obtained from an arbitrary representation by means of orthogonalization.

Lemma. *Simple functions form a subalgebra of the algebra $\mathcal{F}$.*

Proof. Let $f = \sum a_j X_j$, $g = \sum b_k Y_k$ $(a_j, b_k \in \mathbb{L}$, $X_j, Y_k \in Q)$. Then $f + g$, $fg = \sum a_j b_k X_j Y_k \in \mathcal{S}$, since $X_j Y_k \in Q$. $\quad\square$

Let $\mathcal{R}$ denote the set of all simple sets in U. If we identify sets with their indicators, the following inclusions hold: $Q \subseteq \mathcal{R} \subseteq \mathcal{S} \subseteq \mathcal{F}$. The simple sets form a subring of the Boolean ring $\mathcal{P} = \mathcal{P}(U)$, generated by the subring Q. By definition, $\mathcal{R} = \mathcal{P} \cap \mathcal{S}$. For the measure $\mu : Q \to \mathbb{K}$, the integral sum $S(\cdot, \mu) : \mathcal{S} \to \mathbb{K}$ is defined (see Subsection 3.3.1, part 4).

Theorem. *There exists a unique measure $\nu : \mathcal{R} \to \mathbb{K}$ such that ν is an extension of the measure $\mu : Q \to \mathbb{K}$ and $S(\cdot, \nu) = S(\cdot, \mu)$.*

Proof. The measure $\nu : \mathcal{R} \to \mathbb{K}$ is equal to the restriction of the integral sum $S(\cdot, \mu)$ to the ring $\mathcal{R}$. It is an extension of the measure μ because $\nu(X) = S(X, \mu) = \mu(X)$ $(X \in Q)$. Since $\mathcal{R} \subseteq S$, it follows that $L(\mathcal{R}) = S$ and $S(\cdot, \nu) = S(\cdot, \mu)$ because the linear map $S : S \to \mathbb{L}$ that is an extension of μ is unique. $\square$

The above theorem makes it possible to consider measures on rings of sets instead of measures on semirings. This is more convenient because all standard operations on sets are defined for rings.

2. We now consider the integral sum $S : S \to \mathbb{L}$ that is an extension of the measure $\mu : \mathcal{R} \to \mathbb{K}$. By definition,

$$S(f) = \sum a_j \mu(X_j) \quad (f = \sum a_j X_j \in S).$$

We will write $S(\cdot, \mu)$ and $S(f, \mu)$ whenever the measure μ needs to be specified explicitly.

In what follows, the measures under consideration will be usually assumed to be positive and defined on rings of sets. By the term *measure* we will mean a *positive measure on a ring of sets*. Exceptions will be specially mentioned unless they will be obvious from the context.

Remark. The reason for focusing on positive measures is that integration with respect to an arbitrary numeric measure can be reduced to integration with respect to positive measures. Positive measures possess some special properties of great importance.

Throughout this subsection, it is assumed that $\mu \geq 0$.

Lemma. If $f \geq 0$, then $S(f, \mu) \geq 0$.

Proof. Consider a disjoint representation $f = \sum a_j X_j$ of a simple function f. The inequalities $f \geq 0$ and $\mu \geq 0$ mean that $a_j \geq 0$ and $\mu(X_j) \geq 0$ for all j. Therefore, $S(f) = S(f, \mu) \geq 0$. $\square$

Corollary 1. Let $f, g \in S$ be real and $g \geq f$. Then $S(g) \geq S(f)$.

Proof. Since S is a linear functional, the preceding lemma implies that $S(g) - S(f) = S(g - f) \geq 0$ for $g \geq f$. $\square$

Corollary 2. *If a simple function f and numbers a and b are real, $X \in \mathcal{R}$, $a \leq f(x) \leq b$ ($x \in X$), and $f(x) = 0$ ($x \notin X$), then $a\mu(X) \leq S(f, \mu) \leq b\mu(X)$.*

Proof. By assumption, $aX \leq f \leq bX$ and therefore

$$a\mu(X) = S(aX, \mu) \leq S(f, \mu) \leq S(bX, \mu) = b\mu(X).$$

□

The triangle inequality holds for integral sums.

Proposition. $|S(f)| \leq S(|f|)$.

Proof. Take the disjoint representation $f = \sum a_j X_j$. The function $|f|$ with values $|f(x)|$ is also simple and its disjoint representation is $|f| = \sum |a_j| X_j$, since $f(x) = a_j$ for $x \in X_j$. Therefore, $|S(f)| = |\sum a_j \mu(X_j)| \leq \sum |a_j| \mu(X_j) = S(|f|)$. □

Corollary 3. *If $f \in \mathcal{S}$, $c \geq 0$, $X \in \mathcal{R}$, $|f(x)| \leq c$ ($x \in X$), and $f(x) = 0$ ($x \notin X$), then $|S(f, \mu)| \leq c\mu(X)$.*

Proof. From the triangle inequality for S and Corollary 2, it follows that $|S(f, \mu)| \leq S(|f|, \mu) \leq c\mu(X)$. □

Integral sums that are extensions of positive measures possess good properties with respect to the order.

Example. Let $\mathcal{R} = \mathcal{P}(U)$, $a \in U$ and $\mu(X) = \delta_a(X) = 1$ ($a \in X$), $\mu(X) = \delta_a(X) = 0$ ($a \notin X$) for every $X \subseteq U$. Then $S(f) = f(a) = \delta_a(f)$ for every numeric function f on U whose range is finite. The measure δ_a is called the *point measure*, or the *Dirac measure*.

3. Instead of the topological continuity, for measures we introduce the sequential continuity using the monotonic sequences. All the sets considered below are assumed to be subsets of a given set U.

Identifying sets with their indicators makes it possible to define the convergence of a sequence of sets X_n in terms of the convergence of binary sequences $X_n(u)$ ($u \in U$). A sequence X_n is said to *converge* to the set X, written $X_n \to X$, if $X_n(u) \to X(u)$ ($u \in U$). The set X is said to be the *limit* of the sequence X_n, written $\lim X_n = X$.

Lemma. 1) *A decreasing sequence of sets converges to their intersection.*
2) *An increasing sequence of sets converges to their union.* 3) *A sequence of disjoint sets converges to the empty set.*

Proof. 1) Let $X_{n+1} \subseteq X_n$ and $X = \cap X_n$. Then $X(u) = \inf\limits_n X_n(u) = \lim\limits_n X_n(u)$ $(u \in U)$.   2) If $X_{n+1} \supseteq X_n$ and $X = \cup X_n$, then $X(u) = \sup\limits_n X_n(u) = \lim\limits_n X_n(u)$ $(u \in U)$.   3) Let $X_m X_n = O$ $(m \neq n)$. Then, for every $u \in U$, either $X_n(u) = 0$ for all n or $X_{n(u)}(u) = 1$ for some number $n(u)$, and, as a result, $X_n(u) = 0$ for all $n \geq n(u)$. In any case, $\lim\limits_n X_n(u) = \varnothing$. $\square$

Exercise. Prove that $X_n \to X$ is equivalent to $X = \bigcup\limits_m \bigcap\limits_{n \geq m} X_n = \bigcap\limits_m \bigcup\limits_{n \geq m} X_n$.

A measure $\mu : R \to \mathbb{K}$ is said to be (*sequentially monotonically*) *continuous* if $\mu(X_n) \to \mu(X)$ for every monotonic sequence $X_n \to X$ $(X_n, X \in R)$. A measure $\mu : R \to \mathbb{K}$ is said to be *sequentially continuous from above at zero*, if $\mu(X_n) \to 0$ for $X_n \downarrow \varnothing$ $(X_n \in R)$. A measure μ is called *countably additive*, if $\mu(X) = \sum \mu(X_n)$ for every sequence of disjoint sets $X_n \in R$ whose union is $X \in R$. A measure μ is called *countably semiadditive* if $\mu(X) \leq \sum \mu(X_n)$ for any sequence of sets $X_n \in R$, that covers a set $X \in R$.

(If $\sum \mu(X_n) = \infty$, this inequality is true for every $X \in R$.) We denote the property of monotonic continuity of μ by M, and the other three subsequently defined properties by N, A, and D, respectively.

Proposition. $M \Leftrightarrow N \Leftrightarrow A \Leftrightarrow D$.

Proof. Consider the sets $X_n, X \in R$.

1) We now prove that $M \Leftrightarrow N$. Clearly, $M \Rightarrow N$. We will establish that $N \Rightarrow M$. Let $X_n \downarrow X$. Then $Y_n = X_n - X \in R$, $X_n = X + Y_n$, $Y_n \downarrow \varnothing$, and $\mu(X_n) = S(X_n) = S(X) + S(Y_n) = \mu(X) + \mu(Y_n) \to \mu(X)$ for $\mu(Y_n) \to 0$. If $X_n \uparrow X$, then $Z_n = X - X_n \in R$, $X_n = X - Z_n$, $Z_n \downarrow \varnothing$, and $\mu(X_n) = \mu(X) - \mu(Z_n) \to \mu(X)$ for $\mu(Z_n) \to 0$.

2) We now prove that $M \Leftrightarrow A$. Let $X_m X_n = \varnothing$ $(m \neq n)$ and $X = \sum X_n$. Then $Y_p = \sum\limits_{n \leq p} X_n \in R$, $Y_p \uparrow X$, and $\sum\limits_{n \leq p} \mu(X_n) = \mu(Y_p) \to \mu(X)$ if μ is continuous. Therefore, $\mu(X) = \sum \mu(X_n)$ and $M \Rightarrow A$.

Now let $X_n \downarrow \varnothing$. Then $Y_m Y_n = \varnothing$ for $Y_n = X_n - X_{n+1} \in \mathcal{R}$ and $X_1 = X_{n+1} + Z_n = \sum Y_n$, $X_{n+1} = X_1 - Z_n$ for $Z_n = \sum_{m \leq n} Y_m \in \mathcal{R}$, $\mu(Z_n) = \sum_{m \leq n} \mu(Y_m)$. If $\mu(X_1) = \sum \mu(Y_m) = \lim \mu(Z_n)$, then $\lim \mu(X_{n+1}) = \mu(X_1) - \lim \mu(Z_n) = 0$. Consequently, $A \Rightarrow N \Rightarrow M$. Since $M \Leftrightarrow N$ and $M \Leftrightarrow A$, we conclude that $N \Leftrightarrow A$.

3) We now prove that $A \Leftrightarrow D$. Let $X \subseteq \cup X_n$. Then $X = \cup X_n X$ and the orthogonalization of $(X_n X)$ yields a sequence of disjoint sets $Y_n \in \mathcal{R}$ such that $Y_n \subseteq X_n X$ and $\sum Y_n = X$. If the measure μ is countably additive, then $\mu(X) = \sum \mu(Y_n) \leq \sum \mu(X_n X) \leq \sum \mu(X_n)$. As a result, $A \Rightarrow D$.

Now let $X_m X_n = \varnothing$ $(m \neq n)$ and $X = \sum X_n$. Then $Y_n = \sum_{m \leq n} X_n \in \mathcal{R}$ and $\sum_{m \leq n} \mu(X_m) = \mu(Y_n) \leq \mu(X)$ because μ is monotonic. Therefore, $\sum \mu(X_n) \leq \mu(X)$. If the measure μ is countably semiadditive, then the converse inequality holds as well. As a result, $\mu(X) = \sum \mu(X_n)$ and $D \Rightarrow A$.

$\square$

Example 1. The Lebesgue measure dx on the ring of subsets of $\mathbb{R}$ generated by the semiring of intervals $[a, b[$ $(-\infty < a \leq b < \infty)$, is defined by the equality $dx(A) = \sum(b_j - a_j)$ $(A = \sum [a_j, b_j[)$. The measure dx is continuous. If the convergence of a sequence of simple sets A_n to $\varnothing$ is not monotonic, then it is possible that $dx(A_n) \nrightarrow 0$. For example, let $A_n = [n, n+1[$. Then $A_n \to \varnothing$, but $dx(A_n) = 1 \nrightarrow 0$.

Example 2. The Stieltjes measure dg on the same ring of sets for an increasing function $g : \mathbb{R} \to \mathbb{R}$ is defined by the equality $dg(A) = \sum(g(b_j) - g(a_j))$ $(A = \sum [a_j, b_j[)$. The measure dg is continuous if g is continuous from the left.

Exercise. Prove that the Lebesgue measure is continuous and establish the continuity criterion for the Stieltjes measures.

Remark. The convergence of monotonic sequences defines a topology for the algebra of functions $\mathcal{F}$ that is the greatest of all topologies for $\mathcal{F}$ preserving the convergence of monotonic sequences. This topology induces the topology on the Boolean algebra of sets $\mathcal{P}$. Topological continuity of a measure implies its sequential continuity. The converse is not true. There are more sequentially continuous measures than topologically continuous ones.

Different types of sequential convergence, the topologies generated by these types of convergence, and the corresponding classes of continuous measures are described in the paper Savel'ev (1976) and the book Savel'ev (1975) (supplement to Part 4), which contain the proofs of theorems on the extension of measures by continuity.

4. Let $Z \subseteq U$. If for any $\varepsilon > 0$ there is a sequence of sets $X_n \in \mathcal{R}$ that covers Z such that $\sum \mu(X_n) \leq \varepsilon$, then Z is called a μ-zero set. In particular, if $Z \in \mathcal{R}$ and $\mu(Z) = 0$, then Z is a μ@-zero set. We denote by $\mathcal{N} = \mathcal{N}(\mu)$ the class of all μ-zero sets. It is easy to verify that

(1) Every subset of a μ-zero set is a μ-zero set.

(2) The union of a countable collection of μ-zero sets is a μ@-zero set.

Exercise. Prove the above statements.

If $\mathcal{N}(\mu) \subseteq \mathcal{R}$, then the measure μ is said to be *complete*.

It is often required that there exist an increasing sequence of sets $U_n \in \mathcal{R}$ whose union is $\cup U_n = U$. In this case, $\mu(X) \leq \mu(U_n)$ for any $X \in \mathcal{R}$ such that $X \subseteq U_n$. Such rings $\mathcal{R}$ and measures μ on them are called *locally bounded*.

The introduction of μ-zero sets provides a way to extend the class of convergent sequences of functions. A sequence of functions $f_n \in \mathcal{F}$ is said to be μ-*converging* to a function $f \in \mathcal{F}$, written $f_n \rightarrow f(\mu)$, if the set $Z = \{x : f_n(x) \rightarrow f(x)\}$ is μ-zero. Functions $f, g \in \mathcal{F}$ are said to be μ-*equal*, written $f = g(\mu)$, if the set $Z = \{x : f(x) \neq g(x)\}$ is μ-zero. The term μ-*equivalent* is often used instead of μ-equal. In a similar way, we introduce the terms μ-*less* and μ-*greater* and the designations $f \leq g(\mu)$ and $f \geq g(\mu)$. Using μ-zero sets, it is possible to extend the classes of bounded and continuous functions: the μ-*boundedness* of f means that there is a constant $c \geq 0$ such that $|f| \leq c(\mu)$, while the μ-*continuity* of f means that the set of points at which f is discontinuous is μ-zero. If the class of μ-zero sets is large enough, the above generalizations of the corresponding terms are essential. When a measure μ is fixed, for every μ-property we use the name of the property with the term *almost everywhere*.

Proposition. *1) If $f_n \rightarrow f(\mu)$ and $f_n \rightarrow g(\mu)$, then $f = g(\mu)$.*

2) If $f_n \rightarrow f(\mu)$ and $f = g(\mu)$, then $f_n \rightarrow g(\mu)$.

Proof. The proposition follows from the relations

$$\{x : f(x) \neq g(x)\} \subseteq \{x : f_n(x) \nrightarrow f(x)\} \cup \{x : f_n \nrightarrow g(\mu)\},$$
$$\{x : f_n(x) \nrightarrow g(x)\} \subseteq \{x : f_n(x) \nrightarrow f(x)\} \cup \{x : f(x) \neq g(x)\},$$

since the union of two μ-zero sets and any subset of a μ-zero set are also μ-zero sets. □

Examples. 1) The Dirichlet function is equivalent to the constant 0 with respect to the Lebesgue measure. 2) The sequence of functions $f_n(u) = u^n$ $(0 \leq u \leq 1)$ dx-converges to the constant 0. 3) If a measure μ on $\mathcal{P}$ is identically equal to 0, then all functions in $\mathcal{F}$ are μ-equal and μ-bounded, while sequences of such functions μ-converge to functions in $\mathcal{F}$.

Remark. Let $\mathcal{P} = \mathcal{P}(U)$ and $X \in \mathcal{P}$. The equality

$$\mu^*(X) = \inf \left\{ \sum \mu(X_n) : X \subseteq \cup X_n, \ X_n \in \mathcal{R} \right\}$$

defines the *outer measure* $\mu^* : \mathcal{P} \to [0, \infty]$ for any locally bounded measure $\mu : \mathcal{P} \to [0, \infty[$. The outer measure μ^* is countably semiadditive and is an extension of the measure μ (Kolmogorov and Fomin, 1989, Chapter V, Section 3). It follows from the definitions that the μ-zero sets are the sets whose outer measure is equal to zero:

$$\mathcal{N}(\mu) = \{Z \in \mathcal{P} \mid \mu^*(Z) = 0\}.$$

Examples. 1) Prove that the cover sets X_n in the definition of an outer measure μ^* can be taken from the semiring $\mathcal{Q}$ that generates the ring $\mathcal{R}$.

2) Prove that the measure μ^* is countably semiadditive.

5. We will prove a key lemma for the definition of the integral to be given later. A sequence of simple functions $f_n \in \mathcal{S}$ is said to be *integrally bounded* if the sequence of integral sums $S(f_n) \in \mathbb{L}$ is bounded, i.e., there exists a number $c > 0$ such that $|S(f_n)| \leq c$ for all n. The measure μ on the ring $\mathcal{R}$ of simple sets is assumed to be positive and continuous (countably additive).

Levi's lemma. 1) *Every integrally bounded monotonic sequence of simple functions f_n converges almost everywhere to a function f.*

2) *If the limit function f is simple, then $S(f) = \lim S(f_n)$.*

Proof. 1) It suffices to prove the first assertion for increasing sequences of functions $f_n \geq 0$. Indeed, if $g_n \downarrow$, then $f_n = -g_n \uparrow$. Furthermore, $S(f_n) = -S(g_n)$ and therefore $f_n \to f(\mu)$ is equivalent to $g_n \to g = -f(\mu)$. If $h_n \uparrow$, then $f_n = h_n - h_1 \geq 0$; moreover, $S(f_n) = S(h_n) - S(h_1)$ and therefore $f_n \to f(\mu)$ is equivalent to $h_n \to f + h_1(\mu)$.

Let $f_n \geq 0$, $f_n \uparrow$, and $|S(f_n)| \leq c$. Since $\mu \geq 0$, we have $S(f_n) \geq 0$, $S(f_n) \uparrow$, and $S(f_n) \leq c$. Since $f_n \uparrow$, it follows that $f_n(x) \to f(x)$ for some $f(x)$ if and only if $f_n(x) \leq c(x)$ for some $c(x) \geq 0$. Therefore, to prove the first assertion of the lemma, it suffices to show that the sequence (f_n) is μ-bounded.

Consider the set Z of all points $x \in U$ at which (f_n) is not bounded and

$$Z_n(m) = \{x : f_n(x) \geq m\}, \quad Z(m) = \bigcup_n Z_n(m) \quad (m, n \in \mathbb{N}).$$

It is obvious that $Z \subseteq Z(m)$ for all m. Since the functions f_n are simple, it follows that the sets $Z_n(m)$ are also simple. Since $f_n \leq f_{n+1}$, we have $Z_n(m) \subseteq Z_{n+1}(m)$. Let

$$Y_n(m) = Z_{n+1}(m) - Z_n(m), \quad Y_0(m) = Z_1(m).$$

The sets $Y_n(m)$ are simple, $Y_k(m) \cdot Y_n(m) = \varnothing$ $(k \neq n)$, and

$$Z_n(m) = \sum_{0 \leq k < n} Y_k(m), \quad Z(m) = \sum_{n \geq 0} Y_n(m).$$

It follows from the definitions that

$$c \geq S(f_n) \geq S(f_n Z_n(m)) \geq S(m Z_n(m)) \geq m \cdot \mu(Z_n(m)).$$

Hence,

$$\sum_{0 \leq k \leq n} \mu(Y_k(m)) = \mu(Z_n(m)) \leq c \cdot m^{-1}$$

and therefore

$$\sum_{n \geq 0} \mu(Y_n(m)) \leq c \cdot m^{-1}.$$

Since $Z \subseteq Z(m) \subseteq \sum_{n \geq 0} Y_n(m)$, we conclude that Z is a μ-zero set.

2) It suffices to prove the second assertion of the lemma for decreasing sequences of simple functions $f_n \downarrow 0(\mu)$. Indeed, $g_n \uparrow g(\mu)$ implies $f_n = g - g_n \downarrow 0(\mu)$, while $g_n \downarrow g(\mu)$ implies $f_n = g_n - g \downarrow 0(\mu)$. In both cases, $S(f_n) = |S(g) - S(g_n)| \to 0$ is equivalent to $S(g_n) \to S(g)$.

Let $f_n \downarrow 0(\mu)$ and $c \geq f_1 \geq f_n$ (every simple function is bounded). Take an $\varepsilon > 0$ and let

$$X_n(\varepsilon) = \{x : f_n(x) \geq \varepsilon\}, \quad A_n = \{x : f_n(x) > 0\} = \{x : f_n(x) \neq 0\},$$
$$B_n(\varepsilon) = A_n - X_n(\varepsilon) = \{x : 0 < f(x) < \varepsilon\},$$
$$Y_n(\varepsilon) = X_n(\varepsilon) - X_{n+1}(\varepsilon),$$
$$X(\varepsilon) = \cap X_n(\varepsilon) = \{x : f_n(x) \geq \varepsilon \, (n \in \mathbb{N})\}, \quad Z = \{x : f_n(x) \nrightarrow 0\}.$$

The sets $X(\varepsilon)$ and Z are not necessarily simple. We have

$$f_n = A_n f_n = X_n(\varepsilon) f_n + B_n(\varepsilon) f_n,$$

and therefore $B_n(\varepsilon) \subseteq A_n \subseteq A_1$ implies

$$S(f_n) = S(X_n(\varepsilon) f_n) + S(B_n(\varepsilon) f_n)$$
$$\leq c \cdot \mu(X_n(\varepsilon)) + \varepsilon \mu(B_n(\varepsilon)) \leq c \cdot \mu(X_n(\varepsilon)) + \varepsilon \mu(A_1).$$

It remains to prove that $\mu(X_n(\varepsilon)) \to 0$ for $n \to \infty$.

It follows from the definitions that $X(\varepsilon) \subseteq Z$. By assumption, Z is a μ-zero set Let $Z \subseteq \cup Z_q$, $Z_q \in \mathcal{R}$, and $\sum \mu(Z_q) \leq \varepsilon$. We have

$$X_n(\varepsilon) = \sum_{p \geq n} Y_p(\varepsilon) + X(\varepsilon), \quad \sum_{p < n} \mu(Y_p(\varepsilon)) = \mu(\sum_{p < n} Y_p(\varepsilon)) \leq \mu(A_1).$$

Therefore, $\sum\limits_{p \geq n} \mu(Y_p(\varepsilon)) \to 0 \; (n \to \infty)$ and there exists a number $n(0)$ such that $\sum\limits_{p \geq n} \mu(Y_p(\varepsilon)) \leq \varepsilon$ for all $n \geq n(0)$.

Thus, $X_n(\varepsilon) \subseteq \sum\limits_{p \geq n} Y_p(\varepsilon) + \cup Z_q$. The measure μ is continuous and, consequently, countably semiadditive. Hence, $\mu(X_n(\varepsilon)) \leq 2\varepsilon$ and $S(f) \leq (2c + \mu(A_1))\varepsilon$ for $n \geq n(0)$. $\square$

Remark. The proof of the lemma can be substantially simplified if we assume, in addition, that that the measure μ is complete.

Levi's lemma states that every integrally bounded monotonic sequence of simple functions converges almost everywhere and can be integrated term by term if the limit function is also simple.

Consider two integrally bounded, increasing sequences of simple functions f_n and g_n and the sequences of their integral sums $S(f_n)$ and $S(g_n)$. By

Levi's lemma, the equalities $f(x) = \lim f_n(x)$, $g(x) = \lim g_n(x)$ $(x \in U \setminus Z)$, and $f(x) = g(x) = 0$ $(x \in Z)$ define functions $f : U \to \mathbb{R}$ and $g : U \to \mathbb{R}$, where Z denotes a μ-zero set. (Instead of zero value, the functions f and g can be assumed to take any other values $f(x)$ and $g(x)$ for $x \in Z$. The μ-convergence of sequences (f_n) and (g_n) provides only the μ-definiteness of the limit functions f and g.)

Proposition. $\lim f_n \leq \lim g_n \Rightarrow \lim S(f_n) \leq \lim S(g_n)$.

Proof. Take a number m and consider a sequence of simple functions $h_n = f_m - g_n$. Since $g_n \uparrow g(\mu)$, we have

$$h_n \downarrow h = f_m - g \leq f - g \leq 0(\mu).$$

Hence $0 \vee h_n = h_n^+ \downarrow 0(\mu)$. The functions h_n^+ are simple because so are h_n. We have

$$S(f_m) - S(g_n) = S(f_m - g_n) = S(h_n) \leq S(h_n^+).$$

By Levi's lemma, $S(h_n^+) \to 0$. Since the sequences (f_m) and (g_n) increase, the sequences $S(f_m)$ and $S(g_n)$ also increase and are bounded, which ensures the existence of the limits $a = \lim S(f_m)$ and $b = \lim S(g_n)$. As a result, we have $S(f_m) - b \leq 0$, $a - b \leq 0$, and $a \leq b$. $\square$

Corollary. If $\lim f_n = \lim g_n$, then $\lim S(f_n) = \lim S(g_n)$.

Exercise. Prove that $0 \vee h_n = h_n^+ \downarrow 0(\mu)$ if $h_n \downarrow h$ and $h \leq 0(\mu)$ for a sequence of simple functions h_n.

6. Any complex number a is equal to a linear combination of nonnegative numbers:

$$a = b^+ - b^- + ic^+ - ic^-, \quad b = \operatorname{Re} a, \quad c = \operatorname{Im} a,$$

where

$$y^+ = 0 \vee y = 2^{-1}(|y| + y), \quad y^- = 0 \vee (-y) = 2^{-1}(|y| - y)$$

for any real number y. Clearly, $y^+, y^- \geq 0$ and $y = y^+ - y^-$, $|y| = y^+ + y^-$. Therefore, any numeric function f on U is equal to a linear combination of nonnegative functions:

$$f = g^+ - g^- + ih^+ - ih^-, \quad g = \operatorname{Re} f, \quad h = \operatorname{Im} f.$$

By definition, $\varphi^+(x) = (\varphi(x))^+$ and $\varphi^-(x) = (\varphi(x))^-$ for any real function φ on U. As before, $|f|$ stands for the function $|f|(x) = |f(x)|$ $(x \in U)$. It is obvious that the functions g^+, g^-, h^+, h^-, g, h, and $|f|$ are simple if f is simple.

Lemma. *If a sequence of simple functions f_n increases and is integrally bounded, then the sequences (f_n^+), (f_n^-), and $(|f_n|)$ are integrally bounded.*

Proof. If $f_n \uparrow$, then $f_n^+ \uparrow$ and $f_n^- \downarrow$. Hence, $S(f_n^-) \leq S(f_1^-)$ and

$$f_n^+ = f_n + f_n^- \leq f_n + f_1^- \Rightarrow S(f_n^+) \leq S(f_n) + S(f_1^-),$$

$$|f_n| = f_n + f_n^- \Rightarrow S(|f_n|) \leq (S(f_n) + S(f_1^-)) + S(f_1^-). \quad \square$$

Exercise. Give examples of integrally bounded sequences of simple functions f_n such that $(|f_n|)$ are not integrally bounded.

7. Let $\mathcal{A}$ be the set of all increasing integrally bounded sequences of simple functions, and let $\mathcal{C}$ be the algebra of sequences of simple functions convergent almost everywhere on a given set U. By Levi's lemma, $\mathcal{A} \subseteq \mathcal{C}$. Consider the linear span $\mathcal{B} = \operatorname{Lin} \mathcal{A} \subseteq \mathcal{C}$. The sequences in $\mathcal{B}$ will be called *approximating*. By definition, approximating sequences are linear combinations of increasing integrally bounded sequences of simple functions: $(f_n) \in \mathcal{B}$ means that $f_n = \sum a_j f_{jn}$ for some finite collections of numbers a_j and sequences $(f_{jn})_n \in \mathcal{A}$. We denote by $\mathcal{K}$, $\mathcal{L}$, and $\mathcal{M}$ the sets of limit functions for the sequences in $\mathcal{A}$, $\mathcal{B}$, and $\mathcal{C}$, respectively:

$$\mathcal{K} = \operatorname{Lim} \mathcal{A}, \quad \mathcal{L} = \operatorname{Lim} \mathcal{B}, \quad \mathcal{M} = \operatorname{Lim} \mathcal{C}.$$

Functions in $\mathcal{K}$ are called *monotonically integrable*, functions in $\mathcal{L}$ are called *integrable*, and functions in $\mathcal{M}$ are called *measurable*.

Thus, by definition, an integrable function is a function to which a sequence equal to a linear combination of increasing integrally bounded sequences of simple functions (i.e., an approximating sequence) converges almost everywhere. A measurable function is a limit of a sequence of simple functions that converges almost everywhere. Every integrable function is measurable.

Proposition. *Measurable functions form an algebra. Integrable functions form the linear span of the set of monotonically integrable functions:*
$$\mathcal{K} \subseteq \mathcal{L} = \operatorname{Lin} \mathcal{K} \subseteq \mathcal{M}.$$

Proof. The proposition follows from the definitions and the property of the limit to preserve the operations. In particular, $\mathcal{L} = \operatorname{Lim}(\operatorname{Lin} \mathcal{A}) = \operatorname{Lin}(\operatorname{Lim} \mathcal{A}) = \operatorname{Lin} \mathcal{K}$. $\quad \square$

Exercise. Give examples of integrable functions whose product is not integrable.

By combining like terms, every integrable function and every approximating sequence can be reduced to a standard form, which is a representation with coefficients $1, -1, i, -i$:

$$f = g^+ - g^- + ih^+ - ih^- \quad (g^+, g^-, h^+, h^- \in \mathcal{K}),$$
$$(f_n) = (g_n^+) - (g_n^-) + i(h_n^+) - i(h_n^-) \quad ((g_n^+), (g_n^-), (h_n^+), (h_n^-) \in \mathcal{A}).$$

Such standard representations are not unique, since the functions g^+, g^-, h^+, h^- and the sequences (g_n^+), (g_n^-), (h_n^+), and (h_n^-) can be chosen in different ways. In particular, one representation can be obtained from another by adding an arbitrary function $v \in \mathcal{K}$ or sequence $(v_n) \in \mathcal{A}$ to every component.

8. We now describe elementary properties of integrable numeric functions on an abstract set.

Lemma 1. *The function $f = g + ih$ is integrable if its real part $g = \operatorname{Re} f$ and its imaginary part $h = \operatorname{Im} f$ are integrable.*

Proof. Since $\mathcal{L}$ is a vector space, $g, h \in \mathcal{L}$ implies $f \in \mathcal{L}$. (In the real case, we have $h = 0$.) Conversely, if $f \in \mathcal{L}$, then $f = (g^+ - g^-) + i(h^+ - h^-)$ for some $g^+, g^-, h^+, h^- \in \mathcal{K}$ and, consequently, $g = g^+ - g^- \in \mathcal{L}$ and $h = h^+ - h^- \in \mathcal{L}$. $\square$

Lemma 2. *If functions p and q are monotonically integrable, then so is their maximum $p \vee q$.*

Proof. Let $p_n \uparrow p$ and $q_n \uparrow q$ $(p_n, q_n \in \mathcal{S})$. Then $p_n \leq p_{n+1} \leq p_{n+1} \vee q_{n+1}$, $q_n \leq q_{n+1} \leq p_{n+1} \vee q_{n+1}$, $p_n \vee q_n \leq p_{n+1} \vee q_{n+1}$ $(n \in \mathbb{N})$. It is obvious that $p_n \vee q_n \in \mathcal{S}$. If the increasing sequences (p_n) and (q_n) are integrally bounded, then so are $(|p_n|)$ and $(|q_n|)$. Since $|p_n \vee q_n| \leq |p_n| + |q_n|$, we have

$$|S(p_n \vee q_n)| \leq S(|p_n \vee q_n|) \leq S(|p_n| + |q_n|) \leq S(|p_n|) + S(|q_n|).$$

As a result, $(p_n \vee q_n) \in \mathcal{A}$ and $p \vee q \in \mathcal{K}$. $\square$

Proposition. *A function $f = g + ih$ is integrable if and only if the functions $g^+ = \operatorname{Re}^+ f$, $g^- = \operatorname{Re}^- f$, $h^+ = \operatorname{Im}^+ f$, and $h^- = \operatorname{Im}^- f$ are integrable.*

Proof. If the functions g^+, g^-, h^+, and h^- are integrable, then their linear combination f is also integrable. Conversely, if f is integrable, then $g = g^+ - g^- = \operatorname{Re} f$ and $h = h^+ - h^- = \operatorname{Im} f$ are integrable (by Lemma 1). Therefore, $g = p - q$ and $h = r - s$ for some $p, q, r, s \in \mathcal{K}$. Moreover, $p \vee q \in \mathcal{K}$ and $r \vee s \in \mathcal{K}$ by Lemma 2. Since $g^+ = (p \vee q) - q$, $g^- = (p \vee q) - p$, $h^+ = (r \vee s) - s$, $h^- = (r \vee s) - r$, and $p, q, r, s, p \vee q, r \vee s \in \mathcal{K}$, it follows that $g^+, g^-, h^+, h^- \in \mathcal{L}$. □

Corollary. *The maximum and the minimum of a finite collection of integrable real functions are integrable.*

Proof. For the case of two functions f and g, the assertion follows from the equalities $f \vee g = f + (g - f)^+$ and $f \wedge g = f - (g - f)^-$. The assertion for the general case follows by induction. □

Remark. The preceding proposition also implies that for the real integrable function $f = f^+ - f^-$, the function $|f| = f^+ + f^-$ is also integrable.

9. The following two lemmas on the convergence of a sequence of integral sums will be used for the definition of the integral.

Lemma 1. *If (f_n) is an approximating sequence, then the sequence of integral sums $S(f_n)$ converges.*

Proof. Let $f_n = g_n^+ - g_n^- + i(h_n^+ - h_n^-)$ for increasing integrally bounded sequences (g_n^+), (g_n^-), (h_n^+), and (h_n^-). Then the increasing bounded sequences of integral sums $S(g_n^+)$, $S(g_n^-)$, $S(h_n^+)$, and $S(h_n^-)$ converge. Consequently, the sequence $S(f_n) = S(g_n^+) - S(g_n^-) + i(S(h_n^+) - S(h_n^-))$ also converges. □

Lemma 2. *If the limits of approximating sequences (f_n) and (p_n) are equivalent, then $\lim S(f_n) = \lim S(p_n)$.*

Proof. Let $f_n = g_n^+ - g_n^- + i(h_n^+ - h_n^-)$ and $p_n = q_n^+ - q_n^- + i(r_n^+ - r_n^-)$ for some $(g_n^+), (g_n^-), (h_n^+), (h_n^-), (q_n^+), (q_n^-), (r_n^+), (r_n^-) \in \mathcal{A}$. By Levi's lemma, these sequences converge almost everywhere to functions g^+, g^-, h^+, h^-, q^+, q^-, r^+, and r^-, respectively. The sequences (f_n) and (p_n) converge almost everywhere to the functions $f = g^+ - g^- + ih^+ - ih^-$ and $p = q^+ - q^- + ir^+ - ir^-$, respectively. At the same time, the increasing bounded sequences of integral sums $S(f_n)$, $S(g_n^+)$, $S(g_n^-)$, $S(h_n^+)$, $S(h_n^-)$, $S(p_n)$, $S(q_n^+)$, $S(q_n^-)$, $S(r_n^+)$, and $S(r_n^-)$ converge to numbers a, b^+, b^-, c^+, c^-, α, β^+, β^-, γ^+, and γ^-, respectively. If $f = p$ almost everywhere, then the following equalities hold almost everywhere: $g^+ - g^- = \operatorname{Re} f = \operatorname{Re} p = q^+ - q^-$, $h^+ - h^- = \operatorname{Im} f = \operatorname{Im} p = r^+ - r^-$, $g^+ + q^- = q^+ + g^-$, $h^+ + r^- = r^+ + h^-$, $\lim (g_n^+ + q_n^-) = \lim (q_n^+ + g_n^-)$, $\lim (h_n^+ + r_n^-) = \lim (r_n^+ + h_n^-)$. Applying Levi's lemma again and taking into account that the integral sum S is linear, we obtain $b^+ + \beta^- = \beta^+ + b^-$, $c^+ \gamma^- = \gamma^+ c^-$, $b = b^+ - b^- = \beta^+ - \beta^- = \beta$, $c = c^+ - c^- = \gamma^+ - \gamma^- = \gamma$. As a result, $\lim S(f_n) = a = b + ic = \beta + i\gamma = \alpha = \lim S(p_n)$.   □

If an approximating sequence of simple functions f_n converges almost everywhere to f, then the sequence (f_n) is said to *approximate* the function f. The *integral* of a function $f \in \mathcal{L}$ is the limit

$$\int f = \lim S(f_n)$$

of the sequence of integral sums $S(f_n)$ for an approximating sequence of simple functions f_n. This definition is well posed, which is guaranteed by the following theorem on the existence and uniqueness of the integral.

Theorem. *There exists a unique integral for every integrable function.*

Proof. 1) Let f be integrable. Then there exists a sequence of simple functions f_n approximating f, and, by Lemma 1, there exists an integral $\int f = \lim S(f_n)$.

2) Suppose that sequences (f_n) and (p_n) approximate f. By Lemma 2, $\lim S(f_n) = \lim S(p_n)$. It follows that the integral $\int f$ is unique.   □

The functional $\int : \mathcal{L} \to \mathbb{L}$ on the vector space $\mathcal{L}$ of integrable functions that associates every function $f \in \mathcal{L}$ with the number $\int f$ is called the *integral*. If the measure μ needs to be specified explicitly, then the above functional is called the *integral with respect to the measure* μ and write $\int f \, d\mu$ or $\int f(x) \, d\mu(x)$ instead of $\int f$. Furthermore, if it is necessary to specify the set U which is the domain of definition of the functions under consideration, then we will write U under the integral sign.

Proposition. *The integral $\int$ is a linear functional on $\mathcal{L}$ that is an extension of the integral sum S.*

Proof. Any simple function $f \in \mathcal{S}$ is approximated by the constant sequence $f_n = f$. Therefore, $\mathcal{S} \subseteq \mathcal{L}$ and $\int f = S(f)$. Since the limit lim and the sum S are linear, so is the integral:

$$\int (af + dg) = \lim S(af_n + bg_n)$$

$$= a \lim S(f_n) + b \lim S(g_n) = a \int f + b \int g$$

for any numbers a and b, integrable functions f and g and the sequences (f_n) and (g_n) approximating f and g. □

Remark. If sets are identified with their indicators, the measure μ becomes a functional on the Boolean ring $\mathcal{R}$ of simple sets. The measure is uniquely extended to the integral sum S on the algebra $\mathcal{S}$ of simple functions. In its turn, the integral sum is uniquely extended to the integral $\int$ on the subspace $\mathcal{L}$ of integrable functions:

$$\mathcal{R} \subseteq \mathcal{S} \subseteq \mathcal{L}, \quad \mu \subseteq S \subseteq \int .$$

10. Aside from being linear, the integrals possess some important properties that follow from the order properties of integral sums.

Proposition 1. *The integral is a positive linear functional.*

Proof. Let $f \in \mathcal{L}$ and $f \geq 0$. Then $f = g^+ - g^-$ and $g^+ \geq g^-$ for some $g^+, g^- \in \mathcal{K}$ that are the limits of integrally bounded increasing sequences of simple functions g_n^+ and g_n^- converging almost everywhere. By the proposition of Subsection 3.3.2, part 5, the inequality $\lim g_n^- = g^- \leq g^+ = \lim g_n^+$, which holds almost everywhere, implies the inequality

$$\int g^- = \lim S(g_n^-) \leq \lim S(g_n^+) = \int g^+.$$

Hence $\int f = \int (g^+ - g^-) = \int g^+ - \int g^- \geq 0$. □

As for any linear functional, the positivity of the integral implies its monotonicity.

Proposition 2. *The integral increases, i.e., if $f, g \in \mathcal{L}$ such that $f \le g$, then $\int f \le \int g$.*

The monotonicity of the integral implies a useful inequality for the integral $\int f \, d\mu$ of a real integrable function f, real numbers a and b, and an integrable set A. We say that the integral $\int A$ is the measure of A and denote it by $\mu(A)$. Thus, by definition, $\mu(A) = \int A$ ($A \in \mathcal{L}$). In particular, for simple sets $A \in \mathcal{R}$, the value of $\mu(A)$ coincides with its value in the previous sense.

Proposition 3. *If $a \le f(x) \le b$ for $x \in A$ and $f(x) = 0$ for $x \notin A$, then $a\mu(A) \le \int f \, d\mu \le b\mu(A)$.*

Proof. By assumption, $aA \le f \le bA$, which implies $\int aA \le \int f \le \int bA$. Moreover, $\int aA = a \int A = a\mu(A)$ and $\int bA = b \int A = b\mu(A)$. □

The following widely used inequality is called the *triangle inequality for integrals*. It holds for every integrable function f.

Proposition 4. $\left| \int f \right| \le \int |f|$.

Proof. We will prove this inequality for the case of real functions. Take the real integrable function $f = g^+ - g^-$ with positive components g^+ and g^-. From the definition of part 8, it follows that the functions g^+, g^-, and $|f| = g^+ + g^-$ are integrable. Since the integral is positive and linear, we have

$$\left| \int f \right| = \left| \int g^+ - \int g^- \right| \le \int g^+ + \int g^- = \int |f|.$$

□

Let f be an integrable function, c be a positive number, and A be an integrable set. Propositions 3 and 4 imply the following statement.

Proposition 5. *If $|f(x)| \le c$ for $x \in A$ and $f(x) = 0$ for $x \notin A$, then $\left| \int f \, d\mu \right| \le c\mu(A)$.*

Proof. We prove the required inequality in the case of real functions. From the assumptions and Propositions 3 and 4, it follows that $\left| \int f \, d\mu \right| \le \int |f|\mu \le c\mu(A)$. □

Exercise. Prove Propositions 4 and 5 for a complex function f.

3.3.3. Limit theorems

1. Levi's theorem. Consider a sequence of integrable functions f_n. If the sequence of integrals $\int f_n$ is bounded, the sequence of functions f_n is said to be *integrally bounded*, i.e., there exists a number $c > 0$ such that $|\int f_n| \le c$ for all n. If (f_n) converges almost everywhere to an integrable function f and $\int f_n \to \int f$, then the sequence f_n is said to be *integrable term by term*. This property is equivalent to the equality $\lim \int f_n = \int \lim f_n$.

Levi's theorem. *If a monotonic sequence of functions f_n is integrally bounded, then it converges almost everywhere to an integrable function f and is integrable term by term.*

Proof. It suffices to prove the theorem for the case of integrally bounded, increasing sequence of monotonically integrable functions f_n. Every such sequence converges almost everywhere to an integrable function f. Indeed, by definition, each f_n can be approximated by an increasing sequence of simple functions f_{mn} $(m = 1, 2, \ldots)$ and $\int f_n = \lim\limits_{m \to \infty} S(f_{mn})$. We have $f_{mn} \le f_n \le f_p$ for all m if $n \le p$. Consider a sequence of simple functions $g_p = \max\{f_{pn} : n \le p\} \le f_p$. The sequence (g_p) increases and is integrally bounded because the sequences $f_{m1}, \ldots, f_{mp}$ $(m = 1, 2, \ldots)$ increase and f_p $(p = 1, 2, \ldots)$ is integrally bounded.

From the definitions and Levi's theorem, it follows that the sequence (g_p) converges almost everywhere to an integrable function f. Since $f_{pn} \le g_p$ and $f_{pn} \to f_n$, $g_p \to f$ for $p \to \infty$, we conclude that $f_n \le f$. Since $g_n \le f_n$, we have $g_n \le f_n \le f$. This inequality and the convergence $g_n \to f$ imply the desired convergence $f_n \to f$.

It is now easy to prove that the sequence f_n is integrable term by term. The sequence g_n approximates f and therefore $\int f = \lim S(g_n)$. Since $g_n \le f_n \le f$ and $\int g_n = S(g_n)$, it follows that $\int g_n \le \int f_n \le \int f$ and $S(g_n) \le \int f_n \le \int f$. Together with the inequality $\int f = \lim S(g_n)$, the preceding inequalities imply $\lim \int f_n = \int f$, which completes the proof. $\square$

Corollary. *If a monotonic sequence of functions f_n is integrally bounded and converges almost everywhere to a function g, then g is integrable and $\int g = \lim \int f_n$.*

Proof. By Levi's theorem, there exists a function f such that the sequence of functions f_n converges almost everywhere to f and $\int f = \lim \int f_n$. If the sequence f_n also converges almost everywhere to a function g, then g is equivalent to f, integrable as well as f, and its integral is equal to the integral of f. $\square$

2. Fatou's theorem. Let b_n be a sequence of real numbers that is bounded from below. We define the *lower sequence* for b_n as $a_n = \inf\limits_{p \geq n} b_p$. This sequence increases. If it is bounded from above, then it has a limit $a = \lim a_n$. The limit is called the *lower limit* of the sequence b_n and is designated by $\liminf b_n$.

Similarly, for a sequence of real numbers b_n that is bounded from above, we define the *upper sequence* $c_n = \sup\limits_{p \geq n} b_p$. This sequence decreases. If it is bounded from below, then it has a limit $c = \lim c_n$. The limit is called the *upper limit* of the sequence b_n and is designated by $\limsup b_n$.

The following theorem holds for a bounded sequence of numbers $b_n \in \mathbb{R}$ and a number $b \in \mathbb{R}$ (see Savel'ev, 1969b, Section 4.4.5).

Theorem. $\lim b_n = b \Longleftrightarrow \liminf b_n = \limsup b_n = b$.

The lower and upper sequences $f_n = \inf\limits_{p \geq n} g_p$ and $h_n = \sup\limits_{p \geq n} g_p$, as well as the lower and upper limits $f = \lim f_n$ and $h = \lim h_n$ for a sequence of real functions g_n are defined in terms of the sequence of values $g_n(x)$. If $f_n = \inf\limits_{p \leq n} g_p$, $h_n = \sup\limits_{p \leq n} g_p$, $f(x) = \lim f_n(x)$, and $h(x) = \lim h_n(x)$ are defined for almost all x, but not for all x, then for f_n, h_n, f, and h we can take any functions with the specified values. In this case, the lower and upper limits are said to be *defined up to equivalence*.

The theorem on the upper and lower limits of a number sequence implies an analogous theorem for sequences of functions:

$$\lim g_n = g \Longleftrightarrow \liminf g_n = \limsup g_n = g,$$

where the equalities hold almost everywhere and the sequence of real functions g_n is bounded from below and from above by some real functions.

Fatou's lemma. *For every integrally bounded sequence of positive functions g_n, the lower limit $f = \liminf g_n$ is an integrable function and*

$$\int \liminf g_n \leq \liminf \int g_n.$$

Proof. For any n and $q \geq n$, let $f_{qn} = \min\limits_{q \geq p \geq n} g_p$. Clearly, the sequence f_{qn} $(q = n, n+1, \dots)$ decreases. Since $g_p \geq 0$, we have $f_{qn} \geq 0$. Consequently, there exists a limit $f_n = \lim\limits_{q \to \infty} f_{qn} = \inf\limits_{p \geq n} g_p$. The functions $f_{qn} =$

$\min\limits_{p\geq q\geq n} g_p$ are integrable because g_p are integrable. Since $0\leq f_{qn}\leq f_{nn}$ and $0\leq\int f_{qn}\leq\int f_{nn}$, the sequence f_{qn} $(q=n,n+1,\dots)$ is integrally bounded. By Levi's theorem, the limit function f_n is integrable and $\int f_n=\lim\limits_{q\to\infty}\int f_{qn}$.

We have $0\leq f_n\leq f_{qn}\leq g_n$ $(q\geq n)$. By assumption, the sequence of functions g_n is integrally bounded. Hence, there exists a $c>0$ such that $0\leq\int f_n\leq\int f_{qn}\leq\int g_n\leq c$ $(q\geq n)$. At the same time, the sequence $f_n=\inf\limits_{p\geq n}g_p$ $(n=1,2,\dots)$ increases. By Levi's theorem, f_n converges almost everywhere to an integrable function f and $\int f=\lim\int f_n$. By the theorem on the lower and upper limits, $f=\liminf g_n$. On the other hand, since $f_n=\inf\limits_{p\geq n}g_p\leq g_p$ for all $p\geq n$, it follows that

$$\int f_n\leq\int g_p=b_p\ (p\geq n),\quad\int f_n\leq\inf\limits_{p\geq n}b_n=a_n.$$

As a result,

$$\int\liminf g_n=\int f\leq\lim a_n=\liminf b_n=\liminf\int g_n.$$

☐

Fatou's theorem can be obtained from the Fatou's lemma by adding a dual statement for the upper limits and replacing the assumption that g_n is integrally bounded with the assumption that it is absolutely bounded by the function $\bar{g}\geq0$.

Fatou's theorem. *For any sequence of integrable real functions g_n whose absolute values are bounded by an integrable function $\bar{g}\geq0$, the lower limit $f=\liminf g_n$ and the upper limit $g=\limsup g_n$ are integrable functions. Moreover,*

$$\int\liminf g_n\leq\liminf\int g_n\leq\limsup\int g_n\leq\int\limsup g_n.$$

Proof. To prove the theorem, it suffices to apply Fatou's lemma to the sequences of functions

$$u_n=\bar{g}+g_n,\quad v_n=\bar{g}-g_n.$$

The functions u_n, v_n are real and integrable because so are g_n and $\bar{g}$. The assumption $|g_n|\leq\bar{g}$ implies

$$0\leq u_n\leq2\bar{g},\ 0\leq v_n\leq2\bar{g},\ 0\leq\int u_n\leq2\int\bar{g},\ 0\leq\int v_n\leq2\int\bar{g}.$$

Thus, the sequences u_n, v_n satisfy the assumptions of Fatou's lemma. Therefore,

$$\liminf u_n = \bar{g} + \liminf g_n, \quad \liminf v_n = \bar{g} - \limsup g_n$$

are integrable functions and

$$\int \liminf u_n \leq \liminf \int u_n, \quad \int \liminf v_n \leq \liminf \int v_n.$$

Consequently, the functions

$$\liminf u_n = \bar{g} + \liminf g_n, \quad \liminf v_n = \bar{g} - \limsup g_n$$

are also integrable and

$$\int \liminf g_n = \int \liminf u_n - \int \bar{g} \leq \liminf \int u_n - \int \bar{g}$$

$$= \liminf \left(\int \bar{g} + \int g_n \right) - \int \bar{g} = \liminf \int g_n,$$

$$\int \limsup g_n = \int \bar{g} - \int \liminf v_n \geq \int \bar{g} - \liminf \int v_n$$

$$= \int \bar{g} - \liminf \left(\int \bar{g} + \int g_n \right) = \limsup \int g_n.$$

The inequality $\liminf \int g_n \leq \limsup \int g_n$ follows from the definitions of the lower and upper limits of integrable number sequences.   □

3.   The integrability criterion. A real function $f = g^+ - g^-$ is integrable if and only if its positive components g^+ and g^- are integrable. If f is integrable, then so is $|f| = g^+ + g^-$.

Lemma. *Let the absolute values $|f_n|$ of integrable numeric functions f_n be bounded from above by an integrable function $\bar{f} \geq 0$, and let the sequence (f_n) converge almost everywhere to a function f. Then f is integrable and $\int f = \lim \int f_n$.*

Proof. It suffices to prove the assertion for the case of real functions.

Assume that $f_n = g_n$ and $f = g$ are real functions. Then the assumption $\lim f_n = f$ is equivalent to the equalities $\liminf g_n = \limsup g_n = g$. The sequence g_n satisfies the assumptions of Fatou's theorem with $\bar{g} = \bar{f}$. It follows that g is integrable and

$$\int g \leq \liminf \int g_n \leq \limsup \int g_n \leq \int g.$$

Hence $\int f = \int g = \lim \int g_n = \int \lim f_n$.   □

We now prove a convenient integrability criterion, which is widely used.

The integrability criterion. *The numeric function f is integrable if and only if f is measurable and there exists an integrable function $\bar{f} \geq 0$ such that $|f| \leq \bar{f}$.*

Proof. 1. We first prove the formulated conditions are sufficient.

A function $f = g + ih$ is integrable if and only if its real part g and its imaginary part h are integrable. The function f is measurable if and only if g and h are measurable. The inequality $|f| \leq \bar{f}$ implies $|g| \leq \bar{f}$ and $|h| \leq \bar{f}$. Therefore, if the conditions of the criterion hold for f, then they hold for g and h. If g and h are integrable, then so is f. For this reason, when proving the sufficiency of the conditions of the criterion, we can confine ourselves to considering the case of a real function f. Then the measurability of f means that there is a sequence of real simple functions $\bar{f}_n$ that converges almost everywhere to f. Since the condition $|f| \leq \bar{f}$ does not imply $|\bar{f}_n| \leq \bar{f}$ for all n, it is not possible to use the preceding lemma immediately. But the sequence $\bar{f}_n$ can be replaced by the sequence of functions $f_n = \bar{f} \wedge [\bar{f}_n \vee (-\bar{f})]$, which satisfy the assumptions of the lemma.

The functions f_n are integrable, as well as $\bar{f}$ and $\bar{f}_n$. Since $\bar{f}_n \to f$, $\bar{f} \geq 0$, and $-\bar{f} \leq \bar{f}_n \leq \bar{f}$, we have $f_n = \bar{f} \wedge [\bar{f}_n \vee (-\bar{f})] \to \bar{f} \wedge [f \vee (-\bar{f})] = f$. It follows from the definitions that $-\bar{f} \leq f_n \leq \bar{f}$, i.e., $|f_n| \leq \bar{f}$. The assumptions of the preceding lemma are satisfied and therefore f is integrable.

2. We now prove that the formulated conditions are necessary.

It readily follows from the definitions that any integrable function is measurable. The function $f = (g^+ - g^-) + i(h^+ - h^-)$ is integrable if and only if its positive components g^+, g^-, h^+, and h^- are integrable. Then the function $\bar{f} = g^+ + g^- + h^+ + h^-$ is also integrable. The triangle inequality implies that $|f| \leq \bar{f}$. We conclude that the integrable function $\bar{f}$ satisfies the conditions of the criterion. □

Corollary. *A measurable function f is integrable if and only if $|f|$ is integrable.*

Proof. If f is measurable, then $|f|$ is measurable. By the integrability criterion, if f is integrable then there exists an integrable function $\bar{f}$ such that $|f| \leq \bar{f}$. The same criterion implies that $|f|$ is integrable. If f is measurable and $|f|$ is integrable, then the criterion with $\bar{f} = |f|$ readily implies that f is integrable. □

4. The Lebesgue theorem. This theorem supplements Levi's theorem and Fatou's theorem. It is formulated for arbitrary numeric functions, not necessarily real. The Lebesgue theorem is used very often.

The Lebesgue theorem. *Assume that the absolute values $|f_n|$ of measurable numeric functions f_n are bounded almost everywhere by an integrable function $\bar{f} \geq 0$ and the sequence (f_n) converges almost everywhere to a function f. Then f_n and f are integrable and $\lim \int f_n = \int f$.*

Proof. The Lebesgue theorem differs from the lemma of part 3 of this subsection only in the assumption that the functions f_n are measurable and $|f_n|$ are bounded almost everywhere. Therefore, in order to prove the theorem, it suffices to show that the functions f_n are integrable and the lemma of part 3 still holds under the assumption that $|f_n|$ are bounded almost everywhere.

We replace the functions f_n and $\bar{f}$ by equivalent functions g_n and $\bar{g}$ that are equal to zero on the set of measure zero where $|f_n|$ are not bounded. Clearly, $|g_n| \leq \bar{g}$ everywhere. The measurability of f_n implies that of g_n, and the integrability of $\bar{f}$ implies that of $\bar{g}$. Moreover, $f_n \to f$ implies that $g_n \to f$ almost everywhere. By the integrability criterion, since g_n is measurable and $|g_n| \leq \bar{g}$, it follows that g_n is integrable. From the lemma of part 3, it follows that the limit function f is integrable and $\lim \int g_n = \int f$. Since f_n and g_n are equivalent, we conclude that f_n are integrable as well as g_n, and $\lim \int f_n = \lim \int g_n$. Hence $\lim \int f_n = \int f$. □

Corollary. *If functions f_n are integrable and $\sum \int |f_n| < \infty$, then there exists an integrable function f such that $\sum f_n = f$ almost everywhere and $\int f = \sum \int f_n$.*

Proof. The sequence of functions $g_n = \sum_{m \leq n} |f_m|$ satisfies the assumptions of Levi's theorem. Thus, there exists an integrable function g such that $g = \sum |f_n|$ almost everywhere. Since $\left| \sum_{m \leq n \leq p} f_n \right| \leq \sum_{m \leq n \leq p} |f_n|$, by the Cauchy criterion, there is a function f such that $f = \sum f_n$ almost everywhere. The sequence $h_n = \sum_{m \leq n} f_m$ satisfies the assumptions of the Lebesgue theorem: h_n are integrable, $h_n \to f$ almost everywhere, $|h_n| \leq g$, and g is integrable. Hence, f is integrable and $\int f = \lim \int h_n = \sum \int f_n$. □

5. The Fourier transform. We now describe another application of the integrability criterion.

Consider the algebra $\mathcal{R}$ of simple sets consisting of bounded intervals of the real line $\mathbb{R} =]-\infty, \infty[$. Let dx be the Lebesgue measure on $\mathcal{R}$, which is equal to the sum of the lengths of pairwise disjoint intervals that form a simple set:

$$dx\left(\sum A_j\right) = \sum dx(A_j), \quad dx(|a,b|) = b - a,$$

where $|a,b|$ denotes an arbitrary interval with left endpoint a and right endpoint b, i.e., $|a,b| =]a,b[,]a,b], [a,b[, [a,b], -\infty < a \le b < \infty$.

For any real number u, the function $g : \mathbb{R} \to \mathbb{C}$ such that $g(x) = e^{iux}$ $(-\infty < x < \infty)$ is continuous and therefore measurable with respect to the measure dx. The product $h = g \cdot f$ of measurable functions g and f is also a measurable function. Moreover, $|h| = |g| \cdot |f| = 1 \cdot |f| = |f|$ and $|f|$ is integrable together with f. By the integrability criterion, the function h is also integrable.

The function $\varphi : \mathbb{R} \to \mathbb{C}$ with values

$$\varphi(u) = \int e^{iux} f(x)\, dx$$

is called the *Fourier transform* of the function f.

Now consider the Stieltjes measure dF on $\mathcal{R}$, which is defined by the distribution function $F : \mathbb{R} \to \mathbb{R}$ (which is increasing, continuous from the left, and ha slimits $F(-\infty) = 0$ and $F(\infty) = 1$). The constant 1 is integrable with respect to the measure dF. We have $\int 1 \cdot dF = 1$. Since $|g(x)| \le 1$, by the integrability criterion, g is integrable with respect to dF.

The function $\varphi : \mathbb{R} \to \mathbb{C}$ with values

$$\varphi(u) = \int e^{iux}\, dF$$

is called the *Fourier transform* of the measure dF.

3.3.4. Measurable functions

As a rule, in calculus we deal with functions measurable with respect to a given measure.

1. Measurability and integrability. By definition, a measurable function is the limit of a sequence of simple functions that converges almost everywhere. Different algebras of simple sets and different measures can define different classes of measurable functions. As shown above, the measurable functions form an algebra: the sum and the product of measurable functions and the product of a scalar and a measurable function are measurable with respect to the given measure. By definition, an integrable function is the limit of a special sequence of simple functions. Therefore, every integrable function is measurable. The converse is not true. There exist measurable functions that are not integrable.

The converse relationship between measurability and integrability is described by the following statement.

Proposition. *Every measurable function is equal to the limit of a sequence of integrable functions that converges almost everywhere.*

Proof. Indeed, every measurable function is equal to the limit of a sequence of simple functions that converges almost everywhere. At the same time, every simple function is integrable. □

2. Sequences of measurable functions. In what follows, we will assume that there exists an increasing sequence of simple sets U_n whose union is equal to U. For this reason, the measures to be considered will be called *locally bounded*. The above assumption is equivalent to the assumption that there exists a sequence of pairwise disjoint simple sets E_n whose sum is equal to U. Therefore, the measures ot be considered are also called *sigma-finite* $(\mu(A) = \Sigma\mu(AE_n), \mu(AE_n) < \infty)$.

The above assumption ensures that there exists a strictly positive integrable function.

Exercise. Prove the preceding statement (Savel'ev, 1988,1989, Chapter 6, Section 6.2).

Lemma. *If a sequence of measurable functions f_n converges almost everywhere to a function f, then f is measurable.*

Proof. Consider an integrable function $h > 0$ and the sequence of functions $g_n = hf_n(h + |f_n|)^{-1}$. Since h and f_n are measurable, so are g_n. Moreover, $|g_n| \le h$. Finally, $g_n \to g = hf(h + |f|)^{-1}$. From the Lebesgue theorem, it follows that g is integrable and therefore measurable. Hence, the function $f = hg(h - |g|)^{-1}$ is measurable. □

A set of functions is said to be *closed* if for any sequence of its functions that converges almost everywhere, all limit functions of this sequence belong to this set.

Theorem. *The measurable functions form a closed algebra.*

We denote this algebra by $\mathcal{M}$. Since any measure is locally bounded, the algebra $\mathcal{M}$ has the identity element $1 = \lim U_n \in \mathcal{M}$. Therefore, every constant is a measurable function.

Proposition. *A measurable function f is equal to zero almost everywhere if and only if f is integrable and $\int |f| = 0$.*

Proof. If $f = 0$ almost everywhere, then f is integrable and $\int |f| = 0$. Conversely, if f is integrable and $\int |f| = 0$, then the sequence $g_n = n|f|$ satisfies the assumptions of Levi's theorem, i.e., it increases, consists of integrable functions, and integrally bounded (by zero). Hence, g_n converges almost everywhere to an integrable function g, i.e., $n|f(x)| \to g(x)$ for almost all x. We conclude that $f(x) = 0$ for almost all x. $\square$

In particular, if the integral of a positive integrable function is equal to zero, then this function is equal to zero almost everywhere.

3. Measurable and integrable sets. Identifying sets with their indicators makes it possible to define *measurable* and *integrable sets* as sets whose indicators are measurable and integrable, respectively.

The theorem on the algebra of measurable functions implies important corollaries for sets.

Corollary 1. *The measurable sets form a closed algebra of sets.*

A class of integrable sets is said to be *boundedly closed* if, for any sequence of sets in this class that converges almost everywhere and bounded almost everywhere by an integrable set, all limit sets of this sequence belong to this class.

Corollary 2. *The integrable sets form a boundedly closed algebra of sets.*

We denote this algebra by $\bar{\mathcal{R}}$ and give an important definition.

Definition. For any integrable set $A \in \bar{\mathcal{R}}$, the number $\bar{\mu}(A) = \int A\mu$ is called the measure of A.

Theorem. *The function $\bar{\mu} : \bar{\mathcal{R}} \to \mathbb{R}$ is a complete countably additive measure that is an extension of $\mu : \mathcal{R} \to \mathbb{R}$.*

Proof. It follows from the definitions that $\bar{\mu}(A) = \int A\mu = \mu(A)$ for any simple set A. Therefore, $\bar{\mu}$ is an extension of μ.

The additivity and positivity of $\bar{\mu}$ follows from the additivity and positivity of the integral. Hence, $\bar{\mu}$ is a measure. From the corollary of the Lebesgue theorem it follows that the measure $\bar{\mu}$ is complete and countably additive. □

We will call $\bar{\mu}$ the *integral extension* of the measure μ. We will write μ instead of $\bar{\mu}$ whenever it does not lead to a confusion.

The following theorem describes real measurable functions.

The measurability criterion. *A real function f is measurable if and only if the inverse image $f^{-1}[a, \infty[$ of every interval $[a, \infty[$ $(-\infty < a < \infty)$ is a measurable set.*

Remark. The interval $[a, \infty[$ in the criterion can be replaced by any of the intervals $]a, \infty[, \,]-\infty, a], \,]-\infty, a[$ $(-\infty < a < \infty)$.

Corollary. *Any continuous real function is measurable with respect to the Lebesgue measure.*

The following theorem describes the structure of real functions on an interval $[a, b]$ $(-\infty < a < b < \infty)$ that are measurable with respect to the Lebesgue measure λ (Sadovnichii, 1999, Chapter 3).

Luzin's theorem. *If a function $f : [a, b] \to \mathbb{R}$ is measurable with respect to the Lebesgue measure λ, then for any $\varepsilon > 0$ there exists a continuous function $g : [a, b] \to \mathbb{R}$ such that $\lambda(\{x : f(x) \neq g(x)\}) < \varepsilon$.*

3.3.5. The Fubini and Tonelli theorems

The Fubini and Tonelli theorems show the connection between double and repeated integrals.

1. For sets X and Y, let $\mathcal{A}$ and $\mathcal{B}$ be the algebras of simple subsets of X and Y, respectively, and let λ and μ be measures on $\mathcal{A}$ and $\mathcal{B}$, respectively. In the Cartesian product $X \times Y = Z$, we consider the class of simple rectangles $A \times B = C$ with sides $A \in \mathcal{A}$ and $B \in \mathcal{B}$. By simple subsets of $X \times Y$ we mean the unions of finite collections of simple rectangles. Subsets of $X \times Y$ will be called *flat*.

The proofs of the following statements are rather simple (Dunford and Schwartz, 1988, Chapter 7).

Lemma 1. *Every simple flat set is equal to the sum of a finite collection of pairwise disjoint simple rectangles.*

Proposition 1. *The class of all simple sets with the conventional operations forms an algebra.*

The algebra of simple subsets of $X \times Y$ is called the *product of algebras* $\mathcal{A}$ and $\mathcal{B}$. It is designated by $\mathcal{A} \times \mathcal{B}$.

The measure $\lambda \times \mu = \nu$ on the algebra $\mathcal{A} \times \mathcal{B} = \mathcal{C}$ of simple sets $X \times Y = Z$ is defined in a very natural way, but the definition involves a rather lengthy reasoning.

Definition 1. Let $A \in \mathcal{A}$, $B \in \mathcal{B}$, and $A \times B = C$. Then $\nu(C) = (\lambda \times \mu)(A \times B) = \lambda(A) \cdot \mu(B)$.

Lemma 2. *The function ν is additive on simple rectangles.*

Lemma 3. $\sum \nu(C_i) = \sum \nu(Z_i)$ $(\sum C_i = \sum Z_i)$.

Definition 2. The measure $\nu(C)$ of a simple set $C = \sum C_i$ is equal to the sum of the measures of the simple rectangles C_i:

$$\nu(C) = \sum \nu(C_i).$$

As follows from Lemma 3, in this definition, ν is a many-to-one function on the algebra $\mathcal{A} \times \mathcal{B} = \mathcal{C}$ of simple sets.

Proposition 2. *The function ν on the algebra $\mathcal{C}$ of simple sets given by Definitions 1 and 2 is a measure.*

These definitions and lemmas describe the measure $\lambda \times \mu : \mathcal{A} \times \mathcal{B} \to \mathbb{R}$, which is called the *product of measures* $\lambda : \mathcal{A} \to \mathbb{R}$ and $\mu : \mathcal{B} \to \mathbb{R}$. The measure $\lambda \times \mu$ is uniquely defined by its values on rectangles.

Proposition 3. *The product $\lambda \times \mu = \nu$ of countably additive measures λ and μ is a countably additive measure.*

It is easy to verify that the product of locally bounded measures is a locally bounded measure. Thus, the product of measures has all necessary natural properties. The product of measures can be completed.

Remark. The product of complete measures is not necessarily a complete measure (see Halmos, 1974, Section 35, Exercise 2).

2. Double and repeated integrals. As in part 1, we consider a set X, the algebra $\mathcal{A}$ of simple subsets of X, the measure $\lambda = dx$ on $\mathcal{A}$, a set Y and the algebra $\mathcal{B}$ of simple subsets of Y, and the measure $\mu = dy$ on $\mathcal{B}$. The measures $\lambda = dx$ and $\mu = dy$ are assumed to be *countably additive, complete, and locally bounded*. The designations dx and dy are chosen for better clarity of the expressions with integrals. In addition to the measures $\lambda = dx$ and $\mu = dy$, we consider their *completed* product $\lambda \times \mu = dxdy$. This product is a measure on the algebra $\mathcal{A} \times \mathcal{B}$ of simple subsets of $X \times Y$ completed whenever necessary.

Consider a function $h : X \times Y \to \mathbb{C}$ of two variables $x \in X$ and $y \in Y$ with complex values $h(x, y) \in \mathbb{C}$. The function h will be sometimes written as $h(\cdot, \cdot)$. For any $a \in X$ and $b \in Y$, the function h determines the *partial functions* of one variable $h(a, \cdot) : Y \to \mathbb{C}$ and $h(\cdot, b) : X \to \mathbb{C}$ with values $h(a, y)$ and $h(x, b)$ for $y \in Y$ and $x \in X$. The functions $h(a, y)$ and $h(x, b)$ are also called the *sections* of the function h at $a \in X$ and $b \in Y$.

Before formulating the Fubini and Tonelli theorems, we define the double and repeated integrals and introduce the necessary notation.

1. If the function $h(\cdot, \cdot)$ is integrable with respect to the measure $dxdy$, we say that *there exists a double integral*

$$c = \int \int h(x, y)\, dxdy.$$

2. Assume that the measure $h(x, \cdot)$ is integrable with respect to the measure dy for all $x \in X \setminus A$, where $A \subseteq X$ is a set of measure zero. Consider a function f on X that assumes values $f(x) = \int h(x, y)\, dy$ for $x \notin A$ and arbitrary numeric values $f(x)$ for $x \in A$. If the function f is integrable with respect to the measure dx, then we say that *there exists a repeated integral*

$$a = \int f(x)\, dx = \int \left(\int h(x, y)\, dy \right) dx.$$

The existence and the value of this integral do not depend on the choice of the set A of measure $dx(A) = 0$ and the values $f(x)$ for $x \in A$.

3. Assume that the function $h(\cdot, y)$ is integrable with respect to the measure dx for all $y \in Y \setminus B$, where $B \subseteq Y$ is a set of measure $dy(B) = 0$. Let g be a function on Y such that $g(y) = \int h(x, y)\, dx$ for $y \notin B$ and the

values $g(y)$ are arbitrary for $y \in B$. If g is integrable with respect to the measure dy, then we say that *there exists a repeated integral*

$$b = \int g(y)\, dy = \int \left(\int h(x,y)\, dx \right) dy.$$

The existence and the value of this integral do not depend on the choice of the set B of measure $dy(B) = 0$ and the values $g(y)$ for $y \in B$.

The functions f and g are sometimes called the *simple integrals*. They are defined up to equivalence with respect to the measures dx and dy, respectively.

Since the integral is linear and a numeric function is integrable if and only if its positive components are integrable, it suffices to study the relationship between double and repeated integrals for positive functions.

Lemma. *Almost all sections of a flat set of measure zero also have measure zero.*

Exercise. Prove this lemma (see Savel'ev, 1989, Chapter 7, Section 7.2).

The above lemma has many useful corollaries. We now prove one of them under the notation introduced above.

Corollary. *If $h_n(\cdot,\cdot) \to h(\cdot,\cdot)$ almost everywhere with respect to $dxdy$, then $h_n(a,\cdot) \to h(a,\cdot)$ almost everywhere with respect to dy and $h_n(\cdot,b) \to h(\cdot,b)$ almost everywhere with respect to dx for almost all $a \in X$ and $b \in Y$.*
Proof. Let $h_n(x,y) \to h(x,y)$ for $(x,y) \notin C$, $dxdy(C) = 0$. Then $h_n(a,y) \to h(a,y)$ for $x \notin C(a,\cdot)$, $h_n(x,b) \to h(x,b)$ for $x \notin C(\cdot,b)$ and $dy(C(a,\cdot)) = 0$, $dx(C(\cdot,b)) = 0$ for almost all $a \in X$ and $b \in Y$. $\square$

Proposition. *Almost all sections of a measurable function are measurable.*
Proof. Let $h(\cdot,\cdot)$ be measurable. Then, by definition, there is a sequence of simple functions $h_n(\cdot,\cdot)$ that converges to h almost everywhere with respect to $dxdy$. Clearly, the sections $h_n(a,\cdot)$ and $h_n(\cdot,b)$ are also simple functions. From what was proved before, it follows that $h_n(a,\cdot) \to h(a,\cdot)$ almost everywhere with respect to dy for almost all $a \in X$ and $h_n(\cdot,b) \to h(\cdot,b)$ almost everywhere with respect to dx for almost all $b \in Y$. Consequently, for such a and b, the function $h(a,\cdot)$ is measurable with respect to dy and the function $h(\cdot,b)$ is measurable with respect to dx. $\square$

3. We will use the notation introduced in part 2 and the definitions of double and repeated integrals given there.

The Fubini theorem. *If there exists a double integral c for a numeric function h, then for h there exist both repeated integrals a and b. Moreover, a = b = c.*

A short form of the Fubini theorem is represented by the equalities

$$\int\int h(x,y)\,dxdy = \int\left(\int h(x,y)dx\right)dy = \int\left(\int h(x,y)dy\right)dx.$$

Two auxiliary propositions will be used to prove the Fubini theorem.

Lemma 1. *The assertion of the Fubini theorem holds for simple functions.*

Proof. Let $h = \sum_i c_i C_i = \sum_i \sum_j c_{ij}(A_{ij} \times B_{ij})$, where $c_i = c_{ij}$ are numbers, $C_i = \sum_j c_{ij}(A_{ij} \times B_{ij})$ are the sums of simple rectangles. Then

$$c = \int\int h(x,y)\,dxdy = \sum_i\sum_j c_{ij}\,dx(A_{ij})\,dy(B_{ij}),$$

$$h(x,\cdot) = \sum_i\sum_j c_{ij} A_{ij}(x) B_{ij},$$

$$f(x) = \int h(x,y)\,dxdy = \sum_i\sum_j c_{ij} A_{ij}(x)\,dy(B_{ij}),$$

$$a = \int f(x)\,dx = \int\left(\int h(x,y)\,dy\right)dx$$

$$= \sum_i\sum_j c_{ij}\,dx(A_{ij})\,dy(B_{ij}) = c,$$

$$h(\cdot,y) = \sum_i\sum_j c_{ij} A_{ij} B_{ij}(y),$$

$$g(y) = \int h(x,y)\,dx = \sum_i\sum_j c_{ij}\,dx(A_{ij}) B_{ij}(y),$$

$$b = \int g(y)\,dy = \int\left(\int h(x,y)\,dx\right)dy$$

$$= \sum_i\sum_j c_{ij}\,dx(A_{ij})\,dy(B_{ij}) = c.$$

We conclude that $a = b = c$.   □

Lemma 2. *The assertion of the Fubini theorem holds for monotonically integrable functions.*

Proof. Let $h \in \mathcal{K}$. Then there exists an increasing sequence of simple functions h_n that approximates h and

$$c = \int \int h(x,y)\, dxdy = \lim \int \int h_n(x,y)\, dxdy, \tag{1}$$

where $h_n(z) \uparrow h(z)$ $(z = (x,y) \notin C)$, C is a subset of $Z = X \times Y$ such that $dxdy(C) = 0$.

Consider a sequence of simple functions $f_n(x) = \int h_n(x,y)\, dy$. This sequence is increasing and integrally bounded because so is (h_n):

$$\int f_n(x)\, dx = \int \left(\int h_n(x,y)\, dy \right) dx = \int \int h_n(x,y)\, dxdy \le c. \tag{2}$$

Therefore, (f_n) approximates an integrable function f and $\lim \int f_n(x)\, dx = \int f(x)\, dx$. Take a set $A = \{x : dy(C(x,\cdot)) \ne 0\} \cup \{x : f_n(x) \nrightarrow f(x)\}$. Since $dxdy(C) = 0$, we have $dy(C(x,\cdot)) = 0$ for almost all x. At the same time, $f_n(x) \to f(x)$ for almost all x. Hence $dx(A) = 0$.

If $x \notin A$, then $dy(C(x,\cdot)) = 0$. For $y \notin C(x,\cdot)$, we have $(x,y) \notin C$ and $h_n(x,y) \uparrow h(x,y)$, as follows from the definition of the set C. Thus, the increasing sequence of simple functions $h_n(x,\cdot)$ converges almost everywhere to $h(x,\cdot)$ with respect to the measure dy. At the same time, $f_n(x) \uparrow f(x)$ for $x \notin A$ and therefore the sequence $h_n(x,\cdot)$ is integrally bounded, i.e.,

$$\int h_n(x,y)\, dy = f_n(x) \le f(x) < \infty.$$

Hence, for $x \notin A$ the function $h(x,\cdot)$ is integrable with respect to the measure dy and

$$f(x) = \lim f_n(x) = \lim \int h_n(x,y)\, dy = \int h(x,y)\, dy. \tag{3}$$

The function f is integrable with respect to the measure dx. Consequently, there exists a repeated integral

$$a = \int f(x)\, dx = \int \left(\int h(x,y)\, dy \right) dx.$$

It is equal to the double integral, since (1)–(3) imply

$$a = \int f(x)\, dx = \int \left(\int h(x,y)\, dy \right) dx$$

$$= \sum_i \sum_j c_{ij}\, dx(A_{ij})\, dy(B_{ij}) = c.$$

In a similar way, we could prove that there exists a repeated integral b and $b = c$. However, this readily follows by symmetry from what we have already proved if we use the interchange $x \leftrightarrow y$.   □

The proof of the Fubini theorem.   Let h be a numeric function on $X \times Y$ such that h is integrable with respect to the measure $dxdy$ and $h = h_1 - h_2 + ih_3 - ih_4$, where the components h_1, h_2, h_3, and h_4 are monotonically integrable. The components h_j $(j = 1, 2, 3, 4)$ are integrable with respect to the measure $dxdy$. By Lemma 2, the Fubini theorem holds for these components. Indeed, from the existence of repeated integrals

$$c_j = \int \int h_j(x, y) \, dxdy$$

it follows that there exist repeated integrals

$$a_j = \int \left(\int h_j(x, y) \, dy \right) dx, \quad b_j = \int \left(\int h_j(x, y) \, dx \right) dy$$

and $a_j = b_j = c_j$ $(j = 1, 2, 3, 4)$.

Note that for all $x \in X$ and $y \in Y$ the functions $h_j(x, \cdot)$ and $h_j(\cdot, y)$ are the monotonically integrable components of the functions $h(x, \cdot)$ and $h(\cdot, y)$:

$$h(x, \cdot) = h_1(x, \cdot) - h_2(x, \cdot) + ih_3(x, \cdot) - ih_4(x, \cdot),$$
$$h(\cdot, y) = h_1(\cdot, y) - h_2(\cdot, y) + ih_3(\cdot, y) - ih_4(\cdot, y).$$

Since $h_j(x, \cdot)$ is integrable with respect to the measure dy and $h_j(\cdot, y)$ is integrable with respect to the measure dx, it follows that $h(x, \cdot)$ and $h(\cdot, y)$ are integrable with respect to the measures dy and dx, respectively. Therefore, there exist simple integrals f_j and g_j such that

$$f_j(x) = \int h_j(x, y) \, dy, \quad g_j(y) = \int h_j(x, y) \, dx$$

for almost all $x \in X$ and almost all $y \in Y$. By definition, the existence of simple integrals a_j and b_j implies the integrability of f_j and g_j with respect to the measures dy and dx. Hence, f and g are integrable with respect to the measures dy and dx. Consequently, there exist repeated integrals a and b, and

$$a = a_1 - a_2 + ia_3 - ia_4, \quad b = b_1 - b_2 + ib_3 - ib_4.$$

These equalities, together with the equalities $c = c_1 - c_2 + ic_3 - ic_4$ and $a_j = b_j = c_j$ $(j = 1, 2, 3, 4)$, imply that $a = b = c$. Thus, the Fubini theorem is proved in the general case.

4. The Tonelli Theorem. In the Fubini theorem, the existence of the repeated integrals is deduced from the existence of the double integral. In the Tonelli theorem, conversely, the existence of the double integrals is deduced from the existence of the repeated integral, but this is done only for positive measurable functions.

The Tonelli theorem. *If there exists one of the repeated integrals (a or b) for a positive measurable function h of two variables, then the other repeated integral and the double integral c also exist for h, and $a = b = c$.*

Proof. Let a positive measurable function h have a repeated integral

$$a = \int f(x)\,dx = \int \left(\int h(x,y)\,dy \right) dx.$$

There exists an increasing sequence of positive integrable functions h_n that converges almost everywhere to h with respect to the measure $dxdy$:

$$0 \le h_n(z) \uparrow h(z) \quad (z = (x,y) \notin C),$$

where $dxdy(C) = 0$. We also need to prove that the sequence of functions h_n is integrally bounded. (In the Fubini theorem, this readily followed from the integrability of h. In the Tonelli theorem, we only assume that h is measurable.)

Note that the sequence of integrable functions f_n with values

$$f_n(x) = \int h_n(x,y)\,dy$$

increases together with the sequence of functions h_n and is bounded by the integrable function $f(x) = \int h(x,y)\,dy$ for almost all $x \in X$ with respect to the measure dx. Applying the Fubini theorem to the functions h_n, taking into account the existence of the repeated integral a, and integrating the inequalities $f_n \le f$ that hold almost everywhere with respect to the measure dx, we get

$$\int\int h_n(x,y)\,dxdy = \int \left(\int h_n(x,y)\,dy \right) dy$$

$$= \int f_n(x)\,dx \le \int f(x)dx = \int \left(\int h(x,y)\,dy \right) dx = a.$$

Therefore, the sequence of functions h_n is integrally bounded. By Levi's theorem, h is integrable with respect to the measure $dxdy$. There exists a double integral

$$c = \int \int h(x,y)\, dxdy = \lim \int \int h_n(x,y)\, dxdy$$

Then, by the Fubini theorem, the second repeated integral

$$b = \int g(y)\, dy = \int \left(\int h(x,y)\, dx \right) dy$$

also exists, and $a = b = c$.

In a similar way, we could prove that if there exists a repeated integral b for a positive measurable function h, then the double integral c and the repeated integral a also exist for h, and $a = b = c$. However, this readily follows by symmetry from what we have already proved if we use the interchange $x \leftrightarrow y$.   □

It is possible to prove a proposition that generalizes the Fubini and Tonelli theorems to the case of measurable numeric functions h.

As usual, we denote by $|h|$ a function with values $|h(x,y)|$. If h is measurable, then so is $|h|$. Similarly, if h is integrable, then so is $|h|$. Let α, β, and γ denote the two repeated integrals and the double integral of the function $|h|$:

$$\alpha = \int \left(\int |h(x,y)|\, dy \right) dx, \quad \beta = \int \left(\int |h(x,y)|\, dx \right) dy,$$

$$\gamma = \int \int |h(x,y)|\, dxdy.$$

The existence of (the finite) γ is equivalent to the existence of the double integral $c = \int \int h(x,y)\, dxdy$. However, the existence of (the finite) α and β is not guaranteed when a and b exist (and even are equal).

Exercise. Provide a counterexample supporting the preceding statement (see Savel'ev, 1989, Chapter 7, Section 74).

The following theorem combines the Fubini and Tonelli theorems.

The Fubini–Tonelli theorem. *Let h be a measurable function of two variables. Then, if there exist one of the repeated integrals α and β or a double integral γ of the function $|h|$, then there exist repeated integrals a and b and a double integral c of the function h, and $a = b = c$.*

Proof. If h is measurable, then so is $|h|$. By the Tonelli theorem, the measurability of $|h|$ and the existence of one of the repeated integrals α and β for $|h|$ ensures the existence of the double integral of $|h|$ and, consequently, the existence of the double integral c for h. The double integral c of a measurable function h certainly exists if it is assumed that the double integral γ of the function $|h|$ exists. By the Fubini theorem, the existence of the double integral c for h implies the existence of both repeated integrals a and b for h and the equality $a = b = c$. $\square$

5. The Fubini and Tonelli theorems provide the conditions under which it is possible to change the order of integrals. In what follows, we will provide the conditions under which it is possible to interchange the integral and the limit, derivative, or a singular integral.

We will assume that $X \subseteq \mathbb{R}$, and call the variable $x \in X$ the *parameter*. If the function $h(x, \cdot) : Y \to \mathbb{C}$ is integrable with respect to the measure dy for any value of the parameter x, then the function $f : X \to \mathbb{C}$ with values $f(x) = \int h(x, y)\, dy$ will be called the *integral depending on a parameter*. The other designations and terms from the preceding part are assumed to have the same meaning as before.

Let a be an arbitrary point in the closure $\bar{X}$ and $\bar{\mathbb{R}} = [-\infty, \infty]$ (In particular, it is possible that $a = -\infty$ or $a = \infty$.) We now formulate the conditions under which the integral with respect to dx and the limit as $x \to a$ can be interchanged:

(1.1) $h(x, \cdot)$ is measurable with respect to dy for any $x \in X$;

(2.1) $h(\cdot, y)$ converges to the value $g(y)$ of a function $g : Y \to \mathbb{C}$ as $x \to a$ for almost all $y \in Y$;

(3.1) $|h(\cdot, y)|$ is bounded by the value $\bar{g}(y)$ of an integrable function $\bar{g} : Y \to \mathbb{R}$ for almost all $y \in Y$.

The statement "$\lim_{x \to a} \int h(x, y)\, dy$ is well defined" means that the functions $h(x, \cdot)$ are integrable with respect to dy and the function $f(x) = \int h(x, y)\, dy$ has a limit as $x \to a$. The statement "$\int \lim_{x \to a} h(x, y)\, dy$ is well defined" means that there exists a limit function $g : Y \to \mathbb{C}$ with values $g(y) = \lim_{x \to a} h(x, y)$ for almost all $y \in Y$ and g is integrable with respect to dy.

Theorem 1. *Under the conditions* (1.1)–(3.1), *both sides of the equality*

$$\lim_{x \to a} \int h(x, y)\, dy = \int \lim_{x \to a} h(x, y)\, dy$$

are well defined and the equality is true.

Proof. We will reduce the assertion to the case of sequences and apply the Lebesgue theorem. Let $x(n) \in X$ be a sequence converging to $a \in \bar{X}$. If $h(x, y) \to g(y)$ as $x \to a$, then $h(x(n), y) \to g(y)$. The conditions (1.1)–(3.1) ensure that the assumptions of the Lebesgue theorem hold for the functions $g_n = h(x(n), \cdot)$, g, and $\bar{g}$, i.e., the functions g_n are measurable, their absolute values $|g_n| = |h(x(n), \cdot)|$ are bounded almost everywhere by the integrable function $\bar{g}$, and the sequence g_n converges almost everywhere to g. By the Lebesgue theorem, the functions g_n and g are integrable and $\lim \int g_n = \int g$, which means that

$$\lim f(x(n)) = \lim \int h(x(n), y)\, dy = \int g(y)\, dy$$

for any sequence $x(n) \to a$. Hence, the function f has a limit as $x \to a$ and

$$\lim_{x \to a} \int h(x, y)\, dy = \lim_{x \to a} f(x) = \int g(y)\, dy = \lim_{x \to a} \int h(x, y)\, dy.$$

This completes the proof.   □

We now turn to derivatives. Let a be an arbitrary point in an interval X of $\mathbb{R} =]-\infty, \infty[$. (The point a is not necessarily interior, and the interval X is not necessarily open.)

Let $\mathcal{D}_a h(\cdot, y)$ denote the derivative of the function $h(\cdot, y)$ at a, and let $\Delta_a h(\cdot, y)$ denote the relative increment of $h(\cdot, y)$ at a:

$$\Delta_a h(x, y) = \frac{h(x, y) - h(a, y)}{x - a}$$

for $x \neq a$. If $h(\cdot, y)$ is differentiable at a, then the value of $\Delta_a h(\cdot, y)$ for $x = a$ is chosen to be $\mathcal{D}_a h(\cdot, y)$. By definition,

$$\mathcal{D}_a h(\cdot, y) = \lim_{x \to a} \Delta_a h(x, y).$$

We now formulate the conditions under which the integral with respect to dy and the derivative at a can be interchanged.

(1.2) $h(x, \cdot)$ *is measurable with respect to dy for any $x \in X$, and $h(a, \cdot)$ is integrable with respect to dy;*

(2.2) $h(\cdot, y)$ *is differentiable at a for almost all $y \in Y$;*

(3.2) $|\Delta_a h(\cdot, y)|$ *is bounded by the value $\bar{g}(y)$ of an integrable function $\bar{g} : Y \to \mathbb{R}$ for almost all $y \in Y$.*

The statement "$\mathcal{D}_a \int h(x,y)\,dy$ is well defined" means that the functions $h(x,\cdot)$ are integrable with respect to dy and the function $f : X \to \mathbb{C}$ with values $f(x) = \int h(x,y)\,dy$ has a derivative at a. The statement "$\int \mathcal{D}_a h(x,y)\,dy$ is well defined" means that $h(\cdot,y)$ is differentiable at a for almost all $y \in Y$ and the function $\mathcal{D}_a h(\cdot,y)$ for such $y \in Y$ is integrable with respect to dy.

Theorem 2. *Under the conditions* (1.2)–(3.2), *both sides of the equality*

$$\mathcal{D}_a \int h(x,y)\,dy = \int \mathcal{D}_a h(x,y)\,dy$$

are well defined and the equality is true.

Proof. We need to apply Theorem 1 to the function $\Delta_a h(\cdot,\cdot)$.

The conditions (1.2)–(3.2) ensure that the conditions (1.1)–(1.3) hold for the function $\bar{h} = \Delta_a h(\cdot,\cdot)$ with values

$$\bar{h}(x,y) = \frac{h(x,y) - h(a,y)}{x - a} \quad (x \neq a), \quad \bar{h}(a,y) = \Delta_a h(\cdot,y).$$

Indeed, like $h(x,\cdot)$, $\bar{h}(x,\cdot)$ is measurable with respect to dy for all $x \in X$, $\bar{h}(\cdot,y)$ converges to $\mathcal{D}_a h(\cdot,y)$ as $x \to a$ for almost all $y \in Y$ and $|\bar{h}(x,y)|$ is bounded by $\bar{g}(y)$ for almost all $y \in Y$.

By Theorem 1, the functions $\bar{h}(x,\cdot)$ are integrable. By assumption, the function $h(a,\cdot)$ is integrable. Since

$$h(x,\cdot) = h(a,\cdot) + (x-a)\bar{h}(x,\cdot),$$

and since $h(a,\cdot)$ and $\bar{h}(x,\cdot)$ are integrable, it follows that $h(x,\cdot)$ is integrable with respect to the measure dy for all $x \in X$. By Theorem 1, both sides of the equality

$$\lim_{x \to a} \int \frac{h(x,y) - h(a,y)}{x - a}\,dy = \int \lim_{x \to a} \frac{h(x,y) - h(a,y)}{x - a}\,dy$$

are well defined and this equality is true. Furthermore, since $h(x,\cdot)$ is integrable for all $x \in X$, we have

$$\lim_{x \to a} \int \frac{h(x,y) - h(a,y)}{x - a}\,dy$$
$$= \lim_{x \to a} \frac{1}{x - a}\left[\int h(x,y)\,dy - \int h(a,y)\,dy\right] = \mathcal{D}_a \int h(x,y)\,dy,$$
$$\lim_{x \to a} \frac{h(x,y) - h(a,y)}{x - a} = \mathcal{D}_a h(\cdot,y).$$

Hence, both sides of the equality in the assertion of the theorem are well defined and the equality is true. $\square$

The equality in the assertion of Theorem 2 is called the *Leibniz rule* for differentiation of an integral depending on a parameter.

For differentiable functions on an interval, the conditions for integral and derivative to be interchangeable are similar to the assumptions of Theorem 2. Consider an interval X of the real line.

We denote by $\mathcal{D}h(\cdot, y)$ the derivative of $h(\cdot, y)$. We now formulate the conditions for the integral with respect to dy and the derivative with respect to x to be interchangeable.

(1.2′) $h(x, \cdot)$ is integrable with respect to dy for any $x \in X$;

(2.2′) $h(\cdot, y)$ is differentiable for almost all $y \in Y$;

(3.2′) $|\mathcal{D}h(\cdot, y)|$ is bounded by the value $\bar{g}(y)$ of an integrable function $\bar{g} : Y \to \mathbb{R}$ for almost all $y \in Y$.

Theorem 2′ is analogous to Theorem 2.

Theorem 2′. *Under the conditions* (1.2′)–(3.2′), *both sides of*

$$\mathcal{D} \int h(x, y)\, dy = \int \mathcal{D}h(x, y)\, dy$$

are well defined and the equality is true.

Proof. It is required to prove that (1.2′)–(3.2′) imply (1.2)–(3.2) for any $a \in X$. Conditions (1.2) and (2.2) immediately follow from (1.2′) and (2.2′). Condition (3.2) follows from (3.2′) and the Lagrange formula (Schwartz, 1967, Chapter 3, Section 2):

$$\left| \frac{h(x, y) - h(a, y)}{x - a} \right| = |\mathcal{D}_c h(\cdot, y)| \leq \bar{g}(y),$$

where $c = c(a, x)$ is a point of the interval X. □

For singular integrals, we take $X = \mathbb{R} =]-\infty, \infty[$. The Lebesgue singular integrals will be defined by analogy with Riemann integrals.

Consider a function $f : \mathbb{R} \to \mathbb{C}$ that is integrable with respect to the Lebesgue measure dx on every interval $[a, b] \subset \mathbb{R}$. By the Lebesgue *singular integral* of the function f we mean the double limit

$$\int\limits_{-\infty}^{\infty} f(x)\, dx = \lim_{\substack{a \to -\infty \\ b \to \infty}} \int\limits_{a}^{b} f(x)\, dx.$$

If f is integrable with respect to the Lebesgue measure dx on the entire real line $\mathbb{R} =]-\infty, \infty[$, then the singular integral of f is equal to the conventional one. This follows from Theorem 1. Indeed,

$$\int f - \int_a^b f = \int f - \int [a,b] f = \int (1 - [a,b]) f \to 0$$

as $a \to -\infty$ and $b \to \infty$ because $1 - [a,b](x) \to 0$ for any $x \in X$, the function $|f|$, like f, is integrable with respect to dx, and $|(1 - [a,b]) \cdot f| \leq |f|$.

Remark. There exist Lebesgue singular integrals even for some functions that are not integrable with respect to the Lebesgue measure. (For example, for $f(x) = x^{-1} \sin x$ with $x \neq 0$.)

We now formulate the conditions for the integral with respect to dy and the singular integral with respect to dx to be interchangeable.

(1.3) $h(\cdot, \cdot)$ *is measurable with respect to $dxdy$;*

(2.3) *there exists a Lebesgue singular integral of $h(\cdot, y)$ for almost all $y \in Y$;*

(3.3) *for all $a \leq b$ in $\mathbb{R}$, the integrals $\int_a^b |h(x,y)|\, dx$ are bounded by the value $\bar{g}(y)$ of an integrable function $\bar{g} : Y \to \mathbb{R}$ for almost all $y \in Y$.*

The statement " $\int_{-\infty}^{\infty} \left(\int h(x,y)\, dy \right) dx$ is well defined" means that $h(\cdot, \cdot)$ is integrable with respect to dy and there exists a singular integral of $f(x) = \int h(x,y)dy$ with respect to dx. The statement "$\int \left(\int_{-\infty}^{\infty} h(x,y)\, dx \right) dy$ is well defined" means that there exists a singular integral of $h(\cdot, y)$ with respect to dx for almost all $y \in Y$, and $g : Y \to \mathbb{C}$ with values $g(y) = \int_{-\infty}^{\infty} h(x,y)\, dx$ is integrable with respect to dy for the same $y \in Y$.

Theorem 3. *Under the conditions* (1.3)–(3.3), *both sides of the equality*

$$\int_{-\infty}^{\infty} \left(\int h(x,y)\, dy \right) dx = \int \left(\int_{-\infty}^{\infty} h(x,y)\, dx \right) dy$$

are well defined and the equality is true.

Proof. The assertion of Theorem 3 follows from the Fubini–Tonelli theorem and Levi's theorem. Condition (3.3) implies that there exists a singular integral of $|h(\cdot, y)|$

$$\int_{-\infty}^{\infty} |h(x, y)| \, dx = \lim_{\substack{a \to -\infty \\ b \to \infty}} \int_{a}^{b} |h(x, y)| \, dx \leq \bar{g}(y)$$

for almost all $y \in Y$. Consequently, for such $y \in Y$ the function $|h(\cdot, y)|$ is integrable with respect to the measure dx on $\mathbb{R} = \,]{-\infty}, \infty[$. Indeed, from (1.3)–(3.3) it follows that the sequence of functions $\varphi_n = [-n, n] \cdot |h(\cdot, y)|$ with values

$$\varphi_n(x) = |h(x, y)| \ (x \in [-n, n]), \quad \varphi_n(x) = 0 \ (x \notin [-n, n])$$

is an increasing sequence of integrable functions with respect to the Lebesgue measure dx on $\mathbb{R}$, which is also integrally bounded and converges everywhere to $\varphi = |h(\cdot, y)|$. Therefore, as follows from Levi's theorem, the function $|h(\cdot, y)|$ is integrable with respect to dx and therefore so is $h(\cdot, y)$ (as follows from the integrability criterion; since $h(\cdot, \cdot)$ is measurable with respect to $dx\,dy$, $h(\cdot, y)$ is measurable with respect to dx, as was shown in part 2). Moreover, as was shown before, the integral with respect to the measure dx is equal to the singular integral:

$$\int h(x, y) \, dx = \int_{-\infty}^{\infty} h(x, y) \, dx.$$

Since the functions $|h(\cdot, y)|$ are integrable with respect to the Lebesgue measure dx on $\mathbb{R}$, in view of (3.3) it follows that there exists a repeated integral

$$\beta = \int \left(\int |h(x, y)| \, dx \right) dy \leq \int \bar{g}(y) \, dy < \infty.$$

By the Fubini–Tonelli theorem, the measurability of $h(\cdot, y)$ (condition (1.3)) and the existence of the repeated integral β of the function $|h(\cdot, y)|$ ensure that there exist repeated integrals of $h(\cdot, \cdot)$ and they are equal to each other. Since the integrals with respect to the measure dx are equal to the corresponding singular integrals, it follows that both sides of the equality in the assertion of the theorem are well defined and the equality is true.   □

We now prove one more theorem on the interchangeability of integrals. The conditions for this theorem are formulated as follows:

(1.3') $h(x, \cdot)$ *is integrable with respect to* dy *for all* $x \in X$;

(2.3') *there is a Lebesgue singular integral of* $h(\cdot, y)$ *for almost all* $y \in Y$;

(3.3') $|\Delta_u h(\cdot, y)|$ *and* $\int\limits_u^v |h(x, y)| \, dx$ *for all* $u \leq v$ *in* $\mathbb{R}$ *are bounded by the value* $\bar{g}(y)$ *of an integrable function* $\bar{g} : Y \to \mathbb{R}$ *for almost all* $y \in Y$.

The assertions of Theorem 3' are analogous to those of Theorem 3.

Theorem 3'. *Under the conditions* (1.3')–(3.3'), *both sides of*

$$\int\limits_{-\infty}^{\infty} \left(\int h(x, y) \, dy \right) dx = \int \left(\int\limits_{-\infty}^{\infty} h(x, y) \, dx \right) dy$$

are well defined and the equality is true.

Proof. We will verify that both sides of the above equality are well defined. We begin with the right-hand side. By the condition (2.3'), the function $h(\cdot, y)$ is integrable with respect to dx on any $[u, v] \subset \mathbb{R}$ ($-\infty < u \leq v < \infty$) for almost all $y \in Y$. Consider a function g such that $g(u, v, y) = \int\limits_u^v h(x, y) \, dx$ for the same almost all $y \in Y$ and the $g(u, v, y)$ are arbitrary for the remaining $y \in Y$. We now prove that $g(u, v, \cdot)$ is integrable with respect to dy. Note that condition (3.3') implies that $h(\cdot, y)$ is continuous for almost all $y \in Y$. Indeed, if

$$\left| \frac{h(x, y) - h(u, y)}{x - u} \right| \leq \bar{g}(y)$$

for $x \neq u$, then $|h(x, y) - h(u, y)| \leq \bar{g}(u) \cdot |x - u|$. On the interval $[u, v]$, the continuous function $h(\cdot, y)$ is approximated by the step functions

$$h_n(\cdot, y) = \sum_k h(x_{kn}, y) A_{kn},$$

$$A_{kn} = [u + (k - 1)(v - u)/n, u + k(v - u)/n],$$

$$x_{kn} = u + k(v - u)/n, \quad k = 1, \ldots, n.$$

Hence, $g(u, v, \cdot) = \lim\limits_{n \to \infty} \left[\sum_k h(x_{kn}, \cdot) \cdot (v - u)/n \right]$. By (1.3'), the functions $h(x_{kn}, \cdot)$ are integrable with respect to dy and, consequently, are measurable.

Then the sums in square brackets are also measurable, and so is their limit. At the same time, condition (3.3′) implies that

$$|g(u,v,y)| \leq \int_u^v |h(x,y)|\, dx \leq \bar{g}(y)$$

for almost all values of the integrable function $\bar{g} : Y \to \mathbb{R}$. Thus, $g(u,v,\cdot)$ is integrable with respect to dy.

Set

$$\psi(u,x) = \int g(u,x,y)\, dy \quad (x \in [u,v]).$$

We will prove that $\psi(u,\cdot)$ is differentiable and find its derivative. We now verify that $g(u,x,\cdot)$ satisfies the hypothesis of Theorem 2′. Indeed, $g(u,x,\cdot)$ is integrable with respect to dy for all $x \in [u,v]$, and $g(u,\cdot,y)$ is differentiable for $y \in Y$ such that $h(\cdot,y)$ is continuous:

$$\mathcal{D}g(u,\cdot,y) = h(\cdot,y), \quad |\mathcal{D}g(u,\cdot,y)| = |h(\cdot,y)| \leq \bar{h}(y)$$

for almost all values of the integrable function $\bar{h} = |h(\cdot,y)| + g \cdot (v-u)$. This follows from conditions (1.3′) and (3.3′) in view of the inequalities

$$|h(x,y)| - |h(u,y)| \leq |h(x,y) - h(u,y)|$$
$$\leq \bar{g}(y) \cdot |x-u| \leq \bar{g}(y) \cdot (v-u),$$
$$|h(x,y)| \leq |h(u,y)| + \bar{g}(y) \cdot (v-u).$$

By Theorem 2′, the function $\psi(u,\cdot)$ is differentiable and its derivative at $x \in [u,v]$ is written as

$$\mathcal{D}\psi(u,x) = \mathcal{D} \int g(u,x,y)\, dy = \int \mathcal{D}g(u,x,y)\, dy = \int h(x,y)\, dy.$$

We will prove that $\lim\limits_{\substack{u \to -\infty, \\ v \to \infty}} \psi(u,v) = \int \left(\int_{-\infty}^{\infty} h(x,y)\, dx \right) dy.$

The function g satisfies the hypothesis of Theorem 1. Indeed, g has the following properties:

$g(u,v,\cdot)$ is measurable with respect to dy for all $u \leq v$;

$g(u,v,y)$ converges to the singular integral of $h(\cdot,y)$ as $u \to -\infty$ and $v \to \infty$ for almost all $y \in Y$, as follows from (2.3′);

$|g(\cdot,\cdot,y)| \leq \bar{g}(y)$ for almost all values of the integrable function $\bar{g} : Y \to \mathbb{R}$.

By Theorem 1,

$$\lim_{\substack{u \to -\infty, \\ v \to \infty}} \psi(u, v) = \lim_{\substack{u \to -\infty, \\ v \to \infty}} \int \left(\int_u^v h(x, y) \, dx \right) dy$$

$$= \int \left(\lim_{\substack{u \to -\infty, \\ v \to \infty}} \int_u^v h(x, y) \, dx \right) dy = \int \left(\int_{-\infty}^{\infty} h(x, y) \, dx \right) dy.$$

Thus, the right-hand side of the equality in the assertion of Theorem $3'$ is well defined.

We now turn to its left-hand side. By condition $(1.3')$, the function $f : \mathbb{R} \to \mathbb{C}$ with values $f(x) = \int h(x, y) \, dy$ is defined. From $(3.3')$ it follows that f is continuous:

$$|f(x) - f(u)| \le \int |h(x, y) - h(u, y)| \, dy \le |x - u| \int \bar{g}(y) \, dy \le c|x - u|$$

$(c = \int \bar{g}(y) \, dy < \infty)$. Set

$$\varphi(u, v) = \int_u^v f(x) \, dx \quad (u \le v).$$

Since f is continuous, it follows that $\varphi(u, \cdot)$ is differentiable and its derivative at $x \in [u, v]$ is written as

$$\mathcal{D}\varphi(u, x) = f(x) = \int h(x, y) \, dy.$$

We have $\mathcal{D}\varphi(u, x) = \mathcal{D}\psi(u, x)$ $(x \in [u, v])$ and therefore $\varphi(u, x) - \psi(u, x) = c$ $(x \in [u, v])$. Since $\varphi(u, u) = \psi(u, u) = 0$, we have $c = 0$. As a result, $\varphi(u, x) = \psi(u, x)$ $(u \le v)$. Therefore,

$$\int_{-\infty}^{\infty} \left(\int h(x, y) \, dy \right) dx = \lim_{\substack{u \to -\infty, \\ v \to \infty}} \varphi(u, v)$$

$$= \lim_{\substack{u \to -\infty, \\ v \to \infty}} \psi(u, v) = \int \left(\int_{-\infty}^{\infty} h(x, y) \, dx \right) dy.$$

We conclude that the left-hand side of the equality in the assertion of Theorem $3'$ is also well defined and coincides with the right-hand side. This completes the proof of the theorem. □

Theorems 1–3, $2'$, and $3'$ provide the rules for operations under the integral sign. They are widely used and are very effective.

3.3.6. Indefinite integrals

Indefinite integrals are measures whose values are equal to integrals.

1. The Radon-Nikodym theorem. Consider a set U, the algebra $\mathcal{A}$ of subsets of U, a measure μ on $\mathcal{A}$, the algebra $\mathcal{B} \supseteq \mathcal{A}$ of subsets of U, and a measure ν on $\mathcal{B}$. As before, we will assume that the measures under consideration are positive, countably additive, locally bounded, and, in some cases, complete.

We say that *the measure ν is continuous with respect to the measure μ*, written $\nu \ll \mu$, if for any $\varepsilon > 0$ there exists a $\delta > 0$ such that $\nu(A) \leq \varepsilon$ for any $A \in \mathcal{A}$ such that $\mu(A) \leq \delta$.

Let $\bar{\mathcal{A}}$ and $\bar{\mathcal{B}}$ denote the algebras of sets integrable with respect to the measures μ and ν, respectively, and let $\bar{\mu}$ and $\bar{\nu}$ denote the integral extensions of these measures. Assume that $\mathcal{B} \supseteq \mathcal{A}$. It can be proved that the continuity of ν with respect to μ is equivalent to the continuity of $\bar{\nu}$ with respect to $\bar{\mu}$ (see Savel'ev, 1989, Chapter 8, Section 8.2). Furthermore, the continuity of $\bar{\nu}$ with respect to $\bar{\mu}$ is equivalent to the condition that $\bar{\nu}(A) = 0$ for $A \in \mathcal{A}$ such that $\bar{\mu}(A) = 0$. When considering integral extensions below, we will omit the bars in the designations for algebras and measures.

Let f be a function measurable with respect to the measure μ. If the function Af is integrable with respect to the measure μ for all $A \in \mathcal{A}$ and

$$\nu(A) = \int Af \, d\mu = \int_A f \, d\mu,$$

then the measure ν is called the *indefinite integral* of f with respect to the measure μ, and the function f is called *the derivative of the measure ν with respect to the measure μ* and is designated by $d\nu/d\mu$. We also write $d\nu = f d\mu$.

The derivatives of the measure ν with respect to the measure μ form a class of functions equivalent with respect to μ. Since μ and ν are positive, $d\nu/d\mu = f \geq 0$ almost everywhere with respect to μ. Considering linear combinations of measures and their derivatives, it is possible to generalize what was said above to the numeric functions of sets (Halmos, 1974, Section 30).

A function f such that Af is integrable with respect to μ for all $A \in \mathcal{A}$ is said to be *locally integrable* with respect to μ. Since the measure μ is assumed to be locally bounded, from the local integrability of f it follows that f is measurable with respect to μ.

Using the local integrability of derivatives and the Lebesgue theorem, it is easy to verify that indefinite integrals with respect to the measure μ are also continuous with respect to it. (Savel'ev, 1989, Chapter 8, Subsection 8.2.2). The converse statement, which is more essential, is also true: if a measure ν is continuous with respect to μ, then ν is an indefinite integral with respect to μ.

The Radon–Nikodym theorem. *A measure ν is an indefinite integral of a function that is locally integrable with respect to a measure μ if and only if ν is continuous with respect to μ.*

The Radon–Nikodym theorem can be deduced from the Riesz theorem on the representation of linear functionals on a Hilbert space (Savel'ev, 1989, Chapter 8, Section 8.4). There are many useful corollaries of the Radon–Nikodym theorem (see Savel'ev, 1989, Chapter 8, Section 8.5).

2. The change of variables theorem. The change of variables is one of the most effective methods of calculating integrals.

Theorem. *Let f be measurable with respect to μ and integrable with respect to ν, and let ν be continuous with respect to μ. Then the product $f(d\nu/d\mu)$ is integrable with respect to μ and*

$$\int f(d\nu/d\mu)\, d\mu = \int f\, d\nu. \tag{1}$$

Proof. By assumption, the integral on the right-hand side exists. The existence of the integral on the left-hand side needs to be proved. The existence of the derivative $d\nu/d\mu$ follows from the Radon–Nikodym theorem.

From the derivatives of ν with respect to μ that are equivalent with respect to μ, we choose a derivative $d\nu/d\mu = g$ with positive values. The integrability of the product fg and the equality for the integrals will first be established for sets, then for simple functions, then for positive integrable functions, and, finally, for real integrable functions and numeric integrable functions in general.

1. Let $f = X \in \mathcal{A}$. Then the Radon–Nikodym theorem implies that $fg = Xg$ is integrable and

$$\int f\, d\nu = \int X\, d\nu = \nu(X) = \int Xg\, d\mu = \int fg\, d\mu.$$

2. Let $f = \sum c_i X_i$ $(c_i \in \mathbb{R},\ X_i \in \mathcal{A})$. From what was proved in part 1, it follows that fg is integrable and

$$\int f\, d\nu = \sum c_i \int X_i\, d\nu = \sum c_i \int X_i g\, d\nu = \int fg\, d\mu.$$

3. Let $f \geq 0$ be integrable with respect to μ and integrable with respect to ν. Then, since f is positive and measurable with respect to μ, there exists an increasing sequence of simple functions

$$f_n = \sum c_{in} X_{in} \quad (c_{in} \in \mathbb{R},\ X_{in} \in \mathcal{A}),$$

that converges to f almost everywhere with respect to μ. Since $g \geq 0$, $f_n \uparrow f$ implies $f_n g \uparrow fg(\mu)$. From what was proved in the previous part of the proof and from the integrability of f with respect to ν, it follows that $f_n g$ are integrable with respect to μ and

$$\int f_n g\, d\mu = \int f_n\, d\nu \leq \int f\, d\nu < \infty.$$

(Furthermore, since $\mathcal{A} \subseteq \mathcal{B}$ by assumption, it follows that f_n are integrable with respect to ν.)

Applying Levi's theorem to the sequences f_n and $f_n g$, we conclude that the function fg is integrable with respect to μ and

$$\int f\, d\nu = \lim \int f_n\, d\nu = \lim \int f_n g\, d\mu = \int fg\, d\mu.$$

4. Let $f : U \to \mathbb{R}$ be measurable with respect to μ and integrable with respect to ν. Then its positive components f^+ and f^- are also measurable with respect to μ and integrable with respect to ν. As follows from the previous part of the proof, the assertion of the theorem holds for f^+ and f^-. Therefore, $f^+ g,\ f^- g$, and $fg = (f^+ - f^-)g = f^+ g - f^- g$ are integrable with respect to μ and $\int f\, d\nu = \int f^+\, d\nu - \int f^-\, d\nu = \int f^+ g\, d\mu - \int f^- g\, d\mu = \int (f^+ - f^-)\, d\nu = \int fg\, d\mu$.

5. Let $f : U \to \mathbb{C}$ be measurable with respect to μ and integrable with respect to ν. Then so are the real and imaginary parts of f. As follows from the previous part of the proof, the assertion of the theorem is true for the real and imaginary parts of f, which means that it is also true for f. This completes the proof. $\square$

Remark. The use of the assumption that the measures in question are integrally extendable is essential throughout this subsection. In particular, the assertion on the existence of an increasing sequence of simple functions in the third part of the proof of the change of variables theorem does not hold without the said assumption.

3. Consider open subsets X and Y of $\mathbb{R}^m$, the algebras $\mathcal{A}$ and $\mathcal{B}$ generated by bounded open subsets of X and Y, respectively, the Lebesgue measures $\lambda = dx$ and $\mu = dy$ on $\mathcal{A}$ and $\mathcal{B}$, respectively, a smooth homeomorphism T from X onto Y, its derivative T', the absolute value $|\det T'|$ of its determinant, a numeric function g on Y integrable with respect to dy, and the composition $f = gT$ on X.

Since T is a homeomorphism from X onto Y, in addition to the measure μ on $\mathcal{B}$, it defines the measure $\nu = \mu T$ on $\mathcal{A}$ with values $\nu(A) = \mu(T(A))$ $(A \in \mathcal{A})$.

The measure $\nu = \mu T$ is continuous with respect to the measure μ and

$$d\mu T / d\lambda = |\det T'| \tag{2}$$

(see Savel'ev, 1989, Chapter 9, Section 9.3). Moreover, the composition $f = gT$ is integrable with respect to the measure $\nu = \mu T$ and

$$\int_Y g \, d\nu = \int_X gT \, d\mu T. \tag{3}$$

This equality is equivalent to the equality $\nu(B) = \mu(B)$ for $g = B \in \mathcal{B}$. In the case of functions, it is proved successively, in the same manner as equality (1) of the change of variables theorem in part 2 of this subsection. From (1)–(3), it follows that

$$\int_Y g \, d\nu = \int_X gT \, d\mu T = \int_X gT (d\mu T / d\lambda) \, d\lambda = \int_X gT |\det T'| \, d\lambda.$$

Rewriting the equality for the first and the last integrals in classical notation, we obtain the formula for the change of variables for multiple integrals (Savel'ev, 1989, Chapter 9, Section 9.3.4):

$$\int_Y g(y) \, dy = \int_X gT(x) \cdot |\det T'(x)| \cdot dx, \tag{4}$$

where the derivative $T'(x)$ of the map T at $x \in X$ is expressed by the matrix $m \times m$ whose elements are partial derivatives. The condition $\det T'(x) \neq 0$ is not required in equality (4).

Equality (3) is a special case of the general formula for measurable change of variables in an integral. Consider abstract sets X and Y, the algebras $\mathcal{A}$ and $\mathcal{B}$ of subsets of X and Y, respectively, a measure λ on $\mathcal{A}$, a map $T : X \to Y$ such that $T^{-1}(B) = A \in \mathcal{A}$ for any $B \in \mathcal{B}$, the measure $\mu = \lambda T^{-1}$ on $\mathcal{B}$ with values

$$\mu(B) = \lambda(T^{-1}(B)) \quad (B \in \mathcal{B}),$$

a numeric function g on Y that is integrable with respect to μ, and the composition $f = gT$ on X.

Using arguments similar to those in the proof of the theorem of part 2 and the explanation of equality (3), it is easy to verify that

$$\int_B g \, d\mu = \int_A f \, d\lambda \quad (f = gT, \ A = T^{-1}(B), \ \mu = \lambda T^{-1}) \tag{5}$$

for all $B \in \mathcal{B}$. Equality (3) can be obtained from equality (5) with $\lambda = \mu T$ using the local boundedness of the measures λ and μ.

Equality (5) makes it possible to reduce the calculation of integrals over curves and surfaces to the calculation of integrals over domains in finite-dimensional spaces whenever T defines a parametrization of a curve or a surface.

Remark. So far we mostly used countably additive, locally bounded, and complete measures. However, some of the above statements also hold in more general cases.

In Maslov (1987), by analogy with the classical theory of the integral, the theory of the integral for measures with values in ordered metric semirings of idempotents is presented. Measure extension theorems are proved. Simple functions, integral sums, measure functions and their integrals are defined. As in the classical case, the integral of a measurable function is equal to the limit of the sequence of integral sums for the approximating simple functions. However, the corresponding definitions and arguments are more complicated. Analogs of some limit theorems are proved. The theorem on integral representation of functionals is proved and used to define an analog of the Fourier transform. Integration of functions with values in a given semiring is described.

In connection with the problems of pattern recognition and integral geometry, Chapter 9 of Matheron (1975) deals with the theory of integration with respect to measures whose values are convex compact subsets of $\mathbb{R}^m$.

Chapter 10 of Savel'ev (1989) deals with applications of integral calculus in probability theory. It provides a detailed description of the properties of conditional means defined as projections of a special kind using the Radon–Nikodym theorem.

Measures on special algebras of sets of topological vector spaces and integrals with respect to such measures are described in detail in Daletskii and Fomin (1983). There are chapters devoted to evolution equations, continual and stochastic integrals. The conditions for the solution of the Cauchy problem to be well-posed are formulated (Chapter 5, Section 1).

Chapters 6–9 in Maslov (1976) deal with continual integral equations, complex Markov chains, and complex measures in the Feynman integral. Chapter 7 describes a probabilistic model of approximating the trajectories of the Feynman integral by broken lines.

The theory of continual integrals with respect to scalar measures on special algebras of sets in vector spaces is presented in detail in Smolyanov and Shavgulidze (1990). Attention is focused on Feynman integrals and the solution of Schrödinger equations.

3.4. ANALYSIS ON MANIFOLDS

The notion of a manifold generalizes the notions of a curve and a surface. It is the main notion in calculus. Manifolds are extensively studied in the literature. Geometry of surfaces and manifolds is described in detail in Dubrovin, Novikov, and Fomenko (1986). The book provides a brief survey of definitions and theorems related to differential and integral calculus for manifolds.

3.4.1. Manifolds

This subsection deals with topological and smooth manifolds.

1. Consider topological spaces M and N, points $p \in M$ and $q \in N$, and their (open and connected) neighborhoods $U \subseteq M$ and $V \subseteq N$, respectively. If every point $p \in M$ has a neighborhood $U \subseteq M$ homeomorphic to a neighborhood $V \subseteq N$ of a point $q \in N$, then the space M is said to be *locally homeomorphic* to the space N.

Examples. 1) The circle $C(0,1) \subseteq \mathbb{R}^2$ is locally homeomorphic to the interval $[-1,1] \subseteq \mathbb{R}$. 2) The open disk $B(0,1) \subseteq \mathbb{R}^2$ is locally homeomorphic to the plane $\mathbb{R}^2$. 3) The closed disk $\bar{B}(0,1) \subseteq \mathbb{R}^2$ is locally homeomorphic to the lower half-plane $\mathbb{R}^2_- = \{x = (x_1, x_2) : x_2 \leq 0\} \subseteq \mathbb{R}^2$.

Exercise. In each of the above examples, verify that the specified locally homeomorphic spaces with induced standard topologies are homeomorphic.

Local homeomorphisms from the neighborhoods of points $p \in M$ are called *local charts* of M or simply *charts* in N. Collections of charts whose domains of definition cover M are called *atlases* of M in N. The domains of definition of charts are called *coordinate neighborhoods*. The inverse charts are called *parameterizations*.

A topological space M that is locally homeomorphic to the standard space $N = \mathbb{R}^n$, is called an *n-dimensional topological manifold*. It is often assumed in addition that M is separated and there exists a countable base for M.

Examples. 1) The circle $C(0,1)$ is a one-dimensional topological manifold. 2) The open disk $B(0,1)$ is a two-dimensional topological manifold.

A topological space M that is locally homeomorphic to the subspace $N = \bar{\mathbb{R}}^n_- = \{x = (x_1, \ldots, x_n) : x_n \leq 0\}$ of the standard space $\mathbb{R}^n$ is called an *n-dimensional manifold with boundary*. Topological manifolds with boundary, as well as those without boundary (with empty boundary) are called *manifolds*. The type of a manifold is usually clear from the context. The points such that the images of their local charts are included in $\partial \bar{\mathbb{R}}^n_- = \{x = (x_1, \ldots, x_n) : x_n = 0\}$ form the *boundary of the manifold*. The points such that the images of their local charts are included in $\mathbb{R}^n_- = \{x = (x_1, \ldots, x_n) : x_n < 0\}$ form the *interior of the manifold*. The boundary is an $(n-1)$-dimensional manifold, while the interior is an n-dimensional manifold.

Remark. The definition of the dimension is based on the following facts (Postnikov, 1987, Lecture 8). (1) If $m \neq n$, then the open subset $U \neq \varnothing$ of $\mathbb{R}^m$ is not homeomorphic to any open subset $V \neq \varnothing$ of $\mathbb{R}^n$. (2) If U and V are open subsets of $\mathbb{R}^n$ and $h : U \to V$ is a surjective homeomorphism, then $h(U \cap \mathbb{R}^{n-1}) = V \cap \mathbb{R}^{n-1}$. (3) Homeomorphic manifolds have the same dimension.

Example. The closed disk $\bar{B}(0,1)$ is a two-dimensional manifold with boundary $C(0,1)$ and interior $B(0,1)$.

Exercise. Prove that the interior and the boundary of a manifold have no common points.

Remark. Taking the subspaces $\{x = (x_1, \ldots, x_n) : x_{n-m+1} \leq 0, \ldots, x_n \leq 0\}$ $(1 \leq m \leq n)$ instead of $\bar{\mathbb{R}}^n$, we can define manifolds with corner points of different types (Schwartz, 1968). In the case $m = 1$, we obtain manifolds with boundary.

2. Consider an n-dimensional manifold X with atlas $\mathcal{A}$ and arbitrary charts $\alpha : U \to E$, $\beta : V \to F$ in $\mathcal{A}$ $(U, V \subseteq X;\ E, F \subseteq \mathbb{R}^n;\ \alpha(U) = E,$ $\beta(V) = F)$. The composition $\beta\alpha^{-1} : E \to F$, which is a function from $\mathbb{R}^n$ to $\mathbb{R}^n$, is called the *transition* from α to β. If the function $\beta\alpha^{-1}$ is smooth (k times continuously differentiable), then the charts α and β are said to be *smoothly consistent*. The function $\beta\alpha^{-1}$ can also be infinitely differentiable or analytic. If the domains of definition of α and β do not intersect $(U \cap V = \varnothing)$, then these charts are also considered to be smoothly consistent (since the empty map is considered to be smooth). An atlas in which every two charts are smoothly consistent is called a *smooth atlas*. An n-dimensional manifold with smooth atlas is called a *smooth n-dimensional manifold*. Furthermore, there are *infinitely smooth* and *analytic* atlases and manifolds.

Examples. All manifolds in the examples of part 1 are smooth.

Exercise. Prove the preceding statement.

A smooth atlas can be *completed* by adding all charts that are smoothly consistent with every chart of the atlas. (By transitivity, all the added charts will be consistent with each other.)

Atlases obtained using the above procedure are said to be *complete*. If any chart of an atlas is smoothly consistent with any every chart of another atlas, then these atlases are said to be *equivalent*. Every atlas is included in a complete atlas that is equivalent to it.

3. Consider an n-dimensional smooth manifold X with atlas $\mathcal{A}$, a p-dimensional smooth manifold Y with atlas $\mathcal{B}$, and a continuous map $f :$ $X \to Y$. For $\alpha \in \mathcal{A}$ and $\beta \in \mathcal{B}$, the composition $f_{\alpha\beta} = \beta f \alpha^{-1}$ acting from $\mathbb{R}^n$ to $\mathbb{R}^n$ is called the *picture* of f on α and β. A map f is said to be *smooth* if all its pictures $f_{\alpha\beta}$ are smooth. A one-to-one smooth map from X onto Y whose inverse is also smooth is called an *isomorphism between the manifolds* X *and* Y. If such an isomorphism exists for two smooth manifolds, then they are said to be *isomorphic*.

Exercise. Prove that isomorphic smooth manifolds have the same dimension.

A manifold $(X, \mathcal{A})$ is called a *submanifold* of a manifold $(Y, \mathcal{B})$ if $X \subseteq Y$ and every chart $\alpha \in \mathcal{A}$ is a restriction of a chart $\beta \in \mathcal{B}$.

Example. The boundary and the interior of an n-dimensional manifold are its submanifolds of dimensions $n - 1$ and n, respectively.

The images of submanifolds under isomorphisms are sometimes identified with these submanifolds and considered embedded in the corresponding manifolds.

Remark. The abstract definitions given above provide a better understanding of the essence of these notions. Calculus usually deals with manifolds in the space $\mathbb{R}^n$. The Whitney embedding theorem (Postnikov, 1987, Lecture 14) makes it possible to identify any separated smooth manifold with countable base with a manifold in the space $\mathbb{R}^n$.

4. Let X be a smooth manifold with atlas $\mathcal{A}$. Charts $\alpha, \beta \in \mathcal{A}$ such that

$$\det(D(\beta\alpha^{-1})u) > 0 \quad (u \in \mathrm{Dom}(\beta\alpha^{-1})).$$

are be called *positively consistent*. In particular, α and β are positively consistent if $\mathrm{Dom}(\beta\alpha^{-1}) = \varnothing$. An atlas such that any two of its charts are positively consistent is said to be *orientable*. A manifold that has an orientable atlas is said to be *orientable*. The orientation of a smooth manifold X with atlas $\mathcal{A}$ is defined by a function $s : \mathcal{A} \to \{-1, 1\}$ such that $s(\alpha) \cdot s(\beta) = 1$ for $\mathrm{Dom}(\beta\alpha^{-1}) \neq \varnothing$. Every chart $\alpha \in \mathcal{A}$ is assumed to have a sign $s(\alpha) \in \{-1, 1\}$ such that any two charts with intersecting domains of definition have the same sign. For the definition of the orientation, it is convenient to consider the charts defined on open connected sets.

Exercise. Prove that there exist exactly two orientations for any smooth connected orientable manifold (Sulanke and Wintgen, 1972, Section 4).

Obviously, the orientations for different connected components of an orientable manifolds can be chosen independently.

Orientable atlases $\mathcal{A}_1$ and $\mathcal{A}_2$ for a smooth manifold X are said to be *orientable in the same way* if their union $\mathcal{A} = \mathcal{A}_1 \cup \mathcal{A}_2$ is an orientable atlas for X. In this case, the orientations $s_1 : \mathcal{A}_1 \to \{-1, 1\}$, $s_2 : \mathcal{A}_2 \to \{-1, 1\}$ can be defined as restrictions of the orientation $s : \mathcal{A} \to \{-1, 1\}$.

Example 1. Consider the circle $S = \{(x, y) : x^2 + y^2 = 1\}$ in the plane $\mathbb{R}^2$. The circle S is covered by the segments A_-, A_+, B_-, and B_+ defined by the conditions $x < 0$, $x > 0$, $y < 0$, and $y > 0$, respectively. The atlas $\mathcal{A}$ of S consists of the charts α_-, α_+, β_-, and β_+ defined on the above segments and having the values $\alpha_-(x, y) = -y$, $\alpha_+(x, y) = y$, $\beta_-(x, y) = -x$, $\beta_+(x, y) = x$ in $]-1, 1[$. To verify that the atlas $\mathcal{A}$ is orientable, it suffices to verify that the derivatives of the compositions $\beta_-\alpha_-^{-1}$, $\beta_-\alpha_+^{-1}$, $\beta_+\alpha_-^{-1}$, and $\beta_+\alpha_+^{-1}$ are positive on their domains of definition $]0, 1[$, $]-1, 0[$, $]-1, 0[$, and $]0, 1[$. We have $\beta_-\alpha_-^{-1}(u) = \beta_+\alpha_-^{-1}(u) = -\sqrt{1-u^2}$ $(0 < u < 1)$, $\beta_-\alpha_+^{-1}(u) = \beta_+\alpha_-^{-1}(u) = \sqrt{1-u^2}$ $(-1 < u < 0)$. Then $2|u|(1-u^2)^{-1/2} > 0$. There are two orientations for the circle S: one is defined by the equalities $s(\alpha) = +1$ $(\alpha \in \mathcal{A})$, and the other by the equalities $s(\alpha) = -1$ $(\alpha \in \mathcal{A})$.

Exercise. Determine the orientations of the sphere $S^n = \{x : \|x\|^2 = 1\}$ in the Euclidean space $\mathbb{R}^{n+1}$ (see Postnikov, 1987, Lecture 6; Cartan, 1970, Chapter 3, Section 4.12; Dubrovin, Novikov and Fomenko, 1986, Part 2, Section 2).

Example 2. A classical example of a nonorientable manifold is the *Möbius strip*, which is defined as a factor-space obtained from the strip $\mathbb{R} \times [0, 1]$ in $\mathbb{R}^2$ by identifying points (x, y) with $(x+1, 1-y)$. The Möbius strip is described in detail in the following books: Godbillon (1969), Chapter 2, Section 1.6; Chapter 3, Sections 1.5 and 1.9; Dubrovin, Novikov and Fomenko (1986), Part 2, Section 16; Schwartz (1967), Chapter 6, Section 5.

Remark. In Sulanke and Wintgen (1972), Chapter 1, Section 4, the connection of the orientations of a manifold with graphs is described, where the graph vertices represent the charts and the graph edges represent the intersections of the domains of definition of the corresponding charts. The value -1 or $+1$ is assigned to each edge. Paths, cycles, and their values are defined. The orientability of a manifold is equivalent to the absence of cycles with value -1 in the corresponding graph. The graph for the sphere S^n has a particularly simple form.

3.4.2.   The rank theorem

The rank theorem is a generalization of the theorem on reducing the matrix of a linear map to the standard diagonal form.

1. Consider an n-dimensional manifold X with atlas $\mathcal{A}$, a p-dimensional manifold Y with atlas $\mathcal{B}$, a smooth map $f : X \to Y$, a point $x \in X$ and its image $y = f(x) \in Y$, charts $\alpha \in \mathcal{A}$ and $\beta \in \mathcal{B}$ of the neighborhoods of x and y, the chart $f_{\alpha\beta} = \beta f \alpha^{-1}$ of the map f on these charts and its differential $df_{\alpha\beta}(\alpha(x))$ at $\alpha(x) \in \mathbb{R}^n$. It follows from the definition and the differentiation rules for compositions that the rank of the linear map $df_{\alpha\beta}(\alpha(x))$ does not depend on the choice of the charts α and β. This rank is called the *rank of the map* f at the point x. This definition is in agreement with the idea of a smooth map being *approximately locally linear*. If the map f has the same rank at all points x, then this rank is called the *rank of the map* f.

Example. $X = B(0,1) \setminus \{(0,0)\} \subseteq \mathbb{R}^2$, $Y = \mathbb{R}$, $\mathcal{A} = \{\mathrm{id} : X \to \mathbb{R}^2\}$, $\mathcal{B} = \{\mathrm{id} : \mathbb{R} \to \mathbb{R}\}$, $f(x) = (1 - \|x\|^2)^{1/2}$, $f'(x) = x/f(x)$ $(x = (x_1, x_2) \in X)$. The rank of f is equal to 1. The function f has maximum rank at all points.

2. We denote by P_q the projection from $\mathbb{R}^n$ to $\mathbb{R}^p$ such that

$$P_q(x) = (x_1, \ldots, x_q, 0, \ldots, 0) \in \mathbb{R}^p \quad (x = (x_1, \ldots, x_n) \in \mathbb{R}^n),$$

assuming that $1 \leq q \leq n \wedge p$. In addition, consider the zero projection P_0 such that $P_0(x) = 0 \in \mathbb{R}^p$ $(x \in \mathbb{R}^n)$ and set $P = P_q$.

The spaces $X = \mathbb{R}^n$ and $Y = \mathbb{R}^p$ can be viewed as smooth manifolds with atlases $\mathcal{A} = \{\mathrm{id} : \mathbb{R}^n \to \mathbb{R}^n\}$, $\mathcal{B} = \{\mathrm{id} : \mathbb{R}^p \to \mathbb{R}^p\}$. Open sets $U \subseteq X$ and $V \subseteq Y$ with induced topologies and atlases are submanifolds of the manifolds X and Y. Continuously differentiable maps $f : U \to V$ are smooth maps. The rank of f at a point u is equal to the rank of the linear map $df u : \mathbb{R}^n \to \mathbb{R}^p$ or the rank of its matrix $f'(u)$ in standard bases. The map $f : U \to V$ is said to be *locally equivalent* to the projection P if for any point $u \in U$ there exist an open neighborhood $A \subseteq U$ of u, a neighborhood $B \subseteq V$ of the set $f(A)$, and diffeomorphisms $\alpha : A \to X$ and $\beta : B \to Y$ such that $\beta f \alpha^{-1}(x) = P(x)$ for all $x \in \alpha(A)$. The conditions $A \subseteq U$ and $f(A) \subseteq B$ define the composition $\beta f \alpha^{-1}$ on the whole set $\alpha(A) \subseteq X$.

This definition of equivalence makes it possible to formulate the rank theorem for smooth maps the same way as for linear maps.

Theorem. *A smooth map f of rank q is locally equivalent to the projection P_q.*

The proof can be found in Cartan (1970), Part 2, Chapter 3, Section 4.

Exercise. Calculate the rank of a smooth map f that is locally equiva-lent to the projection P_q using the differentiation rules.

The same theorem holds for smooth maps $g : X \to Y$ from the smooth n-dimensional manifold X to the smooth p-dimensional manifold Y. In this case, the definition of the local equivalence of g and P must involve homeomorphisms instead of diffeomorphisms. The general case is reduced to the special case of $X = \mathbb{R}^n$ and $Y = \mathbb{R}^p$ considered above with the use of of g on the charts $k : X \to U$ and $l : Y \to V$ for open $U \subseteq \mathbb{R}^n$ and $V \subseteq \mathbb{R}^p$. By the preceding theorem, f is locally equivalent to P_q if the rank of f is equal to q ($f \sim P_q$ if rank $f = q$). The rank of g is equal to the rank of f by definition. The notion is well defined because it does not depend on the choice of the charts k and l.

Exercise. Using the chain rule, prove that the definition of local equiv-alence does not depend on the choice of the charts k and l.

The local equivalence $g \sim P_q$ means that $(\beta l)g(\alpha k)^{-1}(x) = P_q(x)$ for the corresponding homeomorphisms k and l and diffeomorphisms α and β.

Remark. Changes of variables that preserve the rank make it possible to simplify the systems of equations describing submanifolds of a smooth man-ifold (Godbillon, 1969, Chapter 3, Section 5; Postnikov, 1987, Lecture 13).

3.4.3. Sard's theorem

Sard's theorem has quite extraordinary applications.

1. Let X be a smooth n-dimensional manifold, Y be a smooth p-dimensional manifold, and g be a smooth function from X to Y. The points $x \in X$ at which the function g has maximal rank $n \wedge p$ are said to be *regular*, while the other points are said to be *singular* for g. The same terms are used for the values of g at these points. Let S be the set of all singular points for g and set $T = g(S)$. The set $T \subseteq Y$ *has measure zero* if T can be covered by the domains of definition of a finite collection of charts $l : V \to \mathbb{R}^p$ with images $l(T \cap V)$, of Lebesgue measure zero in $\mathbb{R}^p$.

Example. Let $X = \mathbb{R}^2$, $Y = \mathbb{R}$ and $g(x) = x_1^2 + x_2^2$ ($x = (x_1, x_2)) \in \mathbb{R}^2$). Then $S = \{(0,0)\}$ and $T = \{0\}$ has measure zero.

Sard's theorem. *The set of singular values of a smooth map has measure zero.*

A detailed proof of Sard's theorem is given in Sternberg (1983), Chapter 2, Section 3. A relatively simple case of $n = p$ is considered.

2. Sard's theorem holds, in particular, for smooth maps $f : U \to \mathbb{R}^p$ from open sets $U \subseteq \mathbb{R}^n$. Any set of measure zero in the space $\mathbb{R}^p$ has no nonempty open subsets and its complement is everywhere dense in $\mathbb{R}^p$. The use of charts makes it possible to extend this statement to the case of manifolds.

Corollary. *The set of regular values of a smooth map is everywhere dense.*

In Schwartz (1968), Chapter 1, the Brouwer fixed point theorem is deduced from Sard's theorem. In Postnikov (1987), Lecture 15, the proof of Sard's theorem is associated with the proof of the Whitney theorem on the embedding of a smooth n-dimensional manifold into the space $\mathbb{R}^{2n+1}$.

3. Consider an open set $U \subseteq \mathbb{R}^n$, a smooth injective map $\varphi : U \to \mathbb{R}^n$ and a Riemann integrable function $f : \varphi(U) \to \mathbb{R}^n$ (Spivak, 1965, Chapter 3). In the classical statement of the theorem on the change of variables in the integral, for the equality

$$\int_{\varphi(U)} f = \int_U (f \circ \varphi)|\det \varphi'|$$

to hold, it is necessary that $\det \varphi'(x) \neq 0$ for all $x \in U$. Sard's theorem makes it possible to avoid this requirement. Indeed, the equality $\det \varphi'(x)=0$ means that x is a singular point for the map φ. Let $U = R \cup S$, where R and S are the sets of regular and singular points of φ, respectively. Then $\varphi(U) = \varphi(R) \cup \varphi(S)$, and Sard's theorem implies that $\varphi(S)$ has measure zero. Consequently, the integral of f over the set $\varphi(U)$ is equal to the integral of f over $\varphi(R)$, and, by definition, $\det \varphi'(x) \neq 0$ for $x \in R$. Clearly, the integral over U on the right-hand side of the change of variables formula is equal to the integral over R. Thus, if this formula holds for $U = R$ under the condition that $\det \varphi'(x) \neq 0$, then it also holds without this condition.

3.4.4. Differential forms

The differential forms make up an exterior algebra and are described in detail in the second part of Cartan (1970), Part 2, and in Schwartz (1967), Chapter 6.

1. We will use the definitions and notation of Subsection 2.3.2 and Cartan (1970). Let U be an open set in a Banach space E and let F be a Banach space. (The spaces are assumed to be real, although most of the statements below is also true for complex spaces.) For E and F we define the Banach space $\mathcal{A}_p = \mathcal{A}_p(E, F)$ of antisymmetric continuous multilinear maps $E^p \to F$ that is a closed subspace of the Banach space of all continuous multilinear maps $E^p \to F$. The norm of such maps f is defined as follows:

$$\|f\| = \sup\{\|f(x_1, \ldots, x_p)\| : \|x_1\| \le 1, \ldots, \|x_p\| \le 1\}$$

(see Cartan (1970), Chapter 1, Section 1). A map $\omega : U \to \mathcal{A}_p$ is called a *differential form of degree p* or *p-form* for short. In particular, functions from U to F are 0-forms, while maps from U to $\mathcal{L}(E, F)$ are 1-forms: by definition, $\mathcal{A}_0 = F$ and $\mathcal{A}_1 = \mathcal{L}(E, F)$. A p-form is said to be *nondegenerate* if $p > 0$.

We denote by $\Omega_p^n = \Omega_p^n(U, F)$ the set of all n times continuously differentiable p-forms. They will be called *n-smooth* or *smooth* for short. Smoothness zero means continuity. It is obvious that n-smooth p-forms make up a vector space.

Example. Let $f : U \to F$ be an n-smooth 0-form. Then its differential $df : U \to \mathcal{L}(E, F)$ is an $(n-1)$-smooth 1-form.

2. Since the values of differential forms are antisymmetric multilinear forms, the operations with the latter can be extended to the former. In particular, the exterior product of differential forms can be defined.

Consider Banach spaces E, F, G, and H, the product (a continuous bilinear map) $F \times G \to H$, and differential forms $\xi \in \Omega_p^n(U, F)$ and $\eta \in \Omega_q^n(U, G)$ on an open set $U \subseteq E$. The exterior product of the forms ξ and η is defined to be the form $\xi \wedge \eta \in \Omega_{p+q}^n(U, H)$ such that

$$(\xi \wedge \eta)(u) = \xi(u) \wedge \eta(u) \in \mathcal{A}_{p+q}(U, H),$$

where $u \in U$, $\xi(u) \in F$, $\eta(u) \in G$, and $\xi(u) \wedge \eta(u) \in \mathcal{A}_{pq}(\xi(u) \cdot \eta(u))$ for $\xi(u) \cdot \eta(u) \in H$. The *alternating* of the product $f \cdot g$ of maps $f \in \mathcal{A}_p(E, F)$ and $g \in \mathcal{A}_p(E, G)$ is defined by the equality

$$A_{pq}(f \cdot g)(x_1, \ldots, x_{p+q})$$
$$= \sum_\sigma \operatorname{sign} \sigma \cdot f(x_{\sigma(1)}, \ldots, x_{\sigma(p)}) \cdot g(x_{\sigma(p+1)}, \ldots, x_{\sigma(p+q)}),$$

where the summation is performed with respect to all permutations σ of the indices $1, \ldots, p+q$ such that $\sigma(1) < \cdots < \sigma(p)$ and $\sigma(p+1) < \cdots < \sigma(p+q)$. It can be proved that the *exterior product* is

$$f \wedge g = A_{pq}(f \cdot g) \in \mathcal{A}_{p+q}(E, H)$$

(see Cartan, 1970, Chapter 3, Section 1).

Examples. 1) Let $p = q = 1$. Then $(f \wedge g)(x_1, x_2) = f(x_1) \cdot g(x_2) - f(x_2) \cdot g(x_1)$.

2) Let $p = 0$, $\xi = f$ and $\eta = \omega$. Then $\xi \wedge \eta = f \wedge \omega = f \cdot \omega$: the exterior multiplication by a 0-form coincides with the conventional multiplication.

The exterior product is associative and anticommutative (Cartan, 1970, Chapter 3, Section 1):

$$g \wedge f = (-1)^{pq} f \wedge g.$$

For $f = \xi(u)$ and $g = \eta(u)$, we get the corresponding equalities for the differential forms ξ and η.

3. We now define the operation of exterior differentiation for smooth forms. Consider an n-smooth p-form $\omega \in \Omega_p^n(U, F)$ $(n \geq 1, p \geq 0)$. Its conventional differential at a point u is a linear map $D\omega u \in \mathcal{L}(E, \mathcal{A}_p(E, F))$. Its *exterior differential* $d\omega \in \Omega_{p+1}^{n-1}(U, F)$ is defined by the equality

$$d\omega u(x_0, x_1, \ldots, x_p)$$
$$= \sum_{0 \leq i \leq p} (-1)^i D\omega u(x_i)(x_0, x_1, \ldots, x_{i-1}, x_{i+1}, \ldots, x_p)$$

and is an $(n-1)$-smooth $(p+1)$-form (Cartan, 1970, Chapter 3, Section 2), Here $x_i \in E$ and $d\omega u \in \mathcal{A}_{p+1}(E, F)$. With the appropriate multiplicative representation, the exterior differential is obtained by alternating the conventional one, i.e., $d\omega u = A_{1p}(D\omega u)$.

Examples. If $p = 0$ and $\omega : U \to F$, then $d\omega = D\omega$. If $p = 1$ and $\omega : U \to F$, then $d\omega u(x_0, x_1) = D\omega u(x_0) \cdot x_1 - D\omega u(x_1) \cdot x_0$ in multiplicative notation.

Remark. Let $\omega \in \Omega_1^n(U, F)$, $n \geq 1$. The second example demonstrates that $d\omega = 0$ if and only if the bilinear map $b(x, y) = D\omega u(x) \cdot y$ is symmetric for all $u \in U$.

The rules of exterior differentiation are specified by the equalities

$$d(f \cdot \omega) = (df) \wedge \omega + f \cdot d\omega, \quad d(\xi \wedge \eta) = (d\xi) \wedge \eta + (-1)^p \xi \wedge d\eta.$$

(Cartan, 1970, Chapter 3, Section 2), In the first equality, $f \in \Omega_0^1(U, \mathbb{R})$ and $\omega \in \Omega_p^1(U, F)$, or $f \in \Omega_0^1(U, F)$ and $\omega \in \Omega_p^1(U, \mathbb{R})$. In the second equality, $\xi \in \Omega_p^n(U, \mathbb{R})$ and $\eta \in \Omega_q^n(U, F)$ (with $n \geq 1$).

3.4.5. The Poincare theorem

The Poincare theorem makes it possible to describe solutions to the equation $d\xi = \omega$ for smooth forms.

1. The second differential of a linear function is equal to zero Similarly, by multilinearity, the second exterior differential of a smooth form is equal to zero: $d^2\omega = d(d\omega) = 0$ for $\omega \in \Omega_p^n(U, F)$, $n \geq 0$. This follows from the symmetry of the second differential $d^2\omega$ and the equality $d^2\omega u(v_1, v_2)(x_1, \ldots, x_p) = D^2\omega u(v_1, v_2)(x_1, \ldots, x_p) - D^2\omega u(v_2, v_1)(x_1, \ldots, x_p)$ for all $u \in U$ and $v_1, v_2, x_i \in E$.

Exercise. Prove the preceding equality (see Cartan, 1970, Chapter 3, Section 2).

A smooth form whose exterior differential is equal to zero is said to be *closed*. The equality $d(d\omega) = 0$ means that $d\omega$ is a closed form. External differentials are called *exact* forms. Thus, every exact form is closed. The converse is not necessarily true unless we make additional assumptions on the domain of definition of the forms in question. A classical counterexample is the form $\omega = -x_2(x_1^2 + x_2^2)^{-2}dx_1 + x_1(x_1^2 + x_2^2)^{-2}dx_2$ on the domain $U = \mathbb{R}^2 \setminus \{(0,0)\}$.

Exercise. Prove that this form is closed but not exact. Find its exact restrictions. (See Spivak, 1965, Section 4.10.)

2. An open set $U \subseteq E$ is called a *star domain* if there exists a point $a \in U$ connected with every $u \in U$ by the segment $[a, u] = \{x = (1-t)a + tu : 0 \leq t \leq 1\} \subseteq E$. It is obvious that every star domain is connected and every open convex sets is a star domain. We also say that the domain U can be *shrunk to the point* a or is *homotopic to* a. An example is a five-point star with no boundary in the plane $\mathbb{R}^2$.

The Poincare theorem. *Every nondegenerate closed form on a star domain is exact.*

The proof of this theorem is reduced to solving the equation $d\xi = \eta$ with closed right-hand side $\eta \in \Omega_p^n(U, F)$. The solution $\xi \in \Omega_{p-1}^{n+1}(U, F)$ for the star domain $U \subseteq E$ that can be shrunk into the point $a = 0$ is written as

$$\eta(u)(x_1, \ldots, x_{p-1}) = \int_0^1 t^{p-1}\eta(tu)(x_1, \ldots, x_p)\, dt,$$

$u \in U$, $x_i \in E$, $1 \le i \le p$ (Cartan, 1970, Chapter 3, Section 2).

Consider a star domain $U \in \mathbb{R}^m$ and smooth forms $\omega \in \Omega_p^n(U, F)$, $\xi \in \Omega_{p-1}^{n+1}(U, F)$, $\eta \in \Omega_{p-1}^{n+1}(U, F)$, and $\zeta \in \Omega_{p-2}^{n+2}(U, F)$ $(p \ge 2)$.

Corollary 1. *If $d\xi = \omega$ and $d\eta = \omega$, then $\xi - \omega$ is an exact form.*

Proof. Since $d(\xi - \omega) = d\xi - d\eta = 0$, the form $\xi - \omega$ is closed. By the Poincare theorem, it is exact. $\square$

Corollary 2. *Let $d\xi = \omega$. Then $d\eta = \omega$ if and only if there is a ζ such that $\eta = \xi + d\zeta$.*

Proof. If $\eta = \xi + d\zeta$, then $d\eta = d\xi + d(d\zeta) = d\xi$. Conversely, if $d\xi = d\eta$, then $\eta = \xi + (\eta - \xi)$ and there is a ζ such that $d\zeta = \eta - \xi$ (as follows from Corollary 1). $\square$

Remark. Thus, solutions to the equation $d\xi = \omega$ differ from each other only by exact forms. The general solution is represented as a sum of any special solution and an arbitrary exact form.

In Schwartz (1967), Chapter 6, an integral condition for the solvability of $d\xi = \omega$ for continuous forms ω is given and the connection between the Poincare theorem with the de Rham theorem on the integrals of a form with respect to smooth cycles is demonstrated.

3. Consider a finite-dimensional space $E = \mathbb{R}^m$, its conjugate space E^* of linear functionals, and the standard bases (e_i), (e_i^*) of these spaces. The exterior products $\wedge e_\mu^* = \wedge_i e_{\mu(i)}^*$ for $\mu = (\mu(i))$, $1 \le i \le p \le m$, $1 \le \mu(1) < \cdots < \mu(p) \le m$ make up the basis of $\mathcal{A}_p(E, F)$. Therefore, any $f \in \mathcal{A}_p(E, F)$ is represented as $f = \sum c_\mu(f) \wedge e_\mu^*$, where $c_\mu(f) \in F$ (see Subsection 2.3.2). Every coordinate e_i^* will be identified with the variable x_i. Since $x_i : \mathbb{R}^m \to \mathbb{R}$ is linear and continuous, it is equal to its differential $dx_i u$ at any point $u \in U$. Therefore, $(\wedge dx_\mu)u = (\wedge_i dx_{\mu(i)})u = \wedge_i x_{\mu(i)}^* = \wedge_i x_\mu^*$, the forms $\wedge dx_\mu = \wedge_i dx_{\mu(i)}$ are basic forms, and every form $\omega \in \Omega_p^n(U, F)$ is represented as

$$\omega = \sum f_\mu \wedge dx_\mu,$$

where the components are n-smooth functions $f_\mu = f_{\mu(1)...\mu(p)} : U \to F$. In particular, if $p = 1$ and $\omega = df$ for a smooth function $f : U \to F$ on an open set $U \subseteq \mathbb{R}^m$, then

$$df = \sum f_i' \cdot dx_i \quad (1 \le i \le m).$$

Example. Let $E = \mathbb{R}^3$ and $F = \mathbb{R}$. Consider smooth functions $a, b, c : U \to \mathbb{R}$ on an open set $U \subseteq \mathbb{R}^3$, basic forms dx, dy, dz, and the form $\xi = adx + bdy + cdz$. Its exterior differential is

$$d\xi = (b_x' - a_y')dx \wedge dy + (c_y' - b_z')dy \wedge dz + (a_z' - c_x')dz \wedge dx.$$

In addition, consider smooth functions $f, g, h : U \to \mathbb{R}$ and the differential form $\eta = fdy \wedge dz + gdz \wedge dx + hdx \wedge dy$. Its exterior differential is

$$d\eta = (f_x' + g_y' + h_z')dx \wedge dy \wedge dz.$$

The equality $d\eta = 0$ is equivalent to $f_x' + g_y' + h_z' = 0$. If U is a star domain, then, by the Poincare theorem, the equality

$$\operatorname{div}(f, g, h) = f_x' + g_y' + h_z' = 0$$

implies the vector equality

$$\operatorname{rot}(a, b, c) = (c_y' - b_z', a_z' - c_x', b_x' - a_y') = (f, g, h)$$

for some a, b, and c. The condition $d\eta = 0$ ensures the existence of a solution ξ to the equation $d\xi = \eta$.

Remark. In the finite-dimensional case considered here, the exterior differential d is the only additive map from the exterior algebra of n-smooth forms into the algebra of $(n-1)$-smooth forms for which the differentiation rules for the product and the rules of repeated differentiation hold and coincide with the rules of conventional differentiation D on 0-forms (Sternberg, 1983, Chapter 3, Section 1).

3.4.6. Change of variables

Being locally approximately linear, a smooth change of variables defines a
transition to a new system of local coordinates.

1. Consider Banach spaces $X = E$, F, and Y, open sets $U \subseteq X$ and
$V \subseteq Y$, and a smooth map $\varphi : V \to U$ that defines the change of the variable
$v \in V$ to $\varphi(v) = u \in U$. By definition, $D\varphi v(y) \in \mathcal{L}(Y, X)$, $D\varphi v \in X$ for all
$v \in V$, $y \in Y$. We define the *adjoint map* $\varphi^* : \Omega_p^n(U, F) \to \Omega_p^n(V, F)$ for φ
by the equality

$$\varphi^* \omega(v) \cdot (y_1, \ldots, y_p) = \omega(\varphi(v)) \cdot (D\varphi v(y_1), \ldots, D\varphi v(y_p)),$$

where $v \in V$, $y_i \in Y$ $(1 \leq i \leq p)$, $\omega \in \Omega_p^n(U, F)$, $\varphi(v) = u \in U$, $\omega(u) \in$
$A_p(E, F)$, $D\varphi v(y_i) = x_i \in X$, $\varphi^* \omega(v) \in A_p(Y, F)$, $\varphi^* \omega \in \Omega_p^n(V, F)$. If
$p = 0$, then, by definition, $\varphi^* \omega = \omega \circ \varphi$ for smooth functions $\omega : U \to F$.
The notion of an adjoint map is well defined and the map φ^* is linear, which
was proved in Cartan (1970), Chapter 3, Section 2.

2. Aside from being linear, the adjoint map φ^* possesses other important
properties: it is multiplicative, commutes with the exterior differential, and
is transitive. These properties are expressed by the following equalities:

$$\varphi^*(\xi \wedge \eta) = \varphi^* \xi \wedge \varphi^* \eta, \quad \varphi^*(d\omega) = d(\varphi^* \omega), \quad (\varphi \circ \psi)^* = \psi^* \circ \varphi^*.$$

The forms ξ, η, $\xi \wedge \eta$, and ω in these equalities are defined in the same way as
in Subsection 3.4.4, while $\psi : W \to V$ is a smooth map from the open set W
in a Banach space Z. For $\omega = f : U \to F$, the equality for the differentials
readily follows from the differentiation rules for the composition. In the
general case, it is assumed that φ is two times differentiable.

Exercise. 1) Verify the properties of the adjoint map given above. 2)
Verify the equality for the differentials in the case where φ is once continuously differentiable.

3. Let $E = \mathbb{R}^k$. Then every form $\omega \in \Omega_p^n(U, F)$ on an open set $U \subseteq \mathbb{R}^k$
is defined by the n-smooth coefficients $f_\mu : U \to F$ of the basic forms $\wedge dx_\mu$
with increasing multi-indices $\mu = \mu(1) \ldots \mu(p)$ of numbers $1, \ldots, k$ and is
represented as $\omega = \sum f_\mu \wedge dx_\mu$. In this case, for the adjoint map φ^* and
$\wedge d\varphi_\mu = \wedge_i d\varphi_{\mu(i)}$ with $\varphi_{\mu(i)} = x_{\mu(i)} \circ \varphi$, the following equalities hold:

$$\varphi^* \omega = \sum (f_\mu \circ \omega) \wedge d\varphi_\mu.$$

If the change of variables $\varphi : V \to U$ is defined on an open set V in the space $Y = \mathbb{R}^l$, then $d\varphi_{\mu(i)} = \sum(\varphi_{\mu(i)})'_j dy_j$, where y_j are the basis coordinates for $\mathbb{R}^l$ $(1 \le j \le l)$. Substituting the sums for $d\varphi_{\mu(i)}$ into the products $\wedge d\varphi_\mu$ and combining like terms, we obtain the linear combination of forms $\wedge d\varphi_\nu = \wedge_i d\varphi_{\nu(i)}$ with increasing multi-indices $\nu = \nu(1)\dots\nu(p)$ of numbers $1,\dots,l$. Multiplying the terms of this linear combination by $f_\mu \circ \varphi$ and adding up with respect to μ, we arrive at the general expression for $\varphi^*\omega$.

Example. Let $k = l = p$. Then a smooth p-form on the domain $U \subseteq \mathbb{R}^p$ is written as $\omega = f\,dx$, where $f : U \to F$ is a smooth function and $dx = dx_1 \wedge \cdots \wedge dx_p$ is a basic form. The smooth change of variables $\varphi : V \to U$ such that $x = \varphi(y) \in U$ for $y \in V$ defines the transition from the new coordinates $y = (y_1,\dots,y_p)$ to the original coordinates $x = (x_1,\dots,x_p)$ in $X = Y = \mathbb{R}^p$. The differential $D\varphi : V \to \mathcal{L}(\mathbb{R}^p, \mathbb{R}^p)$ is identified with the $p \times p$-matrix $\varphi' = (\varphi'_{ij})$ whose component for any i, j is the derivatives of the ith coordinate function with respect to the jth variable. The determinant $J(\varphi) = \det \varphi'$ is called the *Jacobian determinant*. For the adjoint function φ^*, we have
$$\varphi^*(f\,dx) = (f \circ \varphi) \cdot \det \varphi' \cdot dy,$$
where $dy = dy_1 \wedge \dots \wedge dy_p$ is the new basic form.

Exercise. Establish the above equality for φ^*.

3.4.7. Integral over a manifold

Based on the general notion of the integral with respect to a measure, it is possible to consider integrals over manifolds with respect to measures of a special kind.

1. First, consider a simple special case of integration of a differential form over a smooth curve defined on an interval. Consider Banach spaces E and F, an open set $U \subseteq E$, a smooth map $\gamma : T \to U$ from a nonempty open interval $T \supseteq [0, 1]$ to $\mathbb{R}$, a form $\omega \in \Omega_1^0(U, F)$, the form $\gamma^*\omega \in \Omega_1^0(T, F)$ obtained by the change of variables such that $h(t) = \omega(\gamma(t)) \cdot \gamma'(t) \in F$ for $t \in T$. The function $h : T \to F$ is continuous and the integral over $[0, 1]$ is defined for it (see Zaidman, 1999, Chapter 5, Section 3). Based on the rule of the change of variables in an integral, it is natural to consider the integral of h over $[0, 1]$ to be the integral of the form $\omega : U \to \mathcal{L}(E, F)$ over the curve $\gamma : [0, 1] \to U$. By definition,

$$\int_\gamma \omega = \int_{[0,1]} \omega(\gamma(t)) \cdot \gamma'(t)\, dt. \tag{1}$$

If we extend the function h to $\mathbb{R}$ setting $h(t) = 0$ for $t \notin T$, then (1) can be rewritten as

$$\int_\gamma \omega = \int \gamma^* \omega.$$

The right-hand side is the integral with respect to the Lebesgue measure on $\mathbb{R}$. The oriented manifold is represented by the map γ. It defines the positive orientation for the image $\gamma[0, 1] \subseteq F$ and the graph $G(\gamma) = \{(t, \gamma(t)) : 0 \le t \le 1\} \subseteq \mathbb{R} \times F$. The change of variables $\varphi(s) = 1 - s$ ($0 \le s \le 1$) that takes γ to $\gamma \circ \varphi$ makes this orientation negative. The sign of the integral changes accordingly:

$$\int_{\gamma \circ \varphi} \omega = \int_{[0,1]} \omega((\gamma \circ \varphi)(s)) \cdot (\gamma \circ \varphi)'(s)\, ds = - \int_\gamma \omega.$$

Let $f : U \to F$ be a smooth function and $\omega = df$ be its differential. Then

$$\int_\gamma df = f(\gamma(1)) - f(\gamma(0)). \tag{2}$$

Exercise. Prove the equalities for the integrals over the curves γ and $\gamma \circ \varphi$ and for the integral of the form df (see Cartan, 1970, Chapter 3, Subsections 3.3 and 3.4).

Example. Let $E = \mathbb{R}^2$, $F = \mathbb{R}$, and let $\gamma : \mathbb{R} \to \mathbb{R}^2$ and $f : \mathbb{R}^2 \to \mathbb{R}$ have values $\gamma(y) = (r \cos 2\pi t, r \sin 2\pi t)$, $f(u, v) = uv$ ($t, u, v \in \mathbb{R}$, $r > 0$). Let x and y denote the standard coordinate functions and let dx and dy denote the corresponding basic forms for $\mathbb{R}^2$. Consider the forms $df = y\,dx + x\,dy$ and $\omega = (1/2)(-y\,dx + x\,du)$. The restriction of γ to the interval $[0, 1]$ can be identified with the oriented circle $S(0, r)$ in $\mathbb{R}^2$. The graph of γ is a spiral line on a horizontal cylinder in $\mathbb{R}^3 = \mathbb{R} \times \mathbb{R}^2$ whose projection onto a vertical plane is the circle $S(0, r)$. Using the rule for the change of variables, we find $\gamma^* df = \gamma^*(y\,dx) + \gamma^*(x\,dy) = (y \circ \gamma) \cdot \gamma^* dx + (x \circ \gamma) \cdot \gamma^* dy = (y \circ \gamma)d(x \circ \gamma) + (x \circ \gamma)d(y \circ \gamma) = h\,dt$ for $h(t) = 2\pi r^2 (\cos^2 \pi t - \sin^2 \pi t)$ ($0 \le t \le 1$). Similarly, $\gamma^* \omega = \pi r^2 dt$. As a result, $\int_\gamma df = 0$, $\int_\gamma \omega = \pi r^2$.

A smooth curve $\gamma : [0,1] \to U$, such that $\gamma(0) = \gamma(1)$, is called a *smooth cycle* in U. The following theorem provides an integral criterion for a form to be exact.

Theorem. *A form $\omega \in \Omega_1^0(U, F)$ is exact if and only if its integral over any smooth cycle γ in U is equal to zero.*

This theorem and its important corollaries are proved in Cartan (1970), Chapter 3, Section 3. Also, integrals over piecewise-smooth curves are described in detail.

Remark. The integral over a curve can be considered the Stieltjes integral of $\omega \circ \gamma : [0,1] \to \mathcal{L}(E, F)$ with respect to the function $\gamma : [0,1] \to U$. In this case, equality (1) is rewritten as

$$\int \omega(\gamma)\, d\gamma = \int \omega(\gamma)(d\gamma/dt)\, dt. \tag{3}$$

The derivative $d\gamma/dt = \gamma'$ can be considered the density of the measure $d\gamma$ with respect to the Lebesgue measure dt. Equality (3) represents the rule for the change of variables.

2. The definition of integral over a manifold involves some technical difficulties. In essence, however, integration in this case is reduced to applying an appropriate change of variables and integrating with respect to the Lebesgue measure in the space of the corresponding dimension. In what follows, we consider surfaces in standard spaces. (Cartan, 1970, Subsection 4.7).

Let $X = \mathbb{R}^k$, $X^\perp = \mathbb{R}^l$, and $Y = \mathbb{R}^k \times \mathbb{R}^l = \mathbb{R}^m$ be standard Euclidean spaces ($k + l = m$). Consider an oriented smooth k-dimensional manifold S embedded in Y and a collection Φ of local parameterizations $\varphi : U \to V$ that map open sets $U \subseteq X$ onto open sets $V \subseteq S$ (traces of open sets in Y on S). Parameterizations $\varphi \in Phi$ are homeomorphisms and their images cover S. The pair (S, Φ) is called a *parameterized surface*. The transition from charts to parameterizations is caused by the convenience of using equations to specify surfaces. For the sake of simplicity, (S, Φ) is denoted by S and called a *surface*. If $k = m$ and $l = 0$, then S is called a *body*. Certain assumptions are made for S when defining the integral. The set S is assumed to be a Borel set (in particular, a compact set or the union of a sequence of compact sets). A smooth surface $S \subseteq \mathbb{R}^m$ is specified using the equation $g(z) = 0$, where $g : A \to \mathbb{R}^m$ is a smooth function on an open set $A \subseteq \mathbb{R}^m$ whose derivative $g'(z)$ has rank l at every point $z \in S = g^{-1}(0)$. This possibility is provided by the implicit function theorem. There exist smooth local parameterizations for For $S = g^{-1}(0)$. (Spivak, 1965, Section 5.2).

3. We now describe an inductive process of defining a measure on the class $\mathcal{B}(S)$ of Borel sets $B \subseteq S$ for a parameterized surface (S, Φ). For any parameterization $\varphi \in \Phi$, we denote by $\mathcal{BC}(\varphi)$ the algebra of all $B \in \mathcal{B}(S)$ included in a compact set $C \subseteq \operatorname{Ran} \varphi$. Let $\mathcal{BC}(S) = \cup \mathcal{BC}(\varphi)$ $(\varphi \in \Phi)$. Every $B \in \mathcal{B}(S)$ is equal to the union of a sequence of disjoint sets $B_n \in \mathcal{BC}(S)$.

Exercise. Prove the preceding statement.

Consider a collection of *consistent* positive measures $\mu_\varphi : \mathcal{BC}(\varphi) \to \mathbb{R}$ $(\varphi \in \Phi)$: if $B \in \mathcal{BC}(\varphi) \cap \mathcal{BC}(\psi)$ for some parameterizations φ and ψ, then $\mu_\varphi(B) = \mu_\psi(B)$. The equality

$$\mu(B) = \mu_\varphi(B) \quad (B \in \mathcal{BC}(\varphi))$$

defines a measure $\mu : \mathcal{BC}(S) \to \mathbb{R}$, while the equality

$$\bar{\mu}(B) = \sum \mu(B_n) \quad (B = \sum B_n, \ B_n \in \mathcal{BC}(\varphi))$$

defines the measure $\bar{\mu} : \mathcal{B}(S) \to \bar{\mathbb{R}}$ that is an extension of μ.

Exercise. Prove the preceding statement and verify that μ and $\bar{\mu}$ are well defined.

The restriction of the measure $\bar{\mu}$ to the algebra $\mathcal{R} = \mathcal{BL}(S, \bar{\mu}) = \{B \in \mathcal{B}(S) \mid \bar{\mu}(B) < \infty\}$ will be also denoted by μ. The sets in $\mathcal{R}$ are said to be μ-integrable. It is usually assumed that the measures μ_φ are *regular*, i.e., for any $B \in \mathcal{BC}(\varphi)$ the value of $\mu_\varphi(B)$ is equal to the supremum of the set of values $\mu_\varphi(A)$ for all compact sets $A \subseteq B$. (The detailed proof of the regularity of the Lebesgue measure is given in Savel'ev (1989), Chapter 9, Subsection 9.2.2.) It is assumed that $S \in \mathcal{BC}(S)$ because this ensures that the measure $\bar{\mu}$ is σ-finite and makes it possible to reduce the general case to the case of finite measure $\bar{\mu} = \mu$ on $\mathcal{R} = \mathcal{BL}(S, \mu)$.

In accordance with the general theory, the algebras $\mathcal{L}(S, \mu)$ of real or vector functions on S that are integrable with respect to the measure μ and the integrals of these functions are defined. Among integrable real functions, we distinguish the Borel functions with supports of finite measure, in particular, continuous functions with compact supports. Larger classes of integrable functions are obtained for measures with compact supports.

Exercise. Prove that any real Borel function $f : S \to \mathbb{R}$ such that $\mu(\operatorname{Supp} f) < \infty$ is integrable with respect to the measure μ on $\mathbb{R}$.

The method used above can be used to define the natural measure ds for a surface S and the integral over this surface. In the classical cases, such integrals express length, area, and volume.

4. We will assume that S is smoothly parameterized, i.e., all local parameterizations $\varphi \in \Phi$ of the surface S are smooth. Surfaces defined by smooth equations satisfy this condition. The map $\varphi : U \to \mathbb{R}^m$ from an open set $U \in \mathbb{R}^k$ is defined by the collection of its coordinates $\varphi_1 : U \to \mathbb{R}, \dots, \varphi_m : U \to \mathbb{R}$. Take the row $\partial_i^* \varphi$ and the column $\partial_j \varphi$ whose elements are the partial derivatives $\partial_i \varphi_1, \dots, \partial_i \varphi_m$ and $\partial_j \varphi_1, \dots, \partial_j \varphi_m$, respectively ($1 \le i, j \le k$). The products $g_{ij}(\varphi) = \partial_i^* \varphi \partial_j \varphi$ are continuous real functions on U. The matrix $G(\varphi) = (g_{ij}(\varphi))$ is called the *Gram matrix* of the parameterization φ. The determinant $g(\varphi) = \det G(\varphi)$ is called the *Gram determinant* of φ. The Gram determinant is also a real function on U. Let $\partial \varphi = (\partial_j \varphi)$ be an $m \times k$-matrix with columns $\partial_j \varphi$ and let $\partial^* \varphi = (\partial_i^* \varphi)$ be a $k \times m$-matrix with rows $\partial_i^* \varphi$ which is the transpose of $\partial \varphi$. Then

$$G(\varphi) = \partial^* \varphi \cdot \partial \varphi, \quad g(\varphi) = \det(\partial^* \varphi \cdot \partial \varphi).$$

The matrix $G(\varphi)$ is a $k \times k$-matrix. Its dimensions do not depend on the difference $l = m - k$. It follows from the definitions that $g(\varphi) \ge 0$, which makes it possible to define the real function $\sqrt{g(\varphi)}$ on U. At every point $x \in U$, the number $\sqrt{g(\varphi)}(x)$ represents the volume of the k-dimensional parallelepiped with edges $\partial_1 \varphi(x), \dots, \partial_k \varphi(x) \in \mathbb{R}^m$.

Let φ and ψ be parameterizations whose images intersect, and put $y = \varphi^{-1} \psi$. The differentiation rules are provided by the equalities

$$G(\psi) = \partial^* h \cdot G(\varphi) \circ h \cdot \partial h, \quad \sqrt{g(\psi)} = \sqrt{g(\varphi)} \circ h \cdot \det(\partial h),$$

where ∂h denotes a $k \times k$-matrix $(\partial_j h_i)$ of partial derivatives of coordinate functions h_i for h. We have $\det(\partial h) > 0$ because the parameterizations are consistent.

Exercise. Prove the rules of transition from φ to ψ specified above for G and $\sqrt{g}$.

The function $\sqrt{g(\varphi)} : U \to \mathbb{R}$ is continuous and therefore locally integrable with respect to the Lebesgue measure dx on $\mathbb{R}^k$. The equality

$$\mu_\varphi(B) = \int_{\varphi^{-1}(B)} \sqrt{g(\varphi)} \, dx \quad (B \in \mathcal{B}(C), \ C \subseteq \operatorname{Ran} \varphi)$$

defines a collection of regular consistent measures. The countable additivity of these measures follows from the Lebesgue limit theorem, while the consistence follows from the transition rules for $\sqrt{g}$ and the rule for the change of variables in the integral.

Exercise. Prove the preceding statement.

The measure defined on the algebra $\mathcal{B}(S)$ of the Borel sets in S using the above collection of measures μ_φ will be called *natural* for the parameterized surface (S, Φ) and denoted by ds. A bounded Borel function $f : S \to \mathbb{R}$ is integrable with respect to the measure ds over any set B of finite measure. It follows from the definitions that

$$\int_B f \cdot ds = \sum_n \int_{\varphi_n^{-1}(B_n)} f \circ \varphi_n \cdot \sqrt{g(\varphi_n)} \cdot dx$$

for $B_n \in \mathcal{BC}(S)$, $\varphi_n \in \Phi$ such that $\bar{B}_n \subseteq \operatorname{Ran} \varphi_n$, $B = \sum B_n$.

Example. Consider the circle $S = S(0, 1)$ in $\mathbb{R}^2$ and $\alpha(t) = (t, \sqrt{1 - t^2})$, $\beta(t) = (t, -\sqrt{1 - t^2})$, $\gamma(t) = (\sqrt{1 - t^2}, t)$, $\delta(t) = (-\sqrt{1 - t^2}, t)$ $(-1 < t < 1)$. The local parameterizations α, β, γ, and δ are the inverses of the charts β_+, β_-, α_+, and α_- of the oriented atlas (see Example 1 in Subsection 3.4.1, part 2). The transitions are defined by the equalities $g(t) = \gamma^{-1}(\alpha(t)) = \gamma^{-1}(\beta(t)) = t$ $(0 < t < 1)$ and $h(t) = \delta^{-1}(\alpha(t)) = \delta^{-1}(\beta(t)) = t$ $(-1 < t < 0)$. The Gram determinants can be easily calculated:

$$g(\alpha) = (1, -t/\sqrt{1 - t^2}) \begin{pmatrix} 1 \\ -t/\sqrt{1 - t^2} \end{pmatrix} = 1/(1 - t^2),$$

$g(\alpha) = g(\beta) = g(\gamma) = g(\delta)$. The circle S is compactly parameterizable, since it is covered by the images of the interval $\bar{T} = [-1/\sqrt{2}, 1/\sqrt{2}]$ under α, β, γ, and δ and is equal to the sum of the arcs $A = \alpha(T)$, $B = \beta(T)$, $C = \gamma(T)$, $D = \delta(T)$, where $T =]-1/\sqrt{2}, 1/\sqrt{2}]$. Therefore,

$$\int_S 1 \cdot ds = 4 \int_T (1/\sqrt{1 - t^2}) \, dt = 4 \cdot \pi/2 = 2\pi.$$

Remark. In Schwartz (1967), Chapter 6, Section 6, Radon measures defined by differential forms on oriented manifolds are described.

5. Consider an oriented compactly parameterized smooth k-dimensional surface (S, Φ) in the space $\mathbb{R}^m$. (The parameterizations $\varphi \in \Phi$ are not assumed to be smooth.) Every parameterization $\varphi \in \Phi$ defines a measure $dx\varphi^{-1}$ with values $dx\varphi^{-1}(B) = dx(\varphi^{-1}(B))$ $(B \in \mathcal{B}(S))$. Such measures are described in detail in Savel'ev (1989), Chapter 10, Subsection 10.3.1. Local measures μ_φ can be specified using a consistent collection of *local densities* $\rho(\varphi) : \operatorname{Ran} \varphi \to \mathbb{R}$ such that their compositions $\rho(\varphi) \circ \varphi$ are locally integrable with respect to the Lebesgue measure dx. These densities are consistent, which is expressed by the equality $\rho(\psi) = \rho(\varphi) \cdot \det h' \circ \psi^{-1}$ for the parameterizations φ and ψ with intersecting images and the parameterization $h = \varphi^{-1}\psi$. The measure μ_φ is defined by the equalities

$$\mu_\varphi(B) = \int_B \rho(\varphi) \cdot dx\varphi^{-1} = \int_{\varphi^{-1}(B)} \rho(\varphi) \circ \varphi \cdot dx$$

for $B \in \mathcal{BC}(S)$, $B \subseteq C \subseteq \operatorname{Ran} \varphi$. By the theorem on the change of variables in the integral, since the densities ρ_φ and ρ_ψ are consistent, so are the measures μ_φ and μ_ψ. Since S is smooth and oriented, it follows that h' exists and $\det h' \geq 0$. From Sard's theorem, it follows that the set of singular values with $\det h' = 0$ is negligible.

Remark. Detailed proofs of the theorems on the change of variables in the integral can be found in Savel'ev (1975), Chapter 9. The properties of densities are also described there.

Example. If the parameterizations $\varphi \in \Phi$ are smooth, then for the densities we can take the compositions $\rho(\varphi) = \sqrt{g(\varphi)} \circ \varphi^{-1}$. Then $\rho(\varphi) \circ \varphi = \sqrt{g(\varphi)}$ and the measures μ_φ are the same as above.

Let μ_φ be a consistent collection of local measures. Let the measure μ_φ be continuous with respect to the measure $dx\varphi^{-1}$ for any parameterization φ. Then the Radon–Nikodym derivatives $\rho(\varphi) = d\mu_\varphi/dx\varphi^{-1}$ form a consistent collection of local densities. (Savel'ev, 1989, Chapter 8).

Exercise. Prove the preceding statement by verifying the equality $dx\varphi^{-1}/dx\psi^{-1} = \det h' \circ \psi^{-1}$ for $h = \varphi^{-1}\psi$ and using the rules for operations with the derivatives of a measure with respect to another measure.

The definition of the measure μ_φ can be written in the differential form

$$d\mu_\varphi = \rho(\varphi) \cdot dx\varphi^{-1}.$$

This form is equivalent to the integral form.

3.4.8. The Stokes formula

The Stokes formula expresses the relationship between the integral of a differential form over a manifold and the integral over the boundary of this manifold. It is a generalization of the Newton–Leibniz formula and is one of the main formulas of integral calculus.

1. Instead of an abstract manifold, we again consider an oriented k-dimensional smoothly parameterized surface (S, Φ) in $\mathbb{R}^m$ described in Subsection 3.4.7, part 4. We will assume that S is *regular*, i.e., the derivative $\varphi'(u) : \mathbb{R}^k \to \mathbb{R}^m$ has rank k at every point $s = \varphi(u) \in S$ ($u \in \text{Dom}\,\varphi \subseteq \mathbb{R}^k$). Sard's theorem makes it possible to reduce the general case to the regular case. Consider the linear manifold

$$T_s(\varphi) = \varphi(u) + \varphi'(u)\mathbb{R}^k = \{\varphi(u) + \varphi'(u)v \mid v \in \mathbb{R}^k\}.$$

Let $\varphi, \psi \in \Phi$, $\varphi(a) = \psi(b) = s$, $h = \varphi^{-1}\psi$. Since $\det h' > 0$ and $\psi'(b) = (\varphi h)'(b) = \varphi'(h(b))h'(b) = \varphi'(a)h'(b)$, it follows that $T_s(\varphi) = T_s(\psi)$. The linear manifold $T_s = T_s(\varphi)$ is called the *tangent manifold*, and the vector space $\varphi'(u)\mathbb{R}^k \subseteq \mathbb{R}^m$ that is parallel to T_s is called the *tangent space* to the surface S at the point s. We emphasize that $s = \varphi(u) + \varphi'(u) \cdot 0 \in T_s$, but it is not necessarily true that $s \in \varphi'(u)\mathbb{R}^k$.

Example. Consider the circle $S = S(0, 1)$ and its local parameterization $\varphi = \alpha$ from the example of Subsection 3.4.7, part 4. Let $u = -1/\sqrt{2}$ and $s = \varphi(u) = (-1/\sqrt{2}, 1/\sqrt{2})$. Then $\varphi'(u) = (1, 1)$ and the line $T_s = (-1/\sqrt{2}, 1/\sqrt{2}) + (1, 1)\mathbb{R}$ is the tangent line to S at the point s, while $(1, 1)\mathbb{R}$ is the bisector parallel to this line.

Remark. The nonconnected union $T = \cup T_s$ ($s \in S$) is called the *tangent bundle over S*. Every T_s in it is open and closed. The parameterizations for S generate the parameterizations for T, making T a $2k$-dimensional manifold. The smoothness of T is one degree lower than that of S which is assumed to be nonzero. The space T_s^* that is conjugate to T_s is called the *cotangent space* to S at the given point s, while the nonconnected union $T^* = \cup T_s^*$

($s \in S$) is called the *cotangent bundle over S*. A detailed description of these bundles can be found in Sulanke and Wintgen (1972), Chapter 1, Section 3. They are used in the definition of differential forms on manifolds. The general theory of smooth bundles (skew products) is presented in Dubrovin, Novikov, and Fomenko (1986), Part 2, Chapter 6.

2. Let E and F be real Banach spaces, $f : E \to F$, $M = f^{-1}(0)$, and $u \in M$. Assume that f is continuously differentiable on a neighborhood of u and the derivative $f'(u)$ maps E onto F. The linear manifold $T_u = u + \operatorname{Ker} f'(u)$ is called the *tangent manifold* to M at the point u. The Lyusternik theorem states that

$$T_u = \{u + v \mid \operatorname{dist}(u + tv, M) = o(t)\}$$

(see Kolmogorov and Fomin, 1989, Section 10.2.3). This theorem makes it possible to generalize the classical Lagrange method for solving extremal problems to the case of Banach spaces (Subsections 3.2.8–3.2.10 of this book; Alekseev, Tikhomirov, and Fomin, 1987).

Example. Let $E = H$ be a real Hilbert space, $F = \mathbb{R}$, $f(x) = \|x\|^2 - 1$, $M = f^{-1}(0) = S(0, 1)$, and $u \in M$. Then $f'(u) \cdot v = 2u \cdot v$, $\operatorname{Ker} f'(u) = \{v : u \cdot v = 0\} = u^\perp$, and $T_u = u + u^\perp$. At the same time, $\operatorname{dist}(u + tv, M) = \min\{\|u + tv - x\| : \|x\|^2 = 1\} = \|(u+tv) - (u+tv)/\|u+tv\|\| = \|u+tv\| - 1 = \sqrt{1 + 2(u \cdot v)t + \|v\|^2 t^2} - 1 = 2(u \cdot v)t + o(t)$ and $\operatorname{dist}(u + tv, M) = o(t)$ is equivalent to $u \cdot v = 0$, $tv \in \operatorname{Ker} f'(u) = u^\perp$, $u + tv \in T_u$. In particular, if $H = \mathbb{R}^2$ and $u = (-1/2, 1/2)$, then $v = (1, 1)$, as in the example of part 1 of this subsection.

3. The integral of a form over a piece of the surface S specified by the parameterization φ can be defined similarly to the integral over a curve γ described in part 1. To combine the integrals over individual pieces into the integral over the whole surface, we use the partition of unity theorem.

Consider a set $M \subseteq \mathbb{R}^m$, its open neighborhood V and a collection of open sets V_i in $\mathbb{R}^m$ that covers M, and a collection of infinitely smooth functions $f_i : V \to \mathbb{R}$ with supports $\operatorname{Supp} f_i \subseteq V_i$ and ranges $f_i(M) \subseteq [0, 1]$. The collection (f_i) is said to be *locally finite on M* if for any $x \in M$ there exists a neighborhood W such that only a finite number of functions f_i are not identically equal to zero on W. The collection (f_i) is said to be a *partition of unity for M inscribed in* (V_i) if it is locally finite and $\sum f_i = 1$ ($x \in M$). The following theorem is proved in Spivak (1965), Section 3.11.

Theorem. For any set $M \subseteq \mathbb{R}^m$ and its open cover (V_i), there exists a partition of unity inscribed in (V_i).

Among all covers (V_i), we distinguish *locally finite* covers with the following property: for any compact set $C \subseteq \mathbb{R}^m$ there is only a finite number of sets V_i such that $C \cap V_i \neq \varnothing$. To construct a partition of unity (f_i), we can use ε-hats $h : \mathbb{R}^m \to \mathbb{R}$ such that $h(x) = c(\varepsilon) \cdot \exp\{-\varepsilon^2/(\varepsilon^2 - \|x\|^2)\}$ for $\|x\| < \varepsilon$ and $h(x) = 0$ for $\|x\| \geq \varepsilon$. The coefficients $c(\varepsilon)$ are chosen to satisfy the normalization condition $\sum f_i = 1$ $(x \in M)$ (see Cartan, 1970, Chapter 3, Section 4; Vladimirov, 1979, Chapter 1, Section 1).

Choose a smoothly parameterized k-dimensional surface (S, Φ) in $\mathbb{R}^m$, a form $\omega \in \Omega_k^0(A, F)$ on an open neighborhood $A \subseteq \mathbb{R}^m$ of the set S, and a set $B \in \mathcal{BC}(\varphi)$ for a parameterization $\varphi \in \Phi$ (see Subsection 3.4.7, part 2). To avoid using the theory of integral for vector functions, we will consider the real line $\mathbb{R}$ instead of the Banach space F. By definition,

$$\int_\varphi B\omega = \int_{\varphi^{-1}(B)} \varphi^* \omega \, dx.$$

As usual, $B\omega(x) = \operatorname{ind} B(x) \cdot \omega(x) = 0$ for $x \notin B$. The function $\varphi^*\omega$ is continuous on the compact set $\varphi^{-1}(\bar{B})$ and is therefore integrable on $\varphi^{-1}(B)$ with respect to the Lebesgue measure dx on $\mathbb{R}^k$. We now write the form ω in the standard form (see Subsection 3.4.5, part 3): $\omega = \sum a_K \wedge dy_K$, where $a_K : A \to F$ are continuous functions, and $\wedge dy_K = dy_{m(1)} \wedge \cdots \wedge dy_{m(k)}$ are basic forms with multi-indices $K = m(1) \ldots m(k)$ $(1 \leq m(1) < \cdots < m(k) \leq m)$. The change of variables φ transforms ω into

$$\varphi^*\omega = \sum (a_K \circ \varphi) \cdot \det \varphi'_K \cdot dx_1 \wedge \cdots \wedge dx_K,$$

where $\varphi_K = (\varphi_{m(j)})_{1 \leq j \leq k}$, $\varphi'_K = (\partial_i \varphi_{m(j)})$ and summation is performed with respect to all the multi-indices K under consideration. As a result,

$$\int_\varphi B\omega = \sum_K \int_{\varphi^{-1}(B)} a_K(\varphi(x)) \cdot \det \varphi'_K(x) \, dx.$$

Since $\varphi, \psi \in \Phi$ are consistent, the rule for the change of variables in the Lebesgue integral yields

$$\int_\varphi B\omega = \int_\psi B\omega \quad (B \in \mathcal{BC}(\varphi) \cap \mathcal{BC}(\psi)).$$

Exercise. Prove the preceding equality (see Spivak, 1965, Section 5.4)

This allows us to define the integral of the form ω over the set B by the equality

$$\int_B \omega = \int_\varphi B\omega \quad (B \in \mathcal{BC}(\varphi)).$$

Assume that S is *compactly parameterized*, i.e., there exists a collection of open sets $B(\nu) \subseteq A$ that covers S and a collection of parameterizations $\varphi_\nu \in \Phi$ such that $B(\nu) \in \mathcal{BC}(\varphi_\nu)$. Take a partition of unity (f_ν) for S inscribed in $(B(\nu))$. Suppose that the collection of integrals of the forms $f_\nu \cdot \omega$ over the sets $B(\nu)$ is summable. The integral of the form ω over the manifold S is given by the equality

$$\int_S \omega = \sum_\nu \int_{B(\nu)} f_\nu \cdot \omega.$$

The sum on the right-hand side does not depend on the choice of the cover $(B(\nu))$ and the partition of unity (f_ν).

Exercise. Prove the preceding statement (Spivak, 1965, Section 3.12)

Remark. If the surface S or the support of the form ω are compact, then the number of terms in the sum in the expression for the integral of ω over S can be made finite.

4. Instead of a form on an open neighborhood A of a manifold S in $\mathbb{R}^m$, it is possible to consider a form ω on S defined as a collection of forms $\omega(s)$ on tangent spaces T_s to S at $s \in S$ (Spivak, 1965, Chapter 5). Alternatively, forms on manifolds are defined as sections of the corresponding bundles (Godbillon, 1969 Chapter 3, Section 7; Sulanke, and Wintgen, 1972, Chapter 2, Section 4). The operations for the forms ω are generated pointwise by the operations for the forms $\omega(s)$. Involving such complex design is justified by the fact that S is usually not a vector subspace of the space $\mathbb{R}^m$ and it is not possible to directly use the standard definition of a form on an open set in a Banach space.

Consider a regular smoothly parameterized oriented k-dimensional manifold (S, Φ) in $\mathbb{R}^m$. By definition, an n-smooth k-form ω on S is a map from S to the union $\cup \Omega_k^n(T_s, F)$ such that $\omega(s) \in \Omega_k^n(T_s, F)$ $(s \in S)$. Under certain conditions, the form $w(s)$ generates a measure on the Borel algebra $\mathcal{B}(S)$. Scalar or vector functions f on S can be integrated with respect to this measure. If $F \neq \mathbb{R}$ and f is a vector function, then the product of values of f and vectors in F must be defined. The easiest way is to put $F = \mathbb{R}$ and consider $f : S \to \mathbb{R}$. The changes of variables defined by parameterizations make it possible to reduce the task to integrating with respect to the Lebesgue measure. Various additional assumptions have to be made in order to implement this process. We use the way of defining the integral described in Subsection 3.4.7, part 4.

Assume that S is compactly parameterized, $S = \sum B_n$ with $B_n \in \mathcal{BC}(\varphi_n)$ for a sequence $\varphi_n \in \Phi$, the form w has a standard representation and $w = \sum a_K \wedge dy_K$ with the basic forms $\wedge dy_K = dy_{m(1)} \cdots \wedge dy_{m(k)}$ for the multi-indices $K = m(1) \ldots m(k)$ $(1 \leq m(1) < \cdots < m(k) \leq m)$. Then the integral of the function f on S with respect to the form w is defined by the equality

$$\int_S f w = \sum_{n,K} \int_{\varphi_n^{-1}(B_n)} f(\varphi_n(x)) \cdot a_K(\varphi_n(x)) \cdot \det \varphi'_{nK}(x)\, dx,$$

where $\varphi'_{nK} = (\partial_i \varphi_{n,m(j)})$ and dx is the Lebesgue measure on $\mathbb{R}^k$. Since S is oriented and regular, we can assume that $\det \varphi'_{nK}(x) > 0$. The function f is integrable on S with respect to the form w if the integrals on the right-hand side and their sum exist. It can be proved that the right-hand side does not depend on the way we choose the sets B_n and the parameterizations φ_n. For the integrable function $f = 1$, we obtain the integral of w over S.

Remark. The detailed presentation of the theory of integration on manifolds can be found in Sulanke and Wintgen (1972), Chapter 3. Absolute forms are introduced there in addition to the conventional forms. Densities are used for the definition of the integral in Sternberg (1983), Chapter 3.

5. Consider a 2-smooth oriented manifold M, its boundary ∂M with induced orientation, and a 1-smooth $(k-1)$-form w on M. Let M or Supp w be compact. Then the following formula holds.

The Stokes formula. $\int_M dw = \int_{\partial M} w.$

It is possible that $\partial M = 0$. It is conventional to write i^*w instead of w in the right-hand side, denoting the identity embedding by $i : \partial M \to M$. The Stokes formula is discussed in detail in Dubrovin, Novikov, and Fomenko

(1986) (4.1, Section 26, and 4.2, Section 8). In Sulanke and Wintgen (1972), Chapter 3, Section 3, several versions of the Stokes formula are deduced from the Fubini theorem for manifolds (Chapter 3, Section 2) and an interesting method is presented for calculating an integral over a sphere and calculating the volume of a ball in $\mathbb{R}^m$, attributed to H. Weyl.

Corollary. *If* $\partial M = \varnothing$, *then* $\int\limits_M d\omega = 0$.

Example 1. Consider the 2-dimensional manifold $\bar{M} = \{(x,y) : 1 \leq x^2 + y^2 \leq 2\}$ with boundary $\partial \bar{M} = S(0,1) \cup S(0,\sqrt{2})$ in $\mathbb{R}^2$ and the form $\omega = (x^2 + y^2)^{-1}(x dy - y dx)$ on $\mathbb{R}^2 \setminus \{(0,0)\}$. It is easy to verify that $d\omega = 0$ and the integrals of the 1-form ω over the circles $S(0,1)$ and $S(0,\sqrt{2})$ are equal to 2π. Hence, in the case of positive orientation we have

$$\int\limits_{\bar{M}} d\omega = 0 = \int\limits_{S(0,\sqrt{2})} \omega - \int\limits_{S(0,1)} \omega.$$

If we replace $\bar{M}$ by a noncompact manifold $M = \bar{M} \setminus S(0,1)$ with boundary $\partial M = S(0,\sqrt{2})$, then the Stokes theorem will not hold for it because

$$\int\limits_M d\omega = 0 \neq 2\pi = \int\limits_{S(0,\sqrt{2})} \omega.$$

Example 2. Consider the forms $\xi = a dx + b dy + c dz$ and $\eta = f dy \wedge dz + g dz \wedge dx + h dx \wedge dy$ described in the example of Subsection 3.4.5, part 3 and their exterior differentials $d\xi$ and $d\eta$. In $\mathbb{R}^3$, take a compact smooth body V with surface S and a compact smooth 2-dimensional surface D with contour C. (In particular, V can be a closed ball, S a sphere, D a disk, and C a circle.) The general Stokes formula implies the classical Stokes formula and the classical Gauss–Ostrogradsky formula:

$$\iint\limits_D (b'_x - a'_y)\, dxdy + (c'_y - b'_z)\, dydz + (a'_z - c'_x)\, dzdx = \int\limits_C a\, dx + b\, dy + c\, dz,$$

$$\iiint\limits_V (f'_x + g'_y + h'_z)\, dxdydz = \iint\limits_S f\, dydz + g\, dzdx + h\, dxdy.$$

For $dz = 0$, the first of the above formulas becomes the classical Green formula.

Example 3. In the proof of the Brouwer fixed point theorem in Subsection 5.1.1, part 2, the Stokes formula is used for calculating the integrals over an n-dimensional ball.

6. In Spivak (1965), Chapter 5, a separate section is devoted to deriving three classical formulas from the general Stokes formula, namely, the Green formula, the Gauss–Ostrogradsky formula, and the Stokes formula. Their applications in the field theory are also described there. A detailed presentation of these applications can be found in Postnikov (1988), Lectures 27 and 28. A brief description of the proof of the Cauchy integral theorem and the Cauchy integral formula for analytic functions with the use of the Stokes formula is given in Spivak (1965), Chapter 4.

In Schwartz (1967), Chapter 6, Section 7, the Stokes formula is derived for manifolds with 1-smooth pseudoboundary and various special cases are considered in detail. It is also noted that integration of vector forms (even with values in an infinite-dimensional space) can be reduced the scalar case under certain assumptions. Application of the theory of differential forms in algebraic topology is described in Chapter 6, Section 8 of the same book.

The Stokes formulas for nonsmooth manifolds are presented in McLeod (1980). Chapter 4 in Federer (1969) is devoted to homological integration theory. In particular, real and integer homologies with common chains are considered, differential forms are associated with distributions, the notion of a *flow* is introduced and the operations with flows are defined. A connection between straightenable flows and oriented manifolds is described. The deformation, closedness, compactness, and approximation theorems are proved. The notion of the *outward normal* is introduced and the natural general formula of Gauss–Green–Ostrogradsky is obtained.

3.4.9. Map degree

We give the differential and the integral definitions of the degree of a map for manifolds. A brief description of the main properties of the degree of a map is provided.

1. Let X and Y be oriented n-dimensional manifolds and $f : X \to Y$ be a smooth map. We assume that the inverse image $f^{-1}(y)$ is finite for every regular value $y \in Y$ of f. Since the manifolds are oriented, for $x \in f^{-1}(y)$ the sign

$$s(f, x) = \operatorname{sgn} \det f'_{\alpha\beta}(\alpha(x))$$

does not depend on the choice of local charts α and β the picture of $f_{\alpha\beta} = \beta f \alpha^{-1}$. If $s(f, x) > 0$, we say that f *preserves the orientation at the point* y.

If $s(f, x) < 0$, f *changes the orientation at* y. The number

$$\deg(f, y) = \sum_{x \in f^{-1}(y)} s(f, x)$$

is called the *degree of a map* f *at the point* y.

Example. Let $X = Y = \mathbb{R}^n$ with standard orientation, $f(x) = x$, and $g(x) = -x$ $(x \in \mathbb{R}^n)$. Then $\deg(f, y) = 1$ and $\deg(g, y) = (-1)^n$ $(y \in \mathbb{R}^n)$.

The degree $\deg(f, y)$ is called the *algebraic number of inverse images* of the point y under the map f.

Remark. The derivative $f'(x)$ can be defined as a linear map of the tangent spaces T_x and T_y to the manifolds X and Y at the points x and y. Then $s(f, x) = \operatorname{sgn} \det f'(x)$.

2. The condition of finiteness for the inverse images $f^{-1}(y) \subseteq X$ of regular values $y \in Y$ is satisfied if the manifold X is compact or if the map f is *proper* (i.e., $f^{-1}(C) \subseteq X$ is compact for any compact set $C \subseteq Y$). Then the inverse image $f^{-1}(y)$ of a regular value y consists of isolated points and therefore the compactness of $f^{-1}(y)$ implies that it is finite. (see Postnikov, 1987, Lecture 13).

From Sard's theorem it follows that the degree $\deg f$ is defined on the everywhere dense set $R(f) \subseteq Y$. If the manifold Y is connected, then the function $\deg f : R(f) \to \mathbb{Z}$ is constant. Its value is called the *degree of the map* f and is also denoted by $\deg f$. Consider a connected manifold Y, a proper map $f : X \to Y$, and a finite n-form ω on Y whose integral is not equal to zero. Then

$$\deg f = \int f^* \omega \Big/ \int \omega.$$

A detailed proof of this can be found in Postnikov (1987), Lecture 26. See also Dubrovin, Novikov, and Fomenko (1986), Part 3, Section 14, and Sulanke and Wintgen (1972), Chapter 3, Section 6.

3. An important property of the degree $\deg$ of a map is its property of being invariant under continuous transformations of the map. This property makes it possible to simplify a given map when calculating its degree.

A continuous map $h : [0, 1] \times X \to Y$ with section $h(0, \cdot) = f$ is called a *homotopy* of the map f. The sections $f_t = h(t, \cdot)$ form the class of maps *homotopic* to f. The relation of homotopy will be denoted by the equivalence symbol $\sim$. For functions of real variable, the graph of h is a continuous surface connecting the curve $f = f_0$ with $g = f_1$. The homotopy h is also called a *continuous deformation* of the map f. We say that h *connects* f with g or *deforms* f into g.

Example. For $X = \mathbb{R}$, the homotopy $h(t, x) = (1 - t)f(x) + tg(x)$ connects the line $f(x) = x$ with the parabola $g(x) = x^2$ (bends the line f).

A homotopy $h : [0, 1] \times X \to Y$ is said to be *smooth* if it is a restriction to $[0, 1] \times X$ of a smooth map from $T \times X$ to Y for an interval T in $\mathbb{R}$ that includes the interval $[0, 1]$. Then the maps $f = f_0$ and $g = f_1$ are called *smoothly homotopic*. Among smooth homotopies h, we distinguish *proper* homotopies. If the manifold X is compact, then all smooth homotopies are proper.

The main result to be formulated here is the theorem on the homotopic invariance of the degree of a map.

Theorem. *If the manifold Y is connected and smooth maps $f, g : X \to Y$ are homotopic, then $\deg f = \deg g$.*

This theorem is proved in Sulanke and Wintgen (1972), Chapter 3, Section 6 using the properties of cohomology groups. In Postnikov (1987), Lecture 26, it is proved for maps connected by a proper homotopy. The continuity of the map $t \to \int f_t^* \omega$ for a finite form ω is used there. Also, smooth approximations are used to define the degree of a smooth map between compact manifolds and the theorem on the equality of degrees is formulated for continuous maps.

In Dubrovin, Novikov, and Fomenko (1986), Part 2, Section 13, the degree is defined for smooth maps between connected oriented compact manifolds with no boundary (i.e., *open*) and the homotopic invariance theorem is proved for this case. For compact manifolds with boundary and maps that preserve boundaries, the degree is defined as the degree of the restriction to the boundary. Maps whose range is an n-dimensional sphere are pointed out.

4. To give an example, we calculate the degree $\deg g$ of the complex polynomial $g(z) = a_n z^n$ ($n > 0$, $a_n \neq 0$). Complex numbers $z = x + iy$ will be identified with points $z = (x, y) \in \mathbb{R}^2$. We assume that it is known that the equation $z^n = c$ has exactly n solutions $z = c^{1/n}$ for any $c \neq 0$. (A detailed elementary proof of the theorem on complex roots can be found in Savel'ev, 1974, Part 2.1, Section 6.) We will prove that $\deg g = n$.

By the binomial formula, $g(z) = a_n(x+iy)^n = g_1(x,y)+ig_2(x,y)$, where $g_1 : \mathbb{R}^2 \to \mathbb{R}$ and $g_2 : \mathbb{R}^2 \to \mathbb{R}$ are polynomials in two real variables. The equalities $g'_{11} = g'_{22}$ and $g'_{21} = -g'_{12}$ for the partial derivatives of g_1 and g_2 are easy to verify. Hence

$$g' = \begin{pmatrix} g'_{11} & g'_{12} \\ -g'_{12} & g'_{11} \end{pmatrix}, \quad \det g' = (g'_{11})^2 + (g'_{12})^2.$$

Since the polynomial $g - c$ has no multiple roots for $c \neq 0$, it follows that $g'(z) \neq 0$ (see van der Waerden, 1991, Section 28) and $\det g'(z) > 0$ for any $z = c^{1/n} \in g^{-1}(c)$. Therefore, $s(g,z) = 1$ and $\deg g = \deg(g,c) = n$.

Remark. In Schwartz (1967), Chapter 6, Section 8, the topological degree of a continuous map is defined using the *cycle index* with respect to a point. The Rouche theorem on the invariance of the topological degree of a map under admissible additive deformations is proved. This theorem is then used to deduce the theorem on the roots of a complex polynomial and the Brouwer fixed point theorem.

3.4.10. Applications

In conclusion, we mention some applications of analysis on manifolds.

1. Vector fields and forms on manifolds are used to describe dynamical systems. They are considered in detail in Dubrovin, Novikov, and Fomenko (1986), Part 2, Chapter 7, and in Godbillon (1969), Chapter 11. Hamiltonian and Lagrangian systems are described and the classical Liouville theorem is proved. Various examples are given.

Complex manifolds are considered in Dubrovin, Novikov, and Fomenko (1986), Part 2, Section 4. Additional literature on the subject can be found therein.

A detailed description of special algebraic, geometric, and analytical structures and concepts associated with manifolds can be found in Nirenberg (2001). They are used in the development of asymptotic methods for solving differential and operator equations.

In addition to conventional derivatives and groups, *Lie derivatives* and *Lie groups* are widely used in the study of dynamical systems. (see Sternberg, 1983, Chapter 5). They have been extensively covered in the literature (see Postnikov, 1988, V). The applications of Lie derivatives and Lie groups to the theory of differential equations are described in detail in Ovsyan-nikov (1978).

Manifolds that describe the global structure of solutions to multidimensional variational problems are studied in connection with the general theory of relativity (Dubrovin, Novikov, and Fomenko, 1986, Part 2, Chapter 4). A detailed presentation of variational calculus on manifolds is given in Sternberg (1983), Chapter 4.

2. Smooth collections of probability distributions can be viewed as smooth surfaces in manifolds which may be infinite-dimensional. A geometric model of this kind is described in Chentsov (1982) and Amari (1985). This model is applied to the general problem of statistical solutions and some important special cases. In Chentsov (1982), Chapter 3, the statistical manifolds considered there are approximated by finite-dimensional manifolds. This compensates for the absence of a well-developed general theory of infinite-dimensional manifolds.

Chentsov (1982) plays a special role in the literature on probability theory and statistics as it features extensive application of algebraic and geometric structures and concepts. The category of solving rules, Markov morphisms, and smooth infinite-dimensional manifolds of probability measures are used already at the stage of formulating the problem of statistical solution. Vector fields, tangent spaces, metrics, and the related notions are described for the mentioned manifolds. A separate chapter is devoted to the theory of exponent collections of probability measures identified with collections of geodesics (surfaces) in an infinite-dimensional manifold equipped with natural polygonal connectedness.

OPERATORS

The second part of this book presents the elements of operator theory. Linear operators are considered first, followed by nonlinear operators. Operators in normed spaces are studied, certain special spaces and classes of operators are described. Considerable attention is given to distributions, their transformations, and operator spectra. Several fixed point theorems and a saddle point theorem are proved for nonlinear operators.

Chapter 4.

Linear operators

An introduction to the theory of linear operators is presented in Chapter 4 of Kolmogorov and Fomin (1989). For a fundamental textbook on this subject, see Dunford and Schwartz (1988). A detailed description of operators in Hilbert spaces is given in Sadovnichii (1999).

4.1.   HILBERT SPACES

Hilbert spaces are complete Euclidean spaces. The description of these spaces that was started in previous chapters is continued in this section.

4.1.1.   Orthogonal projection

Orthogonal projection is one of the main operations in Hilbert spaces.

1. Let $(\mathbb{F}, E)$ be a real or complex space with Hermitian scalar product $p : E \times E \to \mathbb{F}$,

$$p(x, y) = x \cdot y = \langle x, y \rangle \in \mathbb{F} \quad (x, y \in E).$$

It is assumed that $yx = (xy)^* = \overline{xy}$, where the star and the bar mean conjugation in the number field $\mathbb{F}$. The bilinear form p defines a quadratic form $q : E \to \mathbb{R}$ with values $q(x) = p(x, x) \in \mathbb{R}$. We distinguish positive nondegenerate scalar products p such that $q(x) \geq 0$ $(x \in E)$ and $q(x) > 0$ whenever $x \neq 0$.

Such a scalar product defines the Euclidean norm with values $\|x\| = (xx)^{1/2}$ $(x \in E)$. It satisfies the Cauchy inequality

$$|xy| \leq \|x\| \cdot \|y\| \quad (x, y \in E).$$

In the real case, the number $\varphi(x, y) = \arccos(xy(\|x\| \cdot \|y\|)^{-1})$ represents the *angle* between the vectors x and y with nonzero norms.

The triangle identity, the parallelogram law, and the polarization identity hold for the Euclidean norm and the scalar product. Two vectors are orthogonal if their scalar product is equal to zero. The Pythagorean theorem holds for orthogonal vectors. As a rule, we will consider nondegenerate scalar products and separated spaces. If these assumptions are not needed, this will be mentioned explicitly unless it is clear from the context.

2. Only *closed* vector subspaces of H will be called subspaces. Such subspaces are also Hilbert spaces. The orthogonal projection theorem holds for closed subspaces. We now formulate this theorem in a form different from that in Subsection 2.2.6 in order to include an important existence statement.

Theorem. *Let z be a point and A be a subspace of a Hilbert space H. Then*

1) there exists a point $a \in A$ that is closest to z;

2) any point $a \in A$ that is closest to z is an orthogonal projection of z onto A;

3) any orthogonal projection p of the point z onto A is a point of A that is closest to z.

Proof. Since $\|z - x\| \geq 0$ for all $z \in H$ and $x \in A$, there exists $\alpha = \inf\{\|z - x\| : x \in A\} \geq 0$. It follows from the definition of the infimum that for any $n \in \mathbb{N}$ there exists a point $x(n) \in A$ such that $\|z - x(n)\| \leq \alpha + 1/n$.

We now prove that the sequence of points $x(n)$ converges. Consider $z - x(m)$ and $z - x(n)$ for arbitrary $m, n \in \mathbb{N}$. We have

$$(z - x(n)) - (z - x(m)) = x(m) - x(n),$$
$$(z - x(n)) + (z - x(m)) = 2[z - 2^{-1}(x(m) + x(n))],$$
$$x = 2^{-1}(x(m) + x(n)) \in A, \quad \|z - x\| \geq \alpha.$$

Applying the parallelogram law to $z - x(m)$ and $z - x(n)$, we obtain

$$\|x(m) - x(n)\|^2 + 4\alpha^2 \leq \|x(m) - x(n)\|^2 + \|2(z - x)\|^2$$
$$\leq 2[\|z - x(n)\|^2 + L\|z - x(m)\|^2] \leq 2[(\alpha + 1/n)^2 + (\alpha + 1/m)^2],$$

$$\|x(m) - x(n)\|^2 \leq 4\alpha(1/m + 1/n) + 2(1/m^2 + 1/n^2).$$

Hence, the sequence $x(n)$ converges.

Since the space H is complete, there exists a point $a \in H$ to which the sequence $x(n) \in A$ converges. Since the subspace A is closed, it follows that $a \in A$. We now prove that a is a point of A that is closest to z. Since $\alpha \leq \|z - x(n)\| \leq \alpha + 1/n$, we have $\|z - a\| = \lim \|z - x(n)\| = \alpha$. Therefore, $\|z - a\| = \min\{\|z - x\| : x \in A\}$. This proves the first assertion of the theorem. Assertions 2 and 3 follow from the lemma of Subsection 2.2.6, part 7. □

The proposition on orthogonal projections of Subsection 2.2.6, part 7, and the orthogonal decomposition theorem of Subsection 2.2.6, part 8 hold true in the case of Hilbert spaces.

4.1.2.   Continuous linear functionals

A continuous linear functional in a Hilbert space is defined by the scalar product and a coefficient.

1.   A geometric characteristic of continuous linear functionals on a normed space E is provided by the following proposition.

Proposition. *A linear functional is continuous if and only if its kernel is closed.*

Proof. 1) Let $\varphi \in E'$, $z_n \in \mathrm{Ker}\,\varphi$, $z \in E$, and $z_n \to z$. Then $\varphi(z) = \lim \varphi(z_n) = 0$ and $z \in \mathrm{Ker}\,\varphi$. Therefore, $\mathrm{Ker}\,\varphi$ is closed.

2) Let $\varphi \in E^*$ and let $\mathrm{Ker}\,\varphi$ be closed. To prove that $\varphi \in E'$, it suffices to show that the functional φ is continuous at the point 0. Assume the contrary. Let $\varphi(z_n) \not\to 0$ for some $z_n \to 0$, $z_n \in E$. Then $|\varphi(z_{n(k)})| \geq \alpha$ for some $\alpha > 0$ and $n(1) < n(2) < \cdots < n(k) < \ldots$, which implies

$$\|y_k/\varphi(y_k)\| = \|y_k\|/|\varphi(y_k)| \leq \|y_k\|/\alpha \to 0$$

for $y_k = z_{n(k)}$. At the same time, we have $\varphi(y_k/\varphi(y_k)) = \varphi(y_k)/\varphi(y_k) = 1$. Therefore, $\varphi(x_k) = \varphi(y_1/\varphi(y_1) - y_k/\varphi(y_k)) = 1 - 1 = 0$ and $x_k = y_1/\varphi(y_1) - y_k/\varphi(y_k) \in \mathrm{Ker}\,\varphi$.

Consequently, since $\|y_k/\varphi(y_k)\| \to 0$ and $\mathrm{Ker}\,\varphi$ is closed, we have

$$y_1/\varphi(y_1) = \lim[x_k + y_k/\varphi(y_k)]$$
$$= \lim x_k + \lim(y_k/\varphi(y_k)) = \lim x_k + 0 = \lim x_k \in \mathrm{Ker}\,\varphi,$$

i.e., $\varphi(y_1/\varphi(y_1)) = 0$. This contradicts the equality $\varphi(y_1/\varphi(y_1)) = 1$. □

The kernels of linear functionals are also called *hyperplanes*. From what was proved above, it follows that the kernel of every continuous linear functional is a closed hyperplane. By linearity, the kernels of linear functionals are subspaces of the vector space on which they are defined.

Exercise. Prove that any hyperplane in a normed space is either closed or everywhere dense.

2. We now describe continuous linear functionals on a Hilbert space. Consider a Hilbert space H, a vector $c \in H$, and the conjugate space H' of continuous linear functionals on H.

Lemma 1. *The equality* $\varphi(z) = cz$ $(z \in H)$ *defines a continuous linear functional φ on H.*

Proof. Indeed, the above equality associates every vector $z \in H$ with a number $cz \in \mathbb{F}$ and therefore defines a functional φ on H such that $\varphi(z) = cz$ for a vector z. Since the scalar product is linear in the second variable, it follows that φ is a linear functional:

$$\varphi(\alpha_1 z_1 + \alpha_2 z_2) = c(\alpha_1 z_1 + \alpha_2 z_2)$$
$$= \alpha_1 c(z_1) + \alpha_2 c(z_2) = \alpha_1 \varphi(z_1) + \alpha_2 \varphi(z_2)$$

for any $z_1, z_2 \in H$ and $\alpha_1, \alpha_2 \in \mathbb{F}$.

The Cauchy inequality implies $|\varphi(z)| = |cz| \leq \|c\| \cdot \|z\|$ for $z \in H$. This means that the linear functional φ is continuous at the point 0 and is therefore continuous. $\square$

Let $b, c \in H$.

Lemma 2. $bz = cz(z \in H) \Longleftrightarrow \|b - c\| = 0$.

Proof. If $bz = cz$ $(z \in H)$, then $(b - c)z = 0$ $(z \in H)$. In particular, $\|b - c\|^2 = (b - c)(b - c) = 0$ for $z = b - c$ and therefore $\|b - c\| = 0$.

If $\|b - c\| = 0$, then the Cauchy inequality implies $0 \leq |bz - cz| = |(b - c)z| \leq \|b - c\| \cdot \|z\| = 0$ and therefore $bz = cz$. $\square$

Corollary. $bz = cz \,(z \in H) \Leftrightarrow b = c$.

Lemma 3. *Let $c \in H$ and let $\varphi \in H'$, $\varphi(z) = cz$ $(z \in H)$. Then $\|\varphi\| = \|c\|$.*

Proof. By definition, $\|\varphi\| = \sup\{|\varphi(u)| : u \in H, \|u\| \le 1\}$. Using the Cauchy inequality, we obtain $|\varphi(u)| = |cu| \le \|c\| \cdot \|u\| \le \|c\|$ ($u \in H$, $\|u\| \le 1$). Hence $\|\varphi\| \le \|c\|$.

We now prove the reverse inequality. We have $\varphi(c) = cc = \|c\|^2$. If $\|c\| = 0$, then $\|\varphi\| = \|c\| = 0$. If $\|c\| \ne 0$, then $\varphi(u) = \varphi(\gamma c) = \gamma\varphi(c) = \gamma \cdot \|c\|^2 = \|c\|$ for $u = \gamma c$, $\gamma = \|c\|^{-1}$. Since $\|u\| = \gamma \cdot \|c\| = 1$, it follows that $|\varphi(u)| \le \|\varphi\|$. As a result, $\|c\| \le \|\varphi\|$. $\square$

The vector $c \in H$ that defines a continuous linear functional $\varphi \in H'$ with values $\varphi(z) = cz$ $(z \in H)$ is called the *coefficient* of φ.

3. Lemmas 1–3 describe a class of continuous linear functionals on a nondegenerate Hilbert space H. A natural question arises of whether this class contains all continuous linear functionals on H. The Riesz theorem provides the answer.

The Riesz representation theorem. *For any $\varphi \in H'$, there exists a $c \in H$ such that $\varphi(z) = cz$ $(z \in H)$.*

Proof. If $\varphi = 0$, then $c = 0$.

Let $\varphi \ne 0$. Then $A = \operatorname{Ker}\varphi$ is a closed hyperplane in H. By the orthogonal projection theorem, for any $z \in H$ there are $x \in A$ and $y \in B = A^{\perp}$ such that $x + y = z$. At the same time, there are $b \in B$ such that $b \notin A$. (Otherwise, $B \subseteq A$ and $x + y \in A$ for all $x \in A$ and $y \in A$, which implies $A = H$. This contradicts the condition $\varphi \ne 0$.) Hence $H = A + \mathbb{F}b$.

Thus, for any $z \in H$ there are $x \in A$ and $t \in \mathbb{F}$ such that $z = x + tb$, where b is a vector in $B = A^{\perp}$ that does not belong to A. We have $\|b\| \ne 0$. (Otherwise $b = \lim a_n$ for $a_n = 0 \in A$, and $b \in A$ because A is closed.)

We now find the coefficient c for φ. Since $H = A + \mathbb{F}b$, for any $c, z \in H$ there are $a, x \in A$ and $\gamma, t \in \mathbb{F}$ such that $c = a + \gamma b$ and $z = x + tb$. Therefore,

$$\varphi(z) = \varphi(x + tb) = \varphi(x) + t\varphi(b) = t\varphi(b),$$
$$cz = (a + \gamma b)(x + tb) = ax + \bar{\gamma}t \cdot \|b\|^2.$$

(We have $bx = ab = 0$ for $a, x \in A$ and $b \in A^{\perp}$, while $bb = \|b\|^2$.) If $a = 0$ and $\bar{\gamma} = \|b\|^{-2} \cdot \varphi(b)$, then

$$\varphi(z) = t\varphi(b) = t\varphi(b)\|b\|^{-2}\|b\|^2 = \bar{\gamma}t \cdot \|b\|^2 = cz.$$

Therefore, $c = 0 + \overline{\varphi(b)} \cdot \|b\|^{-2} \cdot b = \varphi(b) \cdot \|b\|^{-2} \cdot b$. This completes the proof.
$\square$

The Riesz representation theorem can be briefly stated as follows: *Every continuous linear functional on a Hilbert space has a coefficient.*

The Riesz theorem allows us to view continuous linear functionals on a Hilbert space as its points. This may be convenient in many situations.

4.1.3. The spaces $\mathcal{L}^2 = \mathcal{L}^2(U, \mu)$

The spaces $\mathcal{L}^2 = \mathcal{L}^2(U, \mu)$ are among the most important kinds of Hilbert spaces.

1. We denote by $\mathcal{L}^2 = \mathcal{L}^2(U, \mu)$ the set of functions $f : U \to \mathbb{F}$ measurable with respect to a measure μ whose squares $f^2 : U \to \mathbb{F}$ are integrable with respect to μ. Thus, by definition, if f is measurable with respect to μ, then $f \in \mathcal{L}^2 \Leftrightarrow f^2 \in \mathcal{L}$, where $\mathcal{L} = \mathcal{L}(U, \mu)$ is the set of all numeric functions on U that are integrable with respect to μ.

Since $|fg| \leq 2^{-1}(|f|^2 + |g|^2)$, it follows that $fg \in \mathcal{L}$ for any $f, g \in \mathcal{L}^2$. (It is not necessarily true that $fg \in \mathcal{L}^2$. In addition, the functions f and g are not necessarily integrable.)

Exercise. Provide the corresponding counterexamples.

It is easy to verify that $\mathcal{L}^2$ together with the conventional operations forms a vector space:

$$(\alpha f + \beta g)^2 = \alpha^2 f^2 + 2\alpha\beta fg + \beta^2 g^2 \in \mathcal{L},$$

if $f^2, g^2 \in \mathcal{L}$ and, consequently, $fg \in \mathcal{L}$.

The equality $\langle f, g \rangle = \int \bar{f} g$ ($f, g \in \mathcal{L}^2$) defines a scalar product on $\mathcal{L}^2$. Indeed, we have

$$\langle f, g \rangle = \int \bar{f} g = \overline{\int \bar{g} f} = \overline{\langle g, f \rangle},$$

$$\langle \alpha_1 f_1 + \alpha_2 f_2, g \rangle = \int (\bar{\alpha}_1 \bar{f}_1 + \bar{\alpha}_2 \bar{f}_2) g$$

$$= \bar{\alpha}_1 \int \bar{f}_1 g + \bar{\alpha}_2 \int \bar{f}_2 g = \bar{\alpha}_2 \langle f_2, g \rangle,$$

$$\langle f, \beta_1 g_1 + \beta_2 g_2 \rangle = \beta_1 \langle f, g_1 \rangle + \beta_2 \langle f, g_2 \rangle,$$

$$\langle f, f \rangle = \int \bar{f} f = \int |f|^2 \geq 0.$$

As in all Euclidean spaces, the equality

$$\|f\| = \langle f, f \rangle^{1/2} = \left(\int |f|^2 \right)^{1/2} \quad (f \in \mathcal{L}^2)$$

defines the norm $\| \cdot \|$ on $\mathcal{L}^2$. This norm does not necessarily separate the points of the space.

Examples. 1) $\mathcal{L}^2(\mathbb{R}, dx)$. 2) $\mathcal{L}^2([0, 1], dx)$. 3) $\mathcal{L}^2 = \mathcal{L}^2(\mathbb{N}, dn) = l^2$.

The space $\mathcal{L}^2$ is separated in examples 1 and 2 and is not separated in example 3.

2. We now prove a lemma that describes the connection between convergence in $\mathcal{L}^2$ and convergence almost everywhere. Convergence in $\mathcal{L}^2$ is called *mean convergence (mean-square convergence)*.

By definition, a sequence $f_n \in \mathcal{L}^2$ converges in mean to $f \in \mathcal{L}^2$ if

$$\|f_n - f\| = \left(\int |f_n - f|^2 \right)^{1/2} \to 0.$$

Mean convergence does not imply convergence almost everywhere, and neither does the latter imply the former.

Exercise. Provide the corresponding counterexamples.

The Riesz lemma. *If a sequence of functions in $\mathcal{L}^2$ converges in mean, then it has a subsequence that converges almost everywhere.*

Proof. (1) Let a sequence $f_n \in \mathcal{L}^2$ converge in mean, i.e., $\|f_q - f_p\| \to 0$ $(q, p \to \infty)$. It is required to prove that there exists a subsequence $g_n = f_{r(n)}$ that converges almost everywhere, i.e., $|g_q(x) - g_p(x)| \to 0$ $(q, p \to \infty)$ for almost all $x \in U$ with respect to the measure μ.

Note that for any index n there is an index $r(n)$ such that $\|g_{n+1} - g_n\| \le 2^{-n}$ for $g_n = f_{r(n)}$. The indices $r(n)$ can be chosen to be strictly increasing. Indeed, since f_n converges in mean, there exists an index $r(1)$ such that $\|f_q - f_p\| \le 2^{-1}$ $(p, q \ge r(1))$. Take an arbitrary index n and suppose that strictly increasing numbers $r(m)$ are chosen for $m \le n$ so that $\|f_q - f_p\| \le 2^{-m}$ $(p, q \ge r(m))$. Since f_n converges in mean, there exists an index $r(n+1) > r(n)$ such that $\|f_q - f_p\| \le 2^{-(n+1)}$ $(p, q \ge r(n+1))$. Hence, by induction, there exists a strictly increasing sequence of indices $r(n)$ such that $\|f_q - f_p\| \leq 2^{-n}$ $(p, q \geq r(n))$. In particular, this inequality holds for $p = r(n)$ and $q = r(n+1)$, which is the desired result.

(2) For every index n, consider the function

$$s_n = |g_1| + \sum_{1 \leq m < n} |g_{m+1} - g_m|$$

(in particular, $s_1 = |g_1|$). Since $g_1, g_{m+1} - g_m \in \mathcal{L}^2$, we have $|g_1|, |g_{m+1} - g_m| \in \mathcal{L}^2$ and therefore $s_n \in \mathcal{L}^2$. It is obvious that the sequence s_n increases.

(3) The sequence of functions $h_n = s_n^2$ satisfies the hypothesis of Levi's theorem. Indeed, $s_n \in \mathcal{L}^2$ implies $h_n \in \mathcal{L}$. Since the sequence s_n increases, the sequence h_n also increases. In addition, (h_n) is integrally bounded:

$$\int h_n = \int s_n^2 = \|s_n\|^2 = \left\| |g_1| + \sum_{1 \leq m < n} |g_{m+1} - g_m| \right\|^2$$

$$\leq \left(\|g_1\| + \sum_{1 \leq m < n} \|g_{m+1} - g_m\| \right)^2 \leq \left(\|g_1\| + \sum_{1 \leq m < n} 2^{-m} \right)^2$$

$$\leq (\|g_1\| + 1)^2$$

for any index n. By Levi's theorem, (h_n) converges almost everywhere to a function $h \in \mathcal{L}$. Since $h_n = s_n^2 \geq 0$, it follows that $h_n \to h$ implies $s_n \to s = h^{1/2}$.

(4) As a result, the series $|g_1|, |g_{m+1} - g_m|$ $(m \geq 1)$ has a sum s almost everywhere. Consequently, the series $g_1, g_{m+1} - g_m$ $(m \geq 1)$ also has a sum almost everywhere because the sequence

$$g_n = g_1 + \sum_{1 \leq m < n} (g_{m+1} - g_m)$$

of partial sums of this series converges almost everywhere. We conclude that the subsequence $(f_{r(n)})$ of (f_n) converges almost everywhere. $\quad\square$

3. We now prove the main statement of this subsection.

Theorem. $\mathcal{L}^2$ *is a Hilbert space.*

Proof. It is required to prove that every sequence of functions $f_n \in \mathcal{L}^2$ that converges in itself converges to a function $f \in \mathcal{L}^2$ ($\|f_n - f\| \to 0$).

Take a sequence of $f_n \in \mathcal{L}^2$ that converges in itself. By the Riesz lemma, there exits a subsequence $g_n = f_{r(n)}$ that converges almost everywhere.

Consider a function $f : U \to \mathbb{F}$ such that $f(x) = \lim g_n(x)$ if $g_n(x)$ converges and $f(x) = 0$ otherwise.

We will prove that (1) $f \in \mathcal{L}^2$, (2) $\|f_n - f\| \to 0$.

(1) Since (f_n) converges in itself, the sequence $g_n = f_{r(n)}$ also converges in itself, i.e., for any $\varepsilon > 0$ there exists an index $n(0)$ such that $\|g_q - g_p\| \le \varepsilon$ $(q, p \ge n(0))$.

Take a $p \ge n(0)$ and consider the sequence of functions $h_n = |g_{p+n} - g_p|^2$ $(n \ge 1)$. The sequence (h_n) satisfies the assumption of Fatou's lemma. Indeed, $h_n \ge 0$, and $g_{p+n} - g_p \in \mathcal{L}^2$ implies $h_n \in \mathcal{L}$. The sequence (h_n) is integrally bounded:

$$\int h_n = \int |g_{p+n} - g_p|^2 = \|g_q - g_p\|^2 \le \varepsilon^2$$

for all n. Finally, since $g_{p+n} \to f$ $(n \to \infty)$ almost everywhere, we have $h_n \to h = |f - g_p|^2$ almost everywhere. By Fatou's lemma, $h \in \mathcal{L}$ and

$$\int h = \int |f - g_p|^2 \le \varepsilon^2 \quad (p \ge n(0)). \tag{$*$}$$

Hence $f - g_p \in \mathcal{L}^2$. Since $g_p \in \mathcal{L}^2$, it follows that

$$f = (f - g_p) + g_p \in \mathcal{L}^2.$$

(2) The inequality $(*)$ means that for any $\varepsilon > 0$ there is an index $n(0)$ such that $\|f - g_p\| \le \varepsilon$ $(p \ge n(0))$. At the same time, since $r(p) \ge p$ and the sequence f_n converges in mean, it follows that $\|g_p - f_n\| = \|f_{r(p)} - f_n\| \le \varepsilon$ $(p, n \ge n(1))$ for some $n(1) \ge n(0)$. As a result,

$$\|f - f_n\| \le \|f - g_p\| + \|g_p - f_n\| \le 2\varepsilon \quad (p, n \ge n(1)).$$

We conclude that the sequence (f_n) converges in mean to f. $\square$

Corollary. *Any continuous linear functional φ on $\mathcal{L}^2$ is defined by a function f in $\mathcal{L}^2$ and has values $\varphi(g) = \int \bar{f}g$ $(g \in \mathcal{L}^2)$.*

Proof. Since $H = \mathcal{L}^2$ is a Hilbert space, the Riesz representation theorem for continuous linear functionals holds for it. Therefore, any $\varphi \in H'$ has coefficient $c = f \in H$ and $\varphi(g) = \int \bar{f}g$ $(g \in H)$, which completes the proof. $\square$

The spaces $\mathcal{L}^2$ are widely used in the theory of differential and integral equations.

Remark. Elements of the theory of Hilbert spaces are presented in Halmos (1982) in the form of problems with instructions and solutions. Various kinds of operators are described in detail, namely, multiplication and shift operators, compact, subnormal, Toeplitz operators, and operators represented by infinite matrices.

The second volume in von Neumann (1987) is devoted to a series of classical works on the theory of weakly closed algebras of operators in Hilbert spaces. In the comments, the present condition of the theory is briefly described. The issues related to general measure theory and ergodic theory are discussed.

4.2. FOURIER SERIES

The modern theory of Fourier series is presented in Edwards (1979, 1982).

We describe orthogonal series in a separated Hilbert space. It is usually assumed that the scalar product is nondegenerate and the norm separates the points of the space. The general case can be reduced to this special case using factorization over the subspace of vectors of zero norm. Then equivalences become equalities.

4.2.1. Fourier coefficients

Fourier coefficients are scalar products of a given vector and the vectors of a selected orthonormal basis.

1. Let H be a Hilbert space and let Q be a countable everywhere dense set in H, i.e., for any $z \in H$ there exists a sequence $x_n \in Q$ converging to z. Such spaces are called *separable*. There exist orthonormal basis sequences in them.

Take a finite or infinite number $m = 2, 3, \ldots, \infty$. Choose a maximal collection of linearly independent vectors $x_n \in Q$ ($1 \leq n < m$). Its linear span includes Q and is therefore dense in H. It is easy to verify that the equalities

$$y_n = x_n - \sum_{1 \leq k < n} (x_n e_k) e_k, \quad e_n = \|y_n\|^{-1} \cdot y_n,$$

define the orthonormal collection of vectors e_n ($1 \leq n < m$): $e_k e_n = 0$ ($k \neq n$), $\|e_n\| = 1$. The products of vectors are scalar and equalities mean equivalences in the nonseparated case. Linear independence is also defined up to equivalence. This ensures that $\|y_n\| \neq 0$ (since the equality $\|y_n\| = 0$ implies that $x_1, \ldots, x_n$ are linearly dependent).

Exercise. Prove the above statements.

The inductive method of obtaining the orthonormal collection of vectors e_n from the vectors x_n using the specified equalities is called the *Schmidt algorithm*. The vectors e_n are linear combinations of the vectors x_n and vice versa. Therefore, the linear span of the orthonormal collection of vectors e_n coincides with that of the collection of vectors x_n and is dense in H. This means that the vectors e_n form an *orthonormal basis* of the Hilbert space H: for any $z \in H$ there exists a sequence of linear combinations of e_n that converges to z. Moreover, as we will see below, any vector z is equal to the sum of a series of vectors e_n taken with some numeric coefficients c_n, i.e., $z = \sum c_n e_n$ $(1 \leq n < m)$.

Examples. Take an interval $U = |a, b|$ of the real line $(-\infty \leq a < b \leq \infty)$, the algebra $\mathcal{R}$ generated by bounded subintervals of U, a measure μ on $\mathcal{R}$, the space $\mathcal{L}^2 = \mathcal{L}^2(U, \mu)$ of numeric functions on U that are measurable and square integrable with respect to μ, and the sequence of power functions x^n with values $x^n(t) = t^n$ $(t \in U)$.

It is easy to verify that the functions x^n $(n = 0, 1, 2, \ldots)$ are linearly independent. (If a polynomial is identically equal to zero on a nondegenerate interval, then all its coefficients are equal to zero.) Using the Schmidt algorithm to orthonormalize the sequence x^n, we obtain an orthonormal sequence of polynomials p_n $(n = 0, 1, 2, \ldots)$ such that

$$\int p_j(t) \cdot p_k(t)\, d\mu(t) = 0 \quad (j \neq k), \quad \int p_n^2(t)\, d\mu(t) = 2.$$

1) Put $a = -1$, $b = 1$, and let $d\mu(t) = dt$ be the Lebesgue measure. Then p_n is a Legendre polynomial.

2) Put $a = -1$, $b = 1$, and let $d\mu(t) = (1 - t^2)^{-1/2} dt$ be the measure with the derivative $f(t) = (1 - t^2)^{-1/2}$ $(-1 < t < 1)$ with respect to the Lebesgue measure dt. Then p_n is a Chebyshev polynomial.

3) Put $a = -\infty$, $b = \infty$, and $d\mu(t) = \exp(-t^2) dt$. Then p_n is a Hermite polynomial.

4) Put $a = 0$, $b = \infty$, and $d\mu(t) = \exp(-t) dt$. Then p_n is a Laguerre polynomial.

5) Put $U = [-\pi, \pi]$ and let $d\mu(t) = dt$ be the Lebesgue measure. The collection of functions e_k on $[-\pi, \pi]$ with values $e_k(t) = (2\pi)^{-1/2} e^{-ikt}$ $(k = 0, \pm 1, \pm 2, \ldots)$ forms an orthonormal basis of the space $\mathcal{L}^2$. This basis is

called the *trigonometric basis*, since the Euler formula states that

$$e^{-ikt} = \cos(kt) - i\sin(kt).$$

If $U = [a, b]$ $(-\infty < a < b < \infty)$, then we have the orthonormal basis consisting of functions e_k on $[a, b]$ with values

$$e_k(t) = (2\pi)^{-1/2} \exp\{-i \cdot 2\pi(b - a)^{-1}k(t - 2^{-1}(a + b))\}.$$

Remark. Strictly speaking, the vectors e_k in the trigonometric basis should be numbered properly in order to obtain a series. However this is not necessary, since everything that was said above is true for any orthonormal collection. The proper numeration comes naturally when passing to sines and cosines.

2. If the Hilbert space H with orthonormal basis consisting of vectors e_n $(1 \le n < m)$ is finite-dimensional $(m < \infty)$, then

$$x = \sum_{1 \le n < m} \xi_n e_n, \quad \xi_n = e_n x \quad (x \in H).$$

It turns out that this is also true for $m = \infty$. We have

$$x = \lim s_n, \quad s_n = \sum_{1 \le k \le n} \xi_k e_k, \quad \|x - s_n\| \to 0.$$

The scalar products $\xi_n = e_n x$ are called the *Fourier coefficients* of the vector x with respect to the basis consisting of the vectors e_n or the *coordinates* of x with respect to this basis. The series $\xi_n e_n$ $(\xi_n = e_n x)$ is called the *Fourier series* for x with respect to the basis (e_n).

We will prove the desired assertion about the sum of the Fourier series. Consider a natural number $n < m$ and a subspace A of H generated by the first n basis vectors $e_1, \ldots, e_n$. We now calculate the distance from the point $x \in H$ with coordinates $\xi_k = e_k x$ to the point $a = \sum \alpha_k e_k \in A$ $(\alpha_k \in \mathbb{F}, 1 \le k \le n)$.

Lemma. $\|x - a\|^2 = \|x\|^2 - \sum |\xi_k|^2 + \sum |\xi_k - \alpha_k|^2.$

Proof. Taking into account that the vectors e_k are orthonormal, we have $\|x - a\|^2 = (x - a)(x - a) = xx - xa - ax + aa = \|x\|^2 - \sum \alpha_k \bar{\xi}_k - \sum \bar{\alpha}_k \xi_k + \sum |\alpha_k|^2 = \|x\|^2 - \sum |\xi_k|^2 + \sum |\xi_k - \alpha_k|^2$ because $|\xi_k - \alpha_k|^2 = (\xi_k - \alpha_k)(\bar{\xi}_k - \bar{\alpha}_k) = |\xi_k|^2 - \alpha_k \bar{\xi}_k - \bar{\alpha}_k \xi_k + |\alpha_k|^2$ for any numbers ξ_k, α_k. □

We denote by $B = A^{\perp}$ the orthogonal complement of A with respect to H. Consider the orthogonal projections $P : H \to A$ and $Q : H \to B$.

Corollary. $Px = \sum \xi_k e_k$ $(1 \le k \le n)$.

Proof. By the preceding lemma, for $\alpha_k = \xi_k$ and $a = \sum \alpha_k e_k \in A$ the distance from x to a is the smallest. Therefore, a is a point of A that is closest to x. By the orthogonal projection theorem, it follows that $Px = a$ (Px is equivalent to a in the nonseparated case). $\square$

By the definition of an orthogonal projection, for any numbers α_k we have $(x - \sum \xi_k e_k)(\sum \alpha_k e_k) = 0$. It readily follows from the preceding lemma that

$$\|x\|^2 = \sum |\xi_k|^2 + \left\| x - \sum \xi_k e_k \right\|^2.$$

for $\alpha_k = \xi_k$. This equality also follows from the Pythagorean theorem. Hence $\sum_{1 \le k \le n} |\xi_k|^2 \le \|x\|^2$ for any $n < m$.

Bessel's inequality. $\sum_{1 \le n < m} |\xi_n|^2 \le \|x\|^2$.

Bessel's inequality makes it possible to solve the problem of finding the sum of the Fourier series.

3. The following criterion is convenient in the case of orthogonal series $\alpha_n e_n$ $(\alpha_n \in \mathbb{F})$ in a Hilbert space H with the orthonormal basis consisting of vectors e_n $(1 \le n < m \le \infty)$.

The summability criterion. A series $(\alpha_n e_n)$ is summable in H if and only if $\sum |\alpha_n|^2 < \infty$.

Proof. Let $p > n$ and $s_n = \sum_{1 \le k \le n} \alpha_k e_k$, $s_p = \sum_{1 \le k \le p} \alpha_k e_k$. Then $\|s_p - s_n\|^2 = \left\| \sum_{n < k \le p} \alpha_k e_k \right\|^2 = \sum_{n < k \le p} |\alpha_k|^2$ because the vectors e_k are orthonormal. Applying the Cauchy criterion, we obtain the desired assertion. $\square$

The theorem on the existence and uniqueness of a Fourier series expansion is easily proved using Bessel's inequality and the summability criterion.

Theorem. (1) The sum of the Fourier series $(\xi_n e_n)$, $\xi_n = e_n x$ for the vector $x \in H$ is equal to x.

(2) If the sum of the series $(\eta_n e_n)$ for numbers η_n is equal to x, then $\eta_n = \xi_n$.

Proof. (1) By Bessel's inequality, $\sum |\xi_n|^2 < \infty$. Applying the summability criterion, we conclude that the series $\xi_n e_n$ is summable. We denote its sum by s. It remains to show that $s = x$, i.e., $\|s - x\| = 0$ (s is equivalent to x). Since e_n form an orthonormal basis of the space H, it suffices to show that $e_j(s - x) = 0$ ($1 \le j < m$). Consider the partial sum $s_n = \sum_{1 \le k \le n} \xi_k e_k$. We have $e_j(x - s_n) = \xi_j - \sum_k \xi_k e_j e_k = e_j - e_j = 0$ for $1 \le j \le n$. By the continuity of the scalar product,

$$e_j(s - x) = e_j(x - \lim_n s_n) = \lim_n e_j(x - s_n) = 0, \qquad 1 \le j < m.$$

(2) Let $\sum \eta_n e_n = x = \sum \xi_n e_n$. Then $\sum (\eta_n - \xi_n) e_n = 0$. Hence, $\eta_j - \xi_j = e_j(\eta_j - \xi_j)e_j = e_j \sum(\eta_n - \xi_n)e_n = 0$ for $1 \le j < m$. $\square$

Example. Let $H = \mathcal{L}^2([-\pi, \pi], dt)$ and let (e_k) be the trigonometric basis. Then the Fourier coefficients of the function $x \in H$ are written as

$$\xi_k = e_k x = (2\pi)^{-1/2} \int_{-\pi}^{\pi} e^{ikt} x(t)\, dt.$$

The Fourier series $(\xi_k e_k)$ is mean summable, i.e.,

$$\left\| x - \sum_{|k| \le n} \xi_k e_k \right\|^2 = \int_{-\pi}^{\pi} \left| x(t) - \sum_{|k| \le n} \xi_k e_k \right|^2 dt \to 0 \quad \text{for} \quad n \to \infty.$$

Remark. Convergence of the sequence of partial sums $s_n = \sum_{|k| \le n} \xi_k e_k$ to the function x at every point and especially uniform convergence $s_n \to x$ require additional conditions. Based on the Riesz lemma, we can only assert that there exists a subsequence of the sequence s_n that converges to x at almost every point. L. Carleson proved that the sequence s_n itself converges at almost every point (Carleson, 1966; Edwards, 1982, Subsection 10.4.5).

4. Every vector $x \in H$ and its Fourier coefficients $\xi_n = e_n x$ satisfy Parseval's identity, which is a special case of Bessel's inequality.

Parseval's identity. $\|x\|^2 = \sum |\xi_n|^2$.

Proof. Consider the partial sums $s_n = \sum_{1 \le k \le n} \xi_k e_k$, $\sigma_n^2 = \sum_{1 \le k \le n} |\xi_k|^2$. As was shown above, $\|x\|^2 = \sigma_n^2 + \|x - s_n\|^2$, $\sigma_n^2 \to \sigma^2 = \sum |\xi_k|^2$, $\|x - s_n\| \to 0$. Hence $\|x\|^2 = \sigma^2$. $\square$

Parseval's identity generalizes the Pythagorean theorem to the case of orthonormal collections of vectors and allows expressing the length of a vector x in terms of the Fourier coefficients $\xi_n = e_n x$, i.e.,

$$\|x\| = \left(\sum |\xi_n|^2 \right)^{1/2}.$$

This equality is as widely used as the Pythagorean theorem.

4.2.2. Isomorphism of Hilbert spaces

All separable Hilbert spaces are isomorphic to each other.

1. Consider the space l^2 of sequences of numbers ξ_n ($1 \le n < m$) and a separable Hilbert space H. As follows from what was proved above, every vector $x \in H$ is associated with the sequence of Fourier coefficients $\xi_n = e_n x$ in l^2. Parseval's identity shows that this correspondence preserves the length of vectors. Obviously, it is linear, which means that it is a linear isometry from H to l^2. By proving that any sequence in l^2 is a sequence of Fourier coefficients of a vector in H, we will establish that the Hilbert spaces H and l^2 are isomorphic up to the equivalence of vectors.

2. As before, it is assumed that H has an orthonormal basis consisting of vectors e_n ($1 \le n < m \le \infty$), i.e., H is a separable finite-dimensional or infinite-dimensional Hilbert space. Similarly, l^2 consists of finite or infinite sequences ξ_n ($1 \le n < m \le \infty$) such that $\sum |\xi_n|^2 < \infty$.

The Riesz–Fischer theorem. *Let* $(\xi_n) \in l^2$. *Then there exists an* $x \in H$ *such that* $\xi_n = e_n x$.

Proof. By the criterion of summability of series $\xi_n e_n$ in H, the assumption of the theorem implies that there exists a vector $x = \sum \xi_n e_n \in H$. Since the Fourier series expansion is unique, we have $\xi_n = e_n x$. $\square$

Corollary. *All separable Hilbert spaces of the same dimension are isomorphic to each other up to equivalence.*

Example. The space $\mathcal{L}^2 = \mathcal{L}^2([a, b], dt)$ of functions on an interval $[a, b]$ ($-\infty < a < b < \infty$) is isomorphic (up to equivalence) to the space $l^2 = \mathcal{L}^2(\mathbb{N}, dn)$ of number sequences.

The described isomorphism is widely used in theory and applications.

4.3. FUNCTION SPACES

This Section deals with some popular function spaces (see Dunford and Schwartz, 1988, Chapter 4).

4.3.1. Metric spaces

1. The completion of a metric space. Consider metric spaces (E_1, d_1) and (E_2, d_2) and a map $\varphi : E_1 \to E_2$. If

$$d_1(x_1, y_1) = d_2(\varphi(x_1), \varphi(y_1)) \quad (x_1, y_1 \in E_1),$$

then φ is said to *preserve the distance* or φ is an *isometry*. If (E_1, d_1) is separated, then φ is a one-to-one isometry. Surjective isometries are called *isomorphisms of metric spaces* (up to equivalence).

A metric space $(\bar{E}, \bar{d})$ is called the *completion* of a metric space (E, d) if the space $(\bar{E}, \bar{d})$ is complete and the space (E, d) is isomorphic to an everywhere dense subspace of the space $(\bar{E}, \bar{d})$. Such an isomorphism makes it possible to identify the set E with its image in $\bar{E}$ and consider E a subset of $\bar{E}$. This helps to simplify the terminology.

The completion theorem. *There exists a completion for any metric space. All completions of a given metric space are isomorphic to each other.*

This theorem makes it possible to replace arbitrary metric spaces with complete ones whenever necessary and use all the benefits of completeness.

Examples. 1) If $E = \mathbb{Q}$ is the space of rational numbers, then its completion $\bar{E} = \mathbb{R}$ is the space of real numbers (the metrics are conventional).

2) If $E = \mathcal{P}[a, b]$ is the space of polynomials on the interval $[a, b]$, then its completion $\bar{E} = C[a, b]$ is the space of continuous functions on $[a, b]$ (the metrics are uniform). This follows from the Weierstrass approximation theorem.

3) If $E = \mathcal{S}[a, b]$ is the space of step functions on the interval $[a, b]$, then its completion $\bar{E} = \mathcal{D}[a, b]$ is the space of functions with no essential discontinuities on $[a, b]$ (i.e., functions that have left and right limits at every point). The metrics in this case are uniform.

In the general case, the elements of the completion $\bar{E}$ are convergent sequences of points of the space E. If a sequence converges to a point of E, then it is identified with this point. If the sequence converges in itself but does not converge to any point of E, then it is considered a new element and is added to the completion of E being constructed. The metric $\bar{d}$ on the completion $\bar{E}$ is defined on the basis of the metric d on E by passing to the limit (Kolmogorov and Fomin, 1989, Chapter II, Section 3):

$$\bar{d}(\bar{x}, \bar{y}) = \lim d(x_n, y_n) \quad (\bar{x} = \lim x_n, \ \bar{y} = \lim y_n \in \bar{E}).$$

When passing to the completion, the invariance and absolute uniformity of the metric is preserved. Therefore, the completion of any normed space is a Banach space.

2. Continuous extension. For constructing the completion of metric and normed spaces, it is often necessary to extend functions by continuity. Consider metric spaces E and F, a set $A \subseteq E$ and its closure $\bar{A}$, and a function $f : A \to F$. The general theorems on continuous extensions proved in Subsections 3.1.3 (parts 13 and 14) imply analogous theorems for metric spaces.

The extension theorem. *Let the metric space F be separated and let the function $f : A \to F$ have a limit at every point $\bar{x} \in \bar{A}$. Then the equality*

$$\bar{f}(\bar{x}) = \lim_{x \to \bar{x}} f(x)$$

defines a unique continuous function $\bar{f} : \bar{A} \to F$ that is an extension of f.

In the general case of the theorem of determining a continuous extension $\bar{f}$ of the function f, the equality must be replaced by the condition

$$\bar{f}(\bar{x}) \in \lim_{x \to \bar{x}} f(x)$$

and the part of the assertion stating the uniqueness of a continuous extension must be omitted. Alternatively, all points such that the distance between them is equal to zero can be identified with each other, i.e., the space under consideration can be replaced by the separated space obtained from the former using the above identification of points. Equivalences can be considered instead of equalities.

The extension theorem implies an analogous statement for uniformly continuous functions.

Corollary. *Let the metric space F be separated and complete, and let $f : A \to F$ be uniformly continuous. Then there exists a unique uniformly continuous function $\bar{f} : \bar{A} \to F$ that is an extension of f.*

Another way to formulate the theorem on a uniformly continuous extension is as follows: if F is separated and complete and f is a uniformly continuous function, then its continuous extension $\bar{f}$ exists and is uniformly continuous. In particular, this theorem makes it possible to extend bounded linear operators on Banach spaces from vector subspaces to their closures. This is how generalized solutions to some differential equations are obtained.

4.3.2. Smooth functions

We first consider the largest class of smooth functions, namely, continuous functions, and then turn to some other kinds.

1. **The space of continuous functions.** Consider a metric space (E, d), a Banach space F, a compact set $A \subseteq E$, and the vector space $C = C(A, F)$ of continuous maps from A to F. As usual, we assume that the space F is separated.

For any $f \in C$, the set $f(A) \subseteq F$ is compact and therefore bounded. Then a norm on C can be defined using the equality $\|f\| = \sup\{\|f(x)\| : x \in A\}$. The vector space C with this norm forms a Banach space.

It follows from the definitions that the convergence of a sequence of functions $f_n \in C$ in this space is equivalent to its uniform convergence.

If there exists a countable everywhere dense set of points in the space F, then such a set also exists in C. That is, if F is separable then so is C. In particular, the space of numeric continuous functions is separable as well as the space of finite-dimensional vector functions. This can be easily verified using the property of uniform convergence and the compactness of the domain of definition of functions in C (see Kolmogorov and Fomin, 1989, Chapter 2, Sections 1 and 2).

Compact sets in the space C are described using the notion of uniform equicontinuity.

A set $\mathcal{A} \subseteq C$ is said to be *uniformly equicontinuous* if for any $\varepsilon > 0$ there exists a $\delta > 0$ such that

$$\|f(x) - f(y)\| \le \varepsilon \quad (d(x, y) \le \delta; \; x, y \in A)$$

for all $f \in \mathcal{A}$. In other words, if two values of the argument are sufficiently close, then, for any function in the uniformly equicontinuous set, the corresponding values of this function can be made as close as desired.

We denote by $\mathcal{A}(x)$ the set of values of functions $f \in \mathcal{A} \subseteq C$ at a point $x \in A$:

$$\mathcal{A}(x) = \{f(x) : f \in \mathcal{A}\}.$$

As any set in F, the set $\mathcal{A}(x)$ can be relatively compact.

The Ascoli theorem. *A closed set $\mathcal{A} \subseteq \mathcal{C}$ is compact if and only if $\mathcal{A}$ is relatively compact for every point and uniformly equicontinuous.*

The *if* part is easy to prove. To prove the *only if* part, it is required to show that the specified condition ensures the existence of an ε-net for $\mathcal{A}$ for any $\varepsilon > 0$. This will make it possible to partition $A \times F$ into finitely many rectangles with sides of lengths ε and δ. These rectangles make up step strips over A that cover $\mathcal{A}$. Choosing a function in each of these strips, we obtain an ε-net for $\mathcal{A}$ (see Kirillov and Gvishiani, 1988, Chapter 3, Section 3; Edwards, 1995, Section 0.4).

If $F = \mathbb{R}^m$, then the relative compactness of the set $\mathcal{A} \subseteq \mathcal{C}$ is equivalent to the uniform boundedness of functions in $\mathcal{A}$, ie., there exists an $\alpha > 0$ such that

$$\|f(x)\| \leq \alpha \quad (x \in A; \; f \in \mathcal{A}).$$

Thus, the Ascoli theorem implies the following statement (see Kantorovich and Akilov, 1982, Chapter 1, Section 5).

The Arzelà theorem. *If $F = \mathbb{R}^m$, then a closed set $\mathcal{A} \subseteq \mathcal{C}$ is compact if and only if $\mathcal{A}$ is uniformly bounded and uniformly equicontinuous.*

The uniform boundedness of $\mathcal{A}$ means its boundedness in the normed space $\mathcal{C}$, i.e., there exists an $\alpha > 0$ such that $\|f\| \leq \alpha$ for all $f \in \mathcal{A}$ (cf. Dunford and Schwartz, 1988, Chapter 4, Section 6).

Examples. Let $E = F = \mathbb{R}$ and $A = [0, 1]$. Then the set $\mathcal{A}$ of linear functions on $[0, 1]$ with coefficients in $[-1, 1]$ is closed, uniformly bounded, and uniformly equicontinuous. Therefore, it is compact. At the same time, the set $\mathcal{B}$ of all linear functions on $[0, 1]$ is not compact.

Remark. Different versions of the Ascoli theorem for the sets of maps from a topological space to a uniform space are proved in Kelley (1975), Chapter 7, and Edwards (1995), Section 0.4 (see also Kirillov and Gvishiani, 1988, Problem 337)

The space $\mathcal{V}$ of numeric measures on the algebra generated by closed subsets of A is conjugate to the space $\mathcal{C}$. The norm of the measure is represented by its variation which is equal to the sum of its positive components. Every numeric measure is equal to a linear combination of some positive measures. The correspondence between linear functionals $\varphi \in \mathcal{C}'$ on $\mathcal{C}$ and numeric measures $\mu \in \mathcal{V}$ is established with the use of integration (Dunford and Schwartz, 1988, Chapter 4, Section 6).

The Riesz theorem. *The equality*

$$\varphi(f) = \int f \, d\mu \quad (f \in \mathcal{C})$$

defines an isometric correspondence between the spaces $\mathcal{C}'$ and $\mathcal{V}$.

Example. Let $\mu = \mu_a$ be the measure that describes the unit mass distribution at a point $a \in A$:

$$\mu_a(X) = 1 \ (a \in X), \quad \mu_a(X) = 0 \ (a \notin X)$$

for any $X \subseteq A$. Then the corresponding functional $\varphi = \delta_a$ is written as

$$\delta_a(f) = \int f \, d\mu_a = f(a) \quad (f \in \mathcal{C}).$$

Such functionals are called delta functions.

2. Spaces of smooth functions. Consider an open set U in the space $E = \mathbb{R}^q \ (q \geq 1)$ and the vector space $\mathcal{C}^p = \mathcal{C}^p(U)$ of p times continuously differentiable numeric functions on U $(0 \leq p \leq \infty)$. By definition, 0 times continuously differentiable functions are continuous functions. The points of the space $\mathcal{C}^\infty$ are functions that are differentiable infinitely many times. The functions $\mathcal{C}^p$ are called *p-smooth* or simply *smooth*.

In the notation for derivatives, it is convenient to use multi-indices $\alpha = (\alpha(1), \ldots, \alpha(q))$. Positive integers $\alpha(1), \ldots, \alpha(q)$ determine the number of times the function is differentiated with respect to the variables with indices $1, \ldots, q$, respectively. The corresponding variable is denoted by $\partial^\alpha = \partial^{\alpha(1)} \ldots \partial^{\alpha(q)}$.

The number $|\alpha| = \alpha(1) + \cdots + \alpha(q)$ represents the order of the derivative ∂^α. It is assumed that $|\alpha| \leq p$ for a differentiable function $f \in \mathcal{C}^p$. The derivatives $\partial^\alpha f$ are continuous numeric functions on U. Their values at $u \in U$ are denoted by $\partial^\alpha f(u)$ or $\partial^\alpha f u$. In particular, $\partial^0 f = f$ and $\partial^0 f(u) = f(u)$.

It is conventional to write $\partial/\partial x_1, \ldots, \partial/\partial x_q$ instead of $\partial_1, \ldots, \partial_q$. Some other designations are used as well.

In addition, we introduce the *multifactorial* and the *multidegree*: $\alpha! = \alpha(1)! \ldots \alpha(q)!$ and $v^\alpha = v_1^{\alpha(1)} \ldots v_q^{\alpha(q)}$ for $v = (v_1, \ldots, v_q) \in \mathbb{R}^q$. Then the

Taylor formula for $p < \infty$, $u \in U$, $v \in U - u$, and $f \in C^p$ is written as

$$f(u + v) = \sum_{|\alpha| < p} (\alpha!)^{-1} \cdot \partial^\alpha f u \cdot v^\alpha + r_p f u(v),$$

$$r_p f u(v) = p \sum_{|\alpha| = p} (\alpha!)^{-1} \int_0^1 (1 - t)^{p-1} \partial^\alpha f(u + tv) \, dt \cdot v^\alpha.$$

Consider a collection of smooth numeric functions c_α on U and the polynomial $P(u, \cdot)$ of degree m such that

$$P(u, \xi) = \sum_{|\alpha| \leq m} c_\alpha(u) \cdot \xi^\alpha$$

for $\xi = (\xi_1, \ldots, \xi_q)$. Formally substituting ∂ for ξ, we define the *linear differential operator*

$$P(u, \partial) = \sum_{|\alpha| \leq m} c_\alpha(u) \cdot \partial^\alpha$$

on the space C^p.

Example. If $m = 2$, $c_{jj} = 1$, and $c_{jk} = 0$ for $j \neq k$ ($1 \leq j$, $k \leq q$), then $P(u, \partial) = \partial_1^2 + \cdots + \partial_q^2 = \Delta$ is the Laplace operator.

We now define a metric $C^p = C^p(U)$ in a three-stage process. First, for any $f \in C^p$, positive integer $m \leq p$, and compact set $\mathbb{K} \subseteq U$ we consider the norm

$$\|f\|_{mK} = \sup\{|\partial^\alpha f(u)| : |\alpha| \leq m, \, u \in K\}.$$

Choose an increasing sequence of compact sets $K(n) \subseteq U$ that covers U. For any index n and functions $f, g \in C^p$, consider the number

$$\rho_n(f, g) = \sum_{0 \leq m \leq p} \frac{1}{2^m} \cdot \frac{\|f - g\|_{mK(n)}}{1 + \|f - g\|_{mK(n)}}.$$

Finally, we calculate the distance between f and g in C^p:

$$\rho(f, g) = \sum_{0 \leq n < \infty} \frac{1}{2^n} \cdot \frac{\rho_n(f, g)}{1 + \rho_n(f, g)}.$$

It is relatively easy to verify the triangle inequality for this distance. If $p = \infty$, then the case $m = p$ in the inequality $m \leq p$ is ruled out. The distance does not depend on the choice of the sequence $K(n)$ (see Yosida, 1965, Chapter 1, Section 1).

It should be noted that the distance introduced for $\mathcal{C}^p$ is invariant under translations, but is not absolutely homogeneous. For this reason, the equality $\|f\| = \rho(f, 0)$ does not define a norm for $\mathcal{C}^p$. (Although norms that are not absolutely homogeneous are used in some cases. For example, see Dunford and Schwartz, 1988, vol. 1, Chapter 2, Section 1.)

In particular, consider the space $\mathcal{C}^\infty$. It follows from the definition of the metric on $\mathcal{C}^\infty$ that the convergence $f_n \to f$ in this space means the uniform convergence $\partial^\alpha f_n \to \partial^\alpha f$ on compact sets for the derivatives of any order. Hence, the metric space $\mathcal{C}^\infty$ is complete (see Kirillov and Gvishiani, 1988, Chapter 3, Section 4).

Complete metric vector spaces are called *Fréchet spaces*. Banach spaces are a special case of Fréchet spaces. The spaces $\mathcal{C}^p$ are an example of multi-normed (locally convex) metrizable spaces.

Functions in $\mathcal{C}^p$ that have compact support (i.e., vanish outside a compact set) are often called *finite*. They form the subspace $\mathcal{C}_0^p$ of the space $\mathcal{C}^p$.

Example. The function f on $\mathbb{R}^m$ with values

$$f(x) = \begin{cases} \exp\left\{-(1 - \|x\|^2)^{-1}\right\} & (\|x\| < 1), \\ 0 & (\|x\| \geq 1) \end{cases}$$

belongs to $\mathcal{C}_0^\infty = \mathcal{C}_0^\infty(\mathbb{R}^m)$.

3. Scalar product. Consider a vector space $\mathcal{H}^p = \mathcal{H}^p(U)$ of functions $f \in \mathcal{C}^p = \mathcal{C}^p(U)$ such that

$$\sum_{|\alpha| \leq p} \left(\int_U |\partial^\alpha f(x)| \, dx \right) < \infty.$$

The equality

$$\langle f, g \rangle = \sum_{|\alpha| \leq p} \left(\int_U \partial^\alpha \bar{f}(x) \cdot \partial^\alpha g(x) \cdot dx \right)$$

defines a scalar product on $\mathcal{H}^p$.

The vector space $\mathcal{H}_0^p$ of functions $f \in \mathcal{C}_0^p$ and a scalar product on it are defined in a similar way.

4. Bounded functions. In addition to the space $\mathcal{C}^p = \mathcal{C}^p(U)$ ($0 \leq p < \infty$) of smooth functions on an open subset U of the space $\mathbb{R}^m$, we consider the spaces $\mathcal{C}^p = \mathcal{C}^p(\bar{U})$ of functions that are smooth on U and can be extended to the closure $\bar{U}$ together with all their derivatives to bounded continuous numeric functions. In particular, the restrictions to $\bar{U}$ of smooth functions defined on an open neighborhood of the compact set $\bar{U}$.

A norm on the space $\mathcal{C}^p = \mathcal{C}^p(\bar{U})$ is defined by the equality

$$\|f\|_p = \sup\{|\partial^\alpha f(x)| : |\alpha| \leq p,\ x \in \bar{U})\}.$$

Since the extensions $\partial^\alpha f$ are bounded on $\bar{U}$, the defined norm is finite. The space $\mathcal{C}^p(\bar{U})$ is complete.

4.3.3. Lebesgue spaces

The points of Lebesgue spaces are measurable functions whose powers are integrable with respect to the Lebesgue measure or sets of such functions obtained by factorization in the case where separatedness is required (see Edwards, 1995, Subsection 4.11.10). For the sake of simplicity, such sets are also called functions.

Consider an abstract set X, an algebra $\mathcal{A}$ of subsets of X, and the measure μ on $\mathcal{A}$. As usual, it is assumed that μ is positive, countably additive, complete, and locally bounded (sigma-finite). Take a number $p \in [1, \infty[$. Let $\mathcal{L}^p = \mathcal{L}^p(X, \mu)$ denote the vector space of functions f on X such that they are measurable with respect to the measure μ and their pth power is integrable, i.e.,

$$f \in \mathcal{L}^p \Leftrightarrow \int |f|^p < \infty.$$

1. The equality

$$\|f\|_p = \left(\int |f|^p \, d\mu \right)^{1/p}$$

defines the norm $\|\cdot\|_p$ for $\mathcal{L}^p$. The triangle inequality

$$\|f + g\|_p \leq \|f\|_p + \|g\|_p \quad (f, g \in \mathcal{L}^p)$$

is called the *Minkowski inequality*. It is derived from the Hölder inequality (Dunford and Schwartz, 1988, Chapter 3, Section 4; Edwards, 1995, Subsections 4.11.2–4.11.6)

$$\left| \int fg \, d\mu \right| \leq \|f\|_p \cdot \|g\|_q \quad (f \in \mathcal{L}^p,\ g \in \mathcal{L}^q),$$

which holds for $p > 1$, $q > 1$, $1/p + 1/q = 1$. For $p = 1$, the Minkowski inequality is obtained by integrating the triangle inequality for absolute values.

The normed space $\mathcal{L}^p$ is complete (Dunford and Schwartz, 1988, Chapter 3, Section 6).

The space $\mathcal{L}^\infty$ consists of numeric functions on X that are bounded almost everywhere and measurable with respect to the measure μ. A norm on $\mathcal{L}^\infty$ is defined by the equality

$$\|f\|_\infty = \inf\{\sup_{u \in Z} |f(u)| : Z \subseteq u, \ \mu(Z) = 0\}.$$

This space is also complete (Kirillov and Gvishiani, 1988, Chapter 3, Section 4).

Note that the spaces $\mathcal{L}^p$ are not necessarily separated, since a function is not necessarily equal to zero whenever its integral is equal to zero. For this reason, the space $\mathcal{L}^p$ is often factorized, so the points of the resulting space are represented by classes of functions equivalent with respect to the given measure. Although the resulting space is separated, the operations with its elements are more complicated. It is natural to consider the points of $\mathcal{L}^p$ to be functions and take into account that they are not separated whenever necessary. The spaces $\mathcal{L}^p$ factorized with respect to the equivalence relation are denoted by L^p or L_p.

2. Numbers p and q such that

$$1 \leq p < \infty, \quad 1 \leq q \leq \infty, \quad 1/p + 1/q = 1.$$

are called *conjugate*.

Examples of pairs of conjugate numbers are $p = 1$ and $q = \infty$, $p = 2$ and $q = 2$. It can be proved that the spaces $\mathcal{L}^p$ and $\mathcal{L}^q$ with conjugate indices p and q are also conjugate. This means that the space of continuous linear functionals on $\mathcal{L}^p$ is isomorphic to the space $\mathcal{L}^q$ for $q = p(p-1)^{-1}$ (Dunford and Schwartz, 1988, Chapter 4, Section 8; Kirillov and Gvishiani, 1988, Chapter 3, Section 4).

Among the spaces $\mathcal{L}^p$ ($1 \leq p < \infty$), we point out the space $\mathcal{L}^2$, which is a Hilbert space with scalar product

$$\langle f, g \rangle = \int \bar{f} g \, d\mu \quad (f, g \in \mathcal{L}^p).$$

For all other values of p, it is easy to verify that the parallelogram law does not hold for the norm $\| \cdot \|_p$, which implies that the spaces $\mathcal{L}^p$ with $p \neq 2$ are not Hilbert spaces.

Exercise. Prove the preceding statement.

If X is a compact metric space, then the space $\mathcal{L}^p = \mathcal{L}^p(X, \mu)$ is separable, i.e., there exists a countable everywhere dense set of functions (in particular, simple and continuous functions; cf. Kolmogorov and Fomin, 1989, Chapter 7, Section 1).

Convergence in the space $\mathcal{L}^p$ for $1 \leq p < \infty$ is called *mean convergence* (of *order* p). By definition, if $f_n \to f$ for $f_n, f \in \mathcal{L}^p$, this means that

$$\|f_n - f\|_p = \left(\int |f_n - f|^p \, d\mu \right)^{1/p} \to 0.$$

Obviously, the same condition can be written as

$$\|f_n - f\|_p^p = \left(\int |f_n - f|^p \, d\mu \right) \to 0.$$

3. We now consider several examples of Lebesgue spaces.

Example 1. Put $X = \{1, \ldots, m\} \subseteq \mathbb{N}$ and let $\mu = dn$ be the counting measure on X. The functions on $X = \{1, \ldots, m\}$ are points $f = (f(1), \ldots, f(m))$ of the space $\mathbb{R}^m$. Therefore, $\mathcal{L}^p = \mathbb{R}^m$ for any $p \in [1, \infty]$. A norm is defined by the equality $\|f\|_p = \left(\sum\limits_{1 \leq k \leq m} |f(k)|^p \right)^{1/p}$. In particular, for $p = 1, 2, \infty$, $1 \leq k \leq m$, we have

$$\|f\|_1 = \sum |f(k)|, \quad \|f\|_2 = \left(\sum |f(k)|^2 \right)^{1/2}, \quad \|f\|_\infty = \sup |f(k)|.$$

The space $\mathcal{L}^p = \mathcal{L}^p(X, dn)$ is separated.

Example 2. Put $X = \mathbb{N}$ and let $\mu = dn$ be the counting measure on $\mathbb{N}$. Then the functions on $X = \mathbb{N}$ are number sequences $f = (f(1), \ldots, f(n), \ldots)$. Therefore, the points of $l^p = \mathcal{L}^p(\mathbb{N}, dn)$ for $1 \leq p < \infty$ are number sequences $f = (f(n))$ such that $\|f\|_p = (\sum |f(n)|^p)^{1/p} < \infty$. The points of the space $l^\infty = \mathcal{L}^\infty(\mathbb{N}, dn)$ are bounded number sequences. There is a norm $\|f\|_\infty = \sup\{|f(n)| : 1, 2, \ldots\}$. The space $\mathcal{L}^p = \mathcal{L}^p(\mathbb{N}, dn)$ is separated.

Example 3. Put $X = \mathbb{R}^m$ and let $\mu = dx$ be the Lebesgue measure on $\mathbb{R}^m$. The points of the space $\mathcal{L}^p = \mathcal{L}^p(\mathbb{R}^m, dx)$ for $1 \le p < \infty$ are numeric functions f on $\mathbb{R}^m$ that are measurable with respect to the Lebesgue measure and satisfy the condition $\|f\|_p = \left(\int |f(x)|^p \, dx \right)^{1/p} < \infty$. The space $\mathcal{L}^p = \mathcal{L}^p(\mathbb{R}^m, dx)$ is not separated.

It is possible to formulate a theorem for compact sets in the space $\mathcal{L}^p = \mathcal{L}^p(\mathbb{R}^m, dx)$ with $1 < p < \infty$ that is similar to the Arzelà theorem and is deduced from it with the use of averaging. A set $A \subseteq \mathcal{L}^p$ is said to be *uniformly bounded in mean* if there exists an $\alpha > 0$ such that

$$\|f\|_p = \left(\int |f(x)|^p \, dx \right)^{1/p} \le \alpha \quad (f \in A).$$

This implies that the set A is bounded in the normed space $\mathcal{L}^p$.

Take a point $z \in \mathbb{R}^m$, a function $f \in \mathcal{L}^p$, and the function $\Delta_z f \in \mathcal{L}^p$ such that

$$\Delta_z f(x) = f(x + z) - f(x) \quad (x \in \mathbb{R}^m).$$

A set $A \subseteq \mathcal{L}^p$ is said to be *uniformly equicontinuous in mean* if

$$\|\Delta_z\|_p = \left(\int |f(x + z) - f(x)|^p \, dx \right)^{1/p} \to 0$$

as $z \to 0$ uniformly on A.

Let $B^c(0, n) = \{x \in \mathbb{R}^m : \|x\| \ge n\}$ be the complement of the ball $B(0, n) \subseteq \mathbb{R}^m$ and $h_n = B^c(0, n) \cdot f$ is the product of the indicator of the set $B^c(0, n)$ and a function $f \in \mathcal{L}^p$. Clearly, $h_n \in \mathcal{L}^p$ as well as f. The value of h_n is equal to 1 outside the ball $B(0, n) = \{x \in \mathbb{R}^m : \|x\| < n\}$. A set $A \subseteq \mathcal{L}^p$ is said to be *uniformly equismall at infinity in mean* if

$$\|B^c(0, n) \cdot f\| = \left(\int_{B^c(0,n)} |f(x)|^p \, dx \right)^{1/p} \to 0$$

as $n \to \infty$ uniformly on A.

The Riesz criterion. *A closed set $A \subseteq \mathcal{L}^p(\mathbb{R}^m, dx)$ is compact if and only if A is uniformly bounded, uniformly equicontinuous, and uniformly equismall at infinity in mean.*

The assumption of the Arzelà theorem is supplemented by the requirement that the set in question be uniformly equismall in mean. This additional requirement is not necessary if we consider $\mathcal{L}^p(U, dx)$ for a bounded open set $U \subseteq \mathbb{R}^m$. The proof of the Riesz criterion and also a compactness criterion proposed by Kolmogorov can be found in Kantorovich and Akilov (1982), Chapter 9, Section 1.

4.3.4. Distributions

Distributions are continuous linear functionals on some function spaces. These function spaces are also called *generalized functions*. The choice of the terms can be explained by the natural way to identify such functionals with measures and locally integrable functions. Generalized derivatives were introduced by S. L. Sobolev (see Sobolev, 1991). A comprehensive theory of distributions was developed by L. Schwartz (see Schwartz, 1966). A simple sequential approach to this theory was proposed in Antosik, Mikusinski, and Sikorski (1973).

Distributions can be differentiated infinitely many times. The Fourier transform is defined for them. This makes distributions a convenient tool for analysis. They are especially widely used in the theory of differential equations.

We consider three most often used function spaces that are domains of definition for distributions. These spaces are called *principal* spaces and the functions in them are called *principal* or *test* functions.

As before, in this subsection we consider numeric functions on an open set $U \subseteq \mathbb{R}^m$.

1. The space $\mathcal{E}$. The largest of all function spaces that usually serve as a domain of definition for distributions is the space $\mathcal{E}(U) = C^\infty(U)$ of all smooth functions on U that was described in Subsection 4.3.2 (see also Kirillov and Gvishiani, 1988, Chapter 3, Section 4, part 3).

We point out the space $\mathcal{E} = \mathcal{E}(\mathbb{R}^m) = C^\infty(\mathbb{R}^m)$ of smooth functions defined on the entire space $\mathbb{R}^m$.

2. The space $\mathcal{S}$. One of the widely used spaces is the space $\mathcal{S} = \mathcal{S}(\mathbb{R}^m)$ of smooth functions on $\mathbb{R}^m$ rapidly decreasing at infinity together with all its derivatives. The points of $\mathcal{S}$ are functions $\varphi \in C^\infty(\mathbb{R}^m)$ such that

$$\|\varphi\|_{\alpha\beta} = \sup\left\{|x^\alpha \partial^\beta \varphi(x)| : x \in \mathbb{R}^m\right\} < \infty$$

for any multi-indices $\alpha = (\alpha(1), \ldots, \alpha(m))$ and $\beta = (\beta(1), \ldots, \beta(m))$. As before, $x^\alpha = x_1^{\alpha(1)} \ldots x_m^{\alpha(m)}$, $\partial^\beta = \partial_1^{\beta(1)} \ldots \partial_m^{\beta(m)}$ and $x = (x_1, \ldots, x_n) \in \mathbb{R}^m$. Functions in $\mathcal{C}$ are called *rapidly decreasing*. These functions and all their derivatives decrease at infinity more rapidly than any rational function.

A metric on $\mathcal{S}$ can be introduced with the use of any collection of numbers $c(\alpha, \beta) > 0$ such that $\sum c(\alpha, \beta) = 1$. The distance between any $\varphi, \psi \in \mathcal{S}$ is defined to be

$$\rho(\varphi, \psi) = \sum c(\alpha, \beta) \cdot \frac{\|\varphi - \psi\|_{\alpha\beta}}{1 + \|\varphi - \psi\|_{\alpha\beta}},$$

where the sum is taken with respect to all multi-indices α, β. The general case is usually reduced to the case of series and the distance is defined as follows. For any integer $q \geq 0$, we take the number

$$\|\varphi\|_q = \sup_{|\beta| \leq q} \left(\sup_{x \in \mathbb{R}^m} \{(1 + \|x\|^2)^{q/2} |\partial^\beta \varphi(x)|\} \right)$$

and define the distance between $\varphi, \psi \in \mathcal{S}$ to be

$$\rho(\varphi, \psi) = \frac{1}{2} \sum_{q \geq 0} \frac{1}{2^q} \cdot \frac{\|\varphi - \psi\|_q}{1 + \|\varphi - \psi\|_q}.$$

Since there are $c(\alpha, \beta) > 0$ such that

$$|x^\alpha \partial^\beta \varphi(x)| \leq (1 + \|x\|^2)^{|\alpha|/2} |\partial^\beta \varphi(x)| \leq c(\alpha, \beta) \cdot |x^\alpha \partial^\beta \varphi(x)|,$$

it follows that the above metrics are equivalent (cf. Kirillov and Gvishiani, 1988, Chapter 3, Section 4, part 3). The functions $\varphi \in \mathcal{S}$ can be defined as the functions $\varphi \in C^\infty(\mathbb{R}^m)$ such that $\|\varphi\|_q < \infty$ for all integer $q \geq 0$.

The set $\mathcal{S}$ of rapidly decreasing functions with conventional operations forms an algebra. Indeed, linear combinations and products of rapidly decreasing functions also rapidly decrease. It follows from the definitions that a sequence of functions $\varphi_n \in \mathcal{S}$ converges to a function $\varphi \in \mathcal{S}$ in the space $\mathcal{S}$ if and only if $\|\varphi_n - \varphi\| \to 0$, i.e., when $\|\varphi_n - \varphi\|_{\alpha\beta} \to 0$ for any multi-indices α and β, or, which is equivalent, if and only if $x^\alpha \partial^\beta (\varphi_n - \varphi)(x) \to 0$ uniformly with respect to $x \in \mathbb{R}^m$ for any α and β. The condition that $\|\varphi_n - \varphi\|_{\alpha\beta} \to 0$ for any multi-indices α and β can be replaced by the condition that $\|\varphi_n - \varphi\|_q \to 0$ for any integer $q \geq 0$.

The metric space $\mathcal{S}$ is complete, which means that it is a Frechet space (Vladimirov, 1979, Section 5). The metric on $\mathcal{S}$ does not satisfy the absolute uniformity condition. It is convenient to view the space $\mathcal{S}$ as multinormed space with the specified countable collections of norms that define convergence in $\mathcal{S}$.

3. The space $\mathcal{D}$. The points of the space $\mathcal{D} = \mathcal{D}(U)$ ate functions in $C_0^\infty(U)$, ie., smooth finite functions on U. The collection of norms defining convergence in $\mathcal{D}(U)$ can be specified as follows.

Consider the sequence of compact sets

$$K(n) = \{x \in U : d(x, \bar{U} \setminus U) \geq 1/n, \|x\| \leq n\},$$

that covers U. Note that the sequence of pairwise disjoint sets $A(n) = K(n) \setminus K(n-1)$ $(K(0) = \varnothing)$ also covers U and each point $x \in U$ belongs to exactly one of the sets $A(n)$. The distance from the compact support $\mathrm{Supp}\,\varphi$ of a finite function $\varphi \in \mathcal{D}$ to the boundary $\partial U = \bar{U} \setminus U$ of U is strictly positive if the boundary is nonempty ($U \neq \varnothing$ and $U \neq \mathbb{R}^m$). Therefore, $\partial^\beta \varphi(x) = 0$ ($x \in A(n)$) for all n greater than some number $n(\varphi)$ and for all multi-indices $\beta = (\beta(1), \ldots, \beta(m))$.

For any sequence $\alpha = (\alpha(1), \ldots, \alpha(n), \ldots)$ of integer numbers $\alpha(n) \geq 0$ and any function $\varphi \in \mathcal{D}$, we define the norm

$$\|\varphi\|_\alpha = \sum_{n \geq 1} \alpha(n) \cdot \sup\{|\partial^\beta \varphi(x)| : x \in A(n), |\beta| \leq \alpha(n)\}.$$

(There are finitely many nonzero terms for every $\varphi \in \mathcal{D}(U)$ in the sum on the right-hand side.) This collection of norms is uncountable. It follows from the definitions that the sequence $\varphi_n \in \mathcal{D}(U)$ converges to $\varphi \in \mathcal{D}(U)$ in the space $\mathcal{D}(U)$ if $\|\varphi_n - \varphi\|_\alpha \to 0$ for any index α. It can be proved that this convergence takes place if and only if 1) *the supports of the functions φ_n and φ are included in a compact set $K \subseteq U$; 2) the sequence of functions $\partial^\beta \varphi_n$ uniformly converges to the function $\partial^\beta \varphi$ on K for any multi-index β* (Kirillov and Gvishiani, 1988, Chapter 3, Section 4, part 3). Such convergence cannot be defined by any metric, i.e., the multinormed space $\mathcal{D}(U)$ is not nonmetrizable.

Compare the spaces $\mathcal{D} = \mathcal{D}(\mathbb{R}^m)$, $\mathcal{S} = \mathcal{S}(\mathbb{R}^m)$, $\mathcal{E} = \mathcal{E}(\mathbb{R}^m)$. It follows from the definitions that $\mathcal{D} \subseteq \mathcal{S} \subseteq \mathcal{E}$ for the sets of points of these spaces. Examples can be given to demonstrate that $\mathcal{D} \neq \mathcal{S}$ and $\mathcal{S} \neq \mathcal{E}$. In particular, the function φ on $\mathbb{R}^m$ with values $\varphi(x) = \exp(-\|x\|^2)$ belongs to $\mathcal{S}$, but does not belong to $\mathcal{D}$. Any constant function belongs to $\mathcal{E}$, but does not belong to $\mathcal{S}$. The identity embeddings from $\mathcal{D}$ into $\mathcal{S}$ and from $\mathcal{S}$ into $\mathcal{E}$ are continuous with respect to the types of convergence defined in these spaces: if $\varphi_n \to \varphi$ in $\mathcal{D}$, then $\varphi_n \to \varphi$ in $\mathcal{S}$; if $\varphi_n \to \varphi$ in $\mathcal{S}$, then $\varphi_n \to \varphi$ in $\mathcal{E}$.

Exercise. Prove the preceding statement.

It can be proved that the set $\mathcal{D}$ is dense in the space $\mathcal{S}$ (Yosida, 1965, Chapter 4, Section 1), and the set $\mathcal{S}$ is dense in $\mathcal{E}$ (Kirillov and Gvishiani, 1988, Chapter 3, Section 4). The sets $\mathcal{D}$ and $\mathcal{E}$ are dense in the spaces

$\mathcal{L}^p = \mathcal{L}^p(\mathbb{R}^m, dx)$ for $1 < p \leq \infty$. (The power φ^p of $\varphi \in S$ is integrable with respect to the Lebesgue measure dx, since there exists a $c > 0$ such that

$$(1 + \|x\|^2)^{1/2} |\partial^0 \varphi(x)| \leq c, \quad |\varphi^p(x)| \leq c^p (1 + \|x\|^2)^{-2p};$$

for all $x \in \mathbb{R}^m$ (see Kirillov and Gvishiani, 1988, Chapter 3, Section 4; Bremerman, 1967, Section 3.10).

4. The space $\mathcal{D}'$. The space $\mathcal{D}$ is the smallest of all the spaces considered above. For this reason, the conjugate space $\mathcal{D}'$ is the largest. The elements of $\mathcal{D}'(U)$ are continuous linear functionals on $\mathcal{D}(U)$. They are called *distributions* or *generalized functions*. In particular, $\mathcal{D} = \mathcal{D}(\mathbb{R}^m)$ and $\mathcal{D}' = \mathcal{D}'(\mathbb{R}^m)$ for $U = \mathbb{R}^m$.

It follows from the definitions that a linear functional F on $C_0^\infty(U)$ is a distribution on $\mathcal{D}(U)$ if and only if for any compact set $K \subseteq U$ there exists a number $c > 0$ and an integer number $q \geq 0$ such that

$$|F(\varphi)| \leq c \cdot \sup\{|\partial^\beta \varphi(x)| : |\beta| \leq q, \, x \in K\}$$

for all $\varphi \in C_0^\infty(U)$ with supports included in K. This criterion is convenient for applications.

5. We now consider several examples. In each example, U is assumed to be an open set in $\mathbb{R}^m$.

Example 1. A function f that is measurable with respect to the Lebesgue measure dx on U is said to be *locally integrable* if the restriction of f on every compact set $K \subseteq U$ is integrable with respect to dx. The set of all locally integrable numeric functions on U is denoted by $\mathcal{L}_{\mathrm{loc}}(U)$.

Let $f \in \mathcal{L}_{\mathrm{loc}}(U)$. Then the equality

$$F_f(\varphi) = \int_U \varphi(x) f(x) \, dx \quad (\varphi \in \mathcal{D}(U))$$

defines a functional $F_f \in \mathcal{D}'(U)$. Indeed, since the integral is linear, so is F_f. The continuity of F_f follows from the inequality $|F_f(\varphi)| \leq c(K, f) \cdot \sup\{|\varphi(x)| : x \in K\}$, which holds with $c(K, f) = \int_K |f(x)| \, dx$ for all $\varphi \in C_0^\infty(U)$ whose supports are included in the compact set $K \subseteq U$ because f is locally integrable.

It can be proved that the equality $F_f = F_g$ for $f, g \in \mathcal{L}_{\mathrm{loc}}(U)$ holds if and only if $f = g$ almost everywhere with respect to the measure dx (Yosida, 1965, Chapter 1, Section 8). This makes it possible to identify the functional F_f with the function f and write f instead of F_f whenever it does not lead to a confusion. (More precisely, F_f is identified with a class of functions equivalent to f with respect to the Lebesgue measure dx.) The possibility to identify some functionals in $\mathcal{D}'(U)$ with functions explains the term *generalized functions*. The class of these functionals is larger than the class of locally integrable functions: $\mathcal{L}_{loc}(U) \subseteq \mathcal{D}'(U)$. Then $\mathcal{D}(U) \subseteq \mathcal{D}'(U)$ because $\mathcal{D}(U) \subseteq \mathcal{L}_{loc}(U)$. Generalized functions F_f defined by locally integrable functions f are called *regular*.

Example 2. By numeric measures we mean linear combinations of locally bounded countably additive complete positive measures on the algebra $\mathcal{B}(U)$ of subsets of U generated by the class of compact sets in U. Sets in $\mathcal{B}(U)$ and numeric measures on $\mathcal{B}(U)$ will be called *Baire* sets and measures, as well as sets in the closure $\bar{\mathcal{B}}(U)$ of the algebra $\mathcal{B}(U)$ under simple convergence and numeric measures on $\bar{\mathcal{B}}(U)$.

Integrals with respect to numeric measures are defined to be linear combinations of integrals with respect to positive components of these measures.

Let μ be a Baire measure on $\mathcal{B}(U)$. Then the equality

$$F_\mu(\varphi) = \int \varphi \, d\mu \quad (\varphi \in \mathcal{D}(U))$$

defines a functional $F_\mu \in \mathcal{D}'(U)$. Indeed, since the integral is linear, so is F_μ. The continuity of F_μ follows from the inequality $|F_\mu(\varphi)| \leq c(K, \mu) \cdot \sup\{|\varphi(x)| : x \in K\}$, where $c(K, \mu)$ is the sum of values of the positive components of the measure μ on K. This inequality holds for all $\varphi \in C_0^\infty(U)$ whose supports are included in $K \subseteq U$.

Since Baire measures are defined by their values on compact sets and the indicators of compact sets can be approximated by smooth functions, the equality $F_\mu = F_\nu$ for Baire measures on $\mathcal{B}(U)$ holds if and only if $\mu = \nu$. This makes it possible to identify the functional F_μ with the measure μ whenever it does not lead to a confusion. In this case, the functional F_μ, as well as the Baire measure μ, are called a *distribution*. The other functionals in $\mathcal{D}'(U)$ are also called distributions.

Example 3. We point out the distributions concentrated at isolated points. They are called *delta functions*. Such functionals were already considered in Subsection 4.3.2.

Let $\mu = \delta_a$ be the measure that specifies the unit mass distribution at a point $a \in U$:

$$\delta_a(B) = 1 \; (a \in B), \quad \delta_a(B) = 0 \; (a \notin B)$$

for any $B \in \mathcal{B}(U)$. Then

$$F_\mu(\varphi) = \int_U \varphi \, d\mu = \varphi(a) \quad (\varphi \in \mathcal{D}(U)).$$

The functional F_μ is called the *delta function* (δ-*function*), *concentrated at the point* a, and is denoted by δ_a, as well as the measure that defines it. If $a = 0$, then we write δ instead of δ_0.

By analogy with the regular case, the values of generalized functions are often written in the form of integrals, although these integrals may have no other meaning. In particular, they can be written as

$$F_\delta(\varphi) = \int \varphi(x)\delta(x) \, dx = \varphi(0).$$

The integral in this expression represents the value of the generalized function F_δ for $\varphi \in \mathcal{D}(U)$. The product $\varphi \cdot \delta$ is not an integrable function with respect to the Lebesgue measure dx. In this integral notation, δ can be considered to be the generalized derivative of the point measure $\mu = d^0x$ with respect to the measure dx, written $d^0x = \delta dx$, without giving the formal definition of a generalized derivative. Then the equality $\int \varphi(x)\delta(x) \, dx = \int \varphi(x) \, d^0x$ serves as a formula for the change of variables in the integral.

Thus, functionals in $\mathcal{D}'(U)$ describe, in particular, numeric functions on U that are locally integrable with respect to the Lebesgue measure and Baire measures on $\mathcal{B}(U)$. It is often convenient to combine measures and functions in one model.

6. The space $\mathcal{S}'$. By definition, the elements of the space $\mathcal{S}'$ conjugate to the space $\mathcal{S}$ of rapidly decreasing functions on $\mathbb{R}^m$ are continuous linear functionals on $\mathcal{S}$. They are called *slowly increasing distributions* and are described in detail in Vladimirov (1979), Section 5.

Since $\mathcal{D} \subseteq \mathcal{S}$ and the identity embedding from the space $\mathcal{D} = \mathcal{D}(\mathbb{R}^m)$ into $\mathcal{S} = \mathcal{S}(\mathbb{R}^m)$ is continuous, it follows that the restrictions of slowly increasing distributions to $\mathcal{D}$ belong to $\mathcal{D}'$. This means that the restrictions of slowly increasing distributions to $\mathcal{D}$ are distributions. Furthermore, since $\mathcal{D}$ is dense in the space $\mathcal{S}$, different functionals in $\mathcal{S}'$ have different restrictions to $\mathcal{D}$. We conclude that there is an embedding from $\mathcal{S}'$ into $\mathcal{D}'$ and we can assume that $\mathcal{S}' \subseteq \mathcal{D}'$.

We now give several examples.

Example 4. A measurable numeric function f on $\mathbb{R}^m$ that is measurable with respect to the Lebesgue measure dx is said to be *slowly increasing* if there is a $q \geq 0$ such that the function g_q on $\mathbb{R}^m$ with values $g_q = (1 + \|x\|^2)^{-q} f(x)$ is integrable with respect to dx. The set of all slowly increasing functions on $\mathbb{R}^m$ will be denoted by $\mathcal{M}$. All polynomials on $\mathbb{R}^m$ belong to $\mathcal{M}$. The function h on $\mathbb{R}^m$ with values $h(x) = \exp(\|x\|)$ does not belong to $\mathcal{M}$.

Let $f \in \mathcal{M}$. Then the equality

$$F_f(\varphi) = \int \varphi(x) f(x) \, dx \quad (\varphi \in \mathcal{S})$$

defines a functional $F_f \in \mathcal{S}'$, which is a slowly increasing generalized function. The functional F_f is identified with the function f.

Using the Hölder inequality, it is easy to verify (consider it an exercise) that $\mathcal{L}^P \subseteq \mathcal{M}$ for $1 \leq p \leq \infty$ (see Rudin, 1991, Section 7.12). Thus, $\mathcal{L}^P \subseteq \mathcal{M} \subseteq \mathcal{S}'$.

Example 5. The positive Baire measure μ on the algebra $\mathcal{B} = \mathcal{B}(\mathbb{R}^m)$ generated by compact sets in $\mathbb{R}^m$ is called *slowly increasing* if there is a $q \geq 0$ such that the function h_q on $\mathbb{R}^m$ with values $h_q(x) = (1 + \|x\|^2)^{-q}$ is integrable with respect to μ.

Let μ be a slowly increasing Baire numeric measure on $\mathcal{B}$. Then the equality

$$F_\mu(\varphi) = \int \varphi \, d\mu \quad (\varphi \in \mathcal{S})$$

defines a functional $F_\mu \in \mathcal{S}'$. This follows from the definitions. The functional is identified with the measure μ, which is also called a slowly increasing distribution.

Clearly, point measures $\mu = \delta_a$ are slowly increasing.

Example 6. Let $f \in \mathcal{L}^P$ $(1 \leq p < \infty)$. Then there exists a Baire measure $d\nu = |f| dx$ such that the function $|f|$ is its derivative with respect to the Lebesgue measure dx. This follows from the Radon–Nikodym theorem. The Hölder inequality implies that the measure $d\nu$ is slowly increasing. (Prove this statement.)

Consider a numeric Baire measure $d\mu = f dx$ such that the function f is its derivative with respect to the Lebesgue measure dx. Any function $\varphi \in \mathcal{S}$ is integrable with respect to the measure $d\mu$. The change of variables yields

$$F_f(\varphi) = \int \varphi f \, dx = \int \varphi \, d\mu \quad (\varphi \in \mathcal{S}).$$

The relationship between slowly increasing functions and measures is described in Vladimirov (1979), Section 1.

7. The space $\mathcal{E}'$. By definition, the elements of the space $\mathcal{E}'(U)$ which is conjugate to the space $\mathcal{E}(U)$ of smooth functions on U are continuous linear functionals on $\mathcal{E}(U)$.

Since $\mathcal{D}(U) \subseteq \mathcal{E}(U)$ and the identity embedding from the space $\mathcal{D}(U)$ into the space $\mathcal{E}(U)$ is continuous, it follows that the restriction of every functional $F \in \mathcal{E}'(U)$ to $\mathcal{D}(U)$ belongs to $\mathcal{D}'(U)$. Since $\mathcal{D}(U)$ is dense in the space $\mathcal{E}(U)$, different functionals in $\mathcal{E}'(U)$ have different restrictions to $\mathcal{D}(U)$. Therefore, $\mathcal{E}'(U)$ can be embedded into $\mathcal{D}'(U)$ and we can assume that $\mathcal{E}'(U) \subseteq \mathcal{D}'(U)$.

Similarly, for $U = \mathbb{R}^m$ the space $\mathcal{E}' = \mathcal{E}'(\mathbb{R}^m)$ can be embedded into $\mathcal{S}' = \mathcal{S}'(\mathbb{R}^m)$ and we can assume that $\mathcal{E}' \subseteq \mathcal{S}' \subseteq \mathcal{D}'$.

Let f be a numeric function on U that has compact support and is locally integrable with respect to the Lebesgue measure. Then the equalities

$$F_f(\varphi) = \int \varphi(x)f(x)\,dx = \int \varphi\,d\mu \quad (\varphi \in \mathcal{C}(U))$$

define a functional $F \in \mathcal{C}'(U)$.

The space $\mathcal{E}'(U)$ contains only the functionals in $\mathcal{D}'(U)$ that have compact support (Bremerman, 1967, Sections 3.9–3.12; Kirillov and Gvishiani, 1988, Chapter 3, Section 4). The support of a distribution $F \in \mathcal{D}'(U)$ is defined as follows. A point $x \in U$ does not belong to the support $\operatorname{Supp} F$ if there exists an open neighborhood V of x such that $F(\varphi) = 0$ for any $\varphi \in \mathcal{D}(U)$ with support $\operatorname{Supp} \varphi \subseteq V$. The points that do not possess this property make up the support $\operatorname{Supp} F$ of the distribution F. Distributions in $\mathcal{E}'(U)$ are naturally called *finite* (by analogy with finite functions).

For example, δ-functions are finite since their supports consist of a single point. If a distribution has support consisting of a single point, $\{0\}$, then it can be represented as a linear combination of derivatives of the delta function.

Exercise. Prove the preceding statement.

Remark. For $U = \mathbb{R}^m$, the class $\mathcal{E}'$ of finite distributions is the smallest, the class $\mathcal{S}'$ of slowly increasing distributions is larger, and the largest one is the class $\mathcal{D}'$ of all distributions.

A detailed description of measures and distributions can be found in Sobolev (1991), Chapters 4,5. Major attention is given to their application in the theory of differential equations.

4.3.5. Sobolev spaces

Before describing the Sobolev spaces, we define operations with distributions.

1. Operations with distributions. Linear combinations are defined for distributions, as they are for all linear functionals. The equality

$$fF(\varphi) = F(f\varphi) \quad (\varphi \in \mathcal{D}(U))$$

defines the product $fF \in \mathcal{D}'(U)$ of a distribution $F \in \mathcal{D}'(U)$ and a function $f \in \mathcal{E}(U)$. Note that $f\varphi \in \mathcal{D}(U)$. The product is written as fF in order not to confuse this distribution with the number $Ff = F(f)$ for $f \in \mathcal{D}(U)$.

Example. Since $f\delta(\varphi) = \delta(f\varphi) = f(0) \cdot \varphi(0)$, we have $f\delta = f(0) \cdot \delta$.

For a distribution $F \in \mathcal{D}'(U)$, the equality

$$\partial^\alpha F(\varphi) = (-1)^{|\alpha|} F(\partial^\alpha \varphi) \quad (\varphi \in \mathcal{D}(U))$$

defines the derivative $\partial^\alpha F \in \mathcal{D}'(U)$ with multi-index α.

Examples. 1) Let $F = F_h$ be the distribution identified with the the function $h : \mathbb{R} \to \mathbb{R}$ such that $h(x) = 0$ $(x < 0)$ and $h(x) = 1$ $(x \geq 0)$. Then $\partial F_h = \delta$. Indeed, we have

$$-F_h(\varphi') = \int_{-\infty}^{\infty} \varphi'(x)h(x)\,dx = \int_{0}^{\infty} \varphi'(x)\,dx = \varphi(0) - \varphi(\infty) = \varphi(0) - 0 = \delta\varphi$$

for all $\varphi \in \mathcal{D}(\mathbb{R})$.

2) Let $F = F_f$ be the distribution identified with a smooth function $f : \mathbb{R} \to \mathbb{R}$. Then $\partial F_f = F_{\partial f}$. Indeed, integrating by parts and taking into account that the support of $\varphi \in \mathcal{D}(\mathbb{R})$ is compact, we get

$$-F_f(\varphi') = -\int \varphi'(x)f(x)\,dx = \int \varphi(x)f'(x)\,dx = F_{f'}(\varphi).$$

This example shows that differentiation of distributions is consistent with differentiation of smooth functions.

3) $\partial^\alpha \delta(\varphi) = (-1)^{|\alpha|}\partial^\alpha\varphi(0)$ $(\varphi \in \mathcal{D}(\mathbb{R}))$.

The derivatives of slowly increasing distributions are also slowly increasing distributions: if $F \in \mathcal{S}'$, then $\partial^\alpha F \in \mathcal{S}'$ for any multi-index α. The space $\mathcal{S}'$ is closed under the operation of differentiation.

2. The connection between multiplication by a function and differentiation of distributions is described by the Leibniz formula.

The Leibniz formula. $\partial_j(fF) = \partial_j f \cdot F + f \cdot \partial_j F$.

This formula follows from the definitions and the differentiation rules for the products of smooth functions. (Prove this assertion. For reference, see Vladimirov, 1979, Section 2.) The proof of the general *Hörmander formula* is given in Yosida (1965), Chapter 1, Section 8.

It can be proved that any distribution $F \in \mathcal{D}'(U)$ is equal to the derivative $\partial^\alpha f$ of a function f on U that is locally integrable with respect to dx (see Rudin 1991, Chapter 6). Furthermore, any distribution $F \in \mathcal{S}'(\mathbb{R}^m)$ is equal to the derivative $\partial^\alpha f$ of a slowly increasing function f on $\mathbb{R}^m$ (Vladimirov, 1979, Sections 2 and 5; Kirillov and Gvishiani, 1988, Chapter 3, Section 4).

3. Generalized derivatives. Using the Hölder inequality, it it easy to verify that $\mathcal{L}^p \subseteq \mathcal{L}_{\mathrm{loc}}(U)$ for $1 \le p < \infty$. (Prove this assertion.) Thus, functions $g \in \mathcal{L}^p$ can be identified with distributions $F_g \in \mathcal{D}'(U)$. Take a function $f \in \mathcal{L}_{\mathrm{loc}}(U)$, the distribution $F_f \in \mathcal{D}(U)$ identified with f, and its derivative $\partial^\alpha F_f \in \mathcal{D}'(U)$ with multi-index α. We write f instead of F_f and $\partial^\alpha f$ instead of $\partial^\alpha F_f$ (although f is not necessarily differentiable). If $\partial^\alpha f = F_g \in \mathcal{D}'(U)$ for a function $g \in \mathcal{L}_{\mathrm{loc}}(U)$, then g is called a *generalized derivative* of f, written $\partial^\alpha f = g$. Note that any function $h \in \mathcal{L}_{\mathrm{loc}}(U)$ equivalent to g with respect to the Lebesgue measure is also a generalized derivative of f.

If f is a smooth function, its conventional derivative coincides with its generalized derivative. However, there are differentiable functions that are locally integrable almost everywhere with respect to the Lebesgue measure whose conventional derivatives do not coincide with generalized ones. (For example, functions defined on an interval that are increasing and continuous from the left, but are not absolutely continuous.)

4. The spaces $\mathcal{W}_q^p(U)$. Take integer numbers $p \ge 1$ and $q \ge 0$. The points of the *Sobolev space* $\mathcal{W}_q^p(U)$ are locally integrable functions $f \in \mathcal{L}_{\mathrm{loc}}(U)$ whose generalized derivatives $\partial^\alpha f$ belong to $\mathcal{L}^p$ for $|\alpha| \le q$. A norm on $\mathcal{W}_q^p(U)$ is defined by the equality

$$\|f\|_{p,q} = \left(\sum_{|\alpha| \le q} \int_U |\partial^\alpha f|^p \, dx \right)^{1/p}.$$

If $p = 2$, then this norm is obtained from the scalar product

$$\langle f, g \rangle = \sum_{|\alpha| \leq q} \left(\int_U \overline{\partial^\alpha f(x)} \cdot \partial^\alpha g(x) \, dx \right).$$

The normed space $W_q^p(U)$ is a Banach space, while the normed space $W_q^2(U)$ is a Hilbert space (Yosida, 1965, Chapter 1, Section 9).

The Hilbert space $W_q^2(U)$ is associated with the Euclidean space $\mathcal{H}_0^q(U)$ of smooth functions described in Subsection 4.3.2. It can be proved that the completion of $\mathcal{H}_0^q(U)$ is a subspace of the Sobolev space $W_q^2(U)$ (Yosida, 1965, Chapter 1, Section 10). For $U = \mathbb{R}^m$, the completion of $\mathcal{H}_0^q = \mathcal{H}_0^q(\mathbb{R}^m)$ coincides with $W_q^2 = W_q^2(\mathbb{R}^m)$. For this reason, it is conventional to write $\mathcal{H}_0^q$ instead of W_q^2 Functions in $\mathcal{H}_0^q$ can be approximated by sequence of q-smooth sequences converging in the corresponding norm. The Sobolev lemma establishes a connection between the spaces $W_q^2(U)$ and $\mathcal{C}^p(U)$ for $U \subseteq \mathbb{R}^m$ ($q \geq 0$, $h \geq 1$, $m \geq 1$).

The Sobolev lemma. *If $q > p + m/2$, then for any $f \in W_q^2(U)$ there exists a function $g \in \mathcal{C}^p(U)$ that is equal to f almost everywhere.*

It should be emphasized that the degree of smoothness p of a function g equivalent to f with respect to the Lebesgue measure is strictly less than the order q of the square-integrable derivatives of f. In this sense, the replacement of functions in Sobolev spaces with smooth functions involves a certain loss of smoothness.

Detailed proofs of the Sobolev lemma can be found in Rudin (1991), Chapter 7, and Yosida (1965), Chapter 6.

Remark. Many details related to the spaces of functions and generalized functions can be found in Reed and Simon (1972), Chapter 5. Some important applications are described there as well. More than 60 problems are given at the end of the specified chapter.

Theorems on embeddings from the spaces $W_q^p(U)$ into the spaces of continuous functions are proved in Sobolev (1991), Chapter 1, Section 8,11. The explicit form of the operator that represents these embeddings is presented there. In addition to continuity, the compactness of this operator is proved (i.e., the property to map bounded sets to relatively compact sets). Detailed proofs of the embedding theorems are given in Kantorovich and Akilov (1982), Chapter 11, Section 4.

Remark. Many details related to the spaces of functions and generalized functions can be found in Rudin (1991), Chapter 6, Edwards (1995), Chapter 5, and Reed and Simon (1972), Chapter 5. Some important applications are described there as well. Exercises are given at the end of each chapter. In Bremerman (1967), Chapters 5 and 7, considerable attention is given to applications and to the representation of distributions with the use of analytic functions. Distributions on manifolds are considered in Hörmander (2003), Section 6.3.

4.4. FOURIER TRANSFORM

The theory of Fourier transforms for distributions is a generalization of classical harmonic analysis. The Fourier transform is effectively applied in solving differential equations.

4.4.1. Transforms of rapidly decreasing functions

It is convenient to define the Fourier transform for rapidly decreasing functions first, and then for slowly increasing distributions.

The Fourier transform $g = \Phi(f)$ of a function $f \in S = S\mathbb{R}^m)$ is defined by the equality

$$g(y) = (2\pi)^{-m/2} \int e^{-iyx} f(x)\, dx,$$

where $yx = y_1 x_1 + \ldots + y_m x_m$ is the scalar product of vectors $y = (y_1, \ldots, y_m)$ and $x = (x_1, \ldots, x_m)$ of the Euclidean space $\mathbb{R}^m$, and dx is the Lebesgue measure on $\mathbb{R}^m$. Since functions in S are integrable with respect to dx and $|e^{-iyx} f(x)| \leq |f(x)|$, the integral on the right-hand side exists.

Example. Let $f(x) = \exp(-\|x\|^2/2)$. Then $\Phi(f) = f$.

The Fourier transform is linear, continuous, and one-to-one transformation of the space S. The inverse transformation $g = \Phi^{-1}(f)$ for the function $g \in S$ is defined by the equality

$$f(x) = (2\pi)^{-m/2} \int e^{ixy} g(y)\, dy.$$

This equality is called the *inversion formula*. In it, $xy = yx$ and dy is the Lebesgue measure on $\mathbb{R}^m$. The transform Φ^{-1} is called the *inverse Fourier* transform. It is also a linear, continuous, and one-to-one transformation of the space $\mathcal{S}$. In other words, Φ and Φ^{-1} are linear homeomorphisms from $\mathcal{S}$ onto itself. This can be proved using the rules of differentiation and integration under the integral sign.

Exercise. Use these rules to give detailed proofs of these assertions (see Yosida, 1965, Section 6.1; Rudin, 1991, Sections 7.4 and 7.7).

By the Fourier transform one often means the inverse transform Φ^{-1} rather than Φ One of these transforms is assumed to have coefficient 1, while the other is assumed to have coefficient $(2\pi)^{-m}$. With these definitions, we have

$$\Phi(\bar{f}) = \overline{\Phi^{-1}(f)} \quad (f \in \mathcal{S}).$$

We define the *symmetry operator* S on $\mathcal{S}$ as follows:

$$S\varphi(x) = \varphi(-x) \quad (\varphi \in \mathcal{S}, \ x \in \mathbb{R}^m).$$

It follows from the definition that $S^2 = I$, where I denotes the identity transformation of $\mathcal{S}$. Using the inversion formula, it is easy to verify that $\Phi = \Phi^{-1}S$, $\Phi^2 = S$, and $\Phi^4 = I$.

Exercise. Verify these equalities.

2. In addition to the conventional product, there is another kind of product defined for functions in $\mathcal{S}$. It is called the *convolution* and denoted by $*$. The convolution of functions $g, f \in \mathcal{S}$ is a function $h = g * f$ on $\mathbb{R}^m$ such that

$$h(z) = \int g(z - x)f(x)dx \quad (z \in \mathbb{R}^m).$$

It is easy to verify that $h \in \mathcal{S}$. The convolution is commutative, associative, and distributive over the sum. It is analogous to the rule for calculating the coefficients of the product of polynomials.

Exercise. Prove that the convolution is commutative, associative, and distributive over the sum (Kirillov and Gvishiani, 1988, Chapter 4, Section 1).

The Fourier transform associates the convolution with the conventional product. Using the definitions and the rules of integration, we obtain

$$(2\pi)^{-m/2}\Phi(g * f) = \Phi(g) \cdot \Phi(f), \quad (2\pi)^{-m/2}\Phi(g) * \Phi(f) = \Phi(g \cdot f).$$

Exercise. Prove these equalities.

Remark. The properties of the convolution are described in detail in Yosida (1965), Chapter 6. Mikusinski's operator calculus based on the operation of convolution and its application to solving differential equations is also presented there. Convolutions on a commutative group are considered in Kirillov and Gvishiani (1988).

4.4.2. Transforms of slowly increasing distributions

The Fourier transform ΦF of a distribution $F \in \mathcal{S}'$ is defined to be the composition $F \circ \Phi$ of the Fourier transform $\Phi : \mathcal{S} \to \mathcal{S}$ and the distribution $F : \mathcal{S} \to \mathbb{C}$. By definition,

$$\Phi F(\varphi) = F(\Phi(\varphi)) \quad (\varphi \in \mathcal{S}).$$

Since F and Φ are linear and continuous, we have $\Phi F = F \circ \Phi \in \mathcal{S}'$.

1. The Fourier transform for distributions has the same properties as the Fourier transform for functions. We choose weak convergence in $\mathcal{S}'$, i.e., by definition, $F_n \to F$ for $F_n, F \in \mathcal{S}'$ means that $F_n(\varphi) \to F(\varphi)$ $(\varphi \in \mathcal{S})$. With this convergence, the Fourier transform $\Phi : \mathcal{S}' \to \mathcal{S}'$ turns out to be continuous. It is an extension of $\Phi : \mathcal{S} \to \mathcal{S}$ under the embedding $\mathcal{S} \subseteq \mathcal{S}'$, and we use the same designation for it. The Fourier transform on $\mathcal{S}'$ is linear, continuous, and one-to-one map from $\mathcal{S}'$ onto itself. The inverse Fourier transform Φ^{-1} on $\mathcal{S}'$ is defined by the equality

$$\Phi^{-1}F(\varphi) = F(\Phi^{-1}(\varphi)) \quad (\varphi \in \mathcal{S}).$$

Using the equality

$$SF(\varphi) = F(S\varphi) \quad (\varphi \in \mathcal{S}, \ F \in \mathcal{S}'),$$

we extend the symmetry operator S to $\mathcal{S}'$. The equalities $S^2 = I$, $\Phi = \Phi^{-1}S$, and $\Phi^2 = S$, where I is the identity transformation of $\mathcal{S}'$, remain valid for the extended operator.

Exercise. Prove these equalities.

2. The Fourier transform establishes a connection between the operations of differentiation and multiplication by a function. In addition to the differentiation operator ∂^α, we consider the function $D^\alpha = i^{-|\alpha|}\partial^\alpha$ and denote by M^α the operator of multiplication by the function $f(x) = x^\alpha$. The operators D^α, M^α, and the Fourier transform Φ are related by the equalities

$$D^\alpha \Phi = \Phi M^\alpha, \quad \Phi D^\alpha = (-1)^{|\alpha|} M^\alpha \Phi.$$

These equalities are easy to verify for functions in $\mathcal{S}$. Since $\mathcal{S}$ is dense in $\mathcal{S}'$, the equalities can be extended by continuity to distributions in $\mathcal{S}'$.

Exercise. Provide detailed proofs of the preceding statements.

Example. The Fourier transform for δ is $\Phi(\delta) = (2\pi)^{-m/2} \cdot 1$, where 1 stands for the distribution represented by the function identically equal to unity. Indeed, we have

$$\Phi\delta(\varphi) = \delta(\Phi(\varphi)) = \psi(0) = (2\pi)^{-m/2} \cdot 1(\varphi),$$

where the function $\psi = \Phi(\varphi)$ and the distribution 1 are as follows:

$$\psi(y) = (2\pi)^{-m/2} \int e^{-iyx}\varphi(x)\, dx, \quad 1(\varphi) = \int \varphi(x)\, dx.$$

From the obtained equality for $\Phi(\delta)$, we derive the equality $\Phi(1) = (2\pi)^{m/2}\delta$ for the Fourier distribution of the function identically equal to unity:

$$\delta = S\delta = \Phi^2\delta = \Phi((2\pi)^{-m/2} \cdot 1) = (2\pi)^{-m/2} \cdot \Phi(1).$$

We use $S\delta(\varphi) = \delta(S\varphi) = S\varphi(0) = \varphi(-0) = \varphi(0) = \delta(\varphi)$ ($\varphi \in \mathcal{S}$).

The transition to Fourier transforms often helps to substantially simplify the solution of differential equations Regularizing factors, in particular, exponential ones are effectively used. The Fourier transform of a special product of a slowly increasing distribution and a rapidly decreasing exponential function is known as the *Laplace transform*, which is widely used. (Kolmogorov and Fomin, 1989, Chapter 8, Section 6; Vladimirov, 1979, Section 9; Rudin, 1991, 7.3–4).

4.4.3.   The Fourier–Plancherel transform

The set $\mathcal{L}^2 = \mathcal{L}^2(\mathbb{R}^m, dx)$ can be embedded into the set $\mathcal{S}' = \mathcal{S}'(\mathbb{R}^m)$ by identifying functions in $\mathcal{L}^2$ with the distributions in $\mathcal{S}'$ generated by them. Under the Fourier transform, functions in $\mathcal{L}^2$ are mapped to functions in $\mathcal{L}^2$. Therefore, the restriction of the Fourier transform on $\mathcal{S}'$ to $\mathcal{L}^2$ is a transformation of $\mathcal{L}^2$. It is called the *Fourier–Plancherel transform*. A detailed description of the case $m = 1$ can be found in Kolmogorov and Fomin (1989), Chapter 8, Section 5.

1. For functions $f, g \in \mathcal{L}^2$, we define the scalar product and the norm

$$\langle f, g \rangle = \int \bar{f} g \, dx, \quad \|f\|^2 = \langle f, f \rangle.$$

It is easy to verify that

$$\langle \Phi(f), g \rangle = \langle f, \Phi^{-1}(g) \rangle, \quad \langle \Phi(f), \Phi(g) \rangle = \langle f, g \rangle.$$

This means that the Fourier–Plancherel transform is *unitary*.

Exercise. Verify the above equalities.

Plancherel's theorem. *The Fourier transform maps $\mathcal{L}^2$ onto the entire $\mathcal{L}^2$ and preserves the scalar product and the norm.*

See Yosida (1965), Chapter 8, Section 2; Rudin (1991), Section 7.9.

2. Let $f \in \mathcal{L}^2$ and let $f_n = B_n \cdot f$ be the product of the function f and the indicator B_n of the ball $\bar{B}(0, n) \subseteq \mathbb{R}^m$. Since $B_n \in \mathcal{L}^2$, it follows that $f_n = B_n \cdot f$ are integrable with respect to the Lebesgue measure dx and their Fourier transforms $g_n = \Phi(f_n)$ satisfy the equalities

$$g_n(y) = (2\pi)^{-m/2} \int_{\bar{B}(0,n)} e^{-iyx} f(x) \, dx.$$

From Plancherel's theorem it follows that $g_n \to g = \Phi(f)$ in the space $\mathcal{L}^2$ (i.e., in the mean square sense).

4.4.4.   The Fourier–Stieltjes transform

The Fourier–Stieltjes transform is *the restriction of the Fourier transform to the set of slowly increasing measures* μ on the algebra $\mathcal{B} = \mathcal{B}(\mathbb{R}^m)$ generated by compact sets in $\mathbb{R}^m$. Examples of slowly increasing measures are probability distributions, the Lebesgue measure, and the counting measure. The

Fourier series and integrals are associated with the transforms of the counting and the Lebesgue measures. For probability distributions, we introduce *characteristic functions*.

The characteristic function $\hat{\mu}$ of a probability distribution μ is defined as follows:

$$\hat{\mu}(y) = \int e^{iyx} \, d\mu(x).$$

The following theorem describes characteristic functions.

Bochner's theorem. *A function is characteristic if and only if it is a normed continuous positive definite function on* $\mathbb{R}^m$.

The property of being normed is written as $\hat{\mu}(0) = 1$. The property of being positive definite is expressed by the inequality

$$\sum \hat{\mu}(y_j - y_k) \cdot \bar{z}_j z_k \geq 0 \quad (1 \leq, j, \ k \leq n)$$

for all finite collections of vectors $y_1, \ldots, y_n \in \mathbb{R}^m$ and complex numbers $z_1, \ldots, z_n$. The proof of Bochner's theorem can be found in Loève (1977), Chapter 4, Section 14. A deduction of Bochner's theorem from a more general theorem on positive functionals on Banach algebras can be found in Rudin (1991), Chapter 11.

4.4.5.   The Radon transform

Radon proved that any smooth function on $\mathbb{R}^3$ is defined by the values of its integrals over the planes in $\mathbb{R}^3$. The Radon transforms establish a connection between functions and their integrals over sets of a given class. The Radon transform plays an important role in integral geometry.

Choose a point $a \neq 0$ in the space $\mathbb{R}^m$ and a number c and consider the hyperplane $H = \{x : ax = c\}$ in $\mathbb{R}^m$ and the $(m-1)$-dimensional measure $d_H x$ on H. The equality

$$R_{a,c}(f) = \|a\|^{-1} \int f(x) \, d_H x$$

defines the *Radon transform* $R(f)$ of a function f on $\mathbb{R}^m$ that is integrable over any hyperplane. The expression for the integral on the right-hand side is often written with the delta function δ_H, which is considered the *generalized derivative* of the Lebesgue measure $d_H x$ on the hyperplane H with respect to the Lebesgue measure dx on $\mathbb{R}^m$, without providing a formal definition of the generalized derivative of a measure with respect to another measure. In this case, the equality

$$\int f(x)\delta_H\, dx = \int f(x)\, d_H x$$

serves as the formula for the change of variables $d_H x = \delta_H \cdot dx$ in the integral. If $H = \mathbb{R}^{m-1}$, $d_H x = d^{m-1}x$, and $\delta_H = \delta^{m-1}$, then $d^{m-1}x = \delta^{m-1}(x) \cdot d^m x$. In particular, we can consider rapidly decreasing functions f in $\mathcal{S} = \mathcal{S}(\mathbb{R}^m)$. The Radon transform R is a linear one-to-one map from the space $\mathcal{S}$ onto a vector space $\mathcal{R} = \mathcal{R}(\mathbb{R}^m)$, defined by pairs (a,c) of points of $\mathbb{R}^m$ and numbers (see Helgason, 2000, Chapter 1, Section 2). Using the inverse Radon transform R^{-1}, one can determine a function f from its integrals over hyperplanes.

There is a connection between the Radon transform and the Fourier transform. It is easy to show that

$$R_{a,c}(f) = (2\pi)^{m/2-1} \int_{\mathbb{R}} \Phi^{-1} f(ta) e^{-itc}\, dt.$$

This equality can be used for the definition of the Radon transform.

Exercise. Verify the above equality.

Remark. In Kirillov and Gvishiani (1988), Chapter 4, the Fourier transform is related to characters of commutative groups. More than 100 problems on convolutions, group characters, and Fourier transforms are presented.

4.5. BOUNDED LINEAR OPERATORS

In this section, we describe the main principles of linear analysis. They include the extension, boundedness, and inversion theorems. We formulate these theorems only for the case of normed spaces, although there are more general formulations. (Dunford and Schwartz, 1988, Part 1, Chapter 2; Bourbaki, 1987).

4.5.1. Extensions of functionals

The Hahn–Banach theorem, which was proved in Subsection 2.2.5, part 6, states the principle of a norm-preserving global extension of a bounded linear functional from a subspace to the entire space. This principle can also be formulated in the geometric form.

1. The kernel of any nonzero linear functional on a vector space has codimension 1. At the same time, every subspace of codimension 1 is a kernel of a linear functional (Kolmogorov and Fomin, 1989, Chapter 3, Section 1). A *topological vector space* is a vector space with a topology that ensures the continuity of the operations (Bourbaki, 1987, Chapter 1; Kolmogorov and Fomin, 1989, Chapter 3; Sadovnichii, 1999, Chapter 2). A linear functional on such a space is bounded if and only if its kernel is closed (Bourbaki, 1987, vol. 2, Chapter 2, Section 2). The linear manifold $H = a + Z$ formed by the sums $h = a + z$ of a fixed vector a and vectors $z \in Z$, where Z is a subspace of codimension 1 in a given space Y, is called a *hyperplane parallel to Z*.

The correspondence between linear functionals and hyperplanes yields a geometric form of the Hahn–Banach theorem (Bourbaki, 1987, Chapter 2, Section 3). Let Y be a real topological vector space, $B \subseteq Y$ be a nonempty convex open set, and $A \subseteq Y$ be a liner manifold which does not intersect B. The following theorem holds (see Bourbaki, 1987, Chapter 2, Section 3, Subsection 1).

Theorem. *There exists a closed hyperplane H that includes A and does not intersect B.*

Widely used theorems on separatedness are deduced from this theorem.

2. Let A and B be nonempty sets and H be a hyperplane in a real topological space Y. We say that the hyperplane H (*strictly*) *separates* A from B if A is included in one of the (open) closed half-spaces formed by H and B is included in the other. A hyperplane S in the space Y is called a *support hyperplane* of A if A and S intersect and A is included in one of the half-spaces formed by S (is situated *on one side of S*).

Note that a hyperplane H is defined by the equation $\varphi(y) = \alpha$ for some $\varphi \in Y'$ and $\alpha \in \mathbb{R}$, while the corresponding half-spaces are defined by the inequalities $\varphi(y) > \alpha$, $\varphi(y) \geq \alpha$ or $\varphi(y) < \alpha$, $\varphi(y) \leq \alpha$.

We now formulate several statements on separatedness.

Proposition 1. *For any nonempty open convex set A and any nonempty convex set B such that the intersection of A and B is empty, there exists a closed hyperplane H that separates A and B.*

Corollary. *In a locally convex space Y, for any nonempty closed convex set A and any nonempty convex set B such that the intersection of A and B is empty, there exists a closed hyperplane H that strictly separates A and B.*

The following statement holds for any topological vector space Y.

Proposition 2. *For any nonempty compact set A and any closed hyperplane H, there exists a support hyperplane S of A that is parallel to H.*

A closed convex set $A \subseteq Y$ with nonempty interior is called a *convex body*.

Proposition 3. *Any support hyperplane S of a convex body A is closed and any boundary point of A belongs to a support hyperplane S of A.*

Corollary. *Any convex body A is equal to the intersection of all closed half-spaces that are formed by support hyperplanes of A and include A.*

The above statements are proved in Bourbaki (1987), Chapter 2, Section 3, Subsections 2 and 3.

Exercise. Deduce all the assertions in Subsection 4.5.1 from the Hahn–Banach theorem.

4.5.2.   Uniform boundedness of operators

1. Let E be a Banach space, F be a normed space, and $T_i : E \to F$ be a collection of continuous linear operators with an arbitrary index set. The collection T_i is said to be *bounded at a point* $x \in E$ if the set of values $T_i(x) \in F$ is bounded in F. The collection T_i is said to be uniformly bounded if the union of the sets of values $T_i(x) \in F$ for $\|x\| \leq 1$ is bounded in F (i.e., if the set of numbers $\|T_i\|$ is bounded). In other words, we can say that the collection T_i is *uniformly bounded* on the unit ball. As usual, we assume the spaces in question to be nonempty.

The Banach–Steinhaus theorem. *If a collection of continuous linear operators on a Banach space is bounded at every point, then it is uniformly bounded.*

Proof. The Banach–Steinhaus theorem can be deduced from the Baire theorem (Subsection 3.1.3, part 5).

Consider the sequence of sets $E(n) = \{x : \|T_i(x)\| \leq n \; \forall i\}$. It is easy to verify that these sets are closed and their union is the entire E. Since E has interior points, the Baire theorem implies that some $A = E(n(0))$ has an interior point a. Consequently, A includes a closed ball $B(a, r)$ with center a and radius $r > 0$. Therefore, $\|T_i x\| = \|T_i a + T_i z\| \leq n(0)$, $\|T_i x\| \leq \|T_i a\| + \|T_i x\| \leq c(a) + n(0)$ for $x = a + z$, $\|z\| \leq r$, $\|T_i(a)\| \leq c(a)$. Hence $r\|x\|^{-1} \cdot \|T_i x\| = \|T_i(r \cdot \|x\|^{-1} \cdot x)\| \leq c(a) + n(0)$ for $\|x\| \neq 0$ and $\|T_i x\| \leq c \cdot \|x\|$, where $c = r^{-1}(c(a) + n(0))$. Then $\|T_i x\| \leq c \; (\|x\| \leq 1)$ for all i. □

2. The principle of uniform boundedness given in the Banach–Steinhaus theorem is used to great effect.

Corollary. *The limit T of a sequence of continuous linear operators T_n on a Banach space E that converges at every point $x \in E$ is a continuous linear operator on E.*

Proof. The linearity of T follows from the linearity of T_n and the limit. The continuity of T, which is equivalent to its boundedness on the unit ball, follows from the Banach–Steinhaus. Indeed, since $T_n x \to T x$ for any point $x \in E$, there exists a $c(x) > 0$ such that

$$\|T_n x\| \leq \|T x\| + \|T_n x - T x\| \leq c(x)$$

Hence, by the Banach–Steinhaus theorem, there is a $c > 0$ such that $\|T_n x\| \leq c$ for all indices n and points $x \in B(0, 1) \subseteq E$. Therefore, the limit function T satisfies the inequality $\|T x\| \leq c$ for $\|x\| \leq 1$. We conclude that the operator T is bounded. □

It should be emphasized that if a sequence of nonlinear continuous operators converges at every point, this does not guarantee that the limit operator is continuous.

Exercise. Provide a counterexample to illustrate the preceding assertion.

4.5.3. Inversion of operators

1. The inverse of a one-to-one linear operator is also a one-to-one linear operator. The inverse of a continuous operator, however, is not necessarily continuous.

Example. Consider the Banach space $E = \mathcal{C}[0,1]$ of continuous functions on the interval $[0,1]$ with norm $\|f\| = \sup|f(x)| : x \in [0,1]$.

The indefinite integral $A = \mathcal{I}$ defined by the equalities

$$Af = g, \quad g(y) = \int_0^y f(x)\,dx \quad (y \in [0,1])$$

is a continuous one-to-one linear operator that maps E to E. The inverse of A is the differentiation operator $A^{-1} = \mathcal{D}$ defined on the subspace F formed by smooth functions that vanish at zero.

The operator $\mathcal{D}$ is discontinuous. Indeed, let $g_n(y) = n^{-1}y^n$ $(0 \le y \le 1)$. Then $\|g_n\| = n^{-1}$ and $g_n \to 0$. At the same time, for $f_n = \mathcal{D}g_n$ we have $f_n(x) = x^{n-1}$ $(0 \le x \le 1)$ and $\|f_n\| = 1$. Therefore, $\mathcal{D}g_n = f_n \nrightarrow 0 = \mathcal{D}0$.

Thus, the operator $\mathcal{I}\colon E \to F$ is continuous, whereas the inverse operator $\mathcal{D}\colon F \to E$ is discontinuous.

Note that $A = \mathcal{I}$ viewed as a map from E to E is not an onto map, i.e., $A(E) = F \ne E$. On the other hand, if A is viewed as a map from E onto F, it turns out that the space F is not complete.

2. The following theorem holds for complete spaces.

The Banach theorem. *Let T be a one-to-one continuous linear operator from a Banach space E onto a Banach space F. Then the inverse operator T^{-1} is a one-to-one continuous linear operator from F onto E.*

Proof. From the definition of T^{-1} it follows that it is a one-to-one operator. It is easy to verify that T^{-1} is linear. We need to prove that T^{-1} is continuous. Since T^{-1} is linear, it suffices to show that it is continuous at zero, i.e., that for any $\varepsilon > 0$ there is a $\delta > 0$ such that $T^{-1}B(0,\delta) \subseteq B(0,\varepsilon)$. Since T^{-1} is one-to-one, the preceding inclusion is equivalent to the inclusion $B(0,\delta) \subseteq TB(0,\varepsilon)$.

Consider the sequence of closed sets $F_n = \overline{TB(0, n\varepsilon/8)}$, which are the closures of images of the balls with center 0 and radius $n\varepsilon/8$. Clearly, $E = \cup B(0, n\varepsilon/8)$. By assumption, the operator T maps E onto F, which implies $F = \cup F_n$. Since F is a Banach space, from the Baire theorem it follows that there exists an $n = n(0)$ such that the set $F_n = F_{n(0)}$ has an interior point, i.e., there are $c \in F$ and $r > 0$ such that $B(c,r) \subseteq F_{n(0)} = \overline{TB(0, n(0)\varepsilon/8)}$. It is easy to verify that $\overline{TB(0, n(0)\varepsilon/8)} = n(0) \cdot \overline{TB(0, \varepsilon/8)} = \{z = n(0) \cdot y : y \in \overline{TB(0, \varepsilon/8)}\}$.

Since $V = (c + V) - c$, we have $B(0,r) \subseteq B(c,r) - B(c,r) = \{z_1 - z_2 : z_1, z_2 \in B(c,r)\}$. Furthermore, the inequality $|x_1 - x_2| \leq |x_1| + |x_2|$ implies $B(0, \varepsilon/8) - B(0, \varepsilon/8) \subseteq B(0, \varepsilon/4) \subseteq B(0, \varepsilon/2)$. As a result,

$$B(0, r/n(0)) = (1/n(0)) \cdot B(0,r) \subseteq \overline{TB(0, \varepsilon/8)} - \overline{TB(0, \varepsilon/8)}$$
$$\subseteq \overline{TB(0, \varepsilon/8) - TB(0, \varepsilon/8)} = \overline{T(B(0, \varepsilon/8) - B(0, \varepsilon/8))}$$
$$\subseteq \overline{TB(0, \varepsilon/2)},$$

i.e., $B(0, \delta) \subseteq \overline{TB(0, \varepsilon/2)}$ for $\delta = r/n(0)$.

Considering the sequence of balls $B_n = B(0, \varepsilon/2^n)$, their images $\overline{TB_n}$, and using the properties of the operator T, one can prove that $\overline{TB(0, \varepsilon/2)} \subseteq TB(0, \varepsilon)$. Therefore, $B(0, \delta) \subseteq TB(0, \varepsilon)$, which completes the proof (cf. Kolmogorov and Fomin, 1989, Chapter 4, Section 5).   □

Exercise. Prove the inclusion $\overline{TB(0, \varepsilon/2)} \subseteq TB(0, \varepsilon)$.

3. We now present two corollaries of the Banach theorem. Let $\|\cdot\|_1$ and $\|\cdot\|_2$ be norms on a vector space E such that $E_1 = (E, \|\cdot\|_1)$ and $E_2 = (E, \|\cdot\|_2)$ are Banach spaces. The norm $\|\cdot\|_1$ is said to be *subordinate* to the norm $\|\cdot\|_2$ if there exists a $c_2 > 0$ such that $\|x\|_1 \leq c_2 \cdot \|x\|_2$ ($x \in E$). The norms $\|\cdot\|_1$ and $\|\cdot\|_2$ are said to be *equivalent* if they are subordinate to each other.

Corollary 1. *If one of two norms is subordinate to the other, then these norms are equivalent.*

Proof. Indeed, if $\|\cdot\|_1$ is subordinate to $\|\cdot\|_2$, then the identity operator $T : E_2 \to E_1$ is continuous. By the Banach theorem, the inverse operator $T^{-1} : E_1 \to E_2$ is also continuous, which implies $\|x\|_2 \leq c_1 \cdot \|x\|_1$ ($x \in E$) for $c_1 = \|T^{-1}\|$.   □

Consider the linear equation $Tx = y$, where T is a continuous linear map from a Banach space E to a Banach space F and $x \in E$, $y \in F$.

Corollary 2. *If for any $y \in F$ there exists a unique solution $x = x(y)$ of the equation $Tx = y$, then x continuously depends on the right-hand side y.*

Proof. The assumption of Corollary 2 implies that the operator T is a one-to-one map from E onto F. The assertion of the corollary is equivalent to the assertion that the inverse operator T^{-1} is continuous. The latter follows from the Banach theorem. □

Thus, Corollary 2 provides a condition for the solution of $Tx = y$ to be well defined.

4.5.4. Closedness of the graph of an operator

The closed graph theorem is equivalent to the Banach inverse operator theorem.

1. Consider Banach spaces E and F, their product $E \times F$, a vector subspace A of E which is dense in E ($\bar{A} = E$), and a linear operator T : $A \to F$. The set

$$G(T) = \{(x, Tx) : x \in A\} \subseteq E \times F$$

is called the *graph* of the operator T. An operator is formally identified with its graph.

The product $E \times F$ with norm $\|(x, y)\| = \max(\|x\|, \|y\|)$ is a Banach space and the graph $G = G(T)$ is its vector subspace. If G is closed, then it is also a Banach space. An operator T with closed graph G is said to be a *closed operator*.

Example. Assume that $E = F = \mathcal{C}[0, 1]$, $A \subseteq E$ is a subspace formed by smooth functions, and $T = D$ is the differentiation operator. From the theorem on termwise differentiation of a sequence, it follows that the operator D is closed. We emphasize that D is not everywhere defined. It turns out that an operator T is bounded if it is closed and defined everywhere.

Theorem. *If E and F are Banach spaces, then any closed operator $T : E \to F$ defined on the entire space E is bounded.*

Proof. Let $P : G \to E$ and $Q : G \to F$ be projections defined by the equalities $P(x, Tx) = x$ and $Q(x, Tx) = Tx$ ($x \in E$). They are continuous linear operators. Moreover, P is a one-to-one map from the Banach space G onto E. By the inverse operator theorem, P^{-1} is continuous. Hence, the operator T, which is equal to the composition QP^{-1}, is also continuous and therefore bounded. □

2. The equality

$$\|x\|_G = \|x\|_E + \|Tx\|_F \quad (x \in A)$$

defines another norm on A. This norm is called the *graph norm*. If the operator T is closed, then A equipped with this norm is a Banach space. The operator T is bounded on this space.

Consider the Banach spaces E and F, and let A and B be subspaces of E such that $A \subseteq B$. Let $T : A \to F$ be a closed operator and suppose that an operator $S : B \to F$ has a closed extension. Then there exists a $c > 0$ such that

$$\|Tx\| \leq c(\|x\|^2 + \|Sx\|^2)^{1/2} \quad (x \in A).$$

The proof of this assertion which uses the closed graph theorem can be found in Yosida (1965), Chapter 2, Section 6. In the same book (Chapter 2, Section 7), the closed graph theorem is used to prove the Hörmander theorem on hypoelliptic operators.

Remark. The equality

$$\|x\|_A = \left(\|x\|_E^2 + \|Sx\|^2\right)^{1/2} \quad (x \in A)$$

defines a norm on A. It is also called the *graph norm*.

3. A map $f : X \to Y$ from a topological space $(X, \mathcal{U})$ to a topological space $(Y, \mathcal{V})$ is said to be *open* if f maps every open set $U \in \mathcal{U}$ to an open set $f(U) = V \in \mathcal{V}$.

The following statement can be found in Kolmogorov and Fomin (1989), Chapter 4, Section 5.

Lemma. *The natural map from a Banach space X to the factor space $Y = X \setminus Z$ over a closed subspace Z is open.*

The following theorem is a consequence of the closed graph theorem and the preceding lemma.

Theorem. *A closed operator $T : E \to F$ from a Banach space E onto a Banach space F is an open map.*

The operator T can be assumed to be bounded rather than closed. The corresponding version of the preceding theorem is proved in Kolmogorov and Fomin (1989), Chapter 4, Section 5. It generalizes the inverse operator theorem.

Exercise. Deduce the inverse operator theorem from the open map theorem.

Remark. More general closed graph theorem and open map theorem are proved in Edwards (1995), Chapter 6. Metrizable topological vector spaces and locally convex spaces are considered there. The book contains many examples and references to the literature.

4.5.5. Weak compactness

Continuous linear functionals on normed and locally convex spaces play the role of coordinates.

1. Consider a normed space E, the conjugate space E' of continuous linear functionals on E, and the space E'' conjugate to E', whose elements are continuous linear functionals on E'. Among the elements of E'', we point out the functionals δ_x defined by points $x \in E$ such that

$$\delta_x(x') = x'(x) \quad (x' \in E').$$

It is easy to verify that $\delta_x \in E''$. The correspondence $x \to \delta_x$ is a norm-preserving linear embedding from E into E''. For this reason, functionals δ_x are often identified with points x.

The functionals δ_x $(x \in E)$ represent a convenient coordinate system for E'. The topology for E' defined by this system is called the *E-topology*. It is not to be confused with the E''-topology defined by E''. If $E'' = \{\delta_x : x \in E\}$, then the space E is called *reflexive* and it is necessarily a Banach space.

Exercise. Prove the preceding statement.

For a coordinate system in the original space, we choose the conjugate space E' of all continuous linear functionals x' on E. From the Hahn–Banach theorem, it follows that these coordinates separate the points of E: if $x'(x) = 0$ for all $x' \in E'$, then $x = 0$.

Those properties of sets and sequences in E which are associated with coordinates are called *weak* or *coordinatewise*: weak boundedness, weak convergence, weak compactness. Properties associated with norms are called *strong* or *metric*.

2. Using the embedding into the second conjugate space and the principle of uniform convergence, it is easy to prove the following theorem.

The Mackey theorem. *A set in a normed space is strongly bounded if and only if it is weakly bounded.*

In Bourbaki (1987), Chapter 4, Section 4, this theorem is proved for separated locally convex spaces. The Mackey theorem readily implies the following criterion.

The weak convergence criterion. *A sequence of points $x(n) \in E$ weakly converges to a point $a \in E$ if and only if the sequence $x(n)$ is bounded and $z'(x_n) \to z'(a)$ for any z' in a set $Z' \subseteq E'$ whose linear span is dense in E'.*

This criterion is proved in Yosida (1965), Chapter 5, Section 1 (Theorem 3).

The following statement often appears to be useful.

Mazur's theorem. *If a sequence of points $x(n)$ in a normed space weakly converges to a point a, then there exists a sequence of convex combinations of $x(n)$ that strongly converges to a.*

This theorem can be proved with the use of the separatedness theorems that follow from the Hahn–Banach theorem. The proof can be simplified if we take linear combinations instead of convex combinations. (Prove this assertion.) A proof of Mazur's theorem is given in Yosida (1965), Chapter 5, Section 1 (Theorem 2).

The following theorem, which is analogous to the classical Weierstrass convergent subsequence theorem, states that the conjugate space is weakly compact.

Theorem. *If a sequence of continuous linear functionals on a separable normed space is bounded, then it has a subsequence convergent at every point.*

This theorem is proved in Kolmogorov and Fomin (1989), Chapter 4, Section 3 (Theorem 3).

Remark. The convergence theorems are used for solving linear operator equations. By taking coordinates, it is possible to choose a subsequence of a bounded sequence of approximate solutions that weakly converges to a generalized solution. The existence of a sequence that strongly converges to this generalized solution makes it possible to obtain a solution satisfying the desired conditions.

3. The following statements on compactness, separability, and reflexivity hold for any Banach space E. These statements explain the role of separable and reflexive spaces.

The Alaoglu theorem. *The ball $B' = \{x' : \|x'\| \leq 1\} \subseteq E'$ is compact in the E-topology.*

See Dunford and Schwartz (1988), Chapter 5, Section 4, Theorem 2; Hille and Phillips (1974), Theorem 2.10.2.

Proposition 1. *If the space E is separable, then the ball $B' = \{x' : \|x'\| \leq 1\} \subseteq E'$ is sequentially compact in the E-topology.*

See Hille and Phillips (1974), Theorem 2.10.1. The Alaoglu theorem implies the convergent subsequence theorem of part 2.

Exercise. Provide a detailed proof of these statements.

Proposition 2. *The ball $B' = \{x' : \|x'\| \leq 1\}$ is metrizable under the induced E-topology if and only if E is separable.*

Theorem. *The ball $B' = \{x' : \|x'\| \leq 1\} \subseteq E'$ is weakly compact if and only if E is reflexive.*

Corollary. *A space E is reflexive if and only if its conjugate space E' is reflexive.*

The Eberlein–Smulyan theorem. *A space E is reflexive if and only if every bounded sequence of its points has a weakly convergent subsequence.*

See Dunford and Schwartz (1988), Chapter 5, Section 5, Theorem 1; Dunford and Schwartz (1988), Chapter 5, Section 4, Theorem 1; Hille and Phillips (1974), Theorem 2.10.3; Yosida (1965), Chapter 5, Section 4, respectively.

Remark. Some of the above assertions can be extended to the case of locally convex spaces. (see Bourbaki, 1987, Chapter 4).

4.6.   COMPACT LINEAR OPERATORS

The image of a bounded set under a bounded linear operator is a bounded set. Among all bounded operators, we point out *compact* operators, which take bounded sets to subsets of compact sets. Such operators are also called *completely continuous*. In separable Hilbert spaces, compact operators can be approximated by *degenerate* operators (with finite-dimensional ranges). This explains many good properties of compact linear operators.

A systematic presentation of the theory of compact operators is given in Edwards (1995), Chapter 9. The case of compact operators in locally convex spaces is also considered there.

4.6.1.   Examples of compact operators

Let E and F be normed spaces. A linear operator $T : E \to F$ is said to be *compact* if the image $T(B)$ of any bounded set $B \subseteq E$ is included in a compact set $C \subseteq F$.

1. Note that the condition $T(B) \subseteq C$ is equivalent to the compactness of the closure $\overline{T(B)}$ of the image $T(B)$ of a closed set B. Since T is linear, it suffices to take the unit ball $B(0,1) \subseteq E$ and its image $T(B(0,1)) \subseteq F$ instead of an arbitrary set B. Indeed, the compactness of T means $\overline{TB(0,1)} \subseteq E$.

A linear operator $T : E \to F$ is compact if and only if for any bounded sequence $x_n \in E$ the sequence $Tx_n = y_n \in F$ has a convergent subsequence.

Exercise. Prove the preceding statement.

A linear operator $T : E \to F$ whose range $T(E)$ is finite-dimensional is said to be *degenerate*. By definition, the images of E under degenerate operators are finite-dimensional subspaces of a Banach space F. These subspaces are also called *degenerate*. Since the closure of any bounded set in a finite-dimensional space is compact, it follows that degenerate operators are compact.

It is easy to verify that a normed space is finite-dimensional if and only if the closed unit ball in it is compact, i.e., if the identity operator on this space is compact.

Exercise.   Prove the preceding statement (see Kirillov and Gvishiani, 1988, Chapter 3, Section 3; Kolmogorov and Fomin, 1989, Chapter 4, Section 6).

2. We now give several examples.

Example 1. Set $E = \mathbb{R}^n$ and $F = \mathbb{R}^m$. Then any linear operator $T : E \to F$ is degenerate and therefore compact. Linear maps between finite-dimensional spaces are called *matrix operators*.

Example 2. Let $E = F = l^2$ be the Banach spaces whose elements are square summable sequences of numbers. Any bounded sequence of numbers α_n defines a linear operator $T : E \to F$ that transforms a sequence $x = (\xi_n)$ to the sequence $Tx = y = (\eta_n)$ of numbers $\eta_n = \alpha_n \xi_n$. Such operators are called *diagonal* by analogy with matrices. The operator T is compact if and only if $\alpha_n \to 0$ (Halmos, 1982, Chapter 6 and Problem 132).

Example 3. Set $E = F = C[a, b]$. We define an operator $T : E \to F$ as follows:

$$Tf = g, \quad g(y) = \int_a^b k(y, x) f(x)\, dx \quad (y \in [a, b])$$

for $f \in C[a, b]$, $k(y, x) \in \mathbb{C}([a, b] \times [a, b])$. Clearly, T is linear. It is called an *integral operator with continuous kernel*. Since for any $\varepsilon > 0$ there is a $\delta > 0$ such that

$$|g(y)| \leq (b - a)\|k\| \cdot \|f\|,$$
$$|g(y) - g(z)| \leq \varepsilon(b - a)\|f\| \quad (|y - z| \leq \delta),$$

it follows that $g \in C[a, b]$ and the closure of the set $T(B(0, 1)) = \{Tf = g : \|f\| < 1\}$ is uniformly bounded and uniformly equicontinuous. By the Arzel'a theorem, it is compact and therefore the operator T is compact.

Example 4. Set $E = F = C[a, b]$. We define an operator $T : E \to F$ as follows:

$$Tf = g, \quad g(y) = \int_a^y k(y, x) f(x)\, dx \quad (y \in [a, b])$$

for $f \in C[a, b]$, $k(y, x) \in C([a, b] \times [a, b])$. As in Example 2, it is easy to verify that T is a compact linear operator. It is called an *integral Volterra operator*.

Example 5. Put $E = F = \mathcal{L}^2([a, b], dx)$. We define an operator T by the same equalities as in Example 3 with $f \in \mathcal{L}^2([a, b], dx)$, $k \in \mathcal{L}^2([a, b] \times [a, b], dxdy)$. By the Fubini theorem, $k(y, \cdot) \in \mathcal{L}^2([a, b], dx)$, $k(y, \cdot)f \in \mathcal{L}([a, b], dx)$ for almost all $y \in [a, b]$. We can assume that g is equal to zero at the remaining points of the interval $[a, b]$. Using the Cauchy inequality, it is easy to verify that $g \in \mathcal{L}^2([a, b], dy)$. Obviously, T is linear. Applying the Riesz compactness criterion for the space of integrable functions, it is easy to prove that T is a compact operator. It is called the *integral operator with Hilbert–Schmidt kernel*.

Example 6. The Sobolev embedding operator is compact (see Sobolev, 1991, Chapter 1, Section 11, and Kantorovich and Akilov, 1982, Chapter 11, Section 4.4).

Exercise. Prove that the integral operators in Examples 4 and 5 are compact (see Kolmogorov and Fomin, 1989, Chapter 4, Section 6, and Halmos, 1982, Chapter 15).

4.6.2.   Properties of compact operators

1. We will consider compact operators that map a normed space to itself. They form a *closed two-sided ideal* in the normed ring of bounded operators, which is expressed by the following theorem.

Theorem 1. (1) *A linear combination of compact operators is a compact operator.*

(2) *The product of a compact operator and a bounded operator is a compact operator.*

(3) *The limit of a convergent sequence of compact operators is a compact operator.*

Proof. For example, we will prove (2). Let T be a compact linear operator, L be a bounded linear operator, and B be a bounded set in a normed space E. Then $L(B)$ is bounded and $TL(B)$ is included in a compact set $C \subseteq E$. Similarly, $T(B)$ is included in a compact set, and the same is true for $LT(B)$. Thus, the products TL and LT are compact operators.   □

Exercise. Prove the remaining assertions of Theorem 1 (see Kolmogorov and Fomin, 1989, Chapter 4, Section 6).

The following statement is a consequence of assertion (2) of Theorem 1.

Corollary. *In an infinite-dimensional normed space, the inverse of a one-to-one compact operator is unbounded.*
Proof. Assume the contrary. Then the identity operator on an infinite-dimensional Banach space is compact. A contradiction. □

2. Consider an infinite-dimensional normed space E, a linear compact operator $T : E \to E$, and the equation $Tx = y$. This equation is not well-posed because, as follows from the preceding statements, its solution cannot continuously depend on its right-hand side if its solution is unique.

Compact operators transform weakly convergent sequences into strongly convergent ones. In a reflexive Banach space, every bounded linear operator with this property is compact (Kirillov and Gvishiani, 1988, Chapter 3, Section 3).

Exercise. Prove the preceding statements (see Reed and Simon, 1978, Chapter 5).

Using the property of compact operators to turn weak convergence into strong convergence, one can prove that any compact operator in a separable Hilbert space is the limit of a sequence of degenerate operators. Together with assertion (3) of Theorem 1, this characterizes compact operators in such spaces.

Proposition. *A linear operator T in a separable Hilbert space H is compact if and only if there exists a sequence of degenerate operators S_n in H such that $\|T - S_n\| \to 0$.*

The proof of this proposition is given in Reed and Simon, 1978, Chapter 5. The proposition can be used to verify the compactness condition for the diagonal operator in Example 2. Compact operators can be approximated by degenerate operators in an arbitrary Hilbert space (Kirillov and Gvishiani, 1988, Chapter 3, Section 5, Theorem 43).

4.6.3. Adjoint operators

1. Consider Banach spaces E and F, a bounded operator $T : E \to F$, and the conjugate spaces E' and F'. The operator T takes a point $x \in E$ to a point $Tx = y \in F$. The equalities

$$x' = T'y' = y'T, \quad y'T(x) = y'(Tx)$$

define a bounded linear operator $T' : F' \to E'$ that takes the coordinate $y' \in F'$ of the point $Tx = y$ to the coordinate $x' = T'y'$ of the point x. It is easy to verify that the operator T' is linear and bounded. This operator is called the (Banach) adjoint of T (see examples in Kantorovich and Akilov, 1982, Chapter 9, Section 3).

Example. Let $E = F = l^1$ and $T(\xi_1, \xi_2, \dots) = (0, \xi_1, \xi_2, \dots)$. Then $E' = F' = l^\infty$ and $T'(\eta_1, \eta_2, \dots) = (\eta_2, \eta_3, \dots)$. Note that $\|T\| = \|T'\| = 1$.

Let $E = F$ and $S, T \in \mathcal{B}(E, E)$. Then

$$(ST)' = T'S'$$

(see Yosida, 1965, Chapter 7, Section 1).

Exercise. Prove this equality.

It is easy to prove that $\|T\| = \|T'\|$ (Kolmogorov and Fomin, 1989, Chapter 4, Section 5; Reed and Simon, 1978, Chapter 2). The map $T \to T'$ is a linear isometry from $\mathcal{B}(E, F)$ to $\mathcal{B}(F', E')$. If T is invertible, then so is T'. Moreover, $(T^{-1})' = (T')^{-1}$ (Dunford and Schwartz, 1988, Chapter 6, Section 2).

Exercise. Prove that $\|T\| = \|T'\|$ and $(T^{-1})' = (T')^{-1}$.

The following statements can be found in Kolmogorov and Fomin (1989), Chapter 4, Section 6; Yosida (1965), Chapter 10, Section 4; Rudin (1991), Section 4.19.

Proposition. If an operator T is compact, then the adjoint operator T' is also compact.

Corollary. An operator $T \in \mathcal{B}(E, F)$ is compact if and only if the adjoint operator T' is compact.

2. Let $E = F = H$ be a Hilbert space and $E' = F' = H^*$ be the conjugate space. Then, by the Riesz theorem on the representation of linear functionals, there exists a conjugate linear isometry C from the space H

onto H^* such that $c \in H$ is mapped to a linear functional $x \to c \cdot x$ $(x \in H)$. The operator

$$T^* = C^{-1}T'C : H \to H$$

is called the (*Hermitian, Hilbert*) *adjoint* of $T : H \to H$. Together with T, this operator is bounded and linear (Yosida, 1965, Chapter 10, Section 2).

The adjoint operator $T^* : H \to H$ is defined by the equalities $Tx \cdot y = x \cdot T^*y$ $(x, y \in H)$. Indeed,

$$y \cdot Tx = Cy(Tx) = T'Cy(x) = C^{-1}T'Cy \cdot x = T^*y \cdot x.$$

Examples. 1) Let $H = \mathbb{C}^n$ and let T be a linear operator with matrix $A = (a_{ij})$ with respect to the standard basis. Then the matrix of the conjugate operator T^* is $A^* = (\bar{a}_{ji})$, which is the transpose and complex conjugate of A.

2) Let $H = \mathcal{L}^2([a, b], dx)$ and let T be the integral operator with the Hilbert–Schmidt kernel $k(x, y)$. Then the conjugate operator T^* is the integral operator with the Hilbert–Schmidt kernel $k^*(y, x) = \overline{k(x, y)}$, which is the transpose and complex conjugate of $k(x, y)$.

3. It is easy to verify that for any bounded linear operators S and T in H

$$(ST)^* = T^* \cdot S^*, \quad (T^*)^* = T, \quad \|T^*T\| = \|T\|^2.$$

Moreover, if T is invertible, then so is T^* and $(T^{-1})^* = (T^*)^{-1}$ (Dunford and Schwartz, 1988, Chapter 6, Section 2).

Exercise. Prove the above equalities (see Reed and Simon, 1978, Chapter 3).

The following theorem can be proved using the properties of compact operators, the compactness criterion for the spaces of continuous maps, and the embedding into the second conjugate space (Yosida, 1965, Chapter 10, Section 4).

Theorem. *An operator* $T \in \mathcal{B}(H, H)$ *is compact if and only if the conjugate operator* T^* *is compact.*

Thus, one can pass to the conjugates of given operators without loss of compactness.

Remark. More general definitions for linear operators on dense sets and locally convex spaces are given in Yosida (1965), Chapter 7.

4.6.4.   Fredholm operators

In this subsection, we give an analytic definition of Fredholm operators that does not use the algebraic definition given in Section 2.2. The connection between these definitions is established by the Nikolskii theorem (Kirillov and Gvishiani, 1988, Chapter 3, Section 3). The definition of a Fredholm operator sometimes requires that its range be closed (Kato, 1995, Chapter 4, Section 5), as was the case in Subsection 2.2.4, part 9.

1. Let E and F be normed spaces and $A : E \to F$ be bounded linear operator. The equation

$$Ax = y \quad (x \in E, \ y \in F)$$

is the natural generalization of the system of linear algebraic equations. For some classes of operators, the theory of linear operator equations is well developed. (in particular, for compact operators and Fredholm operators, which are closely related).

Throughout this subsection, we consider a fixed normed space E and linear operators $A : E \to E$. The identity operator on E is denoted by I.

An operator which is represented as a difference of an identity operator and a compact operator is called a *Fredholm operator*, i.e., an operator $A = I - T$ is a Fredholm operator whenever T is a compact operator. Since the operator $-T$ is compact whenever T is compact, the difference can be replaced with the sum $(A = I + (-T))$.

Example 1. Matrix operators in $E = \mathbb{R}^m$ are Fredholm operators. They are described in detail in Kato (1995).

Example 2. Let T be a compact linear operator described in Examples 3–5 of Subsection 4.6.1. Then $A = I - T$ is a Fredholm operator. The equation $Af = h$ for this operator is the integral Fredholm equation

$$f(y) + \int_a^b k(y, x) f(x) dx = h(y).$$

In Example 4, the kernel k is replaced by another kernel with the property $k(y, x) = 0$ for $y < x$. This makes it possible to integrate not only over $[a, y]$, but over the entire interval $[a, b]$.

We will prove the important Riesz theorem on a one-to-one Fredholm operator. Before proving this theorem, we formulate some auxiliary propositions on the properties of Fredholm operators.

2. Let T be a compact operator and $A = I - T$ be a Fredholm operator in a normed space E. Let L and M be closed subspaces of E such that $L \subseteq M$, $L \neq M$, and $AM \subseteq L$.

Lemma 1. $\|Tb - Tx\| \geq 1/2$ for all $x \in L$ and a $b \in M$ with $\|b\| = 1$.

Proof. Since $L \subseteq M$, $L \neq M$, and L is closed, there exists an $a \in M$ such that $d(a, L) = \inf\{\|a - x\| : x \in L\} = \alpha > 0$ and, consequently, there is a $u \in L$ such that $\alpha \leq \|a - u\| \leq 2\alpha$. Take $b = \|a - u\|^{-1}(a - u)$, $\|b\| = 1$. We have

$$\|b - x\| = \|a - u\|^{-1} \cdot \|a - (u + \|a - u\| \cdot x)\| \geq \alpha(2\alpha)^{-1} = 2^{-1} \quad (x \in L),$$

because $u + \|a - u\| \cdot x \in L$ for $x \in L$. Hence,

$$\|Tb - Tx\| = \|(I - A)b - Tx\| = \|b - (Ab + Tx)\| \geq 1/2 \quad (x \in L),$$

because $Ab \in L$ and $Ax \in L$ by assumption, which implies $Tx = x - Ax \in L$ and $Ab + Tx \in L$ for $x \in L$. $\square$

Remark. Using Lemma 1, it is possible to define by induction a sequence of points b_n in the closed unit ball of an infinitely-dimensional normed space so that $\|b_m - b_n\| \geq 1/2$ for $m \neq n$. Hence, this ball is not compact.

Let T be a compact operator and $A = I - T$ be a one-to-one Fredholm operator in a normed space E.

Lemma 2. *The operator A transforms every closed set $X \subseteq E$ into a closed set $AX = Y \subseteq E$.*

Proof. Choose a sequence $y_n \in Y$ that converges to $\bar{y} \in E$. We will show that $\bar{y} \in Y$. To this end, we prove that the sequence $x_n = A^{-1}y_n \in X$ is bounded. Assume the contrary. Then there exists a subsequence $x_{r(n)}$ such that $\|x_{r(n)}\| \geq n$. Note that $Au_n = \|x_{r(n)}\|^{-1} \cdot y_{r(n)} \to 0 \cdot \bar{y} = 0$ for $u_n = \|x_{r(n)}\|^{-1} \cdot x_{r(n)}$. Consider $v_n = Tu_n$. Since the operator T is compact and $\|u_n\| = 1$, there exists a subsequence $v_{s(n)}$ that converges to some $v \in E$. Therefore, $u_{s(n)} = I \cdot u_{s(n)} = (A + T)u_{s(n)} = Au_{s(n)} + v_{s(n)} \to 0 + v = v$. Since T is continuous, $v_{s(n)} = Tu_{s(n)}$, $v_{s(n)} \to v$, and $u_{s(n)} \to v$, it follows that $v = Tv$ and $Av = v - Tv = 0$. Hence $v = A^{-1}0 = 0$. On the other hand, $\|v\| = \|\lim u_{s(n)}\| = \lim \|u_{s(n)}\| = 1$. Thus, the assumption that (x_n) is unbounded leads to a contradiction.

Since x_n is bounded and T is compact, there exist a subsequence $z_{p(n)} = Tx_{p(n)}$ that converges to some $\bar{z} \in E$. Therefore, $x_{p(n)} = I \cdot x_{p(n)} = (A + T)x_{p(n)} = y_{p(n)} + z_{p(n)} \to \bar{y} + \bar{z}$. Since X is closed by assumption and $x_{p(n)} \in X$, we have $x = \bar{y} + \bar{z} \in X$. Consequently, $\bar{y} = \lim y_{p(n)} = \lim Ax_{p(n)} = A(\lim x_{p(n)}) = Ax \in Y$. We conclude that Y is closed. $\square$

Lemma 2 implies the following important statement.

Corollary. *The inverse of a one-to-one Fredholm operator is a continuous operator.*

Proof. As follows from what was proved above, if A^{-1} is the inverse of a Fredholm operator, then the inverse image $(A^{-1})^{-1}X = AX = Y$ of any closed set X is closed. Hence, A^{-1} is continuous. $\square$

3. We now prove the main theorem of this subsection.

The Riesz inverse operator theorem. *If a Fredholm operator A on a normed space E is a one-to-one operator, then A maps E onto itself and the inverse operator A^{-1} is continuous.*

Proof. (1) We will show that $AE = E$. Consider the sequence of operators

$$A^n = (I - T)^n = I - (\alpha_1 T^1 + \cdots + \alpha_n T^n) = I - T_n.$$

Together with $T = I - A$, the operators $T_n = \alpha_1 T^1 + \cdots + \alpha_n T^n$ are compact and therefore A^n are Fredholm operators. In addition, let $A^0 = I$.

The operators A^n define the sets $Y_n = A^n E$. By Lemma 2, Y_n are closed subspaces of E. Moreover, $Y_{n+1} = A^n(AE) \subseteq A^n E = Y_n$ $(n = 0, 1, 2, \ldots)$. There exists an index $n(0)$ such that $Y_n = Y_{n(0)}$ for all $n \geq n(0)$. Indeed, if we assume the contrary, then there exists a strictly decreasing subsequence of subspaces $Z_n = Y_{r(n)}$ and Lemma 1 implies that one can choose a sequence of points $z_n \in Z_n$ such that $\|z_n\| = 1$ and $\|Tz_n - Tz_p\| \geq 1/2$ for $p > n$. This contradicts the compactness of T. (The subspaces $L = Z_{n+1} = Y_{r(n+1)}$ and $M = Z_n = Y_{r(n)}$ and the operator $A^{r(n+1)-r(n)}$ satisfy the hypothesis of Lemma 1.)

There exists a smallest index $m \geq 0$ of all the indices $n(0)$ such that $Y_n = Y_{n(0)}$ for all $n \geq n(0)$. We now prove that $m = 0$. Indeed, if $m > 0$, then

$Y_{m-1} \neq Y_m$. At the same time, we have $Y_{m-1} = A^{-1}Y_m = A^{-1}Y_{m+1} = Y_m$. Hence, $m = 0$ and $E = Y_0 = Y_1 = \ldots$. Therefore, $AE = AY_0 = Y_1 = E$. The operator A maps E onto itself.

(2) By the corollary of Lemma 2, the inverse operator A^{-1} is continuous. □

It should be emphasized that the Riesz theorem on the one-to-one Fredholm operator is proved for the case of arbitrary normed spaces (not necessarily Banach spaces).

4.6.5. Fredholm theorems

The Fredholm theorems provide the conditions under which the Fredholm linear operator equation is well-posed and uniquely solvable. We present the analytic Fredholm theory which does not use the algebraic one presented in Section 2.2. The connection between these theories is described in detail in Kirillov and Gvishiani (1988), Chapter 3, Section 3. Detailed proofs of the Fredholm theorems for Hilbert spaces are given in Sadovnichii (1999), Chapter 3, Section 3. Also, the index of a Fredholm operator is defined and the closedness of the range of a Fredholm operator is proved there. Since the range is closed, it turns out to be equal to its double orthocomplement. (This condition was used in Subsection 2.2.4, part 5 in the proof of the second Fredholm theorem; see also Hutson and Pym, 1980, Section 7.3.)

1. Consider a Fredholm operator $A = I - T$ in a Banach space E and the conjugate operator $A' = I' - T'$ in the Banach space E'. Since T' is compact because so is T, A' is a Fredholm operator because so is A. These operators define the equations

$$Ax = y \quad (1), \qquad A'x' = y' \quad (1'),$$
$$Az = 0 \quad (2), \qquad A'z' = 0 \quad (2'),$$

where $x, y, z \in E$ and $z', y', z' \in E'$. Equation (1) is *well-posed* if for any $y \in E$ there exists a unique solution $x = A^{-1}y \in E$ and the operator A^{-1} is continuous.

The following theorem is a consequence of the Riesz theorem.

Theorem 1. *Either equation* (1) *is well-posed or equation* (2) *has a nonzero solution.*

Proof. Indeed, by the Riesz theorem, equation (1) is well-posed if and only if A is a one-to-one operator. Thus, Theorem 1 is reduced to the trivial statement that A is either one-to-one or not. □

Theorem 1 is called the *Fredholm alternative*. It is equivalent to the Riesz theorem.

2. The second theorem provides the solvability condition.

Theorem 2. *Equation* (1) *is solvable for a given* $y \in E$ *if and only if* $z'(y) = 0$ *for every solution* z' *of equation* (2').

Proof. Indeed, if $Ax = y$ and $A'z' = 0$, then

$$z'(y) = z'(x) - z'(Tx) = z'(x) - T'z'(x) = A'z'(x) = 0.$$

Conversely, if $z'(y) = 0$ for $A'z' = 0$, then, as follows from the Hahn–Banach theorem and the fact that $AE = Y$ is closed, $y \notin Y$ implies the existence of a continuous linear functional φ on E such that $\varphi(y) = 1$ and $\varphi(z) = 0$ for all $z \in Y$. Therefore,

$$0 = \varphi(x - Tx) = \varphi(x) - \varphi(Tx) = \varphi(x) - T'\varphi(x) = A'\varphi(x)$$

for all $x \in E$, $A'\varphi = 0$, and $\varphi(y) = 0$, as follows from the assumption. This contradicts the condition $\varphi(y) = 1$. Hence, $y \in Y = AE$ and equation (1) is solvable.   □

3. The third theorem determines the dimension of the spaces of solutions for the equations in question.

Theorem 3. *Equations* (2) *and* (2') *have the same number of linearly independent solutions.*

Consider the kernels Ker $A = \{z : Az = 0\}$ and Ker $A' = \{z' : A'z' = 0\}$ of the operators A and A', respectively. Theorem 3 states that $\dim(\text{Ker } A) = \dim(\text{Ker } A') < \infty$. The proof of Theorem 3 uses the properties of the spectrum of a compact operator. The solution is unique if and only if the kernels of the operators are zero. All the three theorems are combined in one in Kirillov and Gvishiani (1988), Chapter 3, Section 3, part 3.

4. If the space $E = H$ is a Hilbert space, then equations (1'), (2') must be replaced by

$$A^* x^* = y^* \quad (1^*), \qquad\qquad A^* z^* = 0 \quad (2^*),$$

where $x^*, y^*, z^* \in H$. All three Fredholm theorems will hold true with this substitution. (Sadovnichii, 1999, Chapter 4, Section 2). In this case, the assumption of Theorem 2 means that the right-hand side of equation (1) is orthogonal to every solution of equation (2^*).

Exercise. Use the Fredholm alternative to prove that any complex number $\mu \neq 0$ is either an eigenvalue of every compact operator T or belongs to its resolvent set.

Remark. There are bounded linear operators for which the Fredholm alternative does not hold. For example, the integral Volterra operator with kernel $k = 1$ is a one-to-one map from the Banach space $E = C[a, b]$ onto its proper subset and the inverse of this operator is discontinuous.

The Fredholm alternative for analytic operator functions is proved in Reed and Simon (1978). It is used to deduce the Riesz–Schauder theorem and the Hilbert–Schmidt theorem for compact operators. A detailed description of analytic operator functions is given in Sadovnichii (1999), Chapter 4, Section 3. Idempotent integral operators are introduced and the Fredholm alternative is proved for them in Maslov (1987), Chapter 2, Section 9.

4.7. SELF-ADJOINT OPERATORS

Among operators in a Hilbert space, we point out normal and Hermitian operators. They are defined using adjoint operators. Some properties of Hermitian operators are similar to those of real numbers, while some properties of normal operators are similar to those of complex numbers. As a rule, in this section we consider complex separated spaces and generalize the definitions given in Subsection 4.6.3.

A systematic presentation of the duality theory for locally convex spaces can be found in Edwards (1995), Chapter 8. In it, vector measures and integrals of vector functions are given as examples of adjoint operators.

4.7.1. Banach adjoint operators

Consider normed spaces E and F, a subspace A that is dense in E, a linear operator $T : A \to F$, the algebraic conjugate spaces A^*, E^*, and F^*, the topological conjugate spaces A', E', and F'. Since $\bar{A} = E$, any functional $a' \in A'$ can be extended by continuity to the functional $x' \in E'$ and this extension is unique.

1. The equality

$$S^* y^* = y^* T \quad (y^* \in F^*)$$

defines a linear operator $S^* : A^* \to F^*$. This operator is the algebraic conjugate of the operator T. In addition, consider the subspace $B' = F' \cap$

$(S^*)^{-1}A'$ formed by all $y' \in F'$ such that $S^*y' = a' \in A'$. If $y' \in B'$, then, replacing $a' \in A'$ with its continuous extension $x' \in E'$, we obtain a linear operator $T' : B' \to E'$. This operator is called the *Banach adjoint* of T and is defined by the equalities

$$x' = T'y' \ (y' \in B'), \quad x'(x) = y'(Tx) \ (x \in A).$$

If $A = E$ and the operator T is bounded, then $B' = F'$ and the adjoint operator T' is also bounded (Hille and Phillips, 1974, Section 2.11). Such operators were considered in Subsection 4.6.3.

2. For any densely defined operator T, the adjoint operator T' is always closed (Hille and Phillips, 1974, Section 2.11). However, its domain of definition B' can be trivial, i.e., consist of only the zero functional on F. The properties of adjoint operators are described in detail in Yosida (1965), Chapter 7, Kato (1995), Chapter 3, and Hille and Phillips (1974), Section 2.11. Many examples are given in Kato (1995).

4.7.2. Hilbert adjoint operators

We define Hilbert adjoint operators independently of Banach adjoint operators.

1. Consider Hilbert spaces E and F, their vector subspaces A and B, and linear operators $T : A \to F$ and $S : B \to E$. If

$$Tx \cdot y = x \cdot Sy \quad (x \in A, \ y \in B),$$

then the operators T and S are said to be *adjoint*. For example, the zero operator $T = 0$ on the zero subspace $A = \{0\}$ is adjoint to every linear operator $S : B \to E$. If A is dense in E, then the functional $a' \in A'$ with values $a'(x) = x \cdot Sy \ (x \in A)$ can be extended by continuity to the functional $x' \in E'$ with values $x'(x) = x \cdot Sy \ (x \in E)$. Therefore, it is possible to choose the maximal functional of all functionals adjoint to T. Take the set A^* of all $y^* \in F$ such that

$$Tx \cdot y^* = x \cdot x^* \quad (x \in A)$$

for some $x^* \in E$. Then A^* is a vector subspace of the Hilbert space F. Since $\bar{A} = E$, the desired element is unique. Indeed, $x \cdot x_1^* = x \cdot x_2^* \ (x \in A)$ implies $x \cdot x_1^* = x \cdot x_2^* \ (x \in E)$ and $x_1^* = x_2^*$. Thus, the above equality defines an operator $T^* : A^* \to E$ that maps $y^* \in A^*$ to $x^* = Ty^* \in E$. It is easy to verify that the operator T^* is linear:

$$
\begin{aligned}
Tx \cdot \left(\beta_1 y_1^* + \beta_2 y_2^*\right) &= \beta_1\left(Tx \cdot y_1^*\right) + \beta_2\left(Tx \cdot y_2^*\right) \\
&= \beta_1\left(x \cdot x_1^*\right) + \beta_2\left(x \cdot x_2^*\right) = x \cdot \left(\beta_1 x_1^* + \beta_2 x_2^*\right)
\end{aligned}
$$

for $y_j^* \in A^*$, $x_j^* = T y_j^*$, and arbitrary numbers β_j $(j = 1, 2)$. It follows from the definitions that T^* is adjoint to T and that any operator $S : B \to E$ adjoint to T is a restriction of the operator T.

Thus, the operator $T^* : A^* \to E$ that is Hilbert adjoint to a densely defined operator $T : A \to F$ acts only on those $y^* \in F$ for which there exists an $x^* \in E$ such that $Tx \cdot y^* = x \cdot x^*$ for all $x \in A$, which implies $T^* y^* = x^*$. Note that the kernel of T^* is equal to the orthogonal complement of the range of T:

$$\mathrm{Ker}(T^*) = (\mathrm{Ran}\, T)^{\perp}.$$

If the operator T^* is also densely defined $(\overline{A^*} = F)$, then it is possible to define the adjoint operator $T^{**} : A^{**} \to F$ $(A^{**} \subseteq E)$. It follows from the definitions that T^{**} is an extension of T and $A \subseteq A^{**}$. Furthermore, $A^{**} = E$ whenever T is bounded.

Exercise. Prove the statements above.

2. The properties of the operators T and T^* are closely related to the properties of their graphs (Yosida 1965, Chapter 7, Section 2; Reed and Simon, 1978, VIII.1). It is easy to verify that

$$G(T^*) = W(T)^{\perp}, \quad W(T) = \{(-Tx, x) : x \in A\}.$$

Indeed,

$$W(T)^{\perp} = \{(y, y^*) : (y, y^*) \cdot (-Tx, x) = -y \cdot Tx + y^* \cdot x = 0 \quad (x \in A)\}.$$

Being an orthogonal complement, the graph $G(T^*)$ is closed. If T is a one-to one operator with dense range $(\overline{\mathrm{Ran}\, T} = F)$, then T^* is also a one-to-one operator and $(T^*)^{-1} = (T^{-1})^*$.

Exercise. Prove the preceding statement.

The properties of Hilbert adjoint operators are described in detail in Reed and Simon (1978), VIII.1, and Weidmann (1980), Section 4.4. In what follows, Hilbert adjoint operators will be called *adjoint* for brevity.

Example 1. (The multiplication operator.) Let $E = F = \mathcal{L}^2(\mathbb{R}, dx)$ and $Tf = g$, $g(x) = x \cdot f(x)$ $(x \in \mathbb{R})$. The operator T is called the *operator of multiplication by the independent variable*. Its domain of definition A is specified by the condition $g \in \mathcal{L}^2$. It is easy to verify that $\bar{A} = \mathcal{L}^2$, the operator T is unbounded, and $T^* = T$.

Example 2. (The differentiation operator.) Let $E = F = \mathcal{L}^2([0,1], dx)$ and $Tf = g$, $g(x) = i^{-1} \cdot f'(x)$ ($x \in [0,1]$). The derivatives exist almost everywhere, and the domain of definition A of the operator T is assumed to be the set of all absolutely continuous functions f with derivatives $f' \in \mathcal{L}^2$ satisfying the condition $f(0) = f(1) = 0$. It can be proved that T is densely defined and unbounded, while the operator T^* is an extension of T to the set A^* of all absolutely continuous functions f with derivatives $f' \in \mathcal{L}^2$ (with no boundary conditions): $T^* \supseteq T$, $T^* \neq T$.

If we take $\mathcal{L}^2(\mathbb{R}, dx)$ instead of $\mathcal{L}^2([0,1], dx)$ and do not fix the values of f, then $T^* = T$.

Exercise. Prove the assertions in Examples 1 and 2 (see Yosida, 1965, Chapter 7).

It should be emphasized that points in examples are meant to be classes of functions equivalent with respect to the Lebesgue measure, since the theory is described under the assumption that the spaces are separated. Without this assumption, the theory will still hold if we replace equalities with equivalences and make the necessary adjustments.

4.7.3. Hermitian and normal operators

In this subsection, as well as in 4.7.4 and 4.7.5, we consider linear operators $T : A \to H$ defined on a dense vector subspace A of a Hilbert space H and the adjoint operators $T^* : A^* \to H$ defined on the subspace A^* of the space H.

1. If the operator T^* is an extension of T, then T is said to be *symmetric*. It satisfies the equality

$$Tx \cdot y = x \cdot Ty \quad (x, y \in A),$$

but it is possible that $A^* \neq A$, as in Example 2 of Subsection 4.7.2. If $T^* = T$, then the operator T is said to be *self-adjoint*. It satisfies the same equality and, moreover, $A^* = A$. Any self-adjoint operator T is closed, and so is T^*. The multiplication and differentiation operators in Examples 1 and 2 of Subsection 4.7.2 are closed. If a one-to-one operator T is self-adjoint, then so is the inverse operator T^{-1}. This follows from the properties of the graphs of these operators. If T is closed, then the operators TT^* and T^*T are self-adjoint (see Yosida, 1965, Chapter 7, Section 3).

Exercise. Prove the above assertions.

If $A = H$, then a self-adjoint operator $T : A \to H$ is said to be *Hermitian*. For example, a diagonal matrix operator with real eigenvalues is Hermitian. For an Hermitian operator $T : H \to H$, the equality $Tx \cdot y = x \cdot Ty$ holds for all $x, y \in H$. By the closed graph theorem, any Hermitian operator is bounded.

2. The operator T is said to be *normal*, if it is closed and $TT^* = T^*T$. It follows from the definitions that

$$\|Tx\| = \|T^*x\| \quad (x \in \mathrm{Dom}(T^*T) = \mathrm{Dom}(TT^*)).$$

It can be proved that

$$\mathrm{Ker}\, T = \mathrm{Ker}\, T^* = \mathrm{Dom}(T^*T) = \mathrm{Dom}(TT^*),$$

which implies

$$\|Tx\| = \|T^*x\| \quad (x \in \mathrm{Dom}\, T = \mathrm{Dom}\, T^*).$$

Exercise. Prove these equalities (see Kato, 1995, Chapter 5, Section 3).

Consider Hermitian operators $R : H \to H$ and $S : H \to H$, the operator $R + iS = T$ and the adjoint operator $R - iS = T^*$. It is easy to verify that $TT^* = T^*T$ if and only if $RS = SR$. The analogy between normal operators and complex numbers goes rather far (Weidmann, 1980, Section 5.6).

4.7.4. Unitary operators

1. A bounded linear operator $T : H \to H$ is said to be *isometric* if it preserves the scalar product, i.e.,

$$Tx \cdot Ty = x \cdot y \quad (x, y \in H).$$

The requirement of preserving the scalar product can be replaced by the requirement of preserving the norm:

$$\|Tx\| = \|x\| \quad (x \in H).$$

Indeed, the latter follows from the former for $x = y$. The equivalence of these conditions follows from the polarization identity (Subsection 2.2.6, part 3).

2. An isometric operator $T : H \to H$ is said to be *unitary* if it maps the complex Hilbert space H onto itself, i.e., $T(H) = H$. Since the Hilbert space in question is separated, the property of T to be isometric implies that T is a one-to-one operator, since $Tx_1 = Tx_2$ implies

$$\|x_1 - x_2\| = \|T(x_1 - x_2)\| = \|Tx_1 - Tx_2\| = 0.$$

A bounded linear operator $T : H \to H$ is unitary if and only if the adjoint operator is equal to the inverse of T, i.e., $T^* = T^{-1}$. Indeed, if T is unitary, then T^{-1} is defined on H, and the condition for the scalar product implies $TT^* = T^*T = I$. Hence $T^* = T^{-1}$. Conversely, if $T^* = T^{-1}$, then $T^*T = I$ and the equality for the scalar product holds. Moreover, in this case we have $H = \mathrm{Ran}(T) = \mathrm{Dom}(T^{-1}) = \mathrm{Dom}(T^*)$ and therefore T is a unitary operator.

Since $TT^* = T^*T = I$, it follows that any unitary operator is normal.

4.7.5. Positive operators

1. Every operator $T : H \to H$ defines a functional

$$\varphi(x) = Tx \cdot x \quad (x \in H).$$

An operator T is said to be *positive* if the corresponding functional φ is positive, i.e., $\varphi(x) \geq 0$ ($x \in H$). A positive bounded linear operator $T : H \to H$ is Hermitian. Indeed, if $x \cdot Tx = \overline{Tx \cdot x} = Tx \cdot x$ ($x \in H$), then the polarization identity for the complex case implies $Tx \cdot y = x \cdot Ty$ ($x, y \in H$).

Example. Assume that $H = \mathbb{R}^3$ and an operator T is represented by a matrix $A = (a_{ij})$ with elements $a_{ij} = 1$ ($i \leq j$), $a_{ij} = 0$ ($i > j$) with respect to the standard basis. The operator T is positive, but not Hermitian.

Since
$$T^*Tx \cdot x = Tx \cdot Tx = \|Tx\|^2 \quad (x \in H),$$

the operator T^*T is positive for any bounded linear operator T.

If an operator T is positive, we will write $T \geq O$. This makes it possible to define an order relation for Hermitian operators $S : H \to H$, $R : H \to H$. As in the case of real numbers, $S \geq R$ means that $T = S - R \geq O$, i.e.,

$$(S - R)x \cdot x \geq 0 \quad (x \in H).$$

The order for Hermitian operators has many properties common to other widely used order relations, which is easy to verify.

2. Using the operator Newton's algorithm analogous to the conventional Newton's algorithm for square root, one can prove that for any Hermitian operator $T \geq O$ there is a unique Hermitian operator $S \geq O$ such that $T = S^2$. Moreover, S commutes with every bounded operator R that commutes with T. The operator S is called the *square root* of the operator T and is denoted by $T^{1/2}$. The proof can be found in Sadovnichii (1999), Chapter 4, Section 2, and Trenogin (1993), Section 26.2.

Exercise. Give detailed proofs of the foregoing statements.

The existence of a square root makes it possible to define the *absolute value*

$$|T| = (T^*T)^{1/2}$$

of a bounded operator T. A bounded linear operator $U : H \to H$ is said to be *partially isometric* if

$$\|Ux\| = \|x\| \quad (x \in (\operatorname{Ker} U)^{\perp}).$$

For any bounded linear operator $T : H \to H$, there exists a unique partially isometric operator $U : H \to H$ such that

$$T = U \cdot |T|, \quad \operatorname{Ker} U = \operatorname{Ker} T.$$

This equality is called the *polar decomposition* of the operator T. The proof can be found in Reed and Simon (1978), Chapter 6, Section 4. For the operator T to be normal, it is necessary and sufficient that the operators $|T|$ and U commute, i.e., $|T| \cdot U = U \cdot |T|$.

3. Consider a bounded linear operator $T : H \to H$ and Hermitian operators

$$A = 2^{-1}(T + T^*), \quad B = (2i)^{-1}(T - T^*).$$

It is obvious that $T = A + iB$. The operators A and B are called the *real* and *imaginary parts* of T, respectively, and the equality $T = A + iB$ is called the *Cartesian decomposition* of T.

If T is normal, then

$$AB = (4i)^{-1}(T^2 - T^{*2}) = BA.$$

Conversely, if $AB = BA$, then

$$TT^* = (A + iB)(A - iB) = (A - iB)(A + iB) = T^*T$$

and the operator T is normal. Thus, the normality of a bounded operator means that its real and imaginary parts commute.

Thus, we obtained Cartesian and polar decompositions of bounded linear operators on a Hilbert space with the help of Hermitian operators. The characteristic feature of normal operators is that the terms in their decomposition commute. The polar decomposition theorem is generalized to the case of closed operators in Reed and Simon (1978), Chapter 8, Section 9. In Rudin (1991), Section 12.35, the commutativity theorem is proved using functional calculus for operators.

4.8.   SPECTRA OF OPERATORS

The spectrum of a matrix operator consists of eigenvalues. Many properties of operators can be conveniently described in terms of their spectra. In the general case, the definition and analysis of a spectrum is more complicated than in the matrix case. However, spectral theory is sufficiently productive for some classes of operators.

The central problem of the spectral theory of operators consists in solving the equation $\lambda x - Tx = y$, where λ is a number, x and y are vectors, and T is a linear operator. Spectral theory analyzes the conditions for the existence and uniqueness of solutions x to this equation and the conditions for continuous dependence of solutions x on the right-hand side y.

The spectral theory of operators is presented in detail in Dunford and Schwartz (1988). Two volumes of this fundamental book are devoted to this theory. A brief description of spectral theory for Banach algebras is given in the appendix of Kolmogorov and Fomin (1989). Spectral theory for multinormed algebras is presented in Khelemskii (1989), Chapter 2. In what follows, we will consider only the spectra of linear operators.

4.8.1.   Classification of spectra

Consider a complex Banach space E, its vector subspace X that is dense in E, a linear operator $T{:}X{\rightarrow}E$, the linear operator $I : E \rightarrow E$, and a number $\lambda \in \mathbb{C}$. The restriction of the identity operator I to the subspace X will be also denoted by I. As before, we assume that the space E is nonzero and the norm separates its points. The equation

$$\lambda x - Tx = y \quad (x \in X,\ y \in E) \tag{1}$$

is naturally associated with the operator

$$S(\lambda, T) = \lambda I - T : X \to E$$

and the inverse correspondence

$$R(\lambda, T) = (\lambda I - T)^{-1} : Y \to E,$$

where

$$Y = \operatorname{Ran} S(\lambda, T) = S(\lambda, T)X.$$

By definition, the image

$$R(\lambda, T)y = \{x : \lambda x - Tx = y\}$$

consists of solutions x to equation $\lambda x - Tx = y$. Therefore, the correspondence $R(\lambda, T)$ is called *resolvent*.

1. The set of complex numbers λ is naturally partitioned into classes depending on the character of solutions to equation (1). The first condition forms two main classes. If equation (1) is *well-posed* (i.e., the operator $S(\lambda, T)$ is a one-to-one map from X onto a subspace Y that is dense in E and the inverse operator $R(\lambda, T)$ is continuous), then λ belongs to the *resolvent set* $P(T)$ of the operator T. Otherwise, λ belongs to the *spectrum* $\Sigma(T)$ of the operator T. Thus, by definition,

$$\Sigma(T) = \mathbb{C} \setminus P(T).$$

We emphasize that T (and, consequently, $S(\lambda, T)$) is not assumed to be continuous. It is possible that the operator $R(\lambda, T)$ is continuous and $S(\lambda, T)$ is not.

Here the well-posedness of equation (1) is understood in the *generalized sense*: the equation is solvable not for every right-hand side y, but only for y that belong to a *dense subspace*. Such generalization is justified by the fact that this subspace coincides with the entire space in many important cases.

For $\lambda \neq 0$, equation (1) is equivalent to the equation

$$x - \lambda^{-1}Tx = \lambda^{-1}y \quad (x, y \in E). \tag{2}$$

If T is compact, then this is a Fredholm equation. In this connection, equation (1) with compact operator T is also called a *Fredholm equation*, and $S(\lambda, T) = \lambda I - T$ is called a *Fredholm operator*.

The spectrum $\Sigma(T)$ of the operator T is subdivided into several parts. The first part is the *point spectrum* $\Sigma_p(T)$ consisting of the eigenvalues λ of the operator T such that equation (1) has a solution $x \neq 0$ for $y = 0$ (i.e., for $\lambda \in \mathbb{C}$ such that $Tx = \lambda x$ for some $x \neq 0$). The general definition of an eigenvalue is the same as in the case of matrix operators. From the definitions and properties of matrix operators, it follows that the spectrum of a matrix operator T is a point spectrum, i.e., $\Sigma(T) = \Sigma_p(T)$.

2. A subspace

$$H(\lambda, T) = \{x : Tx = \lambda x\}$$

of a space E is called an *eigenspace* of the operator T defined by its eigenvalue $\lambda \in \Sigma_p(T)$. In some cases, it is convenient to consider $H(\lambda, T) = \{0\}$ for $\lambda \notin \Sigma_p(T)$. The dimension $\dim H(\lambda, T)$ is called the *multiplicity* of the eigenvalue λ of the operator T. The multiplicity of an eigenvalue can be infinite.

Since equation (1) is linear, $\lambda \notin \Sigma_p(T)$ means that equation (1) has exactly one solution for any $y \in Y = \operatorname{Ran} S(\lambda, T)$. If $\bar{Y} = E$, $\lambda \notin \Sigma_p(T)$, and $\lambda \notin P(T)$, then λ is said to belong to the *continuous spectrum* of the operator T, which is written as $\lambda \in \Sigma_c(T)$. Otherwise, if $\bar{Y} \neq E$, then $\lambda \notin \Sigma_p(T)$ is said to belong to the *residual spectrum* $\Sigma_r(T)$ of the operator T.

Thus, the following partitions are considered:

$$P(T) + \Sigma(T) = \mathbb{C}, \quad \Sigma(T) = \Sigma_p(T) + \Sigma_c(T) + \Sigma_r(T).$$

The classification of spectra is illustrated in the table below. In it, the properties of the resolvent correspondence are specified for each spectrum type.

$R(\lambda, T)$	One-to-one		Not one-to-one
	Continuous	Discontinuous	
Densely defined	$P(T)$	$\Sigma_c(T)$	$\Sigma_p(T)$
Not densely defined	$\Sigma_r(T)$		

For example, if $R(\lambda, T)$ is densely defined, one-to-one, and continuous, then $\lambda \in P(T)$. There are other classifications described in the literature (for example, see Reed and Simon, 1978, Weidmann, 1980).

3. We now give several examples.

Example 1. Let $T = O$. Then $S(\lambda, T) = \lambda \cdot I$ and $\Sigma(T) = \Sigma_p(T) = \{0\}$. Let $T = I$. Then $S(\lambda, T) = (\lambda - 1)I$ and $\Sigma(T) = \Sigma_p(T) = \{1\}$.

Example 2. Let $E = \mathbb{C}^n$. Then the operators $T, S(\lambda, T)$ are matrix operators and $\Sigma(T) = \Sigma_p(T) = \{\lambda : \det S(\lambda, T) = 0\}$, since matrix operators are continuous and one-to-one matrix operators with nonzero determinant are surjective along with their inverses.

Example 3. Let $X = E = \mathcal{C}[0, 1]$ and $Tf = g$, where $g(u) = u \cdot f(u)$ ($u \in [0, 1]$). Obviously, the operator T is bounded. It is easy to verify that for $\lambda \notin [0, 1]$ the operator $S(\lambda, T)$ is a homeomorphism from E onto E. For $\lambda \in [0, 1]$, this operator is one-to-one, its range is not dense, and its inverse is discontinuous. This means that for $\lambda \in [0, 1]$ the resolvent correspondence $R(\lambda, T)$ is one-to-one, discontinuous, and is not densely defined. Hence $\Sigma(T) = \Sigma_r(T) = [0, 1]$.

Now let $X = E = \mathcal{L}^2([0, 1], du)$ and T is again the operator of multiplication by the independent variable. As before, it is bounded and has the same spectrum $\Sigma(T) = [0, 1]$. In the new space, however, the resolvent operator $R(\lambda, T)$ is densely defined and the spectrum of the operator T becomes continuous, i.e., $\Sigma(T) = \Sigma_c(T)$.

Let $E = \mathcal{L}^2(\mathbb{R}, du)$ and the operator T of multiplication by the independent variable is defined on the subspace

$$X = \left\{ f : \int |u \cdot f(u)|^2 \, du < \infty \right\},$$

which is dense in E, as in Example 1 in Subsection 4.7.2. The operator T is unbounded. It is easy to verify that the operator $S(\lambda, T)$ is a homeomorphism from X onto E if $\lambda \notin \mathbb{R}$. It can be proved that for $\lambda \in \mathbb{R}$ the operator $S(\lambda, T)$ is one-to-one, its range is dense, and its inverse is discontinuous. This means that the resolvent correspondence $R(\lambda, T)$ is densely defined, one-to-one, and discontinuous for $\lambda \in \mathbb{R}$. As a result, $\Sigma(T) = \Sigma_c(T) = \mathbb{R}$.

Exercise. Give detailed proofs of the assertions of Example 3.

Example 4. Put $E = \mathcal{C}[0, 1]$, $X = \mathcal{C}^1[0, 1]$, and let T be the differentiation operator, i.e., $Tf = g$, $g(u) = f'(u)$ ($u \in [0, 1]$). Let $\lambda \in \mathbb{C}$. Every $h \in E$ under $R(\lambda, T)$ corresponds to the solutions $f \in X$ to the equation $\lambda f - f' = h$ such that

$$f(u) = ce^{\lambda u} - \int_0^u e^{\lambda(u-t)} h(t) \, dt,$$

where $c = f(0)$ is an arbitrary complex number. Consequently, for any $\lambda \in \mathbb{C}$ the resolvent correspondence $R(\lambda, T)$ is defined on the entire E and is not one-to-one. Hence, $\Sigma(T) = \Sigma_p(T) = \mathbb{C}$ and $P(T) = O$. In this case, the operator T has point spectrum and empty resolvent set.

As before, assume that $E = C[0,1]$, but the differentiation operator is defined on the smaller subspace X of functions $f \in C^1[0,1]$ such that $f(0) = 0$. The resolvent correspondence $R(\lambda, T)$ is now one-to-one and is defined on the entire E. Since the exponential function under the integral is bounded for the values of f on the interval $[0,1]$, it follows that the operator $R(\lambda, T)$ is continuous. Thus, we now have $P(T) = \mathbb{C}$ and $\Sigma(T) = 0$. The spectrum of T is empty.

If $E = \mathcal{L}^2([0,1], du)$ and the differentiation operator T is defined on the subspace X of absolutely continuous functions, then everything is the same as in the previous case: the resolvent set $P(T)$ is empty if the condition $f(0) = 0$ does not hold; the spectrum $\Sigma(T)$ is empty if $f(0) = 0$. This means that the solution of the differential equation under consideration is not unique if there are no additional assumptions. With an additional assumption, the equation is well-posed. We emphasize that the condition $f(0) = 0$ is meaningful in the space $E = \mathcal{L}^2([0,1], du)$ because the functions $f \in X$ are continuous and any two continuous functions that are equivalent with respect to the Lebesgue measure are equal, which means that every equivalence class contains no more than one continuous function (see Reed and Simon, 1978, VIII.2).

Let $E = \mathcal{L}^2(\mathbb{R}, du)$ and the differentiation operator T with the factor i^{-1}, as in Example 2 of Subsection 4.7.2, is defined on the subspace X of all functions that are absolutely continuous on every interval. In this case, the spectrum of T is the same as the spectrum of the operator of multiplication by the independent variable, i.e., $\Sigma(T) = \Sigma_c(T) = \mathbb{R}$. Indeed, these operators are unitarily equivalent, since one is obtained from the other by means of the unitary Fourier–Plancherel transform.

Exercise. Prove the preceding assertion (see Yosida, 1965, Chapter 9, Section 6; Weidmann, 1980, Section 10.1).

Example 5. Let $X = E = \mathcal{L}^2([0,1], du)$ and $Tf = g$, where $g(v) = \int_0^v f(u)\, du$ ($v \in [0,1]$). The operator T is one-to-one, compact, and the operator $S(\lambda, T) = \lambda I - T$ is one-to-one if $\lambda \neq 0$. By the Riesz theorem on one-to-one Fredholm operators, the inverse operator $R(\lambda, T)$ is defined on the entire E and is continuous. At the same time, the operator $R(0,T)$, which is the inverse of the compact operator $-T = S(0,T)$, is discontinuous. This is a differentiation operator defined on the subspace of absolutely continuous functions vanishing at 0, which is dense in E. Hence $\Sigma(T) = \Sigma_c(T) = \{0\}$.

For $E = C[0,1]$, the range of $S(0,T)$ is not dense and the continuous spectrum becomes residual, i.e., $\Sigma(T) = \Sigma_r(T) = \{0\}$.

The above examples show that the spectrum depends substantially on the choice of the space on which the operator acts.

4.8.2. The spectrum of a closed operator

1. If an operator T closed, for any number $\lambda \in P(T)$ the resolvent correspondence $R(\lambda, T)$ is a bounded operator defined on the entire E. Then the operator function $R = R(\cdot, T)$ can be defined on the resolvent set $P = P(T)$ with values in the Banach algebra $\mathcal{B} = \mathcal{B}(E, E)$ of bounded linear operators on E.

Lemma. *If T is closed and $\lambda \in P$, then $\operatorname{Ran} S(\lambda, T) = E$.*

Proof. By the definition of the resolvent set P, for $\lambda \in P$ the range of $S(\lambda, T) = \lambda I - T$ is dense in E and therefore it suffices to verify that $\operatorname{Ran} S(\lambda, T)$ is closed. This follows from the closedness of T. ☐

From the lemma and the definition of the resolvent correspondence, it follows that $R(\lambda) = R(\lambda, T) \in \mathcal{B}$ for $\lambda \in P$. Hence, the resolvent R is, indeed, a map from P to $\mathcal{B}$.

2. The function R is said to be *analytic* if the composition $x'R : P \to \mathbb{C}$ of R and every functional $x' \in E'$ is an analytic function. This is equivalent to the condition that the set $P \subseteq \mathbb{C}$ is open and the function R has a power series expansion with coefficients in $\mathcal{B}$ in a neighborhood of any point $\lambda \in P$ (see Reed and Simon, 1978, Theorem VI.4). Many properties of operator analytic functions are the same as those of numeric functions (Yosida, 1965, Chapter 8, Section 2).

Theorem. *The resolvent set $P = P(T)$ of a closed operator T is open and its resolvent $R = R(\cdot, T)$ is an analytic function.*

Proof. Take a $\lambda \in P$ and choose a $\mu \in \mathbb{C}$ such that the norm of the operator $U = (\lambda - \mu) \cdot R(\lambda)$ is strictly less than one ($\|U\| < 1$). For this condition to hold, it is sufficient that $|\lambda - \mu| \|R(\lambda)\| < 1$. Then, in view of the formula for the sum of an operator progression, the operator $I - U$ is invertible $((I - U)^{-1} \in \mathcal{B})$. Therefore,

$$R(\mu) = (\mu I - T)^{-1} = (\mu I - \lambda I + \lambda I - T)^{-1}$$
$$= (\lambda I - T)^{-1}[I - (\lambda - \mu)(\lambda I - T)^{-1}]^{-1} = R(\lambda)(I - U)^{-1} \in \mathcal{B},$$

i.e., $\mu \in P$. Then the disk $B(\lambda, r)$ with center at λ and radius $r \in {]}0, \|R(\lambda)\|^{-1}[$ is included in P. The set P is open.

By the formula for the sum of an operator progression for $|\mu - \lambda| < r$, we have

$$R(\mu) = R(\lambda + (\mu - \lambda)) = \sum (-1)^n R^{n+1}(\lambda)(\mu - \lambda)^n.$$

We conclude that the function R is analytic. □

Since the complement of an open set in the complex plane is closed, the theorem readily implies the following statement.

Corollary. *The spectrum of a closed operator is a closed set.*

In particular, the spectrum of a matrix operator, which is a finite set, is closed.

3. Take numbers λ, μ and closed operators T, U defined on a subspace A that is dense in E. Let I_A and I_E denote the identity transformations of A and E, respectively. We have

$$R(\lambda, T) - R(\mu, U) = R(\lambda, T) \cdot I_E - I_A \cdot R(\mu, U)$$
$$= R(\lambda, T) \cdot [S(\mu, U) \cdot R(\mu, T)] - [R(\lambda, T) \cdot S(\lambda, T)] \cdot R(\mu, T)$$
$$= R(\lambda, T) \cdot [S(\mu, U) - S(\lambda, T)] \cdot R(\mu, U).$$

With $U = T$ and $\mu = \lambda$, the preceding equalities yield the *resolvent equations*

$$R(\lambda, T) - R(\mu, T) = -(\lambda - \mu) \cdot R(\lambda, T) \cdot R(\mu, T),$$
$$R(\lambda, T) - R(\lambda, U) = R(\lambda, T) \cdot (T - U) \cdot R(\lambda, U).$$

The first of these equations is called the *Hilbert identity*. It implies that

$$R(\lambda) \cdot R(\mu) = R(\mu) \cdot R(\lambda)$$

and

$$R'(\lambda) = \lim_{\mu \to \lambda} (\mu - \lambda)^{-1}[R(\mu) - R(\lambda)] = -R^2(\lambda).$$

4. There is a simple connection between the closed operator T and the Banach adjoint T':

$$\Sigma(T') = \Sigma(T), \quad R(\lambda, T') = R(\lambda, T)'$$

for all numbers λ in the resolvent set of T, which is also the resolvent set of T'. If E is a Hilbert space and T^* is the Hilbert adjoint of T, then

$$\Sigma(T^*) = \Sigma(T)^*, \quad R(\lambda, T^*) = R(\lambda^*, T)^*$$

for all $\lambda^* \in P(T)$, where $\Sigma(T)^* = \{\lambda^* : \lambda \in \Sigma(T)\}$ is the reflection of the set $\Sigma(T)$ in the real axis.

Exercise. Prove the above assertions (see Yosida, Chapter 8, Section 6).

Examples of resolvents of differential operators are analyzed in depth in Kato (1995), Chapter 3, Section 6.

4.8.3. The spectrum of a bounded operator

1. A bounded operator is closed. Therefore, everything that has been said about the spectrum of a closed operator also applies to the spectrum of a bounded operator. In addition, the latter is nonempty and bounded.

Let $T \in \mathcal{B} = \mathcal{B}(E, E)$ be a bounded linear operator on a Banach space $E \neq \{0\}$. By the formula for the sum of an operator progression, for $|\lambda| > \|T\|$ we have

$$R(\lambda, T) = (\lambda I - T)^{-1} = \lambda^{-1}(I - \lambda^{-1}T)^{-1} = \lambda^{-1} \sum \lambda^{-n} T^n \quad (n \geq 0).$$

Consequently, $\lambda \in P(T)$ for $|\lambda| > \|T\|$ and the spectrum $\Sigma(T)$ is included in the closed disk with center 0 and radius $\|T\|$. Outside this disk, the resolvent has the Laurent series decomposition specified above.

Using the inversion $z = \lambda^{-1}$ and the formula for the radius of a power series, it is easy to verify that the resolvent series with terms $\lambda^{-n} T^n = T^n z^n$ is summable if $|\lambda| > \rho(T)$ and is not summable if $|\lambda| < \rho(T)$, where $\rho(T) = \overline{\lim}(\|T^n\|^{1/n})$. The number $\rho(T)$ is called the *spectral radius* of the operator T. It is possible that $\rho(T) = 0$, as in the case of an integral Volterra operator It can be proved that the sequence of numbers $\|T^n\|^{1/n}$ converges and therefore

$$\rho(T) = \lim(\|T^n\|^{1/n})$$

(see Sadovnichii (1999), Chapter 4, Section 3).

Exercise. Prove the equalities for $\rho(T)$.

2. The following theorem describes the spectrum of a bounded operator.

Theorem. *The spectrum $\Sigma(T)$ of a bounded operator T is a nonempty bounded closed set included in the disk $\bar{B}(0, \rho(T))$.*

Proof.   We proved that $\Sigma(T)$ is bounded: $\Sigma(T) \subseteq \bar{B}(0, \|T\|)$. The inclusion $\Sigma(T) \subseteq \bar{B}(0, \rho(T))$ follows from the definition of the spectral radius $\rho(T)$. The spectrum $\Sigma(T)$ is closed because T is closed. It remains to prove that $\Sigma(T)$ is nonempty.

We have $\|R(\lambda, T)\| \leq (|\lambda| - \|T\|)^{-1}$ for $|\lambda| > \|T\|$, which implies that $\|R(\lambda, T)\| \to 0$ as $|\lambda| \to \infty$. If $\Sigma(T) = \varnothing$, then $P(T) = \mathbb{C}$ and the resolvent is an entire analytic function. By Liouville's theorem, which holds also for operator functions of a complex variable, it follows that the resolvent $R(\cdot, T)$ is equal to a constant $C \in \mathcal{B}$. Since $\|R(\lambda, T)\| \to 0$ as $|\lambda| \to \infty$, it follows that C is the zero operator on E. By assumption, $E \neq \{0\}$ and the zero operator on E is not one-to-one. Therefore, it cannot be a value of the resolvent of T. We conclude that $\Sigma(T)$ is nonempty. $\square$

3. It is easy to verify that

$$\rho(T) = \inf\{\|T^n\|^{1/n} : n \geq 1\} = \sup\{|\lambda| : \lambda \in \Sigma(T)\}.$$

These equalities are used to determine the spectral radius of an operator.

Exercise. Prove these equalities.

Take numbers $a_0, a_1, \ldots, a_n$ and consider the operator polynomial

$$p(T) = a_0 I + a_1 T + \cdots + a_n T^n.$$

It is easy to verify that

$$\Sigma(p(T)) = p(\Sigma(T)) = \{p(\lambda) : \lambda \in \Sigma(T)\},$$

where

$$p(\lambda) = a_0 1 + a_1 \lambda + \cdots + a_n \lambda^n.$$

This result is a special case of the general theorem on maps of spectra (Reed and Simon, 1978, VII.1).

4.8.4. The spectrum of a compact operator

In many respects, the behavior of compact operators is almost the same as that of matrix operators. The following theorem characterizes the spectra of compact operators.

The Riesz–Schauder theorem. *Let T be a compact operator in a Banach space E. Then*

(1) *every number $\lambda \neq 0$ in the spectrum $\Sigma(T)$ is an eigenvalue of T;*

(2) *there are finitely many points of the spectrum $\Sigma(T)$ outside any disk with center 0 and radius $r > 0$ in the complex plane $\mathbb{C}$;*

(3) *every eigenvalue $\lambda \neq 0$ of the operator T has finite multiplicity.*

The finite set of spectrum points outside a disk in $\mathbb{C}$ can be empty (for example, for matrix operators). The spectrum $\Sigma(T)$ is a countable set.

The special role of the point 0 is evident for infinite diagonal operators in the space of sequences $E = l^2$. Such operators are compact if and only if the diagonal elements tend to zero. Therefore, any nonzero number cannot be repeated in the diagonal infinitely many times, whereas zero can be repeated any number of times.

It follows from the Riesz–Schauder theorem that the only possible non-isolated point of the spectrum $\Sigma(T)$ of a compact operator T is zero. Every point $\lambda \neq 0$ of the spectrum $\Sigma(T)$ has a neighborhood that contains no other points of $\Sigma(T)$.

The Riesz–Schauder theorem is closely related to the Fredholm theorems. Since the eigenspaces $H(\lambda, T)$ are finite-dimensional, it is possible to reduce Fredholm equations to matrix equations if $\lambda \neq 0$.

Proofs of the Riesz–Schauder theorem can be found in Yosida (1965), Chapter 10, Section 5, Reed and Simon (1978), Chapter 6, Section 5, and Kato (1995), Chapter 3, Section 7.

4.8.5. The spectrum of a self-adjoint operator

Consider a nonzero complex Hilbert space H, a vector subspace A that is dense in H, a linear operator $T : A \to H$, the identity operator $I : H \to H$, a number $\lambda \in \mathbb{C}$, the operator $S(\lambda, T) = \lambda I - T : A \to H$, and the adjoint operators $T^* : A^* \to H$ and $S^*(\lambda, T) = S(\lambda^*, T^*) = \lambda^* I - T^* : A^* \to H$. If T is one-to-one and its range $\mathrm{Ran}\, T = T(A)$ is dense, then the operator T^* is also one-to-one and $(T^*)^{-1} = (T^{-1})^*$.

1. The spectrum of a self-adjoint operator is characterized by the following theorem.

Theorem. Let $T = T^*$. Then

$$\Sigma(T) = \{\lambda : \operatorname{Ran} S(\lambda, T) \neq H\} \subseteq \mathbb{R}.$$

Furthermore,

$$\Sigma_p(T) = \{\lambda : \operatorname{Ran} S(\lambda, T) \neq H, \overline{\operatorname{Ran} S(\lambda, T)} \neq H\}, \qquad (1)$$
$$\Sigma_c(T) = \{\lambda : \operatorname{Ran} S(\lambda, T) \neq H, \overline{\operatorname{Ran} S(\lambda, T)} = H\}, \qquad (2)$$
$$\Sigma_r(T) = O. \qquad (3)$$

Proof. If $T = T^*$, $Tu = \lambda u$, and $\|u\| = 1$, we have

$$\lambda = u \cdot \lambda u = u \cdot Tu = Tu \cdot u = (\lambda u) \cdot u = \lambda^*$$

and therefore $\Sigma_p(T) \subseteq \mathbb{R}$.

Let $\Lambda = \{\lambda : \overline{\operatorname{Ran} S(\lambda, T)} \neq H\}$. We will prove that $\Sigma_p(T) = \Lambda$. Indeed, from the orthogonal projection theorem it follows that $\lambda \in \Lambda$ if and only if $B = \overline{\operatorname{Ran} S(\lambda, T)}^{\perp} \neq \{0\}$. Therefore, $S(\lambda, T)x \cdot b = 0$ ($x \in A$) for some $b \neq 0$, $b \in B$. Hence, $x \cdot S^*(\lambda, T)b = 0$ ($x \in A$), $S(\lambda^*, T)b = 0$, $\lambda^* \in \Sigma_p(T)$, and $\lambda = \lambda^* \in \Sigma_p(T)$ because $\Sigma_p(T) \subseteq \mathbb{R}$. Equality (1) of the theorem is proved.

By definition, $\Sigma_r(T) \subseteq \Lambda$ and $\Sigma_r(T) \cap \Sigma_p(T) = \varnothing$. If $\Sigma_p(T) = \Lambda$, this is possible if and only if $\Sigma_r(T) = 0$. This proves equality (3) of the theorem.

We now prove that $\Sigma(T) \subseteq \mathbb{R}$, i.e., that $\mathbb{C} \setminus \mathbb{R} \subseteq P(T)$. Let $\lambda = \mu + i\nu$, where $\mu, \nu \in \mathbb{R}$. Note that if $z = (\mu I - T)x$ and $T = T^*$, then

$$z \cdot x = (\mu I - T)x \cdot x = x \cdot (\mu I - T)x = x \cdot z,$$

$$\|(\lambda I - T)x\|^2 = \|z + i\nu x\|^2$$
$$= \|z\|^2 + i\nu(z \cdot x) - i\nu(x \cdot z) + \nu^2 \cdot \|x\|^2 = \|z\|^2 + \nu^2 \cdot \|x\|^2 \quad (x \in A).$$

Hence, $\|S(\lambda, T)x\| \geq |\nu| \cdot \|x\|$ ($x \in A$). If $\lambda \notin \mathbb{R}$ and $\nu \neq 0$, then $S(\lambda, T)$ is a one-to-one operator. As was proved above, $\Lambda = \Sigma_p(T)$, which implies $\lambda \in \Lambda$ and $\overline{\operatorname{Ran} S(\lambda, T)} = H$. Then the resolvent correspondence is densely defined and one-to-one if $\lambda \notin \mathbb{R}$. Since

$$\|R(\lambda, T)y\| \leq |\nu|^{-1} \cdot \|y\| \quad (y \in S(\lambda, T)x, \ x \in A),$$

the operator $R(\lambda, T)$ is continuous. As a result, $\lambda \in P(T)$.

We now turn to equality (2) of the theorem. Let M denote its right-hand side. We will show that $\Sigma_c(T) \subseteq M$. Let $\lambda \in \Sigma_c(T)$. Then, by the definition of the continuous spectrum, $\overline{\operatorname{Ran} S(\lambda, T)} = H$ and the operator $R(\lambda, T)$ is discontinuous. As was proved above, $\Sigma(T) \subseteq \mathbb{R}$, which implies $\lambda = \lambda^*$ and $R(\lambda, T)^* = R(\lambda, T)$ for $T = T^*$. Consequently, the operator $R(\lambda, T)$ is closed. By the closed graph theorem, $\operatorname{Ran} S(\lambda, T) \neq H$ because $R(\lambda, T)$ is unbounded. Hence, $\lambda \in M$ and $\Sigma_c(T) \subseteq M$.

We now prove the reverse inclusion. Let $\lambda \in M$. As was proved above, $\lambda \notin \Sigma_p(T)$ and $R(\lambda, T)$ is a densely defined operator. It remains to show that $R(\lambda, T)$ is unbounded. The self-adjoint operator $T = T^*$ is closed. Suppose that $R(\lambda, T)$ is bounded. Then $\lambda \in P(T)$ and, by the theorem on the resolvent of a closed operator, $\operatorname{Ran} S(\lambda, T) = H$. This contradicts the assumption that $\lambda \in M$. Thus, $R(\lambda, T)$ is a densely defined discontinuous operator, $\lambda \in \Sigma_c(T)$, and $M \subseteq \Sigma_c(T)$. Equality (2) is proved. $\quad\square$

2. Another characteristic of the spectrum of a self-adjoint operator can be given using the definition of the limit spectrum.

Let $\lambda \in \mathbb{C}$. If there exist vectors $u_n \in A$ such that $\|u_n\| = 1$ and $\lambda u_n - T u_n \to 0$, then λ is called the *limit eigenvalue* of the linear operator $T : A \to H$. Let $\Sigma_l(T)$ denote the set of all limit eigenvalues of the operator T. The set $\Sigma_l(T)$ is called the *limit spectrum* of T. It follows from the definitions that $\Sigma_p(T) \subseteq \Sigma_l(T)$, i.e., In other words, every eigenvalue is a limit eigenvalue.

Using the estimates of the norms of the operators $R(\lambda, T)$ and $S(\lambda, T)$ obtained in the proof of the theorem on the spectrum of a self-adjoint operator, it is easy to prove the following theorem.

Weyl's criterion. *Let $T = T^*$. Then $\Sigma(T) = \Sigma_l(T)$.*

This means that a number belongs to the spectrum of a self-adjoint operator if and only if it is an eigenvalue of this operator.

Exercise. Prove Weyl's criterion (see Reed and Simon, 1978, VII.3).

3. Recall that self-adjoint operators defined on the entire Hilbert space H are called *Hermitian*. By the closed graph theorem, a closed operator $T = T^* : H \to H$ is bounded. It follows that the spectrum $\Sigma(T)$ of a Hermitian operator T is closed.

By applying the Riesz theorem on the representation of linear functionals, it is easy to verify that the formula $B(z, x) = Tx \cdot z$ $(x, z \in H)$ establishes a one-to-one correspondence between Hermitian operators on H and Hermitian forms on $H \times H$. The polarization identity

$$
\begin{aligned}
B(z, x) = 4^{-1}[&B(z + x, z + x) \\
&- B(z - x, z - x) - iB(z + ix, z + ix) + iB(z - ix, z - ix)]
\end{aligned}
$$

gives a one-to-one correspondence between Hermitian forms and quadratic forms

$$
Q(x) = B(x, x) = Tx \cdot x \quad (x \in H)
$$

on the complex Hilbert space H. This makes it possible to identify Hermitian operators with the corresponding quadratic forms and describe the properties of Hermitian operators and their spectra in terms of quadratic forms.

Since $T = T^*$,

$$
Q(x) = Tx \cdot x = x \cdot Tx = \overline{Tx \cdot x} \quad (x \in H)
$$

and all values of the quadratic form Q are real ($Q(x) \in \mathbb{R}$). Note that a quadratic form Q is defined by its values on the unit sphere:

$$
Q(x) = B(x, x) = \|x\|^2 B(u, u) = \|x\|^2 \cdot Q(u)
$$

for $u = \|x\|^{-1} \cdot x$, $x \neq 0$. Moreover, $Q(0) = B(0, 0) = 0$. Since T is bounded, it follows that Q is bounded on the sphere S:

$$
|Q(u)| = |Tu \cdot u| \leq \|Tu\| \leq \|T\|
$$

for $\|u\| = 1$. For this reason, one can consider the bounds

$$
\begin{aligned}
a = \inf\{Tu \cdot u : \|u\| = 1\}, \quad b = \sup\{Tu \cdot u : \|u\| = 1\}, \\
c = \sup\{|Tu \cdot u| : \|u\| = 1\}
\end{aligned}
$$

of the sets of values $Q(u)$ and $|Q(u)|$ on the sphere S. We have

$$
|Q(x)| \leq c \cdot \|x\|^2 \quad (x \in H).
$$

4. In this part, we prove the following lemma.

Lemma. $\|T\| = \max\{|a|, |b|\} = c$.

Proof. It was shown that $|Q(x)| \leq \|T\|$ for $\|u\| = 1$ and, consequently, $c \leq \|T\|$. We will probe the reverse inequality. Let $\|u\| = 1$, $Tu \neq 0$, and $z = \|Tu\|^{-1} \cdot Tu$. Then

$$B(z, u) = Tu \cdot z = \|Tu\|^{-1} \|Tu\|^2 = \|Tu\|.$$

Using the polarization identity, the fact that $Q(x) = B(x, x)$ is real, the inequality $|Q(x)| \leq c \cdot \|x\|^2$, and the parallelogram rule, we obtain

$$\|Tu\| = B(z, u) = 4^{-1} |Q(z + u) - Q(z - u)|$$
$$\leq 4^{-1} (|Q(z + u)| + |Q(z - u)|) \leq 4^{-1} c (\|z + u\|^2 + \|z - u\|^2)$$
$$\leq 2^{-1} c (\|z\|^2 + \|u\|^2) = 2^{-1} c \cdot 2 = c.$$

As a result, $\|T\| = \sup\{\|Tu\| : \|u\| = 1\} \leq c$ and $\|T\| = c$.

The equality $\max\{|a|, |b|\} = c$ follows from the definition of the numbers a, b, and c. This can be easily verified by considering three cases: $0 < a \leq b$, $a \leq b < 0$, and $a \leq 0 \leq b$. We have $c = b$ in the first case, $c = -a$ in the second case, and $c = \max\{-|a|, |b|\}$ in the third case. $\square$

5. The following theorem characterizes the spectrum of an Hermitian operator.

Theorem. $\{a, b\} \subseteq \Sigma(T) \subseteq [a, b]$.

Proof. First, consider a positive Hermitian operator A to prove that $\|A\| \in \Sigma(A)$. Since the spectrum of a self-adjoint operator coincides with its limit spectrum, it suffices to show that $\mu = \|A\|$ is a limit eigenvalue of the operator A.

Since $A \geq O$, we have $|Au \cdot u| = Au \cdot u \geq 0$, and the preceding lemma implies

$$\mu = \sup\{|Au \cdot u| : \|u\| = 1\} = \sup\{Au \cdot u : \|u\| = 1\}.$$

Therefore, there exist $u_n \in H, \|u_n\| = 1$, such that $Au_n \cdot u_n > \mu - 1/n$. Hence,

$$\|(\mu I - A)u_n\|^2 = \mu^2 - 2\mu \cdot Au_n \cdot u_n + \|Au_n\|^2$$
$$\leq \mu^2 - 2\mu(\mu - 1/n) + \mu^2 = 2\mu/n \quad (n \geq 1),$$

$$\|S(\mu, A) \cdot u_n\| \leq (2\mu/n)^{1/2} \to 0 \quad n \to \infty.$$

It follows that $\mu \in \Sigma_l(A)$.

Consider the operators $A = T - aI$ and $B = bI - T$. Since $a \leq Tu \cdot u \leq b$ ($\|u\| = 1$), we have $A \geq O$ and $B \geq O$. It is easy to verify that $\|A\| = \|B\| = b - a$. As a result, $b - a \in \Sigma(A)$ and $b - a \in \Sigma(B)$. It follows from the definitions that $\mu \in \Sigma(A)$ is equivalent to $\lambda = \mu + a \in \Sigma(T)$ and $\nu \in \Sigma(B)$ is equivalent to $\lambda = b - \nu \in \Sigma(T)$. Hence, $a = b - (b - a) \in \Sigma(T)$ and $b = (b - a) + a \in \Sigma(T)$, which implies $\{a, b\} \subseteq \Sigma(T)$.

Since the operators A and B are self-adjoint and bounded, their spectra are real and bounded by the numbers $-(b-a), b-a$. Since $\Sigma(T) = \Sigma(A) + a = b - \Sigma(B)$, the spectrum of T is bounded by the numbers $a = b - (b - a)$ and $b = (b - a) + a$. Indeed, if $\lambda = \mu + a = b - \nu$, $\mu \in \Sigma(A)$, and $\nu \in \Sigma(B)$, then $\lambda \leq (b - a) + a$, $\mu \leq b - a$, $-(b - a) \leq \nu$ and $\lambda = \mu + a \leq (b - a) + a = b$, $\lambda = b - \nu \geq b - (b - a) = a$. We conclude that $\Sigma(T) \subseteq [a, b]$. $\quad\square$

6. The theorem on the spectrum of an Hermitian operator has many useful corollaries.

Corollary 1. *An Hermitian operator is positive if and only if its spectrum contains only positive numbers.*

Proof. Let T be an Hermitian operator and $T \geq O$. By definition, $Tu \cdot u \geq 0$ ($\|u\| = 1$) and therefore $a = \inf\{Tu \cdot u : \|u\| = 1\} \geq 0$. Then, by the preceding theorem, $\Sigma(T) \subseteq [a, \infty[\subseteq [0, \infty[$. Conversely, if $a \geq 0$, then $Tx \cdot x = \|x\|^2 (Tu \cdot u) \geq a \cdot \|x\|^2 \geq 0$ for $u = \|x\|^{-1} \cdot x$, $x \neq 0$, $x \in H$. This means that $T \geq O$. $\quad\square$

Corollary 2. *The spectral radius $\rho(T)$ of an Hermitian operator T is equal to its norm $\|T\|$.*

Proof. Since $\|T^n\| \leq \|T\|^n$, it follows that $\rho(T) = \overline{\lim}(\|T^n\|^{1/n}) \leq \|T\|$. The preceding lemma and theorem imply that $\|T\| = \max\{|a|, |b|\}$ with $a \in \Sigma(T)$ and $b \in \Sigma(T)$. Therefore, $\rho(T) = \sup\{|\lambda| : \lambda \in \Sigma(T)\} \geq \|T\|$. $\quad\square$

Consider the n-dimensional complex space $H = \mathbb{C}^n$, a matrix operator T, the conjugate operator T^*, and their product $S = T^*T$. Note that S is a positive Hermitian operator:

$$S^* = (T^*T)^* = T^* \cdot T^{**} = T^*T = S,$$
$$Sx \cdot x = T^*Tx \cdot x = Tx \cdot Tx = \|Tx\|^2 \geq 0 \quad (x \in H).$$

The eigenvalues of S are called the *singular numbers* of the operator T. They will be denoted by σ_k ($1 \leq k \leq n$). By Corollary 1, $S \geq 0$ implies $\sigma_k \geq 0$. Take the greatest singular number $\sigma = \max\{\sigma_k : 1 \leq k \leq n\}$.

Corollary 3. *The Euclidean norm of a matrix operator is equal to the square root of its greatest singular number*

Proof. We have $\Sigma(S) = \{\sigma_k : 1 \leq k \leq n\}$ and $\sigma_k \geq 0$. By Corollary 2, this means that $\|S\| = \rho(S) = \max\{\sigma_k : 1 \leq k \leq n\} = \sigma$. At the same time, by the preceding lemma, $\|S\| = \sup\{|Su \cdot u| : \|u\|=1\} = \sup\{Su \cdot u : \|u\| = 1\} = \sup\{\|Tu\|^2 : \|u\| = 1\} = \|T\|^2$. As a result, $\|T\|^2 = \sigma$ and $\|T\| = \sigma^{1/2}$. $\square$

In particular, if the matrix operator T is Hermitian, then $T^* = T$, $S = T^2$, and $\sigma_k = \lambda_k^2$, where λ_k are eigenvalues of T. Hence $\|T\| = \max\{|\lambda_k| : 1 \leq k \leq n\}$. The norm of a matrix Hermitian operator is equal to the greatest of the absolute values of its eigenvalues.

Corollary 4. *Vectors from different eigenspaces of an Hermitian operator are orthogonal.*

Proof. Let T be an Hermitian operator, $x \in H(\lambda, T)$, $y \in H(\mu, T)$, and $\lambda \neq \mu$. If $H(\lambda, T) = \{0\}$ or $H(\mu, T) = \{0\}$, then $x = 0$ or $y = 0$ and $x \cdot y = 0$. Suppose that $\lambda \in \Sigma_p(T)$ and $\mu \in \Sigma_p(T)$. Then, since $\Sigma_p(T) \in \mathbb{R}$, we have

$$\lambda(x \cdot y) = Tx \cdot y = x \cdot Ty = \mu(x \cdot y)$$

and therefore $x \cdot y = 0$. $\square$

Thus, for an Hermitian operator T, $H(\lambda, T) \perp H(\mu, T)$ for $\lambda \neq \mu$.

7. From the eigenvectors of a matrix Hermitian operator on a finite-dimensional Euclidean space, it is possible to choose vectors that form an orthonormal basis of this space. The same is true for a compact Hermitian operator on a separable Hilbert space.

In addition, suppose that the Hilbert space H under consideration is separable. Then H has a countable orthonormal basis (h_k) consisting of pairwise orthogonal normalized vectors $h_j \perp h_k$ $(j \neq k)$, $\|h_k\| = 1$. Furthermore, for every $x \in H$ there is a unique coordinate collection $\xi_k \in \mathbb{C}$, such that $x = \Sigma\xi_k h_k$ (we mean the norm convergence of finite sums in the space H).

The following theorem characterizes the spectrum of a compact Hermitian operator.

The Hilbert–Schmidt theorem. *For any compact Hermitian operator T in a complex separable Hilbert space H, there exists a countable orthonormal basis (h_k) consisting of eigenvectors h_k of the operator T.*

Proof. Let $A = \overline{\cup H(\lambda, T)}$ $(\lambda \in \Sigma(T))$ and $B = A^{\perp}$ (where $H(\lambda, T) = \{0\}$ for $\lambda \notin \Sigma(T)$). Since the operator T is continuous,

$$T(A) \subseteq \overline{\cup T H(\lambda, T)} \subseteq \overline{\cup H(\lambda, T)} = A.$$

From $T^* = T$ and $z \cdot y = 0$ $(z \in A, y \in B)$ it follows that $x \cdot Ty = Tx \cdot y = 0$ $(x \in A, y \in B)$ for $TA \subseteq A$. Then $Ty \in B$ for $y \in B$ because $TA \subseteq A$ implies $TB \subseteq B$.

Since the operator T is compact, by the Riesz–Schauder theorem, the dimension of an eigenspace $H(\lambda, T)$ is finite if $\lambda \neq 0$. The dimension of $H(0, T)$ is countable because H is assumed to be separable (in particular, the dimension of $H(0, T)$ can be finite as well) Since T is Hermitian, the subspaces $H(\lambda, T)$ are pairwise orthogonal. Therefore, if we choose an orthonormal basis in every $H(\lambda, T) \neq \{0\}$, we obtain an orthonormal basis for A. It remains to show that $B = \{0\}$ and, consequently, $A = H$.

Consider the restriction $U = T_B$ of the operator T to the subspace B. Since $TB \subseteq B$, it follows that $U : B \to B$ and U is a compact Hermitian operator because so is $T : H \to H$. We now prove that $U = O$. This follows from the fact that U has no eigenvalues: if $\lambda \in \Sigma_p(U)$, then there is a $y \in B$ $(y \neq 0)$ such that $Ty = Uy = \lambda y$, and, consequently, $\lambda \in \Sigma_p(T)$. Hence $y \in H(\lambda, T) \subseteq A$. As a result, we have $y \in AB = \{0\}$ and $y = 0$. On the other hand, we know that $y \neq 0$. The resulting contradiction shows that $\Sigma_p(U) = \{0\}$. Since U is compact, by the theorem on the spectrum of a bounded operator and the Riesz–Schauder theorem, we have $\Sigma(U) = \{0\}$. Since U is Hermitian, it follows that $\|U\| = |0| = 0$ and $U = O$.

Thus, $Uy = Oy = 0 = 0 \cdot y$ for all $y \in B$. Suppose that $y \neq 0$. Then $0 \in \Sigma_p(U)$, which is impossible because, as was shown above, the operator U has no eigenvalues. Hence $B = \{0\}$. □

The Hilbert–Schmidt theorem is widely used to great effect.

8. From the basis vectors h_k, we choose vectors g_j that form a basis of the subspace $H(0, T) = \operatorname{Ker} T$. The remaining vectors will be denoted by f_i. The collection (f_i) of eigenvectors of the operator T is called *fundamental*. The vectors f_i are defined by the equalities $Tf_i = \lambda_i f_i$ for eigenvalues $\lambda_i \neq 0$ of the operator T. Using the Hilbert–Schmidt theorem, the values of a complex Hermitian operator T can be represented in the form of a linear combination of the vectors f_i of the fundamental collection of T. Thus, the Hilbert–Schmidt theorem implies a generalization of the statement that a matrix operator can be diagonalized.

Corollary 1. $Tx = \sum \alpha_i f_i$ $(\alpha_i = f_i \cdot Tx, \ x \in H)$.

Proof. Since the vectors f_i and g_j form a basis of H, we have $x = \sum \xi_i f_i + \sum \eta_j g_j$ for some numbers ξ_i and η_j. Then $Tx = \sum \xi_i T f_i + \sum \eta_j T g_j$ because T is linear and continuous. But $T f_i = \lambda_i f_i$, $T g_j = 0 \cdot g_j = 0$, and $\xi_i \lambda_i = (f_i \cdot x) \lambda_i = (\lambda_i f_i) \cdot x = T f_i \cdot x = f_i \cdot Tx = \alpha_i$, because T is Hermitian and $\lambda_i = \bar{\lambda}_i \in \mathbb{R}$. Hence $\xi_i \cdot T f_i = \xi_i \lambda_i \cdot f_i = \alpha_i f_i$. $\quad\square$

The product $S = T^*T$ of any compact operator T and the adjoint operator T^* is a positive compact Hermitian operator. The eigenvalues σ_k of S are called the singular numbers of the operator T. By the Hilbert–Schmidt theorem, there exists an orthonormal basis of the space H consisting of the eigenvectors b_k of the operator S. Let $\sigma_i > 0$, $\mu_i = \sigma_i^{1/2}$, and $e_i = \mu_i^{-1} \cdot T b_i$. It is easy to verify that $\sigma_i = \|T b_i\|^2$, $\mu_i = \|T b_i\|$, the collection (e_i) is orthonormal, and

$$Tx = \sum \mu_i (b_i x) e_i \quad (x = \sum (b_k x) b_k \in H)$$

is the *canonical form* of the values of a compact operator.

Exercise. Prove the equalities for σ_i, μ_i, and Tx (see Reed and Simon, 1978, VI.5).

The equation $\lambda x - Tx = y$ can be easily solved using the eigenbasis (h_k) of the compact Hermitian operator T.

Corollary 2. Let $\lambda \notin \Sigma_p(T)$, $x = \sum \xi_k h_k$ and $y = \sum \eta_k h_k$. Then the equation $\lambda x - Tx = y$ has a unique solution. The solution is a vector x with coordinates $\xi_k = (\lambda - \lambda_k)^{-1} \eta_k$, $\lambda_k \in \Sigma_p(T)$.

Proof. Taking the scalar product of each side of the equality $\lambda x - Tx = y$ and h_k and using the equalities $Tx = \sum \xi_k \cdot T h_k = \sum \xi_k \cdot \lambda_k h_k$, we obtain

$$\lambda \xi_k - \xi_k \lambda_k = h_k \cdot \lambda x - h_k \cdot Tx = h_k \cdot y = \eta_k.$$

If $\lambda \neq \lambda_k$, we get $\xi_k = (\lambda - \lambda_k)^{-1} \eta_k$. The solution x is unique.

Indeed, the vector x with the specified coordinates is a solution. If $\xi_k = (\lambda - \lambda_k)^{-1} \eta_k$, then $\lambda \xi_k - \xi_k \lambda_k = \eta_k$ and

$$\lambda \xi_k h_k - T(\xi_k h_k) = \lambda \xi_k \cdot h_k - \xi_k \lambda_k \cdot h_k = \eta_k h_k,$$

$$\lambda x - Tx = \lambda \sum \xi_k h_k - T\left(\sum \xi_k h_k\right)$$
$$= \sum (\lambda \xi_k h_k - T(\xi_k h_k)) = \sum \eta_k h_k = y.$$

We conclude that $x = \sum \xi_k h_k$ is the desired solution. $\quad\square$

Corollary 3. *Let $\lambda \in \Sigma_p(T)$, $x = \sum \xi_k h_k$, and $y = \sum \eta_k h_k$. Then the equation $\lambda x - Tx = y$ has a solution only if $\eta_k = 0$ for $\lambda_k = \lambda$. In this case, any vector x with coordinates $\xi_k = (\lambda - \lambda_k)^{-1}\eta_k$ for $\lambda_k \neq \lambda$, $\lambda_k \in \Sigma_p(T)$, is a solution to the equation.*

Proof. Using the same arguments as in the proof of Corollary 2, we obtain the equality $\lambda \xi_k - \xi_k \lambda_k = \eta_k$ for the solution x. If $\lambda_k = \lambda$, then $\eta_k = 0$.

Then, as in the proof of Corollary 2, it can be verified that the vector x with the specified coordinates is a solution to the equation. If $\lambda_k = \lambda$, then ξ_k can be arbitrary. $\square$

Corollaries 2 and 3 are refinements of the corresponding Fredholm theorems for the special case under consideration. It should be noted that the condition $\eta_k = 0$ for $\lambda_k = \lambda$ is equivalent to the condition $y \perp H(\lambda, T)$. Since $\lambda = \lambda^*$ and $T = T^*$, this condition means that y is orthogonal to every solution z^* to the equation $\lambda z^* - Tz^* = 0$.

Remark. The properties of singular numbers of compact operators on Hilbert spaces and the relationship between these numbers and the traces of operators are discussed in detail in Sadovnichii (1999), Chapter 5. Nuclear operators, whose collections of eigenvalues and singular numbers are summable, are pointed out. Estimates of these numbers are given. Theorems on the equality of the matrix trace and the spectral trace of a nuclear operator are proved. Regularized traces of some unbounded operators are described.

Nuclear operators on normed spaces and absolutely summing operators on locally convex spaces, which form a considerably more general class, are described in Pietsch (1972). These operators are used to define nuclear locally convex spaces. Examples of nuclear spaces are the spaces of infinitely differentiable, harmonic, and analytic functions.

The spectra of compact operators on locally convex spaces are analyzed and the Fredholm alternative is formulated for them in Edwards (1995), Chapter 9. Applications of this theory in solving differential and integral equations are described. A general ergodic theorem is proved.

4.9.   SPECTRAL THEOREM

It follows from the Hilbert–Schmidt theorem that the values of a compact Hermitian operator can be represented as sums of vectors from the fundamental collection with numeric coefficients. The spectral theorem generalizes this result to the case of self-adjoint operators. It states that their values can be represented as integrals of certain numeric functions with respect to projection measures.

4.9.1. Projection measures

We will first describe projections, then projection measures, and, finally, integral sums and integrals with respect to these measures.

1. In what follows, we consider only orthogonal projections (Hermitian idempotents) in a complex Hilbert space H. By definition, a linear operator $P : H \to H$ such that $P^2 = P$ and $P^* = P$ is called an *Hermitian projection* (onto the subspace PH). We denote by $\mathcal{P} = \mathcal{P}(H)$ the space of all projections $P : H \to H$. Let $P, Q \in \mathcal{P}$. It is easy to verify that PQ, $P + Q$, and $P - Q$ are projections if and only if $PQ = QP$, $PQ = O$, and $PQ = QP = O$, respectively. If $PQ = O$, then P and Q are said to be *orthogonal*, which is written as $P \perp Q$. (In this case $QP = O$ as well.)

Exercises. 1) Verify the conditions for operations with projections (see Weidmann, 1980, Section 4.6; Trenogin, 1993, Sections 18.6 and 18.7; Sadovnichii 1999, Chapter 4, Section 2).

2) Verify that $P^2 = P$, $Q^2 = Q$, $(PQ)^2 = PQ \Rightarrow (QP)^2 = QP$ and $P^2 = P$, $Q^2 = Q$, $(PQ)^2 = O \Rightarrow (QP)^2 = O$ for linear operators on H (Spivak, 1965, Chapter 7, Section 7).

Since $Pf \cdot f = P^2 f \cdot f = P^* Pf \cdot f = Pf \cdot Pf = \|Pf\|^2 \geq 0$ $(f \in H)$, the projection P is a positive operator $(P \geq O)$. This makes it possible to define a partial order for projections: $P \geq Q$ whenever $P - Q \geq O$ or $PQ = QP = Q$.

For a sequence of projections P_n, weak convergence $P_n f \cdot g \to Pf \cdot g$ $(f, g \in H)$ is equivalent to strong convergence $P_n f \to Pf$ $(f \in H)$.

Exercise. Prove this equivalence.

Theorem. *An increasing sequence of projections strongly converges to its least upper bound, and a decreasing to its greatest upper bound.*

See Weidmann (1980), Theorem 4.32.

Note that O and I are projections. Moreover, $O = \min \mathcal{P}$ and $I = \max \mathcal{P}$. They are, respectively, lower and upper bounds for any sequence of projections $(O \leq P_n \leq I)$.

Corollary. *A series of pairwise orthogonal projections has strong sum equal to its least upper bound.*

2. Consider a set X, the algebra $\mathcal{R}$ of some subsets of X, a Hilbert space H over the field $\mathbb{C}$ of complex numbers, the algebra $\mathcal{P}$ of projections $P : H \to H$. The sets in $\mathcal{R}$ will be called *simple*. Assume that there exist $B_n \in \mathcal{R}$ that form a sequence $B_n \uparrow X$.

Example. We can set $X = \mathbb{R}$ and take the algebra $\mathcal{R}$ of simple sets $A = \sum [a_i, b_i[$ formed by a finite number of pairwise disjoint intervals $[a_i, b_i[$ $(-\infty < a_i \le b_i < \infty)$.

A *projection measure* is an additive map $P : \mathcal{R} \to \mathcal{P}$ that is strongly continuous from above at zero and from below at unity. Thus, a projection measure has the following definitive properties:

$$P(A + B) = P(A) + P(B) \quad (A, B, A + B \in \mathcal{R}),$$
$$P(A_n) \to O \quad (A_n \downarrow 0, A_n \in \mathcal{R}), \quad P(X_n) \to I \quad (X_n \uparrow X, X_n \in \mathcal{R}).$$

For $X = \mathbb{R}$, it is convenient to specify projection measures using projection distribution functions. A *projection distribution function* is a normalized increasing function $F : \mathbb{R} \to \mathcal{R}$ that is strongly continuous from the left. Thus, the definitive properties of the distribution function are

$$F(s) \le F(t) \quad (s \le t), \quad F(t) \to F(u) \quad (t \uparrow u);$$
$$F(s) \to O \quad (s \downarrow -\infty), \quad F(t) \to I \quad (t \uparrow \infty).$$

We emphasize that convergence is assumed to be strong everywhere (at every point).

The requirement of continuity from the left can be replaced by the requirement of continuity from the right, which is often done in practice.

The equalities

$$P([s, t[) = F(t) - F(s), \quad F(t) = \lim_{s \downarrow -\infty} P([s, t[)$$

$(-\infty < s \le t < \infty)$ define a one-to-one correspondence between the projection distribution functions and measures. For $X = \mathbb{R}$, this correspondence makes it possible to consider projection distribution functions instead of projection measures.

Exercise. Prove that the above equalities establish a one-to-one correspondence between projection measures and distribution functions (cf. Loève, 1977, 1978, 4.4).

It is easy to verify that a projection measure is not only additive, but also multiplicative:

$$P(AB) = P(A) \cdot P(B) \quad (A, B \in \mathcal{R}).$$

Since the projection measure is continuous from above at zero, it continuous and countably additive.

Exercise. Prove the above assertions.

3. There is a close analogy between projection and probability measures.

A projection measure $P : \mathcal{R} \to \mathcal{P}$ is naturally associated with collections of numeric and vector measures defined by the equalities

$$\mu_f(A) = P(A)f \cdot f = m_f(A) \cdot f, \quad m_f(A) = P(A)f \quad (A \in \mathcal{R})$$

for every $f \in H$. The necessary and sufficient conditions for such a collection of positive numeric measures $\mu_f : \mathcal{R} \to \mathbb{R}$ $(f \in H)$ to be generated by a projection measure $P : \mathcal{R} \to \mathcal{P}$ are known. They can be used to deduce the properties of projection measures from the known properties of numeric measures.

Remark. The connection between operator measures and numeric measures is described in detail in Berberian (1966), Section 3. This connection is used, in particular, in the proof of an important theorem on the extension of operator measures (Berberian 1966, Section 4, Theorem 7).

The measure μ_f can be normalized in order to use the measure $\rho_f = \|f\|^{-2}\mu_f$ for $f \neq 0$ instead. If $X \in \mathcal{R}$, then $\rho_f(X) = 1$, which means that ρ_f is a probability measure.

In addition to positive measures μ_f, the projection measure P is associated with the collection of numeric measures μ_{fg} $(f, g \in H)$ defined with the help of the polarization identity:

$$\mu_{fg} = 4^{-1}(\mu_{f+g} - \mu_{f-g} - i\mu_{f+ig} + i\mu_{f-ig}).$$

This equality is equivalent to

$$\mu_{fg}(A) = P(A)f \cdot g \quad (A \in \mathcal{R}).$$

Exercise. Prove this equivalence.

Example 1. Let $X = \mathbb{R}$, $H = l^2$, $P_n f = \xi_n e_n$ for $f = (\xi_m) \in H$ and $e_n = (\delta_{mn})$, $\delta_{mn} = 1$ for $m = n$ and $\delta_{mn} = 0$ for $m \neq n$. The operator P_n projects the space $H = l^2$ onto the nth coordinate axis $(n = 1, 2, \ldots)$. The equality

$$P(A) = \sum_{n \in A} P_n \quad (A \in \mathcal{R})$$

defines a *discrete projection measure*.

Example 2. Put $X = \mathbb{R}$, $H = \mathcal{L}^2(\mathbb{R}, dx)$. For every function $f \in H$ and a number $t \in \mathbb{R}$, we define a function $g = \,] - \infty, t[\cdot f$ such that $g(x) = f(x)$ for $x < t$ and $g(x) = 0$ for $t \geq x$. We have $g \in H$. The equality $F(t)f = g$ defines a projection distribution function $F : \mathbb{R} \to \mathcal{P}$ with values $F(t) \in \mathcal{P}$ $(t \in \mathbb{R})$. This can be verified using the Lebesgue theorem on passing to the limit under the integral sign. It is easy to verify that $F(t)f \to 0$ $(t \downarrow -\infty)$, $F(t)f \to f$ $(t \uparrow \infty)$.

The function F generates a projection measure P such that

$$P(A)f = A \cdot f \quad (A \in \mathcal{R}, f \in H).$$

The projection $P(A)$ represents the multiplication of the function f by the set A (i.e., by the indicator of A).

Exercise. Prove the preceding assertion about the function F.

4. To simplify notation, as before, we will identify sets with their indicators and use the same designations for them.

Let

$$u = \sum a_i X_i \quad (a_i \in \mathbb{C}, X_i \in \mathcal{R})$$

be a simple numeric function on X, and let $P : \mathcal{R} \to \mathcal{P}$ be a projection measure. The operator

$$S(u, P) = \sum a_i P(X_i).$$

is called the *integral sum of the function* u *with respect to the measure* P. It is easy to verify that the value of the integral sum does not depend on the representation of the simple function in the form of a linear combination of simple sets

$$\sum a_i P(X_i) = \sum b_j P(Y_j) \quad \left(\sum a_i X_i = \sum b_j Y_j \right),$$

where $a_i, b_j \in \mathbb{C}$ and $X_i, Y_j \in \mathcal{R}$.

Exercise. Verify the preceding statement.

Let $\mathcal{S}$ denote the algebra of simple functions. We will write $S(u)$ instead of $S(u, P)$ whenever the measure P is understood. We denote the algebra of all linear operators $T : H \to H$ by $\mathcal{L} = \mathcal{L}(H)$ and the algebra of bounded operators by $\mathcal{B} = \mathcal{B}(H)$.

It follows from the definitions that the integral sum S with values $S(u)$ is a bounded linear operator:

$$S(\alpha u + \beta v) = \alpha S(u) + \beta S(v) \quad (\alpha, \beta \in \mathbb{C}; \ u, v \in \mathcal{S}).$$

Thus, $S : \mathcal{S} \to \mathcal{B}$. Since the measure P is multiplicative, then the integral sum S is also multiplicative:

$$S(u \cdot v) = S(u) \cdot S(v) \quad (u, v \in \mathcal{S}).$$

The operator $S(u)$ is normal, i.e.,

$$S^*(u) \cdot S(u) = S(u) \cdot S^*(u) \quad (u \in \mathcal{S}).$$

Indeed, since the values of the measure P are Hermitian,

$$S(u^*) = S^*(u) \quad (u \in \mathcal{S}).$$

Then, since S is multiplicative, we obtain

$$S^*(u) \cdot S(u) = S(u^*) \cdot S(u) = S(u^*u) = S(|u|^2) = S(uu^*) = S(u) \cdot S(u^*).$$

5. The following rule for the scalar product of values of S can be readily verified:

$$S(v)g \cdot S(u)f = \int v^*u \cdot d\mu_{gf} \quad (f, g \in H; \ u, v \in \mathcal{S}).$$

The numeric measure μ_{gf} is as follows:

$$\mu_{gf}(A) = P(A)g \cdot P(A)f = P(A)g \cdot f = g \cdot P(A)f \quad (A \in \mathcal{R}).$$

It is equal to a linear combination of positive numeric measures and the integral with respect to μ_{gf} is equal to the corresponding linear combination of integrals with respect to the mentioned measures:

$$\int w \cdot d\mu_{gf} = \frac{1}{4} \left(\int w \cdot d\mu_{g+f} - \int w \cdot d\mu_{g-f} - i \int w \cdot d\mu_{g+if} + i \int w \cdot d\mu_{g-if} \right)$$

for any function w integrable with respect to the measures μ_{g+f}, μ_{g-f}, μ_{g+if}, μ_{g-if}.

Exercise. Prove the above rule for the scalar product.

By analogy with positive numeric measures, the integral of the product v^*u of functions v and u can be considered to be the scalar product of these functions and denoted by $\langle v, u \rangle_{gf}$. Then the rule for the scalar product can be written as

$$\langle S(v)g, S(u)f \rangle = \langle v, u \rangle_{gf} \quad (f, g \in H; \ u, v \in S).$$

For $g = f$ and $v = u$, the rule for the scalar product yields

$$\|S(u)f\|^2 = \|u\|_f^2 = \int |u|^2 d\mu_f \quad (u \in S, \ f \in H),$$

where the positive numeric measure μ_f is as follows:

$$\mu_f(A) = P(A)f \cdot P(A)f = \|P(A)f\|^2 \quad (A \in \mathcal{R}; \ f \in H).$$

Using the equality

$$\|S(u)f\| = \|u\|_f$$

it is easy to prove an important lemma that serves as the basis for the definition of the integral with respect to a projection measure. Let $\mathcal{L}_f^2 = \mathcal{L}_f^2(\mathbb{R}, \mu_f)$ be the Hilbert space of numeric functions on $\mathbb{R}$ that are measurable with respect to μ_f and square integrable with respect to μ_f for a given $f \in H$. The norm in $\mathcal{L}_f^2$ will be denoted by $\| \cdot \|_f$. Clearly, $S \subset \mathcal{L}_f^2$ for any $f \in H$. Therefore, the norm $\|u\|_f$ of $u \in S$ can be considered in every space $\mathcal{L}_f^2$.

Lemma. If a sequence of funcitons $u_n \in S$ convergence in $\mathcal{L}_f^2$, then the sequence of vectors $g_n = S(u_n)f$ converges in H.

Proof. Indeed, from the linearity of the integral sum and the above equality for the norms, it follows that

$$\|g_m - g_n\| = \|S(u_m - u_n)f\| = \|u_m - u_n\|_f \to 0$$

for $m, n \to \infty$. □

Note that the set S of simple functions is dense in the space $\mathcal{L}_f^2$ for any $f \in H$.

The values $S(u)$ of the sum S for real simple functions are Hermitian operators. If $u \geq 0$, then $S(u) \geq 0$. This follows from the fact that values of a projection norm are Hermitian and positive.

Exercise. Give detailed proofs of the preceding statements.

4.9.2. Integrals of bounded functions

Let $\mathcal{M}$ be the algebra of functions $u : X \to \mathbb{C}$ that are measurable with respect to every measure μ_f ($f \in H$). These functions are said to be *measurable with respect to the measure P*, or simply *measurable*. The algebra $\mathcal{M}$ is a Borel algebra, i.e., it is closed under the pointwise convergence of sequences of functions. It follows that $\mathcal{M}$ includes the algebra $\mathcal{S}$ of simple functions together with its Borel closure $\bar{\mathcal{S}}$ and the Borel closure $\bar{\mathcal{R}}$ of the algebra of simple sets.

Example. If $X = \mathbb{R}$ and the algebra $\mathcal{R}$ is generated by intervals, $\mathcal{M}$ includes all Borel functions and, in particular, all continuous functions $f : \mathbb{R} \to \mathbb{C}$.

We denote by $\mathcal{M}_0$ the algebra of functions $u : X \to \mathbb{C}$ that are bounded and measurable with respect to the measure P. We have $\mathcal{S} \subseteq \mathcal{M}_0 \subseteq \mathcal{M}$. We now define the integral with respect to the projection measure for bounded measurable functions and analyze its properties. This integral is a continuous extension of the integral sum S from the algebra $\mathcal{S}$ to the algebra $\mathcal{M}_0$.

1. First, we consider a positive bounded numeric measure $\mu : \mathcal{R} \to \mathbb{R}$, the Hilbert space $\mathcal{L}^2 = \mathcal{L}^2(X, \mu)$ and its subset $\mathcal{L}_0^2 = \mathcal{L}_0^2(X, \mu)$ consisting of bounded functions $u \in \mathcal{L}^2$.

Every function $u : X \to \mathbb{C}$ that is measurable with respect to the measure μ can be associated with a sequence of functions $\varphi_n : X \to \mathbb{R}$ such that

$$\varphi_n(x) = 1 \quad (x : |u(x)| \le n), \quad \varphi_n(x) = 0 \quad (x : |u(x)| > n)$$

and a sequence of functions $\bar{u}_n = \varphi_n u$ such that

$$\bar{u}_n(x) = u(x) \quad (x : |u(x)| \le n) \quad \bar{u}_n(x) = 0 \quad (x : |u(x)| > n).$$

It is obvious that

$$|\bar{u}_n(x)| \le |u(x)|, \quad |\bar{u}_n(x)| \le n \quad (x \in X).$$

The functions φ_n and $\bar{u}_n$ are measurable with respect to μ because so is the function u.

In what follows, we usually assume that the numeric measures under consideration are complete and locally bounded (see Subsection 3.3.2, part 4). These properties will not always be specified explicitly.

The following statements, which are used in the definition of the integral with respect to a projection measure and in the analysis of its properties, can be proved using more or less standard arguments.

Lemma 1. *Any constant* $c : X \to \mathbb{C}$ *is locally integrable with respect to a locally bounded measure* μ.

Lemma 2. *The functions* $\bar{u}_n$ *belong to* $\mathcal{L}^2$ *and* $\bar{u}_n \to u$ *at every point.*

Lemma 3. *Let* $u, v, u_n \in \mathcal{L}^2$ *and* $|u_n| \leq |v|$ $(n = 1, 2, \dots)$. *Then* $u_n \to u(\mu)$ *implies* $u_n \to u(\mathcal{L}^2)$.

Note that $u_n \to u(\mu)$ means convergence almost everywhere with respect to the measure μ, while $u_n \to u(\mathcal{L}^2)$ means $\|u_n - u\| \to 0$. The next lemma follows from Lemmas 1 and 3.

Lemma 4. *Let* $u, u_n \in \mathcal{L}^2$ *and* $|u_n| \leq c$ $(n = 1, 2, \dots)$ *for some* $c > 0$. *Then* $u_n \to u(\mu)$ *implies* $u_n \to u(\mathcal{L}^2)$.

The next statement readily follows from Lemmas 2 and 3.

Proposition 1. *The set* $\mathcal{L}_0^2$ *is dense in the space* $\mathcal{L}^2$.

From this proposition and Lemma 4, it follows that $\mathcal{S}$ is dense in $\mathcal{L}_0^2$.

Proposition 2. *The set* $\mathcal{S}$ *is dense in the space* $\mathcal{L}^2$.

The above statements have several useful corollaries that can be easily deduced.

Corollary 1. *The set* $\mathcal{M}_0$ *is included in* $\mathcal{L}_f^2$ *for any* $f \in H$.

Corollary 2. *For any* $u \in \mathcal{M}_0$ *and* $f \in H$, *there exists a sequence* $u_n^f \in \mathcal{S}$ *that converges to* u *in* $\mathcal{L}_f^2$.

Corollary 3. *Let* $f \in H$, $u \in \mathcal{L}_f^2$, $u_n^f \in \mathcal{S}$, *and* $u_n^f \to u$ *in* $\mathcal{L}_f^2$. *Then there exists a* $g \in H$ *such that* $S(u_n^f)f \to g$ *in* H.

Corollary 4. Let $f \in H$, $u \in \mathcal{L}_f^2$, $u_n^f \in \mathcal{S}$, $v_n^f \in \mathcal{S}$, and $Su_n^f \to u$, $Sv_n^f \to u$ in $\mathcal{L}_f^2$. Then $\lim S(u_n^f)f = \lim S(v_n^f)f$ in the space H.

Exercise. Give detailed proofs of the above statements.

Thus, for any function $u \in \mathcal{M}_0$ and any vector $f \in H$, there exists a sequence of simple functions u_n^f that approximates u in the space $\mathcal{L}_f^2$. Moreover, the sequence of vectors $S(u_n^f)$ has a limit $g \in H$, which is the same for all approximating sequences of simple functions. It is natural to use this limit to define the integral of u with respect to the considered projection measure P.

2. We denote by $\mathcal{F}_0 = \mathcal{F}_0(H)$ the algebra of all (linear and nonlinear) transformations of the space H. These transformations will be called operators acting in H. By the *integral of a function* $u \in \mathcal{M}_0$ with respect to the measure P we mean the operator $E_0(u) \in \mathcal{F}_0(H)$ such that

$$E_0(u)f = \lim S(u_n^f)f$$

for all $f \in H$, where (u_n^f) is a sequence of simple functions that converges to u in $\mathcal{L}_f^2$. By the *integral* of a bounded function that is measurable with respect to the given projection measure we mean the map $E_0 : \mathcal{M}_0 \to \mathcal{F}_0$ that takes a function $u \in \mathcal{M}_0$ to an operator $E_0(u) \in \mathcal{F}_0$. Clearly, the integral E_0 is an extension of the integral sum S:

$$E_0(u) = S(u) \quad (u \in \mathcal{S}).$$

If $u \in \mathcal{S}$, then the constant sequence $u_n^f = u$ $(n = 1, 2 \ldots)$ can be chosen to be the approximating sequence for every $f \in H$.

Example 1. Consider the set $X = \mathbb{N}$ the algebra $\mathcal{R}$ of all subsets of X, the Hilbert space $H = l^2$, and its standard basis $e_n = (\delta_{nm})$. Let P_n denote the projection onto the coordinate axis $\mathbb{C}e_n$, i.e.,

$$P_n f = f(n) \cdot e_n \quad (f = (f(m) \in H)).$$

Take the projection measure

$$P(A) = \sum_{n \in A} P_n \quad (A \in \mathcal{R}).$$

Any bounded sequence $u : \mathbb{N} \to \mathbb{C}$ is measurable with respect to the measure P. For any $f \in H$, it is approximated in $\mathcal{L}_f^2$ by a sequence of simple functions u_n such that $u_n(m) = u(m)$ for $m \le n$ and $u_n(m) = 0$ for $m > n$. Indeed, we have

$$\|u - u_n\|_f^2 = \sum_{m>n} |u(m)|^2 \le c^2 \sum_{m>n} |f(m)|^2 \to 0.$$

Since

$$S(u_n) = \sum_{m \le n} |u(m)|^2 P_m, \quad S(u_n)f = \sum_{m \le n} |u(m)|^2 f(m) \cdot e_m,$$

it follows that

$$E_0(u)f = \lim S(u_n)f = \sum u(m)f(m)e_m$$

or

$$E_0(u)f = uf = (u(m) \cdot f(m))_m.$$

Example 2. Consider a set X, the algebra $\mathcal{R}$ of some subsets of X, a positive numeric measure μ on $\mathcal{R}$ (not necessarily bounded), the Hilbert space $H = \mathcal{L}_f^2 = \mathcal{L}_f^2(X, \mu)$, and a projection measure P on $\mathcal{R}$ defined by the equality

$$P(A)f = Af \quad (A \in \mathcal{R}, \ f \in H),$$

where, as usual, $Af(x) = f(x)$ for $x \in A$ and $Af(x) = 0$ for $x \notin A$. We have

$$S(u)f = \sum a_i P(X_i)f = \sum a_i X_i f = uf \quad (u = \sum a_i X_i \in \mathcal{S}, \ f \in H).$$

Since

$$\mu_f(A) = \|P(A)f\|^2 = \|Af\|^2 = \int A|f|^2 d\mu,$$

it follows that $d\mu_f/d\mu = |f|^2$, $d\mu_f = |f|^2 d\mu$. Then, making the appropriate change of variables, we obtain

$$\|uf - u_n^f f\|^2 = \int |uf - u_n^f f|^2 d\mu$$

$$= \int |uf - u_n^f|^2 |f|^2 d\mu = \int |uf - u_n^f|^2 d\mu_f = |u - u_n^f|_f^2.$$

Hence, $u_n^f \to u$ in H as $u_n^f \to u$ in $\mathcal{L}_f^2$. Therefore,

$$E_0(u)f = \lim S(u_n^f)f = \lim(u_n^f f) = uf.$$

In both examples, integration with respect to the projection measure is reduced to multiplication. The first example is a special case of the second one with $X = \mathbb{N}$ and the counting measure μ.

3. The value $E_0(u)$ of the integral E_0 of a bounded measurable function $u \in \mathcal{M}_0$ is an everywhere defined bounded normal operator in H. The following lemma plays an important role in the proofs of this assertion and other statements about the properties of integrals.

Lemma. *For any $f, g \in H$, there exists an $h \in H$ such that $\mu_f, \mu_g \ll \mu_h$.*

As before, $\mu \ll \nu$ means that the measure μ is continuous with respect to the measure ν. The proof of the lemma is elementary but rather long (cf. the proof of Theorem 7.13 in Weidmann, 1980). The lemma readily implies the following statement.

Corollary. *Let $u \in \mathcal{M}_0$, $|u| < c$, and $f_i \in H$ $(i = 1, \ldots, m)$. Then there exist $u_n \in S$ such that $u_n \to u$ almost everywhere with respect to the measures μ_{f_i}, $u_n \to u$ in $\mathcal{L}^2_{f_i}$ $(i = 1, \ldots, m)$, and $|u_n| < c$ $(n = 1, 2, \ldots)$.*

Exercise. Prove the corollary.

Using this corollary and the corresponding properties of the integral sum S, it is easy to prove the statement about the values of the integral E_0 and the class they belong to that is formulated in the beginning of part 3 of this subsection.

Theorem 1. *The value $E_0(u)$ of the integral E_0 of a function $u \in \mathcal{M}_0$ is an everywhere defined normal linear operator in H.*

Proof. The linearity of $E_0(u)$ follows from the equalities

$$E_0(u)(af + bg) = \lim S(u_n)(af + bg)$$
$$= \lim(aS(u_n)f + bS(u_n)g) = a \lim S(u_n)f + b \lim S(u_n)g$$
$$= aE_0(u)f + bE_0(u)g \quad (a, b, \in \mathbb{C}; \ f, g \in H),$$

which hold for $u \in S$ and $u_n \to u$ in $\mathcal{L}^2_f$, $\mathcal{L}^2_g$, and $\mathcal{L}^2_{af+bg}$.

The linear operator $E_0(u)$ is bounded, since

$$\|E_0(u)\| = \|\lim S(u_n)f\| = \lim \|S(u_n)f\| = \lim \|u_n\|_f \leq c\|f\|$$

for $f \in H$, $u_n \in S$, $u_n \to u$ in $\mathcal{L}^2_f$, and $|u_n| < c$ $(n = 1, 2, \ldots)$.

The operator $E_0(u)$ is normal, since

$$E_0(u^*) = E_0^*(u), \quad E_0(uv) = E_0(u) \cdot E_0(v) \quad (u, v \in \mathcal{M}_0).$$

Indeed, we have

$$E_0(u) \cdot E_0^*(u) = E_0(u) \cdot E_0(u^*)$$
$$= E_0(uu^*) = E_0(|u|^2) = E_0(u^*u) = E_0^*(u) \cdot E_0(u)$$

for any $u \in \mathcal{M}_0$. □

Let $\mathcal{L}_0 = \mathcal{L}_0(H)$ (instead of $\mathcal{B}_0 = \mathcal{B}_0(H)$) denote the algebra of bounded linear operators on H. The integral $E_0 : \mathcal{M}_0 \to \mathcal{L}_0$ is a homomorphism from the algebra $\mathcal{M}_0$ to the algebra $\mathcal{L}_0$. It preserves all operations, namely, addition, multiplication, multiplication by a number, and conjugation.

Theorem 2. *For any* $a, b, \in \mathbb{C}$ *and* $u, v \in \mathcal{M}_0$,

$$E_0(au + bv) = aE_0(u) + bE_0(v), \tag{1}$$
$$E_0(uv) = E_0(u) \cdot E_0(v), \tag{2}$$
$$E_0^*(u) = E_0(u^*). \tag{3}$$

These equalities are obtained by passing to the limits in the analogous equalities for integral sums.

Corollary. *For any real function* $u \in \mathcal{M}_0$, *the value* $E_0(u)$ *of the integral* E_0 *is an Hermitian operator.*

Proof. In view of (3), if $u^* = u$, then

$$E_0^*(u) = E_0(u^*) = E_0(u),$$

which means that $E_0(u)$ is an Hermitian operator. □

Exercise. Prove Theorem 2.

4. Passing to the limits in the rule for the scalar product for integral sums, one can obtain the analogous rule for integrals.

Theorem 3. *For any $u, v \in \mathcal{M}_0$ and $f, g \in H$,*

$$\langle E_0(v), E_0(u) \rangle = \langle u, v \rangle_{fg}, \tag{4}$$

where $\langle u, v \rangle_{fg} = \int v^* u \, d\mu_{fg}$, $\mu_{fg} = \langle P(A)g, P(A)f \rangle$ $(A \in \mathcal{R})$.

Exercise. Prove Theorem 3.

For $f = g$ and $u = v$, equality (4) implies the following statement.

Corollary. *For any $u \in \mathcal{M}_0$ and $f \in H$,*

$$\|E_0(u)f\| = \|u\|_f. \tag{5}$$

Since the projection measure P is assumed to be locally bounded, there exists a sequence of simple sets $X_n \in \mathcal{R}$ such that $X_n \uparrow X$ and $P(X_n) \to I$, where I is the identity transformation of the space H.

Theorem 4. *The integral E_0 is normalized, i.e., $E_0(1) = I$.*

Proof. It is obvious that the constant 1 belongs to $\mathcal{M}_0$ and the sequence of indicators of the sets $X_n \uparrow X$ approximates this constant in $\mathcal{L}_f^2$ for any $f \in H$. Therefore,

$$E_0(1)f = \lim S(X_n)f = \lim P(X_n)f = f.$$

We conclude that $E_0(1) = I$. $\quad\square$

Since the integral E_0 is linear, the theorem readily implies the following corollary.

Corollary. *For any constant c, $E_0(c) = c \cdot I$.*

In addition to the algebraic properties of the integral sum S, the integral E_0 also possesses the order properties of S.

Theorem 5. *The integral E_0 is positive and monotonic, i.e.,*

$$E_0(u) \geq O \ (u \geq 0), \quad E_0(u) \leq E_0(v) \ (u \leq v)$$

for real functions $u, v \in \mathcal{M}_0$.

Proof. By the corollary of Theorem 2, the operator $E_0(u)$ is Hermitian if $u \geq 0$. By equality (4) with $v = 1$ and $g = f$, taking into account that $E_0(1) = I$, we obtain

$$f \cdot E_0(u)f = E_0(1)f \cdot E_0(u)f = \int u \, d\mu_f \geq 0 \quad (u \geq 0).$$

Hence, $E_0(u) \geq O$ for $u \geq 0$. The integral E_0 is positive.

Since E_0 is linear and positive, it follows that it is monotonic:

$$E_0(v) - E_0(u) = E_0(v - u) \geq 0, \quad E_0(u) \leq E_0(v)$$

for $u \leq v$ and $u, v \in \mathcal{M}_0$. □

5. Using the properties of the integral E_0 to be linear and isometric, it is easy to prove a limit theorem for E_0 that describes the convergence of a sequence of its values. Since a sequence of functions $u_n \in \mathcal{M}_0$ may converge in the space $\mathcal{L}_f^2$ to a function $u \notin \mathcal{M}_0$ for some f, the limit theorem includes two assertions: one for the case $u \in \mathcal{M}_0$ and another for the case $u \notin \mathcal{M}_0$. In addition, we prove that the integral does not depend on the choice of an approximating sequence.

Theorem 6. Let $u_n, v_n \in \mathcal{M}_0$, $u \in \mathcal{L}_f^2$, $f \in H$, and $u_n, v_n \to u$ in $\mathcal{L}_f^2$. Then there exists a $g \in H$ such that

$$\lim E_0(u_n) = \lim E_0(v_n) = g.$$

Furthermore, $g = E_0(u)f$ if $u \in \mathcal{M}_0$.

Proof. By equality (5),

$$\|E_0(v_p)f - E_0(u_n)f\| = \|E_0(v_p - u_n)f\| = \|v_p - u_n\|_f.$$

Hence, for $(v_p) = (u_n)$ the sequence $g_n = E_0(u_n)f$ converges to a $g \in H$, and the sequence $h_p = E_0(v)f$ converges to an $h \in H$. Moreover, $\|h - g\| = \|u - u\|_f = 0$ and $h = g$. If $u \in \mathcal{M}_0$, then $E_0(v_p)f \to E_0(u)$ and

$$\|E_0(u)f - E_0(u_n)f\| = \|u - u_n\|_f,$$

which implies

$$g = \lim E_0(u_n)f = E_0(u)f$$

for $u \in \mathcal{M}_0$. □

Theorems 1–6 show that the integrals of bounded measurable functions with respect to projection measures have many favorable properties.

4.9.3. Integrals of unbounded functions

The values of integrals of unbounded functions are unbounded densely defined linear operators. Therefore, the definition of such integrals and the analysis of their properties are more complicated than those of the integrals of bounded functions.

1. Take a measurable function $u \in \mathcal{M}$ and consider the set $D(u) = \{f \mid u \in \mathcal{L}_f^2\} \subseteq H$. We have $f \in D(u) \Leftrightarrow u \in \mathcal{L}_f^2$. This ensures the existence of a sequence of bounded measurable functions $u_n^f \in \mathcal{M}_0$ converging to $u \in \mathcal{M}$ in the space $\mathcal{L}_f^2$ for $f \in D(u)$. For such approximating bounded functions, we can take simple functions $u_n^f \in \mathcal{S}$. By Theorem 6 (Subsection 4.9.2, part 5), the sequence of vectors $g_n = E_0(u_n)f$ has a limit $g \in H$, which is the same for all sequences of functions in $\mathcal{M}_0$ approximating the function $u \in \mathcal{M}$. It is natural to use this limit to define the integral of $u \in \mathcal{M}$ with respect to the given projection measure P.

We denote by $\mathcal{F} = \mathcal{F}(H)$ the set of all partial transformations of the space H (maps from subsets of H to H). Such transformations will be called *partial operators* acting in H, or simply *operators*, as are conventional transformations defined on the entire H.

The *integral of a function* $u \in \mathcal{M}$ with respect to the measure P is the partial operator $E(u) \in \mathcal{F}$ such that there exists a sequence of bounded functions $u_n^f \in \mathcal{M}_0$ that converges to u in $\mathcal{L}_f^2$ and

$$E(u)f = \lim E_0\left(u_n^f\right)f$$

for all $f \in D(u)$.

The *integral* of measurable functions with respect to a projection measure is a map $E : \mathcal{M} \to \mathcal{F}$ that takes a function $u \in \mathcal{M}$ to a partial operator $E(u) \in \mathcal{F}$.

It is obvious that the integral E is an extension of the integral E_0 and the integral sum S, i.e.,

$$E(u) = S(u) \quad (u \in \mathcal{S}), \quad E(u) = E_0(u) \quad (u \in \mathcal{M}_0).$$

In each of these cases, for an approximating sequence we can take the corresponding constant sequence.

Example 1. In Example 1 of Subsection 4.9.2, part 2, take an arbitrary sequence $u : \mathbb{N} \to \mathbb{C}$ instead of a bounded one. It is easy to verify that

$$D(u) = \left\{f : \sum |u(n) \cdot f(n)|^2 < \infty\right\}, \quad E(u)f = uf \quad (f \in D(u)).$$

Example 2. In Example 2 of Subsection 4.9.2, part 2, take any measurable function $u \in \mathcal{M}$ instead of a bounded one. It is easy to verify that

$$D(u) = \left\{ f : \int |uf|^2 < \infty \right\}, \quad E(u)f = uf \quad (f \in D(u)).$$

In both examples, the integration of a function with respect to a projection measure is again reduced to multiplication.

Exercise. Prove the equalities in Examples 1 and 2.

2. For any measurable function $u \in \mathcal{M}$, the value $E(u)$ of the integral E is a densely defined normal linear operator on H. Since the operators under consideration are partial, the proofs of the majority of the statements about the properties of the integral E are complicated. Therefore, they will be formulated only as generalizations of Theorems 1–6 on the properties of the integral E_0. The proofs can be found in Weidmann (1980), Section 7.2.

Lemma. The domain of definition $D(u)$ of the value $E(u)$ of the integral E of a function $u \in \mathcal{M}$ is a dense subspace of H.

Theorem 1. The value $E(u)$ of the integral E of the function $u \in \mathcal{M}$ is a densely defined normal linear operator on H.

Theorem 2. For any $a, b \in \mathbb{C}$ and $u, v \in \mathcal{M}$,

$$E(au + bv) \supseteq aE(u) + bE(v), \tag{1'}$$
$$E(uv) \supseteq E(u) \cdot E(v) \tag{2'}$$

$$E^*(u) = E(u^*). \tag{3'}$$

The inclusions 1′ and 2′ mean that their left-hand sides are extensions of their right-hand sides. We emphasize that, as follows from the rules for operations with partial operators, for a given expression, the domain of definition of the resulting operator is equal to the intersection of the domains of definition of operators occurring in the expression (for example, $D(uv) = D(u) \cap D(v)$).

Corollary. *For any real function $u \in M$, the value $E(u)$ of the integral E is a self-adjoint operator.*

Theorem 3. *For any $u, v \in M$, $f \in D(u)$, and $g \in D(v)$,*

$$\langle E(v)g, E(u)f \rangle = \langle v, u \rangle_{gf}, \tag{4'}$$

where $\langle v, u \rangle_{f,g} = \lim \int \bar{v}_n^* \bar{u}_n d\mu_{gf}$. (Bounded functions $\bar{v}_n$ and $\bar{u}_n$ approximating v and u were defined in Subsection 4.9.2, part 1.)

Corollary. *For any $u \in M$ and $f \in D(u)$,*

$$\|E(u)f\| = \|u\|_f. \tag{5'}$$

Theorem 4. *The integral E_0 is normalized and $E_0(c) = c \cdot I$ for any constant c.*

Theorem 5. *The integral E_0 is positive and monotonic:*

$$E(u) \geq O \quad (u \geq 0), \qquad E(u) \leq E(v) \quad (u \leq v)$$

for any real functions $u, v \in M$.

Theorem 6. *Let $u, u_n \in M$, $f \in D(u) \cap D(u_n)$ $(n = 1, 2, \dots)$, and $u_n \to u$ in $\mathcal{L}_f^2$. Then*

$$\lim E(u_n) = E(u)f.$$

Theorems 4 and 5 are the same as for the integral E_0. They are given in order to preserve the numbering.

The integral $E(u)$ of a function $u \in M$ with respect to a projection measure is often designated the same way as the integral with respect to a numeric measure:

$$E(u) = \int u \, dP.$$

If the measure P is generated by a distribution function F, it is also conventional to write

$$E(u) = \int u \, dF.$$

The variable x and the domain of definition X of the function u are specified whenever necessary.

Remark. Integration with respect to a positive operator measure and its connection with integration with respect to numeric measures are discussed in detail in Berberian (1966). The fact that an operator measure is defined by a collection of numeric measures, as is a projection measure, is used there.

4.9.4. Spectral theorem

The integral of a real measurable function with respect to a projection measure is a self-adjoint operator. A natural question arises of whether every self-adjoint operator is the integral of a function with respect to some
measure. The spectral theorem gives the affirmative answer to this question. Detailed proofs of this theorem can be found in Weidmann (1980),
Section 7.3. See also Reed and Simon (1978), VII, VIII.3; Hutson and
Pym (1980), Chapter 9.

1. The Stieltjes inversion formula, is used in the proof of the spectral
theorem. The formula can be easily verified.

Let $\rho : \mathbb{R} \to \mathbb{R}$ be a function that is continuous from the left and has
bounded variations (i.e., it is the difference of two increasing functions), and
let $\rho(-\infty) = 0$. For any $z \in \mathbb{C}$ such that $\operatorname{Im} z \neq 0$, consider a function
$r(z, \cdot) : \mathbb{R} \to \mathbb{C}$ such that

$$r(z, x) = (z - x)^{-1} \quad (x \in \mathbb{R}).$$

The function $r(z, \cdot)$ is integrable with respect to the numeric measure defined
by the function ρ. If $\rho = \sigma - \tau$, where σ and τ are increasing positive
distribution functions, then

$$\int r(z, \cdot) \, d\rho = \int r(z, \cdot) d\sigma - \int r(z, \cdot) \, d\tau.$$

In addition, consider a function φ such that

$$\varphi(z) = \int\limits_{-\infty}^{\infty} (z - x)^{-1} \, d\rho(x) \quad (\operatorname{Im} z \neq 0).$$

The *Stieltjes inversion formula* is written as

$$\rho(x) = \lim_{\delta \downarrow 0} \lim_{\varepsilon \downarrow 0} (2\pi i)^{-1} \int\limits_{-\infty}^{x+\delta} [\varphi(s - i\varepsilon) - \varphi(s + i\varepsilon)] \, ds.$$

A proof of this formula can be found in Weidmann (1980), Appendix B.
As was shown before, the resolvent

$$R(z, T) = (zI - T)^{-1} \quad (z \in P(T))$$

of a closed operator T is an everywhere defined bounded operator, and the function $R(\cdot, T)$ is analytic on the resolvent set $P(T)$. In addition (see Subsection 4.7.1), a self-adjoint operator is closed and its spectrum is included in $\mathbb{R}$.

2. Consider a complex Hilbert space H, a vector subspace A that is dense in H, the algebra $\mathcal{P}$ of projections in H, the real line $\mathbb{R}$, the algebra $\mathcal{R}$ of simple sets in $\mathbb{R}$ generated by intervals $[a, b[\ (-\infty < a \le b < \infty)$, and a projection measure $P : \mathcal{R} \to \mathcal{P}$. To simplify notation, we denote by x an arbitrary point in $\mathbb{R}$ and the identity function on $\mathbb{R}$.

Theorem. *For any self-adjoint operator $T : A \to H$, there exists a unique projection measure $P : \mathcal{R} \to \mathcal{P}$ such that*

$$T = \int x \, dP.$$

The measure P is defined by the following equality (for all $f \in H$ and $a \le b$ in $\mathbb{R}$):

$$\|P([a, b[)f\|^2 = \lim_{\delta \downarrow 0} \lim_{\varepsilon \downarrow 0} (2\pi i)^{-1} \int_{a+\delta}^{b+\delta} \langle f, [R(x - i\varepsilon, T) - R(x + i\varepsilon, T)]f \rangle \, dx.$$

The expression under the integral sign is the scalar product of the vector f and the vector $g = [R(x - i\varepsilon, T) - R(x + i\varepsilon, T)]f$. Since $\varepsilon > 0$, we have $x - i\varepsilon, x + i\varepsilon \in \mathbb{R}$. Hence, $x - i\varepsilon, x + i\varepsilon \notin P(T)$ and the integrand is defined for any $f \in H$. The measure P defined in the theorem is called the *spectral measure* of the operator T.

The proof of the existence of a spectral measure P of a self-adjoint operator T is complicated (Weidmann, 1980, Section 7.3), while the proof of it uniqueness is simple. We now prove the uniqueness part. This will provide a better understanding of the theorem.

Proof. Let $T = E(x)$ and $D(x) = A$. Then $E(z - x) = zI - T$: $A \to H$ because the integral E is linear and normalized. Take $u = z - x$, $v = r(z, x) = (z - x)^{-1}$ (Im $z \ne 0$). The function v is bounded and $uv = vu = 1$. Therefore, since the integral E is multiplicative, we have

$$E(u) \cdot E(v) = I, \quad E(v) \cdot E(u) = I_A,$$

where I and I_A are the identity transformations of H and A, respectively. Since $E(u) = zI - T$, it follows that

$$E(v) = E^{-1}(u) = R(z, T) : H \to H.$$

By the rule for the scalar product, we have

$$\varphi(z) = f \cdot E(v)f = \int v \, d\mu_f = \int\limits_{-\infty}^{\infty} (z-x)^{-1} \, d\mu_f.$$

Application of the inversion formula yields

$$\mu_f([a,b[) = \lim_{\delta\downarrow0}\lim_{\varepsilon\downarrow0}(2\pi i)^{-1}\int\limits_{a+\delta}^{b+\delta}[\varphi(x-i\varepsilon)-\varphi(x+i\varepsilon)]\,dx$$

for $-\infty < a \le b < \infty$. It remains to note that

$$\mu_f([a,b[) = \|P([a,b[)f\|^2,$$
$$[\varphi(x-i\varepsilon)-\varphi(x+i\varepsilon)] = \langle f, [R(x-i\varepsilon,T)-R(x+i\varepsilon,T)]f\rangle$$

for any $f \in H$ and the projection measure P is uniquely defined by the collection of positive numeric measures μ_f ($f \in H$) and the above equality.

Let $Q : \mathcal{R} \to \mathcal{P}$ be a projection measure such that

$$f \cdot Q([a,b[)f = \|Q([a,b[)f\|^2 = \|P([a,b[)f\|^2 = f \cdot P([a,b[)f$$

for $f \in H$ and $-\infty < a \le b < \infty$. Then, in view of the polarization identity,

$$g \cdot Q([a,b[)f = g \cdot P([a,b[)f \quad (g \in H)$$

and therefore

$$Q([a,b[)f = P([a,b[)f \quad (f \in H),$$
$$Q([a,b[) = P([a,b[) \quad (-\infty < a \le b < \infty).$$

We conclude that $Q = P$. This completes the proof of the uniqueness of a spectral measure. $\square$

Example 1. Consider a compact Hermitian operator $T : H \to H$, the sequences of its eigenvalues $\lambda_n \ne 0$ and projections P_n onto the corresponding eigensubspaces $H_n = \mathrm{Ker}(\lambda_n I - T)$ ($n = 1, 2, \ldots$), and the projection P_0 onto the kernel $\mathrm{Ker}\,T$ of T. Put $\lambda_0 = 0$.

Choose a projection measure $P : \mathcal{R} \to \mathcal{P}$ such that

$$P([a,b[) = \sum_{a\le\lambda_n\le b} P_n \quad (-\infty < a \le b < \infty),$$

where $n = 1, 2, \ldots$ and the sum is strong. Applying the Hilbert–Schmidt theorem, we obtain

$$E(x)f = \sum \lambda_n P_n f = Tf \quad (f \in H),$$

which implies that

$$T = \sum \lambda_n P_n = \int x \, dP$$

and P is the spectral measure of T.

Note that $\lambda_n \to 0$ for $\dim H = \infty$, which follows from the Riesz–Schauder theorem. (Otherwise, the sequence of eigenvalues $\lambda_n \neq 0$ would have a limit $\lambda \neq 0$ or one of the corresponding eigensubspaces would be infinite-dimensional.)

Example 2. Let $H = \mathcal{L}^2(\mathbb{R}, dx)$, $A = \{ f \in H : \int |xf|^2 \, dx < \infty \}$, and T is the operator of multiplication by the independent variable x (the identity function on $\mathbb{R}$), i.e.,

$$Tf = x \cdot f \quad (f \in A).$$

Take a projection measure $P : \mathcal{R} \to \mathcal{P}$ such that

$$P(X)f = X \cdot f \quad (X \in \mathcal{R}, \ f \in A).$$

Repeating the arguments of Example 2 from Subsection 4.9.2, part 2, it is easy to verify that

$$E(x)f = x \cdot f \quad (f \in A).$$

Hence

$$T = \int x \, dP$$

and P is the spectral measure of T.

It follows from the spectral theorem that any self-adjoint operator T is unitarily equivalent to the multiplication operator M:

$$T = U^{-1}MU,$$

where U is a unitary operator. This operator also relates the spectral measures T and M (see Theorem 7.18 in Weidmann, 1980).

Remark. Take the complex plane $\mathbb{C}$, the algebra $\mathcal{R}$ of simple sets in $\mathbb{C}$ generated by rectangles $[a, b[\times[c, d[$ $(-\infty < a \le b, c \le d < \infty)$, the algebra $\mathcal{B}$ of bounded linear operators on H, and a projection measure $P : \mathcal{R} \to \mathcal{B}$. It is possible to define the integral with respect to an operator measure for functions of a complex variable by analogy with the integral with respect to a projection measure and prove the spectral theorem for normal operators (Weidmann, 1980, Theorem 32).

4.9.5. Operator functions

The spectral theorem and integration of measurable functions with respect to projection measures make it possible to define functions of self-adjoint operators.

1. Let $T : H \to H$ be a self-adjoint operator, $P : \mathcal{R} \to \mathcal{P}$ be its spectral measure, and $u : \mathbb{R} \to \mathbb{C}$ be a function measurable with respect to the measure P. Then, by definition,

$$u(T) = \int u \, dP.$$

In particular, this equality holds for any Borel function u.

Example 1. Let

$$u(x) = \sum_{0 \le m \le n} a_m x^m \quad (a_m \in \mathbb{C}).$$

Using the property of the integral to be multiplicative with respect to a projection measure, it is easy to verify that

$$u(T) = \sum_{0 \le m \le n} a_m T^m \quad (T^0 = 1).$$

Example 2. For any natural number $n \ge 1$ and any positive self-adjoint operator T, there exists a unique positive self-adjoint operator $T^{1/n}$ such that $(T^{1/n})^n = T$. The operator $T^{1/n}$ is given by the formula

$$T^{1/n} = \int x^{1/n} \, dP,$$

where P is the spectral measure of the operator T. If T is compact, then so is $T^{1/n}$.

Since the operator T is positive and self-adjoint, its spectrum $\Sigma(T)$ is included in the positive semi-axis of $\mathbb{R}$. Therefore, the spectral measure P of the operator T vanishes on subsets of the negative semi-axis of $\mathbb{R}$ and the values of the integrable function $x^{1/n}$ on this semi-axis can be disregarded and chosen in an arbitrary manner.

2. The spectral theorem for normal operators can be used to prove that for any natural number $n \geq 1$ and any normal operator T there is a normal operator $T^{1/n}$ such that $(T^{1/n})^n = T$.

In some cases, the spectral measure of the operator T is designated by E (as is the integral with respect to it), so we write

$$T = \int \lambda \, dE(\lambda), \quad u(T) = \int u(\lambda) \, dE(\lambda).$$

In particular, a fractional power of a positive self-adjoint operator is written as follows:

$$T^{m/n} = \int \lambda^{m/n} \, dE(\lambda).$$

The integral representation is also used to define fractional powers T^α ($0 < \alpha < 1$) of operators of a more general class (see Kato, 1995, Chapter 5, Section 3).

3. A detailed description of analytic operator functions and the connection between operator functions and spectral theory is given in Dunford and Schwartz (1988), Chapter 7, Section 3.

Consider a complex Banach space E ($\neq \{0\}$, as usual), a bounded linear operator $T : E \to E$, an open set $U \supseteq \Sigma(T)$ with positively oriented piecewise-straightenable boundary $\Gamma = \partial U$, and a complex function f that is analytic in a neighborhood V of the closure $\bar{U}$. The operator function f is defined as follows:

$$f(T) = (2\pi i)^{-1} \int_\Gamma f(\lambda) R(\lambda, T) \, d\lambda.$$

Since the resolvent $R(\cdot, T)$ is an analytic function, the Cauchy integral formula implies that the value $f(T)$ does not depend on the choice of U.

Let a and b be complex numbers and f and g be operator functions defined by the given equality. Then the operator functions $af + bg$ and fg are defined as follows:

$$(af + bg)(T) = af(T) + bg(T), \quad (fg)(T) = f(T)g(T).$$

The composition of functions f and g is defined as

$$(g \circ f)(T) = g(f(T)).$$

Using functional calculus for operator functions, it is easy to prove the theorem on the maps of spectra

$$f(\Sigma(T)) = \Sigma(f(T)).$$

Exercise. Prove this theorem (see Dunford and Schwartz, 1988, Chapter 7, Section 3, part 11).

Remark. Functional calculus for analytic functions is also described in detail in Clément, Heijmans, Angenent, van Duijn, and de Pagter (1987), Chapter 5. The exponential function is pointed out and some estimates are given for it. There are examples and exercises. In Reed and Simon (1978), VII.1, functional calculus for continuous functions and Hermitian operators is considered. Instead of the integral representation, the Weierstrass theorem on approximating continuous functions with polynomials is used.

4.10. OPERATOR EXPONENTIAL

Among all operator functions, we point out exponential operator functions. Using the spectral theorem, they can be defined as the integrals of conventional exponential functions. Operator exponentials are used to great effect in solving differential equations with operator coefficients. The answers to the most essential questions related to the operator exponential can be found in Hille and Phillips (1974).

4.10.1. Problem formulation

1. Consider the complex exponential function $f : [0, \infty[\to \mathbb{C}$ with $a \in \mathbb{C}$ as the coefficient of the exponent, i.e.,

$$f(t) = e^{at} \quad (t \geq 0).$$

The function f is a continuous homomorphism from the additive semigroup of positive numbers onto the multiplicative semigroup of certain complex numbers:

$$f(s + t) = e^{a(s+t)} = e^{as}e^{at} = f(s) \cdot f(t)$$

for any $s, t \geq 0$. In particular, for $a = 2\pi i$ the exponential function maps the interval $[0, \infty[$ onto the unit circle in the plane $\mathbb{C}$.

Every value $f(t)$ of the exponential function f defines a linear transformation $U(t)$ of the plane $\mathbb{C}$ that maps a point $z \in \mathbb{C}$ to the point

$$U(t)z = e^{at}z.$$

Similarly to the values $f(t)$, the transformations $U(t)$ form a multiplicative semigroup:

$$U(s + t) = U(s) \cdot U(t)$$

for any $s, t \geq 0$. Thus, the exponential function can be associated with the semigroup of linear transformations.

If $a = i\alpha$ ($\alpha \in \mathbb{R}$), then

$$U(0) = I, \quad U^*(t) = U(-t) = U^{-1}(t) \quad (t \in \mathbb{R}),$$

the transformations $U(t)$ are unitary, and they make up a group.

2. In addition, note that

$$e^{-at} = \lim_{n \to \infty} \left[\left(1 + \frac{t}{n} a \right)^{-n} \right].$$

The coefficient a of the exponent is equal to its derivative $f'(0)$ at the point 0:

$$a = \lim_{t \downarrow 0} [t^{-1}(e^{at} - 1)].$$

On the other hand, the exponential function f is the only solution to the differential equation

$$\dot{f} = a \cdot f$$

that satisfies the condition $f(0) = 1$.

The exponential function is analytic:

$$e^{at} = \sum (n!)^{-1}(at)^n \quad (n \geq 0).$$

This series is absolutely summable:

$$\sum (n!)^{-1} |at|^n < \infty.$$

3. Problem. *Define the exponential function with operator coefficient whose properties are analogous to those of the main properties of the numeric exponential function.*

The required definition can be easily produced for a bounded operator T in a Banach space X using operator analytic functions:

$$e^{tT} = \sum (n!)^{-1}(tT)^n \quad (n \geq 0).$$

Since $\sum (n!)^{-1}\|tT\|^n < \infty$, it follows that the series on the right-hand side defines a bounded linear operator on X. If the operator T is unbounded, the given definition is not valid, which means that some other properties must be used to define the operator exponential in this case.

4.10.2. Semigroups of operators

1. It appears to be most natural to define the operator exponential as the limit of a special sequence of operators, by analogy with the equality of Subsection 4.10.1, part 2. It is only required to establish the existence of such a limit for operators of a sufficiently large class. Note that negative powers in a chosen sequence make it possible to use the boundedness of the value of the resolvent for bounded operators. It can be proved that that if T is a densely defined closed operator in a Banach space X, the negative semi-axis is included in the resolvent set $P(-T)$ of the operator $-T$, and $\|(\lambda I + T)^{-1}\| \leq \lambda^{-1}$ for $\lambda > 0$, then there exists a limit

$$U(t) = \lim_{n \to \infty} \left[\left(I + \frac{t}{n}T \right)^{-n} \right]$$

in the strong sense for all $t \geq 0$ (see Kato, 1995, Chapter 9, Section 1). The operators $U(t)$ defined by this equality form a semigroup: $U(s+t) = U(s) \cdot U(t)$ for any $s, t \geq 0$.

This semigroup is contracting, i.e., $\|U(t)\| \leq 1$ The function U with values $U(t)$ is strongly continuous. It is called the *exponential with coefficient* $-T$. Its values are written in the usual form

$$U(t) = e^{-tT} \quad (t \geq 0).$$

Note that $U(0) = I$.
   For all $x \in \mathrm{Dom}\, T$,

$$-Tx = \lim_{t \downarrow 0} [t^{-1}(e^{-tT} - I)x].$$

In view of this equality, the operator T is called the *generating operator* for the semigroup $(U(t))$. Different generating operators generate different semigroups. The proofs of the mentioned properties of operator exponentials can be found in Kato (1995), Chapter 9, Section 1. A list of other characteristic formulas is given in Hille and Phillips (1974), Section 11.8.

2. Semigroups of operators have been extensively covered in the literature (Dunford and Schwartz, 1988, Chapter 8; Kato, 1995, Chapter 9; Hille and Phillips, 1974, Chapters 10–16; Reed and Simon, 1978, X.8, 9; Balakrishnan, 1981, Chapter 4; Clément, Heijmans, Angenent, van Duijn, and de Pagter, 1987). The main attention is paid to the theorems on the conditions for a given operator to generate a semigroup of the required class and to the solution of the *abstract Cauchy problem*: for a closed operator T, find a vector function u on $[0, \infty[$ such that $u(t) \in \mathrm{Dom}\, T$ $(t \in [0, \infty[)$ and u is a solution to the equation $\dot{u} = Tu$ with a given value $u(0)$.

Let X be a Banach space. A collection U of bounded linear operators $U(t) : X \to X$ $(t \in [0, \infty[)$ is said to form a *strongly continuous semigroup* if

   (1) $U(0) = I$,
   (2) $U(s + t) = U(s)U(t)$ for all $s, t \in [0, \infty[$,
   (3) $U(t)x \to U(s)x$ as $t \to s$ for all $x \in X$.

In addition, if $\|U(t)\| \leq \gamma$ $(t \geq 0)$ for some $\gamma > 0$, then the semigroup U is said to be *uniformly bounded*. For $\gamma < 1$, U is called *strongly contracting*. Many theorems on strongly continuous semigroups are simple generalizations of analogous theorems for contracting semigroups.

Exercise. Prove that the third condition in the above definition can be relaxed by replacing it with the following one: $(3')$ $U(t)x \to x$ as $t \downarrow 0$ for all $x \in X$, i.e., $(1), (2), (3') \Rightarrow (1), (2), (3)$ (Balakrishnan (1981), Section 4.1).

Example. Let $X = \mathcal{C}([0,1], \mathbb{R})$ and $f(1) = 0$ for $x \in X$. We define U as follows: $U(t)x = y$, where $y(s) = x(s + t)$ for $s + t \in [0, 1]$ and $y(s) = 0$ for $s + t \notin [0, 1]$. The semigroup U is strongly continuous.

Exercise. Find a generating operator T of the semigroup U in the preceding example. Find the spectrum of T.

For operators $U(t)$ that form the strongly continuous semigroup U, there exist $\alpha \in \mathbb{R}$ and $\gamma > 0$ such that

$$\|U(t)\| \leq \gamma e^{\alpha t}$$

(see Yosida, 1965, Chapter 9, Section 1). Therefore, by compensating the exponential growth, U can be turned into the contracting semigroup of operators V with $V(t) = \gamma^{-1}e^{-\alpha t}$. The resolvent set of the generating operator of the semigroup includes the half-plane $\operatorname{Re}\lambda \geq 0$ (Balakrishnan, 1981, Section 4.2).

Remark. The characteristic features of strongly continuous semigroups of operators in real and complex Banach spaces are described in detail in Clément, Heijmans, Angenent, van Duijn, and de Pagter (1987), Section 3.7.

3. If a strongly continuous semigroups U is continuous in the topology generated by the operator norm, then U is said to be *continuous in the uniform topology.*

The semigroup U with $U(t) = e^{-tT}$ $(t \geq 0)$ generated by a bounded operator $T : X \to X$ is continuous in the uniform topology.

Exercise. Prove the preceding statement.

The converse is also true: if U is continuous in the uniform topology, then it is generated by the bounded operator $T = \lim_{t\downarrow 0} t^{-1}(U(t)-I)$. Thus, a semigroup is continuous in the uniform topology if and only if it is generated by a bounded operator (Dunford and Schwartz, 1988, Chapter 8, Section 1).

4. Consider a subspace A that is dense in a Banach space X and a closed linear operator $T : A \to X$ with resolvent set $P(T)$ and resolvent $R(\cdot, T)$. The resolvent set $P(T)$ includes the half-plane $\operatorname{Re}\lambda > \alpha = \inf\{t^{-1}\log\|U(t)\| : t > 0\}$ if T generates a strongly continuous semigroup of operators $U(t)$ (Balakrishnan, 1981, Section 4.1). If the semigroup is uniformly continuous, then $\alpha = 0$ (Balakrishnan (1981), Section 4.2).

The following theorem characterizes closed generating densely defined operators.

The Hille–Yosida theorem. *An operator T generates a strongly continuous semigroup of operators $U(t) = e^{-tT}$ if and only if there are $\alpha \in \mathbb{R}$ and $\gamma > 0$ such that $]\alpha, \infty[\subseteq P(T)$ and*

$$\|R(\lambda, T)^n\| \leq \gamma(\lambda - \alpha)^{-n} \quad (\lambda > \alpha,\ n \in \mathbb{N}).$$

A proof of this theorem and the following corollary can be found in Dunford and Schwartz (1988), VIII.1.

Corollary. *An operator T generates a strongly continuous semigroup of operators $U(t)$ with norms $\|U(t)\| \le e^{\alpha t}$ ($t \ge 0$) for some $\alpha \in \mathbb{R}$ if and only if $]\alpha, \infty[\subseteq P(T)$ and*

$$\|R(\lambda, T)\| \le (\lambda - \alpha)^{-1} \quad (\lambda > \alpha).$$

For $\alpha = 0$, we obtain a criterion for operators generating contracting semigroups:

$$]a, \infty[\subseteq P(T), \quad \|R(\lambda, T)\| \le \lambda^{-1} \quad (\lambda > 0).$$

A proof of the Hille–Yosida theorem is given also in Hille and Phillips (1974), Section 12.3. Some refined and analogous statements for other classes of semigroups are proved there. The generalizations of the theorem to the case of locally convex spaces are described in Yosida (1965), IX.7. A special case of semigroups of compact operators and Hilbert–Schmidt operators is considered in Balakrishnan (1981), Section 4.4. Section 4.6 contains a detailed description of elementary examples of semigroups of operators arising in mathematical physics.

It follows from the definitions that the strongly continuous semigroup U generated by an operator T determines the solution of the Cauchy problem for the equation $\dot{u} = Tu$ so that the problem is well-posed. Therefore, the proof of the well-posedness of this problem is reduced to verifying whether the assumptions of the Hille–Yosida theorem hold. An interesting example of the verification the well-posedness of the system of Maxwell's equations in vacuum is given in Richtmyer (1978, 1981), Chapter 16.

5. It follows from the Hahn–Banach theorem that for any vector $x \in X$ there exists a linear functional $\varphi \in X'$ such that $\|\varphi\| = \|x\|$ and $\varphi(x) = \|x\|^2$. Such a functional is called a *normalized tangent functional* for x. A densely defined operator $T : A \to X$ is called *accretive* if $\operatorname{Re} \varphi(Tx) \ge 0$ for all $x \in A$ and some normalized tangent functional φ for x. In addition, if T has no proper accretive extensions, then it is called a *maximal accretive* (or *m-accretive*) operator. As is proved in Kato (1995), Chapter 5, Section 11, there exists an m-accretive square root of any accretive operator.

For any accretive operator T, the opposite operator $-T$ is called *dissipative*. Dissipative operators are described in detail in Clément, Heijmans, Angenent, van Duijn, and de Pagter (1987), Section 3.3. The notion of dissipativity is generalized to the case of operators in multinormed spaces.

Exercise. Prove that any accretive operator can be extended to a closed accretive operator (Reed and Simon, 1978, Chapter 10, Problem 52; Clément, Heijmans, Angenent, van Duijn, and de Pagter, 1987, Section 3.3)

Proposition. *A closed operator* $T : A \to X$ *generates a contracting semigroup if and only if it is accretive and there is an* $\alpha > 0$ *such that* $\mathrm{Ran}(\alpha I + T) = X$.

A proof of this proposition and the following corollary can be found in Reed and Simon (1978), X.8.

Corollary. *If a closed operator* T *and the conjugate operator* T' *are accretive, then* T *generates a contracting semigroup.*

Examples of application of accretive operators and contracting semigroups in solving the heat conduction equation are given in Reed and Simon (1978), Chapter X, Section 8.

6. Semigroups of operators in *ordered Banach spaces* have important special properties. These spaces are described in detail in Kantorovich and Akilov (1982), Chapter 10, and Clément, Heijmans, Angenent, van Duijn, and de Pagter (1987), Appendix 2. The order on these spaces is consistent with the operations and the norm (*the set of all positive elements is closed*).

Let E be an ordered Banach space with the cone of positive vectors $P = \{x \in E : x \geq 0\}$. By consistency, $P + P \subseteq P$ and $\lambda P \subseteq P$ for $\lambda \geq 0$. Moreover, $\bar{P} = P$. A strongly continuous semigroup U of positive operators $U(t)$ such that $U(t)P \subseteq P$ $(t \geq 0)$ is said to be *positive*. A strongly continuous semigroup U generated by an operator T is positive if and only if there is a β such that the resolvent $R(\lambda, T)$ is a positive operator for all $\lambda > \beta \geq \alpha$ and $\alpha = \inf\{t^{-1} \log \|U(t)\| : t > 0\}$ (Clément, Heijmans, Angenent, van Duijn, and de Pagter, 1987, Section 7.1).

The number $s(T) = \sup\{\mathrm{Re}\,\lambda : \lambda \in \Sigma(T)\}$ is called the *spectral bound* of the semigroup U (if $\Sigma(T) = 0$, then $s(T) = -\infty$). Under some additional assumptions on the set P, for a positive semigroup U generated by T we have $s(T) = \sup\{\lambda \in \mathbb{R} : \lambda \in \Sigma(T)\}$ and

$$R(\mu, T)x = \int_0^\infty e^{-\mu t} U(t)x\, dt \quad (\mathrm{Re}\,\mu > s(T),\ x \in E).$$

Hence, $s(T) \in \Sigma(T)\}$ when $\Sigma(T) \neq \varnothing$ (Clément, Heijmans, Angenent, van Duijn, and de Pagter, 1987, Section 7.1).

Any semiadditive positive homogeneous functional $p : E \to \mathbb{R}$ such that $\|x\|_p = p(x) \vee p(-x)$ $(x \in E)$ defines the norm $\| \ \|_p$ on E. The inequality $|p(x)| \le c\|x\|_p$ which holds for some $c > 0$ and all $x \in E$ ensures that p is continuous. An operator $T : A \to E$ on a subspace A of a real Banach space E such that $p(x) \le p(x - \lambda TX)$ $(\lambda \ge 0, x \in A)$, is called p-*dissipative*. An operator $S : E \to E$ such that $p(Sx) \le p(x)$ $(x \in E)$ is called a p-*contraction*. The strongly continuous semigroup U formed by p-contracting operators $U(t)$ is said to be p-*contracting*.

Proposition. *A strongly continuous semigroup is p-contracting if and only if its generating operator is p-dissipative.*

Under some additional assumptions on the cone P of the ordered Banach space E and the functional p, every p-contraction is a positive operator and the operator T generating a p-contracting semigroup is densely defined, p-dissipative, and there is an $\alpha > 0$ such that $\mathrm{Ran}(\alpha I - T) = E$. Moreover, a theorem holds that is analogous to the Hille–Yosida theorem. The exact formulations and proofs of these statements can be found in Clément, Heijmans, Angenent, van Duijn, and de Pagter (1987), Section 7.2.

The generating operator for a positive semigroup of operators in a σ-complete Banach lattice satisfies the abstract Kato inequality (Clément, Heijmans, Angenent, van Duijn, and de Pagter, 1987, Section 7.4).

4.10.3. The Laplace transform

The result of the Laplace transform of the exponential function with exponent coefficient $-T$ is the resolvent of this operator.

Assume that $f : \mathbb{R} \to \mathbb{C}$ be a Borel function and there exist $c > 0$ and $\alpha \in \mathbb{R}$ such that

$$|f(t)| \le ce^{\alpha t} \ (t \ge 0), \quad f(t) = 0 \ (t < 0).$$

Note that, for any $z \in \mathbb{C}$ with $x = \mathrm{Re}\, z > \alpha$, the function $g : \mathbb{R} \to \mathbb{C}$ with values

$$g(t) = e^{-tx} f(t) \quad (t \in \mathbb{R})$$

is integrable with respect to the Lebesgue measure dt because $x - \alpha > 0$ and

$$|g(t)| \le ce^{-t(x-\alpha)} \ (t \ge 0), \quad g(t) = 0 \ (t < 0).$$

4.10.5. Evolution equations

Using Stone's theorem, it is easy to solve the operator differential equation

$$\dot{U} = iT \cdot U \tag{1}$$

with generating operator iT under the condition $U(0) = I$. The only solution satisfying the specified conditions is the exponential function

$$U(t) = e^{itT} \quad (t \in \mathbb{R}).$$

1. Indeed, from the definition and the rules of integration with respect to a projection measure, it follows that

$$\dot{U}(t) = \lim_{\Delta \to 0} (\Delta^{-1}[U(t + \Delta) - U(t)]) = \lim_{\Delta \to 0} E(\Delta^{-1}[e^{i(t+\Delta)x} - e^{itx}])$$

$$= E(ixe^{itx}) = E(ix) \cdot E(e^{itx}) = iT \cdot U(t) \quad (t \in \mathbb{R}).$$

In addition, $U(0) = I$.

Let V be a solution to equation (1) such that $V(0) = I$. Then $W = U - V$ is also a solution to (1) and $W(0) = O$. Take an $f \in H$ and consider the function $\varphi : \mathbb{R} \to \mathbb{R}$ with values

$$\varphi(t) = \|W(t)f\|^2 = W(t)f \cdot W(t)f \quad (t \in \mathbb{R}).$$

We have

$$\Delta^{-1}[\varphi(t + \Delta) - \varphi(t)]$$
$$= W(t + \Delta)f \cdot \Delta^{-1}[W(t + \Delta) - W(t)]f + \Delta^{-1}[W(t + \Delta) - W(t)]f \cdot W(t)f$$

for $\Delta \neq 0$. Therefore,

$$\dot{\varphi} = 2 \operatorname{Re}[i(g \cdot Tg)], \quad g = W(t)f \quad (t \in \mathbb{R}).$$

Since the operator T is self-adjoint, the product $g \cdot Tg$ is real and $\dot{\varphi}$ is identically equal to zero. Hence, φ is constant. By assumption, $\varphi(0) = 0$. As a result, $\varphi(t) = 0$ and $U(t) - V(t) = W(t) = O$ for any $t \in \mathbb{R}$. We conclude that the solution specified above is unique.

Since $U(A) \subseteq A$, the right-hand side of equation (1) is defined on $A = \operatorname{Dom} T$. Let $f \in A$ and $u(t) = U(t)f$. Then the operator equation (1) is equivalent to the vector equation

$$\dot{u}(t) = iT \cdot u(t), \tag{2}$$

and the condition $U(0) = I$ is equivalent to the condition $u(0) = f$.

Example 1. Let $H = \mathcal{L}^2(\mathbb{R}, dx)$ and let T be the operator of multiplication by the independent variable x, i.e.,

$$Tf = x \cdot f \quad (f, xf \in H).$$

It is easy to verify that

$$e^{itT} f = e^{itx} f \quad (t \in \mathbb{R}, \ f \in H).$$

The exponential e^{itT} is a multiplication operator.

Example 2. Let $H = \mathcal{L}^2(\mathbb{R}, dx)$ and let T be the differential operator, i.e.,

$$Tf = i^{-1} f' \quad (f, f' \in H).$$

It is easy to verify that

$$e^{itT} f(x) = f(x + t) \quad (t, x \in \mathbb{R}, \ f \in H).$$

The exponential e^{itT} is a shift operator.

Exercise. Verify the equalities in Examples 1 and 2.

2. Operator exponentials can be used to solve the Cauchy problem for a large class of evolution equations

$$\dot{U}(t) = T(t) \cdot U(t) \tag{3}$$

with variable operator coefficients $T(t)$ (see Yosida, 1965, Chapter 14; Balakrishnan, 1981, Section 4.12).

Consider a Banach space Y, the interval $[0, 1] \subseteq \mathbb{R}$, a vector subspace A that is dense in Y, and a collection of closed linear operators $T(t) : A \to Y$ ($t \in [0, 1]$). We now rewrite equation (3) in the vector form:

$$\dot{u}(t) = T(t) \cdot u(t) \quad (0 \le t \le 1), \tag{4}$$

where $u(t) = U(t) \cdot f$ for a vector f in the common domain of definition of the operators on the right-hand side of equation (3). In some cases, a solution of equation (4) is sought only for $t \ne 0$. This makes it possible to find solutions that satisfy the condition $u(0) = f$ for any $f \in Y$ (not necessarily from the domain of definition of the operators in question).

We will formulate a theorem on the solution of the Cauchy problem for equation (4).

Let $f \in Y$, $T(0) = T$, and $\mathcal{R}(t) = \mathcal{R}(T(t))$, $R(\lambda, t) = R(\lambda, T(t))$ ($\lambda \in \mathcal{R}(T(t))$). The sum F of iterated kernels K_n ($n = 1, 2, \ldots$) is defined as follows:

$$K_1(t, r) = [T(t) - T(r)]e^{(t-r)T(r)} \ (r \leq t), \quad K_1(t, r) = 0 \ (t < r);$$

$$K_n(t, r) = \int_0^1 K_1(t, s) \cdot K_{n-1}(s, r) \, ds \quad (n \geq 2);$$

$$F(t, s) = \sum K_n(t, s) \quad (n \geq 1).$$

We now make some additional assumptions.

1) There is a $\theta > 2^{-1}\pi$ such that $Z = \{0\} \cup \{z : |\arg z| < \theta\} \subseteq \mathcal{R}(t)$ for any $t \in [0, 1]$.

2) $R(z, \cdot)$ is strongly uniformly continuous with respect to z on every compact set included in the sector Z, and there exist $\alpha > 0$ and $\beta > 0$ such that

$$\|R(z, t)\| \leq \alpha(|z| - \beta)^{-1} \quad (z \in Z, \ |z| > \beta, \ t \in [0, 1]).$$

3) There is a $\gamma > 0$ such that for all $r, s, t \in [0, 1]$ we have

$$\|T(t) \cdot T^{-1}(r) - T(s) \cdot T^{-1}(r)\| \leq \gamma|t - s|.$$

Note that the condition $0 \in \mathcal{R}(t)$ ensures the existence of the inverse operator $T^{-1}(r) = R(0, r) : Y \to A$. The closed graph theorem implies that the operator $S(t, r) = T(t) \cdot T^{-1}(r) : Y \to Y$ is closed. The third assumption means that the function $S(t, \cdot)$ satisfies the Lipschitz condition uniformly with respect to r on $[0, 1]$.

Theorem. *Under the assumptions made above, the equality*

$$u(t) = e^{tT}f + \int_0^t e^{(t-s)T(s)} F(s, 0)f \, ds \tag{5}$$

provides the unique solution to equation (4) for $0 < t \leq 1$ such that $u(0) = f$.

A detailed proof of this theorem is given in Yosida (1965), Chapter 14, Section 5. The proof is based on the Fredholm method and consists in justifying the transition from the differential equation

$$\dot{u}(t) = T(t) \cdot u(t) = T \cdot u(t) + (T(t) - T) \cdot u(t)$$

to the integral equation

$$u(t) = e^{tT}f + \int\limits_0^t e^{(t-s)T}(T(s) - T)u(s)\,ds.$$

This equation is solved using the method of successive approximations to obtain equality (5).

3. Various formulations and methods of solution of the abstract Cauchy problem are presented in Hille and Phillips (1974), Chapter 23, Section 3, and Monna (1970).

A detailed analysis of stochastic evolution differential and integral equations can be found in Clément, Heijmans, Angenent, van Duijn, and de Pagter (1987), Chapter 10. In particular, the direct and inverse Cauchy problems for parabolic Ito equations in Sobolev spaces are solved in Section 4.3.

The questions of the well-posedness of the Cauchy problem and other boundary value problems for operator equations are discussed in detail in Trenogin (1993). Also, various regularization methods are considered and weak well-posedness is analyzed there.

In Balakrishnan (1981), Section 4.8, the nonhomogeneous equation $\dot{u} = Tu + v$ is considered, as well as its homogeneous form. The Cauchy problem is solved in the case where $u(0) \notin \mathrm{Dom}\,T$ (Theorem 4.8.3 and its corollary). The criterion for the system described by this equation to be controllable is formulated in Section 4.9 of the mentioned book. Evolution equations with a variable operator coefficient in Hilbert spaces are considered in Section 4.12. Compact semigroups and Hilbert–Schmidt semigroups that define such coefficients are pointed out.

The *uniform well-posedness* of the abstract Cauchy problem for the *implicit* operator equation $B\dot{u}(t) = Au(t)$ in a Banach space is considered in Zaidman (1999), Chapter 4. In Chapter 5 of the same book, a detailed proof of the theorem on the solution of the Cauchy problem for the equation $\dot{u} = Tu + v$ is proved and the corresponding integral formula is obtained. Chapter 9 contains the proof of the theorem on the uniqueness of weak solutions to the Cauchy problem for the equation $\ddot{u} = Tu$ with symmetric densely defined operator T on a Hilbert space.

Well-posed formulations of a Cauchy problem related to physics are described in Richtmyer (1981), Chapter 16. The Hille–Yosida theorem is used there to prove the well-posedness of the operator form of Maxwell's equations in vacuum and the equation of neutron transport in a slab.

A number of nonstandard methods for solving some equations of mathematical physics are described in Albeverio, Høegh-Krohn, Fenstad, and Lindstrøm (1986). In Chapter 1, a detailed description of an example for the Van Der Pol's equation is given (Section 1.5). It is sown there that the introduction of infinitely small numbers provides a way to properly describe possible forms of the limit cycle. In Chapter 4, considerable attention is given to stochastic differential equations. In particular, a stochastic modification of the Navier-Stokes equation is considered in Section 4.7. In Chapter 6, the questions of the theory of differential operators are discussed. A fair amount of attention is given to a nonstandard approach to the Boltzmann equation (Section 6.5). Among other propositions, the existence theorem is proved in the case of space non-uniformity. In Chapter 7, stochastic evolution of some differential systems on lattices and special semigroups of operators are studied. The concluding Section 7.5 describes a connection between the quantum field theory and stochastic models for polymers. The discrete Laplace operator with Neumann boundary conditions that defines a Markov process is introduced.

Chapter 5.

Nonlinear operators

In this chapter, we prove several fixed point theorems and saddle point theorems. More general formulations of these theorems and their corollaries are given in Warga (1972), Kantorovich and Akilov (1982), Cartan (1970), and Edwards (1995). In addition, we consider nonlinear monotonic operators. They are described in detail in Gajewski, Gröger, and Zacharias (1974). A separate subsection is devoted to degree theory. An introduction to this theory can be found in Hutson and Pym (1980).

5.1.   FIXED POINTS

Fixed point theorems are used in solving operator equations.

5.1.1.   The Brouwer theorem

This theorem serves as a basis for many fixed point theorems.

1. Consider a transformation $f : E \to E$ of a set E. Any element $x \in E$ such that $f(x) = x$ is called the *fixed point* of the transformation f. Thus, by definition, fixed points are solutions to the equation $f(x) = x$. Fixed point theorems can be formulated as theorems on the existence and uniqueness of solutions to such equations. A classic example is the contraction theorem for metric spaces. We say that a set A in a topological space U possesses the *FP-property* if every continuous transformation of A has a fixed point.

Example. (The Brouwer theorem for an interval.) Let φ be a continuous transformation of the interval $E = [-1, 1]$. Suppose that $\varphi(x) \neq x$ for all $x \in E$. Let $f : E \to \mathbb{R}$ be defined as follows:

$$f(x) = (x - \varphi(x))/|x - \varphi(x)| \quad (x \in E).$$

The functions φ and f are continuous and φ maps the interval $E = [-1, 1]$ onto its boundary $\partial E = \{-1, 1\}$: if $\varphi(x) < x$, then $f(x) = -1$; if $\varphi(x) > x$, then $f(x) = 1$. This is impossible because the image of a connected set under a continuous map is also connected. Consequently, there are no continuous transformations φ of the interval $E = [-1, 1]$ without fixed points.

2. The following classic theorem holds.

The Brouwer theorem. *The closed unit ball $B = \{x : \|x\| \leq 1\}$ in the space $\mathbb{R}^n$ possesses the FP-property.*

Proving the theorem takes several stages (Dunford and Schwartz, 1988, Chapter 5, Section 12).

PRELIMINARY STAGE. We will prove the lemma on the vanishing of a special determinant. In what follows, $x_0 = t$, $x_1, \ldots, x_n$ stand for real variables, and f_i, f_{ij} ($0 \leq i, j \leq n$) denote the derivatives of a function $f : \mathbb{R} \times \mathbb{B}^n \to \mathbb{R}^n$ with respect to these variables (the primes are omitted).

Consider a smooth function $f : \mathbb{R}^{n+1} \to \mathbb{R}^n$ (with continuous second-order variables) and the determinants

$$D_i = |f_0 \ldots \tilde{f}_i \ldots f_n|, \quad C_{ij} = |f_{ij} f_0 \ldots \tilde{f}_i \ldots \tilde{f}_j \ldots f_n|$$

of the $n \times n$-matrices whose columns are formed by the partial derivatives, where the exclusion of the variables f_i and f_j is indicated by a tilde. Note that $C_{ij} = C_{ji}$.

Lemma. $\sum\limits_{1 \leq i \leq n} (-1)^i D_i' = 0.$

Proof. Differentiating D_i, we obtain $D_i' = \sum\limits_{j<i}(-1)^j C_{ij} + \sum\limits_{j>i}(-1)^{j-1}C_{ij}$.

Hence, $(-1)^i D_i' = \sum\limits_j (-1)^{i+j}\sigma_{ij}C_{ij}$, where $\sigma_{ij} = \text{sign}(i-j)$ and $S = \sum\limits_i (-1)^i D_i' = \sum\limits_i\sum\limits_j (-1)^{i+j}\sigma_{ij}C_{ij}$. Since $(-1)^{i+j} = (-1)^{j+i}$ and $\sigma_{ij} = -\sigma_{ji}$, we conclude that $S = -S$ and $S = 0$. $\square$

STAGE I. First, take a smooth transformation $\varphi : B^n \to B^n$. Assume that $\varphi(x) \neq x$ for all $x \in B^n$. Consider the quadratic equation $\|x + z \cdot (x - \varphi(x))\|^2 = 1$ or the equivalent equation

$$\|x\|^2 + 2z\langle x, x - \varphi(x)\rangle + z^2\|x - \varphi(x)\|^2 = 1.$$

Let $z = z(x)$ be the greatest root of this equation, i.e.,

$$z = \frac{\langle x, x - \varphi(x)\rangle + \sqrt{\langle x, x - \varphi(x)\rangle^2 + (1 - \|x\|^2)\|x - \varphi(x))\|^2}}{\|x - \varphi(x))\|^2}.$$

By assumption, $\varphi(x) \neq x$ for all $x \in B^n$, which implies that the discriminant is nonzero: $d = \sqrt{\langle x, \varphi(x) - x\rangle^2 + (1 - \|x\|^2)\|x - \varphi(x))\|^2} > 0$. If $\|x\| \neq 1$, then this assertion is obviously true; if $\|x\| = 1$, then $\langle x, x - \varphi(x)\rangle \neq 0$ (because otherwise we have $\varphi(x) = x$, which contradicts the assumption). Indeed, let $\sum x_i^2(x_i - \varphi(x_i))^2 = 0$, $x_j \neq 0$, $x_k = 0$. Then $\varphi(x_j) = x_j$, $\sum \varphi^2(x_j) = \sum x_j^2 = 1$, and $\sum \varphi^2(x_k) = 0$, since $\sum \varphi^2(x_i) \leq 1$, $\varphi(x_k) = 0 = x_k$, and $\varphi(x) = x$. Thus, $d \neq 0$.

Note that $z(x) > 0$ and the function $z(x)$ is smooth. Moreover, $z(x) = 0$ for $\|x\| = 1$ because otherwise the equation for z implies that $0 < 2(1 - \langle x, \varphi(x)\rangle + z\|x - \varphi(x)\|^2 = 0$, since $|\langle x, \varphi(x)\rangle| \leq \|x\| \cdot \|\varphi(x)\| \leq 1$.

Let $f : \mathbb{R} \times B^n \to \mathbb{R}^n$ be defined by the equality $f(t, x) = x + tz(x)(x - \varphi(x))$ ($t = x_0 \in \mathbb{R}; x \in B^n$). The function f is smooth because so are z and φ. For any $t = x_0 \in \mathbb{R}$ we have $f_0(t, x) = z(x)(x - \varphi(x)) = 0$ ($\|x\| = 1$), since $z(x) = 0$ for $\|x\| = 1$. Furthermore, $f(0, x) = x$, $\|f(1, x)\| = 1$ ($x \in B^n$). Consider $D_0(t, x) = |f_1(t, x) \ldots f_n(t, x)|$ and the integral $I(t) = \int_{B^n} D_0(t, x)\, dx$. Note that $D_0(t, x) \equiv 1$ and $I(0)$ is equal to the volume of the ball B^n, which implies $I(0) \neq 0$. Since $\|f(1, x)\| \equiv 1$, it follows that $D_0(1, x) \equiv 0$ and therefore $I(1) = 0$. We will show that $I'(t) \equiv 0$ and therefore the function $I(t)$ is constant, in order to arrive at a contradiction to the fact that $I(0) \neq I(1)$.

Take the derivative $I'(t) = \int_{B^n} D_0'(t, x)\, dx$. It follows from the lemma that

$$I'(t) = \sum_{1 \leq i \leq n} (-1)^{i+1} F_i(t), \quad F_i(t) = \int_{B^n} D_i'(t, x)\, dx.$$

Let $s_i = (x_1, \ldots, \tilde{x}_i, \ldots, x_n)$, $B_i^{n-1} = \{s_i : \|s_i\| \leq 1\}$, $a_i = (1 - \|s_i\|^2)^{1/2}$, and $b_i = -a_i$. We have $\|(a_i, s_i)\| = \|(b_i, s_i)\| = (1 - \|s_i\|^2 + \|s_i\|^2)^{1/2} = 1$. It follows from the general Stokes formula that, after choosing the appropriate orientations for B^n and B_i^{n-1}, the integrals $F_i(t)$ can be replaced with the differences $H_i(t) - G_i(t)$, where $G_i(t) = \int_{B_i^{n-1}} D_i(t, a_i, s_i)\, ds_i$ and $H_i(t) = \int_{B_i^{n-1}} D_i(t, b_i, s_i)\, ds_i$. Since $\|(a_i, s_i)\| = \|(b_i, s_i)\| = 1$, we have $f_0(t, a_i, s_i) = f_0(t, b_i, s_i) = 0$ and therefore $D_i(t, a_i, s_i) = D_i(t, b_i, s_i) = 0$. Hence, $F_i(t) = H_i(t) - G_i(t) = H_i(t) = G_i(t) = 0$ ($1 \leq i \leq n$, $t \in \mathbb{R}$) and $I'(t) \equiv 0$.

STAGE II. Now let $\varphi : B^n \to B^n$ be an arbitrary continuous map. By the approximation theorem (Nirenberg, 2001, Chapter 1, Section 1, Proposition 8) there exists a sequence of smooth functions $\varphi_n : B^n \to B^n$ that uniformly converges to φ. As follows from what was proved above, there exists a sequence of points $y_n \in B^n$ such that $\varphi_n(y_n) = y_n$. Since B^n is a compact set, there is a subsequence $y_{n(k)} \to y \in B^n$. Since $\lim_{k\to\infty} \varphi_{n(k)}(x) = \varphi(x)$ uniformly on B^n, it follows that $\varphi(y) = \lim_{k\to\infty} \varphi_{n(k)}(y_{n(k)}) = \lim_{k\to\infty} y_{n(k)} = y$ and $\varphi(y) = y$. This completes the proof of the Brouwer theorem.

Exercise. Prove the equalities $F_i(t) = H_i(t) - G_i(t)$.

3. In Aubin and Ekeland (1984), Chapter 2, Section 2, the Brouwer theorem is deduced from the theorem on the existence of zero of a continuous vector field on an n-dimensional ball. The proof of this theorem given there is also analytic. A geometric proof of the Brouwer theorem is given in Warga (1972), Chapter 2, Section 2.

The equivalence of the Brouwer theorem to the Fang Zi inequality is proved in Aubin (1978), Chapter 8. In Chapter 9 of the same book, the Brouwer theorem is deduced from the Knaster–Kuratowski–Mazurkiewicz lemma.

In Subsection 5.5.1, part 5, we will easily deduce the Brouwer theorem from the lemma on the degree of a map.

A proof of the Weierstrass theorem on approximation of continuous functions of several variables by Bernstein polynomials with an estimate of the error is given in Savel'ev (1997). Stone's theorem and other generalizations of the classic Weierstrass theorems are considered in Korobeinik (1992).

4. The Brouwer theorem has many corollaries. We give one of them below.

Corollary. *A topological space A homeomorphic to the ball B^n possesses the FP-property.*

Proof. Let $f : A \to A$ be a continuous function. We will prove that there exists an $x_0 \in A$ such that $f(x_0) = x_0$. Consider a homeomorphism $h : B^n \to A$ and the map $g = h^{-1} \circ f \circ h$. The map g is continuous and $g : B^n \to B^n$. By the Brouwer theorem, there exists a $y_0 \in B^n$ such that $g(y_0) = y_0$. Since $h \circ g = f \circ h$, we have $h(g(y_0)) = f(h(y_0))$. Let $h(y_0) = x_0$. Then $f(x_0) = h(y_0) = x_0 \in A$ and x_0 is a fixed point for f. □

5.1.2.   The Tikhonov theorem and the Schauder theorem

The Tikhonov theorem and the Schauder theorem are generalizations of the Brouwer theorem. The sets for which the existence of a fixed point will be proved are assumed to be nonempty.

1. The following is an essential generalization of the Brouwer theorem.

The Tikhonov theorem. *A convex compact set in a multinormed space possesses the FP-property.*

This theorem is proved in Dunford and Schwartz (1988), Chapter 5, Section 10. The proof uses a series of properties of convex sets and multinorms established previously.

The following theorem is a consequence of the Tikhonov theorem.

The Schauder theorem. *A continuous transformation of a convex closed set in a Banach space whose range is relatively compact has a fixed point.*

Proof. Let $f : A \to A$ be a continuous transformation of a convex closed set A in a Banach space E such that the closure $B = \overline{f(A)}$ of the range $f(A)$ is compact. Consider the closed convex hull C of the set B (the intersection of all closed convex sets that include B). By Mazur's theorem (Dunford and Schwartz, 1988, Chapter 5, Section 2), the set $C \subseteq A$ is compact. The restriction of f to C satisfies the assumption of the Tikhonov theorem and therefore has a fixed point. $\square$

Corollary. *A convex compact set in a Banach space possesses the FP-property.*

Proof. If we assume that the set A used in the proof of the Schauder theorem is compact rather than closed, then its image $f(A) \subseteq A$ is a relatively compact and the continuous transformation $f : A \to A$ satisfies the assumption of the theorem. $\square$

Detailed proofs of Mazur's theorem, the Tikhonov theorem, and the Schauder theorem are given in Warga (1972). Application of these theorems to solving nonlinear differential and integral equations is also discussed there. The book also deals with the integrals of functions on metric compact spaces with respect to the measure and the optimization problems related to introducing a control into equations.

A geometric proof of the Schauder theorem is given in Kantorovich and Akilov (1982), Chapter 16. Applications of this theorem to solving differential and integral equations are described.

2. We now formulate two useful fixed point theorems for maps f : $\bar{A} \to E$ from the closure $\bar{A}$ of a bounded convex open nonempty set A in a Banach space E.

Theorem 1. *If f is compact and $f(\partial A) \subseteq A$, then f has a fixed point in A.*

Proof. Since the translations of E are homeomorphisms, it can be assumed that $0 \in A$. Let $p(x) = 1$ $(x \in \bar{A})$, $p(x) = \sup\{\lambda \geq 0 : \lambda x \in A\}$ $(x \notin \bar{A})$ and $g(x) = p(x)x$ $(x \in E)$. The function $g : E \to E$ is continuous, $g(E) = \bar{A}$, and $g(x) = x$ $(x \in \bar{A})$. Take the ball $B = \bar{B}(0,r)$ containing $\bar{A} \vee f(\bar{A})$ and the restriction h of the composition $f \circ g : E \to E$ to B. This restriction is a compact transformation of B, and the Schauder theorem implies that it has a fixed point $h(b) = b \in B$. Note that if $x \notin A$, then $g(x) \in \partial A$ and, by assumption, $h(x) = f(g(x)) \in A$ and $h(x) \neq x$. As a result, we have $b \in A$.   □

In addition, assume that $0 \in A$.

Theorem 2. *If f is continuous and $\|x - f(x)\|^2 \geq \|f(x)\|^2 - \|x\|^2$ $(x \in \partial A)$, then f has a fixed point in A.*

A proof of this theorem is given in Schwartz (1966).

3. Consider a normed space E, a set $A \subseteq E$, and a map $f : A \to E$. By analogy with linear operators, a number λ and a vector $x \in A$ such that $f(x) = \lambda x$ are called an *eigenvalue* and an *eigenvector* of the operator f, respectively. The eigenvectors that correspond to the eigenvalue $\lambda = 1$ are fixed points of the operator f. A continuous operator that maps bounded sets to relatively compact sets is said to be *compact*.

Let $B = \bar{B}(0,r)$ and $S = S(0,r)$ be a closed ball and a sphere of radius $r > 0$ in the space E.

Lemma. *If a compact operator $g : B \to E$ has no eigenvectors $x \in S$ corresponding to eigenvalues $\lambda > 1$, then g has a fixed point.*

Proof. We introduce the transformation $h : B \to B$ as follows:

$$h(x) = g(x) \ (g(x) \in B), \quad h(x) = r \cdot g(x)/\|g(x)\| \ (g(x) \notin B).$$

The operator h is compact because so is g. Indeed, $h = p \circ g$, where the map $p : E \to E$ is given by the formula

$$p(y) = y \ (y \in B), \quad p(y) = r \cdot y/\|y\| \ (y \notin B).$$

Since p is continuous and g is compact, their composition is compact ($X \subseteq B$ is bounded, $Y = g(X)$ is relatively compact, $\bar{Y}$ and $p(\bar{Y})$ are compact, and $h(X) = p(Y) \subseteq p(\bar{Y})$ is relatively compact).

By the Schauder theorem, h has a fixed point $x_0 \in B$. We will prove that $g(x_0) \in B$ and therefore $g(x_0) = h(x_0) = x_0$. Suppose that $g(x_0) \notin B$ and $h(x_0) = (r/\|g(x_0)\|)g(x_0)$. Since $g(x_0) \notin B$, it follows that $\|g(x_0)\| > r$, $\lambda = \|g(x_0)\|/r > 1$, and $g(x_0) = \lambda h(x_0) = \lambda x_0$ for $\lambda > 1$. Since $\|x_0\| = \|h(x_0)\| = r \cdot \|g(x_0)\|/\|g(x_0)\| = r$ and $x_0 \in S$, we arrive at a contradiction. $\square$

Let $g : B \to E$ be a compact operator.

Corollary. *If for any vector $x \in S$ there is a functional $x^* \in E^*$ such that $x^*(x) > 0$ and $x^*(g(x)) \leq x^*(x)$, then the operator g has a fixed point.*

Proof. Let $x \in S$ and $g(x) = \lambda x$. Then $\lambda x^*(x) = x^*(g(x)) \leq x^*(x)$, and $\lambda \leq 1$ because $x^*(x) > 0$. This means that g has a fixed point. $\square$

4. The preceding lemma implies the following theorem.

The Leray–Schauder theorem. *Let $f : E \to E$ be a compact operator. If the set of eigenvectors of f corresponding to its eigenvalues $\lambda > 1$ is bounded, then f has a fixed point.*

Proof. Suppose that $\|x\| \leq c$ for $x \in E$ such that $f(x) = \lambda x$ and $\lambda > 1$. Take $r > c$ and consider the restriction g of the transformation f to the ball $B = \bar{B}(0, r)$. By the preceding lemma, g has a fixed point $x_0 \in B$. Hence $f(x_0) = g(x_0) = x_0$. $\square$

Corollary. *If f is a compact operator and the set of solutions of the collection of equations $\alpha f(x) = x$ ($0 < \alpha < 1$) is bounded, then the equation $f(x) = x$ has a solution.*

Proof. The equality $\alpha f(x) = x$ for $0 < \alpha < 1$ is equivalent to the equality $f(x) = \lambda x$ for $\lambda > 1$. The desired assertion now follows from the Leray–Schauder theorem. $\square$

Remark. Various estimating techniques are often used to find out whether the required set of solutions is bounded. This method of establishing the existence of solutions of nonlinear operator equations is used to good effect.

In Aubin (1978), a *tangent condition* is introduced for operators in Hilbert spaces. Under this condition, a series of fixed point theorems and theorems on the solution of inclusions are proved. In particular, a modification of the Leray–Schauder theorem is proved and two of its corollaries are pointed out in Chapter 9.

5. The Tikhonov theorem employs convex compact sets. An important characteristic of such sets is provided by the Krein–Milman theorem.

Consider a nonempty convex set A in a separated locally convex space E. A point $c \in A$ is called an *extreme point* of A if it is not an interior point of any line segment with endpoints in A.

Example. In the Euclidean space $E = \mathbb{R}^3$, the extreme points of a cube $C \subseteq E$ are its vertices. The extreme points of a ball $B \subseteq E$ are the points of the sphere $S = \partial B$. An open set $U \subseteq E$ has no extreme points. Any cube C and any ball B are equal to the closures of convex hulls of their extreme points.

The Krein–Milman theorem. *Any nonempty convex compact set in a separated locally convex space has extreme points and is equal to the closure of their convex hull.*

A proof of this theorem is given in Yosida (1965), Chapter 12, Section 1, and Edwards (1995). The entire Chapter 10 in Edwards (1995) is devoted to the Krein–Milman theorem and its applications. It also provides a partial inversion of the theorem: *if K is a compact set such that the closure A of its convex hull is also compact, then every extreme point of A belongs to K.* Several sections are devoted to the relationship between the Krein–Milman theorem and the Bochner and Plancherel theorems on the Fourier–Stieltjes and Fourier–Plancherel transforms.

5.2.   SADDLE POINTS

Saddle point theorems are deduced from fixed point theorems for multifunctions. The properties of multifunctions related to continuity are described in detail in Aubin and Ekeland (1984), Chapter 3, and their properties related to measurability are described in Birkhoff (1979) Chapter 1.

5.2.1.   Kakutani's theorem

Kakutani's theorem is a generalization of the Schauder theorem.

1. Consider a Banach space E, a convex compact set $B \subseteq E$, the class $\mathcal{K} = \mathcal{K}(B)$ of all nonempty convex compact subsets B, and a map $F : B \to \mathcal{K}$. The map F is called a *multitransformation* or *multioperator* on B: every point $x \in B$ is mapped to a convex compact set $Y = F(x) \subseteq B$. The set $Y = F(x)$ is called the *multi-image of a point* x, and the union $F(A) = \cup F(x)$ $(x \in A)$ is called the *multi-image of a set* $A \subseteq B$. We will also use the term *multifunction*.

Example. $E = \mathbb{R}^n$, $B = \bar{B}(0,1)$, $F(x) = \bar{B}(x/2, 1/2)$.

Any set V that is a superset of an open set that includes Y is called a neighborhood of the set Y. The definition of continuity for multifunctions is analogous to that for functions. A multifunction F is said to be *continuous at a point* x if for any neighborhood V of the image $y = F(x)$ there exists a neighborhood U of x such that $F(U) \subseteq V$. The continuity of a multifunction on a set means its continuity at every point of the set. The definition of closedness for multioperators is analogous to that for linear operators. A multioperator F is said to be *closed* if from $x_n \to a$, $y_n \in F(x_n)$, and $y_n \to b$ it follows that $b \in F(a)$ $(a, b, x_n, y_n \in B)$.

Remark. The above definitions do not use the properties of convexity or compactness. They can be generalized to larger classes of sets.

2. The following theorem is an analog of the closed graph theorem for linear operators.

Theorem. *A multioperator is continuous if and only if it is closed.*

Proof. 1) We will prove that a multioperator F is discontinuous if it is not closed. Suppose that there exist $a, d, x_n \in B$ and $y_n \in F(x_n)$ such that $x_n \to a$, $y_n \to b$, and $b \notin F(a)$. Since $F(a)$ is compact and $b \notin F(a)$, there is a neighborhood V of the set $F(a)$ and a ball $B(b, \varepsilon)$ $(\varepsilon > 0)$ such that V and $B(b, \varepsilon)$ do not intersect. Since $y_n \to b$, there is a number $n(1)$ such that $y_n \in B(b, \varepsilon)$ for all $n \geq n(1)$, which implies that $F(x_n) \not\subseteq V$ $(n \geq n(1))$. At the same time, $x_n \to a$ and for any neighborhood U of a there is a number $n(2)$ such that $x_n \in U$ for all $n \geq n(2)$. Therefore $F(u) \not\subseteq V$ for $u = x_{n(0)} \in U$ if $n(0) = n(1) \vee n(2)$. Thus, the multifunction F is discontinuous at the point a.

2) We now prove that F is continuous if it is closed. Take a point $a \in B$ and an open neighborhood V of the multi-image $F(a)$. Consider the set

$U = \{x : F(x) \subseteq V\}$. Since $F(a) \subseteq V$, it follows that $a \in U$. We will show that U is open. Consider the complement $U^c = \{x : F(x) \not\subseteq V\}$. Note that $F(x) \not\subseteq V$ if and only if $F(x) \cap V^c \neq \emptyset$. Let $x_n \in U^c$ and $x_n \to x \in B$. Take $y_n \in F(x_n) \cap V^c$. Since V is open, it follows that V^c is closed and compact, as it is closed subset of a compact set B. Consequently, there exists a subsequence $(y_{r(n)})$ of the sequence (y_n) that converges to a point $y \in V^c$. Since $x_n \to x$, we have $x_{r(n)} \to x$. If the multioperator F is closed, from $x_{r(n)} \to x$, $y_{r(n)} \in F(x_{r(n)})$, and $y_{r(n)} \to y$ it follows that $y \in F(x)$. Then $y \in F(x) \cap V^c$, $F(x) \not\subseteq V$, and $x \in U^c$. As a result, U^c is closed, U is open, and U is a neighborhood of a. For any open neighborhood V of the multi-image $F(a)$ (and,consequently, for any neighborhood of $F(a)$), there exists a neighborhood U of the point a such that $F(U) \subseteq V$. We conclude that the multioperator F is continuous.   □

3. A point x is called a *fixed point* of F if $x \in F(x)$.

Kakutani's theorem. *Any continuous multitransformation of a convex compact set in a Banach space has a fixed point.*

Proof. It is natural to reduce the assertion to the Schauder theorem by representing a multifunction F in terms of single-valued functions.

1) For any $n \in \mathbb{N}$, consider the finite collection of balls $B_{ni} = B(x_{ni}, 1/n)$ that covers B. We introduce a positive continuous function $\varphi_{ni} : B \to [0, 1]$ as follows: $\varphi_{ni}(x) = 0 \vee (1/n - \|x - x_{ni}\|)$ $(x \in B)$. Note that $\varphi_{ni}(x) > 0$ for $x \in B_{ni}$ and $\varphi_{ni} = 0$ for $x \notin B_{ni}$. Since $B \subseteq \cup_i B_{ni}$, we have $\sum_i \varphi_{ni}(x) > 0$ $(x \in B)$. We define a positive continuous function $\alpha_{ni} : B \to [0, 1]$ as follows: $\alpha_{ni}(x) = \left(\sum_j \varphi_{nj}(x) \right)^{-1} \varphi_{ni}(x)$ $(x \in B)$ It is obvious that $\sum_i \alpha_{ni}(x) = 1$ $(x \in B)$. The functions α_{ni} make up a continuous partition of unity inscribed in the cover (B_{ni}). It follows from the definitions that $\alpha_{ni}(x) = 0$ for $x \notin B_{ni}$.

2) Consider a point $y_{ni} \in F(x_{ni})$ and a function $f_n : B \to E$ such that $f_n(x) = \sum_i \alpha_{ni}(x) y_{ni}$. Since $F(x_{ni}) \subseteq B$ and B is convex, it follows that $f_n(x) \in B$. Since the functions α_{ni} are continuous, f_n is a continuous transformation of B. It follows from the Schauder theorem that there is a point $x_n \in B$ such that $f_n(x_n) = x_n$.

3) Since B is compact, there exists a subsequence $(x_{r(n)})$ of the sequence (x_n) that converges to a point $a \in B$. We will prove that $a \in F(a)$. Let $\varepsilon > 0$ and consider the open neighborhood $V(\varepsilon) = \{z \in B : d(z, F(a)) < \varepsilon\}$ of the image $F(a)$. By definition, $d(z, F(a)) = \min \|z - y\|$ $(y \in F(a))$. Since $F(a)$ is compact, the specified minimum is attained. The set $V(\varepsilon)$ is

convex. Indeed, if $z_i \in V(\varepsilon)$, then there is a $y_i \in F(a)$ such that $\|z_i - y_i\| = d(z_i, F(a)) < \varepsilon$ and

$$\left\| \sum \lambda_i z_i - \sum \lambda_i y_i \right\| \leq \sum \lambda_i \|z_i - y_i\| < \sum \lambda_i \varepsilon = \varepsilon$$

for $\lambda_i \geq 0$, $\sum \lambda_i = 1$. Since $F(a)$ is convex, $\sum \lambda_i y_i \in F(a)$. Hence, $d\left(\sum \lambda_i z_i, F(a)\right) < \varepsilon$ and $\sum \lambda_i z_i \in V(\varepsilon)$. Note that the closure $\bar{V}(\varepsilon) = \{z \in B : d(z, F(a)) \leq \varepsilon\}$ is also convex.

By assumption, the multifunction F is continuous and there exists a $\delta > 0$ such that $F(U) \subseteq V(\varepsilon)$ for $U = B(a, \delta)$. Since $x_{r(n)} \to a$, there is a number $n(1)$ such that $x_{r(n)} \in B(a, \delta/2)$ for all $n \geq n(1)$. At the same time, $x_n \in B(x_{ni}, 1/n)$ when $\alpha_{ni}(x) > 0$. Consequently,

$$\|x_{ni} - a\| \leq \|x_{ni} - x_{r(n)}\| + \|x_{r(n)} - a\| \leq 1/n + \delta/2 < \delta$$

for $n \geq n(0) \geq n(1) + 2/\delta$ and therefore $x_{ni} \in U$ and $y_{ni} \in F(x_{ni}) \subseteq V(\varepsilon) \subseteq \bar{V}(\varepsilon)$ for $n \geq n(0)$. Since $\bar{V}(\varepsilon)$ is convex and $\alpha_{ni}(x) \geq 0$, $\sum_i \alpha_{ni}(x) = 1$, and $y_{ni} \in \bar{V}(\varepsilon)$, it follows that $x_n = f(x_n) = \sum_i \alpha_{ni}(x_n) y_{ni} \in \bar{V}(\varepsilon)$ for any $\varepsilon > 0$.

Therefore, the limit point a belongs to every set $\bar{V}(\varepsilon)$, which means that $a \in F(a)$. □

Exercise. Prove that $F(a) = \cap \bar{V}(\varepsilon)$ $(\varepsilon > 0)$.

Remark. Any singleton is convex and compact. Therefore, single-valued functions satisfy the definition of multifunctions given above. Kakutani's theorem generalizes the corollary of the Schauder theorem from Subsection 5.1.2, part 1.

In von Neumann (1987), Chapter 9, a modification of Kakutani's theorem and a generalization by Fang Zi are proved for multioperators in Hilbert spaces. They are deduced from the main theorem on the existence of solutions for nonlinear inclusions.

5.2.2. Von Neumann theorem

The von Neumann theorem provides the sufficient conditions for the existence of a saddle point of a function of two vector variables.

1. Consider real Banach spaces X and Y, convex compact sets $A \subseteq X$ and $B \subseteq Y$, and a continuous function $f : A \times B \to \mathbb{R}$. By the Tikhonov theorem, the product $A \times B$ is compact because so are A and B. Moreover, $A \times B$ is convex.

Lemma 1. *The Cartesian product of two convex sets is convex.*

Proof. Let A and B be convex sets in vector spaces X and Y, and let $c = (a, b)$, $z = (x, y) \in A \times B$. By definition, $ta + (t-1)x \in A$ and $tb + (t-1)y \in B$ $(0 \leq t \leq 1)$. Then $tc + (t-1)z = (ta + (t-1)x, tb + (t-1)y) \in A \times B$.

Since $A \times B$ is compact, it follows that f is uniformly continuous. Take partial functions $f(x, \cdot) : B \to \mathbb{R}$ and $f(\cdot, y) : A \to \mathbb{R}$ $(x \in A, y \in B)$. They are defined on compact sets and are continuous because so is f. Therefore, for any $x \in A$ there exists a $y(x) \in B$ and for any $y \in B$ there exists an $x(y) \in A$ such that

$$f(x, y(x)) = \min_{y \in B} f(x, y), \quad f(x(y), y) = \max_{x \in A} f(x, y).$$

We introduce $g : A \to \mathbb{R}$ and $h : B \to \mathbb{R}$ such that

$$g(x) = f(x, y(x)), \quad h(y) = f(x(y), y).$$

It is obvious that $g(x) \leq f(x, y) \leq h(y)$ $(x \in A, y \in B)$.   □

Lemma 2. *The functions g and h are continuous.*

Proof. Since f is uniformly continuous, for any $\varepsilon > 0$ there exists a $\delta > 0$ such that

$$|h(y) - h(b)| = f(x(y), y) - f(x(b), b) \leq f(x(y), y) - f(x(b), y) \leq \varepsilon,$$

or $|h(y) - h(b)| = f(x(b), b) - f(x(y), y) \leq f(x(b), b) - f(x(b), y) \leq \varepsilon$ for all $y, b \in B$ such that $\|y - b\| \leq \delta$. The continuity of g is established in the same way.   □

Since the functions g and h are continuous and their domains of definition are the compact sets A and B, there are points $a \in A$ and $b \in B$ such that

$$\alpha = g(a) = \max_{x \in A} g(x) = \max_{x \in A} \min_{y \in B} f(x, y),$$

$$\beta = h(b) = \min_{y \in B} h(y) = \min_{y \in B} \max_{x \in A} f(x, y).$$

The numbers α and β are called the *maximin* and the *minimax* of f, respectively. Clearly, $\alpha \leq \beta$. If $\alpha = \beta$, then the point (a, b) is called a *saddle point*.

If the partial function $f(x, \cdot)$ is convex and the partial function $f(\cdot, y)$ is concave, then f is said to be *convex-concave*.

Example. Let $X = Y = \mathbb{R}^2$, $A = B = [-1, 1]$, and $f(x, y) = y^2 - x^2$ $(-1 \le x, y \le 1)$. Then $g(x) = f(x, 0) = -x^2$, $h(y) = f(0, y) = y^2$, $a = b = 0$, and $\alpha = \beta = 0$. The point $(a, b) = (0, 0)$ is a saddle point. The function f is convex-concave and its graph is a hyperbolic paraboloid.

Lemma 3. *The equalities* $\alpha = \beta = f(a, b)$ *and* $g(a) = h(b) = f(a, b)$ *are equivalent.*

Proof. We have $\alpha = \beta \Leftrightarrow g(a) = h(b)$. Since $g(x) \le f(x, y) \le h(y)$, it follows that $g(a) = h(b) \Leftrightarrow g(a) = f(a, b) = h(b)$. Consequently, there is a point $(a, b) \in A \times B$ such that $\alpha = \beta \Leftrightarrow \alpha = \beta = f(a, b)$. $\square$

Exercise. Prove that (a, b) is a saddle point for a function f if and only if $f(x, b) \le f(x, y) \le f(a, y)$ $(x \in A, \ y \in B)$.

2. The following theorem provides a sufficient condition for the existence of a saddle point.

The von Neumann theorem. *Any continuous convex-concave function* $f : A \times B \to \mathbb{R}$ *on the product of convex compact sets A and B in Banach spaces X and Y has a saddle point.*

Proof. The proof is based on Kakutani's theorem. For any $x \in A$ and $y \in B$, let

$$B(x) = \{y \in B : f(x, y) = g(x)\}, \quad A(y) = \{x \in A : f(x, y) = h(y)\}.$$

The sets $A(y)$ and $B(x)$ are nonempty convex compact sets. The fact that they are nonempty is obvious from the definition. Being closed subsets of compact sets A and B, the sets $A(y)$ and $B(x)$ are compact. Indeed, since the partial functions $f(x, \cdot)$ and $f(\cdot, y)$ are continuous, it follows that the inverse images $f^{-1}(x, \cdot)(g(x)) = B(x)$ and $f^{-1}(\cdot, y)(h(y)) = A(y)$ are closed. It remains to verify the convexity of these sets. Since $f(x, \cdot)$ is convex by assumption,

$$f(x, tb + (1 - t)y) \le tf(x, b) + (1 - t)f(x, y) \quad (0 \le t \le 1)$$

for all $b, y \in B$. If $b, y \in B(x)$, then $f(x, b) = f(x, y) = g(x)$ and therefore $f(x, tb + (1 - t)y) \le g(x)$. But

$$g(x) = \min_{y \in B} f(x, y) \le f(x, tb + (1 - t)y).$$

Consequently, $f(x, tb + (1 - t)y) = g(x)$ and $tb + (1 - t)y \in B(x)$. The concavity of $A(y)$ is proved in a similar way.

Since $A(y) \times B(x)$ belongs to the class $\mathcal{K} = \mathcal{K}(A \times B)$, we can consider the multifunction $F : A \times B \rightarrow \mathcal{K}$ with values $F(x, y) = A(y) \times B(x)$ ($x \in A$, $y \in B$). We will show that F is continuous. To this end, it suffices to show that F is closed, i.e., that $(x_n, y_n) \rightarrow (x, y)$, $(u_n, v_n) \in F(x_n, y_n)$, and $(u_n, v_n) \rightarrow (u, v)$ imply $(u, v) \in F(x, y)$. Note that $(u_n, v_n) \in F(x_n, y_n)$ implies $u_n \in A(y_n)$, $v_n \in B(x_n)$, and $f(u_n, y_n) = h(y_n)$, $f(x_n, v_n) = g(x_n)$. Since f, g, and h are continuous, taking the preceding equalities as $x_n \rightarrow x$, $y_n \rightarrow y$, $u_n \rightarrow u$, and $v_n \rightarrow v$, we obtain $f(u, y) = h(y)$, $f(x, v) = g(x)$, and $(u, v) \in A(y) \times B(x)$.

By Kakutani's theorem, the continuous multifunction F has a fixed point, i.e., there is a point $(a, b) \in A \times B$ such that $F(a, b) = (a, b)$. It follows from the definitions that $a \in A(b)$ and $b \in B(a)$, which implies $g(a) = f(a, b) = h(b)$. By Lemma 3, $\alpha = \beta = f(a, b)$. We conclude that (a, b) is a saddle point.   □

Remark. The saddle point theorems are widely used in game theory and mathematical economics. They have been extensively studied in the literature (see Aubin, 1978).

3. Consider the spaces $\mathcal{C} = \mathcal{C}(A \times B)$, $\mathcal{A} = \mathcal{C}(A)$, and $\mathcal{B} = \mathcal{C}(B)$ of continuous functions $f : A \times B \rightarrow \mathbb{R}$, $g : A \rightarrow \mathbb{R}$, and $h : B \rightarrow \mathbb{R}$ on compact sets A, B, and $A \times B$; operators $S : \mathcal{C} \rightarrow \mathcal{A}$ and $T : \mathcal{C} \rightarrow \mathcal{B}$, and the functionals $U : \mathcal{A} \rightarrow \mathbb{R}$ and $V : \mathcal{B} \rightarrow \mathbb{R}$ such that $Sf = g$, $Tf = h$, $Ug = \alpha$, and $Vh = \beta$ in the notation of part 1 of this subsection. The compositions $US : \mathcal{C} \rightarrow \mathbb{R}$ and $VT : \mathcal{C} \rightarrow \mathbb{R}$ and the difference $US - VT : \mathcal{C} \rightarrow \mathbb{R}$ are defined. Since the definition of S and T involves maximum and minimum and $U = \max$, $V = \min$, it follows that these operators and functionals are nonlinear. It follows from the definitions that functions f with saddle points are solutions to the equation

$$(US - VT)f = 0.$$

If A and B are convex compact sets in Banach spaces X and Y, then the von Neumann theorem implies that $US = VT$ and every function $f \in \mathcal{C}$ is a solution to this equation.

Remark. With continuous functions replaced by bounded ones and maxima and minima replaced by suprema and infima, the notion of a saddle point and the equation that ensures the existence of a saddle point can be generalized. In the general form, the problem is formulated as follows.

Problem. *Find conditions for a bounded function $f : A \times B \to \mathbb{R}$ on the Cartesian product $A \times B$ to satisfy the equality*

$$\sup_{x \in A} \inf_{y \in B} f(x, y) = \inf_{y \in B} \sup_{x \in A} f(x, y).$$

Functions with values in ordered sets can be considered instead of real functions.

5.3. MONOTONIC OPERATORS

Among nonlinear operators, we point out monotonic operators. They are described in detail in Gajewski, Gröger, and Zacharias (1974), Chapter 3.

5.3.1. Definition and properties

In what follows, E is a real Banach space, E' is the normed conjugate of E, $T : E \to E'$ is an arbitrary map from E to E'. We use multiplicative notation $Tx \cdot y = (T(x))(y)$ for $x, y \in E$. In the case of Hilbert spaces, E' is identified with E and the product becomes a scalar product. In some propositions, the space E is assumed to be separable and reflexive.

1. An operator T is said to be *monotonic* if

$$(Tx - Ty)(x - y) \geq 0 \quad (x, y \in E).$$

If the inequality is strict for $x \neq y$, then T is said to be *strictly monotonic*. This is consistent with the definition in the case $E = E' = \mathbb{R}$.

An operator T is said to be *bounded* if the image of every bounded set $A \subseteq E$ under T is a bounded set $TA = B' \subseteq E'$. If for every point $a \in E$ there is a neighborhood $V \subseteq E$ such that the image of V under T is a bounded set $TV = V' \subseteq E'$, then the operator T is said to be *locally bounded*. (For V we can take the open or closed ball with center a and radius $r > 0$.)

Let $T_i : E \to E'$ be monotonic operators and $a_i \in E$, $b'_i \in E'$ $(1 \leq i \leq n)$. Then the equality $Tx = \Sigma(T_i(a + x) + b'_i)$ $(x \in E)$ defines a monotonic operator T.

Exercise. Prove the preceding statement.

2. Take arbitrary $x, v \in E$ and consider the function $\varphi : [0,1] \rightarrow \mathbb{R}$ with values $\varphi(t) = T(x+tv) \cdot v \ (0 \le t \le 1)$.

Proposition 1. *The operator T is monotonic if and only if every function φ is increasing.*

Proof. Let T be monotonic and $0 \le s < t \le 1$. Then $\varphi(t) - \varphi(s) = T(x+tv) \cdot v - T(x+sv) \cdot v = (t-s)^{-1}(T(x+tv) - T(x+sv))(x+tv-x+sv) \ge 0$. If every φ is increasing, then $(Ty-Tx)(y-x) = \varphi(1) - \varphi(0) \ge 0$ for $v = y-x$. $\square$

Proposition 2. *A weakly continuously differentiable operator T is monotonic if and only if $T'(x)v \cdot v \ge 0$ for all $x, v \in E$.*

Proof. Let T be monotonic and $0 \le s < 1$. Then, by the mean-value theorem for integrals, there is a $r \in [0, s]$ such that

$$0 \le (T(x+sv) - Tx) \cdot sv = \int_0^s T'(x_t v)v \cdot sv \, dt = s^2 T'(x+rv)v \cdot v.$$

Dividing by s^2 and passing to the limit as $s \rightarrow 0$, we obtain $T'(x)v \cdot v \ge 0$. Conversely, if this inequality holds, then $(T(x+sv) - Tx) \cdot sv = \int_0^s T'(x_t v)v \cdot sv \, dt \ge 0$ for all $x, v \in E$. $\square$

The functional and differential criteria of monotonicity formulated above often prove to be useful.

Exercise. Prove that, in the case of a Hilbert space $E = H$, an operator T is monotonic if and only if $S(\lambda) = I + \lambda T$ is a dilation operator for any $\lambda > 0$, i.e., $\|S(\lambda)x - S(\lambda)y\| \ge \|x - y\| \ (x, y \in H)$.

3. The following theorem holds.

Theorem. *Any monotonic operator is locally bounded.*

This theorem is proved in Gajewski, Gröger, and Zacharias (1974), Chapter 3, Section 1. It is based on the Banach–Steinhaus theorem.

Corollary. *Any linear monotonic operator is bounded.*

Proof. Let $T : E \to E'$ be a monotonic linear operator, $x_n \to x$ in E, $v_n = \|x_n - x\|^{-1/2}(x_n - x)$ for $x_n \neq x$, and $v_n = 0$ for $x_n = x$. Since $v_n \to 0$, by the preceding theorem, there is a $c > 0$ such that $\|Tv_n\| \leq c$. Hence

$$\|Tx_n - Tx\| = \|T(x_n - x)\| = \|x_n - x\|^{1/2}\|Tv_n\| \leq c\|x_n - x\|^{1/2} \to 0.$$

A monotonic linear operator is continuous and therefore bounded. $\quad\square$

Thus, the monotonicity of an operator implies local boundedness in the general case and global boundedness in the linear case.

4. In the concluding part, we give an example.

Example. (Gajewski, Gröger, and Zacharias (1974), Chapter 2, Section 2, and Chapter 3, Section 1). Consider a bounded domain G with regular boundary Γ in the space $E = \mathbb{R}^n$; a collection of smooth functions a_{ij} $(1 \leq i, j \leq n)$; the operator $S = -\Sigma \partial_i (a_{ij}\partial_j)$ defined on the selected class of functions on $\bar{G}$; the operators $L : x \to \operatorname{grad} x$, $A : y \to (a_{ij})y^t$, and $M : z \to \operatorname{div} z$ $(x \in X, y \in Y, z \in Z)$. The operator $T = MAL$ is called the *energy extension* of the differential operator S (since it is often associated with energy in physical applications). The domain G may be convex. In some cases, it is required that $u = 0$ on Γ. Let

$$(Sy)(x) = (a_i(x, y(x))), \quad a_i(x, y) = \sum_j b_{ij} y_j \quad (b_{ij} \in \mathcal{L}^\infty(G)).$$

It is easy to prove that the operator S is monotonic if for almost all $x \in G$ and for all $v \in \mathbb{R}^n$ there is a $c \geq 0$ such that $vB(x)v^* \geq c\|v\|^2$ and $B(x) = (b_{ij}(x))$. The case with the monotonicity of the energy extension T is more complicated. Its monotonicity is established under various additional assumptions.

5.3.2. Equations with monotonic operators

In this subsection, we formulate some propositions on solutions of the equation $Tx = f$ with monotonic operator $T : E \to E'$.

1. An operator T is said to be *radially continuous* if every function $\varphi(t) : [0, 1] \to \mathbb{R}$ such that $\varphi(t) = T(x + tv) \cdot v$ $(0 \leq t \leq 1)$ is continuous.

Proposition. *Let a monotonic operator T be radially continuous and $(f - Tv)(x - v) \geq 0$ $(v \in E)$. Then $Tx = f$.*

Proof. If $v = x - tu$ $(t > 0)$, then $t(f - Tv)u = (f - Tv)(x - v) \geq 0$ by assumption. Hence, $t(f - Tv)u \geq 0$ and $(f - Tv)u \geq 0$ for $t \to 0$ and arbitrary $u \in E$. As a result, $Tx = f$.   □

Remark. A more general proposition that includes several more conditions for the equation $Tx = f$ to be solvable is proved in Gajewski, Gröger, and Zacharias (1974), Chapter 3, Lemma 1.3.

2. An operator T is said to be *coercitive* if there is a function $\gamma :$ $[0, \infty[\to \mathbb{R}$ such that $\gamma(t) \to \infty$ as $t \to \infty$ and $Tx \cdot x \geq \gamma(\|x\|)\|x\|$ $(x \in E)$.

Theorem.   *Let a monotonic operator T be radially continuous and coercitive. Then, for any $f \in E'$, the set of solutions of the equation $Tx = f$ is nonempty, weakly closed, and convex.*

This theorem is proved in Gajewski, Gröger, and Zacharias (1974), Chapter 3 (Theorem 2.1). In the same book, a theorem that provides conditions for the problem of solving the equation $Tx = f$ to be well-posed is proved and several corollaries are given.

3. Suppose that $E = H$ is a real Hilbert space, $B = \bar{B}(0,1) \subseteq H$, $T : B \to H$ is a monotonic operator that is continuous on finite-dimensional subspaces, and $S(\lambda) = I + \lambda T$.

Lemma. *There is a point $a \in B$ such that $Ta \cdot (x - a) \geq 0$ for all $x \in B$.*

A proof of this lemma is given in Nirenberg (2001), Subsection 5.1.4.

Theorem. *If $S(\lambda)u \neq 0$ for all $\lambda \geq 0$ and u with $\|u\| = 1$, then the equation $Tx = 0$ has a solution in B.*

Proof. If the inequality of the preceding lemma holds for an interior point a of the ball B, then $Ta = 0$. If $\|u\| = 1$ and $Tu \neq 0$, then $S(\alpha)u = 0$ for some $\alpha \geq 0$.   □

Exercise. Give a detailed proof of the theorem.

Corollary. *If $\|x\|^{-1}Tx \cdot x \to \infty$ as $\|x\| \to \infty$, then the equation $Tx = y$ has a solution for any $y \in H$.*

A proof of the corollary can be found in Nirenberg (2001), Subsection 5.1.4. In Nirenberg (2001), Section 5.2, a general minimax theorem is proved that generalizes the preceding lemma. Section 5.3 of the same book deals with multi-valued monotonic operators.

Remark. A detailed description of monotonic multioperators is given in Aubin and Ekeland (1984), Chapter 6, where they are used for solving variational inequalities. The use of monotonic operators and variational inequalities in solving ill-posed problems is discussed in Bakushinskii and Goncharskii (1989), Chapter 3.

Positive and monotonic operators in partially ordered Banach spaces are the subject of Section 8.3 in Hutson and Pym (1980). The Schauder theorem is used there to prove the fixed point theorem for compact operators in spaces with a normal cone (Theorem 8.3.9). The theory is applied to solving Hammerstein integral equations and some boundary value problems for differential equations (8.3.20–24).

5.4. NONLINEAR CONTRACTIONS

Some semigroups of nonlinear operators can be studied by analogy with the study of semigroups of linear operators.

5.4.1. Contracting semigroups of operators

Contracting semigroups of operators are described in detail in Clément, Heijmans, Angenent, van Duijn, and de Pagter (1987), Chapter 2.

1. Consider a nonempty closed set F in the Banach space E and a collection U of operators $U(t) : F \to F$ ($t \geq 0$). They form a *contracting semigroup* if the following conditions hold:

(1) $U(0) = I$, $U(s+t) = U(s)U(t)$ ($s, t \geq 0$),
(2) $\|U(t)x - U(t)y\| \leq \|x - y\|$ ($t \geq 0$; $x, y \in F$),
(3) $U(t)x \to U(s)x$ ($t \to s$, $x \in F$).

In other words, any strongly continuous semigroup of nondilating operators or (*nonlinear*) *contractions* is said to be a *contracting semigroup*.

Remark. As in the linear case, the condition (3), which requires that U be strongly continuous on $[0, \infty[$, can be replaced by the condition (3′) requiring that U be continuous at the point $s = 0$, so that $(1)(2)(3) \Rightarrow (1)(2)(3')$.

The equality $Tx = \lim_{t\downarrow 0} t^{-1}(U(t)x - x)$ defines a *generating operator* T for U. In the nonlinear case, however, it is possible that $A = \operatorname{Dom} T = \varnothing$. An example of such a situation is given by M. Crandall and T. Liggett (see the corresponding reference in Clément, Heijmans, Angenent, van Duijn, and de Pagter, 1987, Section 2.1). If $A \neq \varnothing$, then

$$\|x - y\| \leq \|(x - y) - \lambda(Tx - Ty)\| \quad (\lambda > 0, \ x, y \in F).$$

Exercise. Prove the preceding inequality (see Clément, Heijmans, Angenent, van Duijn, and de Pagter, 1987, Section 2.1).

Therefore, for $A \neq \varnothing$, the operator $S = I - \lambda T$ is one-to-one for any $\lambda > 0$ and the inverse operator $R = S^{-1}$ is a contraction, i.e.,

$$\|Ru - Rv\| \leq \|u - v\| \quad (u = Sx, \ v = Sy).$$

We will identify correspondences with their graphs and the multifunctions generated by them. Then $R(\lambda, T) = (I - \lambda T)^{-1}$ can be called an operator even if this correspondence is not many-to-one. An operator T is said to be *dissipative* (and $-T$ is said to be *accretive*) if $R(\lambda, T)$ is a contraction for any $\lambda > 0$. A dissipative operator T is called *m-dissipative* (and $-T$ is called *m-accretive*) if $\operatorname{Dom} R(\lambda, T) = E$ for all $\lambda > 0$. The operators in these definitions are assumed to be arbitrary (not necessarily many-to-one). By definition, an operator $R \subseteq E \times E$ is a contraction if

$$\|x - y\| \leq \|u - v\| \quad (x \in Ru, \ y \in Rv, \ u \in \operatorname{Dom} R, \ v \in \operatorname{Dom} R).$$

2. Consider a set $\mathcal{U} = \mathcal{U}(F, E)$ of all contracting semigroups U of operators $U(t) : F \to F$. A sequence U_n is said to converge to U in $\mathcal{U}$ if

$$\sup_{0 \leq s \leq t} \|U_n(s)x - U(s)x\| \to 0 \quad (t > 0, \ x \in F).$$

We denote by $\mathcal{D} = \mathcal{D}(F, E)$ the set of all *m*-dissipative operators $T : A \to E$ densely defined in F. Using $R(\lambda, T) = (I - \lambda T)^{-1}$, we define the convergence $T_n \to T$ in $\mathcal{D}$ as follows:

$$\|R(\lambda, T_n)y - R(\lambda, T)y\| \to 0 \quad (\lambda > 0, \ y \in E).$$

Let $E = H$ be a Hilbert space and $F = B \subseteq H$ be a nonempty closed convex set. This case is described in detail in Brézis (1973). These restrictions make it possible to formulate an analog of the Hille–Yosida for nonlinear operators. Under the specified assumptions, for any contracting semigroup $U \in \mathcal{U}$ there exists a unique operator $T \in \mathcal{D}$ such that $S(t) = t^{-1}(U(t) - I)$ converge to T as $t \downarrow 0$. The map $\mathcal{S} : U \to T$ is a one-to-one map from $\mathcal{U}$ to $\mathcal{D}$; moreover, $U_n \to U$ in $\mathcal{U}$ if and only if $T_n = \mathcal{S}(U_n) \to \mathcal{S}(U) = T$ in $\mathcal{D}$. The ability to formulate this analog of the Hille–Yosida theorem justifies the use of multivalued operators.

Remark. This analog of the Hille–Yosida theorem does not hold in the case of arbitrary Banach spaces, since the operators $S(t)$ do not necessarily converge to the appropriate operator T and the map $\mathcal{S}$ is undefined. Moreover, different m-dissipative operators may generate the same semigroup. Relevant references can be found in Clément, Heijmans, Angenent, van Duijn, and de Pagter (1987), Section 2.1.

5.4.2. Approximation

1. Let $T : A \to E$ be an m-dissipative operator defined on a set A in a Banach space E. Set

$$U_n(t) = R^n(n^{-1}t, T) = (I - n^{-1}tT)^{-n} \quad (t \geq 0).$$

For any $x \in F = \bar{A}$, the sequence $U_n(t)x$ converges to some $U(t)x \in F$ and defines a contracting semigroup of operators $U(t) : F \to F$. Moreover, for any $x \in A$,

$$|Tx| = \sup_{\lambda > 0}(\lambda^{-1}\|R(\lambda, T)x - x\|) < \infty,$$

$$\|U(t)x - U(s)x\| \leq |Tx| \, |t - s| \quad (s, t \geq 0),$$

$$\|U(t) - U_n(t)\| \leq n^{-1/2}|Tx|t \quad (t \geq 0, \, n \geq 1).$$

The above assertions are represented in a theorem that was proved by M. Crandall and T. Liggett (see Clément, Heijmans, Angenent, van Duijn, and de Pagter, 1987, Section 2.2)

2. Consider a sequence of operators $T_n : A_n \to E$ and an operator $T : A \to E$ ($A_n, A \subseteq E$). Assume that these operators are m-dissipative and they generate contracting semigroups of operators U_n and U, respectively. In addition, consider a sequence of vectors $x_n \in F_n = \bar{A}_n$ and a vector $x \in F = \bar{A}$ such that $x_n \to x$ in E.

Theorem. *If there is an $\alpha > 0$ such that $\|R(\alpha, T_n)y - R(\alpha, T)y\| \to 0$ for all $y \in E$, then $\sup\limits_{0 \le s \le t} \|U_n(s)x_n - U(s)x\| \to 0$ $(t > 0)$.*

A proof of this theorem and the following corollary can be found in Clément, Heijmans, Angenent, van Duijn, and de Pagter (1987), Section 2.3. Consider an m-dissipative operator $T : E \to E$ satisfying the Lipschitz condition, a vector $x \in E$, and a function $u : [0, \infty[\to E$ with values $u(t) = U(t)x$.

Corollary. *The function u is continuously differentiable and $\dot{u}(t) = Tu(t)$ $(t \ge 0)$, $u(0) = x$.*

A nonlinear version of the Chernoff estimate for the solution of the Cauchy problem is also established there.

5.5.   DEGREE THEORY

Degree theory is used effectively in the study of nonlinear equations. An introduction to this theory can be found in Hutson and Pym (1980), Chapter 13.

5.5.1.   Finite-dimensional spaces

We begin with the case of finite-dimensional spaces. In what follows, we use the results of Subsection 3.4.9.

1. Assume that U is a nonempty bounded open set in $\mathbb{R}^m$, $f : \bar{U} \to \mathbb{R}^m$ is a continuous map, and $y \notin f(\partial U)$. Take $r = (1/2)\,\text{dist}\{y, f(\partial U)\}$. The image $f(\partial U)$ is compact because so is ∂U. Therefore, $r > 0$ for $y \notin f(\partial U)$. A point $z \in \mathbb{R}^m$ is said to be *close* to y, written $z \approx y$, if $\|z - y\| < r$. Similarly, a map $g : \bar{U} \to \mathbb{R}^m$ is said to be *close* to f, written $g \approx f$, if $\|g - f\|_{\sup} < r$.

If f is smooth, then the *degree of f at a point y with respect to the set U* is the number

$$d(f, y, U) = \sum_{x \in f^{-1}(z)} \text{sign} \det f'(x),$$

where z is a point that is close to y and is neither a singular nor a boundary value of f. In particular, $d(f, y, U) = 0$ for $y \notin f(\bar{U})$.

In the general case, $d(f, y, U) = d(g, y, U)$, where g is a smooth map that is close to f. The designation $\deg(f, y, U)$ is often used instead of $d(f, y, U)$.

Examples. 1) Let $e = \mathrm{id} : U \to \mathbb{R}^m$. Then $d(e, y, U) = 1$ ($y \in U$), $d(e, y, U) = 0$ ($y \notin \bar{U}$). 2) Let $c : \bar{U} \to \mathbb{R}^m$ be constant. Then $d(c, y, U) = 0$ for $y \neq c$.

2. The degree of a map is well defined, since the set of nonsingular values is dense in $f(U)$, the set of smooth functions is dense in $C(\bar{U}, \mathbb{R}^m)$, and $d(f, z_1, U) = d(f, z_2, U)$, $d(g_1, y, U) = d(g_2, y, U)$ for close nonsingular values z_1, z_2 and close smooth maps g_1, g_2. By definition, for all nonsingular $z \approx y$ and smooth $g \approx f$, the degrees $d(g, z, U)$ have the same value $d(f, y, U)$.

We denote by $C(\bar{U}, \mathbb{R}^m)$ the class of all continuous maps from $\bar{U}$ to $\mathbb{R}^m$. Consider its subclass $C^1(\bar{U}, \mathbb{R}^m)$ consisting of smooth (infinitely differential) functions on U. Let $f \in C^1(\bar{U}, \mathbb{R}^m)$ and $S = \{x \in U : \det f'(x) = 0\}$. Then, by Sard's theorem, the Lebesgue measure of the image $f(S)$ of the set of singular points is equal to zero, which implies that there are nonsingular points in the neighborhood of any value of f. Since $\bar{U}$ is compact, it is easy to prove that $f^{-1}(y)$ is finite for $y \notin f(S \cup \partial U)$.

For a given map f and points z and y, we write $z \sim y$ whenever these points belong to the same connected component of the set $\mathbb{R}^m \setminus f(\partial U)$. It is easy to show that $z \sim y$ for $\|z - y\| < 2r$ (and certainly for $z \approx y$).

Exercise. Prove the preceding statements.

Let $f \in C^1(\bar{U}, \mathbb{R}^m)$ and $y \notin f(S \cup \partial U)$. Then $d(f, z, U) = d(f, y, U)$ for $z \sim y$.

Let $f, g \in C^1(\bar{U}, \mathbb{R}^m)$ and $y \notin f(\partial U)$. Then there exist $\delta > 0$ and $y \notin g(\partial U)$ such that $d(f, y, U) = d(g, y, U)$ for all $g \in C^1(\bar{U}, \mathbb{R}^m)$ such that $\|g - f\|_{\sup} < \delta$.

Exercise. Prove that $y \notin g(\partial U)$.

A homotopy $h : [0, 1] \times \bar{U} \to \mathbb{R}^m$ of the map $f \in C(\bar{U}, \mathbb{R}^m)$ is called *admissible* for $y \notin f(\partial U)$ if $y \notin h([0, 1] \times \partial U)$. The maps $f = f_0 = h(0, \cdot)$ and $g = f_1 = h(1, \cdot)$ are called *admissibly homotopic*, written $f \sim g$ (for the given value $y \notin f(\partial U)$). The homotopy invariance theorem states that $d(h_t, y, U)$ does not depend on t. In particular, $d(f, y, U) = d(g, y, U)$ for admissibly homotopic f and g.

The degree of a smooth map satisfies an *integral formula*. Let $f \in C^1(\bar{U}, \mathbb{R}^m)$ and $y \notin f(S \cup \partial U)$. Then

$$d(f, y, U) = \int_U \varphi_\varepsilon(y - f(x)) \det f'(x) \, dx,$$

where $\varphi_\varepsilon : \mathbb{R}^m \to \mathbb{R}$ is the *smoothing factor* with values $\varphi_\varepsilon(x) = c(\varepsilon) \cdot \exp\{-1/(\varepsilon^2 - \|x\|^2)\}$ for $\|x\| \geq \varepsilon$. The normalizing coefficient is chosen to satisfy the condition $\|\varphi_\varepsilon\| = 1$ for the integral norm. Clearly, $\varphi_\varepsilon \in C^\infty(\mathbb{R}^m, \mathbb{R})$ and $\operatorname{supp} \varphi_\varepsilon \subseteq \bar{B}(0, \varepsilon) \subseteq \mathbb{R}^m$.

Proofs of the above statements describing the properties of the degree $d(f, y, U)$ can be found in Hutson and Pym (1980), Section 13.2.

3. Assume that $U \subseteq \mathbb{R}^m$ is a bounded open set, $f : \bar{U} \to \mathbb{R}^m$ is a continuous map, and $y \notin f(\partial U)$. The following theorem describes some properties of the degree $d(f, y, U)$.

Theorem. (1) If $d(f, y, U) \neq 0$, then $y \in f(U)$.

(2) If $g \sim f$, then $d(f, y, U) = d(g, y, U)$.

(3) If $z \sim y$, then $d(f, z, U) = d(f, y, U)$.

(4) If $g_{\partial U} = f_{\partial U}$, then $d(f, y, U) = d(g, y, U)$.

(5) If $y \in U$, then $d(e, y, U) = 1$.

Here $g \in C(\bar{U}, \mathbb{R}^m)$, $z \in \mathbb{R}^m \setminus f(\partial U)$; the maps $g_{\partial U}$ and $f_{\partial U}$ are the restrictions of the maps f and g to ∂U; $e = \operatorname{id} : \bar{U} \to \mathbb{R}^m$.

Proof. We now give a short proof of the theorem. Properties (2) and (3) are generalizations of the assertions formulated in part 2 of this subsection. Property (5) follows from the definitions. It remains to prove only (1) and (4).

We start with property (1). Let $f \in C^1(\bar{U}, \mathbb{R}^m)$ and $y \notin f(\partial U)$. Then $y \notin f(U)$ implies $y \notin f(\bar{U}) = \operatorname{Ran} f$ and $f^{-1}(y) = \varnothing$. Hence, $d(f, y, U) = 0$ for $y \notin f(U)$ and therefore $y \in f(U)$ for $d(f, y, U) \neq 0$. If $f \in C(\bar{U}, \mathbb{R}^m)$, then there exists a map $g \in C^1(\bar{U}, \mathbb{R}^m)$ such that $g \approx f$. The image $g(\partial U)$ is included in the r-neighborhood of the image $f(\partial U)$, which means that $y \notin g(\bar{U})$ for $y \notin f(\bar{U})$ and $r = (1/2) \operatorname{dist}\{y, f(\partial U)\}$. Thus, we have $d(f, y, U) = d(g, y, U) = 0$ for $y \notin f(U)$.

We now turn to (4). If $f(x) = g(x)$ for $x \in \partial U$, then $h(t, x) = (1 - t)f(x) + tg(x)$ defines an admissible homotopy ($t \in [0, 1]$, $x \in \bar{U}$). In view of the homotopy invariance of a degree, $d(f, y, U) = d(g, y, U)$ for $y \notin f(\partial U) = g(\partial U)$. $\quad\square$

Exercise. Deduce (3) from (2).

Remark. Because of the condition $y \notin f(\partial U)$, property (1) can be used to establish the existence of a solution $x \in U$ to the equation $f(x) = y$. It should be emphasized that $d(f, y, U) \neq 0$ is only a sufficient condition for a solution to exist. A solution to the equation may still exist if $d(f, y, U) = 0$.

4. Applying the theorem on the properties of the degree of a map, it is easy to prove the main theorem of the algebra of complex numbers. From the results of Subsection 3.4.9, part 4, it follows that $d(g, z, U) = n$ for the polynomial $g(z) = a_n z^n$ with coefficient $a_n \neq 0$, the disk $U = B(0, r) \subseteq \mathbb{R}^2$ with sufficiently large radius $r > 0$, and a point $z \notin U$. The polynomial $f(z) = a_0 + a_1 z + \cdots + a_n z^n$ and g are related by the admissible homotopy $h(t, x) = (1 - t)g(z) + tf(z)$ $(t \in [0, 1], z \in \bar{U})$. Since $|g(z)| \geq 2|f(z) - g(z)|$ for $|z| \geq r$ and sufficiently large $r > 0$, we have $|h_t(z)| \geq |g(z)| - t|f(z) - g(z)| > 0$ for $|z| = r$ and $0 \notin h_t(\partial U)$ for all $t \in [0, 1]$. Hence $d(f, 0, U) = d(g, 0, U) = n \neq 0$ and $f \in f(U)$. The equation $f(z) = 0$ has a solution $z \in U$.

5. We now prove the Brouwer theorem for the ball $\bar{U} = \bar{B}(0, r) \subseteq \mathbb{R}^m$. The fixed points of the continuous map $f : \bar{U} \to \bar{U}$ are solutions to the equation $g(x) = 0$ with $g = f - e$, where $e = \mathrm{id}$. If $0 \in g(\partial U)$, then there is a point $x \in \partial U$ such that $f(x) = x$, which means that the Brouwer theorem holds. Therefore, it can be assumed that $0 \notin g(\partial U)$ and $f(x) \neq x$ for all $x \in S(0, r) \subseteq \mathbb{R}^m$. The maps e and g are related by the admissible homotopy $h(t, x) = (1 - t)x + tf(z)$ $(t \in [0, 1], z \in \bar{U})$. Indeed, $h_t(x) = 0$ implies $x = tf(x)$ and $r = t\|f(z)\|$ for $x \in \partial U$. Since $0 \leq t \leq 1$ and $\|f(x)\| \leq r$, $r \leq tr$ implies $t = 1$. Then $x = f(x)$, which contradicts the assumption $f(x) \neq x$. Therefore, $d(g, 0, U) = d(e, 0, U) = 1 \neq 0$ and $0 \in g(U)$. The equation $g(x) = 0$ has a solution $x \in U = B(0, r)$ if it has no solutions $x \in \partial U = S(0, r)$.

Remark. The Brouwer theorem can be easily generalized to the case of continuous transformations of convex compact sets. This can be done with the help of the Dugundji theorem, which is proved in Sternberg (1983).

The Dugundji theorem. *In a Banach space X, a continuous map $f : A \to C$ from a closed set $A \subseteq X$ to a convex set $C \subseteq X$ has a continuous extension $\bar{f} : X \to C$.*

Consider a nonempty convex compact set $C \subseteq \mathbb{R}^m$ that includes the ball $\bar{B} = \bar{B}(0, r) \subseteq \mathbb{R}^m$ and a continuous transformation $f : C \to C$. By the Dugundji theorem, there exists a continuous map $\bar{f} : \bar{B} \to C$ which is an extension of f. Since $\bar{f}$ is a continuous transformation of the ball $\bar{B}$, it follows that there exists a point $x \in \bar{B}$ such that $\bar{f}(x) = x$. Since $\bar{f}(\bar{B}) \subseteq C$, we have $x \in C$ and $f(x) = \bar{f}(x) = x$.

5.5.2. The Leray–Schauder degree

Degree theory can be extended from the finite-dimensional case to the infinite-dimensional case only under certain restrictions.

1. If degree theory could be extended from the case of finite-dimensional spaces to the case of arbitrary Banach spaces with no restrictions, then the Brouwer fixed point theorem would hold for Banach spaces. Kakutani's counterexample shows that this is not true.

Consider the Hilbert space $E = l^2$, the ball $\bar{B} = \bar{B}(0,1) \subseteq E$, and the transformation $T : \bar{B} \to \bar{B}$ that maps $x = (x_i) \in E$ to $y = (y_i) \in E$ with coordinates $y_1 = (1 - \|x\|^2)^{1/2}$, $y_{i+1} = x_i$ ($i \geq 1$). The operator T is nonlinear because $T(0) \neq 0$. It is continuous, since the equality

$$\|T(x) - T(a)\|^2 = (1 - \|x\|^2)^{1/2} - (1 - \|a\|^2)^{1/2} + \|x - a\|^2$$

implies that $T(x) \to T(a)$ as $x \to a \in E$. The transformation T has no fixed points. Indeed, $\|T(x)\| = 1 - \|x\|^2 + \|x\|^2 = 1$ and therefore the equality $T(x) = x$ does not hold if $\|x\| < 1$. If $\|x\| = 1$, then $Tx = x$ implies $x = 0$, which is a contradiction. Thus, in order to obtain a more effective theory, it is necessary to choose a class of transformations of Banach spaces that is smaller than the class of all continuous transformations.

2. We now give the necessary definitions. Let E be a Banach space, $U \subseteq E$ be a (nonempty) bounded open set, and $I : \bar{U} \to E$ be the identity embedding. An operator $K : \bar{U} \to E$ is said to be *compact* if it is continuous and the image of any bounded set under K is a relatively compact set, i.e., $\overline{K(B)} \subseteq E$ is compact for any bounded set $B \subseteq \bar{U}$. The operator $F = I - K$ is called a *Fredholm operator* if K is compact. Note that the equality $F(x) = 0$ is equivalent to the equality $K(x) = x$. The study of the fixed points of compact operators is thus reduced to the study of Fredholm equations $F(x) = y$. The latter have been studied in detail in the linear case.

It is easy to prove that $r = 2^{-1} \operatorname{dist}\{y, F(\partial U)\} > 0$ for $y \notin F(\partial U)$ (Hutson and Pym, 1980, Subsection 13.3.1). Then *Schauder projections* can be used to prove that there exist finite-dimensional extensions Q of a Fredholm operator F (Hutson and Pym, 1980, Subsection 8.2.5).

The following lemma is proved in Hutson and Pym (1980), Subsection 13.3.2.

Lemma. *For any* $\varepsilon \in \,]0, r[$ *there is a continuous operator* $Q : \bar{U} \to E$ *of finite rank such that*

$$\|F(x) - Q(x)\| \leq \varepsilon \ (x \in \bar{U}), \quad \|Q(x) - y\| \geq r \ (x \in \partial U).$$

The restriction of Q to the subspace generated by the point y and the image $Q(\bar{U})$ is a continuous map between finite-dimensional spaces. The degree of this map is defined. Different approximations of the operator F that satisfy these conditions have the same degree (Hutson and Pym, 1980, Subsection 13.3.4). The value $d(F, y, U)$ of this degree is called the *Leray–Schauder degree* of the map F at the point y with respect to the set U (Hutson and Pym, 1980, Subsection 13.3.5).

Remark. Using the Dugundji theorem, one can first define the degree of an operator on ∂U and then the degree of its extension on $\bar{U}$ (Hutson and Pym, 1980, Porblem 13.6).

We now refine the notion of the *admissible homotopy* $H : [0, 1] \times \bar{U} \to E$ that relates Fredholm operators $F : \bar{U} \to E$ and $G : \bar{U} \to E$. In addition to the requirements of continuity and the condition $y \notin H([0, 1] \times \partial U)$, we require that H_t be Fredholm operators for all $t \in [0, 1]$. If there is an admissible homotopy between the operators $F = H_0$ and $G = H_1$, this will be written as $G \sim F$. If points y and z belong to the same connected component of $E \setminus \overline{F(\partial U)}$, this will be written as $z \sim y$.

3. A theorem describing the properties of the degree of Fredholm operators can be formulated by analogy with the finite-dimensional case. The proof of the theorem follows the standard procedure: applying the corresponding assertion of the theorem of Subsection 5.5.1 (part 3) and passing to the limit in the equality $F = I - K$, taking into account the compactness of the operator K.

The theorem has a corollary that is especially useful for applications. In it, the condition for a solution of the equation $F(x) = y$ to exist is formulated directly for F.

Let $U \subseteq E$ be a bounded open set, $F : \bar{U} \to E$ be a Fredholm map, and $y \notin F(\partial U)$. The following theorem describes the properties of the degree $d(F, y, U)$.

Theorem. (1) If $d(F, y, U) \neq 0$, then $y \in F(U)$.
(2) If $G \sim F$, then $d(G, y, U) = d(F, y, U)$.

(3) If $z \sim y$, then $d(F, z, U) = d(F, y, U)$.

(4) If $G_{\partial U} = F_{\partial U}$, then $d(G, y, U) = d(F, y, U)$.

(5) If $y \in U$, then $d(I, y, U) = 1$.

Here $G : \bar{U} \to E$ is a Fredholm operator, $z \in E \setminus \overline{F(\partial U)}$, the maps $F_{\partial U}$ and $G_{\partial U}$ are the restrictions of F and G to ∂U, and $I : \bar{U} \to E$ is the identity embedding. A short proof of the theorem can be found in Hutson and Pym (1980), Subsection 13.3.6.

Exercise. Deduce the above theorem from the theorem of Subsection 5.5.1, part 3.

The following theorem is a corollary of the preceding one.

The Leray–Schauder theorem. Let $y \notin H_t(\partial U)$ for $H_t = (1-t)I + tF$ for all $t \in [0, 1]$. Then the Fredholm equation $F(x) = y$ has a solution $x \in U$.

Proof. It follows from the assumption that $I \sim F$. Then, as follows from properties (2) and (5) in the theorem on the properties of the degree, $d(F, y, U) = d(I, y, U) = 1 \neq 0$ for $y \in U$. Hence, property (1) implies $y \in F(U)$. $\square$

Exercise. Deduce the Schauder theorem and Theorem 2 from Subsection 5.1.2, part 2, from the theorem on the degree of a map.

Remark. The application of degree theory in the study of the Chandrasekhar integral equation for radiation transport in stellar coronas is discussed in Hutson and Pym (1980), Section 13.4.

Ill-posed problems

Chapter 6.

Classic problems

6.1. MATHEMATICAL DESCRIPTION OF THE LAWS OF PHYSICS

The mathematical description of physical phenomena consists in assigning a number or a vector to every point in the space (or a domain in the space) at every instant of time to represent the properties of matter at this point. If the properties of matter is described by a number, we have a scalar field. Otherwise, if it is described by a vector, then we have a vector field. Examples of scalar fields are density, temperature, and pressure. Examples of vector fields are the velocity of particles in a flowing fluid, field strength (for example, gravitational, magnetic, or electric field strength).

It follows from the laws of physics that fields and their derivatives are related by certain formulas. In other words, fields satisfy differential equations or systems of differential equations. There is a great multitude of natural phenomena that are described in terms of differential equations. Differential equations describe chemical reactions and various kinds of biological processes.

In what follows, we present equations for some of the most important fields in physics. We will confine ourselves to the fields that are described by equations studied in this book.

The gravitational field. The equation for the gravitational field follows from Newton's law of gravitation, which states that any two particles of matter with masses m and M are attracted to each other with a force $\mathcal{F}$

equal to

$$\mathcal{F} = kmM/r^2, \tag{1}$$

where r is the distance between the particles and k is the gravitational constant.

It follows from (1) that the gravitational field generated by the distribution of mass in space with density ρ has a potential φ, i.e.,

$$\mathcal{F} = \operatorname{grad} \varphi.$$

The potential φ satisfies the Poisson equation

$$\Delta \varphi = 4\pi k\rho. \tag{2}$$

Electric and magnetic fields. Similarly to a gravitational field, a constant electric field has a potential u, i.e.,

$$E = \operatorname{grad} u,$$

and this potential satisfies the Poisson equation

$$\Delta u = \frac{4\pi}{\varepsilon}\rho, \tag{3}$$

where ρ is the density of space charges and ε is the dielectric constant.

A constant magnetic field satisfies the system of equations

$$\operatorname{rot} H = \frac{4\pi}{c}j, \quad \operatorname{div} H = 0, \tag{4}$$

where j is the current density vector.

In the part of space where there are no currents, the potential of the magnetic field satisfies the Laplace equation

$$H = \operatorname{grad} v, \quad \Delta v = 0.$$

Electromagnetic field. A variable electromagnetic field satisfies the system of Maxwell's equations

$$\operatorname{rot} H = \frac{4\pi}{c}j + \frac{1}{c}\frac{\partial}{\partial t}\mathcal{D}, \quad \operatorname{div}\mathcal{D} = 4\pi\rho,$$
$$\operatorname{rot} E = -\frac{1}{c}\frac{\partial}{\partial t}B, \quad \operatorname{div} B = 0, \quad j = \sigma(E + E'), \tag{5}$$

where $B = \mu H$ is the magnetic induction vector, $\mathcal{D} = \varepsilon E$ is the electric induction vector, μ is the magnetic permeability, σ is the electric conductivity, c is the speed of light, and E' is the electric field strength induced by external sources.

Note that equations (3), (4) follow from the system (5) if ε and μ are constant.

In the general case, the system (5) is fairly complex. We consider the case where the system (5) describes an electromagnetic field in vacuum or in the air. We have

$$\mu = 1, \quad \varepsilon = 1, \quad \sigma = 0.$$

In addition, assume that there are no space charges, i.e., $\rho = 0$. In this case, the system (5) is written as follows:

$$\operatorname{rot} H = \frac{1}{c}\frac{\partial E}{\partial t}, \quad \operatorname{div} E = 0,$$
$$\operatorname{rot} E = -\frac{1}{c}\frac{\partial H}{\partial t}, \quad \operatorname{div} H = 0. \tag{6}$$

The last equality in (6) implies that the field H has vector potential A:

$$H = \operatorname{rot} A. \tag{7}$$

Substituting (7) into the third equality in (6), we obtain

$$\operatorname{rot}\left(\frac{\partial A}{\partial t} + cE\right) = 0. \tag{8}$$

It follows from (8) that

$$\frac{\partial A}{\partial t} + cE = -\operatorname{grad}\varphi. \tag{9}$$

Substituting (9) into the second equality in (6), we have

$$\Delta\varphi + \frac{\partial}{\partial t}\operatorname{div} A = 0. \tag{10}$$

As is known, the vector potential A is defined by the field H up to an arbitrary potential vector.

We choose A to satisfy the condition

$$\operatorname{div} A = -\frac{1}{c^2}\frac{\partial}{\partial t}\varphi \tag{11}$$

(which is called the *calibration condition*).

Then the scalar potential φ and the vector potential A satisfy the wave equations

$$\frac{1}{c^2}\frac{\partial^2 \varphi}{\partial t^2} = \Delta\varphi, \quad \frac{1}{c^2}\frac{\partial^2 A}{\partial t^2} = \Delta A. \tag{12}$$

Gas dynamics. Consider the motion of gas or fluid. Let V be the particle velocity field, p be the pressure, and ρ be the fluid density.

If the fluid is ideal (has no viscosity) and the motion is adiabatic (with no heat exchange between separate fluid particles), then the velocity vector field and the scalar fields of density and pressure satisfy the system of equations

$$\frac{\partial V}{\partial t} + (V, \nabla)V = -\frac{1}{\rho}\operatorname{grad} p, \tag{13}$$

$$\frac{\partial \rho}{\partial t} + \operatorname{div}(\rho V) = 0, \tag{14}$$

$$p = a\rho^n. \tag{15}$$

Equation (13) is called the *Euler equation*. It is a consequence of the momentum conservation law. Equation (14) is called the *continuity equation*. It is a consequence of the mass conservation law. Equation (15) is called the *state equation*.

Equation (13) yields an equation that involves only the field V:

$$\frac{\partial}{\partial t}\operatorname{rot} V = \operatorname{rot}[V \times \operatorname{rot} V]. \tag{16}$$

We now describe two cases where the gas dynamics equations are reduced to simpler equations.

1. Assume that the fluid is incompressible, i.e.,

$$\rho = \rho_0 = \text{const},$$

and the velocity field is potential, i.e.,

$$V = \operatorname{grad}\varphi.$$

Then equation (16) holds and (14) implies that

$$\Delta\varphi = 0, \tag{17}$$

i.e., the potential satisfies the Laplace equation.

2. Consider the case of small oscillations in a fluid or gas. Such oscillations are called sound waves.

Let

$$p = p_0 + p_1, \quad \rho = \rho_0 + \rho_1,$$

where p_0 and ρ_0 are constant and p_1 and ρ_1 are sufficiently small.

Moreover, let the velocity field be small as well, so that the squares, the products, and the products of the derivatives of V, p_1, and ρ_1 in equations (13)–(15) can be neglected. Then equations (13)–(15) become as follows:

$$\frac{\partial V}{\partial t} = -\frac{1}{\rho_0} \operatorname{grad} p, \quad \frac{\partial \rho_1}{\partial t} + \rho_0 \operatorname{div} V = 0, \quad p_1 = a\rho_0^{n-1}\rho_1. \qquad (18)$$

It follows from (18) that

$$\frac{\partial \rho_1}{\partial t} + a\rho_0^n \operatorname{div} V = 0.$$

Assume that the velocity field has potential

$$V = \operatorname{grad} \varphi.$$

Then (18) implies that φ satisfies the wave equation

$$\frac{\partial^2 \varphi}{\partial t^2} = c^2 \Delta \varphi,$$

where $c = \sqrt{a\rho_0^{n-1}}$ is the speed of sound.

The vibrating string equation. In classical mechanics, a string is a solid body whose length significantly exceeds its diameter. It is assumed that its resistance to bending is negligibly small compared to its resistance to stretching. The mathematical model of a string that will be considered below provides a fairly good description of strings of musical instruments.

Let $u(x, t)$ denote the deviation of the string from its equilibrium position at an instant t. Let $\alpha(x)$ be the angle between the tangent to the string and the coordinate axis x, $\rho(x)$ be the linear density of the string, and $T(x)$ be the tension force directed along the tangent to the string. Consider points x_1 and x_2 of the string such that $x_1 < x_2$. The components of the tension force at these points are

$$-T(x_1) \sin \alpha(x_1), \quad T(x_2) \sin \alpha(x_2).$$

The angle $\alpha(x)$ is determined from the formula

$$\sin \alpha(x) = \frac{\partial u}{\partial x} \cdot \left[1 + \left(\frac{\partial u}{\partial x} \right)^2 \right]^{-1/2}.$$

If the deviation of the string from its equilibrium position is assumed to be small and the quantity $(\partial u/\partial x)^2$ is neglected, then, taking into account the inertia forces, we obtain the following equation for the function $u(x,t)$:

$$\frac{\partial}{\partial x}\left(T\frac{\partial u}{\partial x}\right) - \rho(x)\frac{\partial^2 u}{\partial t^2} = 0.$$

The so-called vibrating membrane equation is obtained in a similar way. (By a membrane we mean a thin solid body stretched uniformly in all directions.) If $u(x,y,t)$ is the deviation of the membrane from its equilibrium position, then the function $u(x,y,t)$ satisfies the equation

$$\frac{\partial^2 u}{\partial t^2} = a^2 \Delta u,$$

where $a^2 = T/\rho$, T is the tension, and ρ is the density of the membrane.

The heat conduction equation. Let u be the temperature distribution in a medium, c be the specific heat capacity of the medium, and ρ be the density. The variation of the temperature distribution in the medium is based on Fourier's law

$$W = -k\operatorname{grad} u,$$

where W is the heat flow density vector and k is the thermal conductivity coefficient.

From Fourier's law, it follows that u satisfies the equation

$$\frac{\partial u}{\partial t} = \frac{1}{c\rho}\operatorname{div}(k\operatorname{grad} u). \tag{19}$$

The same equation holds for the distribution of mass concentration in the diffusion process.

The eikonal equation. Consider a process of propagation of a perturbation in a continuous medium. The medium may be homogeneous (air, water) or nonhomogeneous (the Earth's crust). As was mentioned above, the deviations of elements of the medium from the equilibrium position are described by wave equations.

Let x be a point with coordinates (x_1, x_2, x_3) in the three-dimensional space, $v(x)$ be the velocity of propagation of the perturbation at the point x. We denote by $\tau(x, x^0)$ the time it takes for the perturbation to travel from x^0 to x. The function $\tau(x, x^0)$ satisfies the differential equation

$$\left[\left(\frac{\partial\tau}{\partial x_1}\right)^2 + \left(\frac{\partial\tau}{\partial x_2}\right)^2 + \left(\frac{\partial\tau}{\partial x_3}\right)^2\right]^{1/2} = \frac{1}{v(x)}.$$

This equation is called the *eikonal equation*. It is a partial differential equation of the first order. Solving this equation is equivalent to solving a system of ordinary differential equations.

6.2.   EQUATIONS OF THE FIRST ORDER

Let $x \in \mathbb{R}^n$ be an n-dimensional vector with coordinates $(x_1, \ldots, x_n)$, and let $u(x)$ be a function defined on a domain $G \subset \mathbb{R}^n$. An equation of the first order for $u(x)$ is an equality of the form

$$F\left(\frac{\partial u}{\partial x_1}, \ldots, \frac{\partial u}{\partial x_n}, u, x_1, \ldots, x_n\right) = 0.$$

The problem of solving this equation under certain conditions imposed on the function F is reduced to the problem of integrating a system of ordinary differential equations. In what follows, we will present only two results related to solving a special case of the first-order equation

$$\sum_{k=1}^{n} a_k(x)\frac{\partial u}{\partial x_k} + b(x, u) = 0. \tag{1}$$

These results can be found in Petrovskii (1984). In this well-known textbook, equations of this form are called semilinear.

Assume that the coefficients $a_k(x)$ have continuous partial derivatives with respect to all arguments in the domain G, the function $b(x, u)$ is defined for all functions u bounded in the domain $\bar{G}$ and has continuous first-order derivatives with respect to all variables x_k. Moreover, let $\sum_{k=1}^{n} a_k^2 > 0$.

Consider the system ordinary differential equations

$$\frac{dx_k}{dt} = a_k(x) \quad (k = 1, \ldots, n).$$

As is known from the theory of ordinary differential equations, for any point in G, there is a unique trajectory of the system passing through it. These trajectories are called characteristics.

The uniqueness theorem. *If $u(x)$ satisfies equation (1) in G and has continuous derivatives, then all values $u(x)$ on any characteristic such that $|u| < M$ are defined by the value of u at a single point $x^0 = (x_1^0, \ldots, x_n^0)$ of this characteristic.*

The existence theorem. *Let S be an $(n-1)$-dimensional surface included in G that has a continuous rotating tangent plane and is not tangent to any characteristic of the equation.*

Let $f(x)$ be a function on S with the following properties:

1) $|f(x)| < M$;

2) *in a neighborhood of any point of S, the function $f(x)$ can be represented as a function of $(n-1)$ coordinates out of $(x_1, \dots, x_n)$ with continuous derivatives with respect to these coordinates.*

Furthermore, suppose that there exists a neighborhood R_0 of the surface S with the following properties:

1) $R_0 \subset G$;

2) *being extended in both directions within R_0, a characteristic passing through any point $x \in S$ does not intersect S; for any $x \in R_0$, there is exactly one characteristic passing through it;*

3) *for any point $x^0 \in S$, the solution $u(x)$ that satisfies the condition*

$$u(x^0) = f(x^0)$$

can be extended to the whole part of the characteristic that is included within R_0; moreover, this solution satisfies the condition

$$|u(x)| < M.$$

Then there exists a function $u(x)$ defined on R_0 such that:

1) $u(x)$ *has continuous first-order derivatives with respect to all variables x_k;*

2) $u(x)$ *satisfies equation (1);*

3) $u(x^0) = f(x^0)$ *for any $x^0 \in S$.*

The problem of finding a function $u(x)$ satisfying the above conditions is called the Cauchy problem for equation (1).

6.3. CLASSIFICATION OF DIFFERENTIAL EQUATIONS OF THE SECOND ORDER

Consider the following linear differential equation of the second order in the n-dimensional space $\mathbb{R}^n$:

$$\sum_{j,k=1}^{n} a_{jk} \frac{\partial^2 u}{\partial x_j \partial x_k} + \sum_{j=1}^{n} b_j \frac{\partial u}{\partial x_j} + cu = f. \tag{1}$$

We change the variables $(x_1, \ldots, x_n)$ for the new independent variables $(y_1, \ldots, y_n)$ using the formulas

$$y_k = \sum_{j=1}^{n} b_{kj} x_j.$$

The matrix $\|b_{kj}\|$ is assumed to be nondegenerate.

After the change of variables, equation (1) becomes as follows:

$$\sum_{p,q=1}^{n} a'_{pq} \frac{\partial^2 u}{\partial y_p \partial y_q} + \sum_{p=1}^{n} b'_p \frac{\partial u}{\partial y_p} + cu = f, \tag{1'}$$

where

$$a'_{pq} = \sum_{j,k=1}^{n} b_{jp} b_{kq} a_{jk}, \quad b'_p = \sum_{j=1}^{n} b_{pj} b_j.$$

Consider the quadratic form

$$\sum_{j,k=1}^{n} a^0_{jk} \xi_j \xi_k, \tag{2}$$

whose coefficients are equal to the coefficients a_{jk} in equation (1) at a point $(x_1^0, \ldots, x_n^0)$. After the change of variables

$$\xi_k = \sum_{j=1}^{n} b_{kj} \eta_j \quad (k = 1, \ldots, n),$$

the quadratic form (2) becomes as follows:

$$\sum_{p,q} a^{01}_{pq} \eta_p \eta_q,$$

where

$$a^{01}_{pq} = \sum_{j,k=1}^{n} b_{jp} b_{kq} a^0_{jk}.$$

Thus, after a linear change of variables, the coefficients of the second-order derivatives in equation (1) change at $(x_1^0, \ldots, x_n^0)$ the same way as the coefficients of the quadratic form (2).

As is known, with a change of variables, a quadratic form can be reduced to the canonical form:

$$a_{pq}^{01} = 0, \quad p \neq q,$$
$$a_{pp}^{01} = 1, \quad p = 1, \ldots, m - 1,$$
$$a_{pp}^{01} = -1, \quad p = m, \ldots, n_1 - 1,$$
$$a_{pp}^{01} = 0, \quad p = n_1, \ldots, n. \tag{3}$$

If $m = n + 1$ or $m = 1, n_1 = n + 1$ in (3), then equation (1) is called *elliptic* at the point $(x_1^0, \ldots, x_n^0)$.

If $m = n$, $n_1 = n + 1$ or $m = 2, n_1 = n + 1$, then equation (1) is called *hyperbolic* at the point $(x_1^0, \ldots, x_n^0)$.

If $2 < m < n, n_1 = n + 1$, then equation (1) is called *ultrahyperbolic*.

If $n_1 \leq n$, then equation (1) is called *parabolic*.

If equation (1) is elliptic (hyperbolic, parabolic) at every point of the domain, then it is said to be elliptic (hyperbolic, parabolic) in the domain.

6.4.   ELLIPTIC EQUATIONS

First of all, elliptic equations describe stationary fields (fields that do not change with time). The Laplace equation

$$\Delta u = 0 \tag{1}$$

and the Poisson equation

$$\Delta u = f. \tag{2}$$

are classic elliptic equations.

Stationary fields in a medium with variable properties satisfy equations with variable coefficients. For example, the concentration field in a stationary diffusion process satisfies the equation

$$\text{div}(k \, \text{grad} \, u) = 0, \tag{3}$$

where k is the diffusion coefficient.

Elliptic equations also describe the fields that represent stabilized oscillating processes. For example, if a solution to the wave equation

$$\frac{\partial^2 u}{\partial^2 t} = \Delta u$$

can be represented in the form

$$u = e^{i\lambda t}v(x, y, z),$$

then the function v satisfies the Helmholtz equation

$$\Delta v + \lambda^2 v = 0. \tag{4}$$

As was noted before, the velocity field potential for an ideal incompressible fluid satisfies the Laplace equation with respect to the space variables at any fixed instant of time.

Boundary value problems are classic problems for elliptic equations. They consist in finding a solution to an elliptic equation in a domain that satisfies a condition on the boundary of this domain. We now give several examples of boundary value problems.

The Dirichlet problem for the Laplace equation. Let $\mathcal{D}$ be a bounded domain with smooth boundary S if the three-dimensional space, and let f be a continuous function defined on S.

Find a solution to the Laplace equation (1) inside $\mathcal{D}$ that satisfies the condition

$$u|_S = f.$$

The Neumann problem. Find a solution to the Laplace equation (1) inside $\mathcal{D}$ that satisfies the condition

$$\left.\frac{\partial u}{\partial n}\right|_S = f,$$

where $(\partial u/\partial n)|_S$ is the derivative of u in the direction of the outward normal n to the surface S.

The general boundary value problem. Let l be a vector and a be a function defined on S. Find a solution to the Laplace equation (1) inside $\mathcal{D}$ that satisfies the following condition on S:

$$(l, \nabla u) + au = f.$$

In the Dirichlet problem, it is assumed that u is continuous in the closed domain $\bar{\mathcal{D}}$. In the Neumann problem and the general problem, it is assumed that u is continuously differentiable in the closed domain $\bar{\mathcal{D}}$. These types of boundary value problems are also considered for other elliptic equations.

The theory of boundary value problems deals with the existence and uniqueness of solutions and continuous dependence of solutions on the data of a boundary value problem.

The uniqueness of solutions to the Dirichlet problem for the Laplace equation follows from the maximum principle, which is a fundamental property of solutions to the Laplace equation.

The maximum principle. *If a solution to the Laplace equation in a bounded domain $\mathcal{D}$ is continuous in the closure $\bar{\mathcal{D}}$, then it attains its maximum (and minimum) value on the boundary of the domain $\mathcal{D}$. If a solution of the Laplace equation attains its maximum (or minimum) value inside the domain $\mathcal{D}$, then the solution is constant.*

As follows from the maximum principle, if a solution to (1) vanishes on the boundary of $\mathcal{D}$, i.e.,

$$u|_S = 0,$$

then the solution is identically equal to zero. Since the problem is linear, this means that the solution is unique.

A slightly modified version of the maximum principle holds for the solutions of elliptic equations of the general form.

Consider the equation

$$\sum_{j,k=1}^{n} a_{jk} \frac{\partial^2 u}{\partial x_j \partial x_k} + \sum_{j=1}^{n} b_j \frac{\partial u}{\partial x_j} + cu = 0 \tag{5}$$

in a bounded domain $\mathcal{D}$ of the n-dimensional space. The coefficients a_{jk}, b_j, and c in (5) are assumed to be continuous.

Moreover, let

$$\sum_{j,k=1}^{n} a_{jk} \xi_j \xi_k \geq \delta \sum_{j=1}^{n} \xi_j^2, \quad \delta > 0, \quad c < 0. \tag{6}$$

The first condition in (6) means that equation (5) is elliptic. The following maximum principle holds for solutions to equation (5).

The maximum principle. *A solution to equation (5) does not attain a positive maximum and a negative minimum inside the domain $\mathcal{D}$.*

As follows from this maximum principle, the solution of the Dirichlet problem for equation (5) is unique.

The question on the uniqueness of solutions to the Neumann problem is answered with the help of Green's formulas.

Green's formulas. *Let functions u and v be twice continuously differentiable in a bounded domain $\mathcal{D}$ of the three-dimensional space and contin-*

uously differentiable in the closure $\bar{\mathcal{D}}$. Then the following equalities hold:

$$\int_{\mathcal{D}} \left(u'_x v'_x + u'_y v'_y + u'_z v'_z\right) dxdydz + \int_{\mathcal{D}} v \cdot \Delta u \, dxdydz = \int_S v \frac{\partial u}{\partial n} \, d\sigma, \qquad (7)$$

$$\int_{\mathcal{D}} (u\Delta v - v\Delta u) \, dxdydz = \int_S \left(u \frac{\partial v}{\partial n} - v \frac{\partial u}{\partial n}\right) d\sigma,$$

where S is the boundary of $\mathcal{D}$.

Now let u be a solution to the Laplace problem. Setting $v = u$ in the first equality in (7), we obtain

$$\int_{\mathcal{D}} \left(u'^2_x + u'^2_y + u'^2_z\right) dxdydz = \int_S u \frac{\partial u}{\partial n} \, d\sigma. \qquad (8)$$

As follows from (8), if

$$\frac{\partial u}{\partial n}\bigg|_S = 0,$$

then

$$\operatorname{grad} u = 0$$

and therefore

$$u = \operatorname{const}.$$

Proposition. If u_1 and u_2 are solutions to the Neumann problem for the Laplace equation and

$$\frac{\partial u_1}{\partial n}\bigg|_S = \frac{\partial u_2}{\partial n}\bigg|_S,$$

then $u_1 - u_2 = \operatorname{const}$.

We now turn to studying the existence of solutions to boundary value problems.

The following existence theorem holds for the Dirichlet problem for the Laplace equation.

Theorem. Let f be a continuous function defined on a smooth surface* S, which is the boundary of a bounded domain $\mathcal{D}$. Then there exists a function u such that u is twice continuously differentiable inside $\mathcal{D}$ and continuous in the closure $\bar{\mathcal{D}}$ and

$$\Delta u = 0, \quad u|_S = f.$$

---

*By a smooth surface we mean a surface that can be represented by a twice continuously differentiable function in a neighborhood of any of its points.

It should be noted that the theorem on the existence of a solution to the Dirichlet problem still holds in the case of weaker conditions imposed on S (more precisely, in the case of the so-called Lyapunov surfaces).

We now formulate the existence theorem for the Neumann problem.

Theorem. *Let f be a continuously differential function defined on S such that*

$$\int f \, d\sigma = 0.$$

Then there exists a function u such that u is twice continuously differentiable inside $\mathcal{D}$ and continuously differentiable in the closure $\bar{\mathcal{D}}$ and

$$\Delta u = 0, \quad \left.\frac{\partial u}{\partial n}\right|_S = f.$$

Explicit formulas for solutions of the Dirichlet problem and the Neumann problem can be obtained in the case of some simple domains. For a ball and a half-space, these formulas are called the Poisson formulas. In these cases, in order to prove the existence theorems, it suffices to verify that the functions defined by these formulas, indeed, satisfy the Laplace equation and the boundary conditions.

In the case of domains of the general form, several methods for proving the existence theorem are known. One of the classical methods consists in reducing boundary value problems to integral equations.

The desired solutions to the boundary value problems are assumed to be represented in the form of potentials. A solution to the Dirichlet problem is assumed to have the form of the double-layer potential:

$$u = \int_S \frac{\partial}{\partial n}\left(\frac{1}{r}\right) \varphi \, d\sigma.$$

A solution to the Neumann problem is assumed to have the form of the simple layer potential:

$$u = \int_S \frac{1}{r}\varphi \, d\sigma.$$

The solution of the boundary value problem turns out to be equivalent to the solution of the integral Fredholm equation of the second kind for the density φ:

$$\varphi + \int_S G\varphi \, d\sigma = \frac{1}{2\pi}f. \tag{9}$$

In addition to equation (9), we consider the so-called *associated homogeneous integral equation*

$$\psi + \int_S G^* \psi \, d\sigma = 0. \tag{10}$$

There are two alternatives.

1. Equation (10) has only the trivial solution

$$\psi = 0.$$

Then equation (9) has a unique solution for any right-hand side f.

2. Equation (10) has finitely many linearly independent solutions

$$\psi_k + \int_S G^* \psi_k \, d\sigma = 0, \quad k = 1, \ldots, n.$$

Then the homogeneous equation (9) (with $f = 0$) has n linearly independent solutions. For equation (9) to have a solution, it is necessary and sufficient that the following conditions hold:

$$\int_S f \psi_k \, d\sigma = 0, \quad k = 1, \ldots, n.$$

The two assertions formulated above are called Fredholm's theorems.

In the study of boundary value problems, it is proved that the associated integral equation is equivalent to a boundary value problem that is called the adjoint of the original problem. The uniqueness of the solution of the adjoint boundary value problem implies the uniqueness of the solution of the associated integral equation and, in view of Fredholm's theorems, the existence of a solution to the integral equation corresponding to the original boundary value problem.

The boundary value problems for the Helmholtz equation (4) are reduced to integral equations with the help of the potentials

$$\int_S \frac{\cos(\lambda r)}{r} \varphi \, d\sigma, \quad \int_S \frac{\partial}{\partial n} \left(\frac{\cos(\lambda r)}{r} \right) \varphi \, d\sigma.$$

Boundary value problems for elliptic equations of the general form with variable coefficients also can be reduced to integral Fredholm equations of the second kind, but the proof of the existence of the corresponding potentials is equivalent to the proof of the existence of a solution to the original boundary value problem.

The classic method of proving the theorems on the existence of solutions to boundary value problems for elliptic equations of the general form consists in reducing these problems to variational ones, i.e., the problems of finding an extremum of certain functionals.

6.5. HYPERBOLIC AND PARABOLIC EQUATIONS

One of the simplest equations of the hyperbolic type is the vibrating string equation, or d'Alembert's equation

$$\frac{\partial^2 u}{\partial t^2} - \frac{\partial^2 u}{\partial x^2} = 0. \tag{1}$$

A solution to this equation characterizes small vibrations of a stretched string: the function $u(x,t)$ represents the deviation of the string from the equilibrium position.

For equation (1), the following Cauchy problem is considered: find a solution to equation (1) that satisfies the conditions

$$u(x,0) = f_0(x), \quad u'_t(x,0) = f_1(x). \tag{2}$$

The theory of solving the Cauchy problem (2) for equation (1) is obtained relatively easily from the general representation of solutions to equation (1). It can be shown that any solution to equation (1) can be represented in the form

$$u(x,t) = \varphi(x - t) + \psi(x + t), \tag{3}$$

where φ and ψ are twice continuously differentiable functions.

The representation (3) implies d'Alembert's formula, which yields the solution of the Cauchy problem

$$u(x,t) = \frac{1}{2}[f_0(x - t) + f_0(x + t)] + \frac{1}{2} \int\limits_{x-t}^{x+t} f_1(\xi)\, d\xi. \tag{4}$$

The uniqueness of the solution also follows from it.

Aside from the Cauchy problem for equation (1), the so-called mixed problems are considered. In mixed problems, boundary conditions are specified in addition to the Cauchy data. Examples of simplest mixed problems are as follows:

1) vibrations of a string with fixed ends

$$u(0,t) = u(l,t) = 0;$$

2) vibrations of a string with loose ends

$$u'_x(0,t) = u'_x(l,t) = 0.$$

Solutions of mixed problems can be obtained using either the representation (3) or the method of separation of variables (the Fourier method).

In the case of two independent variables, the hyperbolic equation of the general form can be reduced to the following form using a change of variables:

$$\frac{\partial^2 u}{\partial t^2} - \frac{\partial^2 u}{\partial x^2} + a\frac{\partial u}{\partial t} + b\frac{\partial u}{\partial x} + cu = 0. \tag{5}$$

The Cauchy problem for equation (5) with variable coefficients can be reduced to a special case of Fredholm integral equations of the second kind, namely, Volterra equations. As was established in the theory of integral equations, a Volterra equation always has a unique solution, which can be obtained using the method of successive approximations.

Solutions to the Cauchy problems for wave equations in the case of two, three, or more space variables are given by certain formulas, which are called the Kirchhoff formulas.

Cauchy problems and mixed problems for hyperbolic equations are reduced to Volterra integral equations of the second kind. The questions concerning the uniqueness of solutions are answered with the help of the so-called energy inequalities.

A simplest equation of the parabolic type is the one-dimensional heat conduction equation

$$\frac{\partial u}{\partial t} - \frac{\partial^2 u}{\partial x^2} = 0. \tag{6}$$

The following Cauchy problem is stated for equation (6): find a solution to equation (6) that satisfies the condition

$$u(x,0) = f(x). \tag{7}$$

For the solution of the problem to be unique, certain conditions need to be imposed on the behavior of $u(x,t)$ as $|x| \to \infty$.

In the simplest case, under the condition

$$\lim_{|x|\to\infty} u(x,t) = 0$$

the uniqueness of the solution follows from the maximum principle.

The solution of the problem (7) is given by the Poisson formula

$$u(x,t) = \frac{1}{2\sqrt{\pi t}} \int_{-\infty}^{\infty} e^{-(x-\xi)^2/4t} f(\xi) \, d\xi. \tag{8}$$

Mixed problems are also stated for equation (6), in which some boundary conditions must be satisfied in addition to the initial condition (7), such as

$$u(0,t) = u(l,t) = 0$$

or

$$u_x(0,t) = u_x(l,t) = 0.$$

The existence, uniqueness, and continuous dependence (stability) theorems are proved for Cauchy problems and mixed problems in the case of parabolic equations of the general form with variable coefficients.

6.6. THE NOTION OF WELL-POSEDNESS

The notion of the well-posedness (properness) of a problem statement for differential equations was first formulated by J. Hadamard in the beginning of the twentieth century.

A problem is said to be well-posed if the following conditions hold:

1) *there exists a solution to the problem,*

2) *the solution is unique,*

3) *the solution continuously depends on the data.*

The first condition represents the requirement that there should not be too much data specified in the problem (i.e., the problem must not be overdetermined).

The second condition requires that there must be sufficient data so that the problem is not underdetermined.

The third condition is explained as follows. If the problem is related to the description of a physical phenomenon, the data of the problem are not exact. It is only an approximation of the data that is known. Thus, if there is no continuous dependence of the solution on the data, the solution is, in fact, undefined.

The well-posedness conditions formulated above need to be refined. In the theory of boundary value problems, the solutions and the data are assumed to be elements of some function spaces and the well-posedness conditions are formulated as follows:

1) *there exists a solution to the problem for any data from a closed subspace of one of the linear normed spaces* $\mathbb{C}^{(K)}, \mathbb{L}_p, \mathbb{W}_p^{(l)}, \ldots,$ *and the solution belongs to one of these spaces;*

2) *the solution of the problem is unique in one of the specified spaces;*

3) *infinitely small variations of the data in the space that the data are assumed to belong to correspond to infinitely small variations of the solution (in the space that solutions are assumed to belong to).*

Theorems on the existence and uniqueness of solutions and on the continuous dependence of solutions on the data constitute a proof of the well-posedness of a given problem.

For elliptic equations, problems whose boundary conditions are specified on the entire boundary of the domain of definition of solutions are well-posed.

For hyperbolic and parabolic equations, Cauchy problems and mixed problems whose data are specified on a part of the boundary of the considered domain are well-posed. From the point of view of the physical process described by the mathematical problem, the Cauchy data are interpreted as the state of the field at a given instant of time. The problem consists in determining the field at the subsequent instants of time.

Having defined the notion of well-posedness, J. Hadamard gave an example of a problem for differential equations that was not well-posed and, in his opinion, did not have any physical meaning. This was the Cauchy problem for the Laplace equation.

This problem and other problems that are ill-posed in the sense of Hadamard will be discussed in detail in the next chapter. At this point we note only that the Cauchy problem for the Laplace equation and a series of other ill-posed problems satisfy the assumption of the classic Cauchy–Kovalevskaya theorem. This theorem implies that there exists a unique solution to the Cauchy problem for the Laplace equation in the class of analytic functions.

Chapter 7.

Ill-posed problems

7.1.   ILL-POSED CAUCHY PROBLEMS

An ill-posed problem can be defined to be a problem that does not satisfy at least one of the well-posedness conditions. In the theory of ill-posed problems, however, the main attention is focused on the third condition.

We now turn to considering examples of ill-posed Cauchy problems for second-order equations.

The Cauchy problem for the Laplace equation. The two-dimensional mixed problem represents the simplest version of an ill-posed problem for the Laplace equation.

It is required to determine a function of two variables $u(x, y)$ in the rectangle $\{0 \leq x \leq \pi, 0 \leq y \leq y_0\}$ that satisfies the condition

$$\Delta u = 0,$$
$$u(0, y) = u(\pi, y) = 0, \quad u(x, 0) = f_0(x), \quad u'_y(x, 0) = f_1(x). \tag{1}$$

A solution to the Laplace equation that satisfies homogeneous conditions on the side boundaries of a half-strip can be represented in the form

$$u(x, y) = \sum_{k=1}^{\infty} \sin kx (a_k e^{ky} + b_k e^{-ky}).$$

The initial conditions yield

$$a_k = \frac{1}{\pi} \left(\int_0^\pi f_0(x) \sin kx \, dx + \frac{1}{k} \int_0^\pi f_1(x) \sin kx \, dx \right),$$

$$b_k = \frac{1}{\pi} \left(\int_0^\pi f_0(x) \sin kx \, dx - \frac{1}{k} \int_0^\pi f_1(x) \sin kx \, dx \right).$$

Thus, the solution of problem (1) is unique.

As follows from the above formulas, for the existence of a solution to problem (1) it is not sufficient that the data $f_0(x)$ and $f_1(x)$ belong to a function space whose norm is defined using finitely many derivatives.

We present Hadamard's example for problem (1).

Consider the following solution to the Laplace equation:

$$u_n(x, y) = a_n \sin(nx)e^{ny}.$$

The Cauchy data for u_n are

$$f_{0n}(x) = a_n \sin nx, \quad f_{1n}(x) = na_n \sin nx.$$

Obviously, n and a_n can be chosen so that the norm of the Cauchy data for u_n is as small as desired while u_n is as large as desired for any fixed y.

We now formulate the Cauchy problem for the Laplace equation in the general form.

Let $\mathcal{D}$ be a bounded domain of the n-dimensional space with smooth boundary S, S_1 be a subset of S, f_0 and f_1 be functions defined on S_1.

It is required to determine a function u satisfying the Laplace equation in the domain $\mathcal{D}$ and boundary conditions on S_1, i.e.,

$$\Delta u = 0, \quad u|_{S_1} = f_0, \quad \frac{\partial u}{\partial n}\bigg|_{S_1} = f_1. \tag{2}$$

Problem (2) has the same properties as problem (1): the solution of the problem is unique; for the existence of a solution, it is not sufficient that the data belong to functional spaces of the mentioned type; there is no continuous dependence of the solution on the data.

The Cauchy problem for elliptic equations of the general form with variable coefficients has similar properties.

The Cauchy problem for the heat conduction equation in reversed time. It is required to determine a function of two variables $u(x, t)$ in the rectangle

$$0 \leq x \leq \pi, \quad 0 \leq t \leq t_0$$

under the following conditions:

$$\frac{\partial u}{\partial t} = -\frac{\partial^2 u}{\partial x^2}, \quad u(0, t) = u(\pi, t) = 0, \quad u(x, 0) = f(x). \tag{3}$$

The solution to problem (3) is given by the formula

$$u(x, t) = \sum_{k=1}^{\infty} a_k \sin(kx) e^{k^2 t},$$

where $a_k = \frac{2}{\pi} \int\limits_0^\pi \sin(kx) f(x)\, dx$. It follows from this formula that the solution of the problem is unique and Hadamard's example holds:

$$u_n(x, t) = a_n \sin(nx) e^{n^2 t}.$$

The Cauchy problem for the heat conduction equation with data specified on a time-like manifold. It is required to determine a function of two variables $u(x, t)$ in the rectangle

$$0 \leq x \leq l, \quad 0 \leq t \leq t_0$$

satisfying the conditions

$$\frac{\partial u}{\partial t} = \frac{\partial^2 u}{\partial x^2}, \quad u(0, t) = f_0(t), \quad u_x'(0, t) = f_1(t). \tag{4}$$

Problem (4) satisfies the assumption of the Cauchy–Kovalevskaya theorem. The solution of the problem is unique, but the proof of uniqueness in this case is not as easy as for problem (3). Hadamard's example for problem (4) is

$$u_n(x, t) = a_n \sin(2n^2 t + nx) e^{nx}.$$

The Cauchy problem for the wave equation with data specified on a time-like manifold. It is required to determine a function of three variables $u(x, y, t)$ in the domain

$$0 \leq x, \quad 0 \leq y \leq \pi, \quad 0 \leq t \leq \pi,$$

that satisfies the conditions

$$\frac{\partial^2 u}{\partial t^2} = \Delta u,$$
$$u(x, y, 0) = u(x, y, \pi) = 0, \quad u(0, y, t) = f_0(y, t), \tag{5}$$
$$u'_x(0, y, t) = f_1(y, t).$$

The solution of problem (5) is unique. Hadamard's example for this problem is

$$u_n(x, y, t) = a_n e^{nx} \sin(\sqrt{5}\, ny) \sin(2nt).$$

7.2.   ANALYTIC CONTINUATION AND INTERIOR PROBLEMS

There are several statements of problems of analytic continuation. We present two statements of the problems of analytic continuation for functions of a complex variable.

The first statement. Assume that $f(z)$ is an analytic function that is regular in a bounded domain $\mathcal{D}$ of the complex plane and continuous on $\bar{\mathcal{D}}$,

$$|f(z)| \le c, \quad z = x + iy, \quad z \in \mathcal{D}, \tag{1}$$

Γ is the boundary of $\mathcal{D}$, Γ_1 and Γ_2 are parts of Γ such that $\Gamma_1 \cup \Gamma_2 = \Gamma$ and $\Gamma_1 \cap \Gamma_2 = \emptyset$. Suppose that the values of $f(z)$ on Γ_1 are known and it is required to determine $f(z)$ in an interior part of $\mathcal{D}$. If $\Gamma = \Gamma_1$, then the solution is given by the Cauchy integral

$$f(z) = \frac{1}{2\pi i} \int_{\Gamma} \frac{f(\xi)}{\xi - z}\, d\xi.$$

For $\Gamma_1 \ne \Gamma$, we will show that the problem of analytic continuation is equivalent to the Cauchy problem for the Laplace equation.

1. Suppose that the values of a harmonic function u and its normal derivative u_n are specified on Γ_1. Put

$$f(z) = u(z) + iv(z),$$

where v is the conjugate of u. As is known,

$$v(z) = \int_{z_0}^{z} \frac{\partial}{\partial n} u(z)\, ds + C_1$$

on the curve Γ_1, where z_0 is one of the endpoints of Γ_1. Thus, if u and u_n are known on Γ_1, it can be assumed that the values of the analytic function $f(z)$ are known on Γ_1.

2. Suppose that a function $f(z) = u(z) + iv(z)$ is specified on Γ_1, where u and v are conjugate harmonic functions. It follows from the Cauchy–Riemann that

$$\frac{\partial u}{\partial n}\bigg|_{\Gamma_1} = \frac{\partial v}{\partial s}\bigg|_{\Gamma_1}$$

(v_s' is a derivative of v along Γ_1). Therefore, taking the derivative of $f(z)$ and $\bar{f}(z)$ along Γ_1, we have

$$\frac{\partial u}{\partial n}\bigg|_{\Gamma_1} = \frac{1}{2}\frac{\partial}{\partial s}(f - \bar{f})\bigg|_{\Gamma_1}.$$

Thus, we obtain the Cauchy data for $u(x, y)$. We conclude that if $\Gamma_1 \neq \Gamma$, then the problem of analytic continuation is equivalent to the Cauchy problem for the Laplace equation and, consequently, it is ill-posed in the classical sense.

The uniqueness of the solution of the above problem of analytic continuation is proved in textbooks on the theory of functions of a complex variable.

The second statement. Suppose that the values of an analytic function $f(z)$ are specified not on Γ_1, but on a set $\mathcal{M}$ included in the interior of the domain $\mathcal{D}$. If $\mathcal{M}$ has at least one limit point inside $\mathcal{D}$, then the solution of the problem of analytic continuation is unique. This assertion is a classic theorem of the theory of analytic functions. Suppose that the set $\mathcal{M}$ is a sequence with one or finitely many limit points inside $\mathcal{D}$. It turns out that the instability of the problem of analytic continuation is, in a sense, even higher than that of the Cauchy problem for the Laplace equation.

The second statement of the problem of analytic continuation is a special case of the class of problems for differential equations that are called *interior problems*.

By interior problems we mean problems of the following type.

Find a solution to an equation (or a system of equations) that is regular in a domain $\mathcal{D}$, provided that the values of the solution (and, possibly, the values of its derivatives) are specified on a set $\mathcal{M}$ included in the interior of $\mathcal{D}$.

We now give an example of an interior problem for the two-dimensional diffusion equation.

Let $u(x, y, t)$ be a solution to the equation

$$\frac{\partial u}{\partial t} = \Delta u$$

that is regular in the cylinder

$$x^2 + y^2 < 1, \quad 0 < t < t_0,$$

and let (x_k, y_k) be a sequence of points inside the unit disk $\lim_{k \to \infty} x_k = \lim_{k \to \infty} y_k = 0$.

It is required to determine the function u if the following functions are known:

$$u(x_k, y_k, t) = f_{0k}(t),$$
$$u'_x(x_k, y_k, t) = f_{1k}(t),$$
$$u'_y(x_k, y_k, t) = f_{2k}(t).$$

7.3.   WEAKLY AND STRONGLY ILL-POSED PROBLEMS. PROBLEMS OF DIFFERENTIATION.

In the above examples of ill-posed problems, there is no continuous dependence of solutions on the data if the solutions and the data are assumed to be elements of function spaces whose norm is defined using finitely many derivatives.

However, problems in which the dependence of solutions on the data is continuous for one pair of spaces and not continuous for another pair of spaces can also be called ill-posed. In the theory of ill-posed problems, those problems in which the dependence of solutions on the data is not continuous for all pairs of functional spaces whose norms are defined using finitely many derivatives are called *strongly ill-posed*.

Problems in which the dependence of solutions on the data is continuous for one pair of spaces and not continuous for another pair of spaces are called *weakly ill-posed*.

The problem of differentiation is a classic weakly ill-posed problem. The problem of differentiation is well-posed if it is considered for the pairs of spaces $\{C, C^1\}$, $\{L_2, W_2^1\}$, and so on, whereas it is ill-posed for the pais of spaces $\{C, C\}$, $\{L_2, L_2\}$.

For a function whose values are obtained in measurements, it is reasonable to consider the problem of differentiation for the last two pairs of spaces because the measurement errors are usually estimated in C or L_2.

7.4. REDUCING ILL-POSED PROBLEMS TO INTEGRAL EQUATIONS

Classic well-posed boundary value problems for differential equations can be reduced to Fredholm integral equations of the second kind. Some boundary value problems are reduced to singular integral equations for which Noether's theorems hold.

Ill-posed problems described in the above examples can be reduced to Fredholm integral equations of the first kind such that the properties of their solutions substantially differ from those of the solutions to Fredholm equations of the second kind and the solutions to classic singular equations. We now give several examples of reducing ill-posed problems to integral equations.

The Cauchy problem for the heat conduction equation in reversed time. Determine a bounded function $u(x,t)$ in the strip

$$-\infty < x < \infty, \quad 0 \le t \le t_0$$

under the conditions

$$\frac{\partial u}{\partial t} = -\frac{\partial^2 u}{\partial x^2}, \quad u(x,0) = f(x). \tag{1}$$

The function $f(x)$ is assumed to be bounded and continuous.

Suppose that there exists a continuous solution of problem (1) in the closed strip. Set

$$\varphi(x) = u(x, t_0).$$

Then, by the Poisson formula,

$$u(x,t) = \frac{1}{2\sqrt{\pi(t_0 - t)}} \int_{-\infty}^{\infty} e^{-\frac{(x-\xi)^2}{4(t_0-t)}} \varphi(\xi)\, d\xi$$

and therefore the solution of problem (1) is equivalent to the solution of the solution of the following integral equation for $\varphi(x)$:

$$\frac{1}{2\sqrt{\pi t_0}} \int_{-\infty}^{\infty} e^{-\frac{(x-\xi)^2}{4t_0}} \varphi(\xi)\, d\xi = f(x).$$

The interior problem for the Laplace equation in the half-plane.
Determine a bounded function $u(x, y)$ in the half-plane $y \geq 0$ that satisfies
the conditions

$$\Delta u = 0,$$

$$u(x, y_0) = f(x), \quad y_0 > 0.$$

(2)

Suppose that there exists a solution to the problem. Put

$$\varphi(x) = u(x, 0).$$

Then, by the Poisson formula,

$$u(x, y) = \frac{1}{\pi} \int\limits_{-\infty}^{\infty} \frac{y}{y^2 + (x - \xi)^2} \varphi(\xi) \, d\xi.$$

As a result, the solution of problem (2) is equivalent to the solution of the
integral equation

$$\frac{1}{\pi} \int\limits_{-\infty}^{\infty} \frac{y_0}{y_0^2 + (x - \xi)^2} \varphi(\xi) \, d\xi = f(x).$$

The Cauchy problem for the Laplace equation. Assume that D
is a domain in the three-dimensional space bounded by a smooth surface S;
S_1 and S_2 are parts of the surface S such that

$$S = S_1 \cup S_2, \quad S_1 \cap S_2 = \varnothing;$$

f_0 and f_1 are continuous functions whose values are specified on S_1.

It is required to determine a function $u(x)$ that is twice continuously
differentiable inside D and continuously differentiable in $\bar{D}$ and that satisfies
the conditions

$$\Delta u = 0,$$

$$u\big|_{S_1} = f_0, \qquad \frac{\partial u}{\partial n}\bigg|_{S_1} = f_1.$$

(3)

Suppose that there exists a solution of problem (3). We will seek a
solution to (3) in the form of simple-layer potentials

$$u = \int\limits_{S_1} \frac{1}{r} \varphi_1 \, d\sigma + \int\limits_{S_2} \frac{1}{r} \varphi_2 \, d\sigma.$$

It follows from the Cauchy conditions that the functions φ_1 and φ_2 satisfy the system of integral equations

$$\int_{S_1} \frac{1}{r} \varphi_1 \, d\sigma + \int_{S_2} \frac{1}{r} \varphi_2 \, d\sigma = f_0, \quad x \in S_1,$$

$$2\pi\varphi_1 + \int_{S_1} \frac{\partial}{\partial n} \left(\frac{1}{r}\right) \varphi_1 \, d\sigma + \int_{S_2} \frac{\partial}{\partial n} \left(\frac{1}{r}\right) \varphi_2 \, d\sigma = f_1, \quad x \in S_1. \tag{4}$$

We set

$$f_1 - \int_{S_2} \frac{\partial}{\partial n} \left(\frac{1}{r}\right) \varphi_2 \, d\sigma = 2\pi f_2.$$

Then the second equality in (4) is written as

$$\varphi_1 + \frac{1}{2\pi} \int_{S_1} \frac{\partial}{\partial n} \left(\frac{1}{r}\right) \varphi_1 \, d\sigma = f_2. \tag{5}$$

Consider (5) as an integral equation for φ_1. It can be shown that there exists a solution of equation (5) that can be represented in the form

$$\varphi_1 = f_2 + \int_{S_1} G f_2 \, d\sigma,$$

where G is a function of two points of the surface S_1.

Substituting the expression for φ_1 into the first equation in (4), we obtain the following integral equation of the first kind for the function φ_2:

$$\int_{S_2} K \varphi_2 \, d\sigma = f_0.$$

The mixed problem for the heat conduction equation with Cauchy data on a time-like manifold. Determine a function $u(x,t)$ in the rectangle

$$0 \le x \le l, \quad 0 \le t \le t_0$$

such that $u(x,t)$ satisfies the conditions

$$\frac{\partial u}{\partial t} = \frac{\partial^2 u}{\partial x^2}, \tag{6}$$

$$u(0,t) = f(t), \quad u_x(0,t) = 0, \quad u(x,0) = 0.$$

We will seek a solution to problem (6) in the form of the heat potential

$$u(x,t) = \frac{1}{2} \int_0^t \left[e^{-\frac{(x-l)^2}{4(t-\tau)}} + e^{-\frac{(x+l)^2}{4(t-\tau)}} \right] \varphi(\tau)\, d\tau.$$

A function that can be represented in this form satisfies the heat conduction equation and the second and the third boundary conditions in (6).

The first boundary condition in (6) yields the following Volterra integral equation for φ:

$$\int_0^t e^{-\frac{l^2}{4(t-\tau)}} \varphi(\tau)\, d\tau = f(t).$$

Chapter 8.

Physical problems leading to ill-posed problems

8.1. INTERPRETATION OF MEASUREMENT DATA FROM PHYSICAL DEVICES

The operation of many devices registering nonstationary physical fields are described by the following scheme:

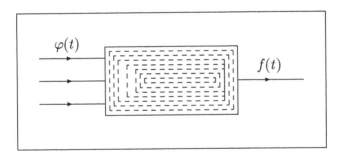

The signal $\varphi(t)$ arrives at the input of the device, and the function $f(t)$ is registered at the output. In the simplest case, the functions $\varphi(t)$ and $f(t)$ are related by the formula

$$\int_0^t g(t-\tau)\varphi(\tau)\,d\tau = f(t). \qquad (1)$$

In this case, $g(t)$ is called the *impulse transition function* of the device. In theory, $g(t)$ represents a function registered by the device in the case where the input of the device is represented by the generalized function $\delta(t)$ (Dirac's delta). In practice, to obtain the function $g(t)$ characterizing the operation of the device, a sufficiently short impulse is provided as an input.

Thus, the problem of interpreting the readings of the device, (i.e., the problem determining the form of the input signal) is reduced to solving the integral equation of the first kind (1).

The relationship between the input signal $\varphi(t)$ and the function $f(t)$ registered at the output can be more complex. In the case of a "linear" device, this relationship is written as

$$\int_0^t K(t,\tau)\varphi(\tau)\,d\tau = f(t).$$

The relationship between the functions $\varphi(t)$ and $f(t)$ can also be nonlinear:

$$\int_0^t K(t,\tau,\varphi)\,d\tau = f(t).$$

For example, this is the case for devices that register variable electromagnetic fields, modes of pressure, and field strength in a continuous medium and seismographs registering the movements of the Earth's crust.

The Fourier and Laplace transforms can be used to solve the simplest equation (1). Let $\tilde{g}(\lambda)$, $\tilde{\varphi}(\lambda)$, and $\tilde{f}(\lambda)$ be the Fourier transforms of the functions $g(t)$, $\varphi(t)$, and $f(t)$:

$$\tilde{g}(\lambda) = \int_0^\infty e^{i\lambda t} g(t)\,dt, \quad \tilde{\varphi}(\lambda) = \int_0^\infty e^{i\lambda t}\varphi(t)\,dt, \quad \tilde{f}(\lambda) = \int_0^\infty e^{i\lambda t} f(t)\,dt.$$

Then the convolution theorem implies that

$$\tilde{g}(\lambda)\tilde{\varphi}(\lambda) = \tilde{f}(\lambda).$$

Hence, inverting the Fourier transform, we obtain the following formula for the solution of (1):

$$\varphi(t) = \frac{1}{2\pi}\int_{-\infty}^\infty e^{-i\lambda t}\tilde{\varphi}(\lambda)\,d\lambda = \frac{1}{2\pi}\int_{-\infty}^\infty e^{-i\lambda t}\frac{\tilde{f}(\lambda)}{\tilde{g}(\lambda)}\,d\lambda. \tag{2}$$

Formula (2) is unstable because the function $\tilde{g}(\lambda)$, which is the Fourier transform of the impulse transition function of the device, tends to zero as $\lambda \to \infty$ in the case of real devices and, consequently, arbitrarily small errors in the measurement of $\tilde{f}(\lambda)$ can lead to large variations in the solution $\varphi(t)$ for sufficiently large λ.

If $g(t)$ is constant, then the problem of solving (1) becomes a differentiation problem.

8.2. INTERPRETATION OF GRAVIMETRIC DATA

Some problems of interpreting gravitational and magnetic fields related to exploration of mineral resources can be reduced to ill-posed problems.

If the Earth were a spherically homogeneous ball, the strength of the gravitational field on the surface would be constant. The inhomogeneity of the land surface and the distribution of density of matter causes the strength of the gravitational field on the surface of the Earth to deviate from the average value. These deviations are small in terms of percentage, but are effectively registered by physical devices, gravimeters*. The results of gravitational measurements are used in prospecting and exploration of mineral resources.

Figures 1 and 2 show the sections of typical geological structures of areas of the Earth's crust and the results of gravitational measurements (the vertical components of the anomalies of gravitational field strength). These anomalies occur because the density of the basement and intrusions[†] is usually substantially higher than the density of sediments[‡]. The interpretation of gravitational data consists in determining the relief of the basement surface and the location and form of intrusions from the gravitational measurements data. Mineral deposits are often accompanied by specific forms of basement and intrusions. For example, oil fields are often accompanied by elevations of the basement surface (domes), while ore deposits are accompanied by intrusions.

We now focus on the problem of determining the location of intrusions from gravitational measurements data. If the distance between two geologi-

---

*The average strength of the gravitational field of the Earth is 1 kg = 980 gal. Typical anomalies are 0,5–10 mgls.

[†] A geological body formed as a result of intrusion of molten magma into the surrounding rock.

[‡] The density of sediments is 2–4 g/cm^3. The density of the basement and intrusions is 4–6 g/cm^3.

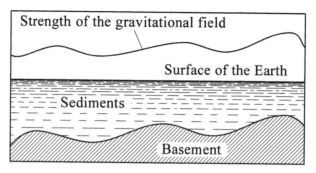

Figure 1

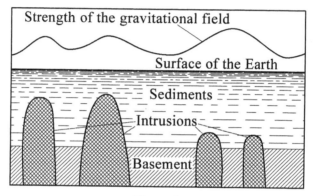

Figure 2

cal bodies (intrusions) is greater than the distance from them to the surface of the Earth, then their locations correspond to the local maxima of the anomalies (see the left side of Fig. 2). Otherwise, if the distance between them is less than the distance from them to the surface, then the two bodies correspond to a single local maximum (see the right side of Fig. 2) Geophysical measurements and the interpretation of their results represent the preliminary stage in prospecting for mineral deposits. For deposits covered by sediments, the main stage consists in drilling the exploratory wells and analyzing the drilling data.

If the shape of the anomaly on the right side of Fig. 2 leads us to the conclusion that it indicates a single body, then the natural decision to drill in the center of the anomaly will result in the well being drilled between the actual bodies that we are interested in. This was often the case in the practice of geological exploration.

Then geophysicists came up with the following proposition: the results of the gravitational measurements performed on the surface of the Earth should be used to calculate the anomalous gravitational field at a certain depth under the surface. If the anomaly at that depth will still have the shape of a single local maximum, then there is a high probability that the anomaly is generated by a single body. Otherwise, if two local maxima appear as a result of the recalculation, then the natural conclusion is that there are two bodies, and the location for drilling the well must be chosen accordingly.

We now consider the statement of a mathematical problem corresponding to the problem of recalculation of the gravitational field described above. We confine ourselves to the case of two variables for the sake of simplicity. This condition often reflects actual geophysical situations, since many geological structures stretch in a certain direction while the geophysical measurements are performed on the path perpendicular to this direction.

As is known, outside of the masses that produce the gravitational field, the field potential and the components of the field strength satisfy the Laplace equation. Let a line on the surface of the Earth be identified with the x axis, and $u(x, y)$ be the vertical component of the gravitational field strength produced by several bodies situated below the level $y = -H$, where $H > 0$. Then the function $u(x, y)$ is a solution to the Laplace equation $\Delta u = 0$ that is regular in the half-plane $y > -H$. The values of $u(x, y)$ are specified on the axis x, i.e., a function $f(x)$ is defined so that

$$f(x) = u(x, 0).$$

It is required to determine the values of the function

$$\varphi(x) = u(x, -h), \quad 0 < h < H.$$

Using the Poisson formula, the solution of this problem is reduced to the integral equation of the first kind

$$\int\limits_{-\infty}^{\infty} \frac{h\varphi(\xi)}{h^2 + (x - \xi)^2} \, d\xi = f(x).$$

The resulting problem is equivalent to the Cauchy problem for the Laplace equation. Hadamard's example for this problem is as follows: if $f(x) = \alpha \sin nx$, then $\varphi(x) = \alpha e^{nh} \sin nx$.

The analogous problem statement arises in the interpretation of anomalies of a constant magnetic field, since its potential and components outside of the magnetic masses also satisfy the Laplace equation.

8.3.   PROBLEMS FOR THE DIFFUSION EQUATION

Consider the equation

$$\frac{\partial u}{\partial t} = \frac{\partial^2 u}{\partial x^2} \quad x \in [0, l], \quad t \in [0, T], \tag{1}$$

which describes the change in temperature in a homogeneous rod or a diffusion process, i.e., the change in the concentration of a substance in a solution. A classic problem for equation (1) is the mixed problem

$$u(x, 0) = f(x), \quad u(0, t) = u(l, t) = 0. \tag{2}$$

The physical meaning of the problem (1), (2) is explained as follows. The temperature distribution in a rod or the concentration of a substance in a cylindrical vessel is known at the initial instant of time. It is required to determine the temperature distribution or the concentration of the substance at the subsequent instants. The boundary conditions mean that constant temperature or concentration is maintained at the endpoints in the given process.

We now turn to the following ill-posed problem for equation (1):

$$u(x, T) = f(x), \quad u(0, t) = u(l, t) = 0. \tag{3}$$

It is obvious that problem (3) for equation (1) is equivalent to problem (2) for the equation

$$\frac{\partial u}{\partial t} = -\frac{\partial^2 u}{\partial x^2}. \tag{4}$$

This problem has a clear physical meaning. Based on physical measurements, the distribution of temperature or concentration is known at a certain instant of time. It is required to determine the history of the process, i.e., the distribution of temperature or concentration at the previous instants of time.

The following ill-posed problem for equation (1) also has a physical meaning:

$$u(0, t) = f_1(t), \quad u'_x(0, t) = f_2(t). \tag{5}$$

Suppose that the concentration of the substance cannot be measured in the whole cylinder, but it is possible to measure the concentration and flow of the substance at one of the endpoints. The following example implies that this problem is ill-posed:

$$f_1(t) = \sqrt{2/n} \sin nt, \quad f_2(t) = \cos nt + \sin nt,$$

$$u(x, t) = \sqrt{2/n} \sin(nt + \sqrt{2/n}\, x) e^{\sqrt{2/n}\, x}.$$

8.4.   DETERMINING PHYSICAL FIELDS FROM THE MEASUREMENTS DATA

In the problem of interpreting the results of gravitational measurements discussed in Section 8.2, in practice, geophysicists know the values of the gravitational field component not on the whole surface of the Earth and not on a segment, but on a finite set of points. Thus, the problem arises of finding a function $\varphi(x)$ from equation (1) of Section 8.1 under the condition that the values of the right-hand side on a finite set of points are known, i.e.,

$$f_k = f(x_k) \quad k = 1, \ldots, n.$$

The solution of this problem is not unique, but it is clear that if the distribution of the points $\{x_k\}$ in an interval $[a, b]$ is sufficiently dense, the function $f(x)$ can be on the entire interval using interpolation techniques. This will introduce an additional error in the right-hand side of equation (1) of Section 8.1.

In applied problems, in particular in the problem of interpreting gravimetric data, measuring a value of a function at a point involves certain expenses. In this connection, the problem arises of choosing a system of measurement points in a rational way. It is obviously not reasonable to choose the grid of measurement points to be so dense that the errors of simple interpolation are considerably less than the measurement errors. It is of practical importance to find out the accuracy of determining a solution to equation (1) in Section 8.1 if the following numbers are known:

$$\{f_k^\varepsilon\}, \quad \left|f_k^\varepsilon - f(x_k)\right| \le \varepsilon, \quad k = 1, \ldots, n.$$

The analogous statement is also meaningful in the case where the set of points $\{x_k\}$ is infinite. If this set has a limit point, then the solution to equation (1) in Section 8.1 is uniquely determined from the sequence of points

$$f_k = f(x_k), \quad k = 1, \ldots, \infty.$$

The uniqueness of determining the solution to equation (1) in Section 8.1 from the data $\{f_k\}$ follows from the well-known theorem of the theory of analytic functions: if an analytic function vanishes on a set that has a limit point inside the domain of regularity for this function, then it is identically equal to zero.

It should be noted that many physical fields are described by analytic functions of space variables.

The components of the variable electromagnetic field strength in media with constant properties satisfy the Helmholtz equation

$$\Delta u + k^2 u = 0.$$

Solutions to this equation with constant coefficient k are analytic functions. Therefore, the problem of determining the structure of an electromagnetic field from the results of measurements performed on a discrete set is reduced to the problem of reconstructing an analytic function (of one or several variables) from its values on this set.

8.5.   TOMOGRAPHY

Tomography has been a fast developing area of science and technology during recent years. The term originates from the Greek words "tomos" (slice, section) and "graphos" (writing, drawing). The term first appeared in the scientific and technical literature in the 1920s in connection with the invention of medical X-ray tomographs. Tomography started to develop rapidly in the 1970s. There are two main aspects that define a modern tomograph. The first one is related to its design including the part that registers the data as a result of scanning the object under study and the part that transforms the data into the image of a section of the object. The second aspect is related to the underlying mathematical theory.

Although, until recently, the term "tomography" was used mostly in connection with the problems of medicine, the same principles have long been used in geophysics for analyzing the sections of areas of the Earth's crust on the basis of the data of geophysical measurements on a profile.

At present, tomographs and tomography are widely used in very diverse areas of science and technology.

Mathematical problems arising in the interpretation of tomographic data (obtaining the image of a section of the object under study) are related to the so-called inverse problems of mathematical physics.

The mathematical methods used in modern medical X-ray tomographs are based on the inversion of the Radon transform. Application of the Radon transform in X-ray tomography involves a certain model of interaction between the X-ray radiation flow and the matter of the object under study. This model does not take into account several factors that can play an important role in various tomographic problems. In the model that leads to the Radon transform, X-ray beams are assumed to be straight, i.e., the effects of geometric divergence and scattering are neglected.

Consider the process of X-ray scanning of an object. Let ψ denote the X-ray flow and a be the absorption coefficient. It is assumed that the scanning is performed within a thin layer, so that the absorption coefficient is a function of two variables that does not vanish in a bounded domain $\mathcal{D}$.

In tomographs, a section of the object under study is scanned with a series of plane-parallel X-ray flows at different angles. Thus, the flow ψ is a function of the variables x and y and the angle α:

$$\psi = \psi(x, y, \alpha).$$

Under the above assumptions, the flow ψ satisfies the simplest transport equation

$$(\cos \alpha)\psi_x + (\sin \alpha)\psi_y + a\psi = 0. \tag{1}$$

For $u = \ln \psi$, equation (1) implies

$$(\cos \alpha)u_x + (\sin \alpha)u_y + a = 0. \tag{2}$$

Assume that the flow is constant before it starts to pass through the object:

$$\psi = c.$$

Then, after the flow passes through the object, the values of the function u will be

$$u(x, y, \alpha) = \int_{\Gamma(x,y,\alpha)} a(\xi, \eta) \, ds, \tag{3}$$

where Γ is a line passing through $\mathcal{D}$.

The initial flow ψ is known, while the flow after passing through the object is to be measured. Thus, we can assume that $u(x, y, \alpha)$ is known and consider (3) to be an integral equation for the distribution of absorption intensity $a(x, y)$.

We now give a more detailed statement of the problem of solving equation (3). It is required to determine a function $a(x, y)$ that vanishes outside the bounded domain $\mathcal{D}$ under the condition that the following function is known:

$$\int_{\Gamma(x,y,\alpha)} a(\xi, \eta) \, ds = u(x, y, \alpha). \tag{4}$$

The transformation that establishes the correspondence between $a(x, y)$ and $u(x, y, \alpha)$ is called *the Radon transform*. The problem of solving equation (4) is the problem of inverting the Radon transform.

There are several known inversion formulas for the Radon transform. If u and a in (4) are viewed as elements of the space $\mathbb{C}$, then the problem of solving (4) is ill-posed. Equation (4) is not an operator equation of the first kind, since the Radon operator is not completely continuous.

The Radon problem is one of the problems of integral geometry. We will formulate the general problem of integral geometry on a plane. Let $\Gamma(p,q)$ be a collection of curves on a plane that depends on two parameters and has the following property: for any point in $\mathcal{D}$, there is a bundle of curves from $\Gamma(p,q)$ that passes through this point. Let $g(x,y,p,q)$ be a given function of four variables. It is required to determine a function $u(x,y)$ that vanishes outside the domain $\mathcal{D}$, provided that the following function is known:

$$\int_{\Gamma(p,q)} g(x,y,p,q)u(x,y)\,ds = f(p,q). \tag{5}$$

We now formulate the geophysical problem of interpreting seismic data, which leads to the problem of integral geometry, i.e., the problem of solving an equation of the form (5). This problem is called the seismic tomography problem.

Let $v(x,y)$ be the velocity of the wave process distribution in a two-dimensional medium and $\tau(x_1,y_1,x_2,y_2)$ be the time it takes for the wave to travel from a point (x_1,y_1) to a point (x_2,y_2).

As is known,

$$\tau = \int_{\Gamma(x_1,y_1,x_2,y_2)} n(x,y)\,ds, \quad n = 1/v, \tag{6}$$

where Γ is a ray connecting the points (x_1,y_1) and (x_2,y_2). The function τ satisfies the eikonal equations

$$\left(\frac{\partial \tau}{\partial x_1}\right)^2 + \left(\frac{\partial \tau}{\partial y_1}\right)^2 = n^2(x_1,y_1),$$

$$\left(\frac{\partial \tau}{\partial x_2}\right)^2 + \left(\frac{\partial \tau}{\partial y_2}\right)^2 = n^2(x_2,y_2). \tag{7}$$

The direct kinematic problem is the problem of determining the function τ from the given function v or n. In the mathematical sense, this is a classical Cauchy problem for the first-order equations (7) or a boundary value problem for the corresponding system of ordinary differential equations.

The inverse problem can be viewed as the problem of solving the operator equation (6). Since the rays Γ depend on the unknown function n, equation (6) is nonlinear.

Suppose that

$$n = n_0 + n_1$$

and the function n_0 is known, while the function n_1 is sufficiently small, so that we can assume that the terms of order $(n_1)^2$ are negligible in our equations.

In geophysics, this assumption is justified either when the distance between the pairs of points is small, in which case n_0 is simply constant (seismic radiography between wells), or when n_0 depends only on one variable and is determined from the solution of the one-dimensional inverse problem (dip shooting of the sedimentary layer). The above hypothesis can also be used in the problem of acoustic stratification of the ocean, where where the deviations of the speed of sound from the average value are small.

Under the above assumptions, equations (7) become as follows:

$$\frac{\partial \tau_0}{\partial x_j}\frac{\partial \tau_1}{\partial x_j} + \frac{\partial \tau_0}{\partial y_j}\frac{\partial \tau_1}{\partial y_j} = n_0(x_j, y_j)n_1(x_j, y_j). \tag{8}$$

From (8) we obtain

$$\tau_1(x_1, y_1, x_2, y_2) = \int\limits_{\Gamma_0(x_1, y_1, x_2, y_2)} n_1(x, y)\, ds, \tag{9}$$

where Γ_0 is the ray that connects the points (x_1, y_1) and (x_2, y_2) in the medium corresponding to the function $n_0(x, y)$. Since n_0 is assumed to be known, the collection of rays Γ_0 can also be assumed to be known.

Chapter 9.

Operator and integral equations

9.1. DEFINITIONS OF WELL-POSEDNESS

In this section we discuss the notions of well-posedness introduced and used by a number of authors. We will consider these notions as applied to operator equations or maps between topological vector spaces. The problems of mathematical physics mentioned above, problems of calculus, algebra, linear programming, and other problems can be represented in these terms relatively easily.

First, we give the general definition of well-posedness, which generalizes *various definitions of this notion.*

Assume that $(X, \mathcal{U})$ and $(Y, \mathcal{V})$ are topological vector spaces, $A \subseteq X$ and $B \subseteq Y$ are sets, $\mathcal{C} = \mathcal{C}(B, X)$ is a selected class of maps from B to X and their restrictions (for example, continuous compact maps or closed maps), $T : X \to Y$ is an arbitrary map from the selected class (for example, continuous or closed), and $T^{-1} : Y \to X$ is the inverse correspondence (not necessarily many-to-one).

Definition. A map (operator) T is called $(A, B, \mathcal{C})$-*well-posed* if the following conditions hold:

(1) $TA \supseteq B$,

(2) T_B^{-1} is many-to-one,

(3) T_B^{-1} is admissible.

Replacing the condition (1) with

(1′) $\overline{TA} \supseteq B$,

we obtain the definition of $(A, B, \mathcal{C})'$-well-posedness in the general sense.

If any of the conditions (1)–(3) does not hold, then the map T is said to be (A, B, C)-*ill-posed* (in the normal or general sense).

As was mentioned above, the notion of well-posedness for the statements of problems for differential equations was introduced by J. Hadamard. For example, in Hadamard (1902), the author writes: "There are two fundamental problems associated with partial differential equations: the Dirichlet problem and the Cauchy problem. In Cauchy problems, the conditions define the values of the function and its first-order derivatives at every point of the boundary. It is required to find sufficiently large classes of the cases where each of these problems is well-posed, i.e., solvable and determined."

It is also stated in Hadamard (1902) that "Any boundary value problem consists in finding a function that

(1°) is a solution of a (undetermined) partial differential equation;

(2°) satisfies certain boundary conditions (or determinedness conditions).

Such a problem is said to be *well-posed* if these conditions ensure the existence and uniqueness of the solution of the given equation." The simplest boundary value problem is the Cauchy problem.

Thus, the original Hadamard's definition of well-posedness includes two requirements: existence and uniqueness.

If the problem is formulated in terms of a map $T : X \rightarrow Y$, then the map corresponding to a well-posed problem is (A, B, C)-well-posed for $A = X$ and $B = Y$, where C represents arbitrary maps.

The requirement of continuous dependence of the solution on the data was introduced by D. Hilbert and R. Courant. The definition of well-posedness in Section 6.6 was originally given in Courant and Hilbert (1962). This definition was further refined. The resulting, more detailed definition is also given in Section 6.6 in the same form as was given in the well-known textbooks Sobolev (1992) and Petrovskii (1984).

The following statement holds for the notion of the well-posedness of a map introduced above.

The map that corresponds to a well-posed problem is (A, B, C)-*well-posed for* $A = X$ *and* $B = Y$, *where* C *represents arbitrary maps.*

We now give examples illustrating the introduced notion.

Examples. 1) A homeomorphism from $A = X$ onto $B = Y$ is an (A, B, C)-well-posed map if $C = C(B, A)$.

2) Assume that X is a Banach space with metric topology $\mathcal{U}$; A is a dense subspace of X $(\bar{A} = X)$; $Y = X$ and $\mathcal{V} = \mathcal{U}$, $B = Y$; $C = C(Y, X)$ is

a space of continuous maps from Y to X; I is the identity operator on X; $S : A \to X$ is a compact operator, $\lambda \in \mathbb{C}$ is a complex number; $T = \lambda I - S$ is a Fredholm operator from A to X.

Then the Fredholm operator $T = \lambda I - S$ is $(A, B, \mathcal{C})$-well-posed in the general sense (satisfies the conditions (1'), (2), (3)) for all $\lambda \notin \Sigma(S)$ (i.e., for λ that do not belong to spectrum of S), and ill-posed in the general sense (and certainly in the normal sense) for $\lambda \in \Sigma(S)$ (for all λ in the spectrum of S).

If the operator S is closed, then the operator T is $(A, B, \mathcal{C})$-well-posed in the normal sense for $\lambda \notin \Sigma(S)$ (see Lavrent'ev and Savel'ev, 1995, Subsection 3.8.2). The same is true if $A = X$ and S is bounded.

3) A compact one-to-one linear operator $S : X \to X$ that transforms an infinite-dimensional Banach space X into itself is $(A, B, \mathcal{C})$-ill-posed if $A = X$, $B = SA$, and $\mathcal{C} = \mathcal{C}(X, X)$ is the class of continuous maps because $S^{-1} : B \to A$ is unbounded (see Lavrent'ev and Savel'ev, 1995, Subsection 3.6.2).

In Lavrent'ev and Savel'ev (1995), one of the authors formulated the notion of Tikhonov well-posedness, or conditional well-posedness, on the basis of concepts described in Tikhonov (1943). Consider the equation

$$Tx = y, \quad x \in X, \quad y \in Y. \tag{1}$$

Definition. The problem of solving equation (1) is said to be *Tikhonov well-posed* if the following conditions hold:

1) It is known *a priori* that there exists a solution to the problem that belongs to a given set A.

2) The solution is unique on the set A.

3) The solution x continuously depends on the right-hand side y for the variations of y that keep the solutions within the set A.

The condition 3) can be alternatively stated as follows:

3') The operator T_{TA}^{-1} is continuous.

As was noted in Chapter 1 of the part "Basic concepts", if A is a compact set, then any of the conditions 3) or 3') follow from the conditions 1) and 2). Thus, the following statements hold.

Proposition. *If the problem of solving (1) is Tikhonov well-posed, then the map T is $(A, B, \mathcal{C})$-well-posed if $\mathcal{C}$ is the class of continuous maps.*

Proposition. *In addition, if A is a compact set and T is continuous, then T is $(A, B, \mathcal{C})$-well-posed if the first two conditions of Tikhonov well-posedness hold.*

Now let X and Y be complete metric spaces and T be a continuous map. We present two theorems from Ivanov, Vasin, and Tanana (2002) and Tikhonov (1943), which provide the sufficient conditions for 1) and 2) to imply 3) or 3').

Theorem 1. (A. M. Tikhonov) *Assume that the solution of equation (1) is unique and $A \subset X$ is a compact set. Then the operator T_{TA}^{-1} is continuous on the set TA.*

Theorem 2. (V. K. Ivanov) *Assume that the solution of equation (1) is unique and the well-posedness set $A \subset X$ is the algebraic sum*

$$A = A_1 + X_1,$$

where A_1 is a compact set and X_1 is a finite-dimensional subspace of the space X. Then the operator T_{TA}^{-1} is uniformly continuous on the set TA.

9.2. REGULARIZATION

In this section we follow the monograph Lavrent'ev (1962). Assume that X and Y are complete metric spaces, $T : X \to Y$ is a continuous map (operator), and the inverse operator T^{-1} is many-to-one and is not continuous.

Consider the equation

$$Tx = y, \quad x \in X, \quad y \in Y. \tag{1}$$

Suppose that the right-hand side of (1) is known only approximately, ie., there is an element y_δ such that

$$\rho(y_t, y_\delta) \le \delta,$$

where y_t is the exact right-hand side of (1), $Tx_t = y_t$, x_t is the exact solution to (1).

It is required to determine an approximate solution to equation (1) from the element y_δ, i.e., find an element x_ε such that

$$\rho(x_t, x_\varepsilon) \le \varepsilon,$$

where ε is sufficiently small for sufficiently small δ.

Definition. An operator $R(y, \alpha) : Y \to X$ depending on a parameter α is said to be *regularizing* for equation (1) in the neighborhood of $y = y_t$ if it has the following properties.

1) There exists a number $\delta_1 > 0$ such that the operator $R(y, \alpha)$ is defined for all $\alpha > 0$ and for $x \in X$ such that

$$\rho(y, y_t) \leq \delta_1.$$

2) There exists a function $\alpha = \alpha(\delta)$ such that for any $\varepsilon > 0$ there is a $\delta(\varepsilon) \leq \delta_1$ such that $\rho(x_\delta, x_t) \leq \varepsilon$ whenever $\rho(y_\delta, y_t) \leq \delta(\varepsilon)$ and $x_\delta = R(y_\delta, \alpha(\delta))$.

If a regularizing operator for equation (1) is determined, then the element

$$x_\alpha = R(y_\delta, \alpha),$$

where the value of α is consistent with the data error δ, can be chosen as an approximate solution constructed from the approximate data. The numeric parameter α is called the *regularization parameter*.

Thus, the problem of determining an approximate solution from approximate data is reduced to finding a regularizing operator and a regularization parameter. It should be noted that the choice of the regularization parameter must be consistent not only with the value of the error δ, but also with certain *a priori* properties of the sought solution x_t. Methods used to ensure this consistency will be considered below.

Theorem. *Suppose that an operator $R(y, \alpha)$ is defined for all $y \in Y$ and $\alpha > 0$, is continuous in y, and for all $y \in Y$*

$$\lim_{\alpha \to 0} R(Tx, \alpha) = x. \tag{2}$$

Then $R(y, \alpha)$ is a regularizing operator for equation (1).

Proof. Let $x_t \in X$, $y_t \in Y$, $y_\delta \in Y$, $Tx_t = y_t$, and $\rho(y_\delta, y_t) \leq \delta$. Then

$$\rho(R(y_\delta, \alpha), x_t) \leq \rho(R(y_\delta, \alpha), R(y_t, \alpha)) + \rho(R(y_t, \alpha), x_t).$$

Since $R(y, \alpha)$ is continuous, for any sufficiently small $\delta > 0$ the inequality

$$\rho(y_\delta, y_t) \leq \delta$$

implies

$$\rho(R(y_\delta, \alpha), R(y_t, \alpha)) \leq \omega(\delta),$$

where $\omega(\delta) \to 0$ as $\delta \to 0$.

As follows from (2), for any $\delta > 0$ there exists an $\alpha_1 = \alpha_1(\delta, x_t)$ such that for $\alpha \leq \alpha_1$ we have

$$\rho(R(y_t, \alpha), x_t) \leq \omega(\delta)$$

and therefore

$$\rho(R(y_\delta, \alpha), x_t) \leq 2\omega(\delta).$$

Since $\omega(\delta) \to 0$ for $\delta \to 0$, it follows that for any $\varepsilon > 0$ there is a $\delta(\varepsilon)$ such that the above inequalities for $\delta \leq \delta(\varepsilon)$ imply

$$\rho(R(y_\delta, \alpha), x_t) \leq \varepsilon.$$

$\square$

We now consider the so-called *variational regularization principle*.

Let $\Omega(x)$ be a nonnegative functional defined on a set X_1 that is everywhere dense in X. Assume that the following conditions hold:

1) $x_t \in X_1$;

2) for any $a > 0$, the set of elements of X_1 such that $\Omega(x) \leq a$ is compact;

3) if a sequence $\{x_n\} \in X_1$ $(n = 1, 2, \dots)$ converges and

$$\lim_{n \to \infty} x_n = x_0, \qquad \Omega(x_{n+1}) \leq \Omega(x_n),$$

then $x_0 \in X_1$ and $\Omega(x_0) \leq \Omega(x_n)$.

Functionals $\Omega(x)$ satisfying these conditions are called *stabilizing functionals* and the corresponding sequences $\{x_n\} \in X_1$ are called *minimizing sequences* for $\Omega(x)$.

Let y_δ be a given element such that

$$\rho(y_\delta, y_t) \leq \delta, \quad y_t = Tx_t.$$

We will seek an approximate solution to (1) on the set $P_\delta \subset X$:

$$x \in P_\delta : \rho(Tx, y_\delta) \leq \delta.$$

Since the inverse of the operator T is not a continuous operator, for any δ as small as desired, the set P_δ can contain elements that deviate from the sought element x_t by a finite value. Set

$$Q_{1,\delta} = P_\delta \cap X_1,$$
$$\tilde{x}_\delta : \Omega(\tilde{x}_\delta) = \inf \Omega(x), \quad x \in Q_{1,\delta}$$

(the existence of $\tilde{x}_\delta$ will be established later).

The element $\tilde{x}_\delta$ can be viewed as the result of application of an operator $\tilde{R}$ to y_δ:

$$\tilde{x}_\delta = \tilde{R}(y_\delta, \delta).$$

We will show that $\tilde{R}(y_\delta, \delta)$ is a regularizing operator for equation (1). First, we will prove that $\tilde{R}(y, \delta)$ is defined for any y_δ such that

$$\rho(y_\delta, y_t) \leq \delta.$$

Since the functional $\Omega(x)$ is nonnegative, there exists a greatest lower bound

$$\inf \Omega(x) = \Omega_0, \quad x \in Q_{1,\delta}.$$

Let x_n be a minimizing sequence for $\Omega(x)$, i.e.,

$$\lim_{n\to\infty} \Omega(x_n) = \Omega_0 \quad and \quad \Omega(x_{n+1}) \leq \Omega(x_n).$$

In view of the property 2), the sequence x_n lies in a compact set, which means, which implies that x_n has a convergent subsequence. Without the loss of generality, this subsequence can be assumed to coincide with the sequence x_n itself, i.e.,

$$\lim_{n\to\infty} x_n = x_\delta \quad and \quad \Omega(x_\delta) = \Omega_0 = \inf_{x\in Q_{1,\delta}} \Omega(x).$$

Since the element x_δ minimizes the functional $\Omega(x)$ on the set $Q_{1,\delta}$, it follows that

$$\Omega(x_\delta) \leq \Omega(x_t)$$

and therefore x_δ belongs to the compact set

$$Q_t = \{x : \Omega(x) \leq \Omega(x_t)\}.$$

Let y_n be a given subsequence such that

$$\rho(y_t, y_n) \leq \delta_n, \quad \delta_n \to 0, \quad n \to \infty.$$

For any δ_n, the sets P_{δ_n} and $Q_{1,\delta_n} = P_{\delta_n} \cap Q_1$ are defined.

As was established above, in any set Q_{1,δ_n} there is an element x_{δ_n} that minimizes the functional $\Omega(x)$ on this set. It follows that the sequence of numbers δ_n is associated with the sequence of elements x_{δ_n} in the compact set Q_t. This sequence has a convergent subsequence. This subsequence can be assumed to coincide with the sequence x_{δ_n} itself. Let $\tilde{x} = \lim\limits_{n \to \infty} x_{\delta_n}$. Since $x_{\delta_n} \in Q_{1,\delta_n} \subset P_{\delta_n}$, the following inequality holds for any x_{δ_n}:

$$\rho(T x_{\delta_n}, y_{\delta_n}) \leq \delta_n.$$

Passing to the limit as $n \to \infty$, we obtain

$$\rho(T\tilde{x}, y_t) = 0,$$

which implies that $\tilde{x} = x_t$ because the solution of (1) is unique. Thus, for the sequence $\delta_n \to 0$, the corresponding sequence x_{δ_n} converges to the exact solution x_t to equation (1).

9.3. LINEAR OPERATOR EQUATIONS

As in the previous section, we consider the operator equation of the first kind

$$Au = f, \quad u \in U, \quad f \in \mathcal{F}, \tag{1}$$

where A is a linear operator and U and $\mathcal{F}$ are Hilbert spaces.

In the case of linear equations, the notions introduced in the previous section are substantially simplified. Furthermore, it is possible to construct regularizing collections that can be explicitly expressed in terms of the operator A. We confine ourselves to discussing several subjects related to regularization under the assumption that the regularizing collection is a collection of linear operators.

Suppose that the solution of (1) is unique and V is a subset of the space U.

Definition. A collection of linear operators $\{R_\alpha\}$ $(0 < \alpha < \alpha_0)$ from $\mathcal{F}$ to U is said to be *regularizing* for equation (1) on V (see Ivanov, Vasin, and Tanana, 2002) if the following conditions hold:

1) for any α such that $0 < \alpha < \alpha_0$, the operator R_α is defined on the entire $\mathcal{F}$ and bounded;

2) for any $u \in V$,

$$\lim\limits_{\alpha \to 0} R_\alpha Au = u.$$

If the convergence in 2) is uniform on V, then R_α is said to be *uniformly regularizing* (Ivanov, Vasin, and Tanana, 2002). Since the operator A^{-1} is unbounded, there is no uniformly regularizing collection if $V = U$.

In some cases, it is convenient to use regularizing collections that depend on an integer parameter n.

Definition. A collection of linear operators $\{R_n\}$ $(n = 1, 2, \dots)$ from $\mathcal{F}$ to U is said to be *regularizing* for equation (1) on V if the following conditions hold:

1) for any n, the operator R_n is defined on the entire $\mathcal{F}$ and bounded;
2) for any $u \in V$,

$$\lim_{n \to 0} R_n Au = u.$$

Suppose that a regularizing solution $\{R_\alpha\}$ for equation (1) is known. Consider the problem of determining an approximate solution to equation (1) from the approximate data (Ivanov, Vasin, and Tanana, 2002; Lavrent'ev, 1962).

Suppose that a given element f_ε is such that

$$\|f - f_\varepsilon\| \le \varepsilon.$$

Let $u_{\alpha\varepsilon}$ denote the result of application of the operator R_α to the approximate right-hand side, i.e.,

$$u_{\alpha\varepsilon} = R_\alpha f_\varepsilon.$$

We now estimate the difference $u - u_{\alpha\varepsilon}$ as follows:

$$\|u - u_{\alpha\varepsilon}\| \le \|u_{\alpha\varepsilon} - R_\alpha Au\| + \|u - R_\alpha Au\| \le \|R_\alpha\|\varepsilon + \|u - R_\alpha Au\|. \quad (2)$$

If $u \in V$, then the second term in the right-hand side of (2) tends to zero as $\alpha \to 0$. If R_α is a uniformly regularizing collection, then

$$\|u - R_\alpha Au\| \le \beta(\alpha),$$
$$\beta(\alpha) = \sup \|R_\alpha Av - v\|, \quad \lim_{\alpha \to 0} \beta(\alpha) = 0,$$

and therefore

$$\|u - u_{\alpha\varepsilon}\| \le \|R_\alpha\|\varepsilon + \beta(\alpha).$$

With the appropriate choice of α, we will have a guaranteed estimate of the accuracy of the approximate solution $u_{\alpha\varepsilon}$. In particular, one can set

$$\alpha = \alpha_0(\varepsilon),$$

where $\alpha_0(\varepsilon)$ is determined from the relation

$$\|R_{\alpha_0}\|\varepsilon + \beta(\alpha_0) = \inf_\alpha [\|R_\alpha\|\varepsilon + \beta(\alpha)].$$

We now give several examples of regularizing collections for equations in a Hilbert space $U = \mathcal{F}$.

Reduction to equations of the second kind. Let the operator A be self-adjoint and positive. In addition to (1), consider the collection of equations

$$\alpha u_\alpha + A u_\alpha = (\alpha E + A) u_\alpha = f. \tag{3}$$

We set

$$R_\alpha = (\alpha E + A)^{-1}.$$

We will show that $\{R_\alpha\}$ is a regularizing collection for equation (1) on the entire space U.

Since the operator A is positive and compact, we have

$$\|R_\alpha\| = 1/\alpha.$$

Now consider the difference

$$u - R_\alpha A u = \alpha(\alpha E + A)^{-1} u.$$

We denote by $\{\varphi_k\}$, $\{\lambda_k\}$, $\lambda_k \geq \lambda_{k+1} > 0$, the complete orthonormal system of eigenelements and eigenvalues of A. Then

$$u = \sum_{k=1}^{\infty} a_k \varphi_k, \quad a_k = (u, \varphi_k),$$

$$u - R_\alpha A u = \alpha \sum_{k=1}^{\infty} \frac{a_k}{\alpha + \lambda_k} \varphi_k,$$

$$\|u - R_\alpha A u\|^2 = \alpha^2 \sum_{k=1}^{\infty} \frac{a_k^2}{(\alpha + \lambda_k)^2}$$

$$= \alpha^2 \sum_{k=1}^{n} \frac{a_k^2}{(\alpha + \lambda_k)^2} + \alpha^2 \sum_{k=n+1}^{\infty} \frac{a_k^2}{(\alpha + \lambda_k)^2} \leq \frac{\alpha^2}{\lambda_n^2} \|u\|^2 + \sum_{k=n+1}^{\infty} a_k^2.$$

$$\tag{4}$$

Since the series $\sum_{k=1}^{\infty} a_k^2$ converges, the second term on the right-hand side of (4) is as small as desired for sufficiently large n. The first term is as small as desired for sufficiently small α and fixed n, which implies

$$\lim_{\alpha \to 0} \|u - R_\alpha A u\| = 0.$$

If the operator A is not positive or self-adjoint, equation (1) can be reduced to an equation with a positive self-adjoint operator by applying the operator A^* to both sides of (1). Thus, we obtain the equation

$$A_1 u = f_1, \quad A_1 = A^* A, \quad f_1 = A^* f.$$

Then the regularizing collection for the original equation (1) has the form

$$R_\alpha = (\alpha E + A^* A)^{-1} A^*.$$

Decomposition with respect to the eigenelements. Let the operator A be self-adjoint, $\{\varphi_k\}$ be its eigenelements and $\{\lambda_k\}$ be its eigenvalues, where $|\lambda_{k+1}| \le |\lambda_k|$. We define operators R_n by the formula

$$R_n f = \sum_{k=1}^{n} \frac{1}{\lambda_k} f_k \varphi_k, \quad f_k = (f, \varphi_k).$$

It is obvious that the operators R_n are continuous and

$$\|R_n\| = 1/\lambda_n,$$

$$\lim_{n \to \infty} R_n A u = \lim_{n \to \infty} \sum_{k=1}^{n} a_k \varphi_k = u, \quad a_k = (u, \varphi_k).$$

Thus, the collection of operators $\{R_n\}$ is a regularizing collection that depends on the parameter n.

The method of successive approximations. Assume that the operator A is positive and self-adjoint, $\{\varphi_k\}$ and $\{\lambda_k\}$ are the sequences of eigenelements and eigenvalues of A, and $\|A\| = \lambda_1 \le 1$. Consider the sequence

$$u_{k+1} = u_k - A u_k + f, \quad u_0 = f, \quad k = 0, 1, \dots.$$

We will show that

$$\lim_{n \to \infty} u_n = u.$$

Indeed, we have

$$u_n = \sum_{k=0}^{n} (E - A^k) A u, \quad u - u_n = (E - A)^{n+1} u.$$

Taking the decomposition of elements u and u_n with respect to the basis $\{\varphi_k\}$, we obtain

$$u - u_n = \sum_{k=0}^{\infty} (1 - \lambda_k)^{n+1} a_k \varphi_k, \quad a_k = (u, \varphi_k),$$

$$\|u - u_n\|^2 = \sum_{k=0}^{m} (1 - \lambda_k)^{2(n+1)} a_k^2 + \sum_{k=m+1}^{\infty} (1 - \lambda_k)^{2(n+1)} a_k^2 \qquad (5)$$

$$\leq (1 - \lambda_m)^{2(n+1)} \|u\|^2 + \sum_{k=m+1}^{\infty} a_k^2.$$

The second term on the right-hand side of the inequality (5) is as small as desired if m is sufficiently large, while the first term is as small as desired for sufficiently large n and fixed m.

The regularizing collection corresponding to the described method of successive approximations is given by the formula

$$R_n = \sum_{k=0}^{n} (E - A)^k, \quad \|R_n\| = n + 1.$$

Regularizing collections can provide existence criteria for solutions of equations.

Theorem. *Suppose that $\{R_\alpha\}$ is a regularizing collection for (1) on the entire space U, $U = \mathcal{F}$ and the operators R_α and A commute:*

$$R_\alpha A = A R_\alpha.$$

Then, if for some f_0 there exists a limit

$$\lim_{\alpha \to 0} R_\alpha f_0 = u_0,$$

then this limit is a solution to (1) with right-hand side f_0, i.e.,

$$A u_0 = f_0.$$

Proof. Since the operator A is continuous, we have

$$\lim_{\alpha \to 0} A R_\alpha f_0 = A u_0,$$

which implies

$$A u_0 = \lim_{\alpha \to 0} R_\alpha A f_0 = f_0.$$

$\square$

9.4.   INTEGRAL EQUATIONS
WITH WEAK SINGULARITIES

Let D be a bounded domain in the space $\mathbb{R}^n$, $x = (x_1, \ldots, x_n)$ and $\xi = (\xi_1, \ldots, \xi_n)$ be elements of $\mathbb{R}^n$, and $u(x)$ be a function defined on D.

Integral equations of the first and the second kind for $u(x)$ are equations of the form

$$\int_D K(x, \xi) u(\xi) \, d\xi = f(x), \tag{1}$$

$$u(x) + \int_D K(x, \xi) u(\xi) \, d\xi = f(x), \tag{2}$$

where $K(x, \xi)$ is such that the integral operator in equations (1), (2) is an integral operator in the special sense defined in Lavrent'ev and Savel'ev (1995), Part I, Subsection 3.6.1. In the equation of the first kind (1), the function $f(x)$ may be defined in a domain D_1 that is different from D. In this case, the kernel K is defined in the domain $D_1 \times D$. The functions $u(x)$ and $f(x)$ may be considered to be elements of different function spaces.

In equations of the second kind, the domains of definition of $u(x)$ and $f(x)$ and the corresponding function spaces coincide.

Among integral equations of the first kind, we point out the class of equations that can be reduced to equations of the second kind by applying differential or pseudo-differential operators. This is the class of integral equations of the first kind that will be considered in what follows. this class. We confine ourselves to considering only the case where the operator that reduces an equation of the first kind to an equation of the second kind is a power of the Laplace operator. The solution $u(x)$ of the equation will be assumed to be an element of the Hilbert space $L_2(D)$.

Assume in (1) that

$$K(x, \xi) = \frac{1}{|x - \xi|^{n-\alpha}} + K_0(x, \xi), \tag{3}$$

where

$$0 < \alpha < n, \tag{4}$$

α is an even integer, $n \geq 3$, and $K_0(x, \xi) \in C^\alpha(\overline{D \times D})$. It is known that

$$\Delta^{\alpha/2} |x - \xi|^{-n+\alpha} = c\delta(x - \xi),$$

where $c = c(n, \alpha)$ is a nonzero constant. Therefore, applying the operator $c^{-1}\Delta^{\alpha/2}$ to (1), we obtain

$$u(x) + \int_D K_1(x, \xi)u(\xi)d\xi = f_1(x), \tag{5}$$

$$K_1(x, \xi) = c^{-1}\Delta^{\alpha/2}K_0(x, \xi), \quad f_1(x) = c^{-1}\Delta^{\alpha/2}f(x).$$

Thus, the problem of solving the equation of the first kind (1) is reduced to the problem of solving the equation of the second kind (5). If the solution of equation (5) is unique, then so is the solution of equation (1). Moreover, the problem of solving (1) is well-posed in the sense of Fichera under the assumption that

$$u(x) \in L_2(D), \quad f(x) \in W_2.$$

Reducing an equation of the first kind with kernel (3) to an equation of the second kind is also possible under the condition (4) without assuming that the number α is even. To this end, it is necessary to define fractional powers of the Laplace operator. In the case where the right-hand side $f(x)$ of equation (1) is defined in the domain $D_1 = \mathbb{R}^n$, this can be easily done with the help of the Fourier transform. This can also be done in the more general case where $D_1 \supset \bar{D}$.

9.5.   SCALAR VOLTERRA EQUATIONS

Volterra integral equations of the first and second kind for a function $u(x)$ are defined to be equations of the form

$$\int_0^x K(x, \xi)u(\xi)\, d\xi = f(x), \tag{1}$$

$$u(x) + \int_0^x K(x, \xi)u(\xi)\, d\xi = f(x). \tag{2}$$

We will consider equations on a finite interval $[0, l]$. Since the integral operator in a Volterra equation is quasinilpotent, there always exists a unique solution to the equation that can be represented in the form of a convergent Neumann series.

Generally speaking, possible cases for Volterra equations of the first kind are almost as diverse as for integral equations of the first kind of the general form. In what follows, we consider Volterra equations of the first kind that belong to the class of integral equations with weak singularities and can be reduced to equations of the second kind. The solutions $u(x)$ will be assumed to belong to the space $C[0, l]$.

Theorem. *Assume that the function $K(x, \xi)$ in (1) can be represented in the form*

$$K(x, \xi) = (x - \xi)^{\alpha-1}/\Gamma(\alpha) + K_0(x, \xi),$$

where $\alpha \in (0, 1)$, $\Gamma(\alpha)$ is the gamma function, and

$$K_0(x, \xi) \in C^1(\Omega), \quad \Omega = \{(x, \xi) \in [0, l] \times [0, l] : 0 < \xi < x < l\}.$$

Then

1) the solution of (1) is unique;

2) for any $f(x) \in C^1[0, l]$ such that $f(0) = 0$, there exists a solution $u \in C[0, l]$.

Proof. The differentiation operator of order α is defined as follows:

$$(D^\alpha f)(x) = \frac{1}{\Gamma(1-\alpha)} \int_0^x \frac{f'(\xi)\, d\xi}{(x-\xi)^\alpha} = \frac{1}{\Gamma(1-\alpha)} \frac{d}{dx} \int_0^x \frac{f(\xi)\, d\xi}{(x-\xi)^\alpha}. \tag{3}$$

Since

$$\int_s^x (x-\xi)^{-\alpha}(\xi-s)^{\alpha-1}\, d\xi = \Gamma(\alpha)\Gamma(1-\alpha),$$

it follows that application of the operator D^α to both sides of (1) yields

$$u(x) + \int_0^x K_1(x, \xi)u(\xi)\, d\xi = f_1(x), \tag{4}$$

where $f_1 = D^\alpha f$ and the continuous kernel $K_1(x, \xi)$ is expressed in terms of the function $K_0(x, \xi)$. It is easy to show that equations (1) and (4) are equivalent. □

Equation (1) is a special case of a general integral equation of the first kind, so general regularization methods presented in Section 9.3 can be applied to it. However, the integral operator in (1) is not self-adjoint. Application of the adjoint operator to (1) decreases the stability of the solution. For this reason and also for the reason that Volterra operators have a number of advantages from the computational point of view, it is reasonable to investigate the possibilities for regularizing equation (1) with the help of Volterra operators. We give an example of Volterra regularization for equation (1) in the case where $K(x, x) = 1$. In addition, we assume that $u(x)$ satisfies the condition $u(0) = 0$.

We denote by $u_\delta(x)$ ($\delta > 0$) a solution to the equation

$$\delta u_\delta(x) + \int_0^x K(x, \xi) u_\delta(\xi) \, d\xi = f(x). \tag{5}$$

There exists a unique solution of equation (5), and the solution can be represented in the form

$$u_\delta(x) = \frac{1}{\delta} f(x) + \frac{1}{\delta} \int_0^x R(x, \xi, \delta) f(\xi) \, d\xi, \tag{6}$$

where $R(x, \xi, \delta)$ is a continuous function.

The right-hand side of (6) with fixed δ is the result of applying a continuous operator to the function $f(x)$. We will show that the collection of these operators depending on the parameter δ is a regularizing collection for equation (1). To this end, it suffices to prove that the collection of functions $u_\delta(x)$ converges to the solution $u(x)$ of (1) as $\delta \to 0$. Differentiating equation (5), we obtain

$$\delta u_\delta'(x) + u_\delta(x) + \int_0^x K_1(x, \xi) u_\delta(\xi) \, d\xi = f_1(x). \tag{7}$$

From (7) it follows that the function $u_\delta(x)$ satisfies the integral equation

$$u_\delta(x) + \int_0^x K_\delta(x, \xi) u_\delta(\xi) \, d\xi = f_\delta(x), \tag{8}$$

$$K_\delta(x, \xi) = \frac{1}{\delta} \int_\xi^x e^{-(x-y)/\delta} K_1(y, \xi) \, dy,$$

$$f_\delta(x) = \frac{1}{\delta} \int_\xi^x e^{-(x-\xi)/\delta} f_\delta(\xi) \, d\xi.$$

Now consider the difference

$$u(x) - u_\delta(x) = f_1(x) - f_\delta(x)$$

$$- \int_0^x [K_1(x,\xi) - K_\delta(x,\xi)]u(\xi)\, d\xi - \int_0^x K_0(x,\xi)[u(\xi) - u_\delta(x)]\, d\xi. \quad (9)$$

It is easy to show that the differences $f_1(x) - f_\delta(x)$ and $K_1 - K_\delta$ satisfy the inequalities

$$|f_1(x) - f_\delta(x)| \le e^{-x/\delta}\mathcal{M}(x),$$
$$|K_1(x,\xi) - K_\delta(x,\xi)| \le e^{-(x-\xi)/\delta}\mathcal{M}_1(x,\xi),$$
$$\mathcal{M}(x) = \max_{\xi}\{|f(\xi)| : \xi \in [0,x]\}, \quad (10)$$
$$\mathcal{M}_1(x,\xi) = \max_{x,y}\{|K_1(x,y)| : y \in [x,\xi]\}.$$

From (10) it follows that the first two terms on the right-hand side of (9) tend to zero as $\delta \to 0$. Then, since the function $K_\delta(x,\xi)$ is uniformly bounded with respect to δ, we obtain

$$\lim_{\delta \to 0} \|u(x) - u_\delta(x)\|_C = 0.$$

9.6.   VOLTERRA OPERATOR EQUATIONS

Let $u(t)$ be a continuous function of a scalar argument t with values in a Hilbert space U, and let $A(t,\tau)$ be a collection of continuous operators whose domain of definition is U and whose values belong to U. It is assumed that the collection $A(t,\tau)$ continuously depends on the variables t and τ. By a Volterra operator equation of the first kind for $u(t)$ we mean an equation of the form

$$\int_0^t A(t,\tau)u(\tau)\, d\tau = f(t). \quad (1)$$

By a Volterra operator equation of the second kind we mean an equation of the form

$$Bu(t) + \int_0^t A(t,\tau)u(\tau)\, d\tau = f(t), \quad (2)$$

where B is a continuous operator and

$$N(B) = 0. \tag{3}$$

Equations (1) and (2) will be considered on a finite interval $t \in [0, T]$.

If the operator B^{-1} is bounded, then the equation of the second kind (2) is obviously equivalent to the equation of the form

$$u(t) + \int_0^t A(t, \tau) u(\tau) \, d\tau = f(t).$$

Since the integral operator in (2) is quasinilpotent in the Banach space $U_T = C([0, T]; U)$ with norm

$$\|u(t)\|_{U_T} = \max \|u(t)\|_U,$$

it follows that there always exists a unique solution to equation (2) and this solution continuously depends on the right-hand side f.

If the Volterra operator equation of the first kind (1) can be reduced to the equation of the second kind (2) with bounded operator B^{-1} by differentiating with respect to the variable t, then the results related to the scalar Volterra equations presented in the previous section can be extended to the case of operator equations without change.

We now consider equation (2) in the case where the operator B^{-1} is not bounded.

Theorem 1. *Assume that the operator B is self-adjoint and commutes with all operators $A(t, \tau)$. Then the solution of equation (2) is unique.*

Proof. Let B_T denote the extension of the operator B to the space U_T, and let A_T denote the integral operator in (2) viewed as an operator from U_T to U_T. Since B is self-adjoint, from (3) it follows that for any $u(t) \subset U_T$ there exists a number $c > 0$ such that

$$\|B_T^n u(t)\|_{U_T} \geq c^n \|u(t)\|_{U_T}$$

(see Lavrent'ev and Savel'ev, 1995, Part I, Section 3.9). On the other hand, since the integral operator in (2) is quasinilpotent, it follows that for any $\delta > 0$ there exists an n_0 such that for all $n \geq n_0$ we have

$$\|A_T^n u(t)\|_{U_T} \leq \delta^n \|u(t)\|_{U_T},$$

which implies the assertion of the theorem. $\quad\square$

It is easy to prove that the norm of the operator A_T^n satisfies the inequality

$$\|A_T^n\| \le a^n T/n!,$$

where $a = \max \|A(t, \tau)\|$.

We will give an example of Volterra operator equations of the first kind that can be reduced to equations of the second kind with unbounded B^{-1}. Suppose that all operators $A(t, \tau)$ commute with each other and are self-adjoint. Then there exists a collection of projection operators $\{E_\lambda\}$ and a scalar function $a(t, \tau, \lambda)$ such that

$$A(t, \tau) = \int a(t, \tau, \lambda)\, dE_\lambda.$$

In this case, the Volterra operator equation (1) is equivalent to the following collection of scalar Volterra equations depending on the parameter λ

$$\int_0^t a(t, \tau, \lambda) u(\tau, \lambda)\, d\tau = f(t, \lambda). \tag{4}$$

If for any λ the function $a(t, \tau, \lambda)$ is continuously differentiable with respect to t and $a(t, \tau, \lambda) > 0$, then equation (4) with fixed λ is equivalent to the Volterra equation of the second kind

$$u(t, \lambda) + \int_0^t a_1(t, \tau, \lambda) u(\tau, \lambda)\, d\tau = f_1(t, \lambda), \tag{5}$$

$$a_1(t, \tau, \lambda) = a_t(t, \tau, \lambda)/a(t, t, \lambda),$$
$$f_1(t, \lambda) = f_t(t, \lambda)/a(t, t, \lambda).$$

When λ is fixed, the solution of equation (5) is unique and its dependence on the right-hand side f is continuous. It follows that the solution of equations (4) and (1) is unique. However, the function $a_1(t, \tau, \lambda)$ may tend to infinity as $\lambda \to \infty$ and the dependence of the solution of (2) on the right-hand side may not be continuous.

If we put

$$\mu(\lambda) = \max |a_1(t, \tau, \lambda)|,$$

it is easy to see that equation (1) is reduced to equation (2), where $B = \int \mu(\lambda)^{-1}\, dE_\lambda$.

Chapter 10.

Evolution equations

10.1. CAUCHY PROBLEM
AND SEMIGROUPS OF OPERATORS

Let $u(t)$ be a function of a scalar argument t with values in a Banach space U and A be a linear operator with domain of definition $D(A)$ that assumes values in U and is everywhere dense in U. Consider the equation

$$\frac{du}{dt} = Au. \tag{1}$$

A *solution to equation* (1) is a strongly continuously differentiable function whose values belongs to the domain of definition of the operator A for all $t \geq 0$ and that satisfies equation (1) (Lavrent'ev and Savel'ev, 1995).

The *Cauchy problem for equation* (1) is the problem of finding a solution to (1) that satisfies the condition

$$u(0) = u_0, \quad u_0 \in D(A).$$

According to the definitions given before, a Cauchy problem for equation (1) is called *well-posed* if the following conditions hold:

1) for any $u_0 \in D(A)$ there exists a solution to the problem;

2) the solution of the problem is unique;

3) the solution of the problem continuously depends on the initial data, i.e., from

$$\lim_{n \to \infty} u_n(0) = 0, \quad u_n(0) \in D(A)$$

it follows that

$$\lim_{n \to \infty} u_n(t) = 0$$

for any t.

In general, we will consider the Cauchy problem on a finite interval $t \in [0, T]$. However, in the case of a well-posed Cauchy problem, its well-posedness on $[0, T]$ implies its well-posedness on any other interval $[0, T_1]$ and, consequently, on the whole semi-axis $[0, \infty)$.

Indeed, consider the interval $[0, 2T]$, and let $u(t)$ be a solution on the interval $[0, T]$. We will construct a solution $v(t)$ to equation (1) with data $v(0) = u(T)$ and introduce a function $W(t)$ such that

$$W(t) = u(t), \quad t \in [0, T],$$
$$W(t) = v(t - T), \quad t \in [T, 2T].$$

Clearly, $W(t)$ is a solution to equation (1) on $[0, 2T]$ with data

$$W(0) = u_0.$$

We denote by $B(t)$ the operator that maps an element $u_0 \in D(A)$ to the value $u(t)$ of the solution to the Cauchy problem for a fixed $t > 0$.

If the Cauchy problem is well-posed, then the operator $B(t)$ is defined on $D(A)$ and is continuous.

Consequently, the operator $B(t)$ can be extended to a linear bounded operator defined on the entire space U. It is clear that the collection of operators $B(t)$ is a semigroup, i.e.,

$$B(t_1 + t_2) = B(t_1) \cdot B(t_2), \quad t_1, t_2 > 0.$$

Indeed, let $u_0 \in D(A)$. Then the function $W(t) = u(t + \tau) = B(t + \tau)u_0$ satisfies equation (1) and the initial condition $W(0) = u(\tau) = B(\tau)u_0$ as a function of t.

The function $v(t) = B(t) \cdot B(\tau)u_0$ is also a solution to equation (1) with initial condition $B(\tau)u_0$. Since the solution is unique, we have $v(t) = W(t)$.

Thus, the operators $B(t + \tau)$ and $B(t) \cdot B(\tau)$ coincide on the set $D(A)$, which is everywhere dense in U; therefore, they coincide on the entire U. If u_0 does not belong to $D(A)$, then the function $B(t)u_0$ is said to be a *generalized solution* to equation (1).

Now consider the nonhomogeneous equation

$$\frac{du}{dt} = Au + f, \tag{2}$$

where $f(t)$ is a given continuous function with values in U.

If the Cauchy problem for the homogeneous equation (1) with operator A is well-posed, then the solution to the nonhomogeneous equation (2) is given by the formula

$$u(t) = B(t)u(0) + \int_0^t B(t-\tau)f(\tau)\,d\tau, \qquad (3)$$

where $B(t)$ is the operator that defines the solution to the Cauchy problem for the homogeneous equation.

Indeed, differentiating (3), we obtain

$$\frac{du}{dt} = B'(t)u(0) + B(0)f(t) + \int_0^t B'(t-\tau)f(\tau)\,d\tau$$

$$= AB(t)u(0) + f(t) + A\int_0^t B(t-\tau)f(\tau)\,d\tau = Au(t) + f(t).$$

10.2. EQUATIONS IN A HILBERT SPACE

Let $u(t)$ be a function with values in a real Hilbert space over the field of real numbers $\mathbb{R}$, and let A be a self-adjoint operator. Consider the equation

$$\frac{du}{dt} = Au. \qquad (1)$$

Theorem 1. *Let $u(t)$ be a solution to equation (1) on an interval $[0, T]$. Then the following inequality holds:*

$$\|u(t)\| \le \|u(T)\|^{t/T}\|u(0)\|^{(T-t)/T}. \qquad (2)$$

Corollary. *The solution of the Cauchy problem for equation (1) is unique.*

Proof. First, suppose that $u(t) \ne 0$, $t \in [0, T]$, and $u(t)$ is twice continuously differentiable.

We introduce the functions

$$\varphi(t) = \|u(t)\|^2 = (u, u), \quad \psi(t) = \ln \varphi(t).$$

Differentiating $\varphi(t)$ and $\psi(t)$, we get

$$\varphi'(t) = 2(u, u_t) = 2(u, Au),$$
$$\varphi''(t) = 2(Au, Au) + 2(u, A^2 u) = 4(Au, Au).$$

Hence, using the Cauchy–Bunyakovsky inequality, we obtain

$$\psi''(t) = \frac{1}{\varphi^2(t)} [\varphi(t)\varphi''(t) - \varphi'^2(t)] \geq 0. \tag{3}$$

From the inequality (3), it follows that

$$\psi(t) \leq \frac{t}{T} \psi(T) + \frac{T-t}{T} \psi(0),$$

which implies the inequality (2).

If $u(t)$ has no second-order derivative, we consider the function

$$u_h(t) = \int_{t-h}^{t+h} g_h(\tau) u(\tau) \, d\tau,$$

where $g_h(t)$ is a smooth function such that

$$g_h(-h) = g_h(h) = 0, \quad \int_{-h}^{h} g_h(t) \, dt = 1.$$

For example, we can put

$$g_h(t) = \frac{1}{2h} \left(1 + \cos \frac{\pi}{h} t\right).$$

The function $u_h(t)$ will obviously be a twice continuously differentiable solution to equation (1) on the interval $[h, T - h]$.

From the inequality

$$\psi_h''(t) \geq 0, \quad \psi_h(t) = \ln \|u_h(t)\|,$$

it follows that for all t_1, t_2, and t such that

$$h \leq t_1 \leq t \leq t_2 \leq T - h, \quad t_1 < t_2,$$

we have

$$\|u_h(t)\| \le \|u_h(t_2)\|^{(t-t_1)/(t_2-t_1)} \cdot \|u_h(t_1)\|^{(t_2-t)/(t_2-t_1)}. \tag{4}$$

Passing to the limit as $h \to 0$, we conclude that the function $u(t)$ satisfies the inequality (4) and, consequently, inequality (2).

From (4) it follows that if there is a $t_0 \in [0, T]$ such that $u(t_0) = 0$, then $u(t) = 0$ for any $t \in [0, T]$.

Thus, the inequality (2) holds in this case too. $\square$

Theorem 2. *For the Cauchy problem for equation (1) to be well-posed, it is necessary and sufficient that the operator A be bounded from above.*

Proof. Let $\{E_\lambda\}$ be a collection of projection operators generated by the operator

$$A = \int_{-\infty}^{\infty} \lambda \, dE_\lambda.$$

(see Lavrent'ev and Savel'ev, 1995, Part I, Section 3.9). If A is semibounded from above, then

$$A = \int_{-\infty}^{a} \lambda \, dE_\lambda.$$

In this case, the solution to the Cauchy problem for equation (1) is

$$u(t) = \left(\int_{-\infty}^{a} e^\lambda \, dE_\lambda \right) u_0, \quad u(0) = u_0. \tag{5}$$

Indeed, the function given by the formula (5) is a solution to the equation. The uniqueness of the solution was established above. It is clear that the solution continuously depends on the data.

Suppose now that the operator A is not semibounded from above. Then for any $a > 0$ there exists a $b > a$ such that the subspace $U_{ab} = (E_b - E_a)U$ is nonempty.

Let $u_0 \in U_{ab}$. Then the norm of the solution of the Cauchy problem with initial condition u_0 satisfies the inequality

$$\|u(t)\| = \left\| \left(\int_{a}^{b} e^\lambda \, dE_\lambda \right) u_0 \right\| \ge e^a \|u_0\|. \tag{6}$$

It readily follows from (6) that for arbitrarily small Cauchy data there are solutions whose norm is as as large as desired, i.e., there is no continuous dependence of the solution on the data.   □

The above theorems admit generalizations to the case of evolution equations with normal operator in a Hilbert space over the field of real numbers. Let A be a normal operator.

Theorem 3. *Let $u(t)$ be a solution to equation (1) on an interval $[0, T]$. Then the inequality (2) holds.*

Proof. As before, we assume that $u(t) \neq 0$, $t \in [0, T]$, and the function $u(t)$ is twice continuously differentiable. Consider the functions φ and ψ (see the proof of Theorem 1).

Differentiating these functions, we obtain

$$\varphi'(t) = (u, (A + A^*)u),$$

where A^* is the adjoint of A, and

$$\varphi''(t) = (Au, (A + A^*)u) + (u, (A + A^*)u) = ((A + A^*)u, (A + A^*)u),$$

$$\psi''(t) = \frac{1}{\varphi^2(t)} [\varphi(t)\varphi''(t) - \varphi'^2(t)] \geq 0.$$

Hence, using arguments very similar to those used above, we arrive at the assertion of the theorem.   □

Recall that a normal operator A can be represented in the form

$$A = X + iY,$$

where X and Y are self-adjoint operators that commute with each other.

Theorem 4. *For the Cauchy problem for equation (1) be well-posed, it is necessary and sufficient that the operator X be semibounded from above.*

The proof of this theorem is completely analogous to that of Theorem 2.

10.3. EQUATIONS WITH VARIABLE OPERATOR

Let $u(t)$ be a function with values in a Hilbert space U over the field of real numbers $\mathbb{R}$ and $A(t)$ be a collection of operators depending on the parameter t, $t \in [0, T]$.

Consider the equation

$$\frac{du}{dt} = Au. \tag{1}$$

Suppose that $A = A_0 + A_1$, where A_0 is a constant self-adjoint operator, the operator $A_1(t)$ has continuous derivative $A_1'(t)$, and the operators $A_1(t)$ and $A_1'(t)$ are bounded.

Theorem. *Suppose that there exist constants $a, b_0 \geq 0$ such that the following inequality holds for any u:*

$$(A_0 u, A_1^* u) - (A_0 u, A_1 u) \geq -a(u, Au) - b_0(u, u). \tag{2}$$

Then any solution to equation (1) satisfies the inequality

$$\|u(t)\| \leq \|u(T)\|^{\omega(t)} \|u(0)\|^{1-\omega(t)} c(t), \tag{3}$$

$$\omega(t) = \frac{1 - e^{-at}}{1 - e^{-aT}},$$

$$c(t) = \exp\left\{\frac{b}{a} \frac{(1 - e^{-at})T - (e^{-aT} - 1)t}{1 - e^{-aT}}\right\},$$

$$b = 2b_0 + 4a_1^2 + 2a_2,$$

$$a_1 = \max \|A_1(t)\|, \quad a_2 = \max \|A_1'(t)\|.$$

Corollary. *The solution of the Cauchy problem for equation (1) is unique.*

Proof. Consider the function $\varphi(t) = \|u(t)\|^2 = (u, u)$. Differentiating $\varphi(t)$, we obtain

$$\varphi'(t) = 2(u, Au) = 2(u, A_0 u) + 2(u, A_1 u),$$

$$\varphi''(t) = 2(Au, Au) + 2(u, A^2 u) + 2(u, A_1' u) = 4(Au, Au)$$

$$+ 2(A_0 u, A_1^* u) - 2(A_0 u, A_1 u) + 2(A_1^* u, A_1 u) - 2(A_1 u, A_1 u)$$

$$+ 2(u, A_1' u) \geq 4(Au, Au) - 2a(u, Au) - b(u, u). \tag{4}$$

It follows from (4) that if $\varphi(t) \neq 0$, then the second-order derivative of the function $\psi(t) = \ln \varphi(t)$ satisfies the inequality

$$\psi''(t) + a\psi'(t) + b \geq 0. \tag{5}$$

As is known from the theory of ordinary differential equations, if $\psi(t)$ satisfies inequality (5), then it also satisfies the inequality

$$\psi(t) \leq \psi_0(t), \tag{6}$$

where $\psi_0(t)$ is a solution to the differential equation $\psi_0''(t) + a\psi_0'(t) + b = 0$ with boundary conditions $\psi_0(0) = \psi(0)$, $\psi_0(T) = \psi(T)$. It is easy to verify that

$$\psi_0(t) = c_1 + c_2 e^{-at} - (b/a)t, \tag{7}$$

$$c_1 = \frac{\psi(T) - \psi(0)e^{-aT} + (b/a)T}{1 - e^{-aT}}, \quad c_2 = \frac{\psi(0) - \psi(T) - (b/a)T}{1 - e^{-aT}}.$$

Finally, the desired inequality (3) follows from (6) and (7).   □

10.4. EQUATIONS OF THE SECOND ORDER

Let $u(t)$ be a function with values in a Hilbert space over the field of real numbers, and let A be a self-adjoint operator. Consider the equation

$$\frac{d^2u}{dt^2} = Au. \tag{1}$$

Theorem 1. *The following inequality holds for any solution to equation (1):*

$$\|u(t)\|^2 \leq c(t)(\|u(T)\|^2 + |a|)^{t/T}(\|u(0)\|^2 + |a|)^{(T-t)/T} - |a|,$$
$$c(t) = e^{2t(T-t)}, \tag{2}$$
$$a = 1/2[(Au(0), u(0)) - (u'(0), u'(0))].$$

Proof. Put $\varphi(t) = (u, u)$. Differentiating $\varphi(t)$, we get

$$\varphi'(t) = 2(u, u'), \quad \varphi''(t) = 2(u', u') + 2(u, Au).$$

We now differentiate the second term in the expression for $\varphi''(t)$:

$$\frac{d}{dt}(u, Au) = 2(u', Au) = 2(u', u'') = \frac{d}{dt}(u', u').$$

Therefore,
$$\varphi''(t) = 4(u', u') + 4a.$$

Now consider the function
$$\psi(t) = \ln(\varphi(t) + |a|).$$

Differentiating $\psi(t)$, we get

$$\psi''(t) = \frac{\varphi''(t)(\varphi(t) + |a|) - (\varphi')^2(t)}{(\varphi(t) + |a|)^2}$$

$$= 4\frac{[(u', u') + a][(u, u) + |a|] - (u, u')^2}{[\varphi(t) + |a|]^2} \geq -4,$$

$$\psi(t) \leq \frac{t}{T}\psi(T) + \frac{T - t}{T}\psi(0) + 2t(T - t),$$

which readily implies the assertion of the theorem. $\square$

Theorem 2. *For the Cauchy problem for equation* (1) *to well-posed, it is necessary and sufficient that the operator* A *be semibounded from above.*

Proof. Let $\{E_\lambda\}$ be a collection of projection operators generated by the operator A:

$$A = \int\limits_{-\infty}^{\infty} \lambda\, dE_\lambda.$$

If A is semibounded from above, then

$$A = \int\limits_{-\infty}^{a} \lambda\, dE_\lambda.$$

For definiteness, assume that $a > 0$. Then the solution to the Cauchy problem is written as

$$u(t) = \left(\int\limits_{-\infty}^{0} \cos(\sqrt{-\lambda}\,t)\, dE_\lambda\right) u_0 + \left(\int\limits_{0}^{a} \mathrm{ch}(\sqrt{\lambda}\,t)\, dE_\lambda\right) u_0$$

$$+ \left(\int\limits_{-\infty}^{0} \frac{\sin(\sqrt{-\lambda}\,t)}{\sqrt{-\lambda}}\, dE_\lambda\right) u_1 + \left(\int\limits_{0}^{a} \frac{\mathrm{sh}(\sqrt{\lambda}\,t)}{\sqrt{\lambda}}\, dE_\lambda\right) u_1,$$

$u_0 = u(0),\ u_1 = u'(0).$ $\square$

If the operator A is not semibounded from above, then, for sufficiently large $a > 0$ and $b > a$, the norm of the solution to the Cauchy problem with data

$$u_0 \in E_b - E_a, \quad u_1 = 0$$

can be as large as desired for arbitrarily small Cauchy data.

10.5. WELL-POSED AND ILL-POSED CAUCHY PROBLEMS

Let D be a bounded domain of the n-dimensional space $\mathbb{R}^n$ with smooth boundary S. Consider the following self-adjoint elliptic differential operator on D with Dirichlet boundary condition:

$$Lu = \sum_{j,k=1}^{n} \frac{\partial}{\partial x_j} a_{jk} \frac{\partial u}{\partial x_k} + cu,$$

$$u|_S = 0,$$

$$\sum_{j,k=1}^{n} a_{jk} \xi_j \xi_k \geq \delta |\xi|^2, \quad \delta > 0,$$

where a_{jk} are continuously differentiable functions and c is a continuous function.

As is known, the operator specified above has complete orthogonal normalized system of eigenfunctions. The sequence of eigenvalues of this operator is bounded from above and tends to minus infinity.

Consider the mixed problems for differential equations in the cylindrical domains $D_T = D \times [0, T]$

$$\frac{\partial u}{\partial t} = Lu, \quad u|_{t=0} = u_0, \tag{1}$$

$$\frac{\partial u}{\partial t} = -Lu, \quad u|_{t=0} = u_0, \tag{2}$$

$$\frac{\partial^2 u}{\partial t^2} = Lu, \quad u|_{t=0} = u_0, \, u'|_{t=0} = u_1, \tag{3}$$

$$\frac{\partial^2 u}{\partial t^2} = -Lu, \quad u|_{t=0} = u_0, \, u'|_{t=0} = u_1, \tag{4}$$

$$u|_S = 0.$$

In accordance with the results presented in Sections 10.2 and 10.4, the problems (1) and (3) are well-posed, whereas the problems (2) and (4) are ill-posed. The solution of each of these problems is unique. The theorems formulated in Sections 10.2 and 10.4 provide the conditional stability estimates for the ill-posed problems (2) and (4).

10.6. EQUATIONS WITH INTEGRO-DIFFERENTIAL OPERATORS

Let $u(t)$ be a continuous function of a scalar argument t with values in a Hilbert space U, and let $A(t, \tau)$ be a collection of continuous linear operators with domain of definition U and with values in U. We assume that the collection $A(t, \tau)$ continuously depends on the variables t and τ. Consider the integro-differential equation

$$\frac{d}{dt} u(t) = Bu(t) + \int_0^t A(t, \tau) u(\tau) \, d\tau + f(t) \tag{1}$$

with initial condition $u(0) = 0$.

Let B be a closed operator such that

$$B^* B - B B^* \geq 0.$$

The operator B is not necessarily bounded, whereas the collection of operators $A(t, \tau)$ is bounded.

Theorem. *The solution of equation* (1) *is unique.*

Proof. We put

$$u_\sigma(t) = e^{\sigma(T-t)^2} \cdot u(t).$$

It is easy to show that the function $u_\sigma(t)$ satisfies the equation

$$\frac{d}{dt} u_\sigma(t) + 2\sigma(T - t) u_\sigma(t) = B u_\sigma(t) + \int_0^t A_\sigma(t, \tau) u_\sigma(\tau) \, d\tau + f_\sigma(t), \tag{2}$$

$$A_\sigma(t, \tau) = e^{\sigma(T-t)^2 - \sigma(T-\tau)^2} A(t, \tau), \quad f_\sigma(t) = e^{\sigma(T-t)^2} f(t).$$

Lemma. *Let $v(t)$ be a twice continuously differentiable function with values in a Hilbert space U, and let B be a closed operator. Then the following identity holds:*

$$\left[-\left(\frac{\partial}{\partial t} + B^* + \sigma t \right) \left(\frac{\partial}{\partial t} - B - \sigma t \right) \right.$$

$$\left. + \left(\frac{\partial}{\partial t} - B - \sigma t \right) \left(\frac{\partial}{\partial t} + B^* + \sigma t \right) \right] v(t)$$

$$= [2\sigma + (B^*B - BB^*)]v(t). \quad (3)$$

To prove that (3) holds, it suffices to perform the multiplication of the operators on the left-hand side of this equality. Then the left-hand side of (3) will be represented as the sum of terms which will coincide with those on the right-hand side after canceling like terms.

Consider the following functional on the solution $u_\sigma(t)$ of equation (2):

$$\mathcal{F}(u_\sigma(t)) = \int_0^T \left\| \left[\frac{\partial}{\partial t} - B + 2\sigma(T - t) \right] u_\sigma(t) \right\|_U^2 dt.$$

Using (3) and integrating by parts with respect to t, we obtain

$$\mathcal{F}(u_\sigma(t)) = \int_0^t \left\| \left[\frac{\partial}{\partial t} + B^* - 2\sigma(T - t) \right] u_\sigma(t) \right\|_U^2 dt$$

$$- \langle (B + B^*)u_\sigma(T), u_\sigma(T) \rangle + 4\sigma T \|u_\sigma(T)\|_U^2$$

$$+ \int_0^T \langle [2\sigma + (B^*B - BB^*)]u_\sigma(t), u_\sigma(t) \rangle \, dt$$

$$= \int_0^T \int_0^t \|A_\sigma(t, \tau)u_\sigma(\tau) + f_\sigma(t)\|_U^2 \, d\tau dt, \quad (4)$$

where $\langle u, v \rangle$ is the scalar product of elements u and v of the space U.

From (4), it follows that

$$4\sigma \int_0^T \|u_\sigma(t)\| \, dt \le \|(B + B^*)u(T)\| \cdot \|u(T)\|$$

$$+ \int_0^T \|A_\sigma(t, \tau)u_\sigma(\tau) + f_\sigma(t)\|_U^2 \, dt. \quad (5)$$

If $f(t) = 0$, then (5) implies

$$(4\sigma - aT) \int\limits_0^T e^{2\sigma(T-t)^2} \|u(t)\|^2 \, dt \leq \|(B + B^*)u(T)\| \cdot \|u(T)\|, \tag{6}$$

$$a = \max \|A_\sigma(t, \tau)\|.$$

Clearly, for sufficiently large σ, inequality (6) holds only if

$$u(t) = 0, \quad t \in [0, T].$$

This completes the proof of the theorem. $\square$

Chapter 11.

Problems of integral geometry

11.1. STATEMENT OF PROBLEMS OF INTEGRAL GEOMETRY

Let $x \in \mathbb{R}^n$, $x = (x_1, \ldots, x_n)$, $y \in \mathbb{R}^m$, $y = (y_1, \ldots, y_m)$, and let $S(y)$ be a collection of manifolds in $\mathbb{R}^n$ depending on the parameter y of dimension m, with $\dim S = p$. Furthermore, let $u(x)$ be a function defined in a domain $D \subset \mathbb{R}^n$, $\rho(x, y)$ be a function of the variables x and y, and $\omega(y)$ be a measure on the manifold $S(y)$.

We introduce the function

$$\int_{S(y)} \rho(x, y) u(x) \, d\omega = f(y). \tag{1}$$

Integral geometry is the area of mathematics that studies various relationships between the elements that appear in (1).

We will assume that $S(y)$, $\rho(x, y)$, and $f(x)$ in (1) are given, and (1) will be viewed as a linear operator equation for $u(x)$.

11.2. THE RADON PROBLEM

Consider the Radon problem for functions with finite support depending on two variables.

Let $u(x, y)$ be a continuous function of two variables that vanishes outside the unit disk

$$D = \{(x, y) : x^2 + y^2 \leq 1\}.$$

It is required to determine the function u from the known integrals of this function along all curves that intersect D.

We point out two representations of the Radon problem in the form of the problems of solving linear operator equations corresponding to the classical parameterization formulas for systems of straight lines on a plane:

$$\int_{-\infty}^{\infty} u(x + s \cos \alpha, y + s \sin \alpha) \, ds = f(x, y, \alpha), \tag{1}$$

$$\iint_D u(\xi, \eta) \delta(x_0 \xi + y_0 \eta - p) \, d\xi d\eta = \varphi(x_0, y_0, p). \tag{2}$$

Since the integrals of u along the lines that do not intersect D are equal to zero, we can assume that the functions f and φ are given for all values of variables (x, y, α), (x_0, y_0, p).

The right-hand sides of equations (1) and (2) include functions of three variables, but they are expressed in terms of functions of two variables.

The function f satisfies the differential equation

$$f'_x \cdot \cos \alpha + f'_y \cdot \sin \alpha = 0,$$

which implies

$$f(x, y, \alpha) = f_0(\sin \alpha \cdot x - \cos \alpha \cdot y, \alpha).$$

In view of the properties of the δ function, the function φ satisfies the relation

$$\varphi(x_0, y_0, p) = \frac{1}{r} \varphi \left(\frac{x_0}{r}, \frac{y_0}{r}, \frac{p}{r} \right) = \frac{1}{r} \psi(\beta, \tau),$$

where $\tau = p/r$, $x_0 = r \cos \beta$, and $y_0 = r \sin \beta$.

The representation (1) is most natural from the point of view of applied problems, whereas the representation (2) yields the simplest inversion formula. Clearly, it is easy to obtain the representation (2) from (1) and vice versa using a change of variables.

From the property of the function φ mentioned above, it follows that equation (2) is equivalent to the equation

$$\iint u(\xi, \eta) \delta(x_0 \xi + y_0 \eta - \tau) \, d\xi d\eta = \psi(\beta, \tau). \tag{3}$$

Since u vanishes outside the disk D, we have

$$\psi(\beta, \tau) = 0, \quad |\tau| > 1.$$

Thus, the operator in (3) takes the function u defined in the disk D to the function ψ defined in the rectangle

$$|\beta| \leq \pi, \quad |\tau| \leq 1.$$

The function ψ is the integral of u along the line defined by the equation

$$\xi \cos \beta + \eta \sin \beta - \tau = 0,$$

the differential being the length element on this line.

We now describe several properties of the solution to equation (3).

1. The problem of solving equation (3) is ill-posed if u and ψ are viewed as elements of the spaces $\mathbb{L}_2$ or $\mathbb{C}$. Let

$$u(x, y) = \begin{cases} 1, & x^2 + y^2 \leq \delta^2 \\ 0, & x^2 + y^2 > \delta^2 \end{cases} \quad (0 < \delta < 1).$$

Then

$$\psi(\beta, \tau) = \begin{cases} 2\sqrt{\delta^2 - \tau^2}, & |\tau| \leq \delta, \\ 0, & |\tau| > \delta, \end{cases}$$

$$\|u\|_{\mathbb{L}_2} = \pi\delta^2, \quad \|\psi\|_{\mathbb{L}_2} = 8\frac{\pi}{\sqrt{3}}\delta^{3/2},$$

$$\|u\|_{\mathbb{C}} = 1, \quad \|\psi\|_{\mathbb{C}} = 2\delta.$$

Thus if δ is sufficiently small, the ratio of the norms of u and ψ is as large as desired. This means that for infinitely small variations of ψ, the corresponding variations of u may not be small.

2. The problem of solving equation (3) is weakly ill-posed.

Applying to (2) the Fourier transform with respect to p, we obtain

$$\int e^{ip}\varphi(x_0, y_0, p)\, dp = \iint e^{i(x_0\xi + y_0\eta)}u(\xi, \eta)\, d\xi d\eta$$

$$= \frac{1}{r}\int e^{ip}\psi\left(\beta, \frac{p}{r}\right) dp = \int e^{i\tau r}\psi(\beta, \tau)\, d\tau = v(\beta, r), \quad (4)$$

where v is the Fourier transform of u in polar coordinates.

Parseval's identity for the functions u and v yields

$$\iint u^2(\xi, \eta)\, d\xi d\eta = \iint |v(\beta, r)|^2 r\, dr d\beta.$$

We denote by $H_{0,1/2}$ the space of functions ψ with the norm

$$\|\psi\|_H = \left(\iint |v(\beta, r)|^2 r\, dr d\beta \right)^{1/2},$$

where v is the Fourier transform of ψ with respect to the variable τ defined by (4). Then

$$\|u\|_{L_2} = \|\psi\|_H.$$

We now present another method for studying the uniqueness and stability of solutions to the Radon problem. This method provides slightly weaker stability estimates, but it admits generalization to a class of problems of integral geometry of the general form.

Consider the problem of solving equation (1). Set

$$F(x, y) = \int_{-\pi}^{\pi} f(x, y, \alpha)\, d\alpha.$$

It is easy to see that the functions F and u are related by the formula

$$F(x, y) = \iint_D \frac{1}{r} u(\xi, \eta)\, d\xi d\eta, \tag{5}$$

$$r = \sqrt{(x - \xi)^2 + (y - \eta)^2}.$$

Equality (5) can be viewed as an integral equation of the first kind with weak singularity for the unknown function u.

We will obtain the inversion formula for equation (5) that involves a fractional power of the Laplace operator. Consider the function

$$W(x, y, z) = \iint_D \frac{1}{R} u(\xi, \eta)\, d\xi d\eta,$$

$$R = \sqrt{(x - \xi)^2 + (y - \eta)^2 + z^2}.$$

The function W is the simple-layer potential with the density of distribution on the disk D.

Using the well-known formulas of potential theory, we get

$$W(x, y, z) = \iint \frac{z}{R^3} F(\xi, \eta) \, d\xi \, d\eta,$$

$$u(x, y) = \frac{1}{2\pi} \frac{\partial}{\partial z} W(x, y, z) \Big|_{z=0}.$$

The Radon problem with incomplete data. From the point of view of applications, the problem of inverting the Radon transform in the case where the integrals of the unknown functions are not known for some lines intersecting the domain.

Consider the problem of solving the operator equation (1) in the case where the values of the right-hand side f are given for all x and y and for all α in the interval

$$|\alpha| \leq \alpha_0 < \pi.$$

It can be proved that this problem is strongly ill-posed. The type of instability in this problem is the same as that in the Cauchy problem for the Laplace equation.

If we reduce the solution of the Radon problem with incomplete data to equation (5), we can assume that the right-hand side $F(x, y)$ is known only for x and y in a domain D_1 such that

$$D_1 \cap D = \emptyset.$$

In the space of (x, y, z), the function W is a solution to the Laplace equation outside the disk D. Clearly,

$$W_z'(x, y, 0) = 0, \quad (x, y) \in D_1,$$

and therefore the Radon problem with incomplete data is reduced to the problem of determining $W(x, y, z)$ from the following data:

$$W(x, y, 0) = F(x, y), \quad W_z'(x, y, 0) = 0, \quad (x, y) \in D_1,$$

which is a Cauchy problem for the Laplace equation.

11.3. RECONSTRUCTING A FUNCTION FROM SPHERICAL MEANS

This section deals with another classic problem of integral geometry that is related to the ill-posed Cauchy problem for the wave equation.

The section is based on the results from Courant and Hilbert (1962) and Lavrent'ev, Romanov, and Shishatskii (1986).

In the n-dimensional space $(x_1, \ldots, x_n)$, where $n \geq 2$, we consider the problem of reconstructing a function $u(x_1, \ldots, x_n)$ from its mean values on spheres (or circles, if $n = 2$) with arbitrary radii r $(0 < r < \infty)$ centered at points of a fixed plane. For convenience, we assume that this plane coincides with the coordinate plane $x_n = 0$ and denote any point $(x_1, \ldots, x_n)$ by (x, y), where $x = (x_1, \ldots, x_{n-1})$. Then $u = u(x, y)$. A sphere of radius r centered at $(x, 0)$ will be denoted by $S(x, r)$. With this notation, the problem formulated above consists in determining the function $u(x, y)$ from the integrals

$$\frac{1}{\omega_n} \int\limits_{S(x,r)} u(\xi, \eta) \, d\omega = v(x, r), \tag{1}$$

where (ξ, η) is a variable point on the sphere $S(x, r)$, $\xi = (\xi_1, \ldots, \xi_{n-1})$, ω_n is the area of the unit sphere in the n-dimensional space $(\omega_n = 2\pi^{n/2}/\Gamma(n/2))$, and $d\omega$ is the solid angle element associated with the surface area element $dS = r^{n-1} d\omega$.

Clearly, any function $u(x, y)$ that is even with respect to y is a solution to the homogeneous equation (1), i.e., the equation with $v(x, r) = 0$. For this reason, it is reasonable to state the problem of determining only the even part of $u(x, y)$ from $v(x, r)$, i.e., the function $u_1(x, y) = 1/2[u(x, y) + u(x, -y)$. This is equivalent to considering the class of functions $u(x, y)$ that are even with respect to y. It is obvious that this statement of the problem is equivalent to the problem of determining $u(x, y)$ that satisfies the condition

$$u(x, y) = 0, \quad y \leq 0.$$

The following uniqueness theorem holds.

Theorem 1. *Any function $u(x, y)$ continuous on the domain $D(x_0, r_0) = \{(x, y) : |x - x_0|^2 + y^2 < r_0^2\}$ and even with respect to y is uniquely determined in this domain by specifying the values of the function $v(x, r)$ in the domain $G_\varepsilon(x_0, r_0) = \{(x, r) : |x - x_0| < \varepsilon, \, 0 < r < r_0 - |x - x_0|\}$, where ε is a fixed positive number such that $\varepsilon \leq r_0$.*

Proof. The proof of the theorem involves calculating all moments of the function $u(x, y)$ corresponding to the variable x on every fixed sphere $S(x, r)$ such that $(x, r) \in G_\varepsilon(x_0, r_0)$. The domain $G_\varepsilon(x_0, r_0)$ resembles a sharpened pencil (Fig. 3). For $\varepsilon = r_0$, this "pencil" degenerates into a part of a cone (Fig. 4).

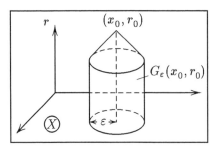
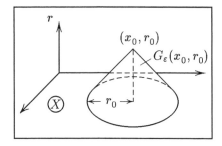

Figure 3   Figure 4

For any point $(x, r) \in G_\varepsilon(x_0, r_0)$, where $0 < \varepsilon \le r_0$, the sphere $S(x, r)$ lies inside the domain $D(x_0, r_0)$.

Consider the result of applying the operator

$$A_i v = \frac{\partial}{\partial x_i} \int_0^r \tau^{n-1} v(x, r) \, d\tau, \quad i = 1, \ldots, n-1,$$

to equation (1). After performing a series of calculations, we obtain

$$A_i v = \frac{1}{\omega_n} \frac{\partial}{\partial x_i} \int_0^r \tau^{n-1} \left[\int_{S(x,\tau)} u(\xi, \eta) \, d\omega \right] d\tau$$

$$= \frac{1}{\omega_n} \frac{\partial}{\partial x_i} \int_{|\xi-x|_i^2 + \eta^2 \le r^2} \left[\int_{x_i - (r^2 - \eta^2 - |\xi-x|_i^2)^{1/2}}^{x_i + (r^2 - \eta^2 - |\xi-x|_i^2)^{1/2}} u(\xi, \eta) \, d\xi_i \right]$$

$$\times \, d\eta d\xi_1 \ldots d\xi_{i-1} d\xi_{i+1} \ldots d\xi_n$$

$$= \frac{1}{\omega_n} \int_{S(x,r)} u(\xi, \eta) \cos(\widehat{n, \xi_i}) \, ds = \frac{r^{n-2}}{\omega_n} \int_{S(x,r)} u(\xi, \eta)(\xi_i - x_i) \, d\omega.$$

In these calculations, $|\xi - x|_i^2$ denotes the expression $[\|\xi - x\|^2 - |\xi_i - x_i|^2]^{1/2}$, which represents the length of the projection of the line segment connecting the points $(\xi, 0)$ and $(x, 0)$ onto the plane $x_i = 0$, $y = 0$. The unit vector of the outward normal to $S(x, r)$ is denoted by n.

We now introduce an operator L_i as follows:

$$L_i v = \frac{1}{r^{n-2}} A_i v + x_i v(x, r), \quad i = 1, \ldots, n-1.$$

From the formula obtained above, it follows that application of L_i to equation (1) is equivalent to calculating the spherical mean for the function $u_i(x, y) = x_i u(x, y)$. Indeed,

$$L_i v = \frac{1}{\omega_n} \int\limits_{S(x,r)} u(\xi, \eta) \xi_i \, d\omega.$$

It is clear that application of the composition $L_j L_i$ to $v(x, r)$ is equivalent to calculating the spherical mean for the function $u(x, y) x_i x_j$. In general, if $P_m(x)$ is a polynomial of degree m with constant coefficients, then application of the operator $P_m(L)$, $L = (L_1, \ldots, L_{n-1})$, to equation (1) yields

$$P_m(L) v = \frac{1}{\omega_n} \int\limits_{S(x,r)} u(\xi, \eta) P_m(\xi) \, d\omega.$$

We now fix (x, r). Let $D_0(x, r) = \{\xi : |\xi - x| \leq r\}$. The preceding equality can be represented on the sphere $S(x, r)$ as follows:

$$P_m(L) v = \int\limits_{D_0(x,r)} \varphi(\xi, x, r) P_m(\xi) \, d\xi, \tag{2}$$

where $\varphi(\xi, x, r)$ (with fixed x and r!) and $u(x, y)$ are related by the formula

$$\varphi(\xi, x, r) = \frac{2}{\omega_n r^{n-2}} \frac{u(\xi, [r^2 - |x - \xi|^2]^{1/2})}{[r^2 - |x - \xi|^2]^{1/2}}.$$

The fact that the function $u(x, y)$ is even with respect to y was taken into account.

The function $\varphi(\xi, x, r)$ is uniquely determined by the system of equalities (2) with different polynomials $P_m(\xi)$. It follows that the same is true for the function $u(x, y)$. This completes the proof of the theorem. $\square$

In order to construct the function $u(x, y)$ from the function $v(x, r)$, for $P_m(\xi)$ we can take the system of polynomials orthogonal inside $D_0(x, r)$. Then the function $\varphi(\xi, x, r)$ can be explicitly represented in the form of a Fourier series. However, more conditions have to be imposed on the function $u(x, y)$ to guarantee the convergence of the Fourier series.

Corollary. *If $u(x, y)$ and its partial derivative $u_y(x, y)$ are continuous in the domain $D(x_0, r_0)$, then $u(x, y)$ is uniquely determined in $D(x_0, r_0)$ by specifying the spherical means of both $u(x, y)$ and its partial derivative $u_y(x, y)$ on all spheres $S(x, r)$ such that $(x, r) \in G_\varepsilon(x_0, r_0)$, $0 < \varepsilon \leq r_0$.*

Indeed, the even part of the function $u(x, y)$ with respect to y and the even part of the partial derivative $u_y(x, y)$ are uniquely determined in this case. Note that any function $u(x, y)$ from the considered class can be represented in the form

$$u(x, y) = \frac{1}{2}[u(x, y) + u(x, -y)] + \frac{1}{2}\int_0^y [u_y(x, y) + u_y(x, -y)]\, dy$$

and, consequently, it can be uniquely determined.

Theorem 1 shows that the function $u(x, y)$ in the domain $D(x_0, r_0)$ is determined by specifying the function $v(x, r)$ in the domain $G_\varepsilon(x_0, r_0)$ for any positive ε as small as desired. Then, if $u(x, y)$ is known in the domain $D(x_0, r_0)$, it is possible to calculate the integrals over all spheres lying inside the domain $D(x_0, r_0)$ and, consequently, determine $v(x, r)$ for $(x, r) \in G_\varepsilon(x_0, r_0)$, $\varepsilon = r_0$. Thus, specifying the function $v(x, r)$ in the domain $G_\varepsilon(x_0, r_0)$ uniquely determines this function in the larger domain $G_{r_0}(x_0, r_0) \supset G_\varepsilon(x_0, r_0)$. This means that the function $v(x, r)$ cannot be specified in an arbitrary way. Moreover, any function $v(x, r)$ that can be represented in the form (1) has the following property analogous to a property of analytic functions: $v(x, r)$ is uniquely determined in $G_{r_0}(x_0, r_0)$ by its values in an arbitrarily thin domain $G_\varepsilon(x_0, r_0)$. At the same time, it is clear that if $u(x, y)$ is not analytic, the corresponding function $v(x, r)$ is also not analytic. As a result, we conclude that providing a constructive description of the class of functions $v(x, r)$ that can be represented in the form (1) is a rather complicated problem.

We will show that the problem of solving equation (1) is ill-posed in the classical sense, which means that it is strongly unstable with respect to small variations of the function $v(x, r)$. The case of the three-dimensional space is the most convenient for this purpose, so we will assume that $n = 3$ in what follows. Furthermore, assume that $u(x, y)$ is twice continuously differentiable in the entire space (x, y) and is even with respect to y. We introduce the function

$$w(x, y, r) = \frac{r}{\omega_3} \int_{|\xi - x|^2 + (\eta - y)^2 = r^2} u(\xi, \eta)\, d\omega, \tag{3}$$

which represents (up to the factor r) the spherical mean on the sphere of radius r centered at (x, y). It is known from courses on equations of mathematical physics that $w(x, y, r)$ satisfies the wave equation

$$\frac{\partial^2 w}{\partial r^2} = \Delta_{xy} w. \tag{4}$$

in the half-space $r > 0$, where Δ_{xy} is the Laplace operator with respect to x and y. (for example, see Courant and Hilbert, 1962). It readily follows from formula (3) that

$$w(x, y, 0) = 0. \tag{5}$$

From (1) it follows that

$$w(x, y, r)|_{y=0} = rv(x, r). \tag{6}$$

Finally, the condition that $u(x, y)$ is even with respect to y yields

$$w_y(x, y, r)|_{y=0} = 0. \tag{7}$$

The problem of finding a function $w(x, y, r)$ that satisfies equation (4) with conditions (5)–(7) is a boundary value problem that is obviously equivalent to the Cauchy problem (4), (6), (7) in the entire space (x, y, r) if we take the even extension of $v(x, r)$ to the domain $r < 0$.

Theorem 2. *The problem* (4)–(7) *is equivalent to the problem of solving equation* (1) *in the class of functions that are even with respect to y.*

Proof. Indeed, if $u(x, y)$ is a solution to (1) such that $u(x, y) = u(x, -y)$, then the function $w(x, y, r)$ given by the formula (3) is a solution of the problem (4)–(7). This solution is unique because the homogeneous problem (4)–(7) has only zero solution. Hence, every function $u(x, y)$ corresponds to a unique solution to the problem (4)–(7).

We will show that the converse is also true, i.e., that any solution to the problem (4)–(7) generates a unique solution to (1) such that $u(x, y) = u(x, -y)$. Let $w(x, y, r)$ be a solution to the problem (4)–(7). We put

$$w_r(x, y, 0) = u(x, y). \tag{8}$$

Then the function $w(x, y, r)$ can be uniquely represented in the domain $r \geq 0$ as a solution to the Cauchy problem with data (5), (8) in the form (3). Comparing formula (3) for $y = 0$ with formula (6), we conclude that $u(x, y)$ is a solution to equation (1). Differentiating both sides of (3) with respect to y, putting $y = 0$, and using (7), we obtain

$$\int_{S(x,r)} w_r(\xi, \eta) \, d\omega = 0. \tag{9}$$

From this equality and Theorem 1, it follows that the even part of the function $u(x, y)$ is equal to zero:

$$u(x, y) + u(x, -y) = 0,$$

i.e., $u(x, y) = -u(x, -y)$.

Thus, the function $u(x, y)$ obtained from the solution to the problem (4)–(7) using formula (8) is even with respect to y. By Theorem 1, if $u(x, y) = u(x, -y)$, we have a one-to-one correspondence. This proves the assertion of the equivalence theorem. ☐

We now illustrate the instability of the problem (4)–(7). Let

$$v(x, r) = \frac{1}{k^s} \frac{\sin kr}{r} \sin kx_1 \sin kx_2.$$

For sufficiently large k and s, the function $v(x, r)$ is small together with finitely many of its derivatives. At the same time, it is easy to verify directly that the solution to the problem (4)–(7) in this case is given by the formula

$$w(x, y, r) = \frac{1}{k^s} \sin(kr) \sin(kx_1) \sin(kx_2) \operatorname{ch}(ky).$$

But this solution tends to infinity as $k \to \infty$ for any fixed y. The function $u(x, y)$, obtained using formula (8), that represents a solution to equation (1) has the same property:

$$u(x, y) = \frac{1}{k^{s-1}} \sin(kx_1) \sin(kx_2) \operatorname{ch}(ky).$$

This means that the solution of the problem (4)–(7) and the equivalent problem (1) is strongly unstable.

11.4. PLANAR PROBLEM OF THE GENERAL FORM

Let D be a bounded open simply connected domain on the plane (x, y) with boundary Γ. Suppose that the boundary is specified by the equation of the form $x = g(s)$, $y = p(s)$, where s is the length of the segment of Γ with a fixed endpoint and direction consistent with the choice of orientation on Γ, $g(s)$ and $p(s)$ are functions of class $C^1[0, l]$, $g(0) = g(l)$, $p(0) = p(l)$, and l is the length of Γ. Let $L(t_1, t_2)$ be a two-parameter collection of curves in D with the following properties.

1. Any two points in the domain $\bar{D}$ are connected by a unique curve from the collection $L(t_1, t_2)$; any curve $L(t_1, t_2)$ intersects Γ at two points (x_1, y_1) and (x_2, y_2), i.e.,

$$x_1 = g(s_1), \quad x_2 = g(s_2),$$
$$y_1 = p(s_1), \quad y_2 = p(s_2),$$

and the other points of $L(t_1, t_2)$ belong to D; the lengths of the curves $L(t_1, t_2)$ are uniformly bounded.

2. The equation of a curve passing through a point (x_0, y_0) in the direction $\nu^0 = (\cos\theta_0, \sin\theta_0)$ is written as follows:

$$x = f_1(s, \theta_0, x_0, y_0) = x_0 + s\cos\theta_0 + s^2\tilde{f}_1(s, \cos\theta_0, \sin\theta_0, x_0, y_0),$$
$$y = f_2(s, \theta_0, x_0, y_0) = y_0 + s\sin\theta_0 + s^2\tilde{f}_2(s, \cos\theta_0, \sin\theta_0, x_0, y_0),$$

(1)

where $\tilde{f}_j(s, \cos\theta_0, \sin\theta_0, x_0, y_0)$ are functions of the length of the segment S with endpoint (x_0, y_0) and the parameters $\theta_0 \in [0, 2\pi]$ and $(x_0, y_0) \in \bar{D}$ that are continuously differentiable and bounded, and so are their derivatives; $f_j(s, 0, x_0, y_0) = f_j(s, 2\pi, x_0, y_0)$ and, moreover,

$$\frac{1}{s}\frac{D(f_1, f_2)}{D(s, \theta_0)} \geq c_0 > 0$$

(2)

in the entire domain of the variables (s, θ_0, x_0, y_0).

If s is close to zero, the validity of inequality (2) obviously follows from (1). Thus, inequality (2) is essential for finite s. In this case, it is equivalent to the positivity of the Jacobian $D(f_1, f_2)/D(s, \theta_0)$.

The following lemma holds for the collection of curves L.

Lemma 1. *Assume that the collection of curves satisfies the conditions listed above and the functions $f_j(s, \theta_0, x_0, y_0)$ have continuous and bounded derivatives of up to the mth order, where $m > 1$. Then (1) defines s and ν^0 as many-to-one m times continuously differentiable functions of points (x_0, y_0), (x, y) for all (x_0, y_0), $(x, y) \in \bar{D}$ such that $(x_0, y_0) \neq (x, y)$; moreover, the following estimates hold in the neighborhood of the set of points $(x_0, y_0) = (x, y)$, $(x_0, y_0) \in \bar{D}$:*

$$|D^\alpha s(x_0, y_0, x, y)| \leq \frac{c}{[(x - x_0)^2 + (y - y_0)^2]^{(|\alpha|-1)/2}},$$
$$|D^\alpha \nu^0(x_0, y_0, x, y)| \leq \frac{c}{[(x - x_0)^2 + (y - y_0)^2]^{|\alpha|/2}},$$
$$|\alpha| \leq m.$$

(3)

Proof. Note that (1) implies the following equalities:

$$s = \frac{[(x - x_0)^2 + (y - y_0)^2]^{1/2}}{[(\cos\theta_0 + s\tilde{f}_1)^2 + (\sin\theta_0 + s\tilde{f}_2)^2]^{1/2}},$$

$$\cos\theta_0 = \frac{x - x_0}{[(x - x_0)^2 + (y - y_0)^2]^{1/2}} [(\cos\theta_0 + s\tilde{f}_1)^2 + (\sin\theta_0 + s\tilde{f}_2)^2]^{1/2}$$

$$+ [(x - x_0)^2 + (y - y_0)^2]^{1/2} \frac{\tilde{f}_1}{[(\cos\theta_0 + s\tilde{f}_1)^2 + (\sin\theta_0 + s\tilde{f}_2)^2]^{1/2}},$$

$$\sin\theta_0 = \frac{y - y_0}{[(x - x_0)^2 + (y - y_0)^2]^{1/2}} [(\cos\theta_0 + s\tilde{f}_1)^2 + (\sin\theta_0 + s\tilde{f}_2)^2]^{1/2}$$

$$+ [(x - x_0)^2 + (y - y_0)^2]^{1/2} \frac{\tilde{f}_2}{[(\cos\theta_0 + s\tilde{f}_1)^2 + (\sin\theta_0 + s\tilde{f}_2)^2]^{1/2}}.$$

Applying the implicit function theorem, we establish that the functions s and ν have the following structure in a sufficiently small domain $(x - x_0)^2 + (y - y_0)^2 < \delta^2$:

$$s = [(x - x_0)^2 + (y - y_0)^2]^{1/2}\{1 + [(x - x_0)^2 + (y - y_0)^2]^{1/2}\varphi\},$$

$$\cos\theta = \frac{x - x_0}{[(x - x_0)^2 + (y - y_0)^2]^{1/2}}\{1 + [(x - x_0)^2 + (y - y_0)^2]^{1/2}\varphi\}$$

$$+ [(x - x_0)^2 + (y - y_0)^2]^{1/2}\psi,$$

$$\sin\theta = \frac{y - y_0}{[(x - x_0)^2 + (y - y_0)^2]^{1/2}}\{1 + [(x - x_0)^2 + (y - y_0)^2]^{1/2}\varphi\}$$

$$+ [(x - x_0)^2 + (y - y_0)^2]^{1/2}\psi,$$

where

$$\varphi = \varphi\Bigg(x_0, y_0, [(x - x_0)^2 + (y - y_0)^2]^{1/2},$$

$$\frac{x - x_0}{[(x - x_0)^2 + (y - y_0)^2]^{1/2}}, \frac{y - y_0}{[(x - x_0)^2 + (y - y_0)^2]^{1/2}}\Bigg),$$

$$\psi = \psi\Bigg(x_0, y_0, [(x - x_0)^2 + (y - y_0)^2]^{1/2},$$

$$\frac{x - x_0}{[(x - x_0)^2 + (y - y_0)^2]^{1/2}}, \frac{y - y_0}{[(x - x_0)^2 + (y - y_0)^2]^{1/2}}\Bigg)$$

are m times continuously differentiable functions of their arguments. In particular, this implies estimates (3). For $(x - x_0)^2 + (y - y_0)^2 > \delta^2$, it is obvious that $s > \delta$ and therefore $D(f_1, f_2)/D(s, \theta_0) \geq c_0\delta > 0$. In this case, the smoothness of s and ν^0 stated by the lemma follows from the smoothness of the functions $f_j(s, \theta_0, x_0, y_0)$, while the condition 1 for the collection of curves L implies that s and ν^0 exist and are many-to-one. □

We denote by $L(x_0, y_0, x, y)$ the segment of a curve from the specified collection that passes through points (x_0, y_0) and (x, y), these points being the endpoints of the segment. Let $\nu = (\cos\theta, \sin\theta)$ be the unit vector tangent to the curve $L(x_0, y_0, x, y)$ at the point (x, y). The vector ν can also be viewed as a function of (x_0, y_0) and (x, y) whose smoothness properties are the same as those of the vector ν^0. Indeed, if $\nu^0 = (h_1(x, y, x_0, y_0), h_2(x, y, x_0, y_0))$, then it is clear that $\nu = (-h_1(x_0, y_0, x, y), -h_2(x_0, y_0, x, y))$, since the points (x_0, y_0) and (x, y) are interchangeable.

Lemma 2. *The following inequality holds:*

$$\frac{\partial}{\partial s_1}\theta(g(s_1), p(s_1), x, y) \geq 0, \quad (x, y) \in D, \ s_1 \in [0, l]. \tag{4}$$

Proof. Note that the level curves of the function $\theta(x_0, y_0, x, y)$ viewed as a function of (x_0, y_0) for fixed $(x, y) \in D$ are the segments of curves from the collection L that pass through (x, y) with one endpoint at (x, y) and the other on the boundary Γ. Consequently, $\mathrm{grad}\,\theta(x_0, y_0, x, y)$ is directed along the normal to the curve $L(x_0, y_0, x, y)$ at (x_0, y_0) whose direction corresponds to increasing θ. Since for fixed x and y the curves $L(x_0, y_0, x, y)$ intersect only at (x, y), on any closed curve with $(x, y) \in D$ enclosed inside it (in particular, on Γ) the positive direction corresponds to the increasing θ. Setting $x_0 = g(s_1)$ and $y_0 = p(s_1)$, we conclude that larger s_1 correspond to larger $\theta(g(s_1), p(s_1), x, y)$. This proves inequality (4). $\square$

For $u(x, y) \in C^1(\bar{D})$ and $\rho(x_0, y_0, x, y) \in C^1(\bar{D} \times \bar{D})$, we consider the function

$$w(x_0, y_0, x, y) = \int_{L(x_0, y_0, x, y)} \rho(x_0, y_0, x_1, y_1)u(x_1, y_1)\,ds, \tag{5}$$

and formulate the following problem: determine $u(x, y)$ if the functions $\rho(x_0, y_0, x, y)$ and $v(t_1, t_2)$ are given, where

$$v(t_1, t_2) = w(g(t_1), p(t_1), g(t_2), p(t_2)), \quad t_1, t_2 \in [0, l]. \tag{6}$$

Theorem 1. *If a collection of curves has the properties 1 and 2 and the weight function* $\rho(x_0, y_0, x, y) \in C^1(\bar{D} \times \bar{D})$ *satisfies the conditions*

$$\rho(x_0, y_0, x, y) \geq \rho_0 > 0, \quad (x_0, y_0) \in \Gamma, \quad (x, y) \in \bar{D}, \tag{7}$$

$$\left|\frac{\partial}{\partial t_1}\ln \rho_0(g(t_1), p(t_1), x, y)\right| \le q\frac{\partial}{\partial t_1}\theta(g(t_1), p(t_1), x, y),$$

$$0 \le q < 1,$$

(8)

then the solution of the specified problem is unique for $u \in C^1(\bar{D})$ and the following stability estimate holds:

$$\|u\|_{\mathbb{L}_2(D)} \le \frac{1}{\rho_0\sqrt{1-q^2}}\frac{1}{2\sqrt{\pi}}\|\mathrm{grad}_{t_1,t_2} v(t_1, t_2)\|_{\mathbb{L}_2(Q)},$$

$$Q = [0, l] \times [0, l].$$

(9)

Proof. First, we establish the estimate (9) in the case where the smoothness of the functions u, ρ, and the functions $\tilde{f}_j$ in (1) is one order higher than that in the assumption of the theorem. In this case, the following lemma holds for $w(x_0, y_0, x, y)$.

Lemma 3. Let the functions $\tilde{f}_j$, ρ, and u be twice continuously differentiable with respect to their arguments. Assume that these functions and their partial derivatives are bounded. Then the function $w(x_0, y_0, x, y)$ is twice continuously differentiable with respect to (x_0, y_0, x, y) everywhere except at the points $(x_0, y_0) = (x, y)$ in whose neighborhoods the following estimates hold:

$$|D^\alpha w(x_0, y_0, x, y)| \le c[(x - x_0)^2 + (y - y_0)^2]^{(1-|\alpha|)/2}, \quad |\alpha| \le 2.$$

(10)

Proof. The assertion of the lemma follows from Lemma 1 and the representation

$$w(x_0, y_0, x, y) = \int_0^{s(x_0,y_0,x,y)} \rho(x_0, y_0, f_1, f_2)u(f_1, f_2)\, ds,$$

where $f_j = f_j(s, \theta_0(x_0, y_0, x, y), x, y)$, $j = 1, 2$. □

We will now obtain a differential equation for $w(x_0, y_0, x, y)$. We have

$$w(x_0, y_0, f_1(s, \theta_0(x_0, y_0, x, y), x, y), f_2(\cdot)) = \int_0^s \rho[x_0, y_0, f_1, f_2]u(f_1, f_2)\, ds.$$

Differentiating this equality with respect to s and taking into account that

$$(f_1(s, \theta_0, x_0, y_0), f_2(\cdot)) = (x, y), \quad (f_{1s}(\cdot), f_{2s}(\cdot)) = \nu(x_0, y_0, x, y),$$

we find

$$(\text{grad}_{x,y} w(x_0, y_0, x, y), \nu(x_0, y_0, x, y)) = \rho(x_0, y_0, x, y) u(x, y). \qquad (11)$$

Dividing both sides of this equality by $\rho(x_0, y_0, x, y)$, putting $x_0 = g(t_1)$ and $y_0 = p(t_1)$, and differentiating the obtained equality with respect to t_1, we arrive at the following equation for the function $w(g(t_1), p(t_1), x, y) = \widetilde{w}(t_1, x, y)$:

$$\frac{\partial}{\partial t_1} \left\{ \frac{1}{\widetilde{\rho}(t_1, x, y)} (\text{grad}_{x,y} \widetilde{w}, \widetilde{\nu}) \right\} = 0, \qquad (12)$$

where

$$\widetilde{\rho}(t_1, x, y) = \rho(g(t_1), p(t_1), x, y), \quad \widetilde{\nu}(t_1, x, y) = \nu(g(t_1), p(t_1), x, y).$$

Equation (12) is an equation of mixed type, namely hyperbolic-parabolic. It holds in the cylindrical domain $[0, l] \times \bar{D}$. The following conditions hold on the boundary of this domain: the periodicity condition

$$\widetilde{w}(0, x, y) = \widetilde{w}(l, x, y), \quad (x, y) \in \bar{D}, \qquad (13)$$

and the condition

$$\widetilde{w}(t_1, g(t_2), p(t_2)) = v(t_1, t_2), \qquad (14)$$

which follows from formula (6).

Thus, the function $\widetilde{w}$ is a solution to the nonclassical problem (12)–(14). Note that formulas (5), (6) imply that $v(t_1, t_1) = 0$ ($t_1 \in [0, l]$). Under this condition, the problem (12)–(14) is equivalent to the problem of integral geometry formulated above. Therefore, it is sufficient to find $\widetilde{w}(t_1, x, y)$ in order to determine $u(x, y)$ from (11). The method of energy estimates can be used to study the uniqueness and stability of solutions to the problem (12)–(14).

For $x \neq g(t_1)$, $y \neq p(t_1)$ and $\beta = (-\sin \widetilde{\theta}, \cos \widetilde{\theta})$, $\widetilde{\theta} = \widetilde{\theta}(t_1, x, y) = \theta(g(t_1), p(t_1), x, y)$, the following obvious identities hold:

$$\widetilde{\rho}(\nabla_{x,y} \widetilde{w}, \beta) \frac{\partial}{\partial t_1} \left[\frac{1}{\widetilde{\rho}} (\nabla_{x,y} \widetilde{w}, \widetilde{\nu}) \right] = (\nabla_{x,y} \widetilde{w}, \beta)(\nabla_{x,y} \widetilde{w}, \widetilde{\nu}_{t_1})$$

$$+ (\nabla_{x,y} \widetilde{w}, \beta)(\nabla_{x,y} \widetilde{w}_{t_1}, \widetilde{\nu}) - (\nabla_{x,y} \widetilde{w}, \beta)(\nabla_{x,y} \widetilde{w}, \widetilde{\nu}) \frac{\partial}{\partial t_1} \ln \widetilde{\rho},$$

$$\tilde{\rho}(\nabla_{x,y}\tilde{w}, \beta)\frac{\partial}{\partial t_1}\left[\frac{1}{\tilde{\rho}}(\nabla_{x,y}\tilde{w}, \nu)\right] = \frac{\partial}{\partial t_1}[(\nabla_{x,y}\tilde{w}, \beta)(\nabla_{x,y}\tilde{w}, \nu)]$$
$$- (\nabla_{x,y}\tilde{w}, \nu)(\nabla_{x,y}\tilde{w}, \beta_{t_1}) - (\nabla_{x,y}\tilde{w}, \nu)(\nabla_{x,y}\tilde{w}_{t_1}, \beta)$$
$$- (\nabla_{x,y}\tilde{w}, \beta)(\nabla_{x,y}\tilde{w}, \nu)\frac{\partial}{\partial t_1}\ln\tilde{\rho}.$$

Adding the first identity to the second and taking into account that

$$\tilde{\nu}_{t_1} = \beta\frac{\partial\tilde{\theta}}{\partial t_1}, \quad \beta_{t_1} = -\tilde{\nu}\frac{\partial\tilde{\theta}}{\partial t_1},$$

$$(\nabla_{x,y}\tilde{w}, \beta)(\nabla_{x,y}\tilde{w}_{t_1}, \nu) - (\nabla_{x,y}\tilde{w}, \nu)(\nabla_{x,y}\tilde{w}_{t_1}, \beta)$$
$$= \frac{\partial}{\partial x}(\tilde{w}_{t_1}, \tilde{w}_y) - \frac{\partial}{\partial y}(\tilde{w}_{t_1}, \tilde{w}_x),$$

we obtain

$$2\tilde{\rho}(\nabla_{x,y}\tilde{w}, \beta)\frac{\partial}{\partial t_1}\left(\frac{1}{\tilde{\rho}}\nabla_{x,y}\tilde{w}, \nu\right)$$

$$= \left\{[(\nabla_{x,y}\tilde{w}, \tilde{\nu})^2 + (\nabla_{x,y}\tilde{w}, \beta)^2]\frac{\partial\tilde{\theta}}{\partial t_1} - 2(\nabla_{x,y}\tilde{w}, \tilde{\nu})(\nabla_{x,y}\tilde{w}, \beta)\frac{\partial}{\partial t_1}\ln\tilde{\rho}\right\}$$

$$+ \frac{\partial}{\partial t_1}[(\nabla_{x,y}\tilde{w}, \tilde{\nu})(\nabla_{x,y}\tilde{w}, \beta)] + \frac{\partial}{\partial x}(\tilde{w}_{t_1}, \tilde{w}_y) - \frac{\partial}{\partial y}(\tilde{w}_{t_1}, \tilde{w}_x). \quad (15)$$

The left-hand side of the obtained identity vanishes on the solutions to equation (12). Under the assumption that $\tilde{w}(t_1, x, y)$ is a solution to equation (12), we integrate the identity (15) over the domain G_ε obtained from the domain $G = [0, l] \times \bar{D}$ by subtracting the set $\{(t_1, x, y) : t_1 \in [0, l], (x, y) \in \bar{D}, (x - g(t_1))^2 + (y - p(t_1))^2 \leq \varepsilon^2\}$ for a sufficiently small positive ε. Then, using the Gauss–Ostrogradsky formula, we have

$$\int_{G_\varepsilon}\left\{[(\nabla_{x,y}\tilde{w}, \tilde{\nu})^2 + (\nabla_{x,y}\tilde{w}, \beta)^2]\frac{\partial\tilde{\theta}}{\partial t_1}\right.$$
$$\left. - 2(\nabla_{x,y}\tilde{w}, \tilde{\nu})(\nabla_{x,y}\tilde{w}, \beta)\frac{\partial}{\partial t_1}\ln\tilde{\rho}\right\} dxdydt_1$$
$$+ \int_{S_\varepsilon}\{[(\nabla_{x,y}\tilde{w}, \tilde{\nu})^2 + (\nabla_{x,y}\tilde{w}, \beta)]\cos(\widehat{n, t_1}) + \tilde{w}_{t_1}[\tilde{w}_y\cos(\widehat{n, x})$$
$$- \tilde{w}_x\cos(\widehat{n, y})]\} dS = 0,$$

where S_ε is the boundary of the set G_ε, n is the outward normal to S_ε, and dS is the area element.

In this equality, we pass to the limit as $\varepsilon \to 0$. In view of the estimates of Lemma 3, the integrals over G_ε and S_ε converge as improper integrals to the integrals over G and S (S is the boundary of G). The integral over the surface S can be represented as the sum of the integral over the top and bottom of the cylinder G and the integral over the lateral surface $[0, l] \times \Gamma$. On the top and bottom of the cylinder, we have $\cos(\widehat{n, x}) = \cos(\widehat{n, y}) = 0$, and $\cos(\widehat{n, t_1})$ at the corresponding points has opposite signs. The periodicity condition (13) implies that the sum of the integrals over the top and bottom of the cylinder is equal to zero. On the lateral surface, we have $\cos(\widehat{n, t_1}) = 0$ and for $x = g(t_2)$ and $y = p(t_2)$

$$\widetilde{w}_{t_1} = v_{t_1}(t_1, t_2),$$

$$\widetilde{w}_y \cos(\widehat{n, x}) - \widetilde{w}_x \cos(\widehat{n, y}) = \frac{\partial}{\partial t_2} \widetilde{w}(t_1, g(t_2), p(t_2)) = v_{t_2}(t_1, t_2),$$

and $dS = dt_1 dt_2$. Finally, we obtain

$$\int\limits_G \left\{ [(\nabla_{x,y}\widetilde{w}, \widetilde{\nu})^2 + (\nabla_{x,y}\widetilde{w}, \beta)^2] \frac{\partial \widetilde{\theta}}{\partial t_1} \right.$$

$$\left. - 2(\nabla_{x,y}\widetilde{w}, \widetilde{\nu})(\nabla_{x,y}\widetilde{w}, \beta) \frac{\partial}{\partial t_1} \ln \widetilde{\rho} \right\} dx dy dt_1$$

$$= - \int\limits_0^l \int\limits_0^l v_{t_1} v_{t_2} \, dt_1 dt_2. \quad (16)$$

By Lemma 2, $\partial \widetilde{\theta}_0 / \partial t_1 \geq 0$. For all points where $\partial \widetilde{\theta}_0 / \partial t_1 = 0$, the expression in braces vanishes because of the condition (8). For the points where $\partial \widetilde{\theta}_0 / \partial t_1 > 0$, we have

$$[(\nabla_{x,y}\widetilde{w}, \widetilde{\nu})^2 + (\nabla_{x,y}\widetilde{w}, \beta)^2] \frac{\partial \widetilde{\theta}}{\partial t_1} - 2(\nabla_{x,y}\widetilde{w}, \widetilde{\nu})(\nabla_{x,y}\widetilde{w}, \beta) \frac{\partial}{\partial t_1} \ln \widetilde{\rho}$$

$$= (\nabla_{x,y}\widetilde{w}, \widetilde{\nu})^2 \left[\frac{\partial \widetilde{\theta}}{\partial t_1} - \left(\frac{\partial}{\partial t_1} \ln \widetilde{\rho} \right)^2 \left(\frac{\partial \widetilde{\theta}}{\partial t_2} \right)^{-1} \right]$$

$$+ \left[(\nabla_{x,y}\widetilde{w}, \widetilde{\nu}) \frac{\partial}{\partial t_1} \ln \widetilde{\rho} \left(\frac{\partial \widetilde{\theta}}{\partial t_1} \right)^{-1/2} - (\nabla_{x,y}\widetilde{w}, \beta) \left(\frac{\partial \widetilde{\theta}}{\partial t_1} \right)^{1/2} \right]$$

$$\geq (\nabla_{x,y}\widetilde{w}, \widetilde{\nu})^2 (1 - q^2) \frac{\partial \widetilde{\theta}}{\partial t_1} \geq \rho_0^2 (1 - q^2) u^2(x, y) \frac{\partial \widetilde{\theta}}{\partial t_1}.$$

Therefore, making inequality (16) stronger, we obtain

$$\rho_0^2(1 - q^2) \int_D u^2(x, y) \, dx dy \cdot \int_0^{2\pi} \frac{\partial \tilde{\theta}}{\partial t_1} \, dt_1 \leq \frac{1}{2} \int_0^l \int_0^l |\nabla_{t_1, t_2} v|^2 \, dt_1 dt_2.$$

The estimate (9) follows.

To conclude the proof of the theorem, we first note that the expression on the right-hand side of equality (9) is meaningful for $u \in \mathbb{C}^1(\bar{D})$, $\rho(x_0, y_0, x, y) \in \mathbb{C}^1(\Gamma \times \bar{D})$, and a collection of curves L satisfying conditions 1 and 2. Therefore, approximating the functions u, ρ, and $\tilde{f}$ that satisfy the assumptions of the theorem by the functions u_n, ρ_n, and $\tilde{f}_n$ for which the estimate (9) has already been established and passing to the limit in this estimate as $n \to \infty$, we obtain the estimate (9) for the functions satisfying the assumptions of the theorem. In particular, this estimate implies the uniqueness of the solution of the problem of integral geometry in the space $\mathbb{C}^1(\bar{D})$. $\square$

11.5. SPATIAL PROBLEMS OF THE GENERAL FORM

Consider the following problem. Let D be a bounded domain in the space $\mathbb{R}^n$, $x = (x_1, \ldots, x_n)$, and Ω be an open domain included in D. If the dimension of the space is even, we will assume that the distance between the boundaries of the domains D and Ω is uniformly minimized by a positive h, so that Ω lies strictly inside D. If the dimension of the space is odd, D and Ω may coincide. Furthermore, assume that for any point $x \in D$ and any unit vector $\nu = (\nu_1, \ldots, \nu_n)$ there exists a unique smooth hypersurface $S(x, \nu)$ that passes through x such that ν is the normal vector to this hypersurface at x.

We denote by $\mathcal{U}$ the class of functions $u(x)$ such that the support of any $u(x) \in \mathcal{U}$ is included in Ω and $u(x) \in \mathbb{L}_2(\Omega)$. Consider the problem of determining a function $u(x) \in \mathcal{U}$ from the equation

$$v(x, \nu) = \int_{S(x,\nu)} \rho(\xi, x, \nu) u(\xi) \, dS, \quad x \in D, \quad |\nu| = 1, \tag{1}$$

where $\rho(\xi, x, \nu)$ is a given smooth function of its arguments and dS is the surface area element.

Generally speaking, the formulated problem is overdetermined because $u(x)$ is a function of n variables, whereas v is a function of $2n - 1$ variables. However, there is an important case where u and v depend on the same number of essential variables. Consider the surface $S(x, \nu)$ and an arbitrary point $x^0 \in S(x, \nu)$. Let ν^0 be the normal vector to $S(x, \nu)$ at the point x^0. Then $S(x^0, \nu^0) = S(x, \nu)$. If $\rho(\xi, x^0, \nu^0) = \rho(\xi, x, \nu)$ for $\xi \in S(x, \nu)$, then $v(x^0, \nu^0) = v(x, \nu)$. Since every point of the hypersurface $S(x, \nu)$ is characterized by $n - 1$ parameter, the preceding equality shows that in this case there are only n essential parameters that $v(x, \nu)$ depends on. Thus, if the weight function depends only on the point $\xi \in D$ and the surface $S(x, \nu)$, then u and v depend on the same number of essential variables. This means that there must be a better form of the statement of the problem of integral geometry for the same collection of surfaces that involves a different parameterization of this collection such that the problem is not overdetermined.

In this section, the problem of solving equation (1) will be studied under the assumption that the diameter of the domain D is sufficiently small. In what follows, we assume that the equation for the surface $S(x, \nu)$ can be specified using a smooth function $\varphi(\xi, x, \nu)$:

$$S(x, \nu) = \{\xi : \varphi(\xi, x, \nu) = 0\}, \tag{2}$$

where $|\nabla_\xi \varphi| \neq 0$ on $S(x, \nu)$.

The assumptions on the parameterization lead to the following necessary conditions to be imposed on the function φ:

$$\varphi(x, x, \nu) = 0, \quad \nabla_\xi \varphi|_{\xi=x} = \nu |\nabla_\xi \varphi|_{\xi=x}. \tag{3}$$

Clearly, one can always assume that

$$|\nabla_\xi \varphi|_{\xi=x} = 1. \tag{4}$$

To satisfy this condition, if necessary, φ can be divided by $|\nabla_\xi \varphi|_{\xi=x}$, so we will assume that equality (4) holds. Then φ can be represented as follows:

$$\varphi(\xi, x, \nu) = (\nu, \xi - x) + \sum_{i,j=1}^n a_{ij}(\xi, x, \nu)(\xi_i - x_i)(\xi_j - x_j), \tag{5}$$

where

$$a_{ij}(\xi, x, \nu) = \int_0^1 \varphi_{\xi_i \xi_j}(x + t(\xi - x), x, \nu)(1 - t)\, dt.$$

We will use this representation in what follows.

First, we will formulate the result and prove it for odd-dimensional spaces, and then for even-dimensional spaces.

Theorem 1. *Assume that n is an odd number, $s = (n-1)/2 \geq 1$, the function $\varphi(\xi, x, \nu)$ and its partial derivatives of order up to $n+2$ are continuous in the domain $G = \{(\xi, x, \nu) : \xi \in \bar{D}, x \in \bar{D}, |\nu| = 1\}$, the weight function $\rho(\xi, x, \nu)$ and its partial derivatives of up to the nth order are continuous in G, and $\rho(\xi, x, \nu)$ satisfies the following condition in G:*

$$\rho(x, x, \nu) \geq \rho_0 > 0. \tag{6}$$

Then there exists a number $d^* > 0$, $d^* = d^*(\varphi, \rho)$ such that if diam $D < d^*$, then the solution of equation (1) in the class of functions $\mathcal{U}$ is unique and the following stability estimate holds:

$$\|u\|_{L_2(\Omega)} \leq C\|\Delta^s v_1\|_{L_2(\Omega)}, \quad v_1 = \int\limits_{|\nu|=1} \frac{v(x, \nu)}{\rho(x, x, \nu)} \, d\omega_\nu, \tag{7}$$

where $d\omega_\nu$ is the surface area element of the unit sphere and Δ^s is the sth power of the Laplace operator.

Proof. Using the delta function and the finiteness of the function $u(x)$, equation (1) can be written as

$$\int\limits_\Omega \rho(\xi, x, \nu) u(\xi) |\nabla_\xi \varphi(\xi, x, \nu)| \delta(\varphi(\xi, x, \nu)) \, d\xi = v(x, \nu). \tag{8}$$

We now divide (8) by $\rho(x, x, \nu)$ and average it over all ν assuming that $x \in D$ is fixed. In view of (7), equality (8) becomes as follows:

$$\int\limits_\Omega K(x, \xi) u(\xi) \, d\xi = v_1(x), \tag{9}$$

where

$$K(x, \xi) = \int\limits_{|\nu|=1} \tilde{\rho}(\xi, x, \nu) \delta(\varphi(\xi, x, \nu)) \, d\omega_\nu,$$

$$\tilde{\rho}(\xi, x, \nu) = \frac{\rho(\xi, x, \nu)}{\rho(x, x, \nu)} |\nabla_\xi \varphi(\xi, x, \nu)|. \tag{10}$$

It is rather easy to study the kernel of equation (10) if the points ξ and x are sufficiently close to each other. Note that we may confine ourselves to considering only this condition because the diameter of the domain D can be assumed to be small. We will use the property of the measure $d\omega_\nu$ to be invariant with respect to orthogonal transformations of the space $\nu_1, \dots, \nu_n$ to perform a transformation with an orthogonal matrix Q. Then the formula for the kernel can be written as

$$K(x, \xi) = \int_{|\nu|=1} \tilde{\rho}(\xi, x, Q\nu)\delta(\varphi(\xi, x, Q\nu)) \, d\omega_\nu. \tag{11}$$

We now introduce the unit vector $\nu^0 = (\xi - x)/|\xi - x|$, $\nu^0 = (\nu_1^0, \dots, \nu_n^0)$. When calculating the integral (11) we use two different orthogonal transformations, depending on the location of the vector ν^0 on the unit sphere. In particular, let $e^1 = (1, 0, \dots, 0)$. Then for $(\nu^0, e^1) \geq 0$ we take the transformation defined by the equality

$$Q\nu = \nu + 2(\nu^0, e^1)\nu^0 - \frac{(\nu, \nu^0) + (\nu^0, e^1)}{1 + (\nu^0, e^1)}(\nu^0 + e^1). \tag{12}$$

It is easy to verify directly that any unit vector ν satisfies the equality $(Q\nu, Q\nu) = 1$. Moreover,

$$Qe^1 = \nu^0. \tag{13}$$

For $(\nu^0, e^1) < 0$, when calculating the integral, we use the transformation (12) with e^1 replaced by $-e^1$, and, consequently, $Qe^1 = -\nu^0$.

Clearly, it suffices to study the case where the position of the points x, ξ is such that $(\nu^0, e^1) \geq 0$. The equation $\varphi(\xi, x, Q\nu) = 0$ with fixed x and ξ defines a surface of dimension $n - 2$ on the surface of a sphere. Indeed, in view of (5), after dividing by $|x - \xi|$, this equation can be written as

$$(\nu^0, Q\nu) + |x - \xi| \sum_{i,j=1}^{n} a_{ij}(x + |x - \xi|\nu^0, x, Q\nu)\nu_i^0 \nu_j^0 = 0. \tag{14}$$

We represent the vector ν as follows:

$$\nu = qe^1 + \sqrt{1 - q^2}\bar{\nu}, \tag{15}$$

where $\bar{\nu}$ is the unit vector orthogonal to e^1. Then (13) implies

$$(\nu^0, Q\nu) = (Q^*\nu^0, \nu) = (e^1, \nu) = q,$$

and equality (14) becomes as follows:

$$q + |x - \xi| \sum_{i,j=1}^{n} a_{ij}(x + |x - \xi|\nu^0, x, q\nu^0 + \sqrt{1 - q^2}Q\bar{\nu})\nu_i^0 \nu_j^0 = 0. \tag{16}$$

The contracting maps principle implies the following lemma.

Lemma 1. If φ satisfies the assumption of the theorem, then for any $q_0 \in (0,1)$ there exists a $d_0 = d_0(\varphi, q_0)$ such that for domains Ω with $\text{diam}\,\Omega < d_0$ the equation $\varphi(\xi, x, Q\nu) = 0$ for $x, \xi \in \Omega$ characterizes q as a many-to-one, continuous, and bounded function of the arguments x, $|x - \xi|$, ν^0, and $\bar{\nu}$ with continuous and bounded derivatives of up to the nth order with respect to these arguments. Moreover,

$$|q(x, |x - \xi|, \nu^0, \bar{\nu})| \leq q_0 < 1. \tag{17}$$

Proof. Since φ satisfies the assumption of the theorem, $a_{ij}(\xi, x, \nu)$ are uniformly bounded for $x, \xi \in \Omega$ and $|\nu| = 1$ by a constant c_0. Therefore, writing (16) in the form $q = A(q)$, we conclude that under the condition $c_0 d_0 \leq q_0$ the operator A maps the set of continuous functions satisfying inequality (17) to itself. The condition for the derivatives of a_{ij} to be uniformly bounded with respect to $\nu_1, \ldots, \nu_n$ implies that A is a contracting operator on this set if $|x - \xi|$ is sufficiently small. Thus, equation (16) characterizes q as a continuous function of the arguments x, $|x - \xi|$, ν^0, and $\bar{\nu}$ that satisfies inequality (17). The existence and boundedness of the derivatives of q It readily follows from (16) that q has bounded derivatives of up to the nth order, since the functions a_{ij} and the matrix $Q = Q(\nu^0)$ given by the formula (12) have the corresponding derivatives. $\quad\square$

It also follows from equality (16) that $q > 0$ for $|x - \xi| \to 0$.

Fix a $q_0 \in (0,1)$ and assume that $\text{diam}\,\Omega < d_0(\varphi, q_0)$. Formula (15), where q is a solution to (16) with arbitrary unit vector $\bar{\nu}$ such that $(\bar{\nu}, e^1) = 0$, defines ν as a smooth many-to-one function of $\bar{\nu}$ (for fixed x and ξ) and thus defines an $n - 2$-dimensional surface on the sphere $|\nu| = 1$. We denote this surface by $\Sigma(x, \xi)$. If $x \to \xi$ so that $(\xi - x)/|\xi - x| \to \nu^0$, then the limit surface $\Sigma(x, \xi)$ coincides with the section of the sphere $|\nu| = 1$ formed by the plane passing through the center of the sphere and orthogonal to the vector e^1.

Let n be the unit normal vector to $\Sigma(x, \xi)$ lying in the plane tangent to the sphere $|\nu| = 1$ at ν. From the above arguments, it follows that formula (11) for the kernel $K(x, \xi)$ of equation (9) can be written as

$$K(x, \xi) = \int_{\Sigma(x,\xi)} \tilde{\rho}(\xi, x, Q\nu) \left|\frac{\partial\varphi}{\partial n}\right|^{-1} d\sigma, \tag{18}$$

where $d\sigma$ is the area element of $\Sigma(x, \xi)$. Since the vector n on $\Sigma(x, \xi)$ is parallel to the vector

$$\nabla_\nu\varphi(\xi, x, Q\nu) - \nu(\nu, \nabla_\nu\varphi(\xi, x, Q\nu)),$$

we have

$$\left|\frac{\partial\varphi}{\partial n}\right| = [|\nabla_\nu\varphi(\xi, x, Q\nu)|^2 - |\nu, \nabla_\nu\varphi(\xi, x, Q\nu)|^2]^{1/2}. \tag{19}$$

From formula (5) we obtain

$$\frac{1}{|x - \xi|}\nabla_\nu\varphi(\xi, x, Q\nu) = e^1 + |x - \xi|\sum_{i,j=1}^n \nabla_\nu a_{ij}(\xi, x, Q\nu)\nu_i^0\nu_j^0.$$

Using (19) and the fact that $\nu = \nu(x, |x - \xi|, \nu^0, \bar\nu)$ on $\Sigma(x, \xi)$, we arrive at the following representation for $|\partial\varphi/\partial n|$ on $\Sigma(x, \xi)$:

$$\left|\frac{\partial\varphi}{\partial n}\right|^{-1} = \frac{1}{|x - \xi|} + b(x, |x - \xi|, \nu^0, \bar\nu), \tag{20}$$

where b and its partial derivatives of order up to $n - 1$ are continuous and bounded functions of their arguments. Similarly, the following representation is valid for the function $\tilde\rho$ on $\Sigma(x, \xi)$:

$$\tilde\rho(\xi, x, Q\nu) = 1 + |x - \xi|c(x, |x - \xi|, \nu^0, \bar\nu), \tag{21}$$

where the function c has the same properties as b.

The vector $\bar\nu$ is orthogonal to e^1 and has unit length. Therefore, it is completely characterized by the angular spherical coordinates $\psi_1, \ldots, \psi_{n-2}$ in the coordinate plane orthogonal to the vector e^1. The surface element $d\sigma$ is given by the formula

$$d\sigma = \left[\Gamma\left(\frac{\partial\nu}{\partial\psi_1}, \ldots, \frac{\partial\nu}{\partial\psi_{n-2}}\right)\right]^{1/2} d\psi, \tag{22}$$

where $d\psi = d\psi_1 \ldots d\psi_{n-2}$ and Γ denotes the Gram determinant for the vectors $\frac{\partial\nu}{\partial\psi_1}, \ldots, \frac{\partial\nu}{\partial\psi_{n-2}}$:

$$\Gamma\left(\frac{\partial\nu}{\partial\psi_1}, \ldots, \frac{\partial\nu}{\partial\psi_{n-2}}\right) = \begin{vmatrix} \left(\frac{\partial\nu}{\partial\psi_1}, \frac{\partial\nu}{\partial\psi_1}\right) & \cdots & \left(\frac{\partial\nu}{\partial\psi_1}, \frac{\partial\nu}{\partial\psi_{n-2}}\right) \\ \vdots & \vdots & \vdots \\ \left(\frac{\partial\nu}{\partial\psi_{n-2}}, \frac{\partial\nu}{\partial\psi_1}\right) & \cdots & \left(\frac{\partial\nu}{\partial\psi_{n-2}}, \frac{\partial\nu}{\partial\psi_{n-2}}\right) \end{vmatrix}.$$

From formulas (18)–(22), it follows that the kernel $K(x, \xi)$ can be represented in the form

$$K(x, \xi) = \frac{\omega_{n-1}}{|x - \xi|} + K_0(x, |x - \xi|, \nu^0), \tag{23}$$

where K_0 and its derivatives of order up to $(n-1)$ are continuous and bounded functions of their arguments. Then, applying the operator Δ^s $(s = (n-1)/2)$ to equation (9) with respect to the variable x and taking into account that

$$\Delta^s \left(\frac{1}{|x - \xi|} \right) = (-1)^s (4\pi)^s (s-1)! \delta(x - \xi), \quad \omega_{n-1} = 2\pi^s / \Gamma(s),$$

we obtain the following equation for $u(x)$:

$$u(x) + \int_\Omega \widetilde{K}(x, \xi) u(\xi) \, d\xi = \frac{(-1)^s}{2^n \pi^{2s}} \Delta^s v_1(x), \tag{24}$$

where

$$\widetilde{K}(x, \xi) = \frac{(-1)^s}{2^n \pi^{2s}} \Delta_x^s K_0(x, |x - \xi|, \nu^0).$$

The kernel $\widetilde{K}(x, \xi)$ of equation (24) has an integrable singularity. Indeed, application of the operator Δ^s involves taking the derivatives of K_0 of order up to $2s$ with respect to x, $|x - \xi|$, and ν^0 and the derivatives of $|x - \xi|$ and ν^0 of order up to $2s$ with respect to x. Since $2s = n - 1$, the derivatives with respect to the arguments of K_0 are continuous and bounded. At the same time, the following estimates hold for the derivatives of $|x - \xi|$ and ν^0 with respect to x:

$$\left| \frac{\partial^k}{\partial x_i^k} (|x - \xi|) \right| \le \frac{c_1}{|x - \xi|^{k-1}}, \quad \left| \frac{\partial^k}{\partial x_i^k} (\nu^0) \right| \le \frac{c_1}{|x - \xi|^k}, \quad 1 \le k \le 2s.$$

It follows that the kernel $\widetilde{K}(x, \xi)$ is continuous everywhere except on the set of points such that $x = \xi$. In the neighborhood of this set, $\widetilde{K}(x, \xi)$ satisfies the estimate

$$|\widetilde{K}(x, \xi)| \le c_2 / |x - \xi|^{n-1}.$$

It is known that equation (24) with a kernel of this type has a unique solution that belongs to $\mathbb{L}_2(\Omega)$ if the diameter of the domain Ω is sufficiently small. In this case, the validity of the estimate (17) for the solution to equation (24) is obvious.   □

We now consider the case of an even-dimensional space. As was mentioned before, in this case we will assume that Ω lies strictly inside D, so that the distance between Ω and the boundary of the domain D is greater than some $h > 0$. In addition, it is assumed that $h > \operatorname{diam} \Omega$. We denote by Ω_h the open set that is the union of all open balls of radius h centered at points $x \in \Omega$.

Theorem 2. *Suppose that $n = 2s$, $s \geq 1$, and the functions φ and ρ satisfy the assumption of Theorem 1 with n replaced by $n + 1$. Then there exists a number $d^* > 0$, $d^* = d^*(\varphi, \rho, h)$ such that if $\operatorname{diam} \Omega < d^*$, then the solution of equation (1) is unique in the class of functions $\mathcal{U}$ and the following stability estimate holds:*

$$\|u\|_{L_2(\Omega)} \leq c_0 \|\Delta^s v_2\|_{L_2(\Omega)}, \tag{25}$$

where

$$v_2(x) = \int\limits_{|x-y| \leq h} \frac{v_1(y)}{|x - y|^{n-1}} \, dy, \tag{26}$$

and $v_1(x)$ is determined from $v(x, \nu)$ using formula (7).

Proof. The outline of the proof is as follows. First, as in the case of odd n, we obtain equation (9). The kernel $K(x, \xi)$ of this equation can be represented in the form (23), the smoothness of K_0 being one degree higher, according to the assumption of the theorem. Namely, K_0 has continuous bounded derivatives of order up to n with respect to the arguments x, $|x-\xi|$, and $\nu^0 = (\xi - x)|x - \xi|^{-1}$. We now apply the averaging operator defined by formula (26) (which is averaging over the ball of radius h) to both sides of the equality. We obtain the equation

$$\int\limits_{\Omega} T(x, \xi) u(\xi) \, d\xi = v_2(x), \quad x \in \Omega, \tag{27}$$

where

$$T(x, \xi) = \int\limits_{|x-y| \leq h} \frac{K(y, \xi)}{|x - y|^{n-1}} \, dy. \tag{28}$$

We will show that $T(x, \xi)$ $(x, \xi \in \Omega)$ can be represented as

$$T(x, \xi) = -\omega_{n-1} \omega_n \ln |x - \xi| + T_0(\xi, |\xi - x|, \nu^0), \tag{29}$$

where the function $T_0(\xi, |\xi - x|, \nu^0)$ and its derivatives of order up to n are continuous and bounded everywhere except on the set $\rho = 0$. In the neighborhood of this set, $T_0(\xi, |\xi - x|, \nu^0)$ satisfies the following inequalities (with $|\alpha| + k \leq n$):

$$\left| \frac{\partial^k}{\partial \rho^k} D_{\xi, \nu^0}^\alpha T_0(\xi, \rho, , \nu^0) \right| \leq c_0 \begin{cases} \ln \rho, & k = 1, \\ \rho^{1-k}, & k \geq 2. \end{cases} \tag{30}$$

Applying the operator Δ^s to equation (27) and taking into account that

$$\Delta_x^s \ln|x - \xi| = (-1)^{s-1} 2^{n-1} \pi^s (s-1)! \delta(x - \xi),$$

we obtain a Fredholm equation analogous to equation (24), with a singularity of the form $|x - \xi|^{n-1} \ln|x - \xi|$, which implies the assertion of the theorem.

Thus, it remains to verify that the kernel T possesses the properties stated above. To this end, we calculate the integral (28) by taking the integral of each term in the representation (23) for $K(x, \xi)$. Calculating the integral of the first term is equivalent to calculating the integral

$$f_1(\rho) = \int\limits_{|x-y| \leq h} \frac{1}{|x - y|^{n-1}} \frac{1}{|y - \xi|} \, dy$$

$$= \int\limits_{|\nu|=1} d\omega_\nu \int\limits_0^h \frac{dr}{[r^2 + \rho^2 + 2r\rho(\nu^0, \nu)]^{1/2}}$$

$$= \int\limits_{|\nu|=1} [\ln|h + \rho(\nu^0, \nu) + [r^2 + \rho^2 + 2r\rho(\nu^0, \nu)]^{1/2}|$$

$$- \ln|1 + (\nu, \nu^0)|] \, d\omega_\nu - \omega_n \ln \rho.$$

For this calculation, we introduced the spherical coordinate system $y = x + r\nu$ and denoted by $(\rho, -\nu)$ the spherical coordinates of ξ in this system, i.e., $\rho = |\xi - x|$ and $\nu^0 = (x - \xi)/|\xi - x|$. As follows from the obtained formula, calculating the integral (28) of the first term in the expression for $K(x, \xi)$ yields the main part of the kernel $T(x, \xi)$, namely $\ln \rho$, and a smooth function (which is even analytic for $\rho < h$) that depends only on ρ. It was essential for the calculation that $\rho \leq \operatorname{diam} \Omega < h$ for $x \in \Omega$ and $\xi \in \Omega$. We now show that calculating the integral (28) of the second term in the expression for $K(x, \xi)$ yields a function that has the same properties as the function T_0. For convenience, we introduce the spherical coordinate system with origin at ξ, $y = \xi + r\nu$, $\nu = (\nu_1, \dots, \nu_n)$, $x = \xi + \rho\nu^0$:

$$f_2(\xi, \rho, \nu^0) = \int\limits_{|x-y| \leq h} \frac{K_0(y, |y - \xi|, \nu^0)}{|x - y|^{n-1}} dy$$

$$= \int\limits_{|\nu|=1} d\omega_\nu \int\limits_0^{r(\rho,(\nu^0,\nu))} \frac{r^{n-1} K_0(\xi + r\nu, r, -\nu)}{[r^2 + \rho^2 - 2r\rho(\nu^0, \nu)]^{(n-1)/2}} \, dr,$$

$$r(\rho, (\nu^0, \nu)) = \rho(\nu^0, \nu) + [h^2 - \rho^2[1 - (\nu, \nu^0)^2]]^{1/2}.$$

As was done before, we introduce an orthogonal transformation of the coordinate system with a matrix Q such that $Q^* \nu^0 = e^1 = (1, 0, \ldots, 0)$ (provided that $(\nu^0, e^1) \geq 0$). Then, setting $K_0(\xi + rQ\nu, r, -Q\nu) = \widetilde{K}(\xi, r, \nu, \nu^0)$, we have

$$f_2(\xi, \rho, \nu^0) = \int\limits_{|\nu|=1} d\omega_\nu \int\limits_0^{r(\rho, \nu_1)} \frac{r^{n-1} \widetilde{K}(\xi, r, \nu, \nu^0)}{(r^2 + \rho^2 - 2r\rho\nu_1)^{(n-1)/2}} \, dr.$$

It readily follows that the derivatives $D^\alpha_{\xi, \nu^0} f_2$ ($|\alpha| \leq n$) are continuous and bounded for $\xi, x \in \Omega$. Indeed,

$$D^\alpha_{\xi, \nu^0} f_2 = \int\limits_{|\nu|=1} d\omega_\nu \int\limits_0^{r(\rho, \nu_1)} \frac{r^{n-1} D^\alpha_{\xi, \nu^0} \widetilde{K}(\cdot)}{(r^2 + \rho^2 - 2r\rho\nu_1)^{(n-1)/2}} \, dr. \tag{31}$$

It remains to deal with the derivatives that involve differentiation with respect to the variable ρ. The change of variables $r = \rho r_1$ in the internal integral shows that for $\rho > 0$ the integral f_2 and its derivatives of order up to n continuously depend on ξ, ρ, and ν^0. However, for $\rho \to 0$, the derivatives that involve differentiation with respect to ρ are not bounded. Indeed, for sufficiently small ρ it is always possible to choose a $\delta > 0$ such that

$$\rho(1 + \delta) < r(\rho, \nu_1), \quad \nu_1 \in [-1, 1].$$

Then, representing the internal integral as the sum of two integrals (one over the interval $[0, \rho(1 + \delta)]$ and the other over the interval $[\rho(1 + \delta), r(\rho, \nu_1)]$) and making the change of variables $r = \rho r_1$ in the first integral, we obtain

$$f_2(\xi, \rho, \nu^0) = \rho \int\limits_{|\nu|=1} d\omega_\nu \int\limits_0^{1+\delta} \frac{r_1^{n-1} \widetilde{K}(\xi, \rho r_1, \nu, \nu^0)}{(r_1^2 + 1 - 2r_1\nu_1)^{(n-1)/2}} \, dr_1$$

$$+ \int\limits_{|\nu|=1} d\omega_\nu \int\limits_{\rho(1+\delta)}^{r(\rho, \nu_1)} \frac{r^{n-1} \widetilde{K}(\xi, r, \nu, \nu^0)}{(r^2 + \rho^2 - 2r\rho\nu_1)^{(n-1)/2}} \, dr. \tag{32}$$

The first integral in this formula and its derivatives of order up to n are continuous and bounded functions. All the singularities in the derivatives are caused by the second integral. Note that the coefficient of $\widetilde{K}$ in the second integral can be represented as follows:

$$\frac{r^{n-1}}{(r^2 + \rho^2 - 2r\rho\nu_1)^{(n-1)/2}} = \Phi\left(\frac{\rho}{r}, \nu_1\right),$$

where $\Phi(z, \nu_1)$, as a function of arguments z and ν_1, and any finite number of its derivatives are bounded in the domain $|z| \le (1+\delta)^{-1}$, $\nu_1 \in [-1, 1]$. As a result, the following estimates hold for the derivatives of Φ with respect to ρ:

$$\left| \frac{\partial^k}{\partial \rho^k} \Phi\left(\frac{\rho}{r}, \nu_1\right) \right| \le \frac{c_0}{r^k}, \quad r \ge \rho(1+\delta). \tag{33}$$

When calculating the derivatives $\frac{\partial^k}{\partial \rho^k} D_{\xi, \nu^0}^\alpha$ of the second term in (32), the symbol D_{ξ, ν^0}^α can be brought under the sign of the internal integral. Calculation of the derivative $\partial^k / \partial \rho^k$ of the internal integral yields several terms resulting from calculating the derivatives with respect to its upper and lower limits and the integral resulting from differentiating the integrand. The former are obviously bounded, since the function $\Phi(\rho/r, \nu_1)$ at the lower limit is bounded and does not depend on ρ, while at the upper limit it coincides with the analytic function $h^{1-n}[r(\rho, \nu_1)]^{n-1}$. The integral obtained as a result of differentiating the integrand is written as follows:

$$\int_{\rho(1+\delta)}^{r(\rho, \nu_1)} \frac{\partial^k}{\partial \rho^k} \Phi\left(\frac{\rho}{r}, \nu_1\right) D_{\xi, \nu^0}^\alpha \tilde{K}(\xi, r, \nu, \nu^0) \, dr.$$

In view of inequality (33), it is estimated by the integral

$$c_1 \int_{\rho(1+\delta)}^{r(\rho, \nu_1)} \frac{dr}{r^k} = c_1 \begin{cases} \ln \frac{\nu_1 + [(h/\rho)^2 - (1-\nu_1^2)]^{1/2}}{1+\delta}, & k = 1, \\ \frac{1}{1-k}\{[r(\rho, \nu_1)]^{1-k} - [\rho(1+\delta)]^{1-k}\}, & k > 1. \end{cases}$$

The estimate (30) follows. This completes the proof of Theorem 2.   □

Remark 1. Let $D(x, \nu)$ denote the set of points of the domain D such that $\varphi(\xi, x, \nu) \ge 0$. The problem of determining $u(x)$ from the equation

$$\int_{S(x,\nu)} \rho(\xi, x, \nu) u(\xi) \, d\xi + \int_{D(x,\nu)} \rho_1(\xi, x, \nu) u(\xi) \, d\xi = v(x, \nu),$$

$$x \in D, \quad |\nu| = 1$$

is more general than the problem of solving equation (1). Theorems 1 and 2 hold for this problem if the function $\rho_1(\xi, x, \nu)$ is at most one degree less smooth than the function $\rho(\xi, x, \nu)$.

Remark 2. Replacing the weight function $\rho(\xi, x, \nu)$ in (1) by a matrix weight function $R(\xi, x, \nu)$ of any finite dimension $m \times m$ and replacing the function $u(\xi)$ by a vector function $\mathbf{u}(\xi)$ with components $u_1, \ldots, u_m$, we obtain the vector problem of integral geometry. The assertions of Theorems 1 and 2 with condition (6) replaced by the condition

$$|\det R(\xi, x, \nu)| \geq \rho_0 > 0$$

and with function ρ^{-1} replaced by the matrix function $R^{-1}(x, x, \nu)$ in (7) are valid for the vector problem of integral geometry.

Remark 3. The following estimate, which holds for any $n \geq 2$, is a consequence of Theorems 1 and 2:

$$\|u\|_{L_2(\Omega)} \leq c_0 \sup_{|\nu|=1} \|v(x, \nu)\|_{W_2^l(\bar{D})}, \quad l = 2\left[\frac{n}{2}\right]. \tag{34}$$

So far in this section we studied the statement of the problem of integral geometry in the case where for any point x and any direction ν there exists a smooth hypersurface $S(x, \nu)$ passing through x, with ν being the normal to this surface at x. In a more general statement of this problem, for any x there are surfaces $S(x, \nu)$ passing through x such that ν belong to a set $\omega(x)$ of points of the unit sphere, where the set $\omega(x)$, in general, does not coincide with the unit sphere. It turns out that the stability of the problem of integral geometry is closely related to the structure of the set $\omega(x)$ (Bukhgeim, 1975; Lavrent'ev, Romanov, and Shishatskii, 1986).

For simplicity, we consider the case where $\omega(x)$ does not depend on x: $\omega(x) = \omega_0$. Take a spherical belt $\omega_{\delta\alpha} = \{\nu : |\nu| = 1, |(\nu, \alpha)| \leq \delta\}$. If there is a unit vector α and a $\delta > 0$ that satisfy the condition $\omega_{\delta\alpha} \cap \omega_0 \neq \emptyset$, then estimates of the form (34) (with the set $|\nu| = 1$ under the supremum sign replaced by the set ω_0) do not hold for any l for the problem of integral geometry. In particular, this implies that there are no finite k, l, and constant $c_0 > 0$ such that

$$\|u\|_{W_2^k(\Omega)} \leq c_0 \sup_{\nu \in \omega_0} \|v(x, \nu)\|_{W_2^l(\bar{D})}. \tag{35}$$

We will prove this statement for the case $n = 2$, putting $\varphi(\xi, x, \nu) = \nu(\xi - x)$ and $\rho = 1$. This example fully illustrates the principle of the general proof. Let $\omega_{\delta\alpha} \cap \omega_0 = \emptyset$ for $\delta > 0$ and $\alpha = (0,1)$. Take a function $u_\lambda \in C_0^\infty(\Omega)$ of the form

$$u_\lambda = \sin(\lambda x_1)\psi(x),$$

where $\psi(x)$ is an infinitely differentiable finite function with support included in Ω that is not identically equal to zero, and λ is a sufficiently large numerical parameter. Then

$$\|u\|^2_{L_2(\Omega)} = \frac{1}{2}\|\psi\|^2_{L_2(\Omega)} - \frac{1}{2}\int_\Omega \cos(2\lambda x_1)\psi^2 \, dx.$$

Integrating by parts the appropriate number of times, we obtain

$$\left|\int_\Omega \cos(2\lambda x_1)\psi^2(x)\, dx\right| \le c_k/|\lambda|^k.$$

On the other hand, for $x \in D$ and $\nu \in \omega_0$ we have

$$v_\lambda(x,\nu) = \frac{1}{|\nu_2|}\int_{-\infty}^\infty u_\lambda\left(\xi_1, x_2 - \frac{\nu_1}{\nu_2}(\xi_1 - x_1)\right) d\xi_1$$

$$= \frac{1}{|\nu_2|}\int_{-\infty}^\infty \sin(\lambda\xi_1)\psi\left(\xi_1, x_2 - \frac{\nu_1}{\nu_2}(\xi_1 - x_1)\right) d\xi_1.$$

Since $|(\nu,\alpha)| = |\nu_2| \ge \delta$ for $\nu \in \omega_0$, it follows that $|\nu_2|^{-1} \le \delta^{-1}$ and, consequently,

$$\left|D^\alpha_{x,\nu} v_\lambda(x,\nu)\right| \le c_{\alpha k}/|\lambda|^k, \quad x \in D, \ \nu \in \omega_0.$$

When $\lambda \to \infty$, $\|v_\lambda\|_{W_2^l(D)} \to 0$ uniformly with respect to $\nu \in \omega_0$ for any finite l, and $\|v_\lambda\|_{L_2(\Omega)} \to (1/\sqrt{2})\|\psi\|_{L_2(\Omega)} \ne 0$. This implies the assertion formulated above.

11.6. PROBLEMS OF THE VOLTERRA TYPE FOR MANIFOLDS INVARIANT WITH RESPECT TO THE TRANSLATION GROUP

In this section, we present a detailed analysis of planar problems of integral geometry Note that the results of this section can be extended to problems of integral geometry of the specified type in a space of any dimension.

Let $u(x, y)$ be a function of two variables. The function $u(x, y)$ will be considered in the half-plane $y \geq 0$ under the assumption that $u(x, y)$ is continuous and has finite support with respect to the variable x, i.e.,

$$u(x, y) = 0, \quad |x| \geq l > 0.$$

Furthermore, suppose that functions $\varphi(y, \eta)$ and $g(y, \eta)$ defined for $y \geq \eta$ satisfy the conditions

1) $\varphi(y, \eta) = \sqrt{y - \eta}\, \varphi_0(y, \eta)$,
2) $g(y, \eta) = (1/\sqrt{y - \eta})g_0(y, \eta)$,

where $\varphi_0(y, \eta)$ and $g_0(y, \eta)$ are continuously differentiable functions such that

$$\varphi_0(y, \eta) \geq \varphi^0 > 0, \quad g_0(y, \eta) \geq g^0 > 0.$$

Consider the following equation for $u(x, y)$:

$$\int_0^y g(y, \eta)[u(x - \varphi, \eta) + u(x + \varphi, \eta)]\, d\eta = f(x, y). \tag{1}$$

The problem of solving equation (1) is a problem of integral geometry, i.e., determining a function from the known integrals of this function over a given collection of curves. In the case being considered, the collection of curves defined by the function $\varphi(y, \eta)$ and the weight function $g(y, \eta)$ are invariant under the group of translations parallel to the x axis.

Two essentially equivalent methods can be applied to study equation (1).

The first method. We apply the Fourier transform to (1) with respect to the variable x. After this, equation (1) becomes as follows:

$$\int_0^y g(y, \eta) \cos(\lambda\varphi)v(\lambda, \eta)\, d\eta = \tilde{f}(\lambda, y),$$

$$v(\lambda, \eta) = \int_{-\infty}^{\infty} e^{i\lambda x} u(x, \eta)\, dx, \quad \tilde{f}(\lambda, y) = \int_{-\infty}^{\infty} e^{i\lambda x} f(x, y)\, dx. \tag{2}$$

Thus, equation (1) became a Volterra integral equation of the first kind. It follows from the properties of the functions $\varphi(y, \eta)$ and $g(y, \eta)$ that the equations (2) satisfy the conditions formulated in the beginning of Section 9.5. Therefore, the solution of (2) is unique, which implies that the solution of (1) is unique.

Applying the operator of fractional differentiation to (2) yields the following equation of the second kind:

$$v(\lambda, \eta) + \int_0^y G(\lambda, y, \eta) v(\lambda, \eta)\, d\eta = \tilde{f}_1(\lambda, y), \tag{3}$$

$$G(\cdot) = \frac{1}{\pi g_0(y, y)} \frac{\partial}{\partial y} \int_\eta^y [g_{0y} \cos(\lambda\varphi) - \lambda g_0 \varphi_y \sin(\lambda\varphi)] \sqrt{\frac{y - \eta}{\xi - \eta}}\, d\xi,$$

$$\tilde{f}_1(\lambda, y) = \frac{1}{\pi g_0(y, y)} \frac{\partial}{\partial y} \int_0^y \frac{1}{\sqrt{y - \eta}} \tilde{f}(\lambda, \eta)\, d\eta.$$

The solution to equation (3) can be represented in the form

$$v(\lambda, y) = \tilde{f}_1(\lambda, y) + \int_0^y R(\lambda, y, \eta) \tilde{f}_1(\lambda, \eta)\, d\eta. \tag{4}$$

In the general case, the norm of the integral operator in equation (3) increases indefinitely with the increase in λ, which means that the problem of solving the equations (3) is ill-posed and so is the problem of solving equation (1).

We now present the complete analysis of the character of instability for the solution of equations (1) and (3) in a simple special case.

In (1), suppose that

$$\varphi_0(y, \eta) = g_0(y, \eta) = 1.$$

Then equation (1) becomes as follows:

$$\int_0^y [u(x - \sqrt{y - \eta}, \eta) + u(x + \sqrt{y - \eta}, \eta)] \frac{d\eta}{\sqrt{y - \eta}} = f(x, y). \tag{5}$$

Application of the Fourier transform to equation (5) yields

$$\int_0^y \cos(\lambda\sqrt{y - \eta}) v(\lambda, \eta) \frac{d\eta}{\sqrt{y - \eta}} = \tilde{f}(x, y). \tag{6}$$

We will show that the solution of equation (6) is given by the formula

$$v(\lambda, y) = \frac{1}{\pi}\frac{\partial}{\partial y}\int\limits_0^y \mathrm{ch}(\lambda\sqrt{y-\eta})\tilde{f}(\lambda, \eta)\frac{d\eta}{\sqrt{y-\eta}}. \qquad (7)$$

Indeed, applying the Volterra operator with kernel $\mathrm{ch}(\lambda\sqrt{z-y})/\sqrt{z-y}$ to (6), we have

$$\int\limits_0^z \mathrm{ch}(\lambda\sqrt{z-y})\tilde{f}(\lambda, y)\frac{dy}{\sqrt{z-y}}$$

$$= \int\limits_0^z\int\limits_\eta^z \cos(\lambda\sqrt{y-\eta})\,\mathrm{ch}(\lambda\sqrt{z-y})\frac{dy}{\sqrt{(y-\eta)(z-y)}}v(\lambda, \eta)\,d\eta.$$

Hence, in view of the well-known formula

$$\int\limits_0^1 \cos(p\sqrt{t})\,\mathrm{ch}(p\sqrt{1-t})\frac{dt}{\sqrt{t(1-t)}} = \pi,$$

we obtain (7).

From formula (7) it follows that the character of instability of the solution of equation (5) is the same as that in the Cauchy problem for the Laplace equation. Analogous conditional stability estimates can be obtained for the general equation (1) and the corresponding equation (2).

In particular, it can be proved that the function R in (4) satisfies the inequality

$$|R(\lambda, y, \eta)| \le a e^{b\sqrt{y-\eta}\lambda}\frac{1}{\sqrt{y-\eta}},$$

where a and b are constants.

The second method. In equation (1), we consider $u(x, y)$ with fixed y to be an element of a Hilbert space W:

$$u(x, y) = w(y) \in W.$$

Then equation (1) can be viewed as a Volterra operator equation defined in Section 9.6 of the present part, i.e.,

$$\int\limits_0^y A(y, \eta)w(\eta)\,d\eta = \psi(y). \qquad (8)$$

The collection of operators $A(y, \eta)$ in (8) is defined as follows:

$$A(y, \eta)w(\eta) = g(y, \eta)[u(x - \varphi, \eta) + u(x + \varphi, \eta)].$$

It is easy to show that the Volterra operator equation (8) defined this way can be reduced to an equation that satisfies the assumption of Theorem 1 in Section 9.6 of the present part. The operator B in this theorem can be represented, for example, by the integral operator

$$Bw(y) = \int_{-l}^{l} p(x, \xi)u(\xi, y)\, d\xi,$$

$$p(x, \xi) = \frac{1}{2l}(l + \xi)(l - x)\xi \leq x, \quad \frac{1}{2l}(l - \xi)(l + x)x \leq \xi.$$

The same results can be obtained for problems of integral geometry in a space of higher dimension.

11.7. PLANAR PROBLEMS OF INTEGRAL GEOMETRY WITH A PERTURBATION

Consider the following operator equation for $u(\xi, \eta)$:

$$\int_{0}^{y} [u(x + h, \eta) + u(x - h, \eta)] \frac{d\eta}{\sqrt{y - \eta}}$$

$$+ \int_{0}^{y} \int_{x-h}^{x+h} K(x, y, \xi, \eta)u(\xi, \eta)\, d\xi d\eta = f(x, y), \quad h = \sqrt{y - \eta}. \quad (1)$$

It is assumed that u is an infinitely differentiable function with finite support included in the rectangle $-l_0 \leq \xi \leq l_0, 0 \leq \eta \leq b$. The function K has continuous derivatives of up to the second order and satisfies the condition $K(x, y, \xi, \eta) = 0$ for $|\xi + x| \leq h$. The function $f(x, y)$ is assumed to be specified in the strip $0 \leq y \leq b$.

Equation (1) represents a problem of integral geometry with a perturbation.

The first term on the left-hand side of (1), namely

$$\int_{0}^{y} [u(x + h, \eta) + u(x - h, \eta)] \frac{d\eta}{\sqrt{y - \eta}} = f_0(x, y),$$

represents the integrals of the unknown function u over the collection of parabolas with vertices at points (x, y). The second term

$$f_1(x, y) = f(x, y) - f_0(x, y)$$

represents the integrals with weight K over the parts of the half-plane that are bounded by the parabolas.

Theorem. *The solution of equation* (1) *is unique.*

Proof. Suppose that the right-hand side of equation (1) is identically equal to zero, i.e.,

$$f(x, y) \equiv 0.$$

Consider the function

$$v(x, y, t) = \frac{\partial}{\partial t} \int_0^y [u(x + gh, \eta) + u(x - gh, \eta)] \frac{d\eta}{\sqrt{y - \eta}},$$

$$g(t) = \sqrt{1 - t}, \quad t \in [0, 1].$$

This function satisfies the integro-differential equation

$$\frac{\partial}{\partial y} v(x, y, t) = -\frac{1}{4} \int_0^t \frac{\partial^2}{\partial x^2} v(x, y, \tau) \, d\tau + \varphi(x, y),$$

$$\varphi(x, y) = -\frac{1}{4} \frac{\partial^2}{\partial x^2} f_0(x, y) = \frac{1}{4} \frac{\partial^2}{\partial x^2} f_1(x, y).$$

(2)

Set $v_\sigma(x, y, t) = e^{\sigma(b-y)^2} v(x, y, t)$. It follows from (2) that v_σ satisfies the equation

$$\frac{\partial}{\partial y} v_\sigma(x, y, t)$$

$$= -\frac{1}{4} \int_0^t \frac{\partial^2}{\partial x^2} v_\sigma(x, y, \tau) \, d\tau - 2\sigma(b - y) v_\sigma(x, y, t) + \varphi_\sigma(x, y),$$

(3)

$$\varphi_\sigma(x, y) = e^{\sigma(b-y)^2} \varphi(x, y).$$

Lemma 1. *There exists a constant M such that*

$$\int_0^b \int_{-l}^l \varphi_\sigma^2(x, y) \, dx \, dy \leq M \int_0^b \int_{-l}^l \int_0^1 v_\sigma^2(x, y, t) \, dx \, dy \, dt,$$

(4)

where $l = l_0 + \sqrt{b}$.

From the definitions of the functions v and f and the properties of K, it follows that

$$2 \int_0^y u(x, \eta) \frac{d\eta}{\sqrt{y - \eta}} + f_1(x, y) = \int_0^1 v(x, y, t) \, dt,$$

$$u(x, y) + \int_0^y \int_{x-h}^{x+h} K_1(x, y, \xi, \eta) u(\xi, \eta) \, d\xi d\eta$$

$$= \frac{1}{2\pi} \int_0^y \int_0^1 v_\eta'(x, \eta, t) \, dt \frac{d\eta}{\sqrt{y - \eta}}, \quad (5)$$

$$K_1(x, y, \xi, \eta) = \frac{1}{2\pi} \int_{\eta+(x-\xi)^2}^y K_{y_1}'(x, y_1, \xi, \eta) \frac{dy_1}{\sqrt{y - y_1}}.$$

The left-hand side of (5) is the result of applying a Volterra operator of the second kind to u. After inverting this operator, we obtain

$$u(x, y) = \frac{1}{2\pi} \int_0^y \int_0^1 v_\eta'(x, \eta, t) \, dt \frac{d\eta}{\sqrt{y - \eta}}$$

$$+ \int_0^y \int_{x-h}^{x+h} K_2(x, y, \xi, \eta) v_\eta'(\xi, \eta, t) \, dt d\xi d\eta, \quad (6)$$

where K_2 is a continuous function.

From (6), transferring the operator of differentiation of v with respect to y to the functions K and K_2, we obtain

$$\varphi(x, y) = \int_0^y \int_{x-h}^{x+h} \int_0^1 K_3(x, y, \xi, \eta) v(\xi, \eta, t) \, dt d\xi d\eta, \quad (7)$$

where K_3 is a continuous function depending on K and its derivatives.

It follows from (7) that

$$\varphi_\sigma(x, y) = \int_0^y \int_{x-h}^{x+h} \int_0^1 K_{3\sigma}(x, y, \xi, \eta) v(\xi, \eta, t) \, dt d\xi d\eta,$$

$$(8)$$

$$K_{3\sigma}(x, y, \xi, \eta) = e^{\sigma(b-y)^2} K_3(x, y, \xi, \eta).$$

Inequality (4) readily follows from (8).

After applying the Fourier transform with respect to x to equations (2) and (3), the equations become as follows:

$$\frac{\partial}{\partial y}w(\lambda, y, t) = \frac{\lambda^2}{4}\int_0^t w(\lambda, y, \tau)\, d\tau + \psi(\lambda, y), \tag{9}$$

$$w(\lambda, y, t) = \frac{1}{\sqrt{2\pi}}\int_{-\infty}^{\infty} e^{i\lambda x} v(x, y, t)\, dx,$$

$$\psi(\lambda, y) = \frac{1}{\sqrt{2\pi}}\int_{-\infty}^{\infty} e^{i\lambda x} \varphi(x, y)\, dx,$$

$$\frac{\partial}{\partial y}w_\sigma(\lambda, y, t)$$

$$= \frac{\lambda^2}{4}\int_0^t w_\sigma(\lambda, y, \tau)\, d\tau - 2\sigma(b - y)w_\sigma(\lambda, y, t) + \psi_\sigma(\lambda, y). \tag{10}$$

We set

$$w_1(\lambda, t) = \begin{cases} w(\lambda, b, t), & t \leq 1, \\ 0, & t > 1. \end{cases}$$

For equation (9), consider the Cauchy problem

$$w(\lambda, b, t) = w_1(\lambda, t) \tag{11}$$

in the half-strip $0 \leq y \leq b$, $0 \leq t < \infty$, and define the function $w(\lambda, y, t)$ for $t > 1$ to be a solution to this Cauchy problem.

Clearly, for $t \leq 1$ the solution of the Cauchy problem (11) for equation (9) coincides with the function w defined earlier. It is easy to verify that the solution of the problem (11) for equation (9) is given by the formula

$$w(\lambda, y, t) = \int_0^t \frac{\cos \lambda\sqrt{(t-\tau)(b-y)}}{\sqrt{t-\tau}} w_2(\lambda, \tau)\, d\tau$$

$$+ \frac{1}{\lambda\sqrt{t}}\int_y^b \sin \lambda\sqrt{t(\eta - y)}\psi_1(\lambda, \eta)\, d\eta, \tag{12}$$

$$w_2(\lambda, t) = \frac{1}{\pi} \frac{\partial}{\partial t} \int\limits_0^t w_1(\lambda, \tau) \frac{d\tau}{\sqrt{t - \tau}},$$

$$\psi_1(\lambda, y) = \frac{1}{2\pi} \frac{\partial}{\partial y} \int\limits_y^b \psi(\lambda, \eta) \frac{d\eta}{\sqrt{\eta - y}}.$$

Indeed, as follows from the properties of the solution of evolution equations, the solution of the problem (11) for equation (9) is unique, and it is easy to verify that the function w in (12) is a solution to this problem.

We introduce the Volterra integral operator J such that

$$J\omega = \int\limits_0^t w(\lambda, y, \tau) \, d\tau$$

and consider the functional

$$F(\omega_\sigma) = \int\limits_0^b \int\limits_0^\infty e^{-\sigma t} \left[\left(\frac{\partial}{\partial y} - \frac{\lambda^2}{4} J + 2\sigma(b - y) \right) \omega_\sigma(\cdot) \right]^2 dt dy$$

$$= - \int\limits_0^b \int\limits_0^\infty e^{-\sigma t} \left[\left(\frac{\partial}{\partial y} + \frac{\lambda^2}{4} J^* - 2\sigma(b - y) \right) \right.$$

$$\times \left. \left(\frac{\partial}{\partial y} - \frac{\lambda^2}{4} J + 2\sigma(b - y) \right) \omega_\sigma(\cdot) \omega_\sigma(\cdot) \right] dt dy$$

$$+ \int\limits_0^\infty e^{-\sigma t} \left[\frac{\partial}{\partial y} w_\sigma^2(\lambda, b, t) - \frac{\lambda^2}{4} J w_\sigma(\lambda, b, t) + 2\sigma b w_\sigma^2(\lambda, b, t) \right] dt,$$

where the operator J^* is the adjoint of J in the Hilbert space of functions defined on the semiaxis $0 \le t < \infty$ with scalar product

$$(g, h) = \int\limits_0^\infty e^{-\sigma t} g(t) h(t) \, dt.$$

From (10) and the lemma of Section 10.6, we obtain

$$
F(\omega_\sigma) = - \int\limits_0^b \int\limits_0^\infty e^{-\sigma t} \left[\left(\frac{\partial}{\partial y} + \frac{\lambda^2}{4} J^* - 2\sigma(b-y) \right) \right.
$$

$$
\times \left. \left(\frac{\partial}{\partial y} - \frac{\lambda^2}{4} J + 2\sigma(b-y) \right) \omega_\sigma(\cdot)\omega_\sigma(\cdot) \right] dt\,dy
$$

$$
+ \int\limits_0^\infty e^{-\sigma t} \left[\frac{\partial}{\partial y} \omega_\sigma^2(\lambda, b, t) - \frac{\lambda^2}{4} J\omega_\sigma(\lambda, b, t)\omega_\sigma(\lambda, b, t) + 2\sigma b \omega_\sigma^2(\lambda, b, t) \right] dt
$$

$$
+ \int\limits_0^b \int\limits_0^\infty e^{-\sigma t} \left\{ \left[2\sigma + \frac{\lambda^4}{16}(J^*J - JJ^*) \right] \omega_\sigma(\cdot) \right\} \omega_\sigma(\cdot)\,dt\,dy
$$

$$
= \int\limits_0^b \int\limits_0^\infty e^{-\sigma t} \left[\left(\frac{\partial}{\partial y} - \frac{\lambda^2}{4} J^* + 2\sigma(b-y) \right) \omega_\sigma(\cdot) \right]^2 dt\,dy
$$

$$
- \int\limits_0^\infty e^{-\sigma t} \left[\frac{\lambda^2}{4}(J + J^*)\omega_\sigma(\cdot)\omega_\sigma(\cdot) \right] dt
$$

$$
+ \int\limits_0^b \int\limits_0^\infty e^{-\sigma t} \left\{ \left[\frac{\partial}{\partial y} + \frac{\lambda^2}{16}(J^*J - JJ^*) \right] \omega_\sigma(\cdot) \right\} \omega_\sigma(\cdot)\,dt\,dy
$$

$$
= \frac{1}{\sigma} \int\limits_0^b \psi_\sigma^2(\lambda, y)\,dy. \quad (13)
$$

Lemma 2. *Let $g(t)$ be a continuous and bounded function on the ray $0 \le t < \infty$ such that $|g(t)| \le g_0$. Then*

$$
(J^*J - JJ^*)g = \frac{1}{\sigma} \int\limits_0^\infty e^{-\sigma(t+\tau)} g(\tau)\,d\tau, \quad (14)
$$

where J^* is given by the formula

$$
J^*g = e^{\sigma t} \int\limits_t^\infty e^{-\sigma\tau} g(\tau)\,d\tau.
$$

It follows from the definition of J^* that

$$J^*Jg = \int_0^\infty G(t,\tau)g(\tau)\, d\tau, \quad G(t,\tau) = \begin{cases} \frac{1}{\sigma}e^{-\sigma t}, & \tau \le t, \\ \frac{1}{\sigma}e^{-\sigma\tau}, & \tau \ge t, \end{cases}$$

$$JJ^*g = \int_0^\infty R(t,\tau)g(\tau)\, d\tau,$$

$$R(t,\tau) = \begin{cases} \frac{1}{\sigma}(1 - e^{-\sigma\tau})e^{-\sigma t}, & \tau \le t, \\ \frac{1}{\sigma}(1 - e^{-\sigma t})e^{-\sigma\tau}, & \tau \ge t, \end{cases}$$

$$G(t,\tau) - R(t,\tau) = \frac{1}{\sigma}e^{-\sigma(t+\tau)},$$

$$(J^*J - JJ^*)g = \frac{1}{\sigma}\int_0^\infty e^{-\sigma(t+\tau)}g(\tau)\, d\tau.$$

Equalities (13) and (14) imply the inequality

$$2\sigma \int_0^b \int_0^\infty w_\sigma^2(\lambda, y, t)\, dt\, dy$$

$$\le \int_0^\infty e^{-\sigma t}\left[\frac{\lambda^2}{4}(J + J^*)w_\sigma(\cdot) - 4\sigma b w_\sigma(\cdot)\right] w_\sigma(\cdot)\, dt$$

$$+ \frac{1}{\sigma}\int_0^b \psi_\sigma^2(\lambda, y)\, dy. \quad (15)$$

Representing the functions w_σ and ψ_σ in (15) in terms of their inverse Fourier images, we have

$$2\sigma \int_0^b \int_0^\infty \int_{-\infty}^\infty e^{-\sigma t}v_\sigma^2(x, y, t)\, dx\, dy\, dt$$

$$\le \int_0^\infty \int_{-\infty}^\infty e^{-\sigma t}\left[\frac{1}{4}(J + J^*)\frac{\partial^2}{\partial x^2}v_\sigma(x, b, t)\right.$$

$$\left. - 4\sigma b v_\sigma(x, b, t)\right] v_\sigma(x, b, t)\, dx\, dt + \frac{1}{\sigma}\int_0^\infty \int_{-\infty}^\infty \varphi_\sigma(x, y)\, dx\, dy. \quad (16)$$

Consider the first term on the right-hand side of (16). It is the integral of the quadratic form including v_σ and $(J + J^*)v''_{\sigma xx}$ over the plane $y = b$ in the space (x, y, t).

From the conditions imposed on the function u and the definition of v_σ, it follows that there are constants M_0 and M_1 such that

$$\int\limits_0^\infty \int\limits_{-\infty}^\infty e^{-\sigma t} \left[\frac{1}{4}(J + J^*) \frac{\partial^2}{\partial x^2} v_\sigma(x, b, t) \right.$$

$$\left. - 4\sigma b v_\sigma(x, b, t) \right] v_\sigma(x, b, t) \, dx dt \le (M_0 + M_1 \sigma) e^{-2\sigma b^2}. \quad (17)$$

By inequalities (4), (16), and (17),

$$(2\sigma - M_2) \int\limits_0^b \int\limits_0^\infty \int\limits_{-\infty}^\infty v_\sigma^2(x, y, t) \, dx dy dt \le (M_0 + M_1 \sigma) e^{-2\sigma b^2}, \quad (18)$$

where M_2 is a constant. It is easy to see that if u and v_σ are not identically equal to zero, there exists a number q such that $0 < q < 1$ and

$$\int\limits_0^b \int\limits_0^\infty \int\limits_{-\infty}^\infty v_\sigma^2(x, y, t) \, dx dy dt \ge M_3 e^{-2\sigma b^2 q}. \quad (19)$$

From (18) and (19) we obtain the inequality

$$M_3(2\sigma - M_2) e^{-2\sigma b^2 q} \le (M_0 + M_1 \sigma) e^{-2\sigma b^2}. \quad (20)$$

This inequality obviously does not hold if the constant σ is sufficiently large. This completes the proof of the theorem. $\square$

Chapter 12.

Inverse problems

12.1. STATEMENT OF INVERSE PROBLEMS

The term "inverse problem" for differential equations or equations of mathematical physics is used to describe a wide variety of statements.

The general rule for using this term is as follows. First, a direct problem is formulated. The direct problem includes the data and the unknown solution. The problem that is inverse to the direct problem can be defined to be the problem in which some part of the data of the direct problem is assumed to be unknown and must determined, provided that some additional information (a collection of functionals) on the solution to the direct problem is given.

Sometimes the inverse problem is defined to be the problem of determining the inverse operator. Differentiation and integration are a classic example of this.

We now give several examples of problems considered to be inverse problems by a number of authors.

1. The inverse problem for the heat conduction equation. This is the Cauchy problem for the heat conduction equation in reversed time formulated in Section 7.1.

2. The one-dimensional inverse kinematic problem of seismology. It is required to determine the propagation velocity of a perturbation of the medium (seismic waves) inside a ball (in geophysics, it is the Earth) if the time of propagation of the perturbation between any two points on the surface of the ball is known. It is assumed that the propagation velocity depends only on one variable, namely the distance between a given point and the center of the ball. This problem was studied in early twentieth century by geophysicists G. Herglotz and E. Wiechert. The solution of this problem and the results of processing the data representing the arrival times for seismic waves generated by earthquakes served as a basis for the model of the internal structure of the Earth presented in school textbooks.

3. The inverse problem for the Sturm–Liouville equation. Consider the following differential operator with a boundary condition:

$$L_q = -\frac{d^2}{dx^2} + q(x), \quad x \in [a, b],$$

$$y'(a) - h_1 y(a) = 0, \quad y'(b) - h_2 y(b) = 0.$$

It is required to determine the coefficients of the differential operator $q(x)$ from the spectral function of the problem $\rho(\lambda)$. This is the simplest problem of this type.

The first results in the study of this problem were obtained by the well-known astrophysicist V. A. Ambartsumian in 1929 and by G. Born in 1945. Various versions of this problem were considered later by V. A. Marchenko, M. G. Krein, I. M. Gelfand, B. M. Levitan, and a number of other authors. It turned out that this problem is related to many applied problems, such as inverse problems of quantum scattering theory, problems of interpreting geophysical measurements data, and some others.

4. The inverse problem of potential theory. It is required to determine the form of a body from the Newton potential generated by this body. It is assumed that the body is homogeneous, i.e., its density is constant and known, while the potential is known in a domain outside the body. If it is a star body with respect to the origin, then the problem is equivalent to the problem of solving the following nonlinear integral equation of the Uryson type:

$$\int_{-\pi/2}^{\pi/2} \int_{-\pi}^{\pi} \int_{0}^{\varphi(\alpha,\beta)} \frac{\rho^2 \sin \beta}{r(x, y)} \, d\rho d\alpha d\beta = u(x),$$

where x and y are points of the three-dimensional space, (ρ, α, β) are polar coordinates of a point, and $r(x, y)$ is the distance between x and y. The function $u(x)$ is assumed to be known in a domain outside the domain of integration, and $\varphi(\alpha, \beta)$ is the unknown function.

The inverse problem of potential theory is related to the problem of interpreting gravimetric data mentioned in Section 8.2.

The theorem on the uniqueness of the solution of this inverse problem in the class of star bodies was proved by S. P. Novikov in 1938. Similar statements of inverse problems of potential theory attracted the attention of a number of researchers, such as V. K. Ivanov, A. I. Prilepko, and one of the authors of the present book.

Problems of determining the coefficients of a differential equation are the most common subject of published studies on inverse problems. In particular, they include problems of the same type as inverse problems for the Sturm–Liouville equation. It would be justified to say that the theory of inverse problems has become an important area of modern mathematics.

The first studies were devoted to one-dimensional inverse problems, i.e., the unknown coefficient was assumed to be a function of one variable.

Probably the first published study of a multidimensional inverse problem (with the unknown coefficient assumed to be a function of several variables) was by Yu. M. Berezanskii (see Berezanskii, 1958). An extensive study of various types of multidimensional inverse problems began in Novosibirsk in 1965 (Lavrent'ev, 1964; Lavrent'ev and Romanov, 1966; Lavrent'ev, Romanov, and Vasiliev, 1970). We confine ourselves to considering four types of multidimensional inverse problems. In the next three sections we will follow Romanov (1987).

12.2. INVERSE DYNAMIC PROBLEM. A LINEARIZATION METHOD

Let L be a uniformly elliptical operator with coefficients depending on the variable $x = (x_1, x_2, x_3)$:

$$Lu = \sum_{i,j=1}^{3} a_{ij}(x) u_{x_i x_j} + \sum_{i=1}^{3} b_i(x) u_{x_i} + c(x) u,$$

$$\mu \sum_{i=1}^{3} \alpha_i^2 \leq \sum_{i,j=1}^{3} a_{ij}(x) \alpha_i \alpha_j \leq \frac{1}{\mu} \sum_{i=1}^{3} \alpha_i^2, \quad 0 < \mu < \infty.$$

Consider the following problem:

$$u_{tt} - Lu = f(x, t), \quad x \in \mathbb{R}^3, \quad t \in \mathbb{R}, \tag{1}$$

$$u|_{t<0} \equiv 0. \tag{2}$$

If the information about the solutions $u(x, t)$ is specified on time-like mani-folds (for example, on a set of lines parallel to the t axis), then the problem of determining the coefficients of equation (1) is called the inverse dynamic problem. This name emphasizes that the information considered in the problem represents a mode of oscillations in time of a set of points in the space $\mathbb{R}^3$. Information of this type can also be effectively used to determine the coefficients of both lower and higher order derivatives of the operator L.

The general statement of the inverse dynamic problem of determining one of the coefficients of the operator L that is unknown inside a domain D bounded by a surface S can be given as follows. The solution to the problem (1), (2) is known at the points of S as a function of time:

$$u(x, t) = g(x, t), \quad x \in S, \quad t \geq 0. \tag{3}$$

It is required to determine the unknown coefficient inside D.

The problem of determining all coefficients of the operator L can be formulated in a similar way. Naturally, the information provided by (3) is not sufficient to determine more than one coefficient of the operator L. Additional information can be obtained, for example, by considering several problems of the form (1), (2) with different functions $f(x, t)$ and specifying data of the form (3) in each of these problems. Also, one can introduce a parameter $\lambda \in \Lambda$ into the function f. Then $f = f(x, t, \lambda)$ and the solution to the problem (1), (2) also depends on λ. In this case, the data (3) depends on the parameter λ. An example of such a parameter λ is the point of application of a concentrated source.

The inverse dynamic problem of determining the coefficients of the operator L is nonlinear. Indeed, the solution to the problem (1), (2) is the result of applying the operator

$$u = A(q, f), \quad q = (a_{ij}, b_i, c),$$

which is linear with respect to f and nonlinear with respect to the coefficients of the operator L, i.e., with respect to the components of the vector q. Using (3), we arrive at the nonlinear operator equation

$$A(q, f) = g. \tag{4}$$

In the analysis of nonlinear equations, the linear equation obtained as a result of linearization plays a very important role. As a rule, this equation reflects the main characteristic features of the original nonlinear equation and provides a better understanding of the essence of the problem. Based on this principle, we present the linearization procedure for the inverse problem and consider some questions of analyzing of the obtained linear problem.

Consider the problem of determining the coefficient $c(x)$ of the operator L in the statement (1)–(3). Suppose that the coefficient $c(x)$ can be represented in the form

$$c(x) = c_0(x) + c_1(x), \tag{5}$$

where the coefficient $c_0(x)$ is known and the absolute value of $c_1(x)$ is small. (As far as the coefficients of the derivatives of the operator L are concerned, in order for linearization to be justified, the additional component of the coefficient must be small in the norm involving the derivatives of the coefficient up to a certain order.) Thus, the inverse problem is reduced to the problem of determining a small additional component of the function $c_0(x)$.

We now describe the essence of the linearization method. Formal parameters λ and $c(x, \lambda)$ are introduced as follows:

$$c(x, \lambda) = c_0(x) + \lambda c_1(x). \tag{6}$$

The solution to the problem (1), (2) with $c = c(x, \lambda)$ is represented as an infinite power series in λ:

$$u(x, t, \lambda) = \sum_{n=0}^{\infty} \lambda^n u_n(x, t). \tag{7}$$

Then the series is substituted into (1), (2). Equations for $u_n(x, t)$ are obtained by equating the expressions containing λ^n. The function u_0 is a solution to the problem

$$\left(\frac{\partial^2}{\partial t^2} - L_0 \right) u_0 = f(x, t), \quad u_0|_{t<0} \equiv 0, \tag{8}$$

where L_0 denotes the operator L with $c = c_0$; the functions u_n $(n \geq 1)$ satisfy the conditions

$$\left(\frac{\partial^2}{\partial t^2} - L_0 \right) u_n = c_1 u_{n-1}, \quad u_n|_{t<0} \equiv 0, \ n \geq 1. \tag{9}$$

These equalities show that u_0 does not depend on c_1, the dependence of u_1 on c_1 is linear, and for $n = 2$ the dependence of u_n on c_1 is nonlinear. Therefore, linearization of the problem consists in replacing the series (7) with its first two terms. Since equality (5) is obtained from (6) by putting

$\lambda = 1$, this means representing the solution to the problem (1), (2) in the form

$$u(x,t) = u_0(x,t) + u_1(x,t). \tag{10}$$

With this approximation, the data of the inverse problem can be written as data for the function $u_1(x,t)$, i.e.,

$$u_1(x,t) = g_1(x,t) \equiv g(x,t) - u_0(x,t) \quad x \in S, \ t \geq 0. \tag{11}$$

We can assume that the function $g_1(x,t)$ is known because the function $u_0(x,t)$ is known from (8).

Thus, the linear approximation of the inverse problem is reduced to the problem of determining the function $c_1(x)$ from the relations

$$\left(\frac{\partial^2}{\partial t^2} - L_0 \right) u_1 = c_1 u_0, \quad u_1|_{t<0} \equiv 0, \tag{12}$$

where $u_0(x,t)$ is known and the data (11) is given. This is a problem of finding the right-hand side of a special form in a differential equation.

Such problems arise in the case where the coefficient is represented by one of the coefficients a_{ij} or b_i. For example, if the coefficient b_i satisfies the assumption

$$b_i = (b_i)_0 + (b_i)_1$$

which is analogous to (5), then we have the following equation for u_1:

$$\left(\frac{\partial^2}{\partial t^2} - L_0 \right) u_1 = (b_i)_1 \frac{\partial}{\partial x_i} u_0, \quad u_1|_{t<0} \equiv 0. \tag{13}$$

Thus, we have the inverse problem of determining $(b_i)_1$ from (13) and (11).

If all coefficients of the operator L are unknown, then, as was mentioned before, it is natural to consider several problems of the form (1), (2) with different functions f. This essentially means that the function $f(x,t)$ must be considered a vector function. Its dimension must coincide with the number of unknown coefficients of the operator L. In this case, the functions u and g also become vector functions. Representing each coefficient of L in the form

$$a_{ij} = (a_{ij})_0 + (a_{ij})_1, \quad b_i = (b_i)_0 + (b_i)_1, \quad c = c_0 + c_1,$$

which is analogous to (5), and denoting by L_k the parts of the operator L corresponding to the coefficients $(a_{ij})_k$, $(b_i)_k$, c_k, $k = 0, 1$, we obtain the inverse problem of determining $(a_{ij})_1$, $(b_i)_1$, and c_1 from the relations

$$\left(\frac{\partial^2}{\partial t^2} - L_0 \right) u_1 = L_1 u_0, \quad u_1|_{t<0} \equiv 0, \tag{14}$$

under the condition (11). As a result, we have a vector version of the inverse problem of determining the right-had side of a special form.

We now return to the problem (12), (11) as a more convenient object for demonstrating the method. The problem of determining $c_1(x)$ can be interpreted as a problem of integral geometry. To show this, we will solve the Cauchy problem (12) using the following formula, which provides the representation of the solution in terms of the fundamental solution:

$$u_1(x,t) = \int_{\mathbb{R}^4} c_1(\xi) u_0(\xi,\tau) H_0(x, t-\tau, \xi) \, d\xi d\tau.$$

The function H_0 corresponds to the operator L_0. Using the structure of the fundamental solution, this formula can be written as

$$u_1(x,t) = \frac{1}{4\pi} \int_{\tau(x,\xi) \le t} c_1(\xi) u_0(\xi, t - \tau(x,\xi)) \frac{d\xi}{\tau(x,\xi)}$$

$$+ \int_{\tau(x,\xi) \le t} c_1(\xi) \int_0^{t-\tau(x,\xi)} u_0(\xi,\tau) v_{-1}(x, t-\tau, \xi) \, d\tau d\xi, \quad t \ge 0, \quad (15)$$

where v_{-1} is the regular part of the function H_0. In view of the condition (11), we obtain the integral equation for the coefficient $c_1(x)$

$$\int_{\tau(x,\xi) \le t} c_1(\xi) \rho(x,\xi,t) \, d\xi = g_1(x,t), \quad x \in S, \quad t \ge 0, \quad (16)$$

where the weight function $\rho(x,\xi,t)$ is given by the equality

$$\rho(x,\xi,t) = u_0(\xi, t - \tau(x,\xi))/4\pi\tau(x,\xi))$$

$$+ \int_0^{t-\tau(x,\xi)} u_0(\xi,\tau) v_{-1}(x, t-\tau, \xi) \, d\tau.$$

The problem of solving equation (16) for the function $c_1(x)$ ($x \in D$) is a problem of integral geometry. The properties of this problem are determined mainly by the weight function ρ, which in turn depends on the function f and the coefficients of the operator L_0.

Consider the simplest version of (16). Let

$$L_0 = \Delta, \quad f(x,t) = \delta(x - x^0, t).$$

Then

$$\tau(x,\xi) = |x - \xi|, \quad u_0(x,t) = \frac{1}{4\pi|x - x^0|}\delta(t - |x - x^0|), \quad v_{-1} \equiv 0,$$

and equation (16) becomes as follows:

$$\frac{1}{(4\pi)^2}\int_{\mathbb{R}^3} c_1(\xi)\frac{\delta(t - |x - \xi| - |\xi - x^0|)}{|x - \xi| \cdot |\xi - x^0|}\,d\xi = g_1(x,t), \tag{16'}$$

$$x \in S, \quad t \geq 0.$$

Since the support of the integrand is concentrated on the ellipsoid

$$S(x, x^0, t) = \{\xi : |x - \xi| + |\xi - x^0| = t\},$$

it follows that the integral on the left-hand side of the equality (16′) can be reduced to an integral over the surface $S(x, x^0, t)$. The most convenient way to achieve this is to use the spherical coordinate system r, θ, φ, with origin at the point x^0 and polar axis passing through the points x and x^0. The equation for the ellipsoid $S(x, x^0, t)$ in this coordinate system is written as follows:

$$r = (t^2 - r_0^2)/2(t - r_0 \cos\theta), \quad r_0 = |x - x^0|, \quad r = |\xi - x^0|.$$

Furthermore,

$$|x - \xi| = (r^2 + r_0^2 - 2rr_0\cos\theta)^{1/2}.$$

After the change of coordinates, equation (16′) becomes as follows:

$$\frac{1}{(4\pi)^2}\int_{\mathbb{R}^3}\frac{c_1(\xi)}{(r^2 + r_0^2 - 2rr_0\cos\theta)^{1/2}}$$

$$\times\,\delta(t - r - (r^2 + r_0^2 - 2rr_0\cos\theta)^{1/2})r\,dr\,d\omega = g_1(x,t), \quad x \in S, \ t \geq 0,$$

where

$$d\omega = \sin\theta\,d\theta\,d\varphi.$$

To calculate the integral with respect to r, we use the following properties of the delta function. Suppose that $r = r^*$ is a simple zero of a differentiable function $\psi(r)$ and there are no more zeros of $\psi(r)$ in a ε-neighborhood of r^*. Then

$$\int_{r^*-\varepsilon}^{r^*+\varepsilon} f(r)\delta(\psi(r))\,dr = \frac{f(r^*)}{|\psi'(r^*)|}.$$

In our case

$$\psi(r) = t - r - \left(r^2 + r_0^2 - 2rr_0\cos\theta\right)^{1/2}, \quad r^* = \left(t^2 - r_0^2\right)/2(t - r_0\cos\theta).$$

Since

$$|\psi'(r^*)| = \left[1 + \frac{r - r_0\cos\theta}{\left(r^2 + r_0^2 - 2rr_0\cos\theta\right)^{1/2}}\right]_{r=r^*}$$

$$= \frac{t - r_0\cos\theta}{|x - \xi|} = \frac{t^2 - r_0^2}{2r^*|x - \xi|}, \quad r^* = |\xi - x^0|,$$

equation (16′) is written as follows:

$$\iint\limits_{S(x,x^0,t)} |x^0 - \xi|^2 c_1(\xi)\, d\omega = 8\pi^2(t^2 - |x - x^0|^2)g_1(x,t), \tag{17}$$

$$x \in S, \quad t \geq 0.$$

Let

$$S = \{x : x_3 = 0\}$$

and let x^0 be a fixed point of S. The following theorem was proved in Romanov (1967).

Theorem. *If the function $c_1(x)$ is even with respect to the plane S and continuous in $\mathbb{R}^3$, then it is uniquely determined from equation (17).*

Since any function can be represented as the sum of its even and odd parts, in particular,

$$c_1(x) = [c_1(x_1, x_2, x_3) + c_1(x_1, x_2, -x_3)]/2$$
$$+ [c_1(x_1, x_2, x_3) - c_1(x_1, x_2, -x_3)]/2,$$

the preceding theorem has the following corollary: if $c_1(x)$ is known in the domain $x_3 < 0$, then it is uniquely determined in the domain $x_3 \geq 0$ by specifying $g_1(x,t)$. To prove the theorem, we first introduce the spherical coordinates $x \in S$ and $\xi \in S(x, x^0, t)$ as follows:

$$x = x^0 + r_0\nu^0, \quad r_0 = |x - x^0|, \quad \nu^0 = (\cos\varphi_0, \sin\varphi_0, 0),$$
$$\xi = x^0 + rQ(\varphi_0)\nu, \quad \nu = (\sin\theta\cos\varphi, \sin\theta\sin\varphi, \cos\theta),$$

$$Q(\varphi_0) = \begin{Vmatrix} 0 & \sin\varphi_0 & \cos\varphi_0 \\ 0 & -\cos\varphi_0 & \sin\varphi_0 \\ 1 & 0 & 0 \end{Vmatrix}. \tag{18}$$

We also introduce the eccentricity ε of the ellipsoid and a polar parameter p:

$$\varepsilon = r_0/t, \quad p = t(1 - \varepsilon^2)/2.$$

Then for $\xi \in S(x, x^0, t)$ we have

$$r = p(1 - \varepsilon \cos \theta)^{-1}. \tag{19}$$

Equation (17) can be written in the form

$$\int_{|\nu|=1} r^2 c_1(x^0 + rQ\nu) \, d\omega = g_2(p, \varepsilon, \varphi_0),$$

$$0 \le p < \infty, \quad 0 \le \varepsilon < 1, \quad 0 \le \varphi_0 \le 2\pi, \tag{20}$$

where r is defined by (19) and the function g_2 is given by the formula

$$g_2(p, \varepsilon, \varphi_0) = \frac{32p^2}{1 - \varepsilon^2} g_1 \left(x^0 + \frac{2\varepsilon p}{1 - \varepsilon^2} \nu_0, \frac{2p}{1 - \varepsilon^2} \right).$$

We put

$$Mg_2 \equiv p \frac{\partial}{\partial \varepsilon} \int_0^p g_2(z, \varepsilon, \varphi_0) \frac{dz}{z}$$

and apply the operator M to both parts of equality (20). The following sequence of equalities shows that the operator is applicable and provides the result:

$$Mg_2 = p \frac{\partial}{\partial \varepsilon} \int_0^p \frac{dz}{z} \int_{|\nu|=1} r^2 c_1(x^0 + rQ\nu) \Big|_{r=z(1+\varepsilon \cos \theta)^{-1}} d\omega$$

$$= p \frac{\partial}{\partial \varepsilon} \int_{|\nu|=1} \int_0^p r^2 c_1(x^0 + rQ\nu) \Big|_{r=z(1+\varepsilon \cos \theta)^{-1}} \frac{dz}{z} \, d\omega$$

$$= p \frac{\partial}{\partial \varepsilon} \int_{|\nu|=1} \int_0^{p(1+\varepsilon \cos \theta)^{-1}} r c_1(x^0 + rQ\nu) \, dr \, d\omega$$

$$= \int_{|\nu|=1} r^3 \cos \theta \, c_1(x^0 + rQ\nu) \, d\omega.$$

Thus, as a result of applying the operator M to equation (20), the factor $r\cos\theta$ appears under the integral sign. Therefore, applying the operator k times produces the factor $(r\cos\theta)^k$:

$$M^k g_2 = \int\limits_{|\nu|=1} r^{k+2}\cos^k\theta c_1(x^0 + rQ\nu)\,d\omega, \quad k = 0,1,2,\ldots. \qquad (21)$$

Put $\varepsilon = 0$ in these equations. Then $r = p = t/2$ and the integrals over the ellipsoids $S(x, x^0, t)$ become the integrals over the spheres $|\xi - x^0| = t/2$. In this case we have

$$r\cos\theta = \left(\xi_1 - x_1^0\right)\cos\varphi_0 + \left(\xi_2 - x_2^0\right)\sin\varphi_0,$$

which means that equations (21) for $\varepsilon = 0$ can be written in the form

$$\int\limits_{|\xi-x^0|=t/2} c_1(\xi)\left[\left(\xi_1 - x_1^0\right)\cos\varphi_0 + \left(\xi_2 - x_2^0\right)\sin\varphi_0\right]^k dS_\xi$$

$$= M^k g_2|_{\varepsilon=0}, \quad k = 0,1,2\ldots, \ t \geq 0, \ 0 \leq \varphi_0 \leq 2\pi. \qquad (22)$$

Hence, since φ_0 is arbitrary,

$$\int\limits_{|\xi-x^0|=t/2} c_1(\xi)\left(\xi_1 - x_1^0\right)^n\left(\xi_2 - x_2^0\right)^m dS_\xi = g_{nm},$$

$$\qquad (23)$$

$$n, m = 0,1,2\ldots.$$

For a fixed sphere $|\xi - x^0| = t/2$, the quantities g_{nm} constitute the complete system of moments of the function $c_1(x)$, which is even with respect to x_3. The assertion of the theorem follows.

The moments g_{nm} given by the formula (23) can be used to construct the system of the Fourier coefficients with respect to the spherical functions on the sphere $|\xi - x^0| = t/2$ and represent $c_1(\xi)$ in the form of a Fourier series.

We now present a modified statement of the problem of determining the function $c_1(x)$ which involves equation (17). Let S be the plane $x_3 = 0$ and x^0 be a variable point in this plane. Consider the case where $x = x^0$. Then the surfaces $S(x^0, x^0, t)$ are the spheres $|\xi - x^0| = t/2$ and equation (17) becomes as follows:

$$\frac{1}{4\pi} \iint\limits_{|\xi-x^0|=t/2} c_1(\xi)\,d\omega = 8\pi g_1(x, t), \quad x \in S, \ t \geq 0. \qquad (24)$$

The problem arises of determining the function $c_1(x)$ from its spherical means, where the centers of the spheres are points of a fixed plane and their radii are arbitrary.

12.3. A GENERAL METHOD FOR STUDYING INVERSE PROBLEMS FOR HYPERBOLIC EQUATIONS

We will consider issues related to the uniqueness and stability of solutions to the inverse dynamic problem. It turns out that these issues are closely related to linear problems of determining the right-hand side of a differential equation.

We will analyze a refined version of the statement of the problem (1)–(3) from Section 12.2. Suppose that all coefficients of the operator L are known outside a closed domain $\Omega \subset D$ and are unknown in Ω. It is required to determine these coefficients from the additional data (3) in Section 12.2. We assume that f is a vector function and its dimension coincides with the number of unknown coefficients of the operator L (there are ten of them in this case, in view of the symmetry of coefficients of the higher order derivatives, i.e., $a_{ij} = a_{ji}$). Accordingly, u and g are also vector functions. We denote the ordered set of coefficients of L by

$$q = (a_{11}, a_{12}, a_{13}, a_{22}, a_{23}, a_{33}, b_1, b_2, b_3, c).$$

Let Q be a set consisting of functions $q(x)$. It is assumed that $q(x)$ and $f(x,t)$ have derivatives of sufficiently high order.

Analyzing the solution stability with respect to the data in the inverse problem, it is necessary to consider a set G of functions $g(x,t)$ and estimate some norm of the difference between the solutions to the inverse problems corresponding to any two elements of G in terms of the norm of the difference between the elements of G. Let $g^{(1)}$ and $g^{(2)}$ be two arbitrary elements of G. Let $(u^{(1)}, q^{(1)})$ and $(u^{(2)}, q^{(2)})$ denote the corresponding solutions to the problem (1)–(3) from Section 12.2, and put

$$u^{(1)} - u^{(2)} = \tilde{u}, \quad q^{(1)} - q^{(2)} = \tilde{q}, \quad g^{(1)} - g^{(2)} = \tilde{g}.$$

Furthermore, let $L^{(i)}$ denote the operator corresponding to $q = q^{(i)}$ $(i = 1, 2)$ and let $\widetilde{L}$ be the operator corresponding to $q = \tilde{q}$. We obtain the following problem for the functions $\tilde{u}$, $\tilde{q}$, and $\tilde{g}$:

$$\tilde{u}_{tt} - L^{(1)}\tilde{u} = \tilde{q}(x)\Phi(x,t), \tag{1}$$

$$\tilde{u}|_{t<0} \equiv 0, \tag{2}$$

$$\tilde{u}(x,t) = \tilde{g}(x,t), \quad x \in S, \ t \geq 0. \tag{3}$$

The matrix $\Phi(x,t)$ is defined by the equality

$$\tilde{q}(x)\Phi(x,t) = \tilde{L}u^{(2)}, \quad \Phi(x,t) = (\Phi_{ij}, \, i,j = 1,\ldots,10). \tag{4}$$

If the problem (1)–(3) of determining $\tilde{q}(x)$ admits the stability estimate that is uniform with respect to the set Q, then the same estimate holds for the inverse problem (1)–(3) from Section 12.2. The uniformity of the estimate with respect to Q is necessary because the operator $L^{(1)}$ depends on $q^{(1)} \in Q$ and the matrix Φ depends on $q^{(2)} \in Q$. Note that the stability estimate in this case is conditional because it is assumed that the solution to the inverse problem belongs to Q. Thus, the necessity to obtain a stability estimate for the problem (1)–(3) in Section 12.2 leads to the problem of determining the right-hand side of a special form which is similar to the problem in Section 12.2 that arose as a result of linearization of the statement of the inverse problem. This shows once again that the linear part of the operator of the inverse problem preserves the main features of the original problem. It should also be noted that the analysis of the uniqueness of solutions to the inverse problem is reduced to the analysis of the uniqueness of solutions to the problem (1)–(3). Indeed, if the function $\tilde{g} \equiv 0$ corresponds to $\tilde{q} \equiv 0$, this means that $g^{(1)} = g^{(2)}$ corresponds to $q^{(1)} = q^{(2)}$.

In the case where $f(x,t)$ is a generalized function, the problem (1)–(3) from Section 12.2 and the corresponding problem (1)–(3) of the present section have been studied only for special function classes (recall that Φ is expressed only in terms of the function $u^{(2)}(x,t)$, which in turn depends on $f(x,t)$).

The complete study of these problems can be carried out in the case of smooth functions $f(x,t)$ satisfying some additional conditions. We now present results obtained in Romanov (1978).

Lemma 1. *Let* $\det \Phi_{tt}(x,0) \neq 0$, $x \in \Omega$. *Then the analysis of the uniqueness and stability of solutions with respect to the data (3) in Section 12.2 for the inverse problem (1)–(3) in Section 12.2 on the set Q is reduced to the analysis of the same properties for the problem of determining the vector function $\varphi(x)$ (supp $\varphi(x) \subset \Omega$) from the relations*

$$v_{tt} - Lv = \varphi(x)R(x,t), \tag{5}$$
$$v|_{t=0} = 0, \quad v_t|_{t=0} = \varphi(x), \tag{6}$$
$$v(x,t) = h(x,t), \quad x \in S, \quad t \geq 0, \tag{7}$$

where L is a given operator, $q \in Q$, $R(x,t)$, and $h(x,t)$ are given functions, and R is a square matrix.

Proof. Note that

$$u^{(2)}\big|_{t=0} = 0, \quad u_t^{(2)}\big|_{t=0} = 0, \quad u_{tt}^{(2)}\big|_{t=0} = f(x,0). \qquad (8)$$

Therefore, the condition $\det \Phi_{tt}(x,0) \neq 0$, $x \in \Omega$, is an easily verifiable condition for the function $f(x,0)$.

After differentiating equalities (1) and (3) three times with respect to t, we set

$$\tilde{u}_{ttt} = v, \quad \varphi(x) = \tilde{q}\Phi_{tt}(x,0),$$
$$R(x,t) = \Phi_{tt}^{-1}(x,0)\Phi_{ttt}(x,t), \quad h(x,t) = \tilde{g}_{ttt}(x,t)$$

and write L instead of $L^{(1)}$. Then we use (8) to obtain (5)–(7), which completes the proof. $\square$

For Q we take the set of infinitely differentiable functions $q(x)$ such that their partial derivatives are bounded by a given constant M and the constant μ defining the uniform positivity of the operator L is fixed. Assume that the collection of characteristic conoids is regular for $q \in Q$ in the sense that any two bicharacteristics with a common point do not intersect each other at any other point. Under these assumptions, if $f(x,t)$ is sufficiently smooth, any solution to the problem (1)–(3) from Section 12.2 is uniformly bounded for $q \in Q$ together with its derivatives in the cylinder $G_T = \{(x,t): x \in D, 0 \leq t \leq T\}$ and the corresponding constant depends only on M, T, μ, and the function $f(x,t)$. For this reason, the functions R and v can be assumed to be uniformly bounded in G_T with respect to the set Q.

We put

$$\|\varphi\|^2 = \int_\Omega |\varphi(x)|^2\,dx, \quad \|h\|_T^2 = \max_{0 \leq t \leq T} \int_S |h(x,t)|^2\,dS,$$

$$\|R\|_T^2 = \max_{(x,t) \in \Omega \times [0,T]} \sum_{i,j} |R_{ij}(x,t)|^2,$$

where dS is the surface area element of S and $R_{ij}(x,t)$ are components of the matrix $R(x,t)$.

We now formulate two theorems on the stability of solutions for the problem (5)–(7).

Theorem 1. *Let*

$$T > \frac{1}{\mu}\operatorname{diam} D. \qquad (9)$$

Then there exists a $\delta > 0$ such that the conditions

$$\operatorname{diam}\Omega < \delta, \quad \|R\|_T < \delta \tag{10}$$

ensure that the solution to the problem (5)–(7) satisfies the estimate

$$\|\varphi\| \leq C\|h\|_T^{1/2} \tag{11}$$

uniformly with respect to the set Q.

Theorem 2. Assume that the condition (9) holds. Then there exists a $\delta > 0$ such that the condition

$$\operatorname{diam} D < \delta \tag{12}$$

ensures that the function $\varphi(x)$ satisfies the estimate (11) uniformly with respect to the set Q.

These theorems are theorems "in the small", since they include smallness conditions on the support of the function to be determined. The proofs of this theorems use the same procedure as before and are based on energy estimates for equation (5).

After taking the scalar product of $2v_t(x,t)$ and both sides of (5), the left-hand side of the resulting equality can be written as

$$(2v_t, [v_{tt} - Lv]) \equiv \frac{\partial}{\partial t}\left[(v_t, v_t) + \sum_{i,j=1}^{3} a_{ij}(v_{x_i}, v_{x_j}) - c(v, v)\right]$$

$$- 2\sum_{i,j=1}^{3} \frac{\partial}{\partial x_i}(v_t, a_{ij}v_{x_j}) - 2\sum_{i=1}^{3}\left(b_i - \sum_{j=1}^{3}(a_{ij})_{x_j}\right)(v_t, v_{x_i}).$$

We denote by

$$I(t) = \int_{t=\text{const}}\left[(v_t, v_t) + \sum_{i,j=1}^{3} a_{ij}(v_{x_i}, v_{x_j}) - c(v, v)\right] dv \tag{13}$$

the integral of the expression in brackets over the section of the domain $G_T = D \times [0, T]$ produced by the plane $t = \text{const}$. The quadratic form under the integral sign in (13) is positive under the condition that $-c > 0$. Assume that the latter is true. Note that this assumption can be made without loss of generality because otherwise it can always be achieved by replacing the function v by $\tilde{v} = ve^{\lambda t}$ with the appropriate numeric parameter λ. On the other hand, as a result of this change of functions, a term that includes the first order derivative with respect to t appears in the operator L. However, the presence of this term is not essential as far as the method of energy estimates is concerned. We now take the scalar product of $2v_t$ and both sides of (5) and integrate the resulting equality over the part of the cylinder G_T contained between the sections $t = t_1$ and $t = t_2$, where $t_2 > t_1$. Applying the Gauss–Ostrogradsky formula, we have

$$I(t_2) - I(t_1) = 2 \int_{t_1}^{t_2} d\tau \int_S \left(v_\tau, \sum_{i,j=1}^3 a_{ij} v_{x_i} \cos(\widehat{n, x_j}) \right) dS$$

$$+ 2 \int_{t_1}^{t_2} d\tau \int_D \left[(v_\tau, \varphi R) + \left(v_\tau, \sum_{i=1}^3 \left(b_i - \sum_{j=1}^3 (a_{ij})_{x_j} \right) v_{x_i} \right) \right] dx,$$

where n is the direction of outward normal to S. Since $v_t|_S = h_t$ and v_{x_i} are bounded on S uniformly with respect to Q, using the Cauchy–Bunyakovsky inequality, we obtain

$$\left| 2 \int_S \left(v_\tau, \sum_{i,j=1}^3 a_{ij} v_{x_i} \cos(\widehat{n, x_j}) \right) dS \right| \le C_1 \|h_t\|_T.$$

Moreover,

$$2|(v_\tau, \varphi R)| \le (v_\tau, v_\tau) + (\varphi R, \varphi R) \le |v_\tau|^2 + \|R\|_T^2 |\varphi|^2.$$

Therefore,

$$2 \left| \int_D (v_\tau, \varphi R)\, dx \right| \le I(\tau) + \|R\|_T^2 |\varphi|^2.$$

Similarly,

$$2 \left| \int_D \left(v_t, \sum_{i=1}^n \left(b_i - \sum_{j=1}^3 (a_{ij})_{x_j} \right) v_{x_i} \right) dx \right|$$

$$\le \int_D \left[|v_t|^2 + \left| \sum_{i=1}^n \left(b_i - \sum_{j=1}^3 (a_{ij})_{x_j} \right) v_{x_i} \right|^2 \right] dx \le C_2 I(t)$$

for a sufficiently large constant C_2. The value of C_2 depends only on the estimate for b_i and $(a_{ij})_{x_j}$ in the domain D and the constant μ.

The obtained estimates readily imply the following estimate for the derivative of I with respect to the parameter t:

$$\left|\frac{d}{dt}I(t)\right| \le C_1\|h_t\|_T + \|R\|_T^2\|\varphi\|^2 + C_3I(t), \quad 0 \le t \le T. \tag{14}$$

Hence

$$\frac{d}{dt}(I(t)e^{-C_3t}) \le (C_1\|h_t\|_T + \|R\|_T^2\|\varphi\|^2)e^{-C_3t},$$

$$I(t) \le [I(0) + (C_1\|h_t\|_T + \|R\|_T^2\|\varphi\|^2)t]e^{C_3t}, \quad t \in [0,T]. \tag{15}$$

The estimate (15) is the usual energy estimate of a solution used in the study of boundary value problems. However, it is of no particular interest to us. More useful for our purposes is the estimate from below for $I(t)$ that follows from the estimate from below for the derivative $I'(t)$. We have

$$\frac{d}{dt}(I(t)e^{C_3t}) \ge -(C_1\|h_t\|_T + \|R\|_T^2\|\varphi\|^2)e^{C_3t}.$$

Hence

$$I(t)e^{C_3t} \ge I(0) - (C_1\|h_t\|_T + \|R\|_T^2\|\varphi\|^2)te^{C_3t}, \quad 0 \le t \le T.$$

Setting $t = T$, we obtain

$$I(0) \le [I(T) + T(C_1\|h_t\|_T + \|R\|_T^2\|\varphi\|^2)]e^{C_3T}. \tag{16}$$

From the data (6) it follows that $I(0) = \|\varphi\|^2$. Therefore, inequality (16) can be written as

$$\|\varphi\|^2(1 - \|R\|_T^2Te^{C_3T}) \le [I(T) + TC_1\|h_t\|_T]e^{C_3T}.$$

Hence, under the condition

$$\|R\|_T^2Te^{C_3T} < 1 \tag{17}$$

we obtain the inequality

$$\|\varphi\|^2 \le C_4(I(T) + TC_1\|h_t\|_T). \tag{18}$$

The condition (17) holds if the condition (10) holds for R with sufficiently small δ or if T in Theorem 2 is small and is in agreement with δ to ensure that (9) holds. From the preceding theorems, it follows that the constant C_4 in (18) can be assumed to be universal, i.e., uniform with respect to the set Q.

We now estimate $I(T)$. To this end, we represent solutions to equation (5) with Cauchy data (6) in the form $v = v_1 + v_2$, where v_1 is a solution to the inhomogeneous equation (5) with zero Cauchy data and v_2 is a solution of the homogeneous equation (5) with data (6). Let $I_1(T)$ and $I_2(T)$ denote the integrals (13) for $t = T$ corresponding to the functions v_1 and v_2. It is obvious that

$$I(T) \le 2[I_1(T) + I_2(T)].$$

The integral $I_1(T)$ can be estimated in the same way using the method of energy estimates of solutions to the Cauchy problem. For this purpose, it is necessary to construct a dome-like domain bounded from above by a space-like surface including the top of the cylinder G_T, and bounded from below by the plane $t = 0$ (see Fig. 5).

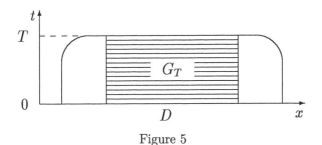

Figure 5

The energy estimates for the sections of this domain produced by the planes $t = $ const have the form of the estimate (15) where the term containing h_t is omitted. In this case, generally speaking, the constant C_3 will be different because the sections of the dome-like domain produced by the planes $t = $ const include the corresponding sections of the domain G_T as subsets. The estimate for $I_1(T)$ is written as follows:

$$I_1(T) \le \|R\|_T^2 \|\varphi\|^2 e^{C_3' T}. \tag{19}$$

In this estimate, we took into account that $I_1(0) = 0$ because of the zero Cauchy data. To estimate $I_2(T)$, we use the representation of the solution v_2 in terms of the fundamental solution $H(x, t - \tau, \xi)$:

$$v_2(x, t) = \int_\Omega H(x, t, \xi)\, d\xi. \tag{20}$$

For $x \in D$, $\xi \in \Omega$, and the values of t that are close to T, the function $H(x, t, \xi)$ is a normal regular function, since under the condition (9) the conoid $t = T - \tau(x, \xi)$, on which the singular part of the function is concentrated, intersects the plane $t = 0$ outside the domain D (the surface $\tau(x, \xi) = T$ includes the domain D for any $\xi \in \Omega$). Therefore, the function $H(x, t, \xi)$ and its partial derivatives of the first order are bounded uniformly with respect to Q for $x \in D$, $\xi \in \Omega$, and t close to T. Using formula (20) to estimate the function v_2 and its derivatives appearing in the integral $I_2(T)$, we obtain

$$I_2(T) \leq C_6 \|\varphi\|^2,$$

where

$$C_6 = \int_D \left\{ \int_\Omega \left(H_t^2 - cH^2 \right) d\xi \right.$$

$$\left. + \sum_{i,j=1}^{3} a_{ij}(x) \left(\int_\Omega H_{x_i}^2 \, d\xi \right)^{1/2} \left(\int_\Omega H_{x_j}^2 \, d\xi \right)^{1/2} \right\}_{t=T} dx$$

$$\leq C_7 (d(D)d(\Omega))^3,$$

$d(D)$ and $d(\Omega)$ are the diameters of the domains D and Ω, respectively. From the estimates obtained for $I_1(T)$ and $I_2(T)$, it follows that

$$I(T) \geq C_8 \|\varphi\|^2,$$

where $C_8 \to 0$ as $\|R\|_T \to 0$ and $d(\Omega) \to 0$, or as $T \to 0$ and $d(D) \to 0$. Then inequality (18) implies the estimate (9).

The estimate (9) also implies the following uniqueness theorem: if $h = 0$, then $\varphi = 0$ for $x \in \Omega$.

12.4. THE CONNECTION BETWEEN INVERSE PROBLEMS FOR EQUATIONS FOR HYPERBOLIC, ELLIPTIC, AND PARABOLIC EQUATIONS

It has long been known that a one-to-one correspondence can be established between solutions to problems of mathematical physics that describe different types of physical processes. Such a correspondence is especially easy to establish in the case of equations with constant coefficients using the Fourier and Laplace transforms. It was first pointed out in Reznitskaya (1974) that the connection between different types of solutions can be useful in the study of inverse problems. When dealing with inverse problems for parabolic or elliptic equations, it turns out to be possible to replace them with some equivalent hyperbolic problems. The opposite transition can also be useful in some cases. We will use some examples of statements of inverse problems to describe this method.

Consider the following mixed problem for a parabolic equation in a domain $D \subset \mathbb{R}^n$ bounded by a surface S:

$$u_t = Lu + f(x,t), \quad u|_{t=0} = \varphi(x), \quad \frac{\partial u}{\partial n}\Big|_S = \psi(x,t). \tag{1}$$

Here L is a uniformly elliptic operator with continuous coefficients that depends only on the variable x; n is the conormal to S. We associate the problem (1) with the following problem for a hyperbolic equation:

$$\tilde{u}_{tt} = L\tilde{u} + \tilde{f}(x,t), \quad \tilde{u}|_{t=0} = \varphi(x), \quad \tilde{u}_t|_{t=0} = 0, \quad \frac{\partial \tilde{u}}{\partial n}\Big|_S = \tilde{\psi}(x,t). \tag{2}$$

Let $\tilde{f}$ and $\tilde{\psi}$ be smooth functions of their arguments that increase as $t \to \infty$ slower than $Ce^{\alpha t}$ and are related to f and ψ by the formulas

$$f(x,t) = \int_0^\infty \tilde{f}(x,t)G(t,\tau)\,d\tau, \quad \psi(x,t) = \int_0^\infty \tilde{\psi}(x,t)G(t,\tau)\,d\tau, \tag{3}$$

where

$$G(t,\tau) = \frac{1}{\sqrt{\pi t}}e^{-\tau/4t}$$

is a solution to the heat conduction equation

$$G_t = G_{\tau\tau}.$$

Then

$$u(x,t) = \int_0^\infty \tilde{u}(x,\tau)G(t,\tau)\,d\tau. \tag{4}$$

Indeed,

$$u_t = \int_0^\infty \tilde{u}(x,\tau)G_t(t,\tau)\,d\tau = \int_0^\infty \tilde{u}(x,\tau)G_{\tau\tau}(t,\tau)\,d\tau$$

$$= (\tilde{u}G_\tau - \tilde{u}_\tau G)|_{\tau=0}^{\tau=\infty} + \int_0^\infty \tilde{u}_{\tau\tau}(x,\tau)G(t,\tau)\,d\tau$$

$$= \int_0^\infty \tilde{u}_{\tau\tau}(x,\tau)G(t,\tau)\,d\tau,$$

$$Lu = \int_0^\infty G(t,\tau)L\tilde{u}(x,\tau)\,d\tau.$$

Therefore, for $t > 0$

$$u_t - Lu - f = \int_0^\infty (\tilde{u}_{\tau\tau} - L\tilde{u} - \tilde{f})G(t,\tau)\,d\tau = 0.$$

On the other hand,

$$u|_{t=0} = \lim_{t\to 0} \int_0^\infty \tilde{u}(x,\tau)G(t,\tau)\,d\tau$$

$$= \lim_{t\to 0} \frac{2}{\sqrt{\pi}} \int_0^\infty \tilde{u}(x, 2\sqrt{t}\,\tau)e^{-\tau^2}\,d\tau = \tilde{u}(x,0) = \varphi(x),$$

$$\left.\frac{\partial u}{\partial n}\right|_S = \int_0^\infty \left.\frac{\partial \tilde{u}}{\partial n}\right|_S G(t,\tau)\,d\tau = \int_0^\infty \tilde{\psi}(x,\tau)G(t,\tau)\,d\tau = \psi(x,t).$$

Thus, formula (4) expresses the solution to the problem (1) in terms of the solution to the problem (2) if the data of these problems are related by (3). It is known that, under rather general assumptions on the operator L and the surface S, the problem (1) has a unique solution, which is given by formula (4). Note that equality (4) is invertible for fixed x, since it can be represented as the Laplace transform with parameter $p = 1/4t$:

$$u(x, 1/4p) = \sqrt{\frac{p}{\pi}} \int_0^\infty e^{-zp}\tilde{u}(x, \sqrt{z})\frac{dz}{\sqrt{z}}. \tag{5}$$

Therefore, formula (4) establishes a one-to-one correspondence between the solutions to the problems (1) and (2).

We now consider the inverse problem for equation (1) that consists in determining $f(x,t)$ or a coefficient q appearing in the operator L from the solution to the problem (1) specified at the points of S:

$$u|_S = g(x,t), \quad t \geq 0. \tag{6}$$

Let the function $g(x,t)$ be associated with the function $\tilde{g}(x,t)$ $(x \in S)$ as a solution to the equation

$$g(x,t) = \int_0^\infty \tilde{g}(x,\tau)G(t,\tau)\,d\tau, \quad x \in S, \ t \geq 0. \tag{7}$$

The function $\tilde{g}(x,t)$ for $x \in S$ is uniquely determined from this equation. Set

$$\tilde{u}|_s = \tilde{g}(x,t), \quad t \geq 0. \tag{8}$$

Then the inverse problem (1), (6) is equivalent to the corresponding inverse problem (2), (8).

It is known from mathematical physics that the solution to the problem (2) satisfies the estimate

$$|\tilde{u}(x,t)| \leq Ce^{\alpha t}$$

under rather general assumptions on the coefficients of the operator L. Formula (4) and the equivalent formula (5) show that $u(x,t)$ is an analytic function of t in the domain $t > 0$. Therefore, in formula (6), it is sufficient to specify the function $g(x,t)$ in a neighborhood of $t = 0$ that is as small as desired, such as the interval $0 \leq t \leq \delta$, $\delta > 0$.

The established connection between inverse problems makes it possible to extend a number of results obtained for hyperbolic equations to the case of parabolic equations. This is possible in all situations where the data of an inverse problem are specified on a cylinder whose directrix is parallel to the t axis.

For example, a simple inverse problem for a parabolic equation that consists in determining the coefficient $q(x)$ in the equation

$$u_t = u_{xx} - q(x)u, \tag{9}$$

from the information on solutions of the Cauchy problem

$$u|_{t=0} = \delta(x), \tag{10}$$

specified by

$$u|_{x=0} = f_1(t), \quad u_x|_{x=0} = f_2(t) \tag{11}$$

is reduced to the following inverse problem:

$$\tilde{u}_{tt} = \tilde{u}_{xx} - q(x)\tilde{u},$$
$$\tilde{u}|_{t=0} = \delta(x), \quad \tilde{u}_t|_{t=0} = 0, \quad \tilde{u}|_{x=0} = \tilde{f}_1(t), \quad \tilde{u}_x|_{x=0} = \tilde{f}_2(t). \tag{12}$$

The functions $\tilde{f}_k$ and f_k are related by the formulas

$$f_k(t) = \int_0^\infty G(t,\tau)\tilde{f}_k(\tau)\,d\tau, \quad k = 1,2.$$

In the inverse problem (9)–(11), the coefficient $q(x) \in C(-\infty, \infty)$ is uniquely determined by specifying the functions f_1 and f_2 on the interval $[0, \delta]$, where δ is an arbitrary positive number.

We now turn to elliptic equations. Let D_0 be a domain in the space of variables x, y ($x \in \mathbb{R}^n$) that is a semi-infinite cylinder with directrix parallel to the y axis:

$$D_0 = D \times \mathbb{R}^+, \quad x \in D, \quad \mathbb{R}^+ = \{y : y \geq 0\}.$$

Consider the equation

$$u_{yy} + Lu + f(x, y) = 0, \tag{13}$$

in the domain D_0, where L is a uniformly elliptic operator with respect to the variables $x_1, \ldots, x_n$ whose coefficients do not depend on y. Let S be the boundary of the domain D. We state the following problem for equation (13): find a solution to equation (13) that decreases as $y \to \infty$ and satisfies the following conditions on the boundary of D:

$$u|_{y=0} = \varphi(x), \quad \left.\frac{\partial u}{\partial n}\right|_{\Gamma} = \psi(x, y), \quad \Gamma = S \times \mathbb{R}^+. \tag{14}$$

In addition to the problem (13), (14), consider the problem (12) in which $\tilde{f}$ and $\tilde{\psi}$ are related to f and ψ by the formulas

$$f(x, y) = \int_0^\infty \tilde{f}(x, t) H(y, t)\, dt, \quad \psi(x, y) = \int_0^\infty \tilde{\psi}(x, t) H(y, t)\, dt, \tag{15}$$

where

$$H(y, t) = \frac{2}{\pi} \frac{y}{y^2 + t^2}$$

is a solution to the Laplace equation

$$H_{tt} + H_{yy} = 0.$$

To ensure that the integrals (15) exist, we assume that $\tilde{f}$ and $\tilde{\psi}$ are bounded. We will also make the same assumption on the solution to the problem (2). Then the unique solution to the problem (13), (14) is given by the formula

$$u(x, t) = \int_0^\infty \tilde{u}(x, t) H(y, t)\, dt. \tag{16}$$

We will show that this formula holds, assuming that all operations performed below are admissible.

We have

$$u_{yy} = \int_0^\infty H_{yy}(y,t)\tilde{u}(x,t)\,dt = -\int_0^\infty H_{tt}(y,t)\tilde{u}(x,t)\,dt$$

$$= (-H_t\tilde{u} + H\tilde{u}_t)|_{t=0}^{t=\infty} - \int_0^\infty H(y,t)\tilde{u}_{tt}(x,t)\,dt = -\int_0^\infty H(y,t)\tilde{u}_{tt}(x,t)\,dt,$$

$$Lu = \int_0^\infty H(y,t)L\tilde{u}(x,t)\,dt.$$

Therefore,

$$u_{yy} + Lu + f = \int_0^\infty (-\tilde{u}_{tt} + L\tilde{u} + \tilde{f})H(y,t)\,dt = 0.$$

At the same time,

$$\frac{\partial u}{\partial n}\bigg|_\Gamma = \int_0^\infty \frac{\partial \tilde{u}}{\partial n}\bigg|_\Gamma H(y,t)\,dy = \int_0^\infty \tilde{\psi}(x,t)H(y,t)\,dt = \psi(x,y),$$

$$u|_{y=0} = \lim_{y\to 0}\int_0^\infty \tilde{u}(x,t)H(y,t)\,dt$$

$$= \lim_{y\to 0}\frac{2}{\pi}\int_0^\infty \tilde{u}(x,y\tau)\frac{d\tau}{1+\tau^2} = \tilde{u}(x,0) = \varphi(x).$$

We conclude that the function $u(x,y)$ given by formula (16) is indeed a solution to the problem (13), (14).

Being an integral transformation with Cauchy kernel, equality (16) is uniquely invertible for fixed x. Therefore, if we consider the problem of determining a coefficient appearing in the operator L in the elliptic equation from the information on the solution to the problem (13), (14)

$$u|_\Gamma = g(x,y), \tag{17}$$

then the inverse problem (13)–(17) can be reduced to the equivalent inverse problem (2), (8), the functions g and $\tilde{g}$ being related by the invertible correspondence

$$g(x,y) = \int_0^\infty \tilde{g}(x,y)H(y,t)\,dt, \quad x \in S,\ y \in \mathbb{R}^+.$$

Another conclusion is that the transition from an elliptic equation to a hyperbolic one is possible only under the following two conditions: 1) the domain in which the equation is considered is a cylindrical domain with directrix parallel to the y axis; 2) the coefficients of the equation do not depend on y. These conditions are rather restrictive, given that the variables x and y are usually equivalent in applied problems.

We now point out another class of inverse problems for elliptic equations that can be studied using reduction to inverse problems for hyperbolic equations. We will confine ourselves to a specific example. For the equation

$$\Delta u + \omega^2 c(x)u = f(x), \quad x \in \mathbb{R}^n, \tag{18}$$

consider the problem of determining the function $f(x)$ or the coefficient $c(x) \geq c_0 > 0$ from the following information. Suppose that, for a set of frequencies ω_k $(k = 1, 2, \dots)$, the solution to equation (18) satisfying the radiation conditions is known on a surface S:

$$u(x, \omega_k) = g(x, \omega_k), \quad x \in S,\ k = 1, 2, \dots. \tag{19}$$

Equation (18) is obtained as a result of the analysis of stabilized harmonic oscillations described by the wave equation

$$c(x)v_{tt} = \Delta v + f(x)e^{i\omega t}.$$

Representing the function v in the form

$$v = u(x)e^{i\omega t},$$

we arrive at equation (18). In the case where $c = 1$, equation (18) becomes the Helmholtz equation. Inverse problems of determining $f(x)$ for the Helmholtz equation were studied in Zapreev (1976) and Zapreev and Tsetsokho (1976). Consider the Cauchy problem

$$c(x)w_{tt} = \Delta w + f(x)\delta(t), \quad w|_{t<0} \equiv 0. \tag{20}$$

If $c(x)$ and $f(x)$ are bounded, the solution to this problem satisfies the estimate $|w| \leq Ce^{\alpha t}$. Therefore, for $\operatorname{Re} p > \alpha$, the Laplace transform $\widetilde{w}(x, p)$ of $w(x, t)$ exists and is an analytic function of the variable p. Under certain conditions (for example, when $c = 1$ and $f(x)$ is a function with finite support), the function $\widetilde{w}(x, p)$ admits an analytic continuation to the imaginary axis $p = i\omega$. Consequently,

$$\widetilde{w}(x, i\omega_k) = g(x, \omega_k), \quad x \in S, \ k = 1, 2, \ldots.$$

If $\omega_k \to \omega_0$, where ω_0 is an interior point of the domain of analyticity of $\widetilde{w}(x, i\omega)$, then $\widetilde{w}(x, p)$ can be determined from the values $\widetilde{w}(x, i\omega_k)$ in the entire domain of analyticity and therefore it is possible to find

$$w|_S = \tilde{g}(x, t). \tag{21}$$

As a result, we arrive at the inverse problem (20), (21), which is equivalent to the problem (18), (19).

12.5. PROBLEMS OF DETERMINING A RIEMANNIAN METRIC

Let D be a bounded simply connected domain in the two-dimensional space (x,y) with piecewise smooth boundary Γ such that

$$x = \xi(z), \ y = \eta(z), \ z \in [0, Z], \ \xi(0) = \xi(Z), \ \eta(0) = \eta(Z), \tag{1}$$

where z is the Euclidean length of the curve Γ.

Furthermore, let the length element in the domain D be defined by the formula

$$d\tau = n(x, y)(dx^2 + dy^2)^{1/2}. \tag{2}$$

The function $n(x, y)$ in (2) belongs to the class $\mathbb{C}^4(\bar{D})$ and $n(x, y) > 0$. The geodesics defined by $n(x, y)$ (with respect to the metric (2)) are specified by the equations

$$x = \varphi(x_0, y_0, \theta; \tau), \quad y = \psi(x_0, y_0, \theta; \tau), \tag{3}$$

where (x_0, y_0) is a fixed endpoint of the geodesic, θ is the outgoing angle of the geodesic at (x_0, y_0), the meaning of the parameter τ of the curve is characterized by formula (2) and the value of τ is given by the formula

$$\tau = \int_{K(x_0, y_0, x, y)} n(x, y)\, ds,$$

where $K(x_0, y_0, x, y)$ is a geodesic connecting the points (x_0, y_0) and (x, y). The geodesics defined by the function $n(x, y)$ must satisfy the following conditions:

a) for any two distinct points in $\bar{D}$, there is a unique geodesic passing through these points; the endpoints of any geodesic $K(\gamma, z)$ belong to Γ and are the points $(\xi(\gamma), \eta(\gamma))$ and $(\xi(z), \eta(z))$, while the other points of $K(\gamma, z)$ do not belong to Γ; the lengths of all geodesics are uniformly bounded;

b) $\dfrac{1}{\tau} \dfrac{D[\varphi(x_0, y_0, \theta; \tau), \psi(x_0, y_0, \theta; \tau)]}{D(\theta, \tau)} \geq c > 0$, where c is a constant and $(x_0, y_0) \in \bar{D}$.

The existence of the derivatives required in the condition b) follows from the conditions $n(x, y) \in \mathbb{C}^4(\bar{D})$, $n(x, y) > 0$.

We consider the problem of determining the function $n(x, y)$ under the condition that the distance between any two points of the boundary of D with respect to the metric (2) is specified. This problem is reduced to solving the equation

$$\tau(\gamma, z) = \int_{K(\gamma, z)} n(x, y)\, ds, \quad ds = (dx^2 + dy^2)^{1/2} \tag{4}$$

for $n(x, y)$, where the function $\tau(\gamma, z) \in \mathbb{C}^1([0, Z] \times [0, Z])$ is known. The geodesic $K(\gamma, z)$ connecting boundary points of the domain D is unknown. Therefore, unlike the problem of integral geometry, the problem (4) is non-linear because the curves $K(\gamma, z)$ are also unknown, their definition involving the unknown function $n(x, y)$.

Theorem 1. *Assume that a bounded simply connected domain D has piecewise-smooth boundary (1), the unknown function $n(x, y)$ belongs to $\mathbb{C}^4(\bar{D})$ and $n(x, y) > 0$. Moreover, let $n(x, y)$ be such that the geodesics defined by the metric (2), which involves $n(x, y)$, satisfy the conditions a) and b). Then the problem (4) has at most one solution and the following stability estimate holds:*

$$\|\tilde{n}\|_{L_2(\bar{D})} \leq (2\pi)^{-1/2} \|\tilde{\tau}(\gamma, z)\|_{L_2([0,Z] \times [0,Z])}, \tag{5}$$

where $\tilde{n}(x, y) = n_1(x, y) - n_2(x, y)$, $\tilde{\tau} = \tau_1(\gamma, z) - \tau_2(\gamma, z)$, and $n_i(x, y)$ is a solution to the problem (4) with data $\tau_i(\gamma, z)$, $i = 1, 2$.

Lemmas on geodesics. We introduce the function

$$\tau(x, y; x_0, y_0) = \int_{K(x_0, y_0, x, y)} n(x, y)\, ds, \tag{6}$$

where $(x, y) \in \bar{D}$, $(x_0, y_0) \in \bar{D}$, and $K(x_0, y_0, x, y)$ is the geodesic connecting the points (x, y) and (x_0, y_0).

Lemma 1. *If* $n(x, y) \in C^4(\bar{D})$ *and* $n(x, y) > C_1 > 0$, *then the paramet- ric equation for geodesics can be represented in the form*

$$x = \varphi(x_0, y_0, \tau p, \tau q), \quad y = \psi(x_0, y_0, \tau p, \tau q), \tag{7}$$

where $(x, y) \in \bar{D}$, p *and* q *are defined by the formulas*

$$p = 2n(x_0, y_0) \cos \theta, \quad q = 2n(x_0, y_0) \sin \theta,$$

θ *is the outgoing angle of the geodesic at* (x_0, y_0), *the parameter* τ *of the geodesic represents the distance between the points* (x, y) *and* (x_0, y_0) *with respect to the metric* (2). *The functions* φ *and* ψ *have continuous bounded derivatives up to the second order with respect to all their arguments* x_0, y_0, τp, *and* τq. *Moreover,*

$$\left.\frac{\partial \varphi}{\partial(\tau p)}\right|_{\tau=0} = \left.\frac{\partial \psi}{\partial(\tau q)}\right|_{\tau=0} = \frac{1}{2n^2(x_0, y_0)}, \quad \left.\frac{\partial \varphi}{\partial(\tau q)}\right|_{\tau=0} = \left.\frac{\partial \psi}{\partial(\tau p)}\right|_{\tau=0} = 0.$$

The proof of this lemma is too long to be presented here and is therefore omitted. A similar statement can be found, for example, in Smirnov (1964).

Lemma 2. *For the condition* $\frac{1}{\tau}\frac{D(x,y)}{D(\tau,\theta)} \geq c_1 > 0$ *to hold, it is necessary and sufficient that* $\frac{D(x,y)}{D(\tau p, \tau q)} \geq c_2 > 0$, *where* c_1 *and* c_2 *are constants.*

Proof. Using Lemma 1, it is easy to show that

$$\frac{1}{\tau}\frac{D(x, y)}{D(\tau, \theta)} = \frac{D(x, y)}{D(\tau p, \tau q)} \cdot 4n^2(x_0, y_0).$$

The desired assertion follows immediately. □

Note that Lemma 2 implies

$$\left.\frac{D(x, y)}{D(\tau p, \tau q)}\right|_{\tau=0} = \frac{1}{4n^2(x_0, y_0)}.$$

For the following arguments, we will need a method for proving the unique- ness theorem for the problem of integral geometry in which the collection of curves $K(\gamma, z)$ is represented by the geodesics (7). If our collection of geodesics is used in the proof of this theorem as the collection of curves $K(\gamma, z)$ and the parameter τ is used as the parameter s of the curves, then all the analogous assertions can be obtained, which is easy to verify. It should be noted that in the proof of Lemma 1 in Section 11.4, in view of the representation (7) for φ and ψ, it is sufficient to require that these functions be differentiable only up to the second order.

As a result, we obtain the following theorem for the problem

$$v(\gamma, z) = \int_{K(\gamma,z)} u \, ds. \tag{8}$$

Theorem 2. *Assume that a bounded simply connected domain D has piecewise-smooth boundary (1), the collection of curves $K(\gamma, z)$ is the set of geodesics defined by a given positive function $n(x, y) \in \mathbb{C}^4(\bar{D})$ with the help of the metric (2), and $n(x, y)$ is such that the geodesics satisfy the conditions a) and b). Then the problem of integral geometry (8) has at most one solution in the class of functions $u(x, y) \in \mathbb{C}^2(\bar{D})$ and the following stability estimate holds:*

$$\|\tilde{u}\|_{\mathbb{L}_2(\bar{D})} \le (2\pi)^{-1/2} \|\tilde{v}_z(\gamma, z)\|_{\mathbb{L}_2([0,Z]\times[0,Z])}.$$

The linearization of the problem (4) yields the problem (8). As a result, Theorem 2 can be viewed as the uniqueness theorem for the linearized statement of the problem (4).

The proof of Theorem 1. At this point it is assumed that the boundary Γ of the domain D is smooth. We introduce the function

$$\tau(x, y, z) = \int_{K(x,y,z)} n \, ds, \tag{9}$$

where $K(x, y, z)$ is a part of the geodesic connecting the points (x, y) and $(\xi(z), \eta(z))$, $(x, y, z) \in \Omega = \bar{D} \times [0, Z]$, and $n(x, y)$ is a solution to the problem (4). The differential properties of the function $\tau(x, y, z)$ are analogous to those of the function $w(x, y, z)$ in the problem of integral geometry in Section 11.4. It is easy to see that $\tau(x, y, z)$ satisfies the eikonal equation

$$\tau_x^2 + \tau_y^2 = n^2(x, y). \tag{10}$$

Differentiating (10) with respect to z, we arrive at the following nonlinear differential equation for $\tau(x,y,z)$:

$$\frac{\partial}{\partial z}(\tau_x^2 + \tau_y^2) = 0. \tag{11}$$

From the data of the problem (4), it follows that $\tau(x,y,z)$ must satisfy the condition

$$\tau(\xi(\gamma),\eta(\gamma),z) = \tau(\gamma,z), \quad \tau(\xi(z),\eta(z),z) = 0. \tag{12}$$

If $n(x,y)$ satisfies the assumptions of the theorem, then the problem (4) is equivalent to the problem (11), (12), provided that $\tau(x,y,z)$ has differential properties analogous to 1) and 2) in Section 11.4.

Equation (11) is a nonlinear hyperbolic-parabolic equation with two collections of bicharacteristics. One collection consists of the geodesics $K(\gamma,z)$, and the other consists of the lines

$$x = x_0, \quad y = y_0, \tag{13}$$

where $(x_0,y_0) \in \bar{D}$. The data of the problem (12) are specified on the entire surface of the torus Ω. It is easy to see that this surface is a characteristic surface formed by the bicharacteristics of the collection (13).

There is a different problem that can be stated for equation (11). To this end, we specify the following data:

$$\tau(x,y,z)|_{z=0} = f(x,y), \quad \tau(\xi(z),\eta(z),z) = 0. \tag{12'}$$

The problem (11), (12') is solved in the usual way. This problem is usually said to be direct, whereas the problem (4) and the equivalent problem (11), (12) are said to be inverse.

Let $\tau_i(x,y,z)$ be solutions to equation (11) that satisfy the following conditions on the boundary of Ω:

$$\tau_i(\xi(\gamma),\eta(\gamma),z) = \tau_i(\gamma,z), \quad \tau_i(\xi(z),\eta(z),z) = 0. \quad i = 1,2.$$

Substituting $\tau_i(x,y,z)$ into (11) and subtracting the corresponding expressions, we obtain

$$Lw \equiv \frac{\partial}{\partial z}\left[\frac{\partial w}{\partial x}\cdot\frac{\partial \tau_1}{\partial x} + \frac{\partial w}{\partial y}\cdot\frac{\partial \tau_1}{\partial y}\right] + \frac{\partial}{\partial z}\left[\frac{\partial w}{\partial x}\cdot\frac{\partial \tau_2}{\partial x} + \frac{\partial w}{\partial y}\cdot\frac{\partial \tau_2}{\partial y}\right] = 0, \tag{14}$$

where $w(x, y, z) = \tau_1(x, y, z) - \tau_2(x, y, z)$. Similarly, from (12) we obtain

$$w(\gamma, z) = w(\xi(\gamma), \eta(\gamma), z) = \tau_1(\gamma, z) - \tau_2(\gamma, z). \tag{15}$$

Using the well-known formulas of variational calculus

$$(\tau_i)_x = n_i \cos \theta_i, \quad (\tau_i)_y = n_i \sin \theta_i, \quad i = 1, 2, \tag{16}$$

where $(\cos \theta_i, \sin \theta_i)$ is the tangent vector at (x, y) to the geodesic connecting (x, y) and $(\xi(z), \eta(z))$, we rewrite (14) in the form

$$\frac{1}{n_1} Lw \equiv \frac{\partial}{\partial z} \left(\frac{\partial w}{\partial x} \cos \theta_1 + \frac{\partial w}{\partial y} \sin \theta_1 \right)$$

$$+ \frac{n_2}{n_1} \frac{\partial}{\partial z} \left(\frac{\partial w}{\partial x} \cos \theta_2 + \frac{\partial w}{\partial y} \sin \theta_2 \right) = 0. \tag{17}$$

From (17), it follows that

$$\frac{2}{n_1} \left(-\frac{\partial w}{\partial x} \sin \theta_1 + \frac{\partial w}{\partial y} \cos \theta_1 \right) Lw \equiv \frac{\partial \theta_1}{\partial z} \left[\left(\frac{\partial w}{\partial x} \right)^2 + \left(\frac{\partial w}{\partial y} \right)^2 \right]$$

$$+ \left[\left(\frac{\partial w}{\partial x} \cos \theta_1 + \frac{\partial w}{\partial y} \sin \theta_1 \right) \left(\frac{\partial w}{\partial y} \cos \theta_1 - \frac{\partial w}{\partial x} \sin \theta_1 \right) \right]_z$$

$$- \left(\frac{\partial w}{\partial x} \frac{\partial w}{\partial z} \right)_y + \left(\frac{\partial z}{\partial x} \frac{\partial w}{\partial y} \right)_x + \frac{1}{2} \left[n_2^2 \sin 2(\theta_1 - \theta_2) \right]_z$$

$$- \left[n_2^2 (\theta_1 - \theta_2) \right]_z = 0. \tag{18}$$

The last two terms in (18) are obtained as follows:

$$2 \left(-\frac{\partial w}{\partial x} \sin \theta_1 + \frac{\partial w}{\partial y} \cos \theta_1 \right) \frac{n_2}{n_1} \frac{\partial}{\partial z} \left(\frac{\partial w}{\partial x} \cos \theta_2 + \frac{\partial w}{\partial y} \sin \theta_2 \right)$$

$$= 2 [-(n_1 \cos \theta_1 - n_2 \cos \theta_2) \sin \theta_1 + (n_1 \sin \theta_1 - n_2 \sin \theta_2) \cos \theta_1]$$

$$\times \frac{n_2}{n_1} \frac{\partial}{\partial z} [(n_1 \cos \theta_1 - n_2 \cos \theta_2) \cos \theta_2 + (n_1 \sin \theta_1 - n_2 \sin \theta_2) \sin \theta_2]$$

$$= 2n_2^2 \sin(\theta_1 - \theta_2) \frac{\partial}{\partial z} \cos(\theta_1 - \theta_2) = -2n_2^2 \sin^2(\theta_1 - \theta_2) \cdot (\theta_1 - \theta_2)_z$$

$$- n_2^2 [1 - \cos 2(\theta_1 - \theta_2)](\theta_1 - \theta_2)_z$$

$$= \frac{1}{2} \left[n_2^2 \sin 2(\theta_1 - \theta_2) \right]_z - \left[n_2^2 (\theta_1 - \theta_2) \right]_z.$$

The remaining terms in (18) are obtained similarly, as in the problem of integral geometry in Section 11.4.

The functions w_{xz}, w_{yz}, w_{xy}, and $[\theta(x, y, z)]_z$ have singularities of type $[(\xi(z) - x)^2 + (\eta(z) - y)^2]^{-1/2}$. We now perform transformations of the formula (18) similar to those in the problem of integral geometry in Section 11.4, i.e., we separate an ε-neighborhood of the singular curve $x = \xi(z)$, $y = \eta(z)$ ($z \in [0, Z]$) from Ω, integrate (18) over the remaining part of Ω using Ostrogradsky's formula, and pass to the limit as $\varepsilon \to 0$. As a result, we have

$$\int\int_\Omega \int \frac{\partial \theta_1}{\partial z} \left[\left(\frac{\partial w}{\partial x} \right)^2 + \left(\frac{\partial w}{\partial y} \right)^2 \right] d\Omega = - \int_0^L \int_0^L \frac{\partial w}{\partial z} \frac{\partial w}{\partial \gamma} dz d\gamma. \tag{19}$$

Using

$$\left(\frac{\partial w}{\partial x} \right)^2 + \left(\frac{\partial w}{\partial y} \right)^2 = (n_1 - n_2)^2 + 4n_1 n_2 \sin^2 \frac{\theta_1 - \theta_2}{2}, \tag{20}$$

$(\theta_1)_z \geq 0$, and the Cauchy inequality, we obtain the estimate (5) from (19).

In the case where the boundary Γ is piecewise-smooth, the proof is similar, as in the problem of integral geometry in Section 11.4. This completes the proof of the theorem.

It is an interesting fact that the estimate (5) coincides with the stability estimate of the problem of integral geometry in Section 11.4.

Formulas for the mean square value of $n(x, y)$ over the domain D. Using (16), we rewrite (11) as follows:

$$L\tau \equiv \frac{\partial}{\partial z} \left(\frac{\partial \tau}{\partial x} \cos \theta + \frac{\partial \tau}{\partial y} \sin \theta \right) = 0. \tag{21}$$

For $2\left(- \frac{\partial \tau}{\partial x} \sin \theta + \frac{\partial \tau}{\partial y} \cos \theta \right) L\tau$ we have an expression analogous to (18) except the last two terms. Then, performing transformations similar to those described above and using equality (10) instead of (20), we obtain

$$2\pi \int\int_D n^2 dx dy = - \int_0^L \int_0^L \frac{\tau(\gamma, z)}{\gamma} \frac{\partial \tau(\gamma, z)}{\partial z} d\gamma dz. \tag{22}$$

The mean square value of $n(x, y)$ over the domain D is easy to obtain from (22). The right-hand side of (22) can be represented in a different form, namely, it can be expressed in terms of the values of $n(x, y)$ on the boundary of D and the angle between the geodesic on the the boundary Γ and the tangent to the boundary of D at the same point.

Let $\beta(z)$ be the angle that defines the direction of the tangent to the boundary Γ of the domain D at the point $(\xi(z), \eta(z))$, its orientation corresponding to the increase in z, and let $\alpha(z, \gamma)$ be the angle that defines the direction of the tangent at $(\xi(z), \eta(z))$ to the geodesic connecting the points $(\xi(z), \eta(z))$ and $(\xi(\gamma), \eta(\gamma))$. Then it is easy to see that

$$\frac{\partial \tau(\gamma, z)}{\partial z} = n(\xi(z), \eta(z)) \cos[\alpha(z, \gamma) - \beta(z)], \tag{23}$$

$$\frac{\partial \tau(\gamma, z)}{\partial \gamma} = n(\xi(\gamma), \eta(\gamma)) \cos[\alpha(z, \gamma) - \beta(\gamma)] \tag{24}$$

and (22) becomes as follows:

$$2\pi \iint_D n^2 \, dx dy = - \int_0^L \int_0^L n(\xi(z), \eta(z)) n(\xi(z), \eta(z))$$

$$\times \cos[\alpha(z, \gamma) - \beta(z)] \cos[\alpha(z, \gamma) - \beta(\gamma)] \, dz d\gamma. \tag{25}$$

In conclusion, we give the formulas of transition from the given function $\tau(\gamma, z)$ to the quantities used in (25). We have

$$n(\xi(z), \eta(z)) = \lim_{\gamma \to z} \frac{\tau(\gamma, z)}{|z - \gamma|},$$

and $\alpha(z, \gamma)$ is determined from (23).

Chapter 13.

Several areas of the theory of ill-posed problems, inverse problems, and applications

The theory of ill-posed problems, inverse problems, and their applications are extensively covered in the literature nowadays. Presenting all achievements in these areas in sufficient detail would take a monograph several times thicker than this one. In this appendix, we characterize research results in several areas related to the topics covered in this book.

Applications. As was noted in Lavrent'ev (1962), Tikhonov (1943), and Tikhonov and Arsenin (1977), the first applied problems leading to ill-posed problems concerned the interpretation of geophysical measurements data.

We pointed out two areas of applied problems that arose in connection with the interpretation of measurements data. The first one is the interpretation of astrophysical data. Research results in this area are presented in Goncharskii, Cherepashchuk, and Yagola (1985). One of the problems described in this monograph is the problem of interpreting the brightness curves for a number of stars. As is known, the luminosity of some stars varies periodically. This phenomenon is due to a satellite of a star passing between the star and the observer on Earth. To determine the parameters of the satellite, a certain integral equation of the first kind is constructed and an algorithm for the numerical solution of this equation is obtained using regularization.

The second area is the interpretation of tomographic data in gas dynamics and plasma physics, which is described in Pikalov and Preobrazhenskii (1987). In Nikolaev (1988), the principles of using tomographic data in seismic prospecting and seismology are presented. In Lavrent'ev, Zerkal, and Trofimov (2001), a classical problem concerning the interpretation of tomographic data, namely the problem of inverting the Radon transform, is analyzed from the point of view of the theory of ill-posed problems. The same monograph also deals with new statements of problems of seismic prospecting, in particular, the problems of determining the structure of geological bodies.

The general theory of ill-posed problems. We now point out the areas of the general theory of ill-posed problems that were left out of the scope of our monograph or were only briefly mentioned.

Variational methods represent one of these areas. These methods are briefly discussed in Tikhonov and Arsenin (1977) and in Chapter 4 of the present book. A detailed discussion of a number of matters related to the application of variational methods is given in Liskovets (1981).

Iterative methods are another example. We confined ourselves to describing one iterative method for linear operator equations. A wide range of matters concerning the application of iterative methods to solving nonlinear ill-posed problems are discussed in Vasin and Ageev (1995) and Tanana (1997). Issues regarding the discretization of ill-posed problems are also considered in these monographs.

Problems of analytic continuation. These problems play an important role in the theory of ill-posed problems. Results of the study of problems of analytic continuation treated as ill-posed problems are presented in Lavrent'ev, Romanov, and Shishatskii (1986). In particular, conditional stability estimates and Carleman formulas that are regularizing for these problems are given there. Substantial developments in this direction were made later. Namely, results were obtained for analytic functions of several variables and for Cauchy problems for elliptic equations in spaces of various dimensions. (Aizenberg, 1990; Ishankulov, 1984; Tarkhanov, 1991; Tsikh, 1992; Yarmukhamedov, 1980).

Volterra equations and evolution equations. In Chapters 9 and 10 we presented research results for Volterra integral operator equations and evolution equations. A considerable number of important results in these areas can be found in Bukhgeim (1983) and Ivanov, Mel'nikova, and Filinkov (1995).

Integral geometry. An original topic in integral geometry is described in Sharafutdinov (1994).

Inverse problems. In Chapter 12, our presentation of the theory of inverse problems mostly followed Romanov (1987). Substantial achievements in the area of coefficient inverse problems were made in the study of the problems of determining the coefficients in the system of Maxwell's equations (see Romanov and Kabanikhin, 1991). A series of important results in the theory of inverse problems are presented in Denisov (1999).

Inverse problems of the theory of the Newton potential, which represent a classical area of the theory if inverse problems, were studied by a number of authors. For example, see Cherednichenko (1996).

Many inverse problems can be reduced to linear or nonlinear Volterra operator equations. The corresponding results of the general theory of Volterra equations in Banach spaces are used to obtain uniqueness theorems and *a priori* estimates for inverse problems. Results in this area are presented in Bukhgeim (1988).

Bibliography

Aizenberg, L. A. (1990). *Carleman Formulas in Complex Analysis. First Applications.* Nauka, Novosibirsk (in Russian).

Albeverio, S., Høegh-Krohn, R., Fenstad, J. E., and Lindstrøm, T. (1986). *Nonstandard Methods in Stochastic Analysis and Mathematical Physics.* Pure Appl. Math., **122**, Academic Press, Inc., Orlando, FL.

Alekseev, V. M., Tikhomirov, V. M., and Fomin, S. V. (1987). *Optimal Control.* Contemporary Soviet Mathematics. Consultants Bureau, New York.

Amari, S. (1985). *Differential-Geometrical Methods in Statistics.* Lecture Notes in Statist., **28**, Springer-Verlag, New York.

Antosik, P., Mikusinski, J., and Sikorski, R. (1973). *Theory of Distributions. The Sequential Approach.* Elsevier, Amsterdam.

Aubin, J.-P. (1978). Analyse fonctionnelle non linéaire et applications à l'équilibre économique. *Ann. Sci. Math. Québec,* **2**(1), 5–47 (in French).

Aubin, J.-P. and Ekeland, I. (1984). *Applied Nonlinear Analysis.* Pure Appl. Math. (N.Y.), Wiley-Interscience, New York.

Bakushinskii, A. B. and Goncharskii, A. V. (1989). *Iterative Methods for Solving Ill-posed Problems.* Nauka, Moscow, (in Russian).

Balakrishnan, A. V. (1981). *Applied Functional Analysis.* Appl. Math. (N.Y.), **3**, Springer-Verlag, New York-Berlin.

Berberian, S. K. (1966). *Notes on Spectral Theory.* Van Nostrand Mathematical Studies, No. 5, D. Van Nostrand Co., Inc., Princeton-Toronto-London.

Berezanskii, Yu. M. (1958). The uniqueness theorem in the inverse problem of spectral analysis for the Schrödinger equation. *Tr. Mosk. Mat. Obs.*, 7, 1–62 (in Russian).

Birkhoff, G. (1979). *Lattice Theory.* Amer. Math. Soc. Colloq. Publ., 25, Amer. Math. Soc., Providence, R.I.

Bourbaki, N. (1987). *Topological Vector Spaces. Chapters 1–5.* Elements of Mathematics, Springer-Verlag, Berlin.

Bourbaki, N. (1998). *General Topology.* Elements of Mathematics, Springer-Verlag, Berlin.

Bourbaki, N. (2003). *Algebra II.* Elem. Math., Springer-Verlag, Berlin.

Bourbaki, N. (2004a). *Theory of Sets.* Elem. Math., Springer-Verlag, Berlin.

Bourbaki, N. (2004b). *Integration. I, II.* Elem. Math., Springer-Verlag, Berlin.

Bremerman, G. (1967). Distributions, complex variables and Fourier transforms. *SIAM J. Appl. Math.*, 15, 929–943.

Brézis, H. (1973). *Opérateurs maximaux monotones et semi-groupes de contractions dans les espaces de Hilbert.* North-Holland Math. Stud., 5, Notas de Matemática (50). North-Holland, Amsterdam-London (in French).

Bukhgeim, A. L. (1975). *Necessary Conditions of Stability for a Class of Integro-differential Equations.* In: Computational methods and programming., Ross. Akad. Nauk Sibirsk. Otdel., Vychisl. Tsentr, Novosibirsk.

Bukhgeim, A. L. (1983). *Volterra Operator Equations.* Nauka, Novosibirsk (in Russian).

Bukhgeim, A. L. (1988). *Introduction to the Theory of Inverse Problems.* Nauka, Novosibirsk (in Russian).

Carleson, L. (1966). On convergence and growth of partial sums of Fourier series. *Acta Math.*, 116, 135–157.

Cartan, H. (1970). *Differential Forms.* Houghton Mifflin Co., Boston, Mass.

Chentsov, N. N. (1982). *Statistical Decision Rules and Optimal Inference.* Transl. Math. Monogr., 53, Amer. Math. Soc., Providence, R.I.

Cherednichenko, V. G. (1996). *Inverse Logarithmic Potential Problem.* Inverse and Ill-posed Problems Series, VSP, Utrecht.

Clément, Ph., Heijmans, H. J. A. M., Angenent, S., van Duijn, C. J., and de Pagter, B. (1987). *One-parameter Semigroups.* CWI Monographs, 5, North-Holland, Amsterdam.

Clifford, A. H. and Preston, G. B. (1967). *The Algebraic Theory of Semigroups. Vol. II.* Mathematical Surveys, 7, Amer. Math. Soc., Providence, R.I.

Courant, R. and Hilbert, D. (1962). *Methods of Mathematical Physics. Vol. II: Partial Differential Equations.* (Vol. II by R. Courant.) Interscience Publ., New York-London.

Daletskii, Yu. L. and Fomin, S. V. (1983). *Measures and Differential Equations in Infinite-Dimensional Spaces.* Nauka, Moscow (in Russian).

Denisov, A. M. (1999). *Elements of the Theory of Inverse Problems.* Inverse and Ill-posed Problems Series, VSP, Utrecht.

Dubrovin, B. A., Novikov, S. P., and Fomenko, A. T. (1986). *Modern Geometry. Methods and Applications.* Nauka, Moscow (in Russian).

Dunford, N. and Schwartz, J. T. (1988). *Linear Operators.* Wiley Classics Lib., Wiley-Interscience, New York.

Edwards, R. E. (1979). *Fourier Series. Vol. 1. A Modern Introduction.* Grad. Texts in Math., 64, Springer-Verlag, New York-Berlin, 1979.

Edwards, R. E. (1982). *Fourier Series. Vol. 2. A Modern Introduction.* Grad. Texts in Math., 85, Springer-Verlag, New York-Berlin.

Edwards, R. E. (1995). *Functional Analysis. Theory and Applications.* Dover Publications, New York.

Engelking, R. (1989). *General Topology.* Sigma Series in Pure Mathematics, 6, Heldermann Verlag, Berlin.

Federer, H. (1969). *Geometric Measure Theory.* Grundlehren Math. Wiss., **153**, Springer-Verlag, New York.

Frölicher, A. and Bucher, W. (1966). *Calculus in Vector Spaces Without Norm.* Lecture Notes in Math., 30, Springer-Verlag, Berlin-New York.

Gajewski, H., Gröger, K., and Zacharias, K. (1974). *Nichtlineare Operatorgleichungen und Operatordifferentialgleichungen. Mathematische Lehrbücher und Monographien, II.* Abteilung, Mathematische Monographien, **38**, Akademie-Verlag, Berlin (in German).

Godbillon, C. (1969). *Géométrie différentielle et mécanique analytique.* Hermann, Paris (in French).

Goncharskii, A. V., Cherepashchuk, A. M. and Yagola, A. G. (1985). *Illposed Problems in Astrophysics.* Nauka, Moscow (in Russian).

Hadamard, J. (1902). Sur les problèmes aux dérivées partielles et leur signification physique. *Princeton University Bulletin*, 49–52.

Halmos, P. R. (1974). *Measure Theory.* Grad. Texts in Math., 18, Springer-Verlag, New York-Heidelberg-Berlin.

Halmos, P. R. (1982). *A Hilbert Space Problem Book.* Grad. Texts in Math., 19, Springer-Verlag, New York-Berlin.

Harrison, J. (1993). Stokes' theorem for nonsmooth chains. *Bull. Amer. Math. Soc. (N.S.)*, **29**(2), 235–242.

Helgason, S. (2000) *Groups and Geometric Analysis. Integral Geometry, Invariant Differential Operators, And Spherical Functions.* Math. Surveys Monogr., 83, Amer. Math. Soc., Providence, RI.

Hille, E. and Phillips, R. S. (1974). *Functional Analysis and Semi-groups.* Amer. Math. Soc. Colloq. Publ., Vol. XXXI. Amer. Math. Soc., Providence, RI.

Hörmander, L. (2003). *The Analysis of Linear Partial Differential Operators. I. Distribution Theory and Fourier Analysis.* Classics Math., Springer-Verlag, Berlin.

Hutson, V. and Pym, J. S. (1980). *Applications of Functional Analysis and Operator Theory.* Mathematics in Science and Engineering, 146, Academic Press, New York-London. Ishankulov, T. (1984). Two analytic continuation problems for functions of several variables. *Siberian Math. J.*, 25, 412–416.

Ivanov, V. K., Mel'nikova, I. V., and Filinkov, A. I. (1995). *Operator-differential Equations and Ill-posed Problems.* Nauka, Moscow (in Russian).

Ivanov, V. K., Vasin, V. V., and Tanana, V. P. (2002). *Theory of Linear Ill-posed Problems and its Applications.* Inverse and Ill-posed Problems Series, VSP, Utrecht.

Kantorovich, L. V. and Akilov, G. P. (1982). *Functional Analysis.* Pergamon Press, Oxford-Elmsford, N.Y.

Kato, T. (1995). *Perturbation Theory for Linear Operators.* Springer-Verlag, Berlin.

Kelley, J. L. (1975). *General Topology.* Grad. Texts in Math., 27, Springer-Verlag, New York-Berlin.

Khelemskii, A. Ya. (1989). *Banach and Multinormed Algebras: General Theory, Representations, Homology.* Nauka, Moscow (in Russian).

Kirillov, A. A. and Gvishiani, A. D. (1988). *Theorems and Problems of Functional Analysis.* Nauka, Moscow (in Russian).

Kolmogorov, A. N. and Fomin, S. V. (1989). *Elements of the Theory of Functions and Functional Analysis.* Nauka, Moscow (in Russian).

Korobeinik, Yu. F. (1992). *The Stone-Weierstrass Theorem.* Rostov. Gos. Univ., Rostov-on-Don (in Russian).

Kostrikin, A. I. and Manin, Yu. I. (1997). *Linear Algebra and Geometry.* Algebra, Logic and Applications, 1, Gordon and Breach, Amsterdam.

Kuratowski, K. and Mostowski, A. (1976). *Set Theory. With an Introduction to Descriptive Set Theory.* Stud. Logic Found. Math., **86**, North-Holland, Amsterdam-New York-Oxford.

Lavrent'ev, M. M. (1962). *Some Ill-posed Problems of Mathematical Physics.* Izdat. Sibirsk. Otdel. Akad. Nauk SSSR, Novosibirsk (in Russian).

Lavrent'ev, M. M. (1964). On an inverse problem for the wave equation. *Soviet Math. Dokl.*, 5, 970–972.

Lavrent'ev, M. M. and Romanov, V. G. (1966). Three linearized inverse problems for hyperbolic equations. *Soviet Math. Dokl.*, 7, 1650–1652.

Lavrent'ev, M. M., Romanov, V. G. and Shishatskii, S. P. (1986). *Ill-posed Problems of Mathematical Physics and Analysis.* Transl. Math. Monogr., 64, Amer. Math. Soc., Providence, RI.

Lavrent'ev, M. M., Romanov, V. G. and Vasiliev, V. G. (1970). *Multidimensional Inverse Problems for Differential Equations.* Lecture Notes in Math., 167, Springer-Verlag, Berlin-New York.

Lavrent'ev, M. M. and Savel'ev, L. Ya. (1995). *Linear Operators and Ill-posed Problems.* With a supplement by A. L. Bukhgeim. Consultants Bureau, New York.

Lavrent'ev, M. M., Zerkal, S. M. and Trofimov, O. E. (2001). *Computer Modeling in Tomography and Ill-posed Problems.* VSP, Utrecht.

Liskovets, O. A. (1981). *Variational Methods for Solving Unstable Problems.* Nauka i Tekhnika, Minsk (in Russian).

Loève, M. (1977). *Probability Theory. I.* Grad. Texts in Math., **45**, Springer Verlag, New York-Heidelberg.

Loève, M. (1978). *Probability Theory. II.* Grad. Texts in Math., **46**, Springer-Verlag, New York-Heidelberg.

Maslov, V. P. (1976). *Complex Markov Chains and the Feynman Path Integral for Nonlinear Equations.* Nauka, Moscow (in Russian).

Maslov, V. P. (1987). *Asymptotic Methods for Solving Pseudodifferential Equations.* Nauka, Moscow (in Russian).

Matheron, G. (1975). *Random Sets and Integral Geometry.* Wiley Ser. Probab. Stat., John Wiley & Sons, New York-London-Sydney.

McLeod, R. M. (1980). *The Generalized Riemann Integral.* Carus Math. Monogr., 20, Math. Assoc. America, Washington, D.C.

Monna, A. F. (1970). *Analyse non-archimédienne.* Ergeb. Math. Grenzgeb., **56**, Springer-Verlag, Berlin-New York (in French).

von Neumann, J. (1987). *Selected Works on Functional Analysis. Vol. I, II.* Classics of Science. Nauka, Moscow (in Russian).

Nikolaev, A. V. (1988). *Seismology: Revolution in Science and Technology and Problems of the XXI Century.* In: Mathematical Modelling in Geophysics. Nauka, Novosibirsk.

Nirenberg, L. (2001). *Topics in Nonlinear Functional Analysis.* Courant Lect. Notes Math., 6, New York.

Ovsyannikov, L. (1978). *Group Analysis of Differential Equations.* Nauka, Moscow (in Russian).

Petrovskii, I. G. (1984). *Lectures on the Theory of Ordinary Differential Equations.* Moskov. Gos. Univ., Moscow (in Russian).

Pietsch, A. (1972). *Nuclear Locally Convex Spaces.* Ergeb. Math. Grenzgeb., **66**, Springer-Verlag, New York-Heidelberg.

Pikalov, V. V. and Preobrazhenskii, N. G. (1987). *Reconstructive Tomography in Gas Dynamics and Plasma Physics.* Nauka, Novosibirsk (in Russian).

Postnikov, M. M. (1986a). *Analytic Geometry. Lectures in Geometry. Semester I.* Nauka, Moscow (in Russian).

Postnikov, M. M. (1986b). *Linear Algebra. Lectures in Geometry. Semester II.* Nauka, Moscow (in Russian).

Postnikov, M. M. (1987). *Smooth Manifolds. Lectures in Geometry. Semester III.* Nauka, Moscow (in Russian).

Postnikov, M. M. (1988). *Differential Geometry. Lectures in Geometry. Semester IV.* Nauka, Moscow (in Russian).

Reed, M. and Simon, B. (1972). *Methods of Modern Mathematical Physics. I. Functional Analysis.* Academic Press, New York-London.

Reed, M. and Simon, B. (1975). *Methods of Modern Mathematical Physics. II. Fourier Analysis, Self-adjointness.* Academic Press, New York-London.

Reed, M. and Simon, B. (1978). *Methods of Modern Mathematical Physics. IV. Analysis of Operators.* Academic Press, New York-London.

Reed, M. and Simon, B. (1979). *Methods of Modern Mathematical Physics. III. Scattering Theory.* Academic Press, New York-London.

Reznitskaya, K. G. (1974). Relationship between the solutions to the Cauchy problem for equations of various types and ill-posed problems. In: *Mathematical problems in geophysics*, 5, Part 1, 55–62, Akad. Nauk SSSR Sibirsk. Otdel., Vychisl. Tsentr, Novosibirsk (in Russian).

Richtmyer, R. D. (1978). *Principles of Advanced Mathematical Physics. Vol. I.* Texts Monogr. Phys., Springer-Verlag, New York-Heidelberg.

Richtmyer, R. D. (1981). *Principles of Advanced Mathematical Physics. Vol. II.* Texts Monogr. Phys., Springer-Verlag, New York-Berlin.

Robinson, A. (1966). *Non-standard Analysis.* North-Holland, Amsterdam.

Romanov, V. G. (1967). The recovery of a function from its integrals over ellipsoids of revolution with one fixed focus. *Soviet Math. Dokl.*, 8, 480–483.

Romanov, V. G. (1978). Inverse problems for hyperbolic equations and energy inequalities. *Soviet Math. Dokl.*, 19, 1154–1158.

Romanov, V. G. (1987). *Inverse Problems of Mathematical Physics.* VNU Science Press, Utrecht.

Romanov, V. G. and Kabanikhin, S. I. (1991). *Inverse Problems of Geoelectrics.* Nauka, Moscow (in Russian).

Rozovskii, B. L. (1990). *Stochastic Evolution Systems. Linear Theory and Applications to Nonlinear Filtering.* Math. Appl. (Soviet Series), 35, Kluwer Acad. Publ., Dordrecht.

Rudin, W. (1991). *Functional Analysis.* Internat. Ser. Pure Appl. Math., McGraw-Hill, New York.

Sadovnichii, V. A. (1999). *Operator Theory.* Vyssh. Shkola, Moscow (in Russian).

Savel'ev, L. (1969a). *Lectures on Mathematical Analysis. Introduction.* Novosibirsk. Gos. Univ., Novosibirsk (in Russian).

Savel'ev, L. (1969b). *Lectures on Mathematical Analysis. Part 1: Numerical Sequences.* Novsibirsk. Gos. Univ., Novosibirsk (in Russian).

Savel'ev, L. (1974). *Lectures on Mathematical Analysis. Parts 2, 3: Elementary Functions.* Novosibirsk. Gos. Univ., Novosibirsk (in Russian).

Savel'ev, L. (1975). *Lectures on Mathematical Analysis. Part 4: The Differential.* Novosibirsk. Gos. Univ., Novosibirsk (in Russian).

Savel'ev, L. (1976). A theorem on the extension of sequential measures. *Soviet Math. Dokl.*, 17, 1031–1034.

Savel'ev, L. (1982). Measures on ortholattices. *Soviet Math.Dokl.*, 25, 837–840.

Savel'ev, L. Ya. (1983). Outer measures and outer topologies. *Siberian. Math. J.*, 24, 263–278.

Savel'ev, L. (1984). *Integration of Uniformly Measurable Functions.* Novosibirsk. Gos. Univ., Novosibirsk (in Russian).

Savel'ev, L. (1987). *Lectures on Differential Calculus.* Novosibirsk. Gos. Univ., Novosibirsk (in Russian).

Savel'ev, L. (1988, 1989). *Lectures on Integral Calculus.* Novosibirsk. Gos. Univ., Novosibirsk (in Russian).

Savel'ev, L. Ya. (1997). Measure spaces. *Dokl. Math.*, **56**(3), 877–879.

Schwartz, J. T. (1968). *Differential Geometry and Topology.* Gordon and Breach. New York-London-Paris.

Schwartz, J. T. (1969). *Nonlinear Functional Analysis.* Notes on Mathematics and its Applications, Gordon and Breach, New York-London-Paris.

Schwartz, L. (1966). *Théorie des distributions.* Publications de l'Institut de Mathématique de l'Université de Strasbourg, No. IX-X. Hermann, Paris (in French).

Schwartz, L. (1967). *Analyse mathématique.* Hermann, Paris (in French).

Segal, I. E. and Kunze, R. A. (1978). *Integrals and Operators.* Grundlehren Math. Wiss., **228**, Springer-Verlag, Berlin-New York.

Sharafutdinov, V. A. (1994). *Integral Geometry of Tensor Fields.* Inverse and Ill-posed Problems Series, VSP, Utrecht.

Smirnov, V. I. (1964). *A Course of Higher Mathematics. Vol. V. Integration and Functional Analysis.* ADIWES International Series in Mathematics, Pergamon Press, Oxford-New York.

Smolyanov, O. G. and Shavgulidze, E. T. (1990). *Path Integrals.* Moskov. Gos. Univ., Moscow (in Russian).

Sobolev, S. L. (1991). *Some Applications of Functional Analysis in Mathematical Physics.* Transl. Math. Monogr., 90, Amer. Math. Soc., Providence, RI.

Sobolev, S. L. (1992). *Equations of Mathematical Physics.* Nauka, Moscow (in Russian).

Spivak, M. (1965). *Calculus on Manifolds. A Modern Approach to Classical Theorems of Advanced Calculus.* W. A. Benjamin, Inc., New York-Amsterdam.

Sternberg, S. (1983). *Lectures on Differential Geometry.* Chelsea Publishing Co., New York.

Sulanke, R. and Wintgen, P. (1972). *Differentialgeometrie und Faserbündel.* Lehrbücher und Monographen aus dem Gebiete der exakten Wissenschaften, Mathematische Reihe, **48**, Birkhäuser Verlag, Basel-Stuttgart (in German).

Tanana, V. P. (1997). *Methods for Solving Operator Equations.* Inverse and Ill-posed Problems Series, VSP, Utrecht.

Tarkhanov, N. N. (1991). *Laurent Series for Solutions of Elliptic Systems.* Nauka, Novosibirsk (in Russian).

Tikhonov, A. N. (1943). On the stability of inverse problems. *C. R. (Doklady) Acad. Sci. URSS (N.S.),* 39, 176–179.

Tikhonov, A. N. and Arsenin, V. Y. (1977). *Solutions of Ill-posed Problems.* Scripta Series in Math., V. H. Winston & Sons, Washington, D.C.

Trenogin, V. A. (1993). *Functional Analysis.* Nauka, Moscow (in Russian).

Tsikh, A. K. (1992). *Multidimensional Residues and their Applications.* Transl. Math. Monogr., 103, Amer. Math. Soc., Providence, RI.

Vasin, V. V. and Ageev, A. L. (1995). *Ill-posed Problems with A Priori Information.* Inverse and Ill-posed Problems Series, VSP, Utrecht.

Vladimirov, V. S. (1979). *Generalized Functions in Mathematical Physics.* Mir, Moscow.

van der Waerden, B. L. (1991) *Algebra. Vol. I, II.* Based in part on lectures by E. Artin and E. Noether. Springer-Verlag, New York.

Warga, J. (1972). *Optimal Control of Differential and Functional Equations.* Academic Press, New York-London.

Weidmann, J. (1980). *Linear Operators in Hilbert Spaces.* Grad. Texts in Math., 68, Springer-Verlag, New York-Berlin.

Yarmukhamedov, Sh. (1980). On the analytic continuation of a holomorphic vector along its boundary values on a boundary piece. *Izv. Akad. Nauk UzSSR, Ser. Fiz.-Mat. Nauk*, 6, 34–40 (in Russian).

Yosida, K. (1965). *Functional Analysis.* Springer-Verlag, Berlin.

Zaidman, S. (1999). *Functional Analysis and Differential Equations in Abstract Spaces.* Chapman & Hall/CRC Monogr. Surv. Pure Appl. Math., 100, Boca Raton, FL.

Zapreev, A. S. (1976). A uniqueness theorem for a planar inverse problem for the Helmholtz equation. In: *Ill-posed problems in mathematics and geophysics*, Akad. Nauk SSSR Sibirsk. Otdel., Vychisl. Tsentr, Novosibirsk, 46–63 (in Russian).

Zapreev, A. S. and Tsetsokho, V. A. (1976). *An Inverse Problem for the Helmholtz Equation.* Preprint No. 22, Akad. Nauk SSSR Sibirsk. Otdel., Vychisl. Tsentr, Novosibirsk.

Index

μ-continuity, 278
μ-convergence, 278
μ-zero set, 278
$\mathbb{K}$-linearity, 89
$\mathbb{K}$-subalgebra, 90
$\mathbb{K}$-submodule, 90

absolute value, 69, 71, 122
absolute value of an operator, 425
absorption, 65
abstract Cauchy problem, 471
accretive operator, 473, 502
additive span, 268
adherent point, 153
Alaoglu theorem, 407
algebraic complement, 51
algebraic conjugate, 102, 124
alternating group, 54
alternation, 148
anti-isomorphism, 110
antimonotonic map, 20
antisymmetric multilinear map, 147
antisymmetric relation, 14
approximating sequence, 283
Archimedean field, 63
Ascoli theorem, 372
atlas, 322
automorphism, 9, 34
Axiom of Choice, 6

Baire measure, 384
Baire theorem, 184
Banach adjoint operator, 412, 420
Banach inverse operator theorem, 401
Banach spaces, 200
base, 152
basis multivector, 144
basis of a vector space, 80
Bessel's inequality, 366

bilinear map, 114
Birkhoff's theorem, 187
Bolzano's theorem, 193
Boolean algebra, 67
Boolean lattice, 67
Boolean ring of sets, 57
boundary of a manifold, 322
bounded functional, 124
broken line, 88
Brouwer theorem, 484

cardinal, 16
cardinal number, 16
Cartesian decomposition of an
    operator, 425
Cartesian product, 6, 21
Cartesian representation, 71
cauchy criterion, 183
centered collection, 161
chain, 26
chain rule, 208
change of variables theorem for
    integrals, 317
characteristic function, 396
chart, 322
closed form, 331
closed multioperator, 491
closed operator, 403
closed subspace, 118, 138
closest point, 138
closure, 118, 153
codimension, 98
coefficient of a linear functional, 358
coercitive, 500
coimage, 98
cokernel, 98
collection, 11
commutative ring, 55
commutative semigroup, 31
compact nonlinear operator, 488

compact operator, 408
compact set, 160
compact space, 160
compactly parameterized, 345
complete atlas, 323
complete measure, 278
complete space, 182
completion theorem, 369
complex conjugate, 71
composition of correspondences, 8
conjugate basis, 110
conjugate operator, 105
connected component of a point, 167
connected component of a set, 168
connected space, 165
continuity, 191
continuity of a measure with respect
        to another measure, 316
continuous differentiability, 217
continuous extension theorem, 195
continuous implicit function
        theorem, 239
continuous inverse function theorem,
        240
continuous measure, 276
continuous multifunction, 491
continuous spectrum, 428
contracting semigroup, 471, 501
contraction, 227
conullity, 98
convergence in itself, 182
convex body, 399
convex hull, 88
convex set, 87
convolution, 86, 392
coordinate neighborhood, 322
coordinate of a vector, 81
coordinate rows, 107
coordinate space, 94
coordinate system, 104
coordinatewise order, 21
countable additivity, 276
countable semiadditivity, 276

countable set, 29
cover, 11
cycle, 53

decomposable tensor, 142
decrement, 54
degenerate operator, 408
degree of a map, 349
delta-function, 90
derivative of a function, 204
derivative of a measure with respect
        to another measure, 316
determinant of a matrix, 149
diagonal operator, 409
diffeomorphism, 237
difference operator, 48
differential form, 329
differential of a function, 202
dilation, 227
dimension of a linear manifold, 100
dimension of a vector space, 83
Dirac measure, 275
direct product of semigroups, 50
direct sum of semigroups, 50
directed set, 19, 171
direction, 171
directional derivative, 257
directional limit, 177
discrete set, 25
discrete space, 151
dissipative operator, 473, 502
distance, 123
distribution, 383
distributive lattice, 66
divergence, 222
domain of definition of a
        correspondence, 7
double integral, 300
double limit, 185
double limit theorem, 185
duality rule, 67

eigenspace, 428
elementary net, 175

embedding, 9, 34
embedding of ordered sets, 20
empty collection, 11
equation for the tangent, 256
equipollent sets, 16
equivalent atlases, 323
Euclidean norm, 131
Euclidean space, 131
Euler equation, 262
exact form, 331
exterior algebra of a space, 148
exterior differential, 330
exterior product, 148
extremal, 261
extreme point, 490

factor algebra, 85
factor class, 15
factor group, 43
factor module, 78
factor set, 16
Fatou's theorem, 291
Fermat's theorem, 253
field, 60
field of p-adic numbers, 74
field of complex numbers, 71
field of rational numbers, 62
field of real numbers, 69
filter, 158
filter base, 158
filter of sections for a directed set, 175
finite distribution, 387
finite function, 375
finite set, 17
First isomorphism theorem, 45
fixed point, 228
fixed point of a multifunction, 492
Fourier coefficient, 137
Fourier coefficients, 365
Fourier series, 365
Fourier series theorem, 366
Fourier transform, 295

Fourier–Plancherel transform, 395
Fréchet filter, 175
Fréchet space, 375
Fredholm alternative, 417
Fredholm equation, 427
Fredholm operator, 100, 116, 414, 427
Fredholm theorem 3, 418
Fredholm theorem 2, 418
Fredholm's theorems 1–3, 117, 118
Fubini theorem, 302
Fubini–Tonelli theorem, 306

general solution, 99
generalized basis, 110
generalized derivative, 389
generalized dimension, 111
generalized function, 383
generalized metric, 157
generalized well-posedness, 427
generating operator, 470, 477, 502
generating system, 82
geometric conjugate, 124
gradient, 257
Gram determinant, 339
graph norm, 404
group, 37
group with operators, 76
groupoid, 31

Hahn–Banach theorem, 397
Hahn-Banach theorem, 126
Hamel basis, 85
Hamilton operator, 258
Hausdorff maximality theorem, 26
Hermitian adjoint operator, 413
Hermitian form, 131
Hermitian operator, 423
Hilbert adjoint operator, 413, 421
Hilbert–Schmidt operator, 410
Hilbert–Schmidt theorem, 441
Hille–Yosida theorem, 472
homeomorphism, 193

Homeomorphism Theorem, 193
homomorphism, 34
Homomorphisms theorem, 44
homothety, 90
homotopic invariance theorem, 350
homotopy, 349
Hyperplane, 398
hyperplane, 125, 357

ideal, 32
ideal of a ring, 59
idempotent, 34
image of an element, 7
image of the set, 7
imaginary unit, 71
implicit function, 226
increment of a function, 198
indefinite integral, 316
index, 11
index of an operator, 100
indicator, 10
infimum, 19
infinite set, 17
injection, 9
inner product, 115
integrability criterion, 292
integrable set, 297
integral boundedness, 279
integral operator, 409
integral sum, 270
integral with respect to a measure,
    459
integral with respect to a projection
    measure, 453
integral with respect to the measure,
    286
intervals, 23
inverse Fourier transform, 392
inversion formula, 391
isometric operator, 423
isometry, 369
Isomorphic sets, 9
isomorphism of ordered sets, 20

isosceles triangle inequality, 74
iterated limit, 185

Jacobian determinant, 335

Kakutani's theorem, 492
kernel of a homomorphism, 40
kernel of an operator, 91
Krein–Milman theorem, 490
Kronecker delta, 81

Lagrange function, 254
Lagrange's increment theorem, 210
Laplace operator, 374
Laplace transform, 476
lattice, 65
lattice order, 171
Lebesgue theorem on term by term
    integrability, 294
Lebesgue's number, 191
Lebesgue's theorem, 165
Leray–Schauder degree, 509
Leray–Schauder theorem, 489, 510
Levi's theorem, 289
lexicographic order, 21
lexicographic product, 21
limit set, 181
limit spectrum, 437
linear bounded operator, 199
linear combination, 79
linear complement relative to a
    basis, 95
linear differential operator, 374
linear independence, 79
linear manifold, 100
linear map, 89
linear operator, 89
linear order, 18
linear representation of a multilinear
    function, 143
linear span, 82
linearly independent collection, 79
Lipschitz condition, 213
local base, 152

local homeomorphism, 321
local integrability, 316
local minimum, 250
locally bounded measure, 278
locally bounded operator, 497
locally compact space, 169
locally constant function, 191
locally convex space, 155
locally finite cover, 344
locally integrable function, 383
locally uniform convergence, 215
lower limit, 290
Luzin's theorem, 298

Mackey theorem, 406
Maclaurin's formula, 245
majorant, 19
manifold, 322
map, 9
map degree, 504
matrix derivative, 205
matrix of an operator, 107
maximal element, 19
maximal ideal, 33
Mazur's theorem, 406
mean convergence, 360, 378
measurability criterion, 298
measurable function, 283
measure, 264
measure of an integrable set, 297
measure space, 264
metric, 123, 154
metric space, 154
minimal element, 19
minorant, 19
module, 77
monotonic map, 20
monotonic operator, 497
monotonically integrable function, 283
multi-index, 144
multidegree, 144
multidirection, 176

multifunction, 491
multilinear map, 141
multimetric, 156
multinorm, 155
multinormed space, 155
multiplication table of an algebra, 86
multiplier, 77
multivalued function, 36
multivectors, 141

natural embedding, 104
natural map, 142
net, 173
Neumann series, 229
neutral element, 31
Newton's algorithm, 70
non-Archimedean norm, 74
noncommutative ring, 55
nondegenerate metric, 154
nondegenerate norm, 123
nondegenerate scalar product, 114, 131
norm, 122
norm of a vector, 123
norm of an operator, 199
normal, 258
normal operator, 423
normed field, 122
normed space, 123
nullity of an operator, 93

open map, 404
Open map theorem, 404
operator exponential, 470
operator with closed range, 120
operators of finite rank, 93
order, 17
order sum, 22
ordered field, 121
ordered pair, 6
ordered set, 18
orientable manifold, 324
orthoadditivity, 265

orthogonal collection, 135
orthogonal complement, 116
orthogonal decomposition, 140
orthogonal projection, 137, 138
Orthogonal projection theorem, 355
orthogonal vectors, 133
orthogonality, 115
ortholattice, 65
orthomodular lattice, 66
orthonormal basis, 115
orthonormal collection, 135
outer measure, 279

parameterization, 322
parameterized surface, 337
Parseval's identity, 367
partial convergence, 188
partial derivatives, 205
partial differential, 203
partial function, 141, 203
partial increment, 203
partial limit, 181
partially isometric operator, 425
partition of unity, 343
Partition of unity theorem, 344
permutation, 9
plane, 255
point spectrum, 428
polar decomposition of an operator, 425
polar representation, 71
polygonally connected component, 88
polygonally connected set, 88
polynomial, 64
positive operator, 424
positive semigroup of operators, 474
precompact set, 189
preorder, 17
prime filter, 159
principal ideal, 59
principle of induction, 24
product, 114

product of matrices, 107
product of operators, 92
projection distribution function, 446
projection measure, 446
proper map, 349
Pythagorean theorem, 133

radially continuous operator, 499
Radon transform, 396
Radon–Nikodym theorem, 317
range of a correspondence, 7
range of an operator, 91
rank of a map, 326
rank of a tensor, 146
rank of an operator, 93
rank theorem, the, 326
rapidly decreasing function, 380
real type field, 109
reflexive relation, 14
reflexive space, 105, 405
regular generalized function, 384
regular integral, 272
regular measure, 338
regular space, 152
regular surface, 342
regular value, 327
relations, 14
relative boundedness, 198
relative complement, 65
relative smallness, 198
relatively compact set, 160
relatively complete lattice, 65
repeated integral, 300
residual spectrum, 428
resolvent correspondence, 427
resolvent equations, 432
resolvent set, 427
Riemannian pair, 172
Riesz compactness criterion, 379
Riesz integral representation theorem, 373
Riesz inverse operator theorem, 416
Riesz lemma, 360

Riesz representation theorem, 358
Riesz–Fischer theorem, 368
Riesz–Schauder theorem, 435
ring, 55
ring of functions, 58

saddle point, 494
scalar, 78
scalar operator, 93
scalar product, 114
Schauder theorem, 487
Schmidt algorithm, 364
Schröder-Bernstein Theorem, 17
Second isomorphism theorem, 46
section of a directed set, 174
section of a function, 300
segment, 87
self-adjoint operator, 121, 422
semigroup, 30
seminorm, 123
separable space, 152
separated space, 151
sequentiable net, 174
sequentially complete space, 182
set partition, 11
sigma-finite measure, 296
simple integral, 301
singular integral, 310
singular number, 440
singular value, 327
slowly increasing distribution, 385
slowly increasing function, 386
slowly increasing measure, 386
smooth curve, 259
smooth cycle, 337
smooth form, 329
smooth function, 217
smooth implicit function theorem,
        230
smooth inverse function theorem,
        237
smooth manifold, 323
smooth surface, 255

Sobolev lemma, 390
special solution, 99
spectral radius, 433
spectrum of an operator, 427
star domain, 331
star product, 68
Stokes formula, 346
Stone's theorem, 477
strong (Fréchet) differential, 213
strongly centered collection, 158
strongly continuous semigroup, 471
subalgebra of a vector algebra, 85
submanifold, 324
submodule, 78
subnet, 173
subsemigroup, 32
subspace of a vector space, 81
substitution, 9
sum of a collection, 178
summation operator, 49
support hyperplane, 398
support of a function, 81
supremum, 19
surjection, 9
symmetric algebra of a space, 148
symmetric difference, 57, 67
symmetric group, 52
symmetric multilinear map, 147
symmetric operator, 121, 422
symmetric relation, 14
symmetrization, 147

tangent bundle, 342
tangent manifold, 342, 343
tangent space, 342
tangent to a surface, 256
Taylor's formula, 245
tensor, 142
tensor of type (p, q), 146
tensor product, 142
tensor product of operators, 145
termwise differentiation theorem, 216
the change of basis theorem, 85

The Fourier transform, 391
The Fourier–Stieltjes transform, 395
The Fredholm alternative, 101
the integral extension of a measure,
        298
The resolvent theorem, 431
Theorem on the completeness of $\mathcal{L}^2$,
        361
theorem on the dimension of a
        vector space, 83
theorem on the interchange of
        derivative and integral, 309
theorem on the interchange of
        integral and limit, 307
theorem on the interchangeability of
        integrals, 313
theorem on the operator regression,
        229
theorem on the solution of an
        evolution equation, 480
theorem on the spectrum of a
        bounded operator, 434
Theorem on the spectrum of a
        self-adjoint operator, 436
Theorem on the spectrum of an
        Hermitian operator, 439
theorem on the symmetry of the
        second differential, 246
theorem on weak and strong
        differentials, 214
Tikhonov product, 163
Tikhonov theorem, 163, 487
Tikhonov topology, 163
Tonelli theorem, 305
topological space, 151
topologically connected set, 165
topology, 150
total differential theorem, 220
totally bounded set, 189
transitive relation, 14
transposition, 53
triangle inequality, 71

ultrafilter, 159

uncountable set, 29
uniform continuity, 194
uniform continuity theorem, 194
uniform convergence, 186
uniform space, 158
uniform topology, 156
uniformly bounded semigroup, 471
uniformly Continuous Extension
        Theorem, 196
uniformly equicontinuous, 371
unital module, 77
unital ring, 56
unitary operator, 423
upper limit, 290

vector, 76
vector algebra, 85
vector space, 78
Volterra operator, 409
von Neumann theorem, 495

weak (Gâteaux) differential, 213
weak topology, 156
Weierstrass's compactness theorem,
        192
Weierstrass's partial convergence
        theorem, 188
well-ordered set, 23

zero ring, 56
Zorn's lemma, 26